# Introduction to
# Marine Biology

## SECOND EDITION

### George Karleskint, Jr.
*St. Louis Community College—Meramec*

### Richard Turner
*Florida Institute of Technology*

### James W. Small, Jr.
*Rollins College*

**THOMSON**

**BROOKS/COLE**

Australia • Canada • Mexico • Singapore • Spain
United Kingdom • United States

Editor-in-Chief: Michelle Julet
Biology Executive Editor: Peter Adams
Development Editor: Elizabeth Howe
Assistant Editor: Elesha Feldman
Editorial Assistant: Lisa Michel
Technology Project Manager: Travis Metz
Marketing Manager: Ann Caven
Advertising Project Manager: Kelley McAllister
Project Manager, Editorial Production: Shelley Ryan
Executive Art Director: Robert Hugel
Art Director: Lee Friedman
Print/Media Buyer: Barbara Britton

Permissions Editor: Sarah Harkrader
Production Service: G&S Book Services
Text Designer: tani hasegawa
Photo Researcher: Kathleen Olson
Copy Editor: Mary Ann Short
Illustrators: G&S Book Services
Cover Designer: tani hasegawa
Cover Image: Corbis
Cover Printer: Transcontinental-Interglobe
Compositor: G&S Book Services
Printer: Transcontinental-Interglobe

Printed in Canada
2  3  4  5  6  7  09  08  07  06  05

For more information about our products, contact us at:
**Thomson Learning Academic Resource Center**
**1-800-423-0563**
For permission to use material from this text or product, submit a request online at
**http://www.thomsonrights.com**.
Any additional questions about permissions can be submitted by email to **thomsonrights@thomson.com**.

**Thomson Higher Education**
10 Davis Drive
Belmont, CA 94002-3098
USA

**Library of Congress Control Number:** 2004112176

ISBN 0-534-42072-9

**Asia (including India)**
Thomson Learning
5 Shenton Way
#01-01 UIC Building
Singapore 068808

**Australia/New Zealand**
Thomson Learning Australia
102 Dodds Street
Southbank, Victoria 3006
Australia

**Canada**
Thomson Nelson
1120 Birchmount Road
Toronto, Ontario M1K 5G4
Canada

**UK/Europe/Middle East/Africa**
Thomson Learning
High Holborn House
50/51 Bedford Row
London WC1R 4LR
United Kingdom

**Latin America**
Thomson Learning
Seneca, 53
Colonia Polanco
11560 Mexico
D.F. Mexico

**Spain (includes Portugal)**
Thomson Paraninfo
Calle Magallanes, 25
28015 Madrid, Spain

# DEDICATION

*To my wife Shaun, who is as enthusiastic about the ocean realm as I, for her love and support during the preparation of this and the first edition.*

*To my children Brad, Kristy, Tim, and Samantha, who share my love of the ocean and its creatures.*

*To my parents, who first introduced me to the wonders of the marine world and continued to encourage me in my scientific pursuits.*

*—GK*

*I dedicate this book to two Massachusetts sisters—my mother Marion Turner and my aunt Casindania Eaton—who in my childhood nurtured my seaside studies in natural history; to my academic mentors John H. Dearborn (University of Maine) and John M. Lawrence (University of South Florida), noble zoologists who guided my research on the puzzling design of echinoderms; to my colleague the late Kerry B. Clark (Florida Institute of Technology), who was for me both Steinbeck and Ricketts; and to my students of the past 29 years who have shared my passion for marine biology.*

*—RT*

*To my students past and present.*

*—JS*

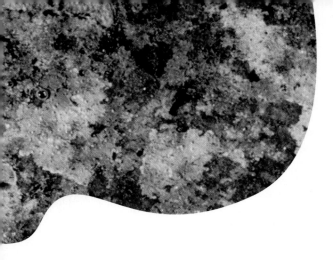

# Contents Overview

# Contents

# PART 4
# HUMANS AND THE SEA

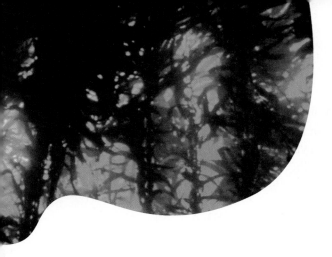

# Preface

*Introduction to Marine Biology* is intended for under-graduate students majoring in a variety of disciplines and biology students interested in learning more about the marine environment. Many such students may already be interested in the field of marine biology. Others, however, are studying marine biology to fulfill a general education requirement, and they may have a fear of science courses. Like many, these students are generally intrigued by the marine environment and especially marine organisms. Having grown up with television programs dealing with the ocean and marine organisms, they have a natural interest in this subject area. This text strives to use this interest as a starting point for teaching biological science. Each of the authors has been teaching marine biology for more than 20 years to both majors and nonmajors. In their lectures and field courses, they stress an ecological approach to the study of marine organisms, and they have used this same approach in preparing this text.

The main focus of the text is the *ecology of the marine environment*—that is, the ways in which marine organisms interact with each other and with their physical environment. The authors believe that there is no better way to teach about the delicate balance of natural systems than within the context of marine ecosystems. This text also strives to educate students about the importance of marine ecosystems to terrestrial ecosystems and to humankind. The authors hope that by studying marine biology, students will be able to make well-informed decisions at the ballot box and when they engage in activities that have an impact on the natural world, especially with respect to the ocean. All citizens need an understanding of topics such as marine pollution and overfishing, and the better informed students are about such issues, the more likely they are to make sound decisions. The authors hope that this text will provide the factual foundation necessary for making such important decisions.

## NEW TO THIS EDITION

The authors have introduced many new topics and improvements to this second edition of *Introduction to Marine Biology*. Chapter 1 contains an improved presentation on the scientific method that takes students step by step through the process with examples from the real world. New boxes titled "The Science of Marine Biology" carry this theme throughout the text by showing how the scientific method is used in actual experiments involving marine organisms. Chapter 2 contains a new section on population dynamics. Chapter 3 has an added description of the evolution of the early earth and sea and an updated presentation on navigation that includes the use of global positioning systems. Chapter 5 now includes coverage of biological chemistry and the cell and cell function. In addition, the section on biological classification has been updated to include domains and some of the new theories on classification of protists. A section on marine viruses has been added to Chapter 6 and the coverage of other marine microbes has been completely updated and reorganized. In response to reviewers' requests, Chapters 7 through 12 contain increased and updated coverage of marine organisms and their ecology, providing more in-depth discussion of physiology, behavior, and ecological interactions. Chapter 12 has increased the coverage of marine mammals, especially cetaceans. New summary tables have been added to the end of each of these chapters, making it easy for students to review the key characters and ecological roles of all major groups of organisms covered. The topics in Chapters 13 and 14 have been switched so that a discussion of intertidal communities (Chapter 13) comes before the discussion of estuaries (Chapter 14). Chapter 13 has been completely revised and now includes a comparison of tropical and temperate rocky shores and a more detailed discussion of rocky shore ecology. Chapter 15 contains more-detailed coverage of coral reef ecology, and a new section on coral diseases has been added. Chapter 16 has been completely reorganized and topics divided into sections on continental shelves, benthic communities, and neritic zone. The content of Chapters 17 and 18 has been updated. Chapters 19 and 20 contain more current information and an expanded discussion of fisheries management and overfishing. Sections dealing with marine pollution have also been reorganized and updated. More than 100 new photos have been added, including some 75 photos selected from the collection of Dr. Richard N. Mariscal, biological science faculty member emeritus of Florida State University. Selected readings now contain some articles from the technical scientific literature as well as the popular scientific literature. Pertinent articles from InfoTrac® College Edition are listed at the end of each chapter along with websites that students can visit to obtain more information and view images and video clips.

# LEARNING AIDS

This edition contains the following learning aids to help students master the text material and successfully use their knowledge.

1. **Key Concepts.** These appear at the beginning of each chapter and identify its main points.

2. **Boldface Terms.** Throughout the text, important terms appear in **boldface** where they are defined.

3. **Pronunciation Guide.** Terms in the text that might pose pronunciation problems for beginning students are accompanied by phonetic transcriptions.

4. **Boxed Readings.** Boxed readings are organized into four categories: "Ecology and the Marine Environment," "Marine Biology and the Human Connection," "Marine Adaptation," and, new to this edition, "The Science of Marine Biology." The first three categories of boxed readings provide students with an interesting and engaging focus on three important aspects of marine biology. The fourth category of boxed readings illustrates for students how science is done in general and specifically in the field of marine biology. The boxed readings also aid instructors in structuring their boxed-reading assignments (by using the four categories) to fit the course emphasis.

5. **In-Chapter Summaries.** A short summary after each major section replaces the end-of-chapter summaries of the previous edition. Each summary is a short synopsis of the section's main points. Students will find it useful for reviewing the content of a section before moving on to the next, as well as in reviewing the entire chapter in preparation for tests.

6. **In Perspective Summary Tables.** Each of the chapters devoted to marine organisms contains a summary table that reinforces for the student the most important biological and ecological aspects of the organisms discussed.

7. **Selected Key Terms.** The Selected Key Terms section includes all of the boldfaced terms in the chapter. This section provides students with a convenient review and helps them with the often daunting task of mastering the language of biology. The boldfaced terms are listed in alphabetical order with page references directing students to the specific locations in the chapter where the terms are defined.

8. **Questions for Review.** There are three levels of these end-of-chapter questions. The first level, Multiple Choice, comprises objective questions that deal primarily with vocabulary, terminology, and some basic concepts. The second level, Short Answer, consists of essay questions that test students' recall of facts and their ability to synthesize information. The last level, Thinking Critically, involves higher-order thinking skills and problem solving. The answers to the first-level Questions for Review are provided in the Appendix of the text. Suggested responses to the other questions are found in the instructor's manual.

9. **Suggestions for Further Reading.** At the end of each chapter is a short list of articles and books that supplement the chapter's content. Most of the readings are taken from popular scientific literature, such as *National Geographic, Natural History, Smithsonian, Discover,* and *Scientific American.* These articles and books are written at a level appropriate for students using this text and are therefore more likely to be enjoyed by them. A few articles from more technical scientific literature are also included for those students and instructors looking to find more in-depth coverage of some of the chapter topics.

10. **InfoTrac College Edition.** New to the second edition is the inclusion of access to InfoTrac College Edition, an online service that provides access to more than 100 journals. At the end of each chapter are short lists of some InfoTrac College Edition articles pertinent to the chapter's content.

11. **Websites.** At the end of each chapter are listed selected websites that students can visit to learn more about the topics in the chapter and to view more pictures and even video clips of marine organisms and habitats.

12. **Separate Glossary.** Located at the end of the text, the Glossary defines all the boldfaced terms that appear in the text, as well as many other biological terms used in the text but not boldfaced.

# ORGANIZATION

*Introduction to Marine Biology* is arranged in four parts that cover the environment, organisms, and ecosystems of the ocean and humans' relationship to the ocean.

## Part 1: The Ocean Environment

Chapter 1 introduces the science of marine biology, presents the scientific method, and orients the student to the rest of the text. Chapter 2 presents the basic principles of ecology as they apply to marine systems. This information is presented early in the text to support the main theme of ecology of the marine environment. Chapters 3 and 4 introduce the physical aspects of the marine environment and discuss their importance to the organisms that live in the ocean.

## Part 2: Marine Organisms

Chapter 5 introduces the student to basic biological concepts such as the chemical basis of life, cell structure and function, energy transfer in biological systems, evolution, and biological classification. Chapters 6 through 12 survey all of the major groups of marine organisms and examine their interrelationships. These chapters are organized on the basis of feeding relationships, proceeding from organisms that produce their own food to those that rely on other organisms for food. Descriptions of animals are presented in a traditional format, beginning with invertebrates and working upward through the vertebrate classes to mammals. The focus of the individual chapters is the role that each group of organisms plays in the overall web of marine life.

## Part 3: Marine Ecosystems

Chapters 13 to 18 examine the major marine ecosystems. Each chapter in this part examines how the interactions of the physical and biological environment make each ecosystem unique and how these factors influence the number and kinds of marine organisms that inhabit a given area. The survey begins at the coastline with the intertidal zone (Chapter 13) and estuaries (Chapter 14) and then progresses to offshore ecosystems, coral reefs (Chapter 15), and the continental shelves and neritic zone (Chapter 16). Chapter 17 deals with the pelagic ecosystem of the open sea, and Chapter 18 addresses life in the ocean's depths.

## Part 4: Humans and the Sea

Part 4 of the text examines the impact that humans have had and continue to have on the marine environment. Chapter 19 addresses fisheries management and the consequences of overfishing as well as problems associated with the extraction of nonliving products from the seas. The chapter also deals with how the use of these natural resources has changed over the last century, and it discusses the effects of this change on the marine environment. Chapter 20 examines marine pollution and habitat destruction and presents ways for students to stem and possibly reverse environmental damage.

## SUPPLEMENTS

*Introduction to Marine Biology* is accompanied by a supplement package that has been designed to aid students in learning and instructors in teaching, including helpful classroom preparation material such as the Instructor's Manual with Test Bank and Exam View computerized testing. In addition, a Multimedia Manager contains all illustrations and most photos from the text. This one-stop lecture tool makes it easy to assemble, edit, publish, and present custom lectures, using Microsoft PowerPoint. The Multimedia Manager brings together text-specific lecture outlines and art from Thomson Higher Education texts, along with video and animations from the web or your own materials—culminating in a powerful, personalized, media-enhanced presentation.

Students will find a robust book-specific website at *http://biology.brookscole.com/marinebio2*. This outstanding site features chapter-by-chapter online tutorial quizzes, a final exam, chapter outlines, chapter review, chapter-by-chapter web links, flashcards, and more!

## ACKNOWLEDGMENTS

The production of a text such as this one involves the collaborative efforts and creative talents of many individuals. Without the help of editors, reviewers, and other professionals this book would never have been published. The authors would like to acknowledge the highly professional and supportive staff at Thomson Higher Education.

The authors would also like to express their gratitude to their colleagues at St. Louis Community College–Meramec, Florida Institute of Technology, and Rollins College for their support and input and to their students, who have provided them with valuable feedback. Lastly, the authors would like to thank Richard N. Mariscal, PhD, for the use of his excellent photographs that aptly capture the beauty and diversity of marine life.

## REVIEWERS

The authors appreciate the diligent work done by the following individuals, who took time from their busy schedules to read and make suggestions concerning the manuscript. Their help and input was invaluable in creating the final text.

Lisa Allen *Gulf Shores High School*

Susan Barrett *Massasoit Community College*

Wayne Bennett *University of West Florida*

Patti Bernard *College of the Albemarle*

Annalisa Berta *San Diego State University*

W. Randy Brooks *Florida Atlantic University*

Douglas G. Capone *University of Southern California*

Suzanne Edmands *University of Southern California*

Austin W. Francis, Jr. *St. Joseph's University*

Kenneth M. Halanych *Auburn University*

Dennis C. Haney *Furman University*

Arthur C. Hulse *Indiana University of Pennsylvania*

Robert E. Knowlton *George Washington University*

Matthew Landau *Richard Stockton College of NJ*

Karen Martin *Pepperdine University*

Marlene Martinez *Pacific University*

W. Linn Montgomery *Northern Arizona University*

John F. Pilger *Agnes Scott College*

Maynard Schaus *Virginia Wesleyan College*

Finally, although every effort was made to produce a product free of errors and inaccuracies, it is inevitable that some will occur. Any errors or oversights are the authors' and not those of the reviewers or the editorial staff at Thomson Higher Education. Comments or suggestions for how we can improve this text in future editions are appreciated and can be sent to the authors in care of the publisher.

# 1

# Science and Marine Biology

There is perhaps no better way to learn about the delicate balance of natural systems than to study marine organisms and the communities they form. This pursuit is not only fascinating in itself but also teaches us how important the ocean and its inhabitants are to humans and why we should try to preserve and conserve this resource. The diversity of marine organisms has attracted naturalists and scientists from the earliest days of human history, and the rich history of marine biology has strongly influenced the modern science we know today. This chapter will introduce you to the importance of oceans and their inhabitants and the developments that have formed the modern science of marine biology.

## IMPORTANCE OF THE OCEANS AND MARINE ORGANISMS

Oceans are the principal physical feature of our planet. They cover nearly 71% of the earth's surface and represent the last great expanse on this planet to be charted and explored. The physical characteristics of these great bodies of water directly and indirectly affect our everyday lives, and the living organisms that inhabit them are an important source of food and natural products.

Oceans act as enormous solar-powered engines that drive the various weather patterns affecting terrestrial environments (Figure 1-1a). Phenomena such as El Niño Southern Oscillation (ENSO), which can cause droughts in Peru, flooding in Texas, and relatively mild winters in the Midwest, are the result of changes originating in the Pacific Ocean. The action of waves and tides changes the contours of continents and affects the lives of people who inhabit coastal areas.

**Ocean productivity**—the amount of food marine organisms can produce and the number of organisms the oceans

## Key Concepts

① Marine and terrestrial environments are interrelated, interactive, and interdependent.

② The ocean is an important source of food and other resources for humans.

③ Marine biology is the study of the sea's diverse inhabitants and their relationships to each other and their environment.

④ The history of marine biology is one of changing perspectives that have shaped the modern science and its applications.

⑤ Marine laboratories play an important role in education, conservation, and biological research.

⑥ It is important to study marine biology in order to make informed decisions about how the oceans and their resources should be used and managed.

⑦ Scientists use an organized approach called the scientific method to investigate natural phenomena.

NOAA; inset, Larry Ulrich /Stone /Getty Images

**Figure 1-1 The Importance of Oceans.** (a) The exchange of heat energy between the atmosphere and the oceans is responsible for creating the weather patterns that affect terrestrial habitats. The white area in this photo is a tropical storm developing in the Pacific Ocean. (b) The oceans supply a significant amount of food in the form of fish, shellfish, and seaweeds.

(a)

(b)

can support—is a leading area of research in marine ecology. The sea has provided, and still provides, a substantial amount of the world's food supply (Figure 1-1b). The United Nations reports that more than 80 million metric tons (1 metric ton = 1.1 tons) of marine fish are harvested annually.

Marine organisms are a main source of food and also provide us with important materials for industry and medicine. The commercial harvest and processing of these organisms and their products provide jobs for millions of people worldwide.

Marine organisms play another significant role in scientific research. Biologists have found that many of the species inhabiting the sea and seashores are ideally suited to the study of such varied fields as ecology, physiology, biochemistry, biogeography, behavior, genetics, and evolution. Experiments using marine organisms have provided biologists with considerable information not only about the organisms studied but also the workings of biology in general. Discoveries based on experiments with marine organisms have greatly advanced our understanding of biology.

The full extent of what the sea and its inhabitants have to offer has yet to be discovered, and only by learning more about the sea and its creatures will we be able to realize its full potential.

## In Summary

The world's oceans play an important role in everyday life because they affect weather patterns, provide food, and provide vital resources. ●

## STUDY OF THE SEA AND ITS INHABITANTS

**Oceanography** is the study of the oceans and their phenomena, such as waves, currents, and tides. The science of oceanography draws from many different disciplines, including chemistry, physics, geology, geography, meteorology, and even biology. The study of the living organisms that inhabit the seas and their interactions with each other and their environment is **marine biology** (Figure 1-2). These two areas of study are not completely distinct from each other, and they frequently overlap. It is necessary to combine elements from both fields to form a complete picture of the oceans and their inhabitants.

With this in mind, the theme of this book is the ecology of marine organisms, the interplay and interdependence of organisms with each other and their environment. There is tremendous diversity among the organisms in the world's oceans, and we know very little about many of them. Some of these organisms may be important new sources of food, useful commercial products, or chemicals that could be used as medicines (Figure 1-3). Millions of visitors are drawn yearly to seaside resorts to admire the beauty and variety of marine organisms. This is a key factor in industries such as ecotourism and sports such as scuba diving. Unless we learn more about these organisms, we will never know all of the substantial contributions that they make to our lives now and could make to our lives in the future.

We hear a great deal today about the impact humans have on the environment, including the sea. There are even some who claim that the sea is doomed and that we can-

Richard N. Mariscal

**Figure 1-2 Marine Biology.** The marine organisms in this tidal pool interact with and are influenced by each other and their physical environment. The study of marine organisms and these interactions is the science of marine biology.

Stuart Westmorland/Stone/GettyImages

**Figure 1-3 Marine Organisms and Medicine.** The cartilage that makes up the skeletons of sharks is an important source of antiangiogenesis factor, a chemical that prevents tissues from establishing a blood supply. This chemical may be useful in the fight against cancer by depriving tumors of blood, thus killing them.

not reverse the damage that has been done. Although the media play a prime role in informing us about human impact on the sea, they interpret much of this information for us as well. The attitude that people take toward such topics as the dumping of trash, disposal of radioactive and industrial wastes, oil spills, and overfishing is greatly influenced by the media. To make intelligent decisions about the ocean, as well as the rest of the environment, responsible citizens need to have a background of facts when analyzing these complex issues. This is perhaps the most important reason for a concerned citizen to learn more about marine biology.

## In Summary

The science of oceanography is the study of the oceans and their phenomena. Marine biology is the study of the organisms that inhabit the sea, their interrelationships, and their interactions with their environment. A basic knowledge of marine biology is necessary to understand how marine organisms relate to us and how human activities affect the marine environment. A basic knowledge also helps conscientious citizens make prudent decisions about activities that involve and affect the sea. ●

## MARINE BIOLOGY: A HISTORY OF CHANGING PERSPECTIVES

Human interest in the sea probably dates back to the time we first set eyes on it, fished its waters, and sailed across it. The great expanse of water with its variety of strange and wonderful creatures has inspired awe, wonder, curiosity, myth, and at times, fear. This interest in the sea and its creatures was the beginning of the sciences of oceanography and marine biology.

Like living organisms, the science of marine biology has evolved over time. Early investigations of the sea's creatures centered on simple observations of marine organisms that were easily accessible from the shore. As ships and equipment became available, humans set out to conquer the sea and attempted to control its awesome magnitude and power. In their quest for discoveries, explorers and merchants alike eagerly set about emptying the sea of its contents, often overexploiting its resources. Improvements in shipbuilding and navigation opened up new frontiers in marine exploration. These same technical improvements also allowed the science of marine biology to expand as a body of knowledge. In more recent times, advances in submersibles, robotics, computers, and other technology have broadened our view of the marine environment even more. This new knowledge has led marine biology to a more global perspective of not only the interrelatedness of ocean habitats but also interactions between ocean and terrestrial habitats.

### Early Studies of Marine Organisms

Early attempts to study the sea's creatures can be traced back to the ancient Greeks and Romans. The Greek philosopher Aristotle, who was an accomplished naturalist, was one of the first to develop a scheme of classification, which he called the "ladder of life." His writings contain descriptions of more than 500 species, almost one third of which are marine. He also studied fish gills, proposing that they functioned in gas exchange, and made detailed observations of the anatomy of the cuttlefish (*Sepia*).

Pliny the Elder was the foremost of the Roman naturalists. His only surviving work is the 37-volume *Natural History,* which contains mostly information about terrestrial animals, but it does include references to marine fishes and bivalves (shellfish such as clams and mussels). During the Middle Ages, the Catholic Church became the primary overseer of scholastic pursuits but gave most of its attention to matters of theology and philosophy. The study of natural history still depended on the works of the early Greek and Roman naturalists. Arabian philosophers of the Middle Ages involved themselves with interpreting and explaining the works of the ancient naturalists, rather than engaging in their own studies and observations. It would be the late eighteenth century before biologists would again conduct studies of the marine environment based on original observations.

## Renewed Interest in Marine Organisms

During the late eighteenth and early nineteenth centuries, biology expanded into several disciplines. This was a time of great discovery, fueled in part by the development of better sailing ships and navigational instruments. The exploration of new lands and new sea routes provided information for the various branches of science, including biology. Scientists, such as the French naturalist Lamarck and the anatomist Cuvier, studied and described many marine organisms during this time.

In December 1831 the vessel HMS *Beagle* set sail on a 5-year voyage of exploration that would take it around the globe. Among the members of the expedition was Charles Darwin. During the voyage, Darwin, father of the theory of evolution by natural selection and an early marine biologist, was able to observe marine life firsthand and collect many specimens of marine organisms. On the basis of his observations of atolls (ring-shaped coral reefs that enclose a lagoon) in the Pacific, he proposed an explanation of atoll development that is still widely accepted. It was during this voyage that he began to formulate what would eventually become his theory for the process of evolution. In 1859 Charles Darwin published his landmark work *On the Origin of Species by Means of Natural Selection*. This work stimulated many scientists to investigate the causes of adaptations observed in marine organisms. It also sparked study of the interrelationships among marine organisms and between marine organisms and their environment. In the years following the voyage of the *Beagle,* Darwin engaged in a detailed study of the barnacles that inhabit the rocky coasts of England and produced a monograph on the subject that is still used today (Figure 1-4).

In the early part of the nineteenth century, it was generally agreed that living organisms could not survive in the cold and darkness of the ocean depths, an idea proposed by the English naturalist Edward Forbes. Evidence to the contrary was produced when a transatlantic telegraph cable linking the United States and England failed shortly after it was laid in 1858 and had to be retrieved. The cable was located at a depth of approximately 1.7 kilometers (1 mile) in the northern Atlantic Ocean. As the repair ship retrieved the cable, it was found to be covered with all sorts of marine organisms that had never been seen before. These organisms apparently flourished in the depths of the ocean, and their discovery sparked investigations by several countries into life on the ocean floor. Subsequent dredging expeditions recovered animals from as deep as 4.42 kilometers (14,500 feet, or more than 2¾ miles).

Copyright Library, Academy of Natural Sciences, Philadelphia

**Figure 1-4  Charles Darwin.** Although better known for his theory of evolution by natural selection, Darwin was an accomplished marine biologist. This page is from Darwin's monograph on barnacles, a reference work that is still used by marine biologists today.

## Beginnings of Modern Marine Science

By the end of the nineteenth century, many countries, including the United States, were spending significant amounts of time and money to learn more about the marine environment and its creatures.

### *Challenger* Expedition

As a result of increasing interest in the marine environment, the British Admiralty organized the *Challenger* expedition, which lasted 3½ years. The expedition was named after the research vessel HMS *Challenger,* a ship containing state-of-the-art equipment and research facilities. When the *Challenger* returned to England in May 1876, it had crisscrossed the major oceans of the world and brought back enough information to fill 50 volumes of scientific reports. During this expedition, more than 4,700 new species of marine organisms were collected and described. Many of these new species were dredged from great depths (Figure 1-5).

The *Challenger* expedition gave birth to the modern sciences of marine biology and oceanography, and even today, marine scientists continue to refer to the *Challenger* reports in their research. Later expeditions by Great Britain

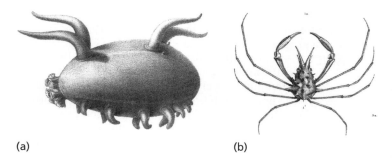

(a)                                    (b)

and other countries would add to and build on the broad base of information accumulated by this groundbreaking expedition.

Charles Wyville Thomson, the driving force behind the *Challenger* expedition and its chief scientist, collected samples of the microscopic organisms floating in the water (Figure 1-6). Previously, biologists had paid little attention to these organisms, but now they were coming under close scrutiny. In 1887 Victor Hensen coined the term **plankton** to describe all of the different organisms that float or drift in the sea's currents. These tiny organisms are at the base of the ocean's complex food webs, and only in the last 50 years have marine biologists started to understand the specific roles that these organisms play.

## Marine Studies in the United States

American scientists were also contributing to the ever-growing body of knowledge that was marine biology. Through expeditions and the founding of marine laboratories, the United States became a leader in the emerging field of marine studies.

***Expeditions of Alexander Agassiz***    In 1877 the American naturalist Alexander Agassiz began the first of several expeditions he would direct to investigate the organisms of the sea. Agassiz collected samples of animals from hundreds of locations and dredged animals from depths of 180 to 4,240 meters (600 to 14,000 feet). The amount of material he collected rivaled that brought back from the *Challenger* expedition (Figure 1-7).

In addition to collecting and cataloging marine organisms, Agassiz studied coloration in marine animals. He noted that the most brightly colored animals were found in surface waters and that, as one proceeded deeper, brilliant colors gave way to blues and greens and ultimately reds and blacks. He theorized that the colors were related to the absorption of different wavelengths of light at different depths, a theory later proved correct. He also noted a great deal of similarity in the deepwater organisms on the east and west coasts of Central America and hypothesized that the Pacific and the Caribbean were at one time connected. Agassiz spent much of the latter part of his life studying the structure and formation of coral reefs.

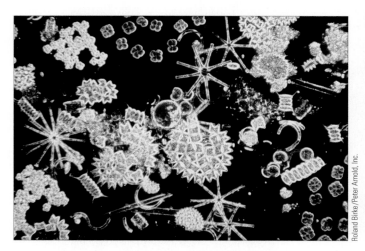

**Figure 1-6  Plankton.** Some examples of marine plankton, organisms that float or drift in the sea's currents. Charles Wyville Thomson, the chief scientist of the *Challenger* expedition, was one of the first scientists to seriously investigate the role of plankton in marine communities.

**Figure 1-7  Alexander Agassiz.** Alexander Agassiz was one of the foremost U.S. marine biologist of the nineteenth century. He is pictured here in his laboratory with jars containing some of the marine specimens that he collected during his many expeditions.

***Founding of the First Marine Biological Laboratory***
Alexander's father, Louis Agassiz, founded the Museum of Comparative Zoology at Harvard University. He also founded the first marine biology laboratory in the United States, in July 1873. Originally located on Penikese Island off the Massachusetts coast and called the Anderson Summer School of Natural History, it was founded to help teachers at all levels improve their methods of teaching natural history. Agassiz was a firm believer in learning through observation and hands-on acquisition of knowledge. It was Agassiz's hope that the Summer School would allow him the opportunity to teach his methods to others using marine organisms as subjects of study. A primary reason for locating his school by the sea was that all the major groups of organisms (phyla) are represented in the ocean and many groups are found only in the ocean. The animals would be within easy reach and could be studied in their natural habitats. Agassiz's goal of teaching biology through direct observation is still a major objective of the courses that are offered at marine laboratories around the world.

Agassiz's school was the inspiration and predecessor of the Marine Biological Laboratory at Woods Hole (Figure 1-8), which was founded in 1888 in a little fishing village on Cape Cod, Massachusetts. In 1922 the Woods Hole Oceanographic Institution was constructed down the street from the Marine Biological Laboratory, and their proximity allowed a great deal of interchange between the two institutions. The Marine Biological Laboratory is today one of the foremost institutions of its kind in the United States and is a major, internationally recognized research institution. Louis Agassiz's influence on the field of marine biology can still be seen at the Marine Biological Laboratory at Woods Hole. One of his favorite sayings, "Study nature, not books," is prominently posted in the library to remind students and researchers alike of this important concept.

***Other U.S. Marine Laboratories***   During the twentieth century, other notable research institutions were founded, such as the Scripps Institution of Oceanography in Califor-
nia, the University of Miami's Rosenstiel School of Marine and Atmospheric Science in Florida, the Harbor Branch Oceanographic Institution in Florida, the Friday Harbor Laboratories of the University of Washington, and Duke University Marine Laboratory in North Carolina, to name a few.

Understanding commercial fishery production was a driving force in the establishment of many marine laboratories and continues to be a chief goal. In addition to research on ocean productivity and the interrelationships of marine organisms, modern marine laboratories focus on the use of marine organisms to solve fundamental problems in biology, such as the control of cell division, the process of embryological development, the functioning of the nervous system, and the applications of biotechnology.

## Marine Biology in the Twentieth Century

Early in the twentieth century, expeditions were mounted to study the Arctic and Antarctic seas. Individuals such as the Norwegian Fridtjof Nansen and the Englishman Sir Alistair Hardy (Figure 1-9) led expeditions that collected information and organisms from these two areas. Nansen was interested in reaching the magnetic North Pole as well as charting the waters around the Pole. He was not successful in his attempt to reach the Pole, but his expedition to gather information about the polar seas was quite successful. Hardy was interested in the biology of whales for the purpose of commercial exploitation, and it was this interest that took his expeditions to the Antarctic Sea. In addition to making new observations concerning whales, he also increased the amount of information available concerning the Antarctic Sea. His book *The Open Sea: Its Natural History* remains a classic in the field of marine biology.

**Figure 1-8  Marine Biological Laboratory at Woods Hole.** This Massachusetts facility is one of many institutions worldwide whose goal is to learn more about the organisms that inhabit the oceans.

**Figure 1-9  Sir Alistair Hardy.** The Englishman Sir Alistair Hardy led expeditions to the Antarctic Sea to study whales.

Since the latter part of the nineteenth century, advances in the field of marine biology have literally given us both a broader look and a deeper view into the sea itself. With this new perspective came a broader and deeper understanding about global ecology. For years our technology for exploiting the sea has outpaced our understanding of marine biology, but advances in the science have led to new attitudes about conservation and resource management. Starting in the last century, some investigators began to center their attention on not only the relationships of marine organisms to each other but also the impact of human activities, such as fishing and pollution, on the marine environment.

## Marine Biology Today

We now live in the information age. Each day new findings that increase our comprehension of the living world that surrounds us are added to the world's data banks. Knowledge of the inner workings of the sea as well as its inhabitants is increasing at a breakneck speed, aided by the many advances in technology. Today, deep-sea submersibles such as *Alvin* (Figure 1-10) can take marine scientists to the very floor of the ocean to view and collect organisms that live in the deepest recesses of the sea, and researchers can live in underwater habitats while they observe the activities of these organisms. The advent of computers with connections to the Internet and the "information superhighway" allows scientists and nonscientists around the world to share the enormous amounts of information collected on a variety of subjects, including marine biology. With the aid of these new tools, marine biologists are discovering the multifaceted interrelationships among marine organisms and marine systems and their ties to terrestrial environments.

In the following chapters, you will learn more about the sea and the organisms that inhabit it. You will learn how both physical and biological factors interact to produce the complex ecosystems of the marine environment. You will be introduced to the methods of science as they apply to the study of marine biology, and you will learn what impact humans have had on this truly wondrous realm.

## In Summary

The science of marine biology has changed over the years as new technologies have been developed. Human interest in the sea and its creatures can be traced back to the ancient Greeks and Romans. It was not until the nineteenth century, however, that the foundations for the modern sciences of marine biology and oceanography were formed. The research expedition of HMS *Challenger* was a landmark in the study of the sea. Information collected during this expedition and subsequent expeditions laid the foundations for modern marine science. The late nineteenth century saw the first marine laboratories founded in the United States, beginning with the Marine Biological Laboratory at Woods Hole. These laboratories continue to play a vital role in both marine and basic biological research. ●

## PROCESS OF SCIENCE

A particular endeavor or study becomes a science when the principles on which it is based can be presented as hypotheses, explanations that can be tested by experiments. A good **hypothesis** can explain past events and predict the outcome of current or future experiments. All branches of science seek to organize observations so that hypotheses can be formed that suggest relationships between the observations. These hypotheses are then tested by experiments. The data gathered from these experiments are evaluated, and logical conclusions are drawn from the information at hand.

## Scientific Method

Not all researchers follow the same approach to doing science. Some believe that science should be done inductively. **Inductive reasoning** involves looking at individual observations and proposing a general explanation for them. For instance, a marine biologist might observe that the octopus and squid, both cephalopods, have arms with suckers and conclude, using inductive reasoning, that all cephalopods have arms with suckers. Other researchers believe that good science should be done deductively. In **deductive reasoning**, observations suggest some general principle or idea from which specific statements can be derived. Thus a marine biologist might note that all cephalopods have arms with suckers, and because a cuttlefish is a cephalopod, it must have arms with suckers. Most scientists feel free to use either method and sometimes the best science is accomplished by using both methods. For example, if a body of facts suggests some general principle (inductive reasoning), then specific points can be drawn from that principle (deductive reasoning) and those points tested.

Scientists use an organized, commonsense approach to their work that begins with making observations of the natural world. On the basis of these observations, the scientist asks questions, proposes hypotheses, designs experiments

**Figure 1-10  The Submersible *Alvin*.** Submersibles like Alvin allow marine scientists to investigate life in the ocean's deepest recesses.

to test the hypotheses, gathers results, and draws conclusions. This orderly pattern of gathering and analyzing information is called the **scientific method.** The scientific method represents what might be called a scientific approach to interpreting the natural world. It should be emphasized that the steps of the scientific method are simply an idealized approach to problem solving in science. Individual researchers tend to bring their own variations to the plan as they construct and conduct their experiments.

## Step 1: Making Observations

Generally, the first step in the process of science is to gather observations that stir the curiosity of scientists, causing them to ask questions concerning the subject that is being studied. Making observations is a substantial aspect of the scientific process, but observations alone are not very helpful without some thread or pattern that ties them together. It is not enough just to note facts. A good researcher is not only a good observer but also a creative thinker. For instance, like many visitors to a rocky shore, a biologist named J. H. Connell observed that barnacles grew on rocks in groups forming definite zones. Connell also noticed that two different types of barnacle occupied distinctly different zones on the rocks. These observations set him wondering about the possible causes for the separation. Were the two types separated because of some characteristic of their environment, such as type of rock or amount of moisture, or was it due to some biological interaction, such as competition? To answer these questions, Connell would have to form hypotheses and test them experimentally. In another case researchers found that night-feeding fishes, such as moray eels, ocean catfish, and some shark species, are unable to locate their food by sight. These observations led the researchers to ask how these fishes were finding their prey. Again, finding the answer would involve forming a hypothesis and testing it experimentally.

## Step 2: Using Inductive Reasoning to Form a Hypothesis

After noticing a recurring pattern or a relationship between observations, a researcher uses inductive reasoning to make an educated guess as to the probable answer to the question asked. This process is called formulating a hypothesis. On the basis of his observations of the barnacles, Connell hypothesized that biological interactions between species were responsible for the pattern he observed. In the example of the night-feeding fishes, researchers hypothesized that the fishes rely on their sense of smell to locate their prey. A hypothesis may be as simple as proposing a cause-and-effect relationship between observed events, or it may be more complex and propose a model for the way a particular process may work. There can be more than one method of formulating a hypothesis. Insights can arise from accident and intuition as well as from methodical observations.

## Step 3: Using Deductive Reasoning to Design Experiments

Regardless of how the hypothesis is stated, to be valid it must be testable. It is this step of testing the hypothesis that sets science apart from other disciplines and enables scientists to produce accurate explanations of natural phenomena. Once the hypothesis has been formed, it can be used to predict what the consequences might be if the hypothesis is correct. This process of reasoning from a general statement to predicting consequences involves the use of deductive reasoning (the if–then process). For instance, Connell hypothesized that the two types of barnacles were restricted to their particular zones by biological interaction. *If* that were true, *then* removing one of the types of barnacle would allow the other to move into its zone. Connell could then design experiments, such as scraping one type of barnacle off the rocks, to test his hypothesis. In the example of the night-feeding fishes, researchers hypothesized that the fishes used their sense of smell to find their prey. On the basis of this hypothesis, *if* the fishes could not smell their prey, *then* they would not be able to locate it. Again, researchers could then design experiments to test this hypothesis, such as plugging the fishes' nostrils.

Hypotheses are tested by performing experiments or sometimes, in fieldwork, by making systematic and detailed observations. For this step to be effective, scientists must design experiments that will disprove incorrect hypotheses. A well-designed experiment involves running two trials at the same time. In these two trials only one factor, the **experimental variable,** is altered. The trial that contains the experimental variable is called the **experimental set.** The trial without the experimental variable is known as the **control set.**

Experiments sometimes produce results that have occurred purely by chance and cannot be duplicated. Therefore, before accepting a hypothesis, a researcher should repeat the experiment several times to ensure that the results are valid. Another scientist working independently must be able to repeat the same experiments and get essentially the same results for a hypothesis to be valid and accepted. Alternative experiments should also be performed to ensure that other variables have not been overlooked. Only if many different approaches fail to disprove a hypothesis will it be considered accurate, and even then the acceptance of the hypothesis is subject to change if new information comes to light. Thus the aim of a scientist is not to prove something but to test it many times to achieve a more accurate explanation of the phenomenon being studied. Whenever possible, the data that are collected in an experiment are quantitative, or numerical. Numerical data are easier to compile and subject to statistical analysis.

## Steps 4 and 5: Gathering Results and Drawing Conclusions

After all of the results of the experiments are gathered, the researcher draws conclusions and either accepts or rejects the hypothesis. In the research with the night-feeding fishes, researchers found that the fishes could not find their prey with their nostrils plugged. They concluded that these fishes locate their prey with their sense of smell and accepted their hypothesis.

If the results of an experiment do not support the hypothesis, the investigator will then alter or modify the hypothesis in light of the new information and begin a new round of testing. This process may be repeated several times before a researcher arrives at a hypothesis that is consistently supported by experimentation. Connell found that removing one type of barnacle from the rocks did not always allow the other type to move into the newly vacated space. Through a series of several experiments, Connell was able to show that the zonation of barnacles on the rocks was due to a combination of physical and biological factors (for more information about Connell's work see the box in Chapter 13).

Over time, some hypotheses are expanded upon and eventually become generally accepted as theories. A scientific **theory** represents a body of observations and experimental support that have stood the test of time.

## Plant Growth in a Salt Marsh: A Case Study of the Scientific Method

Ivan Valiela is an ecologist who has spent many years studying salt marshes. While conducting studies on salt marshes in New England, he observed that marsh grasses that grew in areas receiving higher levels of nutrients seemed to grow taller than marsh grasses in areas where nutrient input appeared to be limited (Figure 1-11). Analysis of soil samples from different areas suggested that the levels of nitrogen, an essential nutrient in plant growth, were higher where the plants were taller. These observations led him to hypothesize that the growth of marsh grass was limited by the availability of nitrogen. On the basis of this hypothesis, he predicted that if nitrogen were added to the soil, then marsh grass would grow larger or faster or both.

To test his hypothesis, Valiela picked several plots of salt marsh that were as identical as possible in several respects. In each plot the types of plants, plant density, soil type, freshwater input, and height above the average tide level were as similar as possible. To his experimental plots, Valiela added nitrogen-containing fertilizer. The control plots received no nitrogen-containing fertilizer but otherwise were subjected to the same environmental conditions as the experimental plots.

All through the growing season, Valiela monitored the plots, measuring the growth rate of the plants, chemical composition of the leaves, and rate at which the leaves decayed when they were removed from the plant. He analyzed his data statistically and, among other things, found that several well-known species of marsh grasses grew taller and larger than the controls when they were supplemented with nitrogen-containing fertilizer. Further tests of the hypothesis continued to support these findings, thus leading to the conclusion that a major limiting factor in the growth of salt marsh plants is the availability of nitrogen.

## Alternative Methods of Science

Although the process of forming hypotheses and testing them with experimentation is central to scientific study, it is not the only approach that scientists use. There are times when scientists must depend on observation alone. For instance, certain behaviors of humpback whales may occur more frequently while mating, leading to the hypothesis that these behaviors are important for attracting a mate. In this situation, it would not be possible to conduct controlled experiments to test this hypothesis. In this case the weight of observations alone would support or deny the hypothesis. This approach to science is sometimes referred to as **observational science.**

## The Science of Marine Biology

As you read the text, watch for boxes titled "The Science of Marine Biology" They will introduce you to experiments biologists have performed to increase our knowledge of marine organisms and shape our understanding of the ocean environment. These boxes will help you better understand how the scientific method is applied to marine studies and how scientists go about formulating hypotheses and designing experiments to test their hypotheses.

## In Summary

Scientists use an orderly method of investigation known as the scientific method to learn about the natural world. The scientific method begins with observations of natural phenomena. On the basis of these observations, hypotheses are formed using inductive reasoning. Deductive reasoning is then used to design experiments to test the hypotheses. The conclusions drawn from the experimental results lead scientists to either accept or reject their hypotheses. ●

## SELECTED KEY TERMS

control set, *p. 8*

deductive reasoning, *p. 7*

experimental set, *p. 8*

experimental variable, *p. 8*

hypothesis, *p. 7*

inductive reasoning, *p. 7*

marine biology, *p. 2*

observational science, *p. 9*

oceanography, *p. 2*

ocean productivity, *p. 1*

plankton, *p. 5*

scientific method, *p. 8*

theory, *p. 9*

## QUESTIONS FOR REVIEW

### Multiple Choice

1. One of the first naturalists to systematically study marine organisms was

   a. Darwin
   b. Forbes
   c. Agassiz

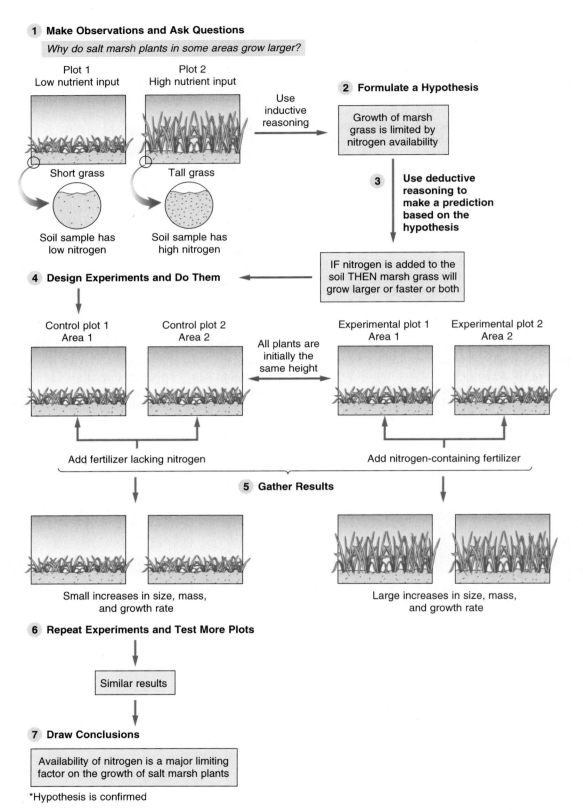

**1** **Make Observations and Ask Questions**

*Why do salt marsh plants in some areas grow larger?*

Plot 1
Low nutrient input

Plot 2
High nutrient input

Use inductive reasoning

**2** **Formulate a Hypothesis**

Growth of marsh grass is limited by nitrogen availability

Short grass

Tall grass

**3** **Use deductive reasoning to make a prediction based on the hypothesis**

Soil sample has low nitrogen

Soil sample has high nitrogen

IF nitrogen is added to the soil THEN marsh grass will grow larger or faster or both

**4** **Design Experiments and Do Them**

Control plot 1
Area 1

Control plot 2
Area 2

All plants are initially the same height

Experimental plot 1
Area 1

Experimental plot 2
Area 2

Add fertilizer lacking nitrogen

Add nitrogen-containing fertilizer

**5** **Gather Results**

Small increases in size, mass, and growth rate

Large increases in size, mass, and growth rate

**6** **Repeat Experiments and Test More Plots**

Similar results

**7** **Draw Conclusions**

Availability of nitrogen is a major limiting factor on the growth of salt marsh plants

*Hypothesis is confirmed

**Figure 1-11 The Scientific Method.** The scientific method is an approach that researchers use in doing science. Shown here are the steps of this method as they apply to Ivan Valiela's experiments on salt marsh plants.

d. Aristotle

e. Pliny

2. Publication of Darwin's book *On the Origin of Species by Means of Natural Selection* sparked an interest in the study of
   a. physical oceanography
   b. animal and plant adaptations
   c. plankton
   d. barnacles
   e. polar seas

3. Marine biology is the study of
   a. the physical characteristics of the ocean
   b. the organisms that inhabit the sea and their relationships to each other and their environment
   c. marine animals but not marine plants
   d. the organisms found in the open sea but not along the shoreline
   e. marine fishes and mammals only

4. The modern science of marine biology can trace its beginnings to
   a. the invention of steam-powered ships
   b. the founding of the first marine laboratory at Woods Hole
   c. the *Challenger* expedition
   d. the voyage of HMS *Beagle*
   e. the development of modern fishing techniques

5. The term *plankton* refers to
   a. all kinds of marine plants and algae
   b. microscopic animals only
   c. organisms that float or drift in the sea's currents
   d. animals that are active swimmers
   e. all of the marine organisms that can produce their own food

## Short Answer

1. What event proved that some organisms could live in the dark recesses of the ocean's depths?

2. What was Louis Agassiz's goal in founding the first marine laboratory in the United States?

3. Explain the significance of the *Challenger* expedition.

4. Describe how the focus of marine biology has changed from early times to the present day.

5. Describe the kinds of research performed at marine biology laboratories today.

6. Compare inductive and deductive reasoning.

## Thinking Critically

Apply the scientific process to a scenario of your own (real or imagined). Develop a hypothesis to explain something you have observed, and design an experiment to test your hypothesis. Try to relate the steps in your process to the generalized steps of the scientific method presented in the chapter.

## SUGGESTIONS FOR FURTHER READING

Edmunds, P. J. 1996. Ten Days under the Sea, *Scientific American* 275(4):88–94.

Hartwig, E. O. 1990. Trends in Ocean Science, *Oceanus* 33(4):96–100.

Orange, D. L. 1996. Mysteries of the Deep, *Earth* 5(6):42–45.

Valiela, I., and J. Teal. 1979. The Nitrogen Budget of a Salt Marsh Ecosystem, *Nature* 280:652–56.

### InfoTrac College Edition Articles

Dunn, R. J. 2001. William Francis Thompson (1888–1965) and the Dawn of Marine Fisheries Research in California, *Marine Fisheries Review* 63(2).

Heinonen, K. C. 1995. New Frontiers in Marine Career Opportunities, *Sea Frontiers* 41(4).

### Websites

**http://www.mbl.edu/** Marine Biological Laboratory website offers opportunities for research and education in biology, biomedicine, and ecology.

**http://sio.ucsd.edu/** Scripps Institution of Oceanography website provides current news, education, and resources concerning marine science research.

**http://depts.washington.edu/fhl/** Offers general information about University of Washington's Friday Harbor Laboratories.

**http://podaac-www.jpl.nasa.gov/edudoc.html** California Institute of Technology's Jet Propulsion Laboratory provides educational resources in oceanography and earth science.

**http://www.env.duke.edu/marinelab/** Provides general information about Duke University Marine Laboratory.

**http://www.nws.bnl.gov/marine.html** National Weather Service Marine Page provides educational information and marine forecasts for the New England area.

**http://www.oceanfund.org/gallery-1.html** Ocean Fund supports marine conservation organizations. The website provides grant information, current activities, and links to grant recipients.

# 2

# Fundamentals of Ecology

## Key Concepts

**1** Ecology is the study of relationships among organisms and the interactions of organisms with their environment.

**2** An organism's environment consists of biotic (biological interactions) and abiotic (physical characteristics of the environment) factors.

**3** An organism's habitat is where it lives, and its niche is the role the organism plays in its community.

**4** All organisms expend energy to maintain homeostasis.

**5** Physical factors of the environment, such as sunlight, temperature, salinity, exposure, and pressure, will dictate where organisms can live.

**6** Species interactions that influence the distribution of organisms in the marine environment include competition, predator–prey relationships, and symbiosis.

**7** Marine ecosystems consist of interacting communities and their physical environments.

**8** Most populations initially grow at an exponential rate, but as they approach the carrying capacity of the environment, the growth rate levels off.

**9** Energy in ecosystems flows from producers to and through consumers.

**10** The average amount of energy passed from one trophic level to the next is approximately 10%, and this ultimately regulates and limits the number and biomass of organisms at different trophic levels.

**11** With the exception of energy, everything that is required for life is recycled.

The term *ecology* is derived from the Greek word *oikos* meaning "home," in reference to nature's household and the economy of nature. The science of ecology deals with the interactions of organisms with each other and with their environment and how these interactions affect survival and reproduction. Frequently, the popular press uses the term *ecology* to refer to environmental science, or the study of the effects of human activity on the environment. In this text we will use the term *ecology* in its scientific sense.

## STUDY OF ECOLOGY

The organisms that inhabit the seas are integrated components of a living network that encompasses the globe. Just as cells are parts of living organisms, organisms are parts of **ecosystems,** systems composed of living organisms and their nonliving environment. All of the earth's ecosystems taken together compose the **biosphere.** The structure of the biosphere is determined by the basic principles of life: the capture of energy, the cycling of nutrients, survival and reproduction, and the process of evolution that has shaped the natural world. The world's ecosystems are all interconnected, and what happens to one ultimately affects the others. Each organism and each ecosystem on the planet play a crucial role in the function of the biosphere. Energy flows through the biosphere, and within the biosphere nutrients are cycled. The oceans play an important role in all of this as they produce food and recycle wastes.

## Environment

An organism's environment consists of all the external factors acting on that organism. These factors can be physical (abiotic factors) or biological (biotic factors). The physical environment consists of the nonliving aspects of an organism's surroundings. For marine organisms, the physical en-

vironment includes temperature, salinity, pH, the amount of sunlight, ocean currents, wave action, and the type and size of sediment particles. The biological environment consists of living organisms and their interactions with each other. For example, some feed on plants, some feed on animals, and others are parasites. Although we speak of the **biotic** and **abiotic environments** as two separate aspects of an organism's surroundings, in reality the two are difficult to separate.

## Habitat: Where an Organism Lives

The specific place in the environment where an organism is found is called its **habitat.** Marine habitats are characterized primarily by their abiotic features, the physical and chemical characteristics of the environment. Some examples of marine habitats are rocky shores, sandy shores, mangrove swamps, coral reefs, and deep sea vents. Each of these habitats is characterized by its own set of physical and chemical characteristics, and these characteristics dictate what types of organisms can live in that habitat. Each habitat can be divided into smaller subdivisions called **microhabitats.** For instance, the sandy shore habitat contains several different microhabitats for microscopic organisms in the spaces between the sand granules. These habitats are characterized by the size of the sand particles, the amount of space between them, and the ability of these spaces to hold water during intervals between the tides. As a general rule, the more complex the habitat, the more microhabitats it contains. Coral reefs contain literally thousands of microhabitats for the many organisms that live in this community (Figure 2-1).

## Niche: An Organism's Environmental Role

What an organism does in its environment (in a sense, its occupation) is its **niche.** For example, mussels stick to rocks

**Figure 2-1 The Coral Reef Habitat.** Large habitats, such as the coral reef, can contain many smaller microhabitats. Microhabitats in the coral reef include the crevices in the coral, the sediments surrounding the coral stands, and even the tissues of the organisms themselves.

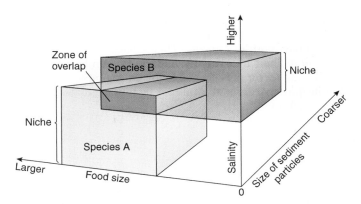

**Figure 2-2 A Niche.** An organism's niche is determined by a variety of abiotic and biotic factors acting together on the organism. This three-dimensional graph shows how several factors (food size, salinity, and size of sediment particles) interact to form niches for two species of burrowing worm. On the basis of this graph, we can see that species A prefers to burrow in substratum composed of smaller sediment particles where the salinity of the water is low and prefers to feed on medium- to large-sized food items. Species B, on the other hand, prefers coarser sediments where the salinity of the water is higher and prefers smaller food items. The zone of overlap indicates the combination of sediments, salinity, and food that would meet the requirements of both organisms.

and filter seawater for food, crabs scavenge, and some worms burrow into bottom sediments, extracting organic material as they do so. A full description of an organism's niche would include the range of environmental and biological factors that affect its ability to survive and reproduce. Because a niche is so complex and involves so many different factors, it is not possible to show a picture of a niche as you could a habitat, but it is possible to examine different aspects of the niche separately to see how each of these factors affects an organism (Figure 2-2).

If we were to examine the organisms on a rocky shore, we would notice that their distribution from high tide line to low tide line is determined by factors such as available moisture, the length of time exposed to air during low tide, their ability to withstand the force of waves, and the characteristics of the rock itself. This would be an abiotic view of this particular niche.

Biological factors that describe a niche include predator–prey relationships, parasitism, competition for the same resources, and organisms that provide shelter for other organisms. If we return to the example of a rocky shore, we would find that some organisms such as blue mussels (*Mytilus edulis*) are distributed in their particular zone because of the zone's abiotic characteristics. Predators, such as sea stars, are found in an overlapping zone because of the abundance of prey, namely the mussels. The seaweed that grows on the rocks provides food and shelter for a variety of small crustaceans, and the distribution of snails is determined by the amount and distribution of seaweed.

An organism's behavior also plays an important role in defining its niche. Behavioral factors, such as when and where an organism feeds, how it mates, where it bears its young, and social behaviors, influence an organism's niche. For instance, two very similar species that require the same kind of food can coexist if one feeds at night and the other during the day, provided that the amount of available food will support both organisms. This is because each species' niche is defined by so many factors. Remember, however, that no one factor is responsible for an organism's distribution in the environment. Multiple factors regulate organismal distribution.

## In Summary

An organism's environment consists of all the external factors acting on that organism. The specific place in the environment where an organism is found is called its habitat. Marine habitats are characterized primarily by their physical features. The organism's role in the environment, in a sense its profession, is its niche. ●

# ENVIRONMENTAL FACTORS AFFECT THE DISTRIBUTION OF MARINE ORGANISMS

Did you ever wonder why some rocky shores are covered with seaweeds whereas others are not, or why butterflyfishes are found on coral reefs but not in the shallow water off rocky coasts? As mentioned previously, the distribution of organisms in the marine environment is determined by a variety of factors, both abiotic and biotic. Living organisms must expend energy to survive and reproduce. The demands of their environment determine how much energy is necessary to survive and whether there will be enough energy to reproduce. This in turn determines the distribution of organisms in the marine environment.

## Maintaining Homeostasis

One of the greatest challenges faced by living organisms, whether they are microscopic and consist of a single cell or are larger and multicellular, is to maintain a stable internal environment. Factors such as temperature, the levels of waste products, and the amount of water, salts, and nutrients all have to be maintained within narrow limits for an organism to survive. These factors can be constantly changing in the external environment, but the internal environment of living organisms must be maintained within the boundaries required for that organism's survival. When any one of these factors changes in the external environment, the organism must make the proper adjustments internally to reestablish a balanced state. This internal balancing of factors that occurs in the face of changes in the external environment is called **homeostasis,** and the means of maintaining homeostasis is vital to the life of all organisms. Different types of organisms can tolerate different levels of fluctuation in environmental factors, and some organisms may be able to maintain homeostasis with respect to some factors but not others. An organism's ability to maintain homeostasis limits the areas of the world in which it can survive and reproduce.

## Homeostasis and the Distribution of Marine Organisms

If there is a range of environmental conditions and animals are able to move freely within the range, the animals will preferentially occupy those areas that offer the best set of conditions. If, however, animals are forced to occupy habitats that have a less than optimal range of environmental factors, they may fail to reproduce or, worse, die. The same is true of organisms that rely on sunlight to provide energy for food production (a process called photosynthesis). These organisms will thrive in environments with the proper amounts of sunlight and nutrients, such as nitrogen and phosphorus, but will fail to reproduce or will die if the environment is too deficient in these factors.

For every species there is an **optimal range** for each environmental factor that affects its life. As long as the factors remain within the optimal range, the organism should be able to thrive and reproduce. When one or more of the environmental factors is outside the optimal range, an organism's chances of survival are decreased (Figure 2-3). **Zones of stress** are regions above or below the optimal range of an environmental variable. An organism may be able to exist in a stress zone if the stress is not too great. If the stress is high, the organism may have to expend so much energy maintaining homeostasis that it will not have enough left to reproduce. Beyond the stress zones are zones where an environmental variable is so far from the optimal range that the organism cannot survive. These areas are called **zones of intolerance.** Consider, for example, the crabs in Figure 2-3. For a particular species, the optimal environmental temperature might be 20° to 30°C, and this is where most of the crabs are found. The crabs may be able to tolerate temperatures slightly warmer (30° to 35°C) or cooler (15° to 20°C), but in these stress zones the amount of energy that the crab has to expend to cool down or warm up is so great that very little remains for reproduction. Consequently, these animals have fewer offspring and the numbers of crabs in these areas are fewer. Temperatures that are even higher (above 35°C) or lower (below 15°C) constitute zones of intolerance. At these temperatures the crabs would have to expend so much energy trying to maintain a reasonable internal temperature (maintain homeostasis) that there would not be enough energy left over for other life processes, much less reproduction. Thus the crabs could not exist in water with these temperatures. The same principle applies to all of the other environmental factors that affect an organism.

## Physical Environment

A marine organism's physical environment includes such factors as sunlight, temperature, salinity, pressure, nutrients, and wastes. The ability of an organism to tolerate changes in these and other environmental factors plays a major role in determining that organism's distribution in the marine environment. If an organism must expend too much energy to maintain homeostasis when any of these environmental factors changes, it will not be able to survive in that area of the sea.

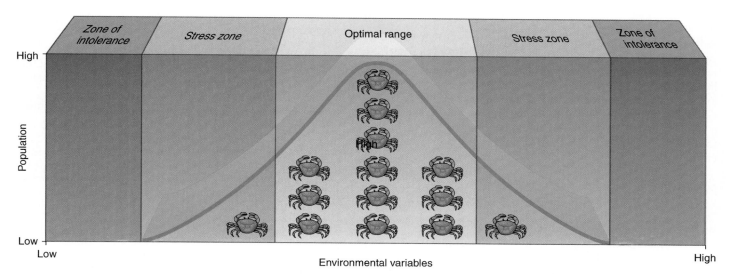

**Figure 2-3  Optimal Ranges.** An organism survives and reproduces best when environmental factors affecting it fall with an optimal range. Although organisms can live outside of their optimal ranges, they expend more energy maintaining homeostasis, leaving less energy available for reproduction.

## Sunlight

Sunlight plays an essential role in the marine environment. It powers the process of photosynthesis that, directly or indirectly, provides energy to nearly all forms of life on earth (this process is described in detail on page 23). The largest group of photosynthetic organisms in marine environments are **phytoplankton,** the tiny plantlike organisms and bacteria that float in ocean currents. Phytoplankton, together with seaweeds and plants, are the primary sources of nutrients and energy for marine animals. The distribution of these leading food producers is determined by the available sunlight and nutrients. In cloudy coastal waters, as in those of some North Atlantic bays and estuaries, phytoplankton can survive only in the shallowest areas because sunlight penetrates to a depth of less than 1 meter (3.3 feet). In the very clear water of the South Pacific, there may be enough sunlight for photosynthesis at depths of 200 meters (660 feet).

Sunlight is also necessary for vision. Many animals rely on their vision to capture prey, avoid predation, and communicate with each other. The distribution of these animals is affected by the depth that the sunlight can penetrate and allow for accurate vision. Fishes and other animals that live in areas of the ocean that do not receive enough light for good visibility must rely on other senses, such as taste and smell, to find food, avoid predation, and find mates.

Excessive sunlight can be a problem as well. Organisms that live in the harsh environments along shorelines that are periodically covered with water then exposed by the changing tides are subject to the intense heat of the sun and **desiccation** (drying out) that results from overexposure to sunlight (solar energy). Many algae suffer pigment destruction when exposed to intense sunlight, limiting their ability to photosynthesize.

## Temperature

The majority of marine animals are **ectotherms,** which means they obtain most of their body heat from their surroundings. Ectotherms become sluggish when the temperature drops and more active when temperatures rise. Marine mammals and birds, on the other hand, are endotherms. An **endotherm** can maintain a constant body temperature because its **metabolism** (the chemical reactions within its cells) can generate sufficient heat internally. Because these animals maintain temperatures that are usually higher than those of their surroundings, they have to be very well insulated so that they do not lose too much heat to the water that surrounds them (for more information on marine endotherms see Chapter 12).

Temperature often influences the distribution of organisms in shallow water and in the **intertidal zone,** the region that is covered at high tide and exposed at low tide. Shallow water like that in tide pools can experience large temperature changes in a short time because of hot summer sun or freezing nights. Organisms that survive in this habitat must be able to adapt quickly to a wide range of temperatures. The same is true of organisms that live in intertidal zones. The temperatures in these environments change quickly and substantially, and only the hardiest organisms are able to survive in this type of habitat. Large bodies of water like oceans do not change temperature rapidly (for more information on the role of oceans in moderating temperature see Chapter 4), and consequently, organisms that live in the open ocean away from the shore experience relatively constant temperatures at a given depth.

Most organisms can tolerate only a specific range of environmental temperatures. Temperatures above or below this critical range disrupt metabolism, resulting in decreased ability to reproduce, injury, or even death.

## Salinity

**Salinity** is a measure of the concentration of dissolved inorganic salts in the water. Substances that are dissolved in water are generally referred to as **solutes**. The membranes of living cells are almost always permeable to water, but they are not always permeable to the substances that are dissolved or suspended in water. All organisms must maintain a proper balance of water and solutes in their bodies (maintain homeostasis) to keep their cells alive. When a solute cannot move across the cell membrane to reach a balanced state on both sides (equilibrium), water moves instead to achieve the balance. The movement of water across a membrane in response to differences in solute concentration is called **osmosis** (Figure 2-4). The process of osmosis is vital to the life of a cell. If a cell loses too much water, it will become dehydrated and die. On the other hand, if a cell takes in too much water, it will swell and, in the case of animal cells, possibly burst. Living organisms invest both time and energy in maintaining the proper amount of salts and water in their cells and the fluid that surrounds them. This is a particularly important challenge for many marine organisms, because they live in water that has a high concentration of solutes. Some marine animals, such as the spider crab (*Macrocheira*) and other inhabitants of the open ocean, cannot regulate the salt concentration of their body fluids. Because their body surfaces are permeable to both salts and water, the concentration of solutes in their body fluids rises and falls with changes in the concentration of solutes in seawater. This is not a problem for these animals because the open ocean is a very stable environment, and they are seldom exposed to fluctuations in salt and water content. They do, however, have very little ability to withstand changes in salinity, and if they are moved to a less-stable environment, they will die.

Along coastal areas, the concentration of salts in seawater can vary greatly, especially in bays, estuaries, and tide pools. In these areas, pockets of water can lose moisture to evaporation, thus concentrating the salt content, or gain moisture through rain and freshwater runoff, thus diluting the normal salt concentration of the seawater. Animals that thrive in this type of environment, such as the fiddler crab (*Uca pugnax*), must be able to adjust the salt content of their body tissues by regulating salt and water retention. As we will see in the following chapters, salinity is critical in determining the distribution and types of organisms in many marine habitats.

## Pressure

The pressure at sea level is 760 mm Hg, or 1 atmosphere (14.7 lb per square inch). Because water is so much denser than air, for every 10 meters (33 feet) below sea level in the ocean, the pressure increases by 1 atmosphere. For instance, the pressure at an average ocean depth of 3,700 meters is 370 atmospheres (2.7 tons per square inch). A wig head lowered to a depth of 4,000 meters, where the pressure is 400 atmospheres, is compressed to approximately one third of its original size (Figure 2-5). As you might imagine, very few surface-dwelling organisms are able to survive at these depths (although pressure may not be the only factor limiting the number of organisms at great depths). The pressure of the water not only affects organisms that inhabit the deep regions of the seas but it poses problems for animals that sometimes frequent these depths, such as whales and some species of deep-sea fishes. These animals possess

**Figure 2-4 Osmosis.** Water tends to move from areas of lower solute concentrations to areas of higher solute concentrations. (a) An isotonic solution contains the same concentration of solute molecules (green) and water molecules (blue) as a cell. Cells placed in isotonic solutions do not change since there is no net movement of water. (b) A hypertonic solution contains a higher concentration of solute than a cell. A cell placed in a hypertonic solution will shrink as water moves out of the cell to the surrounding solution by osmosis. (c) A hypotonic solution contains a lower solute concentration than a cell. A cell placed in a hypotonic solution will swell and possibly rupture as water moves by osmosis from the environment into the cell.

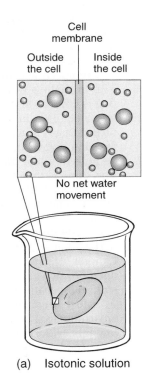

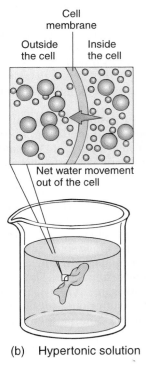

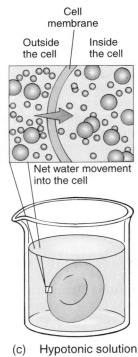

(a)   Isotonic solution          (b)   Hypertonic solution          (c)   Hypotonic solution

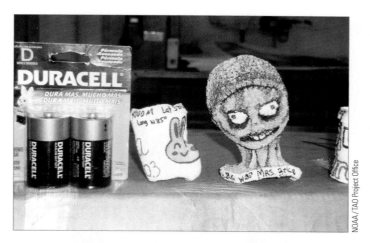

NOAA/TAO Project Office

**Figure 2-5 Pressure.** To demonstrate the pressure in the sea's depths, this wig head, normally the size of a human head, was lowered to a depth of 4,000 meters (13,200 feet). The pressure at this depth is so great that it compressed the Styrofoam to the size you see in this photograph.

specialized adaptations that allow them to survive at great depths (see Chapters 10, 12, and 18).

## Metabolic Requirements

The availability of nutrients strongly influences the distribution of organisms in the marine environment. The term **nutrient** refers not just to food but also to all of the organic and inorganic materials that an organism needs to metabolize, grow, and reproduce. The chemical composition of seawater supplies many mineral nutrients, such as nitrogen and phosphorus, required by phytoplankton, seaweeds, and plants, which in turn supply other marine organisms with nutrients (for more information on the composition of seawater see Chapter 4). For instance, the mineral calcium, essential for the synthesis of mollusc shells, coral skeletons, and the exoskeleton of crustaceans, is readily available in seawater. The availability of a key nutrient may play a significant role in the distribution of organisms in the marine environment. For instance, marine plants may have sufficient amounts of sunlight, but if they do not have enough nitrogen in their environment, they will not be able to thrive. Nutrients that limit the number or distribution of marine organisms are referred to as **limiting nutrients.**

A particularly important requirement for metabolism is oxygen. Oxygen is produced as a by-product of the photosynthesis performed by phytoplankton, seaweeds, and plants. Life evolved in an environment that lacked free oxygen, and when free oxygen entered the environment, it was probably the first harmful chemical, or pollutant, released into the earth's atmosphere. Although toxic to most early life forms, the presence of free oxygen revolutionized life on earth by producing an environment that would allow the evolution of multicellular organisms.

Oxygen in seawater dissolves at or near the surface, coming from photosynthesis or the higher concentration in the atmosphere. The ability of water to dissolve oxygen depends on its temperature and salinity. Cooler, less-salty water of the open sea contains more oxygen than the warm, saline water in a tide pool. Not all organisms require oxygen for life. Some bacteria are **anaerobic organisms** (anaerobes), meaning that they can survive and even thrive in the absence of oxygen. These organisms are often found in areas where the amount of oxygen is very limited, such as the ocean's depths, salt marshes, sand and mud flats, and between sediment particles, and they are responsible for the familiar odor of rotten eggs associated with many of these areas. **Aerobic organisms** (aerobes), such as plants, algae, animals, and the majority of marine microbes, require oxygen for their survival and are limited to regions of the ocean that contain sufficient quantities of oxygen.

Although nutrients are necessary for life, too much nutrient material in seawater can be a problem. Water runoff from land can greatly increase the nutrient levels in some coastal waters. This process of nutrient enrichment is known as **eutrophication.** Eutrophication leads to population explosions of certain types of photosynthetic plankton, an event referred to as an **algal bloom.** When these excess plankton die, they are decomposed by bacteria. The process of decomposition robs the water of oxygen. As the oxygen is depleted, the water can no longer support other forms of life. When these organisms die, their remains are decomposed, robbing the water of more oxygen. A vicious cycle begins that leads to massive die-offs of marine organisms.

## Metabolic Wastes

All organisms produce waste products when they metabolize. Most living organisms release carbon dioxide as a product of respiration. Animals excrete nitrogen-rich waste products, and plants release oxygen when they photosynthesize. Most of the time, waste products of metabolism are either removed from the environment or broken down and recycled by a variety of organisms, especially bacteria. In some environments, waste products can accumulate to toxic levels and prohibit the growth of all but the hardiest organisms. Certain small tide pools and coastal marsh areas are especially susceptible to the problems of accumulating metabolic wastes because the exchange of waste-laden water with new, uncontaminated water is limited.

## Biological Environment

An organism's biological environment is composed of all of the different species with which the organism comes in contact and the ways that the organism interacts with those species. Some of the more notable species interactions are competition, predator–prey relationships, and symbioses.

## Competition

When organisms require the same limited resource, such as food, living space, or mates, competition occurs. Competition may occur between similar species (**interspecific competition**) or between members of a single species (**intraspecific competition**), and it prevents two groups of organisms from occupying the same niche. In other words, usually no two groups of organisms can use exactly the same resources in exactly the same place at exactly the same time. If this were to occur, chances are good that one group will be more efficient at what it does than the other and will have better success at survival and reproduction. Possible results of competition are local extinction of the less-successful competitor (a process that ecologists call **competitive exclusion**), displacement of the less-successful competitor to another area where there is less or no competition, or selection for specializations that would lessen the competition.

Those individuals that are successful at using resources not in demand by other individuals experience less competition and usually produce more offspring. For example, several species of angelfish (*Holocanthus*) feed almost exclusively on sponges, a food source utilized by few other species. Many times, two groups of organisms competing for the same resource have developed anatomical and behavioral specializations that enable them to use that resource more efficiently. If a variety of similar, yet distinct, resources are available, a single niche can be subdivided into two or more smaller niches that have only a minimal amount of overlap. This process that allows organisms to share a resource is called **resource partitioning**. For instance, plankton feeders on coral reefs divide their food species in a number of different ways. Some plankton-feeding fishes, such as sea bass (*Anthias*), feed close to the reef, whereas damselfish (*Chromis*) go on short forays away from the reef to feed on plankton. The sea bass and damselfishes swarm after plankton during the day, whereas soldierfish (Holocentridae) hunt for plankton at night.

Many similar species can occupy similar small niches if the anatomy, feeding behavior, and preferred territory of each species are just a little different from another's. Many butterflyfishes (Chaetodontidae) have elongated jaws that allow them to reach deep into cracks and crevices for small invertebrates. Some species, however, have a blunt snout and must feed on the surface of coral heads, where they suck food-laden mucus from the coral animals (Figure 2-6). These different characteristics allow several similar species to thrive in the same coral reef community.

## Predator–Prey Relationships

Another important interaction between species is that between predators and their prey. The number of **herbivores** (plant-eating animals) that a given area can support is relative to the amount of vegetation that is available for food. The number of herbivores, in turn, supports a certain number of predators that use the herbivores as food. If the number of herbivores in an area increases so that there is not enough food for all of them, some will starve and the number of individuals will decline. As the number of individuals declines, the vegetation may be able to grow back, and as a result of more food, the population of herbivores may increase again.

A similar situation can exist between **carnivores** (meat-eating animals) and their prey. If the number of carnivores increases too quickly, there will not be enough prey to support them, and the carnivores will starve and begin to die off. As the carnivore population declines, the number of prey species may start to increase again as the result of less predation.

In some habitats, the presence and activities of a particular organism prevent one or a few aggressively coloniz-

Hal Beral / Visuals Unlimited

**Figure 2-6 Competition.** Competition among butterfly fishes is limited by the shape of their mouths, which dictates where they can find food and the types of food they can eat. This saddled butterfly fish (*Chaetodon ulietensis*) has a blunt mouth that restricts it to feeding on the surface of corals.

ing species from multiplying and crowding out others and thus dominating a community. For example, in rocky intertidal communities along the northern Pacific coast of the United States, the ochre sea star (*Pisaster ochraceus*) is a dominant predator. The ochre sea star feeds on a variety of prey but seems to prefer mussels (*Mytilus*). In experiments performed by Robert Paine in the early 1970s, the ochre sea stars were removed from some rocky areas, and those areas were kept relatively free of sea stars for 5 years. In the absence of the sea stars, mussels quickly colonized more of the rocky habitat, crowding out many other species in the process. The predation of mussels by sea stars keeps these highly competitive animals in check, allowing many other species, such as sea anemones, chitons, snails, and seaweeds, to survive in this habitat. Animals, such as the ochre sea star, whose presence in a community makes it possible for many other species to live there are called keystone predators or keystone species (Figure 2-7). By definition, the effect of a **keystone predator** on the biological diversity of an area is disproportionate to its abundance, because only a few predators have a large effect on the community's diversity.

If something happens to upset the natural balance of predator–prey relationships, a boom-or-bust cycle frequently occurs. For instance, in the early part of the nineteenth century, sea otters along the Pacific coast (a keystone predator in this case) were hunted nearly to extinction for their furs. Sea otters are predators, and one of their favorite foods is the sea urchin, an animal that feeds on a variety of

Adam Jones / Dembinsky Photo Associates

**Figure 2-7  Keystone Predator.** This ochre sea star limits the size of the mussel population in this community. This prevents the mussels from crowding out other species of rock dwellers.

seaweeds. As the number of sea otters decreased, the number of sea urchins increased, and they began to overgraze on the forests of kelp, a giant seaweed (brown alga) that dominates their habitat. In localized areas, the kelp was almost eradicated. Because many fish species relied on the kelp for cover, the loss of kelp led to a reduction in the local fish populations, which in turn depressed local populations of eagles. Fortunately, sea otters were eventually protected by the International Marine Mammal Protection Act, and the sea urchin population was thus again brought under control. Populations of sea urchins still thrive, however, in sewage-polluted areas where sea otters cannot survive and in some coastal areas of the North Pacific where killer whales feed on sea otters. Human intervention can also disrupt the delicate interactions between predators and their prey, as in southern California, where commercial harvest of sea urchins for food threatens the balance of this community once again by reducing the size of the sea urchin population on which sea otters depend.

## Symbiosis: Living Together

Some organisms have developed very close relationships with each other, to the extent that one frequently depends on the other to survive. Any change in environmental factors that affects one partner will invariably affect the other as well. This arrangement is called **symbiosis,** a term that means "living together." Types of symbiotic relationships are distinguished by the nature of the relationship between the two organisms. There are three main types: mutualism, commensalism, and parasitism. Although in theory each of these symbiotic relationships is clearly defined, determining where one ends and the other begins can be difficult. In nature the distinctions between commensalism and mutualism, and commensalism and parasitism are frequently vague and open to interpretation.

***Mutualism***  In **mutualism,** both organisms benefit from the relationship. Sometimes, as in the case of coral animals and their zooxanthellae (a type of single-celled, photosyn-

thetic organism), the two organisms are so dependent that they appear and function as a single organism. The coral animal provides the zooxanthellae with nitrogen, phosphate, and carbon dioxide—nutrients needed for photosynthesis. In return, the zooxanthellae provide the polyp with food in the form of carbohydrates. Also, the removal of carbon dioxide from the polyp by the photosynthesizing zooxanthellae makes it easier for the polyp to form its stony skeleton (for more information regarding coral and zooxanthellae see Chapter 15).

This type of mutualistic relationship is at one extreme of the spectrum. Not all mutualistic relationships are equally beneficial to both organisms involved. In some mutual relationships, two organisms can be separated but at the risk that one or both may not thrive or may die, as in the case of Pacific clownfishes (*Amphiprion*) and their mutualistic anemones (Figure 2-8a). The body of the clownfish is coated with a special mucus that protects it from the anemone's toxic stings. The fish acclimates the anemone to its presence by rubbing against it so that the fish's mucous covering picks up the anemone's scent. Thus the anemone doesn't recognize the fish as food or foe. The clownfish gains protection by living within the stinging tentacles of the anemone. In return, the clownfish defends the anemone from other fishes that might eat the anemone's tentacles, rendering it defenseless and unable to feed. In this case, both organisms benefit from the relationship, but the benefit is greater for the clownfish. Although the anemone can often survive without the clownfish, the clownfish cannot survive very long without its symbiotic partner.

***Commensalism***  In **commensalism,** one organism benefits from the relationship, whereas the other partner is neither harmed nor benefited. An example of commensalism is the relationship between the remora fish (*Echeneis*) and some species of sharks and rays (Figure 2-8b). Remoras have a flattened suction cup on top of their heads that allows them to attach to the shark's body. In addition to getting a "free ride" on the shark, they eat the scraps of food left over when a shark feeds. Remoras are also less likely to be attacked by predators because they are attached to the shark. The shark, on the other hand, does not appear to benefit from the relationship, but is not harmed either, although a certain amount of hydrodynamic drag presumably results from the attached remoras.

***Parasitism***  In **parasitism,** one organism, the **parasite,** lives off another organism, the **host.** The parasite benefits from the relationship, whereas the host is harmed. Many fishes and marine mammals are infected by parasitic worms, such as tapeworms (Figure 2-8c). Tapeworms live in the intestines of their hosts and derive their nourishment from them (for more information regarding tapeworms see Chapter 8). As a result, the host is weakened and becomes more vulnerable to disease and predation.

## In Summary

Many environmental factors influence the distribution of marine organisms. Abiotic factors include sunlight, temperature, salinity,

# Amphipod and Sea Butterfly

Not all symbiotic relationships fit neatly into one of the three categories of mutualism, commensalism, or parasitism. For instance, marine biologists working in the ocean around Antarctica observed sea butterflies (*Clione antarctica*) being carried around on the backs of amphipods (*Hyperiella dilatata;* Figure 2-A). The observations started them wondering about the relationship and raised several questions, such as which organism was doing the carrying and how does the association form? Both field and laboratory observations clearly confirmed that the amphipods were doing the swimming while the sea butterflies were attached to their backs. When solitary amphipods and sea butterflies were placed in a laboratory aquarium, the amphipods were observed to swim quickly to a sea butterfly and grab it with their anterior appendages. They then used their sixth and seventh appendages to position the sea butterfly on their backs and hold it there.

**Figure 2-A** *Amphipod and Sea Butterfly. The amphipod and sea butterfly exhibit an interesting symbiotic relationship in which the sea butterfly is kidnapped by the amphipod and provides protection against fish predators.*

James McClintock, Bill Baker, American Scientist, Vol. 86, No. 3, P. 254, DOI: 10.1511/1998.3.254, used with permission.

From the standpoint of traditional symbiosis, the relationship did not make good sense in that both partners seemed to be negatively affected. Studies revealed that amphipods with sea butterflies attached could swim only half as fast as they could without their passenger. Presumably this would make them easier prey for their predators and make it more difficult for them to capture their prey. Other studies suggested that while attached to the amphipod the sea butterflies were unable to feed. The relationship appeared to be a disadvantage for both partners, but if that were true, why did it form? This question prompted researchers to ask if one or the other of these organisms derived a benefit from the relationship. Because the amphipods purposely "kidnapped" the sea butterflies, it was hypothesized that advantages for amphipods outweighed the disadvantages.

Analysis of stomach contents from several predatory fishes revealed that the plankton-feeding fish *Pagothenia,* as well as at least five other species, were major predators of the amphipods. This observation led researchers to wonder if the presence of the sea butterfly somehow deterred predators. To answer this question, researchers offered captive *Pagothenia* codfish muscle to determine whether the fish were hungry. They then presented the fish with amphipods, and the amphipods were quickly eaten. Another group of hungry *Pagothenia* were given sea butterflies to eat. The fish began to feed on the sea butterflies, but immediately spit them out with a violent head shake. The same response was observed when hungry *Pagothenia* were offered amphipods carrying sea butterflies.

Having established that fish rejected both sea butterflies and amphipods with attached sea butterflies, researchers wanted to know if the fishes' response was due to a chemical cue or a visual one. To find out, researchers homogenized some sea butterflies and added the extract to fish meal to form food pellets. Control fish were fed food pellets without the homogenate, whereas the experimental group was fed fish pellets with the homogenate. The control *Pagothenia* readily ate their pellets without the homogenate, but the experimental fish rejected the pellets with the homogenate. The results of this experiment suggested that some chemical from the sea butterfly was responsible for the avoidance response.

To identify the specific chemical, researchers homogenized more than 4,000 sea butterflies and separated the extracted compounds into two groups: fat soluble and water soluble. Food pellets were made containing each of the two extracts and fed to hungry *Pagothenia*. The fish readily ate pellets made with the water-soluble extract but rejected the pellets containing the fat-soluble extract. This suggested that one or more chemicals in the fat-soluble extract were distasteful to the fish. The fat-soluble fraction was further processed, and five pure organic compounds were isolated. Each of these compounds was tested, and one was determined to be the compound of interest. Using several chemical techniques, the compound was identified as a previously unknown chemical that investigators named pteroenone. Apparently the sea butterflies manufacture this compound to deter predators.

The results of these experiments revealed that this may be a unique case of symbiosis in which one species "kidnaps" another as a way of protecting itself against predators. Biologists are tentatively referring to this as antagonistic symbiosis. ●

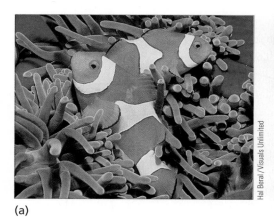
(a)

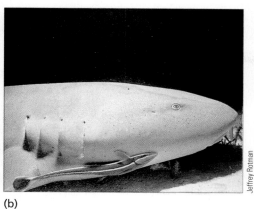
(b)

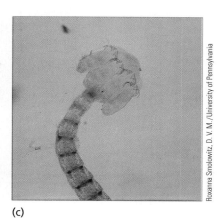
(c)

**Figure 2-8  Symbiotic Relationships.** (a) Mutualism: a clownfish and sea anemone. (b) Commensalism: a remora fish and shark. (c) Parasitism: a portion of a tapeworm taken from the intestine of a dogfish shark.

pressure, and the levels of nutrients and wastes. Biotic factors include competition, predator–prey relationships, and symbioses. ●

## POPULATIONS AND COMMUNITIES

To a biologist, a **population** is a group of the same species that occupies a specified area. Members of a population interact with each other and are able to breed with each other. In nature, populations are separated from one another by barriers that prevent organisms from interacting or breeding (for more information regarding isolating mechanisms see Chapter 5). The population, rather than the individual, is the basic unit that many ecologists study.

A biological **community** is composed of populations of different species that occupy one habitat at the same time. The species that make up a community are linked to some degree by competitive relationships, predator–prey relationships, and symbiosis. For instance, we can talk about the populations of barnacles, mussels, seaweeds, sea stars, and snails that inhabit a rocky shore on the Pacific coast. This assemblage of populations would make up a rocky shore community (see Figure 2-7). Communities can be large or small, depending on the area that is being discussed.

### Population Growth

There are many different ways by which a population can increase in size. The number of new individuals added through reproduction can increase while the death rate remains constant, the rate at which individuals die can decrease while the birth rate remains constant, or the birth rate can rise while the death rate declines. Populations can grow as the result of immigration of new individuals from other populations or decline because of emigration of members to neighboring populations. The addition of new members to a population through reproduction or immigration is frequently termed **recruitment.** The majority of marine organisms living on rocks and bottom sediments have

planktonic larval stages. The size of these populations can be affected by the rate at which the larvae leave the water column and settle to the bottom, a process called **larval settlement.**

When populations of organisms have sufficient food or nutrients and are not affected greatly by predators or disease, they will grow rapidly. For example, phytoplankton will undergo large population increases called **blooms** when there is an increase in available nutrients. In a short time the phytoplankton reproduce so quickly that they exploit all of the available nutrients or space and then begin to die back. If we were to graph the growth of the phytoplankton population as a function of time, it would be a J-shaped curve (Figure 2-9a). Initial growth would be slow, represented by the bottom of the J, but the doubling of the population again and again by cell division, represented by the nearly vertical top of the J, would produce ever-bigger numbers. This pattern of growth is called **exponential,** or **logarithmic, growth** and is characteristic of rapidly growing populations. Although some populations can grow exponentially for short periods, none can sustain this type of growth for long. Eventually, some resource needed for growth or survival will become scarce and the growth rate will begin to decline.

A more typical growth pattern for a population is one in which the original growth is exponential, but then the rate of population increase reaches an equilibrium point and the population growth is zero. This type of population growth is called **logistic growth** (Figure 2-9b). A population reaches its equilibrium point when it contains as many organisms as its environment can sustain for an extended period. That point is called the carrying capacity of the environment. Growth rate slows as a population reaches its carrying capacity for any of three reasons: the birth rate drops, the death rate rises, or birth rates and death rates change at the same time.

The carrying capacity inhibits the tendency of populations to grow exponentially. Any of the physical or biological factors discussed earlier in the chapter can limit population size. Insufficient amounts of sunlight can limit the number

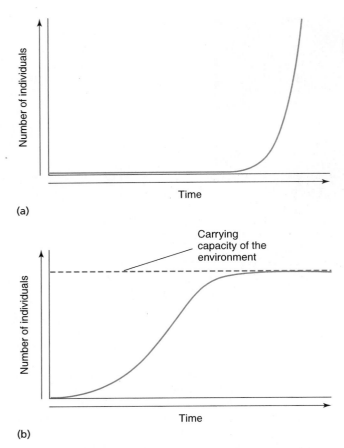

**(a)**

**(b)**

**Figure 2-9  Population Growth.** (a) Under proper conditions, some marine organism populations can grow exponentially. A graph of exponential growth gives a characteristic J-shaped curve. (b) As the growth of a population approaches its carrying capacity, the graph flattens out. As a result, the logistic growth curve shown here has somewhat of an S-shape.

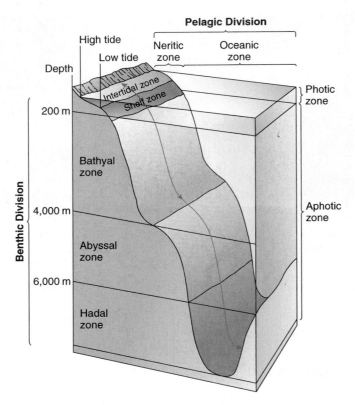

**Figure 2-10  Ocean Divisions and Zones.** Ecologists frequently divide the ocean into two major divisions: the pelagic division, consisting of the water column, and the benthic division, consisting of the sea bottom. The pelagic division can be subdivided based on the availability of sunlight (photic zone and aphotic zone) or distance from the shore (neritic zone and oceanic zone). The benthic division can be subdivided on the basis of depth (intertidal zone, shelf zone, bathyal zone, abyssal zone, and hadal zone).

of photosynthetic organisms, whereas insufficient numbers of prey can limit the number of predators. Although it may appear that the carrying capacity of a specific environment for a particular species is constant, it often is not. The physical and biological factors that determine the carrying capacity can vary, for instance, with time of year, and thus the carrying capacity will vary as well. The typical pattern shown in Figure 2-9b predicts the growth of laboratory populations and some populations in nature, but most natural populations exhibit many different growth patterns that are variations of the typical pattern.

## Distribution of Marine Communities

Marine communities can be designated by the regions of ocean that they occupy. Ecologists frequently divide the marine environment into two major divisions: the **pelagic division,** composed of the ocean's water (the **water column**), and the **benthic division,** the ocean bottom. These divisions can be subdivided into zones on the basis of three characteristics: distance from land, light availability, and depth (see Fig-

ure 2-10). We will examine the physical characteristics of these zones in more detail in later chapters.

### Pelagic Division

The pelagic division is subdivided into the **neritic zone** and the **oceanic zone.** The neritic zone is composed of the water that overlies the continental shelves. The larger oceanic zone consists of the water that covers the deep ocean basins. The pelagic division can also be divided into the **photic zone,** where sunlight is present to support photosynthesis; the **disphotic zone,** or **twilight zone,** where light is dim and photosynthesis cannot take place; and the **aphotic zone,** where sunlight is absent. The photic zone contains not only the greatest number of photosynthetic organisms but also the largest number of animals. As we saw earlier, in typical predator–prey relationships, large populations of photosynthesizers can support correspondingly large populations of herbivores and the predator species that feed on them. The organisms that inhabit the pelagic division exhibit one of two different lifestyles. **Plankton** drift with the currents, whereas **nekton** are active swimmers that can move against currents. The term **neuston** refers to small plankton that float at or near the surface of the ocean.

### Benthic Division

The benthic division begins at the shore with the intertidal zone. This region of ocean bottom is covered with water only during high tide. During low tide it is exposed to air. The **shelf zone** extends from the line of lowest tide to the edge of the continental shelf. From the edge of the continental shelf to a depth of 4,000 meters (13,200 feet) is the **bathyal zone.** The **abyssal zone** extends from 4,000 to 6,000 meters. At an ocean bottom depth greater than 6,000 meters (19,800 feet) lies the **hadal zone.**

Benthic organisms live primarily in or on the bottom sediments. Benthic organisms that live on the bottom are called **epifauna;** those that live in the bottom sediments are called **infauna.**

## In Summary

In ecological terms, a population is a group of the same species that occupies a specific area. Populations grow when more organisms are added through reproduction and immigration than are lost through death and emigration. Initially, populations grow quickly, a process known as exponential growth. Such growth cannot be maintained indefinitely. Characteristics of the environment, such as space and available food, limit the number of organisms an area can support. This limit is called the carrying capacity of the environment. Ecologists divide the marine environment into two major divisions: the pelagic division, composed of the ocean's water (the water column), and the benthic division, the ocean bottom. These divisions can be subdivided into zones on the basis of distance from land, light availability, and depth. •

## ECOSYSTEMS: BASIC UNITS OF THE BIOSPHERE

Communities in nature are rarely isolated from each other, resulting in interactions among different communities. As indicated earlier in the chapter, biological communities both influence and are influenced by their abiotic environment. Biological communities and the physical environment interact to produce a relatively stable system, the ecosystem. Some examples of marine ecosystems are estuaries, salt marshes, mangrove swamps, rocky shores, sandy beaches, kelp forests, coral reefs, and open ocean.

Estuaries occur where partially enclosed areas of the sea receive freshwater runoff to produce an area of mixed salinities. Because estuaries are such changeable environments, they are home to opportunistic species that can cope with environmental variation.

The intertidal zone is the area of shore defined by the high tide and low tide mark. In the northeastern United States and along much of the Pacific coast, this area is rocky shoreline. In the southeastern part of the country, sandy beaches predominate. These areas offer particular chal-

lenges to organisms in dealing with the extremes of temperature, moisture, and salinity that occur between the tides.

Kelps are cold-water brown algae that are found off the coasts of North America, Japan, Siberia, South America, Great Britain, Scandinavia, and the Atlantic coast of South Africa. These algae form an amazing undersea forest that provides food and shelter for many marine organisms.

Coral reefs are complex ecosystems formed by living organisms, including coral animals and coralline algae. They are found in subtropical and tropical waters and are home to thousands of species. The great diversity of life on coral reefs reflects the thousands of niches that are available.

As with communities, the different ecosystems are not independent, and there is a great deal of interaction among them. We will study the characteristics of these ecosystems in more detail in Part 3 of this book.

## Energy Flow through Ecosystems

All living organisms require energy to live, grow, and reproduce. The source of this energy for practically all life on earth is the sun. Organisms that are capable of photosynthesis convert the radiant energy of the sun into the chemical energy of food molecules. These molecules in turn serve as a source of nutrition for not only photosynthesizers but also the organisms that feed on them. As each organism in turn feeds on another, energy is funneled through levels of the ecosystem.

## Producers

Some organisms contain special pigment molecules like **chlorophyll** that capture the sun's energy. This energy is then stored in organic molecules that can serve as a source of energy for the organisms that produce them or as food for other organisms. The process by which the energy of sunlight is captured and stored in organic molecules is called **photosynthesis** (Figure 2-11). In photosynthesis light energy is used to combine relatively low-energy molecules of

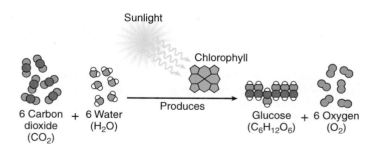

**Figure 2-11 Photosynthesis.** In the process of photosynthesis, carbon dioxide and water combine to form a sugar called glucose. Oxygen is a byproduct of the reaction. The energy for the process is supplied by sunlight. Special molecules, like the green pigment chlorophyll, absorb light energy and make it available to power the photosynthetic process. The glucose produced by photosynthesis can be used by the photosynthetic organism as food or to make other important molecules.

carbon dioxide and water to form high-energy carbohydrate molecules like the sugar glucose. In advanced photosynthetic organisms, such as plants, oxygen gas is also released in the process. In the marine environment, the primary photosynthetic organisms are phytoplankton, seaweeds, and plants. Because these organisms are able to produce their own food, as well as food for other organisms, they are called **autotrophs** (from *auto,* meaning "self," and *troph,* meaning "feed"), or **producers.** More than one half of the photosynthesis (measured in kilocalories per year) on earth occurs in the oceans.

In the marine environment, not all producers are photosynthetic. Some are **chemosynthetic,** using the energy from chemical reactions, rather than sunlight, to form organic molecules from carbon dioxide and other compounds. For example, bacteria that inhabit regions of the ocean floor where water heated by the earth's core seeps through (deep-sea vents) and no light is available produce their food by chemosynthesis (for more information on the process of chemosynthesis see Chapter 6).

## Measuring Primary Productivity

The amount of food being produced in an area of the ocean can be determined by measuring primary productivity. **Primary productivity** refers to the rate at which energy-rich food molecules (organic compounds) are being produced from inorganic materials. If you look at the equation for photosynthesis in Figure 2-11, you will notice that it would be possible to measure the amount of photosynthesis that is taking place by measuring either how much carbon dioxide and water are being used in the process or how much carbohydrate (glucose) and oxygen are being produced. This is the basis for methods used to measure primary production.

The traditional method employed by marine biologists is the **light–dark-bottle method.** In this procedure, two bottles that are identical in all respects except that one is clear (light can penetrate) and the other opaque (light cannot penetrate) are used. Equal volumes of ocean water taken from the same area and depth are placed in each bottle. The water sample in each bottle contains the plankton present in the water source. Another sample of seawater taken from the same area is analyzed separately to determine the water's initial concentration of oxygen. The light and dark bottles are stoppered and lowered on a line to the depth from which the water sample was taken. The sample in the clear bottle is able to receive sunlight and the producers photosynthesize, releasing oxygen, and they and the zooplankton respire, consuming oxygen. The sample in the opaque bottle does not receive any sunlight, so there is no photosynthesis, but the organisms respire, using the oxygen in the sample. After a time determined by the researcher, the samples are retrieved and the oxygen content of each bottle is measured. It is assumed that the same amount of respiration occurred in both bottles, so that value is subtracted from the oxygen content in the bottle in which photosynthesis occurred. The oxygen remaining is assumed to have come from photosynthesis and is thus an estimate of the primary production in the water sample. A problem with this method is that the technique for measuring oxygen is not sensitive enough to pick up changes in water samples containing small amounts of phytoplankton.

Another method for measuring primary production involves measuring the amount of carbon incorporated into the organic products of photosynthesis. A radioactive form of carbon, $^{14}C$, is used in this measurement. The $^{14}C$ is introduced into the water in the form of bicarbonate ion ($HCO_3^-$). After allowing the phytoplankton in the sample to photosynthesize for a known time, they are filtered out of the water and a radiation counter is used to measure the amount of radioactive carbon in the sample. The amount of radioactive carbon present in the sample represents carbon that has been incorporated into food molecules and is thus proportional to the amount (rate) of primary production. To correct for the possible uptake of $^{14}C$ by nonphotosynthetic organisms, a dark-bottle control may also be used. This method tends to give a low estimate of productivity, especially in water that is low in nutrients or has a low density of phytoplankton.

Other methods for determining primary production are being developed that use satellite images to estimate the amount of chlorophyll in the water, and then, on the basis of estimates of how much radiant energy is being absorbed by the chlorophyll, primary productivity can be estimated.

## Consumers

Organisms that rely on other organisms for food are collectively called **heterotrophs** (from *hetero,* meaning "other," and *troph,* meaning "feed"), or **consumers. First-order consumers** (also known as **primary consumers**) are those that feed directly on producers. These organisms are called herbivores. **Second-order consumers** (also known as **secondary consumers**) are carnivores that feed on herbivores. **Third-order consumers** (also known as **tertiary consumers**) are carnivores that feed on other carnivores, and so on. **Omnivores** are consumers that feed on both producers and other consumers. **Detritivores** are organisms that feed on **detritus,** organic matter such as animal wastes and bits of decaying tissue. **Decomposers** are organisms that break down the tissue of dead plants and animals and help to recycle nutrients. Detritivores and decomposers are also considered consumers because they cannot produce their own food and they rely on other organisms for their organic nutrients.

## Food Chains and Food Webs

In every ecosystem, producers and consumers are linked by feeding relationships called **food chains** (Figure 2-12). For instance, in tropical regions, sea urchins feed on seagrass and seaweeds and helmet snails feed on the urchins. Humans, fishes, and marine mammals, in turn, prey on the urchins and helmet snails. This is an example of a food chain. The concept of food chains, although helpful in establishing feeding relationships, is too simplified for what actually occurs in an ecosystem. If we were to study the dynamic feeding relationships in a particular community or ecosystem, we would notice that the relationships are interconnecting and much more complex than the simple food chain implies.

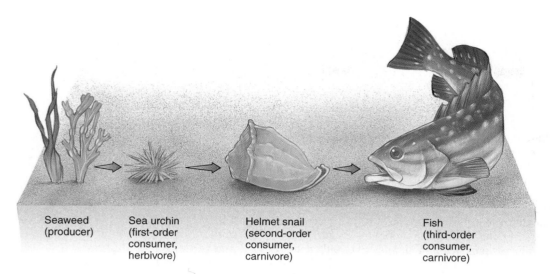

**Figure 2-12 A Food Chain.**
Food chains depict the feeding relationships among a group of organisms as a linear sequence from producers to higher level consumers.

Seaweed
(producer)

Sea urchin
(first-order
consumer,
herbivore)

Helmet snail
(second-order
consumer,
carnivore)

Fish
(third-order
consumer,
carnivore)

These complex feeding networks are referred to as **food webs.** Figure 2-13 shows an example of a simplified marine food web.

## Other Energy Pathways

Not all energy pathways in marine ecosystems involve one organism consuming another. For instance, it is difficult for one-celled organisms to trap small molecules within their cells, and some inevitably leak into the environment. In addition, phytoplankton may release some of their photosynthetic products into the surrounding seawater. These organic molecules that are lost or released into the water column are referred to as **dissolved organic matter (DOM).** These energy-rich organic molecules can be taken in by plankton and other small organisms and used as a nutrient source. These smaller organisms are eaten by larger animals, and thus the energy of the organic compounds is funneled through the oceanic food web.

Detritus represents an enormous supply of energy for marine organisms. The major sources of detritus are decaying plant and algal matter that is not consumed by grazing herbivores, animal wastes, and bits and pieces of animal tissue. All of this material contains energy trapped in the organic compounds that compose them. As the detritus rains down through the water column, it serves as an important food source for pelagic bacteria and some zooplankton. When detritus settles to the bottom, it is consumed by animals, such as worms or clams, that in turn can channel the energy to larger animals when they are consumed. Although the formation of detritus represents an initial loss of energy to pelagic organisms, the detritivores feeding on the detritus return the energy to food chains.

## Trophic Levels

Energy flows from the sun through producers to the various orders of consumer organisms. The energy that producers receive from photosynthesis or that consumers get from eating other organisms is temporarily stored until the organism is consumed by another or decomposed. Each of these energy storage levels is called a **trophic level,** with the producers making up the first trophic level. Theoretically, at least, there is no limit to the number of trophic levels, but in reality there are limits because only a fraction of the energy available on one level is passed along to the next when an animal feeds.

## Ecological Efficiency

Energy transfer from one trophic level to the next is not efficient. First-order producers capture and store less than 1% of the available solar energy. The percentage of energy that is taken in as food by one trophic level and passed on as food to the next higher trophic level is called **ecological efficiency.** The **ten percent rule** of ecology states that, on average, only approximately 10% of the energy available at one trophic level is passed along to the next, although in reality the amount can be quite variable. This small amount of transferred energy is what remains after energy is expended on finding, capturing, eating, and digesting food, as well as the heat lost during metabolism. The higher the trophic level, the lower the amount of available energy. This controls the number of trophic levels possible. Because energy is lost going from one trophic level to the next, most food chains are short (two or three links), although open-water food chains may be five or six links long because nutrients are so scarce and the primary producers are very small.

## Trophic Pyramids

The flow of energy from one trophic level to another can be pictured in the form of an **energy pyramid** (Figure 2-14a). Energy pyramids not only indicate that the amount of available energy decreases with each trophic level but also that each trophic level supports fewer organisms as a result. It is for this reason that herbivores are generally more numerous in a community than carnivores.

These patterns are only generalizations. The actual energy relationships in ecosystems are much more complicated than this. Although energy pyramids give the best

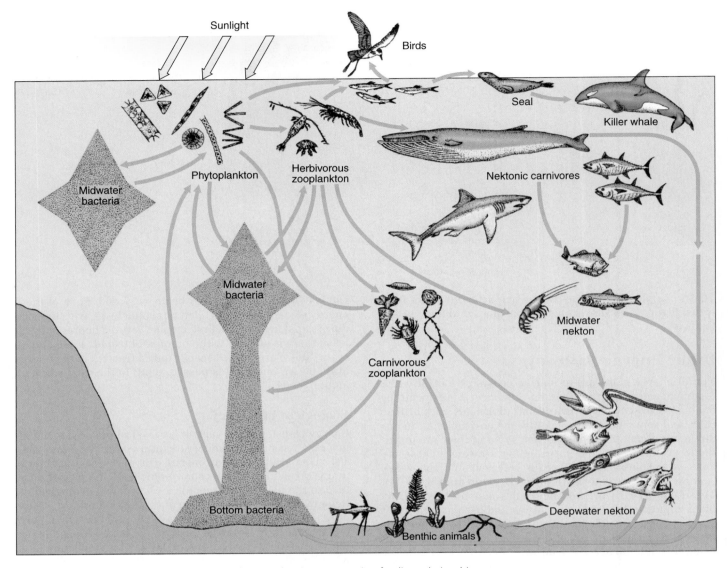

**Figure 2-13  A Food Web.** Food webs show the complex, interconnecting feeding relationships among members of a community or ecosystem.

overall picture of community structure, ecologists also use ecological **pyramids of biomass** (the amount of all living tissue in an area) and **pyramids of numbers** to indicate relationships among trophic levels (Figure 2-14b).

## In Summary

The energy for most life on earth comes from sunlight. Producers capture the energy of sunlight in the chemical bonds of organic molecules. The rate at which these energy-rich molecules are formed is called primary production. Consumer organisms rely on these molecules as a source of food, because they cannot synthesize their own. In every ecosystem, producers and consumers are linked by feeding relationships called food chains. In reality the interactions among most living organisms are more complex than simple food chains. These complex feeding networks are known as food webs. The energy that an organism receives from photosynthesis or from feeding on other organisms is temporarily stored until the organism is consumed by another or decomposed. These energy storage levels are called trophic levels. Energy transfer from one trophic level to the next is not efficient. The ten percent rule of ecology states that the average amount of energy passed from one trophic level to the next is approximately 10%. This energy flow from one trophic level to another can be pictured in the form of an energy pyramid.  ●

## BIOGEOCHEMICAL CYCLES

Unlike energy, the nutrients that are necessary for life are available in limited supply and are recycled. The cycling of these various nutrients involves biological, physical, and

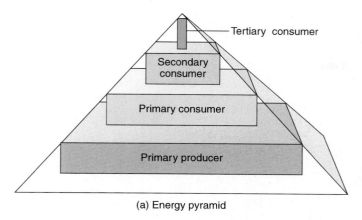

- Tertiary consumer
- Secondary consumer
- Primary consumer
- Primary producer

(a) Energy pyramid

**Figure 2-14 Ecological Efficiency.** (a) The flow of energy from one trophic level to another can be represented as an energy pyramid, emphasizing the decrease in available energy from one level to the next. This diagram also represents a pyramid of numbers. The size of each box indicates the relative number of organisms that can be supported by each level. (b) Typically a pyramid of biomass resembles a pyramid of numbers. A pyramid of biomass indicates the mass (kilograms) of living organisms that each level can support. For instance, 10,000 kilograms of phytoplankton are required to support 1,000 kilograms of herbivorous zooplankton and when channeled through several layers of consumers, would be enough to support 0.1 kilogram (100 grams) of tuna or about enough to make one tuna sandwich. (The numbers have been rounded to illustrate the general principle.)

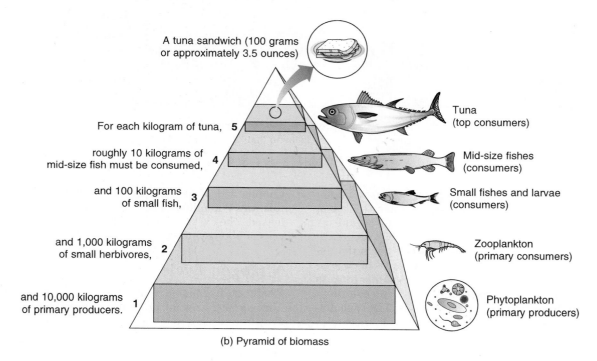

A tuna sandwich (100 grams or approximately 3.5 ounces)

For each kilogram of tuna, **5**    Tuna (top consumers)

roughly 10 kilograms of mid-size fish must be consumed, **4**    Mid-size fishes (consumers)

and 100 kilograms of small fish, **3**    Small fishes and larvae (consumers)

and 1,000 kilograms of small herbivores, **2**    Zooplankton (primary consumers)

and 10,000 kilograms of primary producers. **1**    Phytoplankton (primary producers)

(b) Pyramid of biomass

chemical processes. For this reason they are frequently referred to as **biogeochemical cycles.** A cycle exists for each of the nutrients needed for life, such as water, carbon, nitrogen, and phosphorus. In this section we will examine a few of the more important cycles.

## Hydrologic Cycle

The most abundant compound in all living organisms is water. Obviously, the cycling of water is of primary importance to all living things as well as to the marine environment (Figure 2-15). Because the regions of the earth around the equator receive the most heat energy and sunlight, this is the area of the ocean that supplies the greatest amount of water to the atmosphere via evaporation.

Water vapor is carried north and south from the equator and from west to east within each hemisphere. As the air masses rise and cool, the water falls to earth as precipi-

tation. Although most sea salt remains in the ocean, some enters the air owing to waves crashing on the beach and similar action. Sea salt acts as **precipitation nuclei,** airborne particulates that attract water droplets. When these get heavy enough, they fall back to earth as precipitation.

Precipitation over the sea returns a great deal of water to the ocean. Precipitation that falls on land collects in the many rivers and streams that carry water back to the sea. On the way, minerals and organic substances dissolve in the water. Thus fresh water that enters the marine environment returns not only water but also large amounts of organic and inorganic nutrients.

## Carbon Cycle

Compounds containing the element carbon are essential components of all living organisms. Carbon makes up the backbone of carbohydrates, proteins, lipids, and nucleic

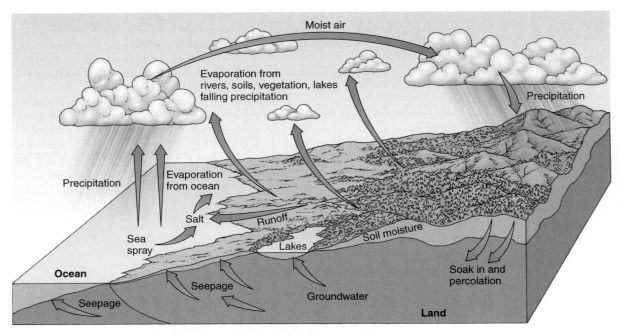

**Figure 2-15  The Hydrologic Cycle.** Water leaves the oceans by way of evaporation and returns in the form of precipitation. Rivers and streams collect the precipitation that falls on land and return it to the sea.

acids, the basic molecules of life (for more information on these molecules and their role in life see Chapter 5). Carbon cycling, then, is vital to the scheme of life. Living organisms produce carbon dioxide when they respire. When an organism dies, a variety of decay organisms break down the tissues (for more information regarding the process of decomposition see Chapter 6), and a major product of this process is carbon dioxide. Marine producers use the carbon dioxide in photosynthesis to make carbohydrates. The carbohydrates can then be used to make other organic molecules (Figure 2-16).

Carbon dioxide ($CO_2$) reacts with seawater to form carbonic acid ($H_2CO_3$), which in turn forms hydrogen ions ($H^+$) and bicarbonate ions ($HCO_3^-$), as shown in the following equation:

$$CO_2 + H_2O \rightarrow H_2CO_3 \rightarrow H^+ + HCO_3^-$$

The bicarbonate ions can then be taken up by some marine organisms and combined with calcium to form the calcium carbonate needed for their shells and skeletons. The calcium carbonate from shells, corals, animal skeletons, and coralline algae eventually collects in bottom sediments and becomes limestone. Geological upheavals sometimes bring the limestone to the surface, where it is weathered by wind, rain, and other physical forces. The calcium and carbonate released from the limestone then washes back into the sea by way of rivers and streams. Another source of calcium and carbonate is the recycling of dead shells and skeletons by organisms, such as boring sponges.

When some marine plants and animals die, their tissues are trapped in sediments and decay, ultimately forming deposits of fossil fuel (for more information on fossil fuels see Chapter 19). Most of the world's oil reserves were formed in this way. When these fuels are burned, carbon dioxide is released. The carbon dioxide can then be recycled by the process of photosynthesis.

## Nitrogen Cycle

Producers like plants, seaweeds, and phytoplankton require nitrogen fertilizer for protein synthesis and thus for proper growth and reproduction. The nitrogen they need is usually in a simple form, such as ammonia ($NH_3$), ammonium ($NH_4^+$), nitrite ($NO_2^-$), or nitrate ($NO_3^-$). The producers use energy from photosynthesis to concentrate the nitrogen in their tissues and assemble it into amino acids and then proteins. The nitrogen is passed in the form of amino acids and proteins to consumers when they eat producers or other consumers. Animals process the amino acids and proteins and excrete nitrogen in the form of ammonia, urea, and uric acid. Certain bacteria convert some of the ammonia into nitrites and nitrates that can be used again by the producers to make more amino acids and proteins (Figure 2-17). When an organism dies, decomposers release nitrogen into the environment from the tissue, and various groups of bacteria then convert the nitrogen into forms used by producers.

The atmosphere consists of almost 79% nitrogen and represents a major reservoir for this critical nutrient. Electrical discharges during thunderstorms produce nitrates, which enter the oceans during precipitation. Some microorganisms are capable of converting atmospheric nitrogen into forms that are usable by producers in a process called nitrogen fixation). The major nitrogen-fixing organisms in the marine environment are cyanobacteria, and we will ex-

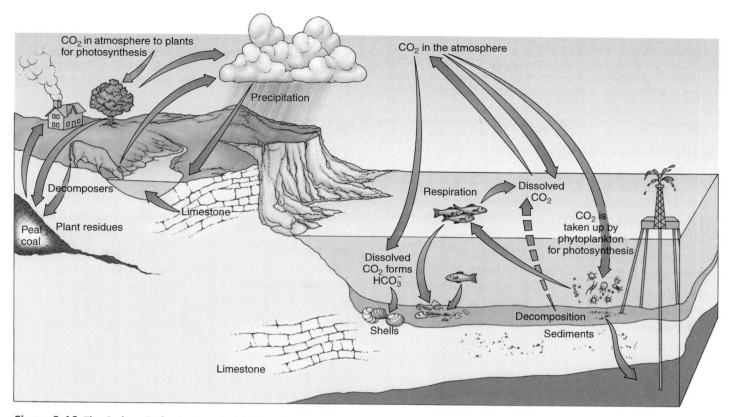

**Figure 2-16  The Carbon Cycle.** Carbon dioxide from the atmosphere that is dissolved in seawater is used by producers to make food through the process of photosynthesis. When the food is metabolized in respiration, the carbon dioxide is returned to the environment. Some carbon dioxide is converted into bicarbonate ions and incorporated into the shells of marine organisms. When these organisms die, their shells sink to the bottom. Some of the shells are compressed to form limestone, some of which is moved by geological process to the surface, where erosion will return the carbon to the sea.

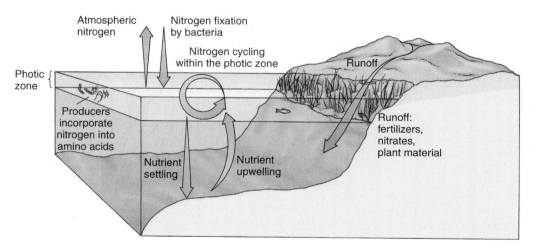

**Figure 2-17  The Nitrogen Cycle.** Upwellings and runoff from the land bring nitrogen into the photic zone, where producers can incorporate it into amino acids. Nitrogen-fixing bacteria in the photic zone can convert atmospheric nitrogen into forms that can be used by producers. Nitrogen is returned to the enviroment when organisms die or animals eliminate wastes.

amine the role they play in this process more closely in Chapter 6.

Runoff from the land may contain nitrogen from fertilizers, sewage, and dead plants and animals, as well as from animal wastes. This nitrogen can collect in shallow coastal waters and support a large amount of phytoplankton growth. In the open ocean, dead organisms and the nutrients they contain sink from the photic zone and are unavailable for the producers in those areas. Thus large quantities of nutrient-laden material end up on the ocean floor.

In certain areas, such as the coasts of California and Peru, the combination of winds and ocean currents brings large quantities of this nutrient-laden material from the bottom back into the photic zone, a phenomenon known as upwelling (for more information regarding upwellings see Chapter 4), where it can support the growth of phytoplankton.

## In Summary

Although energy constantly flows through ecosystems, nutrients necessary for life do not. These nutrients are constantly recycled from one generation to the next through biogeochemical cycles. ●

## SELECTED KEY TERMS

abiotic environment, *p. 13*

abyssal zone, *p. 23*

aerobic organism, *p. 17*

algal bloom, *p. 17*

anaerobic organism, *p. 17*

aphotic zone, *p. 22*

autotroph, *p. 24*

bathyal zone, *p. 23*

benthic division, *p. 22*

biogeochemical cycle, *p. 27*

biosphere, *p. 12*

biotic environment, *p. 13*

bloom, *p. 20*

carnivore, *p. 18*

chemosynthetic, *p. 24*

chlorophyll, *p. 23*

commensalism, *p. 19*

community, *p. 20*

competitive exclusion, *p. 18*

consumer, *p. 24*

decomposer, *p. 24*

desiccation, *p. 15*

detritivore, *p. 24*

detritus, *p. 24*

disphotic zone, *p. 22*

dissolved organic matter (DOM), *p. 25*

ecological efficiency, *p. 25*

ecosystem, *p. 12*

ectotherm, *p. 15*

endotherm, *p. 15*

energy pyramid, *p. 25*

epifauna, *p. 23*

exponential growth, *p. 20*

eutrophication, *p. 17*

first-order consumer, *p. 24*

food chain, *p. 24*

food web, *p. 25*

habitat, *p. 13*

hadal zone, *p. 23*

herbivore, *p. 18*

heterotroph, *p. 24*

homeostasis, *p. 14*

host, *p. 19*

infauna, *p. 23*

interspecific competition, *p. 17*

intertidal zone, *p. 15*

intraspecific competition, *p. 17*

keystone predator, *p. 18*

larval settlement, *p. 20*

light–dark-bottle method, *p. 24*

limiting nutrient, *p. 17*

logistic growth, *p. 20*

metabolism, *p. 15*

microhabitat, *p. 13*

mutualism, *p. 19*

nekton, *p. 22*

neritic zone, *p. 22*

neuston, *p. 23*

niche, *p. 13*

nutrient, *p. 17*

oceanic zone, *p. 22*

omnivore, *p. 24*

optimal range, *p. 14*

osmosis, *p. 16*

parasite, *p. 19*

parasitism, *p. 19*

pelagic division, *p. 22*

photic zone, *p. 22*

photosynthesis, *p. 23*

phytoplankton, *p. 15*

plankton, *p. 22*

population, *p. 20*

precipitation nuclei, *p. 27*

primary consumer, *p. 24*

primary productivity, *p. 24*

producer, *p. 24*

pyramid of biomass, *p. 26*

pyramid of numbers, *p. 26*

recruitment, *p. 20*

resource partitioning, *p. 18*

salinity, *p. 16*

secondary consumer, *p. 24*

second-order consumer, *p. 24*

shelf zone, *p. 23*

solute, *p. 16*

symbiosis, *p. 19*

ten percent rule, *p. 25*

tertiary consumer, *p. 24*

third-order consumer, *p. 24*

trophic level, *p. 25*

twilight zone, *p. 22*

water column, *p. 22*

zone of intolerance, *p. 14*

zone of stress, *p. 14*

## QUESTIONS FOR REVIEW

### Multiple Choice

1. The specific place in the environment where an organism is found is called its
   a. community
   b. habitat
   c. niche
   d. ecosystem
   e. trophic zone

2. The role that an organism plays in its living and nonliving environment defines its
   a. community
   b. habitat
   c. niche
   d. ecosystem
   e. trophic level

3. The area of the ocean that receives the greatest amount of sunlight for photosynthesis is the
   a. coastal zone
   b. benthic zone
   c. photic zone

d. nutrient zone
e. trophic zone

4. According to Figure 2-2, species B prefers or tolerates _____ better than species A.
   a. higher salinity, coarse sediments, larger food
   b. lower salinity, fine sediments, smaller food
   c. higher salinity, fine sediments, larger food
   d. lower salinity, coarse sediments, smaller food
   e. higher salinity, coarse sediments, smaller food

5. The upside-down jellyfish depends on algae in its body for certain nutrients. The algae are protected by the jellyfish and supplied with nutrients. This relationship would be an example of
   a. mutualism
   b. commensalism
   c. parasitism

6. According to the ten percent rule, how many kilograms of phytoplankton would be needed to produce 10 kilograms of fish that were second-order consumers?
   a. 1 kilogram
   b. 10 kilograms
   c. 100 kilograms
   d. 1,000 kilograms
   e. 10,000 kilograms

7. The ultimate source of energy for most life in the ocean is
   a. photosynthesis
   b. the sun
   c. thermal vents
   d. predation
   e. phytoplankton

8. The most important primary producers in marine ecosystems are
   a. seaweeds
   b. plants
   c. phytoplankton
   d. detritivores
   e. filter feeders

9. The abiotic factor that most influences life in the ocean is
   a. temperature
   b. pressure
   c. salinity
   d. sunlight
   e. predation

10. An organism that feeds on organic wastes and decaying material would be a(n)
    a. herbivore
    b. carnivore
    c. omnivore
    d. detritivore
    e. piscivore

## Short Answer

1. What are six abiotic factors that affect the distribution of organisms in an ecosystem?

2. What would probably happen to the natural balance in a community if the population of predators dramatically decreased or was wiped out?

3. What is the difference between a community and an ecosystem?

4. Describe the three main types of symbiotic relationships found in nature.

5. Describe the marine nitrogen cycle.

6. Why are there fewer marine organisms in the ocean's depths?

7. Why are so many marine organisms ectotherms?

8. Why is energy transfer between trophic levels inefficient?

9. How can groups of similar species avoid competition?

10. How can primary production be measured?

11. Why can't most populations continually grow at an exponential rate?

12. Describe a detritus-based food chain.

## Thinking Critically

1. Does the area of overlap in Figure 2-2 indicate that species A and species B have overlapping niches and will be in direct competition under those conditions? Why or why not?

2. Why does it make good ecological sense for whales to feed on plankton?

3. Although the open ocean receives plenty of radiant energy and has a larger area, it is not nearly as productive as the shallow coastal seas. Why?

4. Why do organisms that live in tide pools have to be more tolerant of changes in salinity than organisms that live in the open sea?

5. Which type of competition would you think would be more intense: interspecific or intraspecific? Why?

6. Would it be possible to have a pyramid of numbers where the trophic level above was larger than the trophic level below? Explain using an example with marine organisms.

## SUGGESTIONS FOR FURTHER READING

Bertness, M. D., S. D. Gaines, and M. E. Hay. 2001. *Marine Community Ecology*. Sunderland, Mass: Sinauer Publishing.

Grall, G. 1992. Pillar of Life, *National Geographic* 182(1): 95–114.

Paine, R. T. 1974. Intertidal Community Structure: Experimental Studies on the Relationship between a Dominant Competitor and Its Principal Predator, *Oecologica* 15: 93–120.

Rennie, J. 1992. Living Together, *Scientific American* 266(1):122–33.

### InfoTrac College Edition Articles

Boyd, I. L. 1996. Temporal Scales of Foraging in a Marine Predator, *Ecology* 77(2).

Downing, J. A., C. W. Osenberg, and O. Sarnelle. 1999. Meta-Analysis of Marine Nutrient Enrichment Experiments: Variation in the Magnitude of Nutrient Limitation, *Ecology* 80(4).

Gillis, A. M. 1993. Sea Dwellers and Their Sidekicks: The Private Lives of a Squid and Its Bacteria and Fish Family Trees Provide Clues about Symbiosis, *Bioscience* 43(9).

McClintock, J. B., and B. J. Baker. 1998. Chemical Ecology in Antarctic Seas: Chemical Interactions Can Lead to Unusual Arrangements between Species, *American Scientist* 86(3).

Poulin, R., and A. S. Grutter. 1996. Cleaning Symbioses: Proximate and Adaptive Explanations, *Bioscience* 46(7).

### Websites

**www.ecosystemworld.com**  A good general site with articles on many aspects of ecology, such as physical and biological habitats, population dynamics, and energy flow.

**http://www.ots.duke.edu/tropibiojnl/TROPIWEB/enlaces/marec.htm**  This website has links to marine journals that publish articles relating to marine ecology as well as websites related to the topic.

**http://www.epa.gov/bioiweb1/aquatic/marine.html**  An introduction to marine ecosystems with links to other sites.

# Geology of the Oceans

The physical characteristics of the environment play an important role in determining the kinds of organisms that can live in a given area and the traits that they will exhibit. Before we begin our study of the ocean's inhabitants and their interactions with each other and their environment, we need to gain a basic understanding of the physical characteristics of the ocean itself. In this chapter you will learn how oceans were formed and about the forces that shaped the physical features of the ocean in the past and continue to shape them today. You will be introduced to the methods scientists and navigators use to locate their position when they are at sea with no landmarks for reference points and how this information is used to produce charts that accurately represent the geography of the ocean. We will also examine the role that living organisms play in forming some of the physical features of the ocean environment.

## Key Concepts

1. The world ocean has four main basins: the Atlantic, Pacific, Indian, and Arctic.

2. Life first evolved in the ocean.

3. The earth's crust is composed of moving plates.

4. New seafloor is produced at ocean ridges and old seafloor is removed at ocean trenches.

5. The ocean floor has topographical features similar to those found on continents.

6. The seafloor is composed of sediments derived from living as well as nonliving sources.

7. Latitude and longitude determinations are particularly necessary for precisely locating positions in the open sea, where there are no features at the surface.

## WORLD OCEAN

Current theories hold that our solar system was formed more than 5 billion years ago and the planet earth about 400 million years later. The primitive earth and surrounding atmosphere were quite different in the beginning than they are now. At that time the surface of the earth was so hot that water could not remain there, and there was no free oxygen gas in the atmosphere as there is today. It was in this type of environment that the ocean began to form about 4.2 billion years ago. It continued to form for the next 200 million years, and it was in the newly formed ocean that life first evolved. Today the **world ocean** covers 70.8% of the planet's surface and is the most visible physical feature of our planet when viewed from space.

# Primitive Earth and Formation of the Ocean

It is generally believed that for the first billion years of its existence the earth was mainly composed of silicon compounds, iron, magnesium oxide, and small amounts of other elements. Geologists hypothesize that originally the earth was composed of cold matter and that over time several factors, such as energy from space and the decay of radioactive elements, contributed to raising its temperature.

The process of heating continued for several hundred million years, until the temperature at the center of the earth was high enough to melt iron and nickel. As these elements melted, they moved to the earth's core, displacing lighter, less-dense elements, and eventually raised the core temperature to approximately 2,000°C. Molten material from the earth's core moved to the surface and spread out, creating some of the features of the early earth's landscape. It is probable that the melting and solidifying happened repeatedly, ultimately separating the lighter elements of the earth's crust from the deeper, denser elements. This separation of elements eventually produced the various layers of the planet that will be discussed in a later section.

In these early times, any water present on the planet was probably locked up in the earth's minerals. As the cycles of heating and cooling took place, water, in the form of water vapor, was carried to the surface, where it cooled, condensed, and formed the ocean.

## Ocean and the Origin of Life

The earth's early atmosphere is thought to have been formed by some of the hot gases escaping from deep within the planet. Initially, the atmosphere contained gases such as carbon monoxide, carbon dioxide, nitrogen, methane, and ammonia. Because oxygen is so chemically active, any free oxygen gas originally present would have combined with other elements to form compounds called **oxides**. Gaseous oxygen did not begin to accumulate in the atmosphere until the evolution of modern photosynthesis (for more information regarding photosynthesis see Chapter 5).

In the early twentieth century biologists, such as the Russian A. I. Oparin and the Englishman J. B. S. Haldane, theorized that heat, lightning, and radiation, forms of energy that were prevalent in the early days of earth, may have provided enough energy to produce the organic molecules needed for life from the gases present in the primitive atmosphere. In 1953 while a graduate student at the University of Chicago working under Harold Urey, Stanley Miller did some experiments in which he tested this hypothesis. Using an apparatus like the one shown in Figure 3-1, he attempted to replicate the conditions present on the primitive earth. The results of Miller's experiments showed that organic molecules could form from the gases present in the primitive atmosphere, including some of the basic molecules necessary for life. Since Miller's initial experiments, other scientists have done similar experiments and achieved similar results. Biologists theorize that as these molecules formed they accumulated in the ocean, and that over time

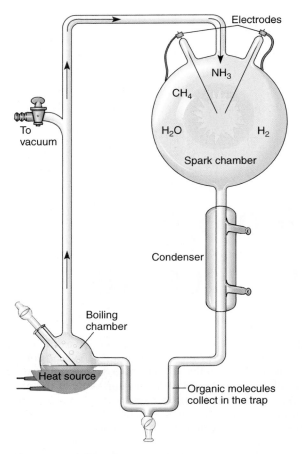

**Figure 3-1  Miller's Apparatus.** Using an apparatus like the one shown here, Stanley Miller was able to demonstrate that simple organic compounds, including some necessary for life, could have formed under the conditions found on the primitive earth.

the ocean became a highly concentrated nutrient soup. It was in an environment such as this that the molecules became organized and the first cells evolved. Because there was no free oxygen in the atmosphere, the first cells must have been anaerobic, and they were most likely heterotrophs that gained nutrition from the organic molecules in the surrounding water. It is not surprising then that the oldest known fossils are of marine bacteria (Figure 3-2). The fossils were found in northwestern Australia and are between 3.4 and 3.5 billion years old.

## The Ocean Today

Today 1.37 billion cubic kilometers (approximately $362 \times 10^{18}$ gallons) of water covers the earth's surface and forms the ocean, the largest habitat on the planet. The ocean, which is frequently referred to as the world ocean, is a continuous mass of water that covers most of the planet. Traditionally, humans have divided the world ocean into artificial compartments called oceans and seas using the boundaries

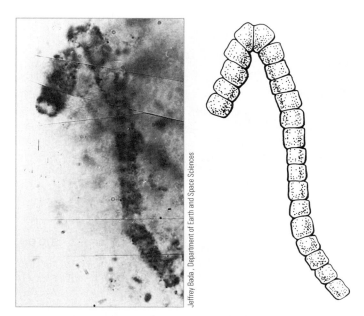

Jeffrey Bada., Department of Earth and Space Sciences

**Figure 3-2 Oldest Known Fossils.** These fossils of marine bacteria represent some of the earth's earliest life forms.

of continents and imaginary lines such as the equator. In reality, there are few natural divisions that can be used to separate the world ocean into smaller bodies of water. Beneath the surface of the world ocean are four main basins, and water flows freely among them. These major ocean basins are the Pacific, Atlantic, Indian, and Arctic basins. The body of water in each ocean basin is referred to as an ocean (Figure 3-3). The Pacific Ocean is the largest, and the Arctic Ocean is the smallest. Each ocean has its own characteristic surface area, volume, and depth, and these are shown in Table 3-1. Oceanographers also refer to the water surrounding the continent of Antarctica as an ocean, the Southern or Antarctic Ocean.

Other divisions of the world ocean, such as seas and gulfs, are so named for our convenience and are in reality only temporary features of the single world ocean. A **sea** is a body of salt water that is smaller than an ocean and is more or less landlocked. The Mediterranean Sea, Red Sea, and Caribbean Sea are a few examples. A **gulf** is a smaller body of water that is mostly cut off from the larger ocean or sea by land formations. Some examples are the Gulf of California, Gulf of Mexico, and Persian Gulf.

## In Summary

The world's oceans are believed to have formed approximately 4.2 billion years ago when water vapor escaping from minerals in the earth cooled and condensed on the earth's surface. It was in this early ocean environment that the first cells evolved. Today the world ocean covers nearly 71% of the planet's surface. The body of water in each of the four major ocean basins is referred to as an ocean. These four oceans are the Atlantic, Pacific, Indian, and Arctic Ocean. Smaller subdivisions of the oceans are seas and gulfs. ●

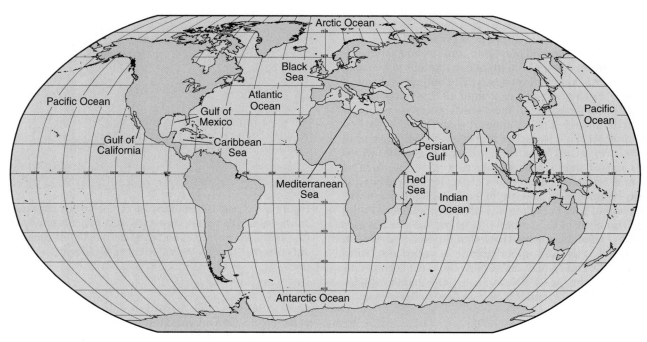

**Figure 3-3 The World Ocean.** There are four main ocean basins and the body of water in each basin is traditionally referred to as an ocean. These are the Pacific, Atlantic, Indian, and Arctic oceans. Other common smaller divisions of the world ocean, such as seas and gulfs, are temporary features that are named for convenience.

| Table 3-1 **Characteristics of Major Ocean Basins** | | | |
|---|---|---|---|
| **Ocean** | **Surface Area (km²)** | **Volume (km³)** | **Average Depth (m)** |
| Atlantic | 82,400,000 | 323,600,000 | 3,575 |
| Pacific | 165,200,000 | 707,600,000 | 4,282 |
| Indian | 73,400,000 | 291,000,000 | 3,962 |
| Arctic | 12,257,000 | 13,702,000 | 1,117 |

## CONTINENTAL DRIFT

The features of the earth are constantly changing, and the seafloor is no exception. New seafloor is continually being added as old seafloor is removed. This process reshapes the features of the ocean beneath the surface of the water and influences marine organisms, especially those that live in and on the bottom. The movement of the seafloor also causes movement of the continents that rest on it, a process known as continental drift.

### Layers of the Earth

The earth is made up of several layers (Figure 3-4). At the center of the planet is an inner core, which is solid (because of the extremely high pressure), very dense, very hot, and rich in iron and nickel. Surrounding the inner core is an outer core that consists of a transition zone and a thick layer of liquid material. The liquid material has the same composition as the inner core, only it is under less pressure and thus cooler. The next layer, called the **mantle,** is the thickest layer and contains the greatest mass of material. The mantle is mainly composed of magnesium–iron silicates. Although the outer region of the mantle is thought to be rigid, the inner layer is able to flow slowly. The outermost layer of the earth is the thinnest and coolest of all, the crust.

Because the crust is the most easily accessible portion of the earth, more is known about it than the other layers. The continental crust is thicker and slightly less dense than the oceanic crust and is mainly composed of granite, which contains mostly lightweight silicate-rich minerals, such as quartz and feldspar. The oceanic crust, by comparison, is primarily composed of basalt-type rock, which has a lower silicate content and is higher in iron and magnesium than granite. The layer of the mantle just below the crust is also composed of rigid, basalt-type rock that is fused to the crust. The region of crust and upper mantle is called the **lithosphere,** and the region of mantle below the crust is known as the **asthenosphere** (as-THEN-uh-sfeer). The asthenosphere is thought to be liquid, and it is able to flow under stress.

### Moving Continents

With the advent of more accurate world maps in the later seventeenth century, several scientists became intrigued with the observation that the continents fit together like pieces of a jigsaw puzzle (Figure 3-5). For instance, the east-

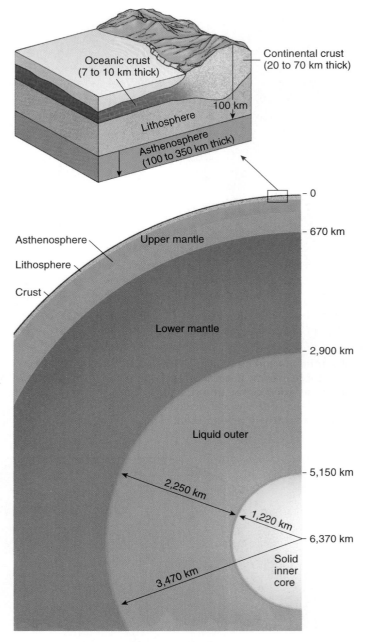

**Figure 3-4 Composition of the Earth.** The earth is composed of several layers. The surface layer is the crust, which is fused to the lithosphere, the outermost portion of the mantle just beneath it. The asthenosphere lies beneath the lithosphere and is able to flow under stress.

ward bulge of South America would fit well with the west coast of the African continent. Also, the comparative study of rock formations, the placement of mountain ranges, and the distribution of fossils all suggested that the continents were at one time connected.

This idea was not new. As early as the 1600s, the English scholar Sir Francis Bacon suggested that the continents

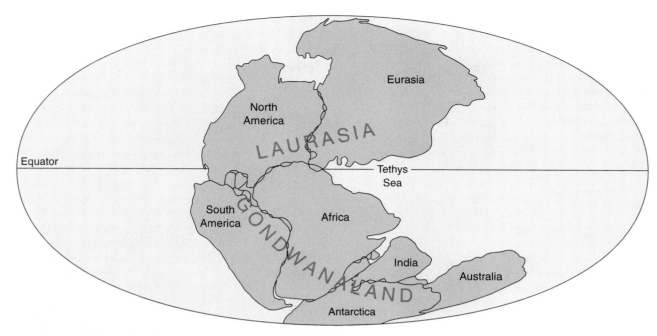

**Figure 3-5  The Supercontinent Pangaea.** This map of Pangaea shows how today's continents were once connected to form a single supercontinent approximately 400 million years ago.

may have once been connected to each other. In the latter part of the nineteenth century, the Austrian geologist Edward Suess even proposed a name, Gondwanaland, for the fusion of southern continents. In 1915 the German meteorologist Alfred Wegener proposed that at one time there was only one supercontinent, which he named **Pangaea.** He suggested that forces associated with the earth's rotation were responsible for breaking the supercontinent into a northern portion, **Laurasia,** composed of Europe, Asia, and North America, and a southern portion, **Gondwanaland,** composed of India, Africa, South America, Australia, and Antarctica. Following World War II, there was renewed interest in the theory of moving continents. With more sophisticated equipment, studies were made during the 1950s to determine what forces might be acting to produce the continental movement.

### Forces that Drive Continental Movement

In the early 1960s, H. H. Hess proposed that molten rock, called **magma,** located deep in the earth's mantle, moved by convection currents to the region below the solid upper mantle and crust (Figure 3-6). The convection currents are driven by the heat of the mantle and the cooling of the magma. The molten magma would then flow laterally under this region, cooling as it did so, and then sink back toward the core. Occasionally, when the upward-moving magma breaks through the crust of the ocean floor, volcanoes are formed. Over time, a long mountain range called a **midocean ridge** will form along the crack produced by the erupting magma, and the magma that oozes out of these mountain ranges, or **ridge systems,** will cool and form new crust,

known as **oceanic basaltic crust.** In some areas of the range, a **rift valley** runs along the length of a portion of the mountain crests. These are areas of high volcanic activity. Steep-sided **fracture zones,** linear regions of unusually irregular ocean bottom, may also occur, running perpendicular to the ridges and rises and separating sections of the range.

Because new crust is being formed and the earth is not growing larger, there must be some area where old crust is removed. In regions called **subduction zones,** such as the deep recesses of ocean trenches, old crust at the bottom of the trenches sinks and eventually reaches the mantle, where it is liquefied and recycled by convection currents into the earth's core (Figure 3-7).

Most of the magma that rises from the deep mantle is blocked by the rigid lithosphere and moves horizontally under it, ultimately descending along the sides of ascending convection currents and carrying pieces of the lithosphere with it. This activity produces the movement of the crust called **seafloor spreading.** This is the force that causes **continental drift.** Seafloor spreading causes movement of the basaltic crust of the seafloor, and because the continents rest on this crust, they also move, much in the same way that boxes move on a conveyor belt.

### Evidence for Continental Drift

Along with the fit of the continental boundaries, other evidence that supports the theory of continental drift includes observations on the distribution of earthquakes, the temperature of the sea bottom, the age of rock samples from the seafloor, and the analysis of core samples drilled through the ocean sediments. Earthquakes are known to occur

**Figure 3-6 Formation of Oceanic Crust and Mountains.** Molten material called *magma* rises from the earth's core to the upper mantle. The hot magma (red arrows) rises because it is less dense than the surrounding material. As magma reaches the mantle it cools (blue arrows), becomes more dense, and sinks back toward the core. This cycling of magma from the core to the mantle and back that results from changes in temperature and density is called convection. Occasionally, the heated magma breaks through the earth's crust, forming volcanos. Lines of volcanos form mountain ranges called *midocean ridges.*

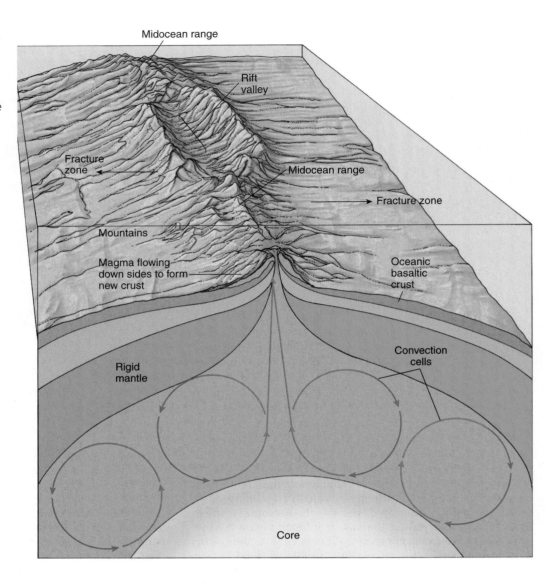

around the globe in narrow zones that correspond to areas along ridges and trenches, the most active areas of crustal movement (Figure 3-8a). The highest seafloor temperatures occur in the regions of the ridges; temperatures decrease with distance from the ridge. These observations are consistent with heated magma oozing from the ridges and cooling as the seafloor spreads away from the ridges.

From 1968 to 1983, a specially constructed drilling ship, the *Glomar Challenger,* gathered core samples from the ocean's bottom. Their analysis revealed that the crust was thinnest and the rock youngest near the ridges, evidence for the hypothesis that new crust is being formed in these areas. The farther from the ridges, the older the rocks and the greater the amount of sediment overlying them. This is to be expected because new crust replaces older crust as convection moves the crust away from the ridges. Because old crust was in place longer, it accumulated more sediment. No rock was found that was older than 180 million years, much

younger than the age of continental rocks. All of these data support the theory of continental drift.

## Theory of Plate Tectonics

All of the information concerning the movement of the earth's crust is combined to form the **theory of plate tectonics.** In this theory, the lithosphere is viewed as a series of rigid plates (see Figure 3-8b) that are separated by the earthquake belts of the world—that is, the trenches, ridges, and faults. There are seven major lithospheric plates: the Pacific, Eurasian, African, Australian, North American, South American, and Antarctic plates. Each plate is composed of continental or oceanic crust or both.

At the midocean ridges, where plate boundaries move apart as new lithosphere is formed, **divergent plate boundaries** occur. **Convergent plate boundaries** occur at trenches, where plates move toward each other and old lithosphere is

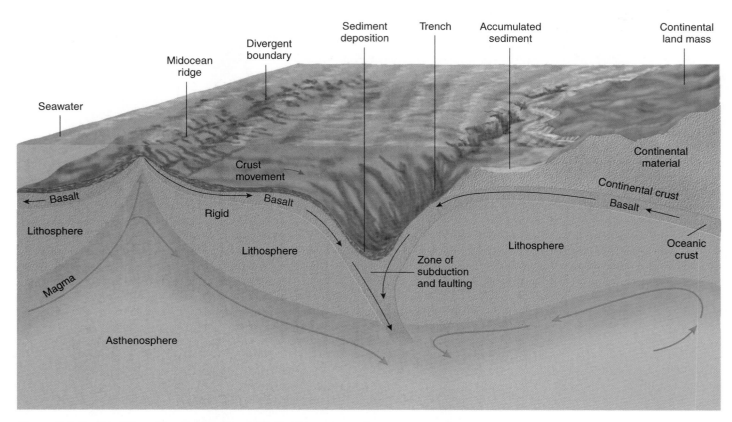

**Figure 3-7 Seafloor Spreading and Continental Drift.** Rising magma forms new oceanic crust that moves away from the midocean ridges. At subduction zones, old crust sinks and is ultimately returned to the mantle, where it melts and forms new magma. Since the continents rest on the basaltic crust, as the crust moves, the continents are carried along.

destroyed. The plates move past each other at regions known as **faults** (see Figure 3-8b).

A **transform fault** is a special kind of fault that is found in sections of the midocean ridge. Each side of a transform fault is formed by a different plate, and the plates scrape against each other as they move in opposite directions. The fault zone produced by this movement is the site of frequent earthquakes. The motion of the plates along these faults produces a sharp demarcation and often a nearly continuous line of cliffs with sharp vertical drops, known as **escarpments.** In these regions, there are sudden changes in the ocean depth.

Regions where the lithosphere splits, separates, and moves apart as new crust is formed are called **rift zones.** The midocean ridge and rise systems represent the major rift zones. It is generally thought that rifting occurs when rising magma stretches the overlying crust, creating a sunken rift zone. As this process continues, the rift spreads and the fault deepens and cracks, allowing the magma to seep through and eventually form a ridge. When this happens to the lithosphere under the continents, the rift zone can eventually fill with seawater. The Red Sea is an example of this type of formation.

As a plate moves away from the rift zone, it cools and thickens. At the rift, thinning of the crustal plate and increased flow of magma into the rift cause the land mass to separate. A low-lying region of oceanic basalt is formed, and a new ocean basin and ridge system are formed as well.

## Rift Communities

**Rift communities,** also known as **deep-sea vent communities,** are thriving communities of marine organisms that depend on the specialized environments found at divergence zones in the ocean floor. The first of these communities was discovered in 1977 by Dr. Robert Ballard and J. F. Grassle of Woods Hole Oceanographic Institution. It was found at a depth of 2,500 meters (8,250 feet) in the Galápagos Rift, which lies between the East Pacific Rise and the coast of Ecuador. Rift communities consist of a variety of animals, such as clams, worms, and crabs, some of which are relatively large (Figure 3-9). These organisms depend on the chemosynthetic activity of bacteria for their nutrients. Rift communities are unique because they represent food webs that exist in the absence of sunlight. We will examine the

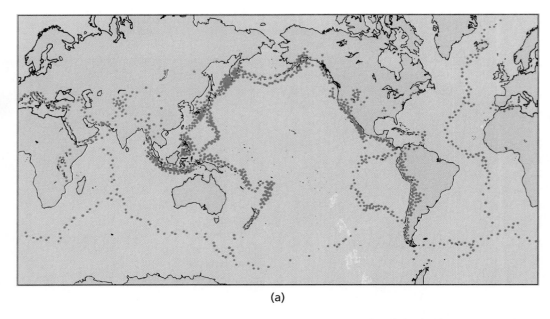

(a)

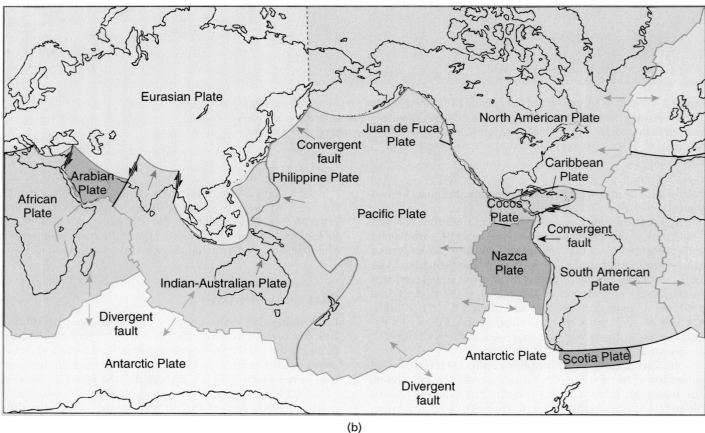

(b)

**Figure 3-8 Earthquake Zones and Tectonic Plates.** (a) This map shows the earth's earthquake zones. Each red dot represents an earthquake epicenter (the area of the earth's surface directly above the origin of an earthquake). Notice that the earthquake zones correspond closely to the location of the major plate boundaries (see part b). The distribution of volcanos also follows along these lines. (b) This map shows the location of the major tectonic plates and their general direction of movement. Compare the plate boundaries to the earthquake zones shown in (a).

# Magnetic Evidence for Continental Drift

Over the past 76 million years, there have been 170 reversals of the earth's magnetic field, times when its north–south polarity switched. No one knows what causes these reversals, but their existence can be demonstrated by measuring the magnetic poles of magnetized rocks that were formed at different times. For example, when iron particles in a magnetic field are heated, they become magnetized as they cool, forming miniature compass needles. The north and south magnetic poles of these particles are lined up with the direction of the north and south magnetic poles of the earth at the time they formed. Examination of the north and south poles of magnetized iron found in rocks of various ages in the earth's crust indicates that reversals in the earth's magnetic field have occurred in the past.

In 1960 oceanographers from the Scripps Institution of Oceanography towed magnetometers (devices that measure magnetic fields) over the ocean floor and recorded a series of band patterns (Figure 3-A) that represented changes in the polarity of magnetic components in the earth's crust. No one was sure of the significance of these patterns until 1963, when F. J. Vine and D. H. Matthews of Cambridge University hypothesized that the band patterns corresponded to changes in the earth's magnetic field, which had been frozen in the rock of the seafloor. They suggested that, as magma containing iron particles rose to the surface at ridge systems, the iron particles oriented their north and south magnetic poles in the same direction as the north and south poles of the earth. As the magma reached the crust, cooled, and hardened, the direction of the earth's field was frozen in the magnetized iron particles in the rocks. As new crust formed and the plates moved apart, these magnetized rocks were carried away from the ridge system. Over time, this process would produce strips of crust parallel to the ridges that recorded alternating reversals in the polarity of the magnetic field. This, then, would account for the band patterns observed by the research team from the Scripps Institution.

Vine and Matthews deduced that if their hypothesis were correct, there would be symmetrical, mirror-image band patterns on the two sides of a ridge and that the oldest rocks would be found farthest away from the ridge. Furthermore, the rocks' polarity should be the same as that in similarly dated rocks found on land.

To test their hypothesis, they measured the polarity and age of rock samples taken from various sites relative to ridges on the ocean bottom. The results confirmed their hypothesis. The band patterns on either side of the ocean ridge were mirror images of each other, and the ages of the rocks matched their predictions. The conclusions drawn, that new crust was being formed on the seafloor and was spreading, added strong evidence for the theory of plate tectonics. ●

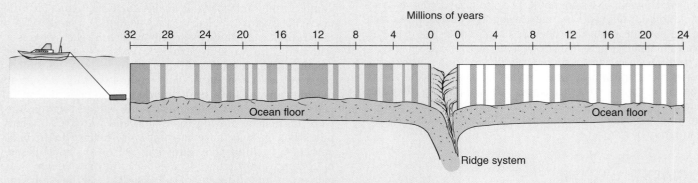

*Figure 3-A  Magnetometer Data. This pattern of bands was produced when a magnetometer was towed across the floor of the ocean. The alternation of dark and light bands indicates reversals in the earth's magnetic field over millions of years. Notice that the older rock is farther away from the site of crust formation and that the newer rock is closer to it.*

WHOI/D. Foster/Visuals Unlimited

**Figure 3-9 Vent Community.** Thriving communities are found in some rift valleys such as this one in the Galápagos Rift. Since sunlight does not penetrate to this depth, these organisms rely on chemosynthetic bacteria for food.

biology and ecology of deep-sea vent communities in detail in Chapter 18.

## In Summary

The continental masses are not locked into position but are constantly moving at a very slow rate. Geologists have determined that in the past there was only one large continent that fragmented one or more times to form the continents that we have today. Continents move when the plates on which they rest move. The crustal plates move horizontally when molten magma from the earth's core moves to the crust and breaks through, pushing the plates apart and forming new crust. Old crust is removed in the deep trenches and other subduction zones of the ocean, preventing the earth from getting larger.  ●

## OCEAN BOTTOM

The ocean bottom has geological features similar to those found on land. Mountain ranges, canyons, valleys, and great expanses are all part of the vast underwater landscape (Figure 3-10). These physical features of the ocean bottom are called **bathygraphic features,** and unlike their counterpart topographic features on land, they change relatively slowly. Erosion is slow in the relatively calm recesses of the ocean, and changes mainly involve sedimentation, uplift, and subsidence. We can divide the ocean bottom into two major regions. The region that lies beneath the neritic zone is called

the **continental margin.** Beyond the continental margin lies the ocean basin.

## Continental Margins

The continental margin is composed of the continental shelf and the continental slope. **Continental shelves** (Figure 3-11) are the shallow, submerged extensions of continents. They are composed of granite that is covered by sediments and have physical features similar to the edge of the nearby continent. Continental shelves are generally flat areas, averaging 68 kilometers (40 miles) wide and 130 meters (430 feet) deep, that slope gently toward the bottom of the ocean basin. The width of a continental shelf is frequently related to the slope of the land it borders. Mountainous coasts like the west coast of the United States usually have a narrow continental shelf, whereas low-lying land like the east coast of the United States usually has a wide one. They are produced by waves that constantly erode the land mass and by natural dams, such as reefs, volcanic barriers, and rock, that trap sediments between the dam and the coastal land mass (Figure 3-12).

The transition between the continental shelf and the deep floor of the ocean is called the **continental slope.** At the point where the continental shelf ends and the continental slope begins there is an abrupt change in the landscape called the **shelf break.** From this point there is an increase in the slope and a rapid change in depth to the seafloor. The extent of the sloping can vary from a gradual decline to a steep drop into an ocean trench, like the slope that occurs off the western coast of South America. Because of its steep angle, the continental slope usually has less sediment.

### Submarine Canyons

Some continental slopes have **submarine canyons** (see Figure 3-11) that are similar to canyons found on land. Many of these submarine canyons are aligned with river systems on land and were probably formed by these rivers during periods of low sea level. The Hudson River canyon on the east coast of the United States is an example of this. Other submarine canyons have ripple marks on the floor, and at the ends of the canyons sediments fan out, suggesting that they were formed by moving sediments and water. Oceanographers believe that these canyons were formed by turbidity currents. **Turbidity currents** are swift avalanches of sediment and water that erode a slope as they sweep down and pick up speed. At the end of the slope, the current slows and the sediments fan out. Turbidity currents can be caused by earthquakes or the collapse of large accumulations of sediments on steep slopes. Oceanographers are currently engaged in producing computer models of this process, and one such model is being tested in the submarine Scripps Canyon in California.

### Continental Rise

At the base of a steep continental slope there may be a gentle slope called a **continental rise** (see Figure 3-11). A continental rise is produced by processes such as landslides that carry sediments to the bottom of the continental slope. Most continental rises are located in the Atlantic and Indian Oceans and around the continent of Antarctica. In the Pacific Ocean, it is more common to find trenches located at the bottom of the continental slopes.

### Shaping the Continental Shelves

During the glacial periods, much of the ocean water was incorporated into the polar ice sheets, and the continental shelves were at times uncovered. Erosion carved valleys in the shelves, waves altered their contours, and rivers deposited sediments. When the glaciers melted and the sea level rose, the continental shelves were again covered by water, retaining the surface features that were established during the time they were uncovered.

Some continental shelves still experience changes today. The Mississippi and Amazon Rivers, for instance, carry mud, silt, and sand out to sea to be deposited on the continental shelves. These thick deposits of sediment have a significant effect on the productivity of these areas, as we shall see in Chapter 16. However, not all shelves have this thick layer of sediment. For instance, the continental shelf around the eastern tip of Florida is bare of sediment. The swift Florida current sweeps the sediments northward to deeper water, leaving the shelf sediment free.

## Ocean Basin

The floor of the ocean is referred to as the **ocean basin,** and it covers slightly more of the earth's surface than the continents. As noted earlier in this chapter, there are four main ocean basins: the Pacific, Atlantic, Indian, and Arctic basins. Ocean basins are composed of basaltic rock that is covered by a thick blanket of sediments that have either washed down from the continental shelves or settled there from the surface waters. Some of the planet's most spectacular geological formations are found in the ocean basins.

### Abyssal Plains and Hills

The bottom of many ocean basins is a flat expanse called the **abyssal plain** (Figure 3-13). These plains extend from the seaward side of the continental slope and are formed by sediments deposited by turbidity currents as well as sediments falling from above. Dotting the ocean floor are **abyssal hills** rising as high as 1,000 meters (3,300 feet). Abyssal hills cover as much as 50% of the Atlantic seafloor and as much as 80% of the seafloor of the Pacific and Indian Oceans. They are formed by volcanic action and the movements of the seafloor described earlier.

### Seamounts

Another feature of ocean basins is the **seamount,** a steep-sided formation that rises sharply from the bottom. All seamounts are formed from underwater volcanoes and are most prevalent in the Pacific Ocean. Some seamounts show evidence of coral reefs and surface erosion, suggesting that at one time they were above the surface. Plate tectonic movement of the ocean floor, the natural process of compaction that volcanic material undergoes, subsidence due to cooling of the ocean floor, erosion, and the increased weight of sediments at the top may be the reasons for the sinking of these structures.

### Ridges and Rises

A continuous series of large, underwater, volcanic mountains runs through every ocean of the world and stretches some 68,000 kilometers (40,000 miles) around the earth. The ridges and rises of these mountain ranges separate the ocean basins into a series of smaller, deepwater sub-basins.

### Trenches

Other impressive features of the ocean floor are the trenches. Deepwater trenches are more common and more spectacular in the Pacific Ocean and are usually associated with chains of volcanic islands called **island arcs.** For instance, the deepest of all ocean trenches is the Mariana Trench. It is associated with the Mariana Islands chain (Figure 3-14). A portion of the Mariana Trench, the Challenger Deep, is 11,020 meters (6.85 miles) deep, the deepest spot on earth. The Peru-Chile Trench extends for more than 6,120 kilometers (3,600 miles) along the coast of South America, making it the longest of the ocean trenches. By comparison, the Atlantic has only two, relatively short trenches, the South Sandwich Trench and the Puerto Rico–Cayman Trench. Table 3-2 lists the major ocean trenches and their depths.

## Life on the Ocean Floor

The water above the continental shelf is comparatively shallow, and sunlight can sometimes reach the bottom. Because the coastal seas are next to the continental land masses, they receive high levels of nutrients washed from the land. The combination of plentiful sunlight and high nutrient levels makes these areas very productive, and marine life on the continental shelf is abundant.

Although abyssal plains may contain nutrients, they do not receive any sunlight. Thus there is no photosynthesis and food production is limited to the chemosynthetic activity of bacteria in vent communities. The organisms that live on the ocean floor rely heavily on food delivered from the sunlit surface waters. Because primary production is so limited, life is not as abundant in this environment compared with the continental shelf.

## In Summary

The seafloor can be divided into the continental margins, composed of the continental shelf and the continental slope, and ocean basins. Sloping away from the edge of the continental masses are the continental shelves. Continental shelves have been alternately submerged and exposed over the history of the earth. During the times they have been exposed, forces, such as rivers and turbidity currents, have carved submarine canyons. The transition between

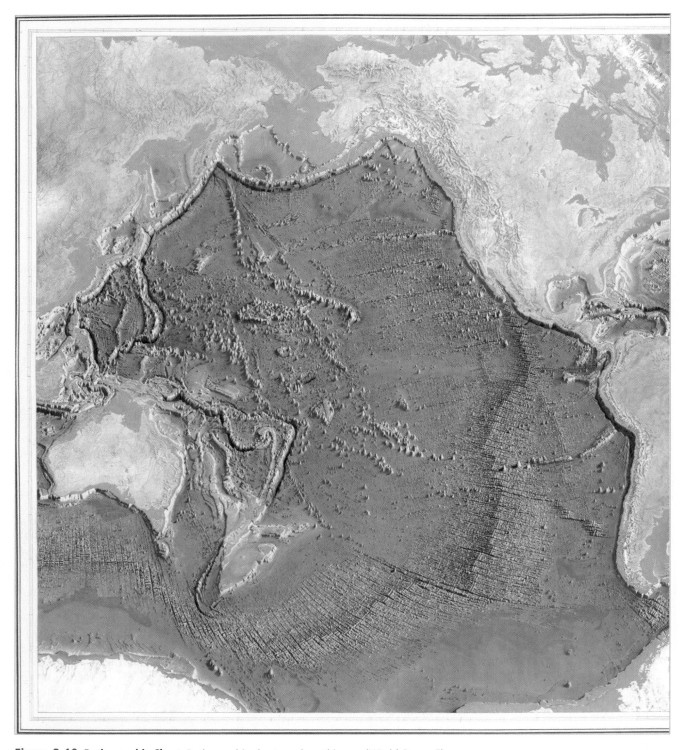

**Figure 3-10 Bathygraphic Chart.** Bathygraphic charts such as this one (*World Ocean Floor Panorama* by Bruce C. Heezen and Marie Tharp) depict the physical features of the ocean bottom. *(Copyright by Marie Tharp 1977/2003. Reproduced by permission of Marie Tharp, Oceanic Cartographer, One Washington Ave., South Nyack, New York, 10960.)*

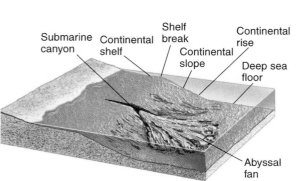

Submarine canyon

Continental shelf

Shelf break

Continental slope

Continental rise

Deep sea floor

Abyssal fan

**Figure 3-11 Continental Shelf.** Continental shelves are generally flat areas that extend from the continental land mass into the sea. Many physical features of continental shelves were formed during glacial periods when these areas were exposed. Submarine canyons are thought to be formed by turbidity currents. As the current slows at the end of the continental slope, sediments fan out to form an abyssal fan.

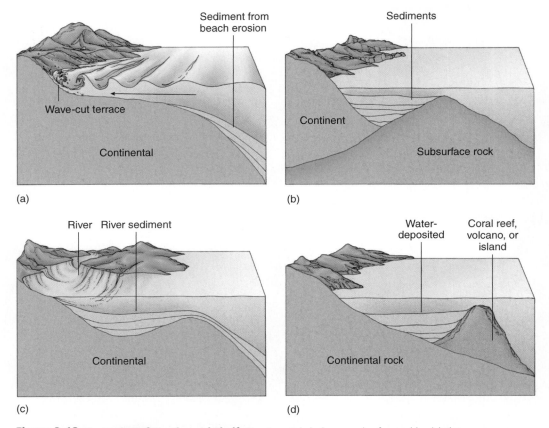

(a)

Sediment from beach erosion

Wave-cut terrace

Continental

(b)

Sediments

Continent

Subsurface rock

(c)

River   River sediment

Continental

(d)

Water-deposited

Coral reef, volcano, or island

Continental rock

**Figure 3-12 Formation of Continental Shelf.** Continental shelves can be formed by (a) the eroding action of waves, (b) the trapping of sediments by natural dams, (c) the accumulation of sediments deposited by rivers, or (d) the trapping of sediments by coral reefs, islands, or offshore volcanos.

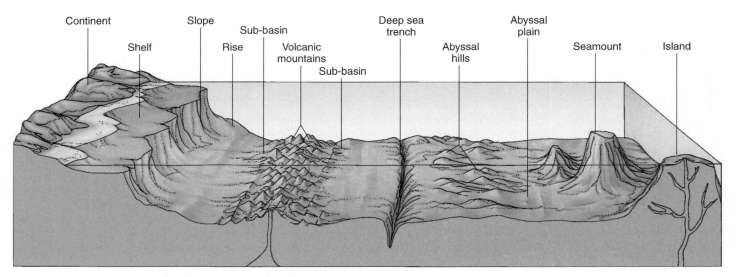

Continent   Shelf   Slope   Rise   Sub-basin   Volcanic mountains   Sub-basin   Deep sea trench   Abyssal hills   Abyssal plain   Seamount   Island

**Figure 3-13 Landscape of the Ocean Floor.** Like the surface of the continents, the ocean floor displays a variety of physical features, including mountains, hills, trenches, and expansive plains.

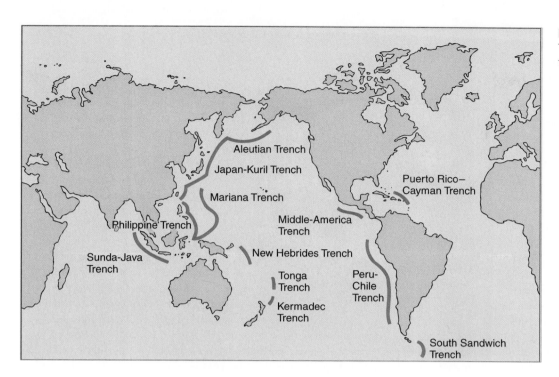

**Figure 3-14  Ocean Trenches.** This map shows the location of the major ocean trenches.

| Table 3-2  Depth of Major Ocean Trenches | | |
|---|---|---|
| **Trench** | **Depth in Meters** | **Depth in Feet** |
| Puerto Rico–Cayman Trench | 8,605 | 28,232 |
| Peru-Chile Trench | 8,065 | 26,460 |
| Middle America Trench | 6,669 | 21,880 |
| South Sandwich Trench | 7,235 | 23,736 |
| Aleutian Trench | 7,635 | 25,194 |
| Kuril Trench | 10,542 | 34,587 |
| Japan Trench | 8,513 | 27,929 |
| Mariana Trench | 11,020 | 36,366 |
| Tonga Trench | 6,000 | 19,800 |
| Sunda-Java Trench | 7,450 | 24,440 |
| Kermadec Trench | 10,047 | 32,962 |
| Philippine Trench | 10,190 | 33,431 |
| New Hebrides Trench | 8,100 | 26,730 |

make up these sediments may come from living organisms, the land, the atmosphere, or even the sea itself and accumulate over time on the ocean floor. The amount and type of sediments that are found on continental shelves, continental slopes, and abyssal plains are particularly important to the organisms that inhabit these areas. The sediments provide a habitat for many organisms and a source of nutrients for others.

The many different types of sediments in the ocean can be classified on the basis of particle size or according to their source. A scheme of classifying sediments on the basis of particle size was devised in the late nineteenth century. In this scheme of classification the finest particles are clay measuring less than 0.004 mm (0.00016 inches) and the largest are boulders measuring greater than 256 mm (10 inches). Table 3-3 lists the different types of sediment particles found in the ocean and along its shores classified according to particle size.

the continental shelf and the deep seafloor is the continental slope. At the bottom of the continental slope is the ocean basin. Some features of the ocean basin are the abyssal plain, mountain ranges, ridges, valleys, trenches, and seamounts. ●

## COMPOSITION OF THE SEAFLOOR

The seafloor is composed of basaltic rock that is covered by a coating of sediment. **Sediment** is composed of loose particles of inorganic and organic material. The particles that

| Table 3-3  Sediment Types | |
|---|---|
| **Sediment** | **Size of Sediment Particle (diameter)** |
| Clay | Less than 0.004 mm ($1.6 \times 10^{-4}$ inch) |
| Silt | 0.004 to 0.062 mm ($1.6 \times 10^{-4}$ to $2.5 \times 10^{-3}$ inch) |
| Sand | 0.062 to 2 mm (0.0025 to 0.08 inch) |
| Granule | 2 to 4 mm (0.08 to 0.16 inch) |
| Pebble | 4 to 64 mm (0.16 to 2.5 inches) |
| Cobble | 64 to 256 mm (2.5 to 10 inches) |
| Boulder | Greater than 256 mm (10 inches) |

# Animal Sculptors of the Seafloor

In the late 1970s while surveying the floor of the Bering Sea for geologic hazards to offshore oil platforms, C. Hans Nelson discovered pits and furrows that could not be attributed to any known geologic process. Nelson was aware that many marine mammals either live in or frequent these relatively shallow waters between Alaska and Siberia, and he wondered whether any of these animals might be responsible for the disturbances he discovered. In investigating the disturbances, Nelson and his colleagues discovered that California gray whales (*Eschrichtius gibbosus*) produce the pits and that Pacific walruses (*Odobenus rosmarus*) produce the furrows. Both animals alter the seafloor in the course of feeding on the continental shelf and, in the process, introduce more sediment into the water of the region than the Yukon River, which annually dumps 60 million metric tons of sediment into the sea.

The gray whales are not permanent residents of this region. They leave their breeding grounds off Baja California in March and migrate up the Pacific coast to feed in the waters of the northeastern Bering Sea. When the winter ice melts, the water is relatively calm and teems with life. The whales spend May to November feeding in these rich waters before returning to their breeding grounds, and while feeding, they dig up huge amounts of sediment searching for the small, bottom-feeding crustaceans known as amphipods (Figure 3-B), which are their preferred food.

Initially, the evidence for gray whales disturbing the seafloor was circumstantial. As far back as the nineteenth century, biologists had suspected that whales feed on animals in the sea bottom. Whalers frequently reported whales coming to the surface with muddy water streaming from their mouths, and when the stomachs of captured animals were opened, they were full of bottom-dwelling animals that have

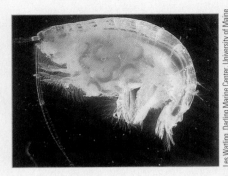

**Figure 3-B** *An Ampeliscid Amphipod.* Crustaceans such as this are filtered from the sediments by gray whales.

since been identified as amphipod crustaceans. When the whale feeds, it rolls to one side so that its mouth is parallel to the seafloor. It then retracts its tongue, creating a suction that draws a large amount of food-laden sediment into the mouth, where a series of fibrous plates, called *baleen,* growing from the jaw, filter out the water.

Two other findings suggested that whales were responsible for the seafloor disturbances. In 1979 aerial observers who were tracking feeding gray whales by following emitted mud plumes found them to be concentrated in the Chirikov Basin, the area where the bottom disturbances were found. Nelson had sampled bottom sediments from several sites on the continental shelf of the Bering Sea and found that a sheet of sand covering the bottom of the Chirikov Basin was inhabited by the amphipods that were the whales' favorite food. In addition, the extent of this sand sheet matched the area where the whales were observed to be feeding. Kirk Johnson and his colleagues confirmed that the pits Nelson found had been caused by the gray whales. Using direct measurements of the pits and data obtained from sonograms (impressions of the seafloor derived from sound waves), they were able to show that

the pits could indeed have been formed by whales and that the area where the pits are located corresponded to the whales' feeding grounds.

It was later realized that walruses were also involved in changing the shape of the seafloor in the region. Pacific walruses disturb the bottom as they forage for clams and other types of bottom-dwelling prey, producing grooves or furrows as they feed. The walruses are year-round inhabitants of the area. In 1972 Samuel Stoker of the University of Alaska saw the furrows from a submarine and noticed walruses feeding nearby. He suggested that the furrows were the feeding tracks of the walruses. Ten years later John Oliver of Moss Landing Marine Laboratories showed that clam shells that appeared to have recently been excavated and emptied were found along the furrows. These observations confirmed the findings of sand, gravel, and clam shells in the stomachs of walruses by Eskimos, who concluded that they were bottom feeders. Researchers noted that the furrows were generally in clam beds and that the width of the furrows was approximately the same as a walrus snout. This finding was consistent with Oliver's suggestion that walruses unearth their food with their lips and not with their tusks, as was previously assumed.

The feeding activities of whales and walruses seem to be beneficial to the ecosystem and enhance the area's productivity. The feeding activity of the whales separates mud deposited by the Yukon River from the sand, thus preserving the sandy bottom as an ideal habitat for amphipods. During feeding, only adult amphipods are trapped on the baleen, whereas the smaller juveniles escape to replenish the supply. The feeding activity also releases nutrients from the bottom sediment and moves them into the water column where they stimulate the growth of plankton. ●

Sediments can also be classified according to their origin. Sediments formed by the precipitation of dissolved minerals from seawater are called **hydrogenous sediments.** Sediments formed from the hard parts of dead organisms are called **biogenous sediments. Terrigenous sediments** come from land and are washed into the sea. **Cosmogenous sediments** are formed from dust and debris from outer space.

## Hydrogenous Sediments

Some sediments form from seawater as a result of a variety of chemical processes. Carbonates, phosphorites, and manganese nodules are examples of the sediments that may form in salt water and accumulate on the seafloor. In shallow areas, calcium salts, sulfates, and rock salt can form on the bottom of pools when enough water evaporates to cause these minerals to precipitate.

## Biogenous Sediments

Sediments from living organisms (biogenous sediments) are mostly particles of corals, mollusc shells, and the shells of microscopic, planktonic organisms. In the deep sea almost all of the biogenous sediments are composed of the remains of single-celled organisms, such as diatoms, radiolarians, foraminiferans, and coccolithophores, and of molluscs known as pteropods, or sea butterflies (Figure 3-15). (For more information on single-celled marine organisms see Chapter 6). These organisms produce shells of calcium carbonate (foraminiferans, coccolithophores, and molluscs) or silica (diatoms and radiolarians). When they die their remains settle on the sea bottom. If more than 30% of an area's sediment is made up of these fine biogenous particles, then the sediment is termed an **ooze: calcareous ooze** (calcium) or **siliceous ooze** (glass), depending on which of the two shell types is dominant. The amount of biogenous sediment in an area depends on the mass of organisms contributing to the

sediment and the ability of the water to dissolve the minerals in the shells.

## Terrigenous Sediments

Some sediments are produced from continental rocks by the various actions of wind, water, freezing, and thawing. These particles are carried from the land by water, wind, ice, and gravity and are deposited primarily on the continental shelves. **Mud** is found in ocean basins throughout the world wherever marine life is too scarce to form biogenous sediments. Mud is composed of clay and silt. Clay is composed of fine powdered rock, and its color depends on the type of minerals that it contains. For instance, red clay contains large amounts of iron compounds. It is believed that this fine powder is blown from land by the wind or rinsed by rain from the atmosphere. The particles are so fine that they may remain suspended in the water for many years before settling.

## Cosmogenous Sediments

Iron-rich particles from outer space strike the surface of the ocean and slowly drift to the seafloor. Many dissolve before reaching the bottom. Although not as numerous as other sediments, small amounts are found scattered on the bottom of all seas.

## In Summary

The seafloor is covered with a variety of sediments that can be classified according to size or on the basis of their origin. Sediment particles can be formed by the action of physical processes on rocks, from the seawater itself, and from space. A substantial source of ocean sediment is small organisms that have shells of silica or calcium carbonate. When these organisms die, their shells are deposited on the seafloor and make up biogenous sediments.  ●

# FINDING YOUR WAY AROUND THE SEA

Sometimes it is necessary for marine biologists to return to the same point in the ocean to take samples or to make observations. This can be quite challenging on the open sea, out of the sight of land, because everything looks the same. There are no obvious features such as the landmarks on land or along coasts that can be used to determine your exact location. Under these circumstances, marine biologists must have a basic knowledge of navigational techniques so they can find their way around the ocean accurately. Navigators use maps and charts as well as sophisticated modern technology to locate specific areas of the ocean.

## Maps and Charts

A two-dimensional representation of the earth's three-dimensional surface is called a map or chart. **Maps** usually

Dr. Leslie Sautter/Ocean Explorer/NOAA

**Figure 3-15 Biogenous Sediments.** Biogenous sediments are primarily composed of the shells of organisms such as foraminiferans, molluscs, and corals.

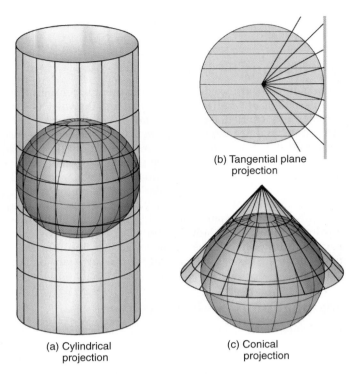

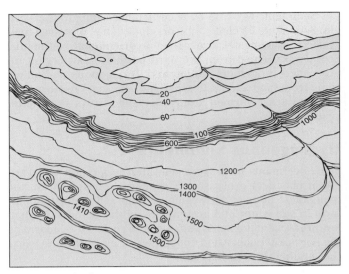

(b) Tangential plane projection

(a) Cylindrical projection

(c) Conical projection

**Figure 3-16  Map Projections.** If a light were placed inside a transparent globe, the pattern of light and shadow falling on a piece of paper placed in a particular position would form a map. The three most common arrangements for producing a map are the (a) cylindrical projection, (b) tangential plane projection, and (c) the conical projection.

show the land features of the earth, whereas **charts** display the oceans and their features. Maps and charts are made by projecting the features of the earth's surface together with reference lines onto a surface to produce a chart projection (Figure 3-16). If you could imagine a clear glass globe with reference lines drawn on the surface and a light bulb placed inside, you could get an idea of how a projection is made. As the light shines through the globe, it would project an image onto a surface, such as a piece of paper that is placed in a particular position with reference to the globe. Depending on the position of the paper, one could make three basic types of chart projections: cylindrical, conical, or tangential plane.

The most familiar maps and charts, such as the ones found in texts (including this one) and in classrooms, are modifications of the cylindrical type called a **Mercator projection** (for examples see Figures 3-3 and 3-8). Even though the pole regions are greatly distorted and the poles themselves cannot be shown with this projection, it has the advantage that a straight line drawn on it is a line of true direction (constant compass heading), and as such is useful for navigation. Charts of the ocean that show lines connecting points of similar depth are called **bathymetric charts** (Figure 3-17a); these are similar to topographic maps showing land elevations. A chart that uses perspective drawing

(a) Bathymetric chart

(b) Physiographic chart

**Figure 3-17  Bathymetric and Physiographic Charts.** (a) A bathymetric chart indicates variations in ocean depth by lines that connect areas of similar depth. (b) A physiographic chart shows the same information as a bathymetric chart, but it uses coloring or shading rather than lines to indicate areas of similar depth.

(picturing the sea as it appears to the eye with reference to a relative depth), coloring, or shading instead of lines to show the varying depths of the oceans is called a **physiographic chart** (Figure 3-17b).

## Reference Lines

To locate a specific spot on the surface of the earth or navigate from one place to another, we need a frame of reference. For this we use a grid composed of two sets of lines, called **latitude** and **longitude.** The grid is superimposed on the earth's surface and divides it into sections.

## Latitude

The earth is essentially a sphere, and a circle drawn around the center of the earth perpendicular to its axis of rotation is the **equator** (Figure 3-18a and b). This line is designated 0 degrees latitude. Each half of the earth is then divided again by other latitude lines at equal intervals ending at the poles. The North Pole is 90 degrees north with reference to the equator, and the South Pole is 90 degrees south. Notice that each line of latitude forms a progressively smaller circle, and that it is necessary to state whether the latitude is north (N) or south (S) of the equator. Latitude lines are sometimes referred to as **parallels** because they are parallel to the equator as well as each other.

In addition to the equator, other well-known parallels include the **Tropic of Cancer** (23.5 degrees N) and the **Tropic of Capricorn** (23.5 degrees S), which set the boundaries of what is called the tropics, or tropical zone, and the **Arctic Circle** (66.5 degrees N) and the **Antarctic Circle** (66.5 degrees S). The odd latitudes (numbers that are not multiples of ten) for these parallels are determined by the position of the sun and radiant energy during the June and December solstices (times of the year when the overhead position of the sun is farthest from the celestial equator). The Tropic of Cancer lies at the point where the sun shines directly overhead during the summer (June) solstice. Likewise, the Tropic of Capricorn lies where the sun shines directly overhead during the winter solstice (December). The Arctic Circle marks the northernmost limit of sunlight at the December solstice.

In the area north of the Arctic Circle the sun does not rise for one or more days on approximately December 21 and does not set for one or more days on approximately June 21. The Antarctic Circle marks the southernmost limit of sunlight at the June solstice. The area south of the Antarctic Circle experiences constant darkness for one or more days on approximately June 21 and constant sunlight for one or more days on approximately December 21.

## Longitude

Lines of longitude, also known as **meridians,** are at right angles to the latitude lines, extending from the North Pole to the South Pole. The primary line of longitude is an arbitrary one that passes through the Royal Naval Observatory in Greenwich, England. This line is known as the **prime, or Greenwich, meridian** and is designated 0 degrees longitude. Directly opposite the prime meridian at 180 degrees is a line that approximates the **international date line.** Longitudinal lines are identified by the angle that they form with respect to the prime meridian and their direction from it, east (E) or west (W). Unlike latitude lines, longitude lines all form circles of the same size, because each passes through the earth's poles and equator.

## Divisions of Latitude and Longitude

Using latitude and longitude one can exactly locate any point on the earth's surface. For instance, the Canary Islands lie at 15 degrees W and 28 degrees N (see Figure 3-18c). A de-

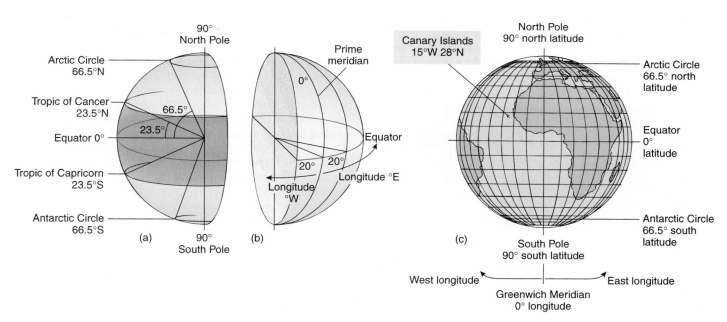

**Figure 3-18  Latitude and Longitude.** (a) The point of reference for latitude is the equator (0 degrees latitude). Latitude lines are then drawn at equal intervals from the equator to each pole (90 degrees latitude). (b) The prime, or Greenwich, meridian is the reference point for lines of longitude (0 degrees longitude). Other lines of longitude are designated by the angle they form with respect to the prime meridian and by direction (east or west). (c) Together lines of latitude and longitude form a grid that, when superimposed on the earth's surface, can be used to locate the precise position of any point on the globe.

gree of latitude or longitude covers a large area, so each degree is subdivided into 60 minutes and each minute into 60 seconds for giving a more precise location. A common unit of distance used in navigation is the **nautical mile,** which is equal to 1 minute of latitude (1.85 kilometers, or 1.15 land miles).

## Navigating the Ocean

Early navigators had a difficult time accurately determining their position. Charts and maps were not particularly accurate, and they exhibited a great deal of artistic license. To compensate, early sailors learned to use the sun and stars to navigate. Today sophisticated electronic devices and satellites help navigators find their position at sea.

### Principles of Navigation

From the earliest days of sailing the ocean, humans have used celestial guides, the stars and sun, to navigate. For instance, sailors knew that the North Star, Polaris, was positioned approximately over the North Pole. By measuring the angle of the North Star with respect to the horizon, using an instrument called a **sextant** (Figure 3-19), a seaman could get a good idea of his latitude, as long as he was in the Northern Hemisphere. In the early days of sailing, the standard procedure for navigation, once out of sight of land, was to sail north or south to the desired latitude, then sail east or west until reaching land again. When land was sighted, north–south adjustments to the course could be made with reference to landmarks along the coast.

Determining longitude was much more difficult, because the earth rotates and so do the longitudinal lines. In the 1500s, the Flemish astronomer Gemma Frisius proposed that longitude could be determined by setting a clock

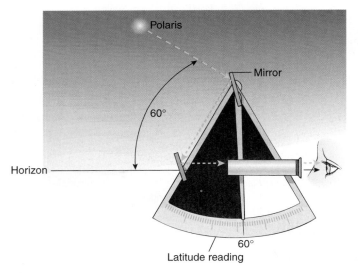

**Figure 3-19 A Sextant.** Sextants like this one are used to measure the angle of the North Star with respect to the horizon. This information can then be used to determine a ship's latitude.

to exactly noon when the sun was at its zenith (highest point above a reference longitude), then moving to a different longitude and taking the time when the sun reached its zenith at the new position. Because the earth rotates 15 degrees each hour, one could determine the new longitude by finding the difference in time: If the zenith at the new longitude occurred at 1 p.m. (one hour later), the ship would be 15 degrees to the west. If the zenith at the new longitude occurred at 11 a.m. (23 hours later), the ship would be 15 degrees to the east. Despite such early theories, however, clocks were not accurate enough on sailing vessels to be of any great use. In 1735 an Englishman named John Harrison built the first seagoing **chronometer,** a clock that could keep precise time, and he used it to determine longitude. However, it was not until 1761, with his fourth model, that the instrument proved itself, losing only 51 seconds during an 81-day voyage.

Today the reference longitude for time is the prime meridian, and the earth is divided into time zones (Figure 3-20), each one 15 degrees of longitude wide. You will notice as you look at Figure 3-20, however, that time zones have become greatly modified by humans, and they often differ dramatically from the ideal 15 degrees. For example, even though China spans four actual time zones, the entire country is on a single time.

### Global Positioning System

Modern electronic navigational devices are very important to marine scientists because they help them determine the exact position of their research vessels in any type of weather, permitting them to return again and again to the same position for more sampling or study. The most sophisticated of these modern systems is the **global positioning system,** or GPS. GPS utilizes a system of satellites that can be used to find an exact position anywhere on the earth at any time and in any weather. There are three parts to GPS: the space component, the user component, and the control component. The space component consists of 24 satellites, each in its own orbit 11,000 nautical miles above the earth. The satellites are spaced in such a way that at any time at least four of them will be above the horizon from any point on earth. The user component consists of receivers, which you can hold in your hand or mount in your vessel (or other vehicle). The control component consists of five ground stations located around the world that make sure the satellites are working properly.

Each of the GPS satellites takes 12 hours to orbit the earth and is equipped with an atomic clock, radio transmitter, and computer. The user-component unit receives the satellite signal and determines the amount of time it took the signal to reach the receiver. The difference between the time the signal was sent and the time it was received enables a computer in the receiver to calculate the distance to the satellite. By calculating this information from at least three different satellites, the receiver can calculate a position (latitude and longitude) that is accurate to less than 1 meter (3.3 feet), depending on the quality of the receiver being used.

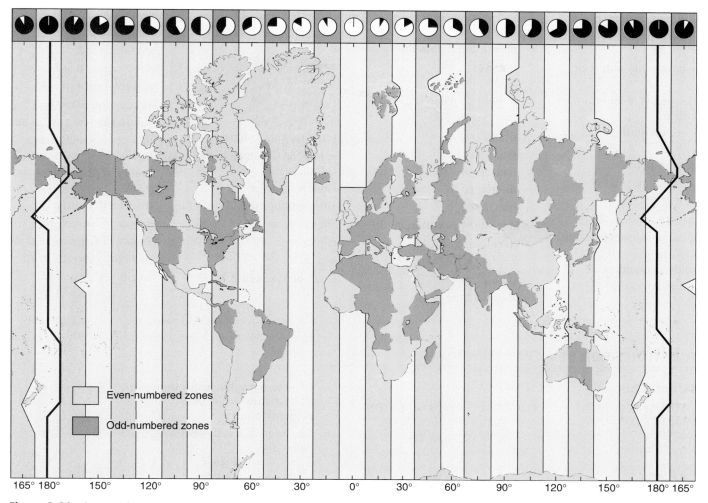

**Figure 3-20  The Earth's Time Zones.** The reference longitude for time is the prime meridian. Because the earth rotates 15 degrees each hour, each of the world's time zones is 15 degrees wide. Yellow bands represent even-numbered hours and pink bands represent odd-numbered hours. Countries shown in yellow are in the closest yellow time zone, whereas countries shown in red are in the closest pink time zone. Notice that some countries like China span more than one time zone but the country is on one time. Countries in blue or purple set their own time independent of the times zones.

## In Summary

Navigators use charts and maps with lines of latitude and longitude to locate their position while at sea. Lines of latitude run perpendicular to the long axis of the earth, which extends from pole to pole. Lines of longitude run perpendicular to the lines of latitude. Modern navigational techniques use sophisticated electronic equipment and satellites to help marine biologists find their location accurately any time, anywhere, and in any weather. ●

## SELECTED KEY TERMS

abyssal hill, *p. 43*

abyssal plain, *p. 43*

Antarctic Circle, *p. 51*

Arctic Circle, *p. 51*

asthenosphere, *p. 36*

bathygraphic features, *p. 42*

bathymetric chart, *p. 50*

biogenous sediment, *p. 48*

calcareous ooze, *p. 48*

chart, *p. 50*

chronometer, *p. 52*

continental drift, *p. 37*

continental margin, *p. 42*

continental rise, *p. 42*

continental shelf, *p. 42*

## QUESTIONS FOR REVIEW

### Multiple Choice

1. What portion of the earth's surface is covered by oceans today?
   a. 25%
   b. 50%
   c. 66%
   d. 71%
   e. 88%

2. How many major ocean basins are there?
   a. 1
   b. 2
   c. 3
   d. 4
   e. 5

3. A line drawn around the center of the earth perpendicular to the earth's north–south axis is the
   a. prime meridian
   b. Tropic of Cancer
   c. Tropic of Capricorn
   d. equator
   e. international date line

4. One minute of latitude is equal to
   a. one minute of time
   b. one degree of angle
   c. one nautical mile
   d. one minute of longitude
   e. one meridian

5. Charts that use lines connecting points of similar depth to show the physical characteristics of the ocean bottom are
   a. topographic charts
   b. bathymetric charts
   c. physiographic projections
   d. Mercator projections
   e. depth charts

6. The edge of a continent where land meets water is called the
   a. continental shelf
   b. continental ridge
   c. continental rise
   d. continental trench
   e. subduction zone

7. The bottom of many ocean basins is a flat
   a. ridge
   b. seamount
   c. trench
   d. valley
   e. plain

8. Biogenous sediments are formed from
   a. sediments washed into the sea by rivers
   b. chemical reactions between elements in seawater
   c. meteor particles showering down on oceans
   d. the shells of plankton and small animals
   e. all of the above

9. The outermost layer of the earth is called the
   a. mantle
   b. lithosphere
   c. asthenosphere
   d. atmosphere
   e. crust

10. Regions where the lithosphere splits, separates, and moves apart as new crust is formed is called a
   a. rift zone
   b. subduction zone

c. deep-sea vent
d. trench
e. continental rise

## Short Answer

1. How do oceans differ from seas?

2. What was the earth's early atmosphere like and how did it influence the evolution of cells?

3. What evidence supports the theory of continental drift?

4. What processes are responsible for the formation and sculpting of the continental shelves?

5. What is a seamount and how is it formed?

6. Explain how the oceans were originally formed.

7. Explain how time can be used to help determine a ship's longitude.

8. Explain how biogenous sediments are formed.

9. Describe how the continents are thought to move apart.

10. What is the global positioning system?

## Thinking Critically

1. How would you expect continental drift to affect the distribution of bottom-dwelling marine organisms?

2. Would you expect to find more actively swimming fish species above the continental shelves or in the large expanses of open sea? Explain.

3. While doing research along a coastal area, you discover that the bottom sediments are predominantly calcareous ooze. What does this imply about the local conditions?

## SUGGESTIONS FOR FURTHER READING

Allegre, C. J., and S. H. Schneider. 1994. The Evolution of the Earth, *Scientific American* 271(4):66–75.

Ballard, R. 1975. Dive into the Great Rift, *National Geographic* 147(5):604–15.

Bloxham, J., and D. Coubbins. 1989. The Evolution of the Earth's Magnetic Field, *Scientific American* 261(6): 68–75.

Bonati, E. 1994. The Earth's Mantle below the Oceans, *Scientific American* 270(3):44–51.

*Continents Adrift and Continents Aground: Readings from Scientific American,* 1976. San Francisco: Freeman.

Curtsinger, B. 1996. Realm of the Seamount, *National Geographic* 190(5):72–80.

Marine Geology and Geophysics, *Oceanus* 35(4), Winter 1992–1993.

### InfoTrac College Edition Articles

Hutchison, D. 1992. Continental Margins, *Oceanus* 35(4).

Taylor, B. 1992. Island Arcs, Deep Sea Trenches, and Back Arc Basins, *Oceanus* 35(4).

### Websites

**http://pubs.usgs.gov/publications/text/dynamic.html**
A concise history and explanation of plate tectonics.

**http://www.sfos.uaf.edu/msl111/notes/bottom.html**
A good overview of the different types of ocean sediments and their distribution.

# 4

# Water, Waves, and Tides

## Key Concepts

1. The polar nature of water accounts for many of its physical properties.

2. Seawater contains a number of salts, the most abundant being sodium chloride.

3. Salts are constantly being added to and removed from the oceans.

4. The exchange of energy between oceans and the atmosphere produces winds that drive ocean currents and weather patterns.

5. The density of seawater is mainly determined by temperature and salinity.

6. Vertical mixing of seawater carries oxygen to the deep and nutrients to the surface.

7. Waves are the result of forces acting on the surface of the water.

8. The gravitational pull of the moon and the sun on the oceans produces tides.

Living cells contain about two thirds water by mass and have a concentration of salts similar to that of seawater. The similarity between the salt concentration in cells and seawater reflects the fact that life evolved in the oceans. From the beginning, water has played a central role in the evolution of living organisms on this planet and continues to play a vital role in the maintenance of life. Marine organisms are constantly influenced by water and the forces associated with it, such as waves and tides. In this chapter you will learn about the characteristics of water, especially seawater, that make it unique and vital to life. You will also learn about currents, waves, and tides so that you can better understand the role these forces play in the biological communities that we will study in later chapters.

## NATURE OF WATER

Not only is water necessary for life but it is also the most abundant component of living things. For instance, most marine organisms are between 70% and 80% water by mass. By comparison, terrestrial organisms are approximately 66% water by mass. The importance of water to living organisms is related to its unique physical and chemical properties (Table 4-1).

### Physical Properties of Water

Water is an excellent **solvent** (medium for dissolving other substances). It has a high boiling point and freezing point compared with similar chemical compounds. It is denser in its liquid form than it is as a solid. Water helps to support marine organisms by buoyancy and provides a medium for the different chemical reactions that are necessary for life. Many of these physical properties are related to the structure of the water molecule itself.

NOAA; inset, NOAA

| Table 4-1  **Physical Properties of Water** | |
| --- | --- |
| Boiling point | 100°C |
| Freezing point | 0°C |
| Heat capacity | 1.00 cal/g/°C |
| Density (at 4°C) | 1.00 g/cm³ |
| Latent heat of fusion | 80 cal |
| Latent heat of vaporization | 540 cal |

## Structure of a Water Molecule

Water molecules are composed of two atoms of hydrogen bonded to one atom of oxygen. Chemists refer to the water molecule as being **polar** because different parts of the molecule have different electrical charges. Although the water molecule as a whole is electrically neutral, the electrons that form the chemical bond between the hydrogen and oxygen atoms are attracted more toward the large, positively charged oxygen atom. This causes the oxygen end of the molecule to have a slightly different charge than the two hydrogen ends (Figure 4-1a). The oxygen atom carries a slight negative electrical charge, and by the same token, the hydrogen atoms carry a slight positive electrical charge.

## Freezing Point and Boiling Point

Because water molecules are polar, they have a tendency to come together and form **hydrogen bonds** with each other. Hydrogen bonds are weak attractive forces that occur between the slightly positive hydrogen atoms of one molecule and the slightly negative oxygen ends of a neighboring molecule (see Figure 4-1b). It is the ability of water molecules to form hydrogen bonds that accounts for many of water's special properties. For instance, the relatively high boiling point of water (100°C) reflects the substantial amount of energy

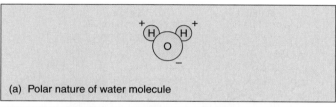

(a)  Polar nature of water molecule

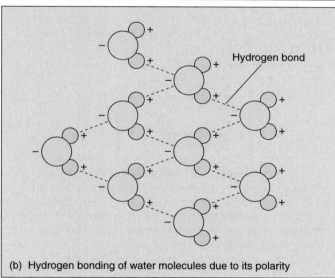

(b)  Hydrogen bonding of water molecules due to its polarity

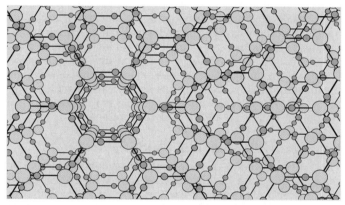

(c)  Structure of water molecules in a solid state (ice)

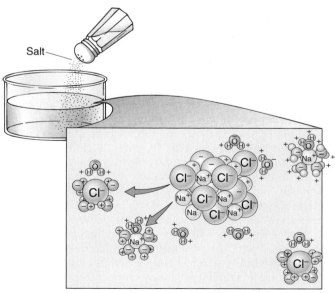

(d)  Salt crystals dissolving in water

**Figure 4-1  Water Molecules.** (a) A water molecule contains two atoms of hydrogen and one atom of oxygen. Water molecules are polar because the hydrogen atoms have a slight positive charge and the oxygen has a slight negative charge. (b) The polar nature of water molecules causes them to be attracted to each other and form hydrogen bonds. The hydrogen bonding in water accounts for its relatively high freezing point and boiling point. (c) Because of the polar nature of water molecules, when water freezes the molecules spread out. As a result solid water (ice) is less dense than liquid water and floats. (d) The ions in salts are attracted to polar water molecules and readily move into solution. In solution, the ions are surrounded by water molecules so there is less chance of their being attracted to each other and reforming crystals.

that is required to overcome the attractive forces of the hydrogen bonds so that individual water molecules can separate from each other and go into the gaseous state. The comparatively high freezing point of water (0°C) is also due to the presence of hydrogen bonds. Because the molecules have a natural attraction to each other, less energy is required to fix them into position so that they will form a solid. When water does freeze, the molecules move away from each other because of repulsive electrical forces between electrons of neighboring atoms. As a result, solid water (ice) is less dense than liquid water, and it floats (see Figure 4-1c). This characteristic is significant because a body of water freezes at its surface, forming an insulating layer, and the water below stays liquid. This keeps the oceans from freezing solid and allows living organisms to survive, even when the ocean surface is frozen.

## Water as a Solvent

The polar nature of water molecules also accounts for water's remarkable solvent properties. Other polar substances, such as salts, readily dissolve in water. Salt crystals are made up of individually charged particles called **ions,** which are attracted to the polar water molecules. As ions separate from the crystal, the crystals dissolve in the water. The attraction of polar water molecules for charged ions then helps to keep the ions in solution once the crystals have dissolved (see Figure 4-1d). Water is not able to dissolve molecules that are nonpolar, such as oil and petroleum products.

## Cohesion, Adhesion, and Capillary Action

The hydrogen bonds in water cause water molecules to be very cohesive, that is, they stick together. This cohesiveness gives water a high **surface tension.** The water molecules at the surface of the liquid have a greater attraction for other water molecules than they do for molecules in the air. As a result, the molecules at the surface form a tight, closely packed layer held by the hydrogen bonds of the water molecules beneath the surface. The high surface tension contributes to the high boiling point of water, since water must take in enough energy (the latent heat of evaporation) to overcome the surface tension before the molecules move into the vapor state. Many small organisms such as water striders (*Halobates*) take advantage of water's surface tension to give them support (Figure 4-2).

Water is also attracted to the surface of objects that carry electrical charges. This property is called **adhesion** and accounts for the ability of water to make things wet. It also accounts for the ability of water to rise in narrow spaces, a property known as **capillary action.** Water adheres to the surface of sediment particles and can become trapped in the spaces between them. The smaller the sediment particles, the smaller the spaces between them, which increases the movement of water by capillary action. Microscopic organisms that live in the spaces between sediment particles rely on adhesion and capillary action to supply the water necessary to support life (Figure 4-3). Without adhesion and capillary action, sediments could not trap or move sufficient amounts of water for these organisms to survive.

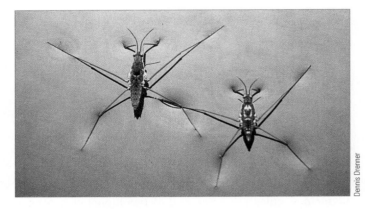

**Figure 4-2  Surface Tension.** Since water molecules have a greater attraction for each other than for molecules in the air, they form a tight layer at the surface that is able to support small organisms like this water strider.

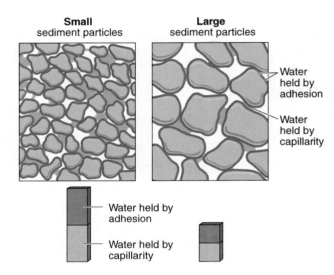

**Figure 4-3  Adhesion and Capillary Action.** Water moves through the tiny spaces between sediment particles and adheres to the surface of the particles, supplying moisture for numerous organisms that live in bottom sediments. The smaller the particles, the smaller the spaces between them and the more capillary water they can hold.

## Specific Heat

Because the presence of hydrogen bonds greatly restricts the movement of water molecules, it takes significantly more energy to raise the temperature of 1 gram of water 1°C than other common liquids. The amount of heat energy required to change the temperature of 1 gram of a substance 1°C is called the **specific heat,** or **thermal capacity.** Large bodies of water such as oceans maintain a more or less constant temperature because of the relatively large amount of energy they can absorb without changing temperature. Because the temperature of the ocean is relatively stable, marine organisms, with the exception of intertidal plants and animals, have not evolved mechanisms for adapting to rapid fluctuations in temperature, as have terrestrial plants and animals.

### Water and Light

The ability of light to penetrate water has important consequences for marine life. As you learned in Chapter 2, light (solar radiation) powers the process of photosynthesis, which produces food for most life in the sea, and is required for vision. Only a small portion of the light that reaches the sea, however, is available for driving photosynthesis. Much of the solar radiation is reflected by the water into the atmosphere, and wave action increases the amount of light that is reflected. The ability of water to selectively absorb certain wavelengths of light is another notable physical property of water. Low-energy red, orange, and yellow light are quickly absorbed, while high-energy blue, green, and violet light can penetrate deeper. This explains why objects appear more blue and green when viewed underwater. Clear water, such as the water around tropical coral reefs, appears blue because blue light predominates. Coastal seas appear greener because the water contains more sediments and plankton that absorb more of the blue, allowing the wavelengths of green light to dominate. In the turbid water of the Atlantic Ocean, almost 65% of the light that enters the water is absorbed within the first meter (3.3 feet). In clear ocean water, such as the tropical South Pacific Ocean, long-wavelength, low-frequency infrared light is absorbed within the first few centimeters, whereas short-wavelength, high-energy ultraviolet light can penetrate as deep as 200 meters (660 feet), the limit of the photic zone. In most clear water, however, less than 1% of the light entering the ocean can penetrate deeper than 100 meters (330 feet). The absorbed light energy is converted to heat that helps to warm the water. Heat energy, however, cannot power photosynthesis, so photosynthetic organisms are restricted to the upper, sunlit surface waters where there is sufficient radiant energy.

## Chemical Properties of Water

Chemical compounds are classified as acids or bases on the basis of how they ionize in water. **Acids** are compounds that release hydrogen ions ($H^+$) when they are added to water, whereas **bases,** or **alkaline** substances, can bind hydrogen ions and remove them from a solution. To indicate whether a solution is acidic or alkaline (basic), biologists use a shorthand notation called the **pH scale** (Figure 4-4). The pH is an indicator of the number (concentration) of hydrogen ions in a volume of a solution, and it is equal to the negative logarithm of the concentration of hydrogen ions in solution ($pH = -\log [H^+]$). The pH scale ranges from 0 to 14, with numbers below 7 indicating an acidic solution (higher concentration of $H^+$) and numbers above 7 indicating a basic, or alkaline, solution (lower concentration of $H^+$). The lower the pH, the more hydrogen ions there are in solution and the stronger the acid. The higher the pH, the fewer hydrogen ions there are in solution and the stronger the base. Since the pH scale is logarithmic, a change of one pH unit indicates a 10-fold change in the concentration of hydrogen ions. The

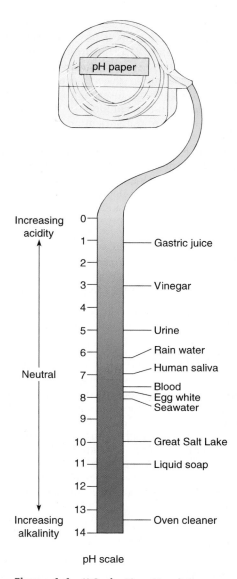

**Figure 4-4  pH Scale.** The pH scale is a convenient way of expressing the acidity of a solution. A pH of 7 is neutral. The more hydrogen ions a solution contains, the more acidic it is and the lower the pH. The more hydroxide ions a solution contains, the more alkaline it is and the higher the pH.

pH of pure water is 7, which is considered the neutral point. At a pH of 7, the number of acidic hydrogen ions in solution is equal to the number of basic hydroxide ions; thus pure water is neither acidic nor basic.

The pH of a body of water, such as the ocean, is determined by the substances that are dissolved in it. If more compounds are dissolved that release hydrogen ions, then the water will be acidic. On the other hand, if there are more ions that can bind to hydrogen ions and remove them from the solution, then the water will be alkaline. Seawater is slightly alkaline, with an average pH of 8. The reason for this is that seawater contains a large amount of alkaline ions, mainly bicarbonate ($HCO_3^-$) and carbonate ($CO_3^{2-}$), that can bind hydrogen ions and remove them from solution (see the following section on gases in seawater).

The pH of an organism's internal and external environment is a vital factor in determining the distribution of marine life. Changes in an organism's internal pH can affect the function of vital molecules, such as enzymes, and interfere with metabolism. Such a change could injure or even kill an organism. Changes in external pH can also affect metabolism and can interfere with growth. For example, corals cannot grow in water that is acidic because the low pH inhibits their ability to form external skeletons.

## In Summary

The unique physical and chemical properties of water make it a critical component of all living cells. Many of water's properties are due to the polar nature of the water molecules and their ability to form hydrogen bonds. ●

## SALT WATER

Ocean water is referred to as salt water because of the high quantity of dissolved salts compared with freshwater. As you learned in Chapter 2, the salinity of ocean water is an environmental factor that strongly influences marine organisms as they work to maintain appropriate levels of salts and water in their bodies. The variety of salts and other substances found dissolved in seawater are also important in determining the kinds and distribution of organisms in the marine environment.

## Composition of Seawater

Most of the salts that are present in seawater are present in their ionic form (Table 4-2). A total of six ions are responsible for 99% of the dissolved salts in the ocean: sodium ($Na^+$), magnesium ($Mg^{2+}$), calcium ($Ca^{2+}$), potassium ($K^+$), chloride ($Cl^-$), and sulfate ($SO_4^{2-}$). Other elements that are

**Table 4-2  Major Ions in Seawater**

| Ion | g/kg of Seawater | Percentage by Weight |
|---|---|---|
| Chloride ($Cl^-$) | 19.35 | 55.07 |
| Sodium ($Na^+$) | 10.76 | 30.62 |
| Sulfate ($SO_4^{2-}$) | 2.71 | 7.72 |
| Magnesium ($Mg^{2+}$) | 1.29 | 3.68 |
| Calcium ($Ca^{2+}$) | 0.41 | 1.17 |
| Potassium ($K^+$) | 0.39 | 1.10 |
| Bicarbonate ($HCO_3^-$) | 0.14 | 0.40 |
| Total | | 99.76 |

dissolved in seawater but present in concentrations of less than one part per million are called **trace elements.** The proportions of these major salts in seawater are relatively constant, and even though salinities can change, the relative proportions do not over much of the world ocean. This helps marine organisms because it makes this aspect of their environment predictable.

The ions and minerals in seawater are vital to the lives of marine organisms. Snails, clams, corals, and various crustaceans, for instance, require calcium to form their shells. Even minerals that are present in trace quantities are necessary to many marine organisms. Several marine organisms can concentrate trace minerals in their tissues at much higher levels than exist in seawater. For example, kelps (*Macrocystis*) and shellfish (shrimp, oysters, and so on) are able to concentrate high levels of iodine in their tissues. Kelps have even been commercially harvested as a source of this element, an essential nutrient in the human diet.

## Salinity

Seawater is approximately 3.5% salt and 96.5% water by mass. Sodium chloride (NaCl), or table salt, is the most common salt present. Since **salinity** (the concentration of salt in a given volume of water) is usually expressed in either grams of salt per kilogram of water or **parts per thousand** (ppt or symbolically represented as ‰), seawater would have an average salinity of 35 grams per kilogram, or 35 parts per thousand. The salinity of surface seawater varies with latitude and the topographical features of an area. These variations are the result of evaporation, precipitation, freezing, thawing, and freshwater runoff from nearby land masses. As you might expect, areas of ocean that are close to the mouths of rivers (fresh water) have comparatively low salinity, whereas surface waters in subtropical areas of the world where evaporation is high and precipitation is low have comparatively high surface salinities (Figure 4-5).

The processes of evaporation and precipitation contribute heavily to the salinity of surface water in the mid-ocean. Between 10 degrees N and 10 degrees S of the equator, rainfall is heavy and the surface water has a relatively low salinity. The regions of ocean at approximately 30 degrees N and S, the latitudes that correspond with many of the world's deserts, are characterized by evaporation exceeding precipitation. As a result, they have a relatively high salinity. Farther north and south, from 50 degrees latitude, precipitation is again heavy and the surface water is relatively less salty. Near the poles, freezing removes water from the sea, leaving the salt behind. Thus, the water beneath the ice has a relatively high salinity.

## Cycling of Sea Salts

The original sources of sea salts were rocks and other constituents of the earth's crust and interior. It is currently thought that these salts entered the ocean dissolved in water that seeped up through the ocean floor. Several processes continue to contribute salts to the ocean. Rocks previously formed on the seafloor release their ions as they are broken down by physical and chemical processes. Volcanic

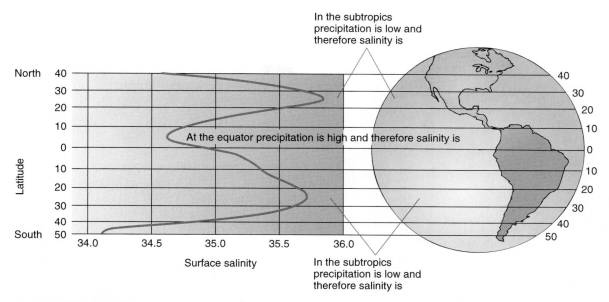

**Figure 4-5  Ocean Salinity.** The salinity of the surface water of the ocean varies with latitude as a result of regional differences in evaporation and precipitation.

eruptions produce gases, such as hydrogen sulfide and chlorine, which then dissolve in rainwater and enter the ocean with precipitation. River water carries large amounts of ions, formed by the weathering of rocks, into the sea.

Geologists believe that the composition of the oceans has been the same for about the last 1.5 billion years, despite runoff from the land annually adding $2.5 \times 10^{12}$ kilograms (about 250 million dump-truck loads) of salt to the sea. For the salinity of the ocean to remain the same over time, the amount of salt added to the oceans by runoff must be balanced by salt removal. Salt is removed from seawater in several ways. Some salt ions can react with each other to form insoluble complexes that precipitate to the ocean floor. Alternatively, when waves strike a beach, the sea spray is carried off by air currents and forms a layer of salt on surrounding rocks, land, buildings, and vehicles.

During the development of the earth, shallow extensions of the ocean became cut off from the sea, and the water evaporated. This process produced salt deposits called **evaporites** (the salt flats of southern California are an example). Similar situations exist today when small, shallow bodies of water become separated from the ocean and the water evaporates, leaving the salt behind. Biological processes also play a role in salt cycling. Certain ions absorbed by marine organisms are later returned to the environment when the organisms excrete them or die and decompose. Salts concentrated in the body tissues of marine organisms are permanently removed if the organisms are harvested for food.

The process responsible for removing the largest amount of salt, however, involves ions sticking to the surface of fine particles, a process known as **adsorption**. These particles then settle to the bottom and become trapped in the sediment. Clay particles formed from weathered rock and carried to the oceans by rivers attract ions and attach them to their surfaces by the process of adsorption. Ions and min-

erals deposited in this manner in the ocean's sediments can later be moved by geological processes to areas well above sea level where weathering can remove the minerals and return them to the oceans (Figure 4-6).

## Gases in Seawater

The gases of the atmosphere, primarily oxygen ($O_2$), carbon dioxide ($CO_2$), and nitrogen ($N_2$), are also found in seawater (Table 4-3). These gases dissolve at the surface of the sea from the atmosphere and are introduced into the water by biological processes. All of the gases are distributed throughout the water by mixing processes and currents.

### Gases from Biological Processes

Some bacteria, phytoplankton, algae, and plants use carbon dioxide in the process of photosynthesis, producing oxygen as a by-product. Most living organisms require oxygen and produce carbon dioxide. The process of decomposition, which requires respiration by bacteria or other organisms, also uses oxygen and produces carbon dioxide. Oxygen as a by-product of photosynthesis is added to seawater only near the surface. Just below the sunlit surface waters is the **oxygen-minimum zone.** This is an area in which oxygen is depleted by the resident animal life but not replaced by photosynthesis. Carbon dioxide, on the other hand, is added at all depths by the processes of decomposition and respiration. The levels of these two gases in seawater have a profound effect on the kinds of organisms that can live in a given area and the number of organisms a region can support.

### Solubility of Gases in Seawater

The proportion of oxygen and carbon dioxide in surface seawater is greater than it is in the atmosphere. The amount of nitrogen in seawater is less. Oxygen is far more soluble in

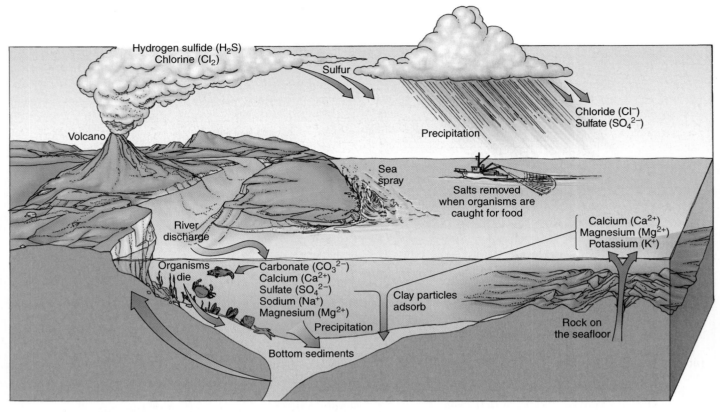

**Figure 4-6  Sea Salt Cycling.** Salts are constantly being added to and removed from seawater by a variety of processes.

seawater than nitrogen, and carbon dioxide is more soluble than oxygen. Carbon dioxide dissolves in seawater more readily than oxygen because it combines chemically with the water to form a weak acid called carbonic acid, which then dissociates into hydrogen ions and bicarbonate ions (Figure 4-7a). The chemical combination of carbon dioxide with water allows the seawater to hold much more carbon dioxide than the other two gases, which do not chemically combine. The amount of gas that water can hold depends on the temperature, salinity, and pressure of the water. For instance, cold water holds more gas than warm water, and more gas will dissolve if the salinity of the water is low or the

gas pressure is high. For these reasons warm tropical waters contain less oxygen than the cold water of polar seas.

### Role of Bicarbonate as a Buffer

The bicarbonate that is formed from the solution of carbon dioxide plays the vital role of a buffer. A **buffer** is a substance that can maintain the pH of a solution at a relatively constant point. Bicarbonate ions are buffers because they remove excess hydrogen ions if the water becomes too acidic or release hydrogen ions if the water becomes too alkaline (Figure 4-7b). By counteracting changes in the hydrogen ion concentration, the bicarbonate helps to maintain the pH of the water at a more or less constant value, maintaining a stable environment for marine organisms. The large amount of bicarbonate ion in the ocean is the main reason seawater has a slightly alkaline pH.

## In Summary

Seawater contains a number of salts; sodium chloride is the most abundant. The salinity of seawater varies with latitude because of evaporation, precipitation, freezing, thawing, freshwater runoff, and other processes. Evaporation and surf spray return salts to the land, and erosion processes return the salt to the sea. Seawater also contains gases, such as oxygen and carbon dioxide. Oxygen levels are highest and carbon dioxide levels lowest in the upper re-

| Table 4-3   Gases Found in Seawater | | | |
|---|---|---|---|
| Gas | Percentage by Volume in Atmosphere | Percentage by Volume in Surface Seawater | Percentage by Volume in Ocean Total |
| Nitrogen ($N_2$) | 78.08 | 48 | 11 |
| Oxygen ($O_2$) | 20.99 | 36 | 6 |
| Carbon dioxide ($CO_2$) | 0.03 | 15 | 83 |
| Other gases | 0.95 | 1 | |

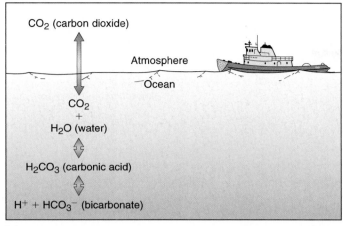

**(a)**

In acid, bicarbonate can act as a base.

$$H^+ + HCO_3^- \longrightarrow H_2CO_3 \longrightarrow H_2O + CO_2$$

In base, bicarbonate can act as an acid.

$$HCO_3^- + OH^- \longrightarrow H_2O + CO_3^{2-}$$

**(b)**

**Figure 4-7  Gases in Seawater.** (a) Carbon dioxide combines more readily with seawater than oxygen because it combines chemically with the water to form a weak acid, carbonic acid, which then dissociates into a hydrogen ion and a bicarbonate ion. (b) Bicarbonate ion acts as a buffer to help maintain the pH of seawater. If there is excess acid, it can combine with the hydrogen ion to form carbonic acid, which will break down to carbon dioxide and water. If the water is too alkaline, it can donate a hydrogen ion to form a carbonate ion and water.

gions of the ocean, where physical processes and photosynthesis add and remove the two gases. Deeper in the ocean, there is more carbon dioxide as the result of respiration and decay processes. Carbon dioxide is more soluble in seawater than oxygen because it can chemically combine with the water. The bicarbonate ion that forms from the chemical reaction between carbon dioxide and water is a buffer and accounts for seawater's slightly alkaline pH. •

# OCEAN HEATING AND COOLING

As mentioned earlier in this chapter, it takes a large amount of heat energy to change the temperature of a body of water as big as the ocean. The ability of the ocean to store and release heat has substantial implications for life on land as well as life in the sea.

## Earth's Energy Budget

The earth and the atmosphere continually receive energy from space, primarily in the form of radiant energy. They return energy to space in the form of heat. For the earth to maintain a relatively constant annual average temperature, the amount of heat gained has to be roughly equal to the amount lost. An imbalance in these processes can produce global warming or global cooling.

## Energy Input

The sun's radiant energy is responsible for warming the earth's surface. The latitudes between the Tropic of Cancer and the Tropic of Capricorn receive the greatest amount of radiant energy, whereas the middle latitudes receive moderate amounts, and the poles receive the least. This difference in the amount of radiant energy striking different parts of the earth's surface is caused by the earth's spherical shape and the presence of the atmosphere (Figure 4-8). The greatest amount of radiant energy is received where the sun's rays strike the earth at a right angle, around the equator. As you move to the north or the south, the angle of the sun's rays relative to the earth's surface increases because of the earth's curved surface. Consequently, at the middle latitudes and the poles, the same amount of sunlight falls on a larger area of the earth. Thus, these regions receive proportionately less radiant energy. The earth's atmosphere also absorbs some of the radiant energy before it strikes the earth. Atmospheric absorption causes the amount of radiant energy reaching the earth's surface to decrease with increasing latitude. The decrease is due to the longer path through the atmosphere that the sun's rays must travel to reach the earth's surface at higher latitudes. The combined effect of the earth's curved surface and atmosphere is a larger amount of radiant energy being received and transformed to heat around the equator, less in the temperate regions, and least at the poles.

## Energy Output

Because the earth's temperature is not constantly increasing over the short term, heat must be lost as well as gained. Some of the incoming solar radiation is reflected into space by the atmosphere and the surface of the earth and plays no

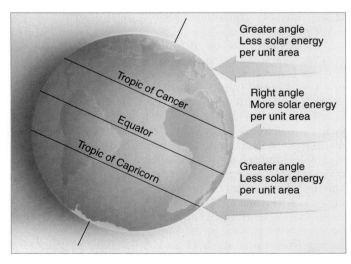

**Figure 4-8  Distribution of Solar Energy.** Because of the earth's curved surface, more radiant energy reaches the tropics than the temperate regions or the regions surrounding the poles.

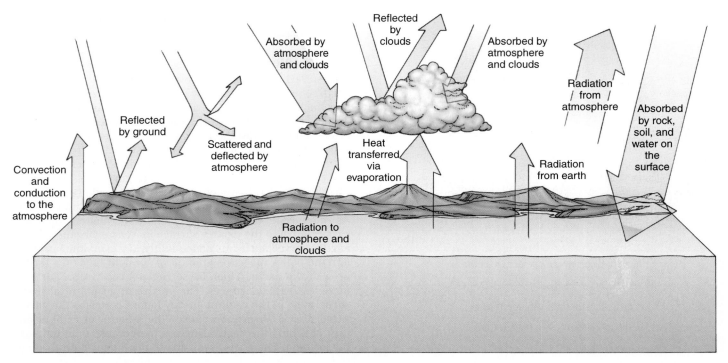

**Figure 4-9** **The Earth's Energy Budget.** Only about 47% of incoming solar radiation is absorbed by the earth's surface. The remainder is absorbed by the atmosphere or reflected into space. Heat is lost from the surface of the earth by the processes of evaporation, radiation, conduction, and convection. The width of the arrows in the diagram indicates the relative amount of energy being transferred.

role in heating the earth's surface. Of the remaining energy, most is absorbed by the earth's surface, and the rest is absorbed by the atmosphere. Ultimately, an equivalent amount of heat energy is reradiated back into space. The atmosphere loses more energy back to space than it gains from the sun, whereas the earth loses less energy back to space. In this way the atmosphere cools, and the surface of the earth warms.

To maintain the earth's temperature, the excess energy must be transferred from the earth to the atmosphere, and this is accomplished by evaporation and radiation (Figure 4-9). Energy from the earth's surface produces water vapor from liquid water by evaporation. This energy is then transferred to the atmosphere when the vapor condenses to form rain or freezes to form hail or snow. The heat that radiates from the earth's surface is absorbed by the water content of the atmosphere. Approximately two thirds of the heat transferred to the atmosphere is transferred by evaporation, and the other one third is transferred by radiation and other processes. The accumulation of certain gases, called **greenhouse gases,** such as carbon dioxide, methane, and chlorofluorocarbons (CFCs), in the atmosphere can prevent heat energy from radiating back to space. Like the glass in a greenhouse, these gases tend to block the escape of heat energy, causing an increase in the earth's average temperature known as **global warming.** The increase in average sea

temperature that has occurred because of global warming may be pushing some marine organisms into their stress zones, discussed in Chapter 2.

## Sea Temperature

When considering the amount of heat energy in an area of ocean, we must take into account several factors, such as the amount of energy absorbed at the sea's surface, the loss of heat energy by evaporation, the transfer of energy into or out of the area by ocean currents, the warming or cooling of the overlying atmosphere by heat from the sea's surface, and heat lost from the sea back to space by radiation. Since all of these processes change with time, regions of the ocean show both daily and seasonal variations in temperature (Figure 4-10). More heat is gained than lost in the region around the equator, whereas more heat is lost than gained at the higher latitudes. Winds and ocean currents are responsible for removing the excess heat from the tropics and moving it to the higher latitudes, so that the overall pattern of surface temperatures across the earth is maintained.

The amount of solar radiation reaching the earth also undergoes an annual cycle of seasonal variations. These variations are most pronounced between the latitudes 40 degrees and 60 degrees N and 40 degrees and 60 degrees S because the angle of the sun's rays changes so dramatically at

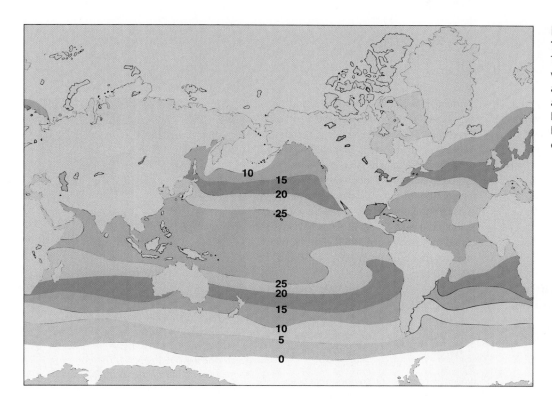

**Figure 4-10** **Average Surface Temperatures of the World Ocean.** The surface temperature of the sea varies from one region to the next as the result of differences in the amount of solar radiation received, loss of heat via evaporation, and heat energy transferred by ocean currents and the atmosphere.

these latitudes with changes in season. This seasonal change in radiant energy produces seasonal changes in the surface temperature of the ocean. Between the Tropic of Cancer and the Tropic of Capricorn, the amount of solar radiation remains relatively constant, and the surface temperature of the ocean also remains fairly uniform. As the sun makes its annual migration between the Tropic of Cancer and the Tropic of Capricorn, its rays cross this area twice, producing only a small, semiannual variation in the intensity of solar radiation. As mentioned previously in this chapter, oceans have a very high heat capacity, absorbing and releasing large amounts of heat at their surface with very little change in ocean temperature. During the summer, when the ocean receives more radiant energy, heat that is absorbed at the surface is moved deeper by the action of winds, waves, and currents, helping to stabilize the surface temperature.

## In Summary

The earth and the atmosphere receive energy from space in the form of radiant energy and return energy to space in the form of heat. An imbalance in the amount of heat gained and lost can produce global warming or cooling. Excess energy must be transferred from the earth to the atmosphere through evaporation and radiation. The accumulation of greenhouse gases can prevent heat from radiating back to space. These gases tend to reflect heat energy back to earth, causing an increase in the earth's average temperature known as global warming. Regions of the ocean show both daily and seasonal variations in temperature. Winds and ocean cur-

rents are responsible for removing the excess heat from the tropics and moving it to the higher latitudes, so that the overall pattern of surface temperatures across the earth is maintained. ●

## WINDS AND CURRENTS

Approximately 10% of ocean water is involved in surface currents. Driven by winds, the currents move water in the upper 400 meters (1,320 feet) of the ocean. These currents carry plankton in predictable paths through the surface waters, changing the distribution of planktonic food and dispersing planktonic larvae. Surface currents also influence the distribution of nekton, because predators follow the plankton and plankton feeders on which they prey. The factors that influence ocean currents have a dramatic impact on marine life as evidenced by changes caused by El Niño and similar phenomena.

### Winds

Winds are the result of horizontal air movements that are caused by factors such as temperature and density. As air heats, its density decreases and it rises; and as it cools, its density increases and it falls toward the earth. The density of air is determined by its temperature, water vapor content, and atmospheric pressure. Air density decreases when it is warmed, when its vapor content increases, or when the atmospheric pressure decreases. Air becomes denser when

# El Niño Southern Oscillation

Not only do oceans supply the water that allows life to exist on land they also influence other aspects of terrestrial climate. In 1983 it appeared that a dramatic change was occurring in the earth's climate. These changes were manifested in a variety of ways. Some areas, such as California and the coasts of South America, were drenched with rains that caused severe flooding. Other areas, such as Australia and Indonesia, suffered terrible droughts, whereas parts of Polynesia were hammered by severe typhoons. The effects of these changes were also seen in marine organisms. Off the coast of South America, large numbers of fish and sea birds died for no apparent reason, and many marine mammals, such as whales, dolphins, and seals, disappeared from their usual feeding grounds.

Scientists studying these events found that they were the result of changes in wind patterns and ocean currents that occur occasionally in winter. Farmers and fishermen in Peru had been aware of the phenomenon for years and had named it El Niño ("the Child"), in honor of the Christ child, since the phenomenon usually occurred around Christmas. An El Niño is usually a short-term, local phenomenon, but in 1983 it was particularly severe.

The coastal waters of Peru and Ecuador are highly productive fishery areas because of the nearly constant upwelling of deep, nutrient-rich water. The upwelling is driven by the Pacific trade winds that move large volumes of water westward along the ocean's surface. Upwelling cold, nutrient-rich water along the coast replaces the water moving out to sea. Sometimes the trade winds lessen, typically around Christmas, and warm tropical water moves eastward across the Pacific to accumulate along the coasts of North and South America.

The warm water kills cold-water organisms that are the basis of food chains for many marine fishes, mammals, and birds. The loss of food causes the death of these animals. In severe cases the number of decaying organisms is so great that the surface water contains large quantities of hydrogen sulfide. Although the increased temperature of the coastal waters usually ends by April, sometimes it may persist for as long as a year. Severe El Niños have occurred frequently since 1953, with the two most severe occurring in 1982–1983 and 1997–1998.

No one is certain of the exact cause of El Niños, but certain processes have been correlated with the phenomenon's appearance. One process is the Southern Ocean Oscillation, in which the atmospheric pressure on one side of the Pacific increases while the pressure on the opposite side decreases. The pressure changes then reverse. Normally, there is a high pressure system over Easter Island in the eastern Pacific and a low-pressure system over Indonesia in the western Pacific. Under these conditions the east-to-west trade winds are strong and constant, and upwellings occur along the coasts of Peru and Ecuador. When the normal pressure system reverses, the trade winds break down and warm water from the western Pacific moves eastward, ultimately warming the surface waters along the coast of South America and depressing the upwelling (Figure 4-A). The elevated surface temperature is the result of both warm water moving in and colder, denser water not being able to rise. The warm surface water carries the low-pressure zone of rising air and precipitation along with it, producing large amounts of precipitation in normally dry areas. The effects of El Niño usually lessen by midsummer. Another slight warming is often observed in November and December before the Southern Ocean Oscillation reverses again and the trade winds return to normal. A severe El Niño can last for 15 months.

**Figure 4-A** *El Niño Southern Ocean Oscillation.* *The orange and yellow band in the lower right of this picture is warm water of the 1991 El Niño. The orange and yellow band further north is a sea temperature change caused by the 1982–1983 El Niño. El Niños are thought to be the result of changes in atmospheric pressure over the eastern and western Pacific Ocean that produce changes in the tradewinds, causing a mass of warm water to move east across the Pacific from Asia. This changes the position of the atmospheric jet streams and results in climate shifts across the globe.*

Attempts have been made to forecast El Niños on the basis of changes in the Southern Ocean Oscillation and the strength of the trade winds, both of which appear to precede El Niños. The models used to study previous El Niños did not predict some events that occurred and predicted others that did not occur. As researchers continue to search for more clues that might increase their forecasting accuracy, the phenomenon of El Niño continues to remind us of the close interrelationship between marine and terrestrial ecosystems. ●

G. Jacobs, Stennis Space Center/SPL/Photo Researchers, Inc.

it is cooled, its vapor content decreases, or the atmospheric pressure increases. These forces produce two sets of winds that circulate in opposite directions: surface winds at the earth's surface and upper winds such as the jet stream.

## Wind Patterns

As noted previously, the region of the earth around the equator receives large amounts of the solar energy that heats the air and water. This combination of warm air with a high water vapor content produces less-dense air that rises. At the poles, on the other hand, the air is cooling, moisture has been removed by precipitation, and there is increased atmospheric pressure. This combination of factors causes the air to become denser and sink. The denser air from the poles is then sucked toward the equator to replace the less-dense air that has moved away. These processes establish a basic pattern of upper air flow (winds) from the equator toward the north in the Northern Hemisphere and from the equator toward the south in the Southern Hemisphere (Figure 4-11).

## Coriolis Effect

If the atmosphere were attached tightly to the earth's surface and the earth and its atmosphere moved in unison, then these north and south movements would be the primary air movements. The atmosphere, however, is not rigidly attached to the earth's surface, and the amount of friction between the moving atmosphere and the moving earth is so

small that the movement of the atmosphere is somewhat independent of the movement of the earth's surface. For instance, if a mass of air lies over the equator and appears to be stationary, it is actually moving eastward with the rotating earth.

As the earth rotates on its axis, a point at the equator moves faster than a point at a higher latitude. Both points make one full revolution per day, but the point at the equator has to travel a longer distance, therefore it must be moving faster. For this reason, air moving north or south away from the equator toward the poles moves progressively faster than the earth beneath it. This causes the track the air mass follows to appear to curve relative to the earth's surface. In the Northern Hemisphere this effect causes the air to deflect to the right of the direction of the air movement, and in the Southern Hemisphere the deflection of the air is to the left of the direction of air flow.

This apparent deflection of the path of the air is called the **Coriolis effect** (Figure 4-12a and b). The Coriolis effect is named after Gaspard Coriolis, who developed a mathematical model for deflection in frictionless motion when the motion is referenced to a rotating body. The Coriolis effect is applicable to any rotating body, and describes not just air movements but also movement of surface water and currents.

## Surface Wind Patterns

Because of the Coriolis effect, air that rises at the equator does not move straight to the north or the south but is deflected as it moves (see Figure 4-12c). At 30 degrees N and S, the equatorial air sinks and either moves back along the surface of the water to the equator or to 60 degrees N or S. The air that reaches the poles cools and sinks and begins to move back toward the equator. As the air flows back toward the equator, it warms and picks up water vapor, and by the time it reaches 60 degrees N or S, it rises again. The net result of these processes is the division of the earth's surface into three convection cells in each hemisphere.

Between the equator and 30 degrees N, the surface winds are deflected to the right, producing the **northeast trade winds.** In the Southern Hemisphere, the winds are deflected to the left, producing the **southeast trade winds.** (Winds are designated with reference to the direction from which they are coming, not to which they are blowing.) Between 30 degrees N and 60 degrees N, the deflected surface air produces winds that blow from the south and the west, producing the **westerlies.** In the Southern Hemisphere, the westerlies blow from the north and the west. Between 60 degrees N and the North Pole, the winds blow from the north and east, forming the **polar easterlies.** Between 60 degrees S and the South Pole, the easterlies blow from the south and east.

Low-density air rises at 0 degrees and 60 degrees N and S, whereas high-density air descends at 30 and 90 degrees N and S. Areas where the air rises are zones of low atmospheric pressure and usually exhibit clouds and rain. Areas where the air descends are associated with high atmospheric pressure, clear skies, and low precipitation. Surface winds move from areas of high pressure to areas of

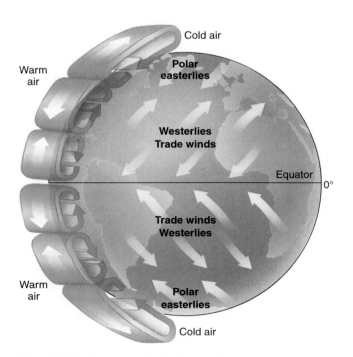

**Figure 4-11 North–South Air Flow.** As warm air from the equator moves north and south, it displaces colder, denser polar air that then flows toward the equator. This produces the basic pattern of upper air currents.

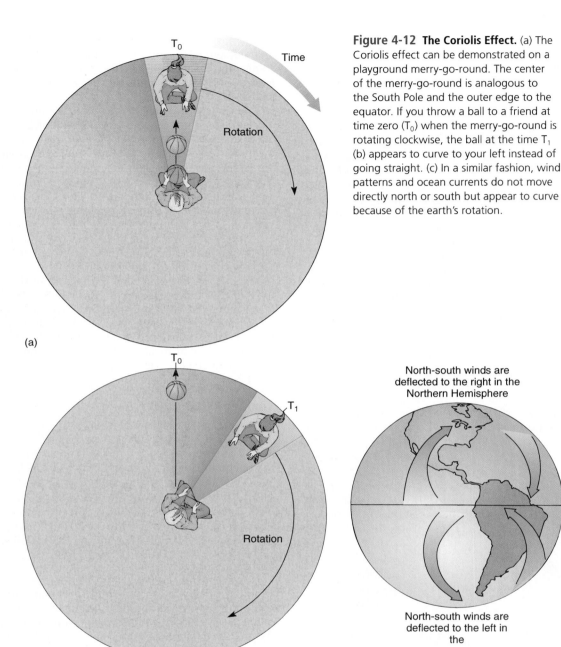

(a)

(b)

(c)

North-south winds are deflected to the right in the Northern Hemisphere

North-south winds are deflected to the left in the

**Figure 4-12  The Coriolis Effect.** (a) The Coriolis effect can be demonstrated on a playground merry-go-round. The center of the merry-go-round is analogous to the South Pole and the outer edge to the equator. If you throw a ball to a friend at time zero ($T_0$) when the merry-go-round is rotating clockwise, the ball at the time $T_1$ (b) appears to curve to your left instead of going straight. (c) In a similar fashion, wind patterns and ocean currents do not move directly north or south but appear to curve because of the earth's rotation.

low pressure (Figure 4-13). In the areas of vertical movement between the wind belts, wind movement is unsteady and unreliable. In the early days of sailing ships, these areas were a cause of great concern to sailors because they could be stranded for days without a breeze to move them. The area of rising air at the equator is known as the **doldrums,** and the areas of descending air at 30 degrees N and S are known as the **horse latitudes.** It is said that the horse latitudes received their name from early sailing ships carrying horses. If the ship became becalmed for a long time, there would not be enough drinking water to support both the crew and the horses, so the horses were thrown overboard.

## Ocean Currents

Ocean currents are produced when wind blowing across the surface of the ocean pushes and pulls the water, causing it to move. Like rivers on land, the ocean currents move water in predictable patterns within the ocean basins.

### Surface Currents

Winds transfer energy from the air to water by friction. This action produces waves that transfer energy to the water, causing a mass of water beneath the moving air to flow and form a surface current. The main driving forces for surface

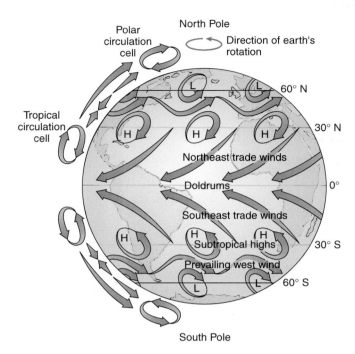

**Figure 4-13 Surface Wind Patterns.** Differences in air temperature and density are responsible for producing global wind patterns.

currents are the trade winds (easterlies and westerlies) in each hemisphere. As the water moves, it begins to accumulate in the direction that the wind is blowing. Gravity acts on the mass of accumulating water and tries to pull it back against the pressure gradient, but the water moves according to the Coriolis effect and continues moving in a circular pattern.

***Coriolis Effect***    Just as with air, the friction between the moving water and the earth is very small, and the water's motion is described by the Coriolis effect. Since water moves more slowly than air, it takes the water longer to move the same distance. During this longer time, the earth rotates farther out from under the water than it would from under air. Therefore, the slower-moving water is deflected to a greater degree than air. In the Northern Hemisphere, the current is deflected to the right of the prevailing wind direction, and in the Southern Hemisphere, the deflection is to the left. In the open sea, this deflection can be as much as a 45-degree angle from the wind direction.

***Gyres***    The position of the continents and the features of the ocean basins interfere with the continuous flow of water and contribute to the deflection of currents, so that water flows in a circular pattern around the edge of an ocean basin. This type of flow pattern is called a **gyre** (Figure 4-14a). There are five major gyres in the world ocean, three in the Southern Hemisphere and two in the Northern Hemisphere (Figure 4-14b), and each for the most part is independent of the others. A sixth major surface current is the Antarctic Circumpolar Current, which is driven by the strong westerly

winds. This is the largest of the ocean currents and flows continuously eastward around the continent of Antarctica. Since it is not deflected by a continental mass, it is technically not considered a gyre.

## Classification of Currents

Although the water within a gyre flows continuously, oceanographers divide each gyre into interconnected currents. Each of the currents can be distinguished on the basis of distinct temperature and flow characteristics. The currents can be classified by position as eastern-boundary currents, western-boundary currents, or transverse currents. The water in a current can move long distances along a well-defined path, and frequently the flow of water in these currents is compared to the flow of water in a river. With few exceptions these currents have a nearly constant pattern (Figure 4-15).

***Western-Boundary Currents***    The fastest and deepest currents are the **western-boundary currents** found along the western boundaries of ocean basins. They move warm water toward the poles in each gyre. The **Gulf Stream** is the largest of these western-boundary currents, moving at an average speed of 2 meters per second (5 miles per hour). At this rate the Gulf Stream can travel more than 160 kilometers (100 miles) per day. As you might imagine, the amount of water that a western-boundary current can move is quite large. The volume of water transported in an ocean current is measured in **sverdrups** (sv), a unit named in honor of the oceanographer Harald Sverdrup. A sverdrup equals a flow of 1 million cubic meters of water per second. The flow of the Gulf Stream averages 55 sverdrups.

***Eastern-Boundary Currents***    Eastern-boundary currents are exactly opposite western-boundary currents in most of their characteristics. Eastern-boundary currents carry cold water toward the equator. They are slow moving compared with their western counterparts, moving at rates of only 10 to 20 kilometers (6–12 miles) per day. The volume of water transported by eastern-boundary currents is also small, averaging 10 to 15 sverdrups. Unlike western-boundary currents, which frequently have sharp boundaries and are narrow, eastern-boundary currents lack sharp boundaries and are frequently very wide.

***Transverse Currents***    The eastern- and western-boundary currents in each gyre are connected by **transverse currents**. Although gyres are divided into currents, remember that each current flows uninterrupted into the next, such that the flow of water within a gyre is continuous.

***Biological Impact***    The characteristics of the individual ocean currents have significant biological implications. Because western-boundary currents move so quickly and move such large volumes of water, they tend to carry little in the way of nutrients, and are thus not very productive. The rapid movement, however, does increase the amount of oxygen mixed into the water. On the other hand, eastern-boundary currents, because they move less water and move more slowly, tend to be more productive and contribute to

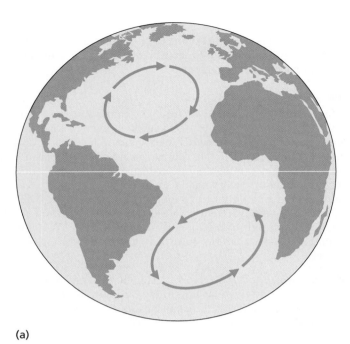

(a)

**Figure 4-14 Gyres.** (a) A combination of surface winds, the Coriolis effect, position of continental land masses, and the shape of ocean basins causes surface currents to form circular currents of water called gyres. (b) This map shows the location of the five major gyres.

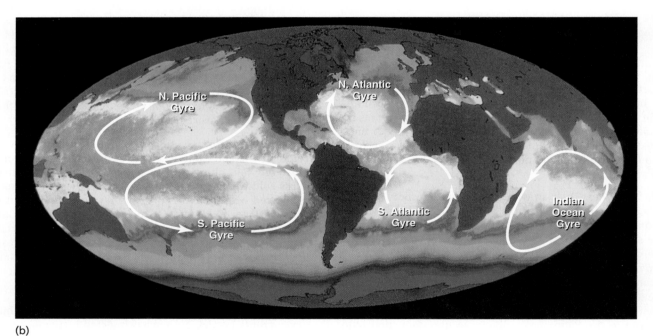

(b)

the mixing of nutrients into surface waters that is discussed later in this chapter.

## Currents below the Surface

The energy transferred from winds to water at the surface continues down to other layers of water. The movement of water at the surface causes a layer of water below it to move as a result of friction. The movement of this deeper layer of water is deflected to the right in the Northern Hemisphere and to the left in the Southern Hemisphere as a result of the Coriolis effect. This process continues to affect successively deeper layers of water to a depth of about 100 meters (330 feet). Each layer slides horizontally over the layer beneath, with each layer moving at an angle to the one above and moving slower than the layer above because of frictional loss of energy (Figure 4-16). The spiral flow of water that results is called the **Ekman spiral** after the Swedish oceanographer who developed a mathematical model to describe the process. The net movement of water to the 100-meter depth is called **Ekman transport.**

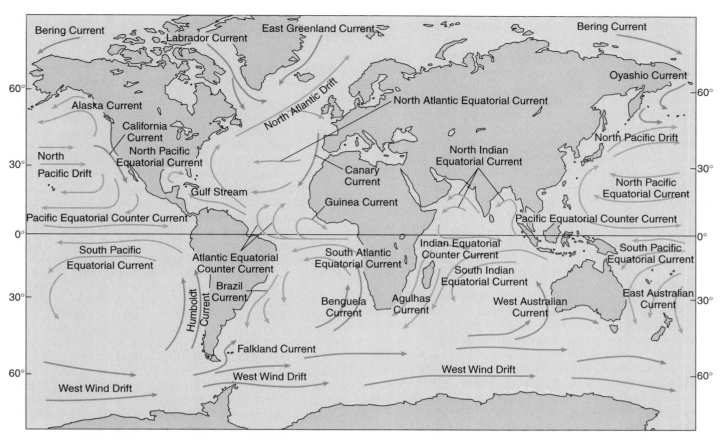

**Figure 4-15  Major Ocean Currents.** Ocean currents are produced by the driving force of wind. Notice the similarities in the patterns of ocean currents and the surface winds shown in Figure 4-13.

## In Summary

Winds are produced by differences in the density of air. Warmer air at the equator rises and moves toward the poles, whereas colder air at the poles sinks and returns to the equator. Air masses in the Northern Hemisphere move to the right, and air masses in the Southern Hemisphere move to the left. This apparent deflection of air masses is called the Coriolis effect. Wind patterns are responsible for driving ocean currents. Like air masses, ocean currents are affected by the Coriolis effect, causing them to deflect. The combination of wind, gravity, and the Coriolis effect produces gyres, circular patterns of water flow at the ocean's surface. Within each gyre specific currents are identified based on their physical properties. Although several currents make up a gyre, the surface current is one continuous flow of water.  ●

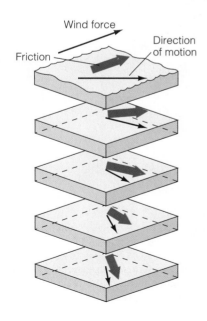

**Figure 4-16  Ekman Spiral.** The movement of surface water is passed along to deeper layers of water causing them to move in a characteristic pattern known as the Ekman spiral. As you go deeper, each successive layer of water moves at a slower speed and an angle to the layer above, producing the spiral movement.

## OCEAN LAYERS AND OCEAN MIXING

**Density** is a physical property of a substance: the mass of the substance in a given volume. It is generally measured in grams per cubic centimeter (g/cm³). For instance, pure water has a density of 1 gram per cubic centimeter. Because seawater contains dissolved salts, it has a higher density than pure water (1.0270 g/cm³). Variations in the surface temperature and salinity of the oceans control the water's density. Density increases when salinity increases or temperature decreases (recall that water reaches its maximum density at a temperature of 4°C). Density decreases when salinity decreases or temperature increases. Salinity increases when evaporation occurs or when some of the seawater comes out of solution to form ice. Precipitation, influx of river water, melting ice, or a combination of factors can all contribute to a decrease in salinity. Pressure can also affect the density of seawater, but pressure effects at the surface are minor.

In the open ocean, surface temperature is more decisive than salinity in determining the water's density. For instance, surface water in the tropics has the highest salinity in the open ocean, but the water is so warm that it remains less dense than the water below it and does not sink. On the other hand, water in the North Atlantic has a lower salinity but is much colder, and thus the surface water is denser and sinks. Salinity becomes a more important factor in when determining the density of water close to shore. This is especially evident in semi-enclosed bays that receive a large amount of freshwater runoff. Wave action contributes to the density and mixing of ocean water close to shore, where the water is relatively shallow. The combination of different surface temperatures and salinities produces regions of ocean with different densities (Figure 4-17). Denser water sinks

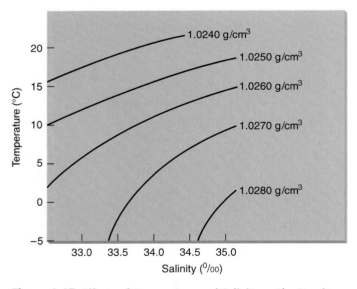

**Figure 4-17 Effects of Temperature and Salinity on the Density of Seawater.** Increases in salinity or decreases in water temperature result in an increase in the density of seawater. This graph shows the combined effects of salinity and temperature on seawater densities.

until it reaches water of similar density, whereas less-dense water rises. This situation is responsible for layered water in the ocean.

### Characteristics of Ocean Layers

The characteristics of temperature, salinity, and density change with depth. The surface layer of the ocean extends down to about 100 meters (330 feet), is warmed by solar heating, and is well mixed by a variety of processes. From 100 meters to 1,000 meters (3,300 feet), temperatures decrease, creating layers of water of increasing density (Figure 4-18a). This zone of rapid temperature change is called a **thermocline.** Similarly, below the surface waters in the temperate zone, the salinity increases with depth to about 1,000 meters (Figure 4-18b). This zone is called a **halocline.** The changes in temperature and salinity in the region from 100 meters to 1,000 meters produce a **pycnocline** (PIK-nuh-klyn), a zone where density increases rapidly with depth (Figure 4-18c). Below the thermocline, temperatures are relatively stable, with only small decreases in temperature toward the ocean bottom. Likewise, below the halocline, salinities are constant down to the ocean floor. As a result, water deeper than 1,000 meters has relatively constant density.

The thermocline, halocline, and pycnocline divisions are permanent features of the ocean because they are too deep to be affected by the mixing action of winds. In polar and temperate seas, however, there is also a shallow-water, seasonal thermocline that is generally not deeper than 70 meters (231 feet). During the summer months, the surface water in these seas is warm, the deep water is cold, and the water column is relatively stable. The seasonal thermocline traps nutrients in deeper water and limits productivity during the summer months. In the fall, as the surface water cools, its density increases, the water column becomes unstable, and the surface water sinks. The mixing process continues with the aid of winter storms and the continued cooling of surface waters. The thermocline developed during the previous summer is ultimately eliminated, and the density of the water becomes uniform from top to bottom and is easily mixed (Figure 4-19). Storms at this time of year produce strong winds that disrupt the water column and help to mix nutrients into the surface waters. The seasonal influx of nutrients supports high levels of primary production during these months. With spring comes warmer temperatures, and the thermocline begins to reestablish itself, stabilizing the water column. This process of stabilization then continues through the summer.

### Horizontal Mixing

Seawater that is not very dense, such as the warm, low-salinity water around the equator (1.0230 g/cm³), remains at the surface. Surface water at 30 degrees N and 30 degrees S is also warm, but it has a higher salinity and is denser (1.0267 g/cm³) than the equatorial waters. This difference in density causes the water at 30 degrees N to form a curved layer that sinks below the less-dense surface water at the equator and then rises to rejoin the surface at 30 degrees S. The combination of colder temperatures and higher

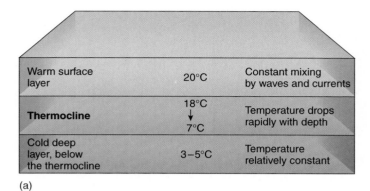

(a)

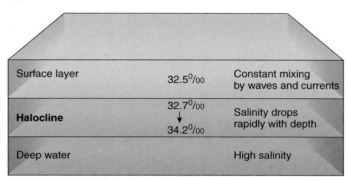

(b)

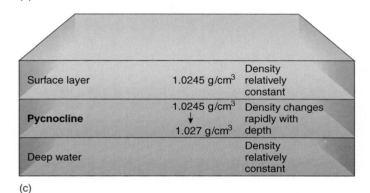

(c)

**Figure 4-18 Changes in Temperature, Salinity, and Density of Seawater with Depth.** (a) Changes in water temperature with depth. (b) Changes in salinity with depth. (c) Changes in density with depth.

salinities produces even denser surface water at 60 degrees N and 60 degrees S (1.0276 g/cm³). This water extends from the surface in one hemisphere below the other surface waters to the surface of the other hemisphere. During winter at the poles, the lower water temperature and increased salinity due to formation of sea ice result in very dense water that sinks toward the floor of the ocean.

## Vertical Mixing

When the density of water increases with depth, the water column from the surface down is said to be stable. If the top water in a water column is more dense than the water be-

low it, then the water column is unstable. Unstable water columns do not persist, because the denser water at the top sinks and the less-dense water below rises to the surface. This change-over in the water column produces a **vertical overturn** (Figure 4-20). If the water column has the same density from top to bottom it is neutrally stable, or **isopycnal** (eye-soh-PIK-nuhl). Neutral stability means there is no tendency for water in the column to sink or rise. A water column that is neutrally stable can be easily vertically mixed by wind, wave action, or currents.

Any process that increases the density of surface water will cause vertical movement of the water, or **vertical mixing**. Vertical mixing is one of the principal processes for the exchange of water from top to bottom throughout the world ocean. Since bottom water usually contains abundant nutrients from the settling of organic matter from above and the process of decomposition, vertical mixing provides a means of exchanging this nutrient-rich bottom water with oxygen-rich surface water. Since density is generally controlled by temperature and salinity, this type of circulation is also known as **thermohaline circulation**.

## Upwelling and Downwelling

Sometimes the horizontal movement of water that is produced by the winds can cause surface water to move vertically. This phenomenon is known as **wind-induced vertical circulation**. Since the quantity of water in the oceans is essentially fixed, any movement of water from one place to another would cause a similar but opposite movement to replace the water that leaves an area. For instance, when dense surface water sinks to a depth at which it is no longer denser than the water beneath it, it stops sinking. At this point, the water begins to move horizontally, making room for denser water that is sinking behind it. Eventually, this water rises back to the surface, replacing the surface water that has been sinking. At the same time, surface water is also moving horizontally into areas where the surface water is sinking. If the vertical movement is upward, it is called an **upwelling**, and if it is downward, it is called a **downwelling**. Upwellings and downwellings mix usually cold, nutrient-rich bottom water with the warmer, sunlit surface waters.

### Equatorial Upwelling

In most tropical seas, productivity is low because of the pycnocline that separates the nutrient-rich bottom water from the sunlit surface waters. The tropical waters along the equator are by contrast quite productive regions. This is due to water from currents on either side of the equator deflecting toward the poles, pulling surface water away that is replaced by deeper water. This movement of water is known as the **equatorial upwelling**. The equatorial upwelling brings nutrients to the surface and supports a high level of productivity all along the equator, which in turn provides food for large schools of fish, such as tuna.

### Coastal Upwelling

Upwellings also occur along coastal areas (Figure 4-21a). Winds blowing parallel to the coast produce surface currents that deflect according to the Coriolis effect. The Ekman

**Figure 4-19 Seasonal Changes and Vertical Mixing.** Seasonal changes in the temperature and salinity of surface waters of temperate and polar seas produce a mixing effect that brings nutrient-rich bottom water closer to the surface and oxygen-rich surface water closer to the bottom.

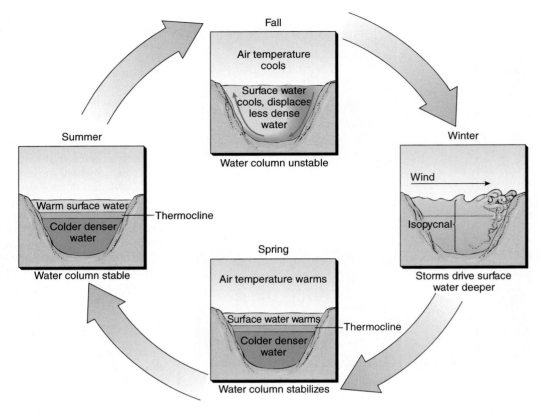

**Figure 4-20 Vertical Overturn.** (a) In a stable water column, water density increases with depth. (b) In an unstable water column, the water at the top of the column is denser than the water beneath. As a result, the denser surface water sinks and the less dense water beneath is displaced to the surface. This process is called vertical overturn. (c) An isopycnal water column has the same density from top to bottom and is stable.

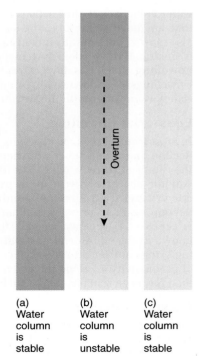

transport that results causes the water to move offshore. As the water moves offshore, deeper water along the shore rises to take its place. The deeper water is rich in nutrients from sedimentation and decay processes, and it supplies the needs of photosynthetic organisms and the zooplankton that are vital parts of oceanic food chains. Along coastlines where prevailing winds blow more or less continuously, such as the west coast of South America, the constant upwelling supports high levels of biological productivity. In other areas, upwellings are seasonal. In the areas where upwelling occurs, phytoplankton thrive and rapidly reproduce. This forms the basis for productive food webs that may include anchovies, herring, lobster, cod, hake, bluefish, tuna, and other commercial species. As much as 50% of the world's commercial fish catch comes from upwelling areas that amount to only about 0.1% of the ocean's surface.

## Coastal Downwelling

When winds blow water toward a coastline, it tends to be forced downward and then returns to sea along the continental shelf (see Figure 4-21b). Downwelling does not directly affect ocean productivity, but it does force gases and nutrients from the surface into deeper waters and plays a significant role in the distribution of marine organisms. Because downwelling carries oxygen-rich surface water to deeper areas, many organisms can live in the deep water where downwellings occur. This process is especially crucial for organisms that live below the photic zone.

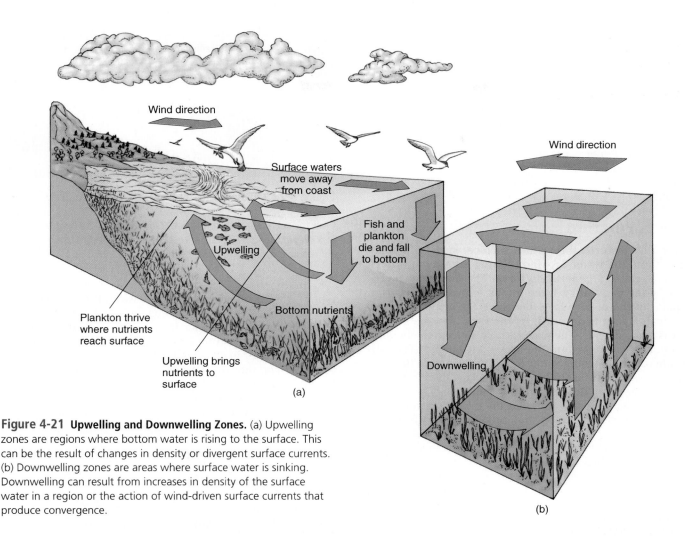

**Figure 4-21  Upwelling and Downwelling Zones.** (a) Upwelling zones are regions where bottom water is rising to the surface. This can be the result of changes in density or divergent surface currents. (b) Downwelling zones are areas where surface water is sinking. Downwelling can result from increases in density of the surface water in a region or the action of wind-driven surface currents that produce convergence.

## Deepwater Circulation

The currents that we have discussed so far have primarily been in the surface waters, but there are also horizontal and vertical currents in deep ocean water below the pycnocline. In the ocean's depths, it is a difference in densities that causes the water to move, rather than wind energy. As previously mentioned, water density is most affected by salinity and temperature. The densest water in the ocean is Antarctic Bottom Water ($1.0279$ g/cm³). The majority of Antarctic Bottom Water forms in the winter in an area known as the Weddell Sea. As the water freezes, the majority of salt remains in solution, producing water that is extremely dense because of the cold temperature and high salinity. This dense water sinks and mixes with water from the Antarctic Circumpolar Current. The mixture of water continues to sink to the bottom and then begins to spread north. The movement of Antarctic Bottom Water is extremely slow. In the Pacific basin it may take as long as 1,600 years to reach the Arctic Sea. Since the Atlantic basin is much smaller, the water moves faster and may reach the North Atlantic in 800 years.

Dense water forms and sinks to the bottom in the Arctic Ocean as well, but the characteristics of the ocean basin prevent most of it from moving southward. A deep channel east of Greenland does allow some North Atlantic Deep Water to flow into the North Atlantic.

Another deep-water mass forms in the Mediterranean Sea. Here evaporation exceeds the input of freshwater, producing very dense water that sinks. The density is most pronounced in the winter when the water salinity reaches 38 parts per thousand. The Mediterranean Deep Water flows through the Strait of Gibraltar and into the Atlantic Ocean where it is found along the bottom of most of the central Atlantic. Some Mediterranean Deep Water has also been found as far south as the Antarctic. The salinity of Mediterranean deep water is higher than other deep water but it is warmer and so it tends to sit on top of other deep water.

## In Summary

In some areas of the ocean, seasonal turnover mixes the deeper nutrient-rich, oxygen-poor water with the oxygen-rich, nutrient-poor surface waters. Wave action and currents play leading roles in mixing deep and surface waters. Evaporation, precipitation, and temperature affect the density of seawater. Changes in density contribute to vertical mixing. ●

# WAVES

Waves are the result of forces acting on the surface of the water. A wave in the ocean does not represent a flow of water but a flow of energy or motion. The energy that is added to the water to cause the wave is either dissipated at sea or transferred to a beach or structures on the beach when the wave strikes.

## Wave Formation

A force that disturbs the water's surface, such as a stone dropped into the water or wind blowing across the water's surface, is called a **generating force.** The disturbance produced by the generating force moves outward, away from the point of the disturbance. Consider, for instance, water into which a rock is dropped. As the rock hits, the surface water is pushed aside. Then, as the rock sinks to the bottom, the water moves back into the space left behind. The momentum of the returning water forces it upward so that it is raised above the surface. The raised surface then falls back down and creates a depression in the surface. The depression is filled with more water, and the process is repeated, producing a series of waves that ripple outward from the point of disturbance.

The force that causes the water to return to the undisturbed level is called the **restoring force.** If the amount of water that is displaced is small, the restoring force is the surface tension of the water (surface tension was discussed earlier in this chapter), and the small waves are referred to as **capillary waves.** When the amount of water displaced is sizable, the restoring force is gravity and the waves are referred to as **gravity waves.** A wave, then, results from the interactions between generating forces and restoring forces. Generating forces can be any event that adds energy to the surface of the sea. Geological events, such as earthquakes and volcanic eruptions, objects dropped into the water, and the movement of ships, are all possible generating forces, but the most common generating force is the wind.

As wind blows across the surface of still water, it creates drag (friction) that lifts some of the water away from the surface (Figure 4-22). If the amount of water displaced is very small, the surface tension of the water pulls it back to restore a smooth surface, and a series of ripples are formed. If the force of the air is greater than a small breeze, more friction is created. As the water is stretched by the wind, more of it is pulled away from the surface, and a combination of surface tension and gravity acts to pull the water back to the surface. As the surface becomes rougher, it becomes easier for the wind to add more energy. The frictional drag between the air and the water is increased, and the waves become progressively larger.

## Types of Waves

Most ocean waves are generated by wind and restored by gravity, and they progress in a particular direction. This

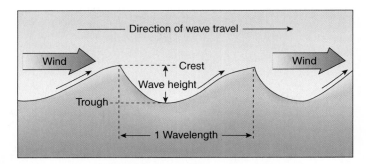

**Figure 4-22 Characteristics of Waves.** The drag produced by wind blowing across the surface of still water lifts some of the water away from the surface, producing a wave. Principal wave characteristics include wavelength (the distance between the crests of two consecutive waves) and wave height (amplitude).

type of wave is called a **progressive wave.** Progressive waves can be formed by local storm centers or by the prevailing winds of the wind belts, such as the trade winds or westerlies. Waves produced by storms at sea move outward from the storm center in all directions. As wind waves are formed by the storm, they are forced to increase in size and speed by the input of energy from the storm; for this reason they are also known as **forced waves.** When the energy from a storm or other generating force no longer has an effect on the waves, they become **free waves,** moving at speeds that are determined by the wave's length and **period,** the time required for one wavelength to pass a fixed point (a wave's speed is calculated by dividing its length by its period).

The various parts and characteristics of waves are shown in Figure 4-22. Waves with long periods and long wavelengths move faster than those with short periods and short wavelengths. Eventually, they escape a storm and form a pattern of wave crests on the ocean's surface. These long-period, uniform waves are called **swells.** Swells carry a considerable amount of energy and can travel for thousands of kilometers.

## Deepwater and Shallow-Water Waves

Waves that occur in water that is deeper than one half of a wave's wavelength are called **deepwater waves.** The height of a deepwater wave increases with wind speed, the duration that the wind blows, and the fetch, or distance over the water that the wind blows. If the wind speed is low, the waves will be small, regardless of the duration or the fetch. A strong wind blowing for a short time does not produce large waves even if the fetch is large. If a strong wind blows for a long time over a short fetch, the waves are again small. Large waves are produced only when all three factors (wind speed, wind duration, and fetch) are of high magnitude.

## Breakers

When a deepwater wave enters shallow water it becomes a **shallow-water wave** (the depth of the water is less than half the wavelength). The **surf zone** is the area along a coast where waves slow down, become steeper, break, and disappear (Figure 4-23). As the waves approach the shore, the decreasing depth begins to affect their shape and speed, and

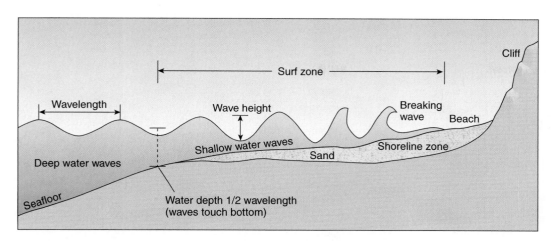

**Figure 4-23 Breakers.** As a wave approaches shallow water, the lower part of the wave may become slowed by friction with the bottom, although the top of the wave is not affected much. The result is a wave known as a breaker.

their crests become flatter. The friction from the lower part of the wave dragging on the bottom slows the forward movement of the wave. Whereas the length and speed of a deep-water wave are determined by the wave period, the length and speed of a shallow-water wave are determined by the depth of the water.

**Breakers** form in the surf zone when the bottom of the wave slows but its crest continues moving toward the shore at a speed faster than that of the wave. As a result, the crest overtakes the base of the wave in front of it and eventually falls into the preceding trough and breaks up. The two most common types of breakers are called **plungers** and **spillers.**

Plunging breakers form when the beach slope is steep. The crest curls and curves over and outruns the rest of the wave. It then breaks with a sudden loss of energy and a splash. Spilling breakers are more common and are found on flatter beaches, where the energy is dissipated more gradually as the wave moves over the shallow bottom. This action produces waves that are less spectacular and consist of turbulent water and bubbles that flow down the face of the wave. Spilling breakers last longer than plungers because they lose energy more gradually. If you were a surfing enthusiast, you would look for spillers to get a longer ride and plungers to get a more exhilarating ride.

As waves enter shallow water they almost always approach the shore at an angle. The portion of the wave that reaches shallow water first slows, but the portion still in deeper water continues at the original speed. The wave bends, a process called **wave refraction.**

### Tsunamis

Sudden movements of the earth's crust produce earthquakes, which may produce large **seismic sea waves,** or **tsunamis.** These waves are sometimes called tidal waves, but oceanographers use that term to describe a totally unrelated event. If a large area of the seafloor rises or falls, it will cause a proportional movement in the surface of the sea. The disruption of the surface forms waves with long wavelengths, long periods, and low height. Because tsunamis are such long waves, they behave like shallow-water waves when they move out from the point of the seismic disturbance. As the wave approaches a coast or an island, the wave energy is compressed into a smaller volume of water

as the depth decreases. The sudden increase in energy causes the height of the wave to increase dramatically, and the energy rapidly dissipates as the water races over the land mass. The wave's surge over land can cause mass destruction, wrecking buildings and docks and depositing vessels high on dry land (Figure 4-24).

## TIDES

The periodic changes in water level that occur along coastlines are called **tides.** Tides play a considerable role in the life of many marine organisms, especially those that live in the area exposed at low tide and covered by high tide (the intertidal zone). The adaptations that allow intertidal organisms to survive and thrive in this potentially hostile environment will be explored later in the text.

### Why Tides Occur

Tides are the result of the gravitational pull exerted on the water of the oceans by the moon and the sun. The easiest way to explain the effects of the sun and the moon that result in tides is to imagine the earth as being completely covered with water and the tides behaving as ideal waves, ones that are not influenced by friction. In this model, the moon is held in orbit around the earth by the earth's gravitational force. There is also an apparent force called centrifugal force acting in the opposite direction and pulling the moon away from the earth (Figure 4-25). Together, the earth and moon are held in their orbit around the sun by the gravitational attraction between the sun and the center of mass of the earth–moon system. An opposing centrifugal force pulls the center of mass of the earth–moon system away from the sun. For the earth–moon system to remain in orbit around the sun, the gravitational force must equal the centrifugal force. Because the moon also exerts a gravitational pull on the earth, there must be an opposing centrifugal force to prevent the earth from moving toward the moon. This centrifugal force is created by the earth's center of mass rotating around the center of mass of the earth–moon system.

Calculations using Newton's law of gravity, which describes the gravitational force between two or more bodies,

**Figure 4-24  Tsunami.** This damage to the town of Aonae on Okushiri Island, west of Japan's main northern island of Hokkaido, was caused by a tsunami that struck following an earthquake on July 14, 1993. Tsunamis form when activity along fault lines causes large areas of seafloor to rise or fall, producing proportional movement at the surface of the ocean.

Reuters/Bettmann /Corbis

**Figure 4-25  Tides.** Tides occur when the gravitational force of the moon pulls ocean water toward it, while the centrifugal force of the earth–moon system forces a mass of ocean water to move in the opposite direction.

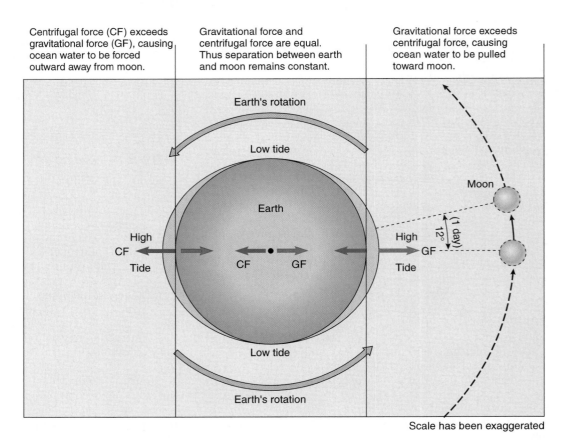

Centrifugal force (CF) exceeds gravitational force (GF), causing ocean water to be forced outward away from moon.

Gravitational force and centrifugal force are equal. Thus separation between earth and moon remains constant.

Gravitational force exceeds centrifugal force, causing ocean water to be pulled toward moon.

Earth's rotation

Low tide

Earth

High

CF

Tide

CF    GF

High

GF

Tide

Moon

(1 day) 12°

Low tide

Earth's rotation

Scale has been exaggerated

indicate that the opposing forces are in balance at the earth's center, but at the earth's surface, the forces are not in balance. Although the mass of the moon is much less than the mass of the sun, it is much closer to the earth. When the forces are calculated for the effects of the sun and the moon on the water at the earth's surface, the moon's gravitational pull is found to have the greater effect.

The water on the side of the earth facing the moon is acted on by a gravitational pull from the moon that is larger than the opposing centrifugal force. Because water is fluid, it moves to a point directly under the moon and produces a bulge. On the opposite side of the earth, the centrifugal force acting on the water is larger than the moon's gravitational pull, producing another bulge on the side of the earth opposite the moon. At the same time that the bulges form, areas of low water form between the two bulges. Because the earth rotates on its axis, a location on the earth's surface would experience a high tide when it is under one of the bulges and a low tide when it is not under one. Therefore most locations experience two high tides and two low tides each day. Although one rotation of the earth takes 24 hours, the moon also moves slightly in its orbit each day, so that a location on the earth's surface needs an extra 50 minutes to come in line with the moon again. For this reason a full tidal cycle takes 24 hours and 50 minutes.

## Spring and Neap Tides

Although the moon plays the greater role in producing tides, the sun also participates. Even though the sun's mass is much larger than that of the moon, it is so far away from the earth that its tide-producing gravitational force is only about 46%

that of the moon. The bulges that are produced by the sun are smaller and usually masked by the moon's.

Twice a month, at the full moon and new moon, the earth, moon, and sun are all in a straight line. As a result, the gravitational pull of the sun is added to that of the moon. These are the times of the highest and lowest tides, referred to as **spring tides,** not for the season of the year but for the water "springing up" at the shore. During the first and last quarter of the moon, the sun and moon are at right angles to each other. The pull of the sun cancels some of the moon's pull and produces **neap** (from an Anglo-Saxon word meaning "napping") **tides,** which have the smallest change between the high and low tide (Figure 4-26).

## Tidal Range

Tidal range is greatly influenced by geographic factors, such as the contours of shorelines and the depth of coastal waters. Remember that the model for tides assumes the water is moving as an ideal wave. As the earth rotates, continents' positions interfere with tidal crests, altering their size and movement. The shape of ocean basins can also affect tidal patterns and height. For these reasons, some coastal areas have only one high tide and one low tide each day. This condition is referred to as a **diurnal tide.** Most areas experience two high tides and two low tides each day, a condition called a **semidiurnal tide.** In the typical semidiurnal tide, the two high tides are the same height and the two low tides are at about the same level. If the high and low tides are at different levels, the tide is referred to as a **mixed semidiurnal tide** (Figure 4-27).

In a tidal system, the greatest height to which the high

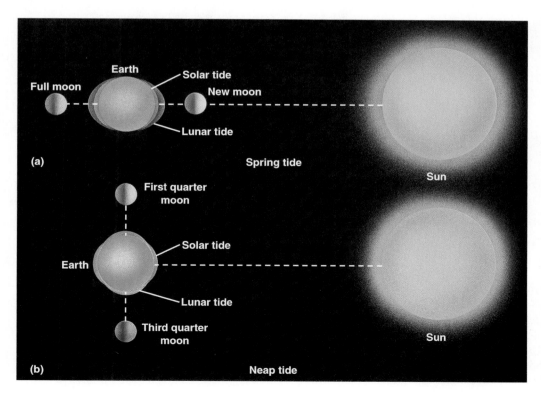

**Figure 4-26 Spring and Neap Tides.** (a) Spring tides occur when the moon and the sun are in line with each other and the gravitational pull of the two are working together. (b) Neap tides occur when the sun is at a right angle to the position of the moon. The small gravitational pull of the sun cancels some of the moon's gravitational pull, producing smaller tides.

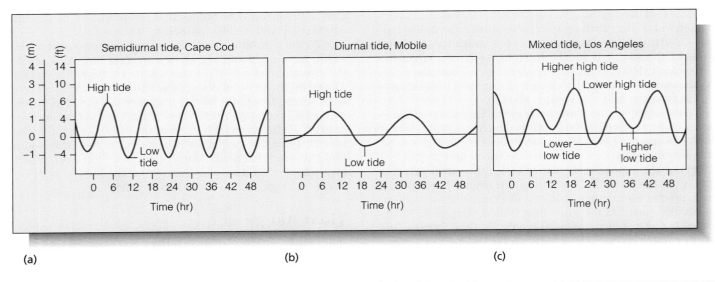

(a)          (b)          (c)

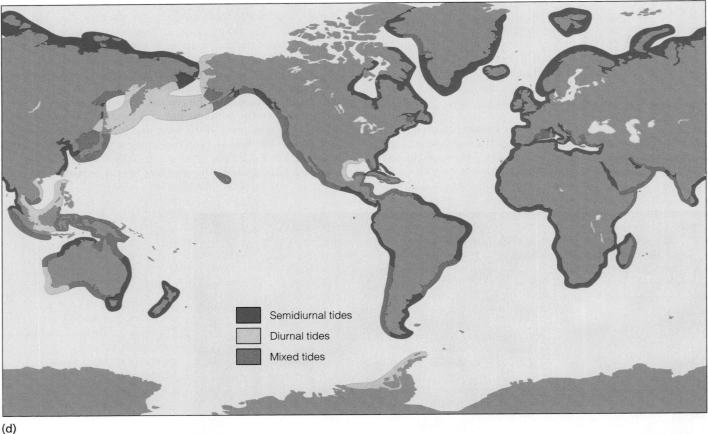

(d)

**Figure 4-27 Tidal Patterns.** Tidal patterns are influenced by geographic factors, resulting in three main types of tide: (a) semidiurnal tide, (b) diurnal tide, and (c) mixed tide. The map (d) shows the geographic distribution of the three types of tides.

tide rises is called **high water** and the lowest point is called **low water.** Observations of tides over a long time are used to determine the average tide. A rising tide is called a **flood tide,** because the water floods the coast, and a falling tide is referred to as an **ebb tide.** Currents called **tidal currents** are associated with the rising and falling of the tides. Tidal currents can be very swift and sometimes dangerous as they move water onshore and offshore. During the change of tide from high to low or low to high, there is a period called **slack water,** during which the tidal currents slow down and then reverse.

## In Summary

Waves are produced by forces acting at the surface of the water, most commonly wind. The force raises the water, and capillary action or gravity restores the water to its original position. This combination of events creates a wave. The energy transferred to the water remains with the wave until it can be dissipated, as when it crashes against the shore.

The periodic rise and fall in the level of coastal waters is known as tides. Tides are the result of the gravitational pull of the moon and the sun acting on the waters of the ocean. ●

## SELECTED KEY TERMS

acid, *p. 59*

adhesion, *p. 58*

adsorption, *p. 61*

alkaline, *p. 59*

base, *p. 59*

breaker, *p. 77*

buffer, *p. 62*

capillary action, *p. 58*

capillary wave, *p. 76*

Coriolis effect, *p. 67*

deepwater wave, *p. 76*

density, *p. 72*

diurnal tide, *p. 79*

doldrums, *p. 68*

downwelling, *p. 73*

eastern-boundary current, *p. 69*

ebb tide, *p. 80*

Ekman spiral, *p. 70*

Ekman transport, *p. 70*

equatorial upwelling, *p. 73*

evaporites, *p. 61*

flood tide, *p. 80*

forced wave, *p. 76*

free wave, *p. 76*

generating force, *p. 76*

global warming, *p. 64*

gravity waves, *p. 76*

greenhouse gas, *p. 64*

Gulf Stream, *p. 69*

gyre, *p. 69*

halocline, *p. 72*

high water, *p. 80*

horse latitudes, *p. 68*

hydrogen bond, *p. 57*

ion, *p. 58*

isopycnal, *p. 73*

low water, *p. 80*

mixed semidiurnal tide, *p. 79*

neap tide, *p. 79*

northeast trade winds, *p. 67*

oxygen-minimum zone, *p. 61*

parts per thousand (ppt), *p. 60*

period, *p. 76*

pH scale, *p. 59*

plunger, *p. 77*

polar, *p. 57*

polar easterlies, *p. 67*

progressive wave, *p. 76*

pycnocline, *p. 72*

restoring force, *76*

salinity, *p. 60*

seismic sea wave, *p. 77*

semidiurnal tide, *p. 79*

shallow-water wave, *p. 76*

slack water, *p. 80*

solvent, *p. 56*

southeast trade winds, *p. 67*

specific heat, *p. 58*

spiller, *p. 77*

spring tide, *p. 79*

surf zone, *p. 76*

surface tension, *p. 58*

sverdrup, *p. 69*

swell, *p. 76*

thermal capacity, *p. 58*

thermocline, *p. 72*

thermohaline circulation, *p. 73*

tidal current, *p. 80*

tide, *p. 77*

trace elements, *p. 60*

transverse currents, *p. 69*

tsunami, *p. 77*

upwelling, *p. 73*

vertical mixing, *p. 73*

vertical overturn, *p. 73*

wave refraction, *p. 77*

westerlies, *p. 67*

western-boundary currents, *p. 69*

wind-induced vertical circulation, *p. 73*

## QUESTIONS FOR REVIEW

**Multiple Choice**

1. The relatively high boiling and freezing points of water are due to
   a. the size of the water molecules
   b. the ability of water molecules to form hydrogen bonds
   c. the low thermal capacity of water
   d. the neutral pH of water
   e. the shape of water molecules

2. Compounds that release hydrogen ions when they dissolve in water are called
   a. acids
   b. bases
   c. salts
   d. alkaline compounds
   e. organic compounds

3. Carbon dioxide gas is produced by living organisms during the process of
   a. photosynthesis
   b. digestion
   c. respiration
   d. chemosynthesis
   e. anabolism

4. A zone that exhibits a rapid change in temperature with depth is called a
   a. pycnocline
   b. halocline
   c. barocline
   d. thermocline
   e. mesocline

5. The Coriolis effect refers to
   a. the process by which ocean currents are formed

b. the movement of nutrient-rich bottom water to the surface in exchange for oxygen-rich surface water

c. the apparent deflection of the path of air currents and ocean currents that results from the earth and atmosphere and ocean water moving at different speeds

d. the vertical exchange of water due to differences in density

e. the horizontal exchange of water between the poles and the equator that occurs as the result of changes in the temperature and salinity of the water

6. Ocean currents are produced by
   a. earthquakes
   b. winds
   c. differences in water density
   d. upwellings and downwellings
   e. tides

7. The restoring force for large waves is
   a. wind
   b. capillary action
   c. adhesion
   d. friction
   e. gravity

8. Long-period, uniform waves that can travel for thousands of kilometers are called
   a. plungers
   b. spillers
   c. breakers
   d. tsunamis
   e. spoilers

9. _____ are periodic changes in sea level due to the gravitational pull of the moon and the sun.
   a. waves
   b. ocean currents
   c. winds
   d. tides
   e. tsunamis

10. When the sun, moon, and earth are all in a line, _____ tides occur.
    a. ebb
    b. neap
    c. spring
    d. seasonal
    e. no

## Short Answer

1. What roles do photosynthesis and respiration play in the distribution of gases in seawater?

2. What factors are responsible for the prevailing wind patterns of the earth?

3. What combination of factors produces neap tides?

4. Explain how the polar nature of water molecules influences water's physical characteristics.

5. Explain how salt from the sea is returned to the land.

6. Explain how vertical mixing of seawater occurs.

7. Describe how winds are produced.

8. Explain how waves are formed.

9. Explain how breakers are formed.

10. Describe how upwellings occur.

11. What is the biological importance of upwelling and downwelling zones?

12. What are gyres and how are they formed?

13. Why is carbon dioxide more soluble in seawater than oxygen?

14. List three different processes that circulate water in the oceans.

15. Why is the shallow-water thermocline in polar and temperate seas seasonal?

## Thinking Critically

1. Would it be easier for a planktonic organism to float in water with a high salinity or a low salinity? Explain.

2. Why do upwelling zones and downwelling zones support more biomass than areas of the open sea where these zones don't exist?

3. Why do coastal cities usually experience cooler summers and warmer winters than cities of the same latitude that are inland?

4. Sometimes the discharge of organic wastes from land into coastal waters results in a population explosion of phytoplankton. How would this affect organisms that live in the water below the surface and on the bottom?

5. What effect do you think boundary currents have on coastal communities of organisms?

## SUGGESTIONS FOR FURTHER READING

Canby, T. Y. 1984. El Niño's Ill Wind, *National Geographic* 165(2):144–183.

Curtsinger, B. 1996. Realm of the Seamount, *National Geographic* 190(5):72–80.

Garrison, T. 2002. *Oceanography,* 4th ed. Pacific Grove, Calif.: Thomson Learning.

MacLeish, W. H. 1989. The Blue God, Tracking the Mighty Gulf Stream, *Smithsonian* 19(11):44–59.

McCredie, S. 1994. When Nightmare Waves Appear Out of Nowhere to Smash the Land, *Smithsonian* 24(12): 28–39.

Open University. 1989. *Ocean Circulation.* Oxford, U.K.: Pergamon Press; Milton Keynes, U.K.: Open University.

Open University. 1989. *Seawater: Its Composition, Properties, and Behavior.* Oxford, U.K.: Pergamon Press; Milton Keynes, U.K.: Open University.

Weller, R. A., and D. M. Farmer. 1992. Dynamics of the Ocean Mixed Layer, *Oceanus* 35(2):46–55.

### InfoTrac College Edition Articles

Garrett, C., L. R. M. Maas. 1993. Tides and Their Effects, *Oceanus* 36(1).

Hogg, N. 1994. The Deep Basin Experiment: How Does the Water Flow in the Deep South Atlantic? *Oceanus* 37(1).

Pickart, R. S. 1994. Where Currents Cross: Intersection of the Gulf Stream and the Deep Western Boundary Current, *Oceanus* 37(1).

### Websites

**http://seawifs.gsfc.nasa.gov/OCEAN_PLANET/HTML/ oceanography_currents_1.html** Good information on ocean currents and their importance.

**http://co-ops.nos.noaa.gov/tide_pred.html** National Oceanographic and Atmospheric Administration website with information on tides and tidal predictions.

**http://www.saltwatertides.com/** Tide tables for U.S. coasts.

**http://www.pmel.noaa.gov/tsunami/** Research on predicting tsunamis.

# 5 Basics of Life: Molecules, Cells, Evolution, and Biological Classification

## Key Concepts

1. To understand living organisms, one must have a basic understanding of the variety of compounds from which organisms are built.

2. Four groups of macromolecules are necessary for life: carbohydrates, lipids, proteins, and nucleic acids.

3. All living organisms are composed of cells.

4. Cells can be either prokaryotic or eukaryotic.

5. Cells produce new cells by the process of cell division.

6. Evolution is the process by which the genetic composition of populations of organisms changes over time.

7. Natural selection favors the survival and reproduction of those organisms that possess variations that are best suited to their environment.

8. A species is a group of physically similar, potentially interbreeding organisms that share a gene pool, are reproductively isolated from other such groups, and are able to produce viable offspring.

9. The binomial system of nomenclature uses two words, the genus and the species epithet, to identify an organism.

10. Most biologists classify organisms into one of three domains, categories that reflect theories about evolutionary relationships.

L iving organisms are composed of cells. These basic units of life contain all of the chemicals necessary to support life and pass along hereditary information. To understand the wide variety of organisms that inhabit the sea, it is necessary to know something about the chemicals that make up cells and the cells themselves. Through the process of cell division, cells produce more cells and pass along genetic information that allows their offspring to continue functioning in the marine environment. If you spend time observing marine organisms, you will surely notice that all of their activity seems to be directed toward two objectives: survival and reproduction. In animals, for instance, behaviors and body structures are specialized to gather food, produce reproductive cells, and use those reproductive cells to produce new individuals. Other behaviors, such as defense against predators, enhance an organism's ability to survive so that it can continue to feed and reproduce. All organisms that are alive today descended from ancestors that were well adapted to their environments and survived long enough to reproduce successfully.

By studying and comparing the fossilized remains of extinct organisms and modern organisms, biologists try to piece together the evidence for evolution and discover the natural processes that have produced and continue to produce the diversity of species that we find today. To add order to this vast array of diversity, biologists employ a common system of naming and classification that facilitates communication among scientists throughout the world.

## BUILDING BLOCKS OF LIFE

Living organisms are composed of chemical compounds, and to understand living organisms one must have a basic understanding of the variety of compounds from which or-

NOAA, inset, Fred McConnaughey 1985/Photo Researchers, Inc.

ganisms are built. Some of the most important chemicals in living organisms are large molecules called **macromolecules**. There are four major classes of macromolecules in living organisms: carbohydrates, lipids, proteins, and nucleic acids.

## Carbohydrates

The molecules known as **carbohydrates** contain the elements carbon, hydrogen, and oxygen, frequently in a ratio of $1:2:1$, or $CH_2O$, thus the name carbohydrate (carbon water). The most common carbohydrates in living organisms are sugars and polysaccharides.

### Sugars

The most common sugars in nature are monosaccharides and disaccharides. **Monosaccharides,** or simple sugars, are small molecules usually containing five or six carbon atoms (Figure 5-1a). The five-carbon sugars ribose and deoxyribose are essential to the structure of nucleic acids (discussed later) as well as other molecules in cells. The six-carbon sugar **glucose** is the basic fuel molecule for living cells.

**Disaccharides** are composed of two bonded monosaccharides (see Figure 5-1b). **Sucrose,** common table sugar, contains a molecule of glucose and a molecule of fructose (another six-carbon sugar). Marine plants and algae use sucrose to transport sugars within their bodies. **Maltose** is a disaccharide formed from two molecules of glucose. Maltose is produced from starches during digestion in animals and is a source of energy. Another disaccharide is lactose. **Lactose** is known as milk sugar because it is found in mammalian milk. The lactose molecule contains a molecule of glucose and another six-carbon sugar called galactose. Lactose supplies much of the energy that newborn mammals require.

### Polysaccharides

The carbohydrates known as **polysaccharides** are examples of polymers. A **polymer** is a large molecule that consists of the same basic units linked together. **Starches** store polysaccharides produced by plants, algae, and some microorganisms. The basic unit of a starch polymer is the sugar glucose. Starches are an efficient way for cells and organisms to store large amounts of glucose for future use. Starches

(a)

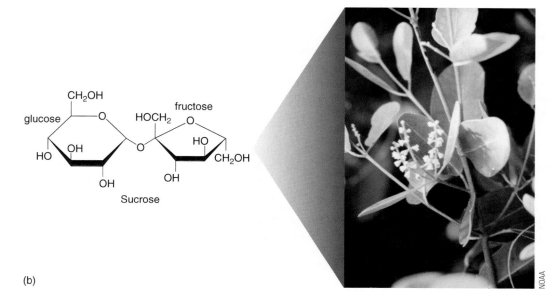

(b)

**Figure 5-1  Sugars.** (a) Glucose is an important six-carbon monosaccharide that is used by marine organisms, such as this fish, as a source of energy. (b) Marine plants, like this mangrove, use the disaccharide sucrose to transport sugars within their bodies.

**Figure 5-2 Cellulose and Chitin.**
(a) The polysaccharide cellulose makes up the structural parts of this cordgrass. (b) This crab's hard covering is composed of the polysaccharide chitin.

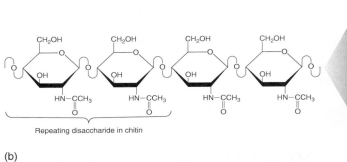

Repeating disaccharide in chitin

(b)

are also important food sources for many animals. A polysaccharide similar to starch, **glycogen,** or animal starch, is produced by animals and some unicellular organisms to store glucose for future use. **Cellulose** is an example of a structural polysaccharide. It is found in the cell walls of plants and algae as well as their stiff and supportive structures (Figure 5-2a). Like starches and glycogen, cellulose is a polymer of glucose molecules. Cellulose differs from starch and glycogen in the way the glucose molecules are arranged, which accounts for its strength and durability. Another structural polysaccharide is **chitin.** Chitin is a polymer composed of molecules of *N*-acetylglucosamine and is also strong and durable. Chitin is in the cell walls of fungi and the hard exterior skeletons of some marine animals such as crabs and lobsters (see Figure 5-2b).

## Lipids

Fats, oils, and waxes are all examples of **lipids,** macromolecules composed primarily of carbon and hydrogen. Many lipids contain molecules called fatty acids, which are made up of long hydrocarbon chains containing an acid group (Figure 5-3a). Simple fats, or **triglycerides,** are composed of three fatty acids attached to a carbohydrate molecule known as glycerol. Marine organisms use triglycerides to store energy, to cushion vital organs, and to increase buoyancy. Ho-

meothermic animals use triglycerides as insulation to trap heat. Another kind of lipid, **phospholipid,** is similar to triglycerides and is in the membranes that surround cells and some of the internal components of cells. **Steroids** are lipids that have complex ring structures (see Figure 5-3b). They function as chemical messengers within the bodies of animals. Waxes are lipids that are composed of very long hydrocarbon chains (see Figure 5-3c). They coat the exposed surfaces of some marine plants and algae and act as a water barrier. Waxes are also found in the body coverings of some marine animals and in the ear openings of some marine mammals.

## Proteins

**Proteins** are polymers made up of basic units called **amino acids** (Figure 5-4). Twenty different amino acids make up the various proteins found in living organisms. The number and type of amino acids determine the characteristics of protein molecules. Within cells, individual amino acids are assembled into chains called **polypeptides.** The polypeptide chains are then coiled and folded into complex, three-dimensional structures, protein molecules. The complex structures of protein molecules allow them to serve several functions. For instance, the primary structural components of animals, muscles and connective tissues, are composed

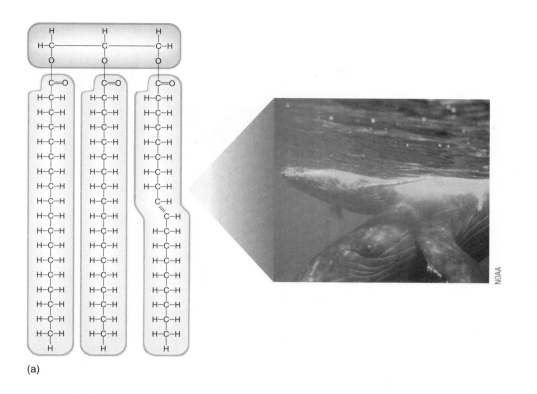

(a)

**Figure 5-3  Lipids.** (a) This young whale is insulated from the cold by a thick layer of blubber that is composed of triglyceride fat. (b) Steroids, like the hormone testosterone, control this fish's reproduction. (c) Lipids known as waxes waterproof the cells of this alga.

(b)

(c)

**Figure 5-4 Proteins.** Proteins are composed of molecules called amino acids. The muscles that allow this bird to fly are composed of protein.

of protein. Proteins known as **enzymes** are essential for life. Enzymes are biological catalysts that speed up the rate of chemical reactions, allowing metabolism to function efficiently. Some proteins, such as hemoglobin, transport chemicals within organisms, whereas others store chemicals. Table 5-1 lists the many functions of proteins and some examples.

## Nucleic Acids

**Nucleic acids** are polymers of molecules called nucleotides (Figure 5-5a). Each **nucleotide** is composed of a five-carbon sugar, a nitrogen-containing base, and a phosphate group. There are two types of nucleic acids found in living organisms: **deoxyribonucleic acid (DNA)** and **ribonucleic acid (RNA).**

### DNA

DNA is a large molecule with the shape of a helix (see Figure 5-5b). Each nucleotide found in DNA contains the sugar deoxyribose and one of the following four nitrogen-containing bases: adenine, guanine, cytosine, or thymine. The nucleotides are linked by phosphate groups to form long chains, or strands, of DNA. In the DNA molecule two strands of nucleotides are wound around each other to form the characteristic double helix (except for some viruses that have a single strand of DNA). DNA contains an organism's genetic material, or **genes,** and is capable of copying itself so the genes can be passed from one generation to another. Encoded in the genes are directions for synthesizing the various proteins that an organism needs. The proteins are responsible for an organism's appearance and control its function, growth, and reproduction. In many cells the DNA is located in a specialized structure called the nucleus.

### RNA

RNA is composed of nucleotides that contain the sugar ribose and the same bases found in DNA, with the exception that in RNA the base uracil takes the place of thymine. RNA

molecules are usually single-stranded (although some viruses have double-stranded RNA) and can come in different forms that have evolved for specific roles in the cell (see Figure 5-5c). RNA functions in **protein synthesis.** In this process the information carried by individual genes is copied, or transcribed, from DNA onto a molecule of RNA known as **messenger RNA (mRNA).** The message carried by the mRNA is decoded, or translated, by a cellular structure known as a **ribosome,** composed of protein and another type of RNA called **ribosomal RNA (rRNA).** The ribosome synthesizes protein by connecting the appropriate amino acids that are brought to the ribosome by a third type of RNA called **transfer RNA (tRNA).** The proteins produced by this process make new cell parts and enzymes that give the cells their characteristics.

### Table 5-1  Proteins and Their Functions

| Types of Proteins | Function |
|---|---|
| Enzymes | Biological catalysts that speed up the rate of chemical reactions in cells |
| Structural Proteins | Make up body parts of animals such as hair, skin, scales, tendons, cartilage |
| Contractile Proteins | Make up muscle |
| Messenger Proteins | Send signals from one cell to another and from organ to organ |
| Transport Proteins | Transport important substances such as oxygen and fatty acids |
| Storage Proteins | Store important materials in cells, such as iron |
| Antibodies | Protect animals from foreign proteins and disease-causing microbes |
| Toxins | Help to capture prey and protect animals from predators |

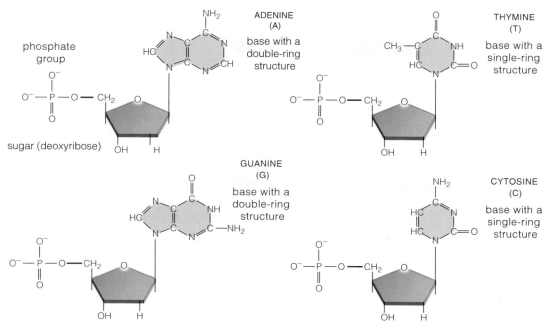

(a) Nucleotides

phosphate group

sugar (deoxyribose)

**ADENINE (A)**
base with a double-ring structure

**THYMINE (T)**
base with a single-ring structure

**GUANINE (G)**
base with a double-ring structure

**CYTOSINE (C)**
base with a single-ring structure

covalent bonding in carbon backbone

hydrogen bonding between bases

(b) DNA

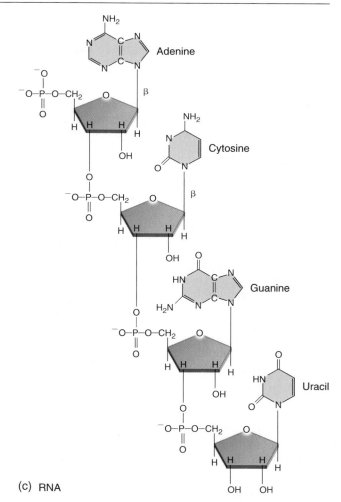

Adenine

Cytosine

Guanine

Uracil

(c) RNA

**Figure 5-5 Nucleic Acids.** (a) Nucleotides are the building blocks of nucleic acids. A nucleotide is composed of a five-carbon sugar, a nitrogen base, and a phosphate group. (b) The nucleic acid DNA is composed of two strands of nucleotides wound into a double helix. (c) The nucleic acid RNA is normally a single strand of nucleotides. There are three different types of RNA, and they all function in protein synthesis.

## In Summary

Four main groups of large molecules are necessary for life. Carbohydrates supply energy and some contribute to the structure of plants, fungi, algae, and some invertebrates. Lipids function in the storage of energy, in insulation, buoyancy, and as chemical messengers. Proteins provide structural materials for making many body parts. Enzymes are proteins that function as catalysts, regulating the metabolism of cells, and as chemical messengers to coordinate the activities of many cells. Nucleic acids are molecules that carry genetic information and direct the synthesis of proteins. ●

## CELLS

All living organisms are composed of basic units called **cells** (Figure 5-6), and every cell is capable of the basic processes of life: metabolism, growth, and reproduction. Each cell is surrounded by a plasma membrane that is referred to as a cell membrane. The cell membrane separates the fluid contents of the cell from the external environment. The cell membrane is a selective barrier, regulating what goes in and out of the cell. Some cells also have a **cell wall** that gives them protection and support.

The fluid content of a cell is called **cytosol.** Cytosol is a complex mixture of chemicals but mostly water. In addition to cytosol, some cells contain tiny internal structures called **organelles** that perform specific jobs within the cell. The combination of cytosol and organelles is called **cytoplasm.**

### Types of Cells

There are two major types of cells in nature: prokaryotic cells and eukaryotic cells. **Prokaryotic cells** (see Figure 5-6a) lack a nucleus and do not have any membrane-bound organelles. Organisms with prokaryotic cells are called **prokaryotes,** and they are always unicellular. Examples of prokaryotes are the many different types of marine bacteria and archaeons. **Eukaryotic cells** (see Figure 5-6b) have a well-defined nucleus and many membrane-bound organelles. Organisms with eukaryotic cells, or **eukaryotes,** can be either unicellular or multicellular. Seaweeds, fungi, plants, and animals are some examples of marine eukaryotes.

### Organelles

Eukaryotic cells contain several different organelles. Just as organs in a multicellular organism specialize in certain functions, organelles have specific functions within a cell.

### Nucleus and Ribosomes

The most obvious organelle in a eukaryotic cell is the nucleus (see Figure 5-6b). The **nucleus** is a large structure surrounded by a nuclear membrane composed of two layers of plasma membrane. It contains the cell's DNA and acts as the cell's control center. The DNA in the nucleus is combined with protein to form structures called **chromosomes,** which contain an organism's genes. Eukaryotic cells usually contain two copies of each different chromosome. Exceptions to this rule are cells that function in reproduction. In addition to being the repository of a cell's genes, the nucleus is where ribosomes are formed. Ribosomes are assembled in a specific area of the nucleus called the **nucleolus.** As we noted earlier, ribosomes are organelles that function in the synthesis of proteins. Ribosomes are not surrounded by a plasma membrane and are in both prokaryotic and eukaryotic cells.

### Organelles Involved in Synthesis, Processing, and Storage

In eukaryotic cells a series of membranes called **endoplasmic reticulum (ER)** winds through the cytoplasm (see Figure 5-6b). Some endoplasmic reticulum has ribosomes attached to its surface (**rough endoplasmic reticulum,** or **RER**) and some does not (**smooth endoplasmic reticulum,** or **SER**). Rough endoplasmic reticulum functions in the modification of proteins as they are synthesized. Smooth endoplasmic reticulum functions in the synthesis of lipids and carbohydrates and the detoxification of harmful substances. Organelles called **Golgi apparatuses** function in the modification of molecules and place plasma membranes around them. These packages of chemicals may be stored for later use or released from the cell in the process of secretion. **Lysosomes** are an example of the membrane-bound sacs produced by the Golgi apparatus. Lysosomes contain enzymes that function in digestion. **Vacuoles** are structures surrounded by a plasma membrane and may contain food, wastes, or water.

### Organelles Involved in Energy Conversion

The most complex organelles found in eukaryotic cells are those involved in energy conversions. **Chloroplasts** are organelles that function in converting the radiant energy of light into chemical energy. They are found in photosynthetic organisms such as phytoplankton, seaweeds, and marine plants. Organelles known as **mitochondria** transfer the chemical energy in food to molecules of **adenosine triphosphate,** or **ATP.** ATP supplies the energy for most of the metabolism and other activities that occur in living cells. All eukaryotic cells contain mitochondria, even if they are photosynthetic. Unlike other cellular organelles, chloroplasts and mitochondria are not synthesized by the cell but reproduce themselves.

### Organelles of Movement

Some cells have organelles called **flagella** (singular, flagellum) and **cilia** (singular, cilium) that are used to move the cell. Although the basic structures of flagella and cilia are the same, there are some differences. Flagella are long, hairlike structures, and cells that have flagella usually have one, two, or three of these structures. Cells use flagella to

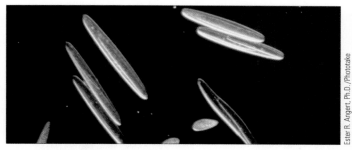

**Figure 5-6 Cells.** (a) Prokaryotic cells, like this bacterial cell, do not have a nucleus or membrane-bound organelles. (b) Eukaryotic cells like this plant and animal cell have a nucleus, and many membrane-bound organelles. Plant cells have a cell wall in addition to their cell membrane, organelles called chloroplasts, and a large central vacuole. Animal cells lack these features.

Ester R. Angert, Ph.D./Phototake

(a) Prokaryotic cell

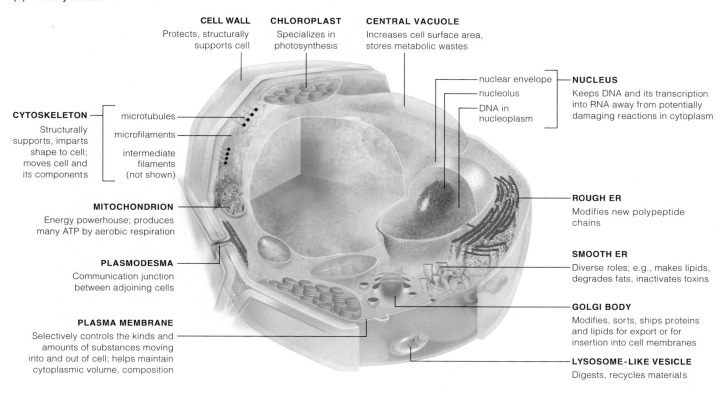

**CELL WALL**
Protects, structurally supports cell

**CHLOROPLAST**
Specializes in photosynthesis

**CENTRAL VACUOLE**
Increases cell surface area, stores metabolic wastes

nuclear envelope
nucleolus
DNA in nucleoplasm

**NUCLEUS**
Keeps DNA and its transcription into RNA away from potentially damaging reactions in cytoplasm

**CYTOSKELETON**
Structurally supports, imparts shape to cell; moves cell and its components

microtubules
microfilaments
intermediate filaments (not shown)

**ROUGH ER**
Modifies new polypeptide chains

**MITOCHONDRION**
Energy powerhouse; produces many ATP by aerobic respiration

**SMOOTH ER**
Diverse roles; e.g., makes lipids, degrades fats, inactivates toxins

**PLASMODESMA**
Communication junction between adjoining cells

**GOLGI BODY**
Modifies, sorts, ships proteins and lipids for export or for insertion into cell membranes

**PLASMA MEMBRANE**
Selectively controls the kinds and amounts of substances moving into and out of cell; helps maintain cytoplasmic volume, composition

**LYSOSOME-LIKE VESICLE**
Digests, recycles materials

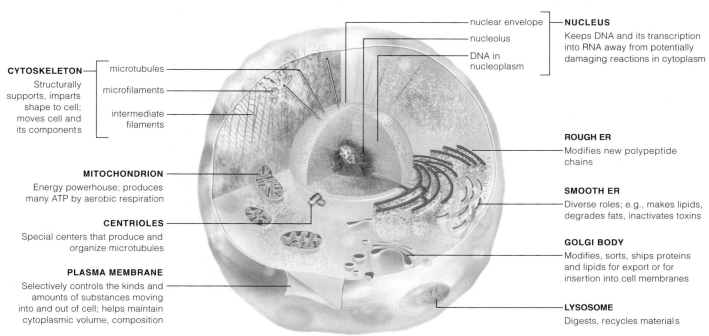

nuclear envelope
nucleolus
DNA in nucleoplasm

**NUCLEUS**
Keeps DNA and its transcription into RNA away from potentially damaging reactions in cytoplasm

**CYTOSKELETON**
Structurally supports, imparts shape to cell; moves cell and its components

microtubules
microfilaments
intermediate filaments

**ROUGH ER**
Modifies new polypeptide chains

**MITOCHONDRION**
Energy powerhouse; produces many ATP by aerobic respiration

**SMOOTH ER**
Diverse roles; e.g., makes lipids, degrades fats, inactivates toxins

**CENTRIOLES**
Special centers that produce and organize microtubules

**GOLGI BODY**
Modifies, sorts, ships proteins and lipids for export or for insertion into cell membranes

**PLASMA MEMBRANE**
Selectively controls the kinds and amounts of substances moving into and out of cell; helps maintain cytoplasmic volume, composition

**LYSOSOME**
Digests, recycles materials

(b) Eukaryotic cell

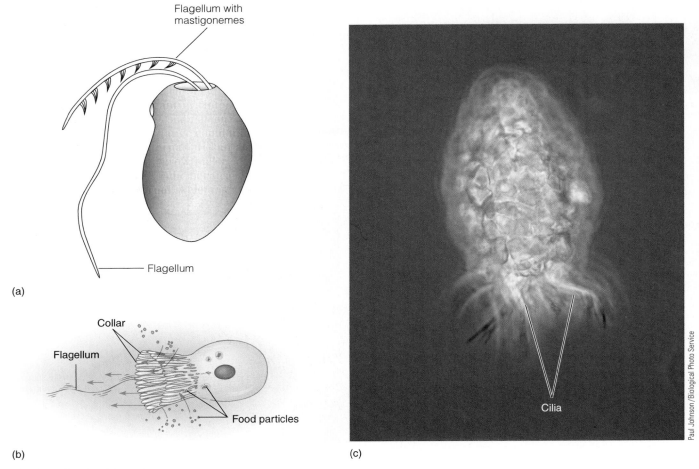

Flagellum with
mastigonemes

Flagellum

(a)

Collar

Flagellum

Food particles

(b)

Cilia

Paul Johnson / Biological Photo Service

(c)

**Figure 5-7  Organelles.** (a) Some cells, like this stramenopile, use flagella for locomotion. (b) Others, like this collar cell from a sponge, use the flagellum to move materials, in this case water, through the animal's body. (c) Cilia are more numerous than flagella and can be used for locomotion, as in this marine ciliate or for moving material across the surface of the cell.

propel themselves through their watery environment (Figure 5-7a and b). Cilia are short, hairlike structures that are quite numerous and can cover large areas of the cell surface (see Figure 5-7c). Cilia are used by single cells to move through their aqueous environment. In multicellular animals, cilia may cover the surface of some cells and move material along the cell's surface.

## Energy Transfer in Cells

As we noted in Chapter 2, energy is necessary to power all of life's activities, and without energy life would be impossible. Autotrophs use energy from their environment to synthesize food molecules, which can then be used as a source of energy to form ATP. Heterotrophs rely on other organisms

for energy. The food they take in is broken down and some of the energy transferred to ATP molecules for the organism's use. The rest is released as heat energy (Figure 5-8).

### Photosynthesis

You learned in Chapter 2 that in the process of photosynthesis low-energy molecules, such as carbon dioxide and water, combine to form high-energy food molecules like carbohydrates. In photosynthetic prokaryotes this occurs in areas of the cell where the membrane has folded in to form a surface to attach the molecules and enzymes required for this vital process. In eukaryotic cells photosynthesis occurs in organelles called chloroplasts. Chloroplasts are small, oval structures surrounded by two membranes. Within the chloroplasts are membrane-bound discs called **thylakoids**

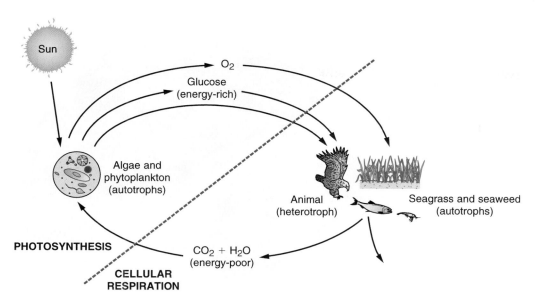

**Figure 5-8 Energy Transfer.** In photosynthesis low-energy inorganic compounds such as water and carbon dioxide are combined using the radiant energy of sunlight to form high-energy organic compounds that can be used for food. In the process of respiration, the high-energy organic compounds produced by photosynthesis are broken down to low-energy compounds like carbon dioxide and water. Some of the energy released by the process is transferred to adenosine triphosphate (ATP) and used to power cellular processes.

(Figure 5-9a) that contain pigment molecules, such as chlorophyll, that are needed to trap the radiant energy of light. The thylakoids are arranged in stacks, like stacks of poker chips, called **grana** (singular, granum). Surrounding the grana is a fluid material called the **stroma** that contains the enzymes necessary to convert the carbon in carbon dioxide into an organic form, a process called **carbon fixation**.

## Cellular Respiration

**Cellular respiration** releases energy from food molecules. Most of this energy-conversion process occurs in mitochondria. Like chloroplasts, mitochondria are complicated organelles surrounded by two membranes. The inner membrane is folded many times to form the mitochondrial cristae (see Figure 5-9b). This folding increases the total surface area available for the attachment of molecules that play key roles in energy conversion. In a stepwise series of complex reactions that begin in the cell's cytoplasm, food molecules are broken down and some of the energy released is used to synthesize molecules of ATP. During cellular respiration the carbon from the food molecules combines with oxygen, producing carbon dioxide as a waste product. Organisms exchange gases to supply oxygen for respiration and remove the waste carbon dioxide.

## Cellular Reproduction

Cells reproduce by cell division. Cell division in prokaryotic cells is relatively simple. In eukaryotic cells, however, the process is more complex and the nucleus divides before the cell.

### Cell Division in Prokaryotes

Prokaryotic cells have only a single, circular chromosome. The chromosome must be duplicated before the cell divides so that each new daughter cell will receive a complete copy of the necessary genetic information. Once duplicated, the two chromosome copies migrate to opposite sides of the cell. The cell then splits in two, forming two new daughter cells (Figure 5-10a). This is known as **binary fission**.

### Cell Division in Eukaryotes

Before a eukaryotic cell can divide, its nucleus divides, a process called **mitosis.** Before the nucleus can divide, all its chromosomes must be duplicated. During mitosis the nuclear membrane disappears and the two sets of chromosomes separate. A new nuclear membrane then forms around each of the two identical sets of chromosomes. Mitosis ensures that each new cell will receive a complete set of chromosomes containing the necessary genetic information. Once the nucleus has divided, the cell itself divides. In cells with cell walls, a new wall forms between the two nuclei, producing two new cells. In animal cells, which lack a cell wall, the cytoplasm pinches off between the two new nuclei, forming two new cells (see Figure 5-10b). The actual division of the cell into two new cells is called **cytokinesis.**

## Levels of Organization

Although many marine organisms are composed of only a single cell, others are made up of many cells that work together at various levels of organization to form an organism

**Figure 5-9  Photosynthesis.** (a) Photosynthesis occurs in the chloroplasts of eukaryotic cells. Pigments, such as chlorophyll, carotenoids, and xanthophylls in the thylakoid membranes trap the light energy that drives the process. (b) In eukaryotes, respiration occurs in an organelle called the mitochondrion. The highly folded internal membrane provides a large surface area for the attachment of enzymes that are necessary for the process.

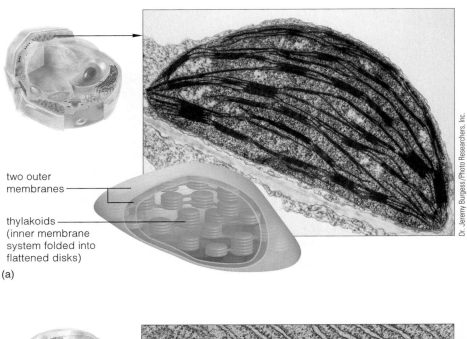

two outer membranes

thylakoids (inner membrane system folded into flattened disks)

(a)

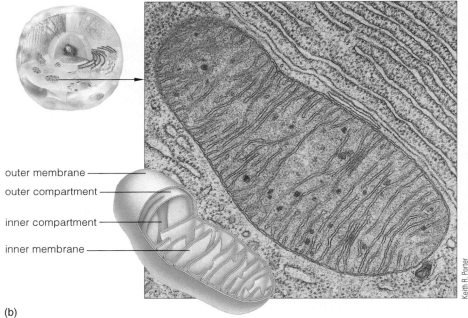

outer membrane

outer compartment

inner compartment

inner membrane

(b)

(Figure 5-11). Cells that serve one particular function are grouped into types of tissues. For instance, many muscle cells join to form muscle tissue in animals, or tubular cells make up vascular tissue to carry liquids through plants. Several different tissues can combine to form more complex structures called organs. Organs also specialize in performing certain functions. The stomach is an example of an animal organ whose function is digestion. It is composed of tissues that line the inside and outside of the organ and nerve and muscle tissue. Groups of organs can combine to form organ systems. For instance, an animal digestive system can be made up of an oral cavity, pharynx, stomach, and intestines, all working together to fully digest the food an animal eats. The most complex organisms are made up of several organ systems all working together to maintain the organism's life.

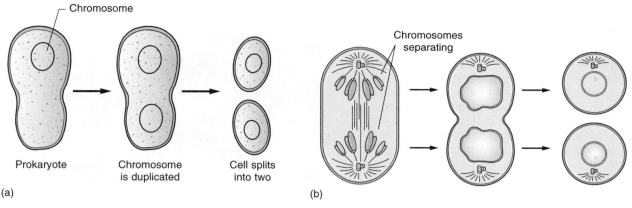

(a)                                                              (b)

**Figure 5-10  Cell Division.** (a) Cell division in prokaryotes is called binary fission. In this process the cell duplicates its single chromosome, the two chromosomes are separated, and the cell splits into two. (b) In eukaryotes cell division (cytokinesis) is preceded by mitosis, division of the nucleus. Once the nucleus has been duplicated and the two new nuclei separated to opposite sides of the cell, the cell divides. In organisms with cell walls, cell division is accomplished by forming a new cell wall between the two nuclei. In animals, cell division involves the formation of furrows that deepen until the cell is divided into two.

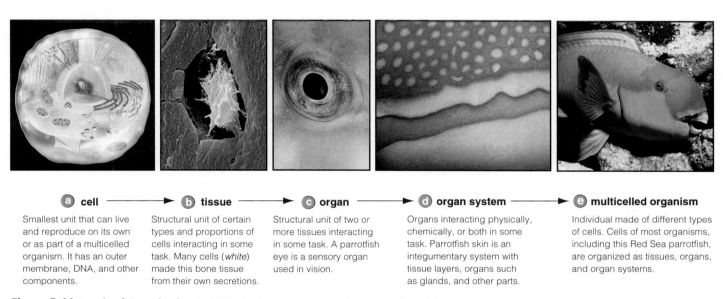

**Figure 5-11  Levels of Organization.** Individual cells can group together as a unit and form tissues that perform a single function. Several types of tissue come together to form organs, and groups of organs performing the same function make up and organ system. Organ systems working together form an organism. *(Lisa Starr with PDB file courtesy of Dr. Christina A. Bailey, Department of Chemistry and Biochemistry, California Polytechnic State University, San Luis Obispo, CA.)*

## In Summary

All living things are composed of cells. Prokaryotic cells lack a nucleus and membrane-bound organelles. Eukaryotic cells have a nucleus and many membrane-bound organelles. The organelles found in eukaryotic cells perform specific functions within the cell, much like organs within a multicellular organism. All cells reproduce by a process of cell division, giving rise to new cells. In multicellular organisms, groups of like cells form tissues. Different tissues combine to form organs, and organs can combine to form organ systems. ●

# EVOLUTION AND NATURAL SELECTION

The sea is home to large numbers of living organisms, and each one seems well suited for surviving and reproducing in its particular habitat. How and when did these organisms evolve, and what role does the environment play in determining the characteristics of organisms that can live in any given area? The answers to these and other questions lie in **evolution,** the process by which populations of organisms change over time.

## Darwin and the Theory for Evolution

In 1831 Charles Darwin set sail aboard HMS *Beagle* for a voyage of discovery around the world. During the next 5 years, he observed and collected specimens from many parts of the globe and began to formulate his ideas on the mechanism by which evolution operates.

### Voyage of Discovery

In the years before Darwin's great voyage, geologists were finding that the earth was much older than most people believed and that layers of rock of different ages contained fossils of life forms that were different from, but obviously related to, living species. Observations such as these laid the groundwork for the theory of evolution by natural selection that Darwin would eventually propose. Darwin was aware of these fossil findings as he set sail on the *Beagle*. As he traveled, he read books by the geologist Charles Lyell in which Lyell outlined his hypotheses for geological change. Lyell proposed that the physical features of the earth, such as mountains and valleys, were formed over long periods by the same slow geological processes, such as uplifting and erosion, that occur today. Darwin was greatly influenced by two conclusions drawn from the observations of geologists such as Lyell. First, if geological change is slow and continuous, then the earth must be very old. Second, slow and subtle changes that occur over centuries and millennia can produce substantial changes. Darwin realized that the age of the earth allowed ample time for gradual changes to occur in different populations of organisms and for new forms of organisms to arise.

As he observed and collected specimens from around the world, Darwin was amazed at the diversity of living organisms. He marveled at how each organism he observed seemed to have what he referred to as a "perfection of structure" for doing whatever was necessary to survive and produce offspring. He wondered what natural processes might be responsible for the evolution of the diverse life forms and their beneficial characteristics (Figure 5-12).

### Formulating a Theory for Evolution

Following his return to England, Darwin began to document his ideas on the process of evolution. In 1838 he read a 40-year-old essay by the English mathematician Thomas

**Figure 5-12 Beneficial Adaptations.** This triggerfish is adapted for the aquatic life on a coral reef. Fins help the fish maneuver in water, and the laterally compressed body allows the fish to easily navigate among the corals. Powerful jaws allow the triggerfish to feed on small animals with protective body coverings that are found on the reef.

Malthus. Malthus had observed that the human population was continuing to grow and, if unchecked, would eventually run out of space and food. Malthus believed that the consequences of uncontrolled population growth would be famine, disease, and war and that these consequences were external controls that would keep the human population from growing too large. If this were true of humans, who have relatively few offspring, Darwin reasoned, it must be even more true of plants and animals that produce numerous offspring. After years of work, Darwin came up with a scientific explanation for why populations generally do not exhibit unchecked growth and how they can change over time. He called his hypothesis "evolution by natural selection."

For many years Darwin discussed his ideas with only a few scientific colleagues. It was not until 1858, when he read a paper by the naturalist Alfred Russel Wallace, that he was encouraged to present his views. Wallace had been working in the country now known as Malaysia and had formulated a theory for evolution similar to the one Darwin had been working on for years. Wallace's paper, as well as a paper by Darwin, was presented at a meeting of the Linnaean Society in 1858. A year later, Darwin published his classic work, *On the Origin of Species by Means of Natural Selection.*

### Theory of Evolution by Natural Selection

Darwin was quite familiar with the process known today as **artificial selection,** which is practiced by farmers and animal breeders. In artificial selection, only animals and plants with certain desirable traits are selected for breeding in an effort to produce more animals or plants with the same traits. Darwin believed that a similar process was occurring in nature as the result of what he called **natural selec-**

tion, a process that favors the survival and reproduction of those organisms that possess variations best suited to their environment.

Unlike artificial selection, natural selection acts without the purpose of a farmer or a breeder. In fact, the process of evolution operates without any ultimate goal, selecting those forms of organisms that are best able to survive and reproduce under the environmental conditions in which they live. Note that natural selection does not have the ability to cause variations that are better suited than others. The variations that occur are due to chance mutations. Only after the variations appear can they be affected by natural selective forces. **Selective forces** are the physical and biological characteristics of the environment (such as temperature, salinity, predation, availability of food, and so on) that favor the survival of one species over another. According to Darwin's theory, selection occurs over many generations. During this time, organisms become better adapted by a process that can involve all aspects of an organism, such as anatomy, physiology, and behavior.

Darwin's theory of evolution by natural selection contains four basic premises.

1. All organisms produce more offspring than can possibly survive to reproduce.

2. There is a great deal of variation in traits among individuals in natural populations. Many of these variations can be inherited.

3. The amount of resources necessary for survival (food, light, living space, and so on) is limited. Therefore organisms must compete with each other for these resources.

4. Those organisms that inherit traits that make them better adapted to their environment are more successful in the competition for resources. They are more likely to survive and produce more offspring. The offspring inherit their parents' traits, and they continue to reproduce, increasing the number of individuals in a population with the adaptations necessary for survival.

Not all of an organism's traits are necessarily advantageous. Some modern evolutionary biologists believe that many traits become fixed in a population without necessarily being the best traits. Some traits may remain not because they are beneficial but because they confer no disadvantage. Others may be linked to traits that are beneficial, and the linked trait is "extra baggage" that is not harmful. To put it another way, the combination of traits exhibited by a successful organism represents a balance among several selective forces. Table 5-2 summarizes some key points concerning the evolutionary process.

## Genes and Natural Selection

At the time Darwin proposed his theory of evolution by natural selection, cell division, genes, and chromosomes had not yet been discovered. Since that time, the discovery of DNA and its role in heredity has completely revolutionized

the field of biology. Modern discoveries in the fields of **genetics** (the study of heredity) and **molecular biology** (the study of the structure and function of nucleic acids like DNA) have not contradicted Darwin's theory but have expanded and deepened our understanding of the evolutionary process.

### Modern Evolutionary Theory

Today's view of how evolution occurs, the **modern synthetic theory of evolution,** is essentially Darwin's 1858 idea as it has been refined by modern genetics. Recall from our previous discussion that a gene is a unit of hereditary information. It contains the chemical instructions for making proteins, and these proteins are largely responsible for the structure and function of organisms (in other words, their traits). Genes produce traits, such as body color, longer fins, or better vision, when the genetic information is translated into proteins by protein synthesis. The protein molecules can then form parts of the organism (structural proteins), direct the organism's biological activities (chemical messengers), or aid in the production of other molecules necessary for the organism's structure and function (enzymes).

Genes can exist in different forms called **alleles.** We now know that the variation in natural populations is the result of the alleles present in individual organisms. For instance, a snail can have a gene for shell color (the trait) and there can be several alleles (alternative forms) of the gene, so that some snails would have red shells, others yellow or brown (Figure 5-13). Each snail would have two of each kind of chromosome in its cells (one from each parent) and thus would have two alleles for each gene. The alleles could be the same, two yellow alleles, for instance, or different, one yellow and one red. There would be many possible combinations and variations in shell color from the same genes. Because chromosomes from the two parents are randomly mixed in their offspring, several offspring from the same parents can receive different combinations of alleles. This accounts for the variations among offspring of the same parents.

### Role of Reproduction

Genes are transmitted from one generation to another when DNA in the form of chromosomes is passed from parents to offspring in the process of reproduction. Reproduction can involve a single parent (asexual reproduction) or two parents (sexual reproduction).

***Asexual Reproduction*** In asexual reproduction (Figure 5-14a), offspring receive their genes from only one parent, and as a result, they are identical to the parent in all of their characteristics (clones). This type of reproduction introduces no variety into the population. The only source of variation is mutation. Asexual reproduction is common in organisms that live in stable environments where there is a benefit to reproducing many individuals in a short time. Asexual reproduction is very efficient because every member of the population can reproduce and mating is not required.

## Table 5-2  Evolution: What, Who, How

| | Random Genetic Changes | Natural Selection | Evolution |
|---|---|---|---|
| What Is It? | The raw material acted upon by natural selection | Difference in reproductive success between genetically different individuals of a population | Genetic change in a population from generation to generation |
| Whom Does It Affect? | Only individuals | Individuals and populations | Only populations |
| How Does It Work? | Produces an organism with an individual genetic makeup that:<br>1. is unique<br>2. may be inherited by the organism's offspring | Produces an organism with genetic differences that:<br>1. are advantageous under the existing environmental conditions<br>2. make it more likely to survive and reproduce, passing these advantages on to offspring | Natural selection affects generations and produces a population that is better adapted to existing environmental conditions |

Wrong: "Individuals can evolve."

Right: "Individuals do not evolve."

The genes that determine the anatomical features of the mouth cannot be modified in an individual. Therefore, this seahorse has no way to change its mouth to bite or chew prey if the small organisms that it feeds on were to become scarce or disappear completely.

Wrong: "Evolution produces perfection."

Right: "Organisms are not perfectly adapted."

Almost all adaptations represent a compromise in which the adaptation has advantages and disadvantages. For example, a turtle's shell provides good protection against the attack of sharks and other predators, but this same shell slows locomotion, decreases maneuverability, and makes reproduction challenging.

Wrong: "Evolution has a purpose."

Right: "Evolution involves chance."

Selective factors are unpredictable, as is the course of evolution. For example, when shoreline rocks are continuously submerged, this rock anemone survives and reproduces. If geological changes or human intervention causes these same rocks to be raised above the water line, the same genes that allowed the anemone to survive previously will not help it survive in the dry environment, and the anemone will die.

Wrong: "More evolved equals better."

Right: "New species are not better than older species."

Any species alive today is successful (so far), whether it originated 200 years ago or 200 million years ago. Is a bacterium that evolved a billion years ago better than a blue whale? It is less beautiful to the human eye, perhaps, but the bacterium is just as successful. Indeed, bacteria may very well outsurvive the blue whale and all of its relatives.

Bacterium magnified 1,000,000 times

***Sexual Reproduction***  In **sexual reproduction** chromosomes from two parents are combined when specialized cells called **sex cells,** or **gametes** (eggs, sperm, pollen, and so on) unite during **fertilization** (see Figure 5-14b). Recall that each cell in a eukaryotic organism normally contains two copies of each of the organism's different chromosomes. This number of chromosomes is called the **diploid number** and is represented by **$2N$,** where $N$ is the number of different chromosomes. Unlike the other cells in a sexually reproducing organism, gametes have only a single copy of each

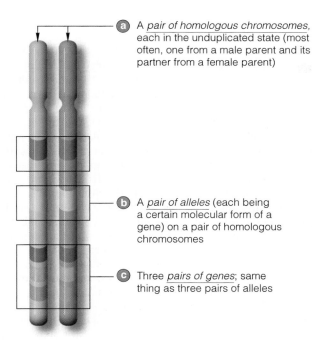

(a) A *pair of homologous chromosomes*, each in the unduplicated state (most often, one from a male parent and its partner from a female parent)

(b) A *pair of alleles* (each being a certain molecular form of a gene) on a pair of homologous chromosomes

(c) Three *pairs of genes*; same thing as three pairs of alleles

**Figure 5-13 Genes.** Chromosomes contain an organism's genes. Each gene can exist in more than one form, and each form of a gene is called an allele.

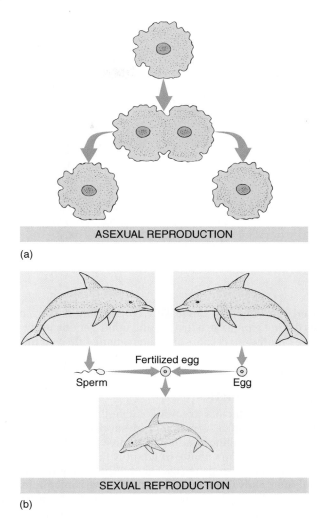

**ASEXUAL REPRODUCTION**

(a)

Fertilized egg

Sperm          Egg

**SEXUAL REPRODUCTION**

(b)

**Figure 5-14 Asexual and Sexual Reproduction.**
(a) Asexual reproduction requires only one parent and produces offspring that are genetically identical to the parent (clones). (b) Sexual reproduction requires two parents to contribute genes to the next generation. This results in offspring that have new combinations of genes (half from each parent) and are not identical to their parents.

chromosome, the **haploid number (*N*)**. This is necessary because, when sexually reproducing organisms reproduce, a cell from each parent fuses to form the next generation. If the number of chromosomes were not reduced to one copy of each, then each generation would have cells with twice as many chromosomes as the last, and eventually there would be no room in the cell for anything but chromosomes and the cell would die. Sexual reproduction is not as efficient as asexual reproduction because large amounts of resources and energy are wasted in forming gametes that will not be fertilized or survive. Also, only those organisms successful in having a mate are able to reproduce. This limits the number of individuals in a population that can reproduce.

***Meiosis***    Gametes are formed by a special kind of cell division called **meiosis, or reduction division.** Meiosis is called reduction division because it produces cells with nuclei containing only one half the number of chromosomes as the parent cell. The main difference between meiosis and mitosis (mitosis was discussed earlier in the chapter) is that chromosomes are duplicated once in meiosis and then the cell divides twice. This produces twice as many cells with half as many chromosomes.

Meiosis not only reduces the number of chromosomes in the cell's nucleus it also increases the amount of variation in populations. During the initial phase of meiosis, all like chromosomes connect with each other for a short time. During this phase, alleles of the same gene on neighboring chromosomes can switch from one chromosome to another. This **crossing over** produces chromosomes with combina-

tions of alleles that did not exist before the meiotic division (Figure 5-15). This **recombination** of alleles increases the amount of variation within populations. As meiosis continues, more variety is added as the chromosomes are randomly assorted into new combinations in the gametes that did not exist in the cells of the previous generation. During fertilization, a male gamete fuses with a female gamete to produce a new individual that has inherited half its chromosomes from one parent and half from the other.

As you can see, the sexual reproduction that occurs in each generation produces a constant reshuffling of alleles in the species' **gene pool** (all of the alleles in a given population that exist at a given time). This random mixing of alleles by

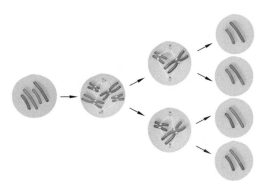

**Figure 5-15  Meiosis.** In sexually reproducing organisms, sex cells are formed by meiosis. In meiosis the number of chromosomes in a cell is reduced by one half.

sexual reproduction along with the random sorting of alleles during gamete production and other genetic processes, such as mutation, create variations among different individuals of a species. The variations are acted upon by natural selection.

## Population Genetics

Environmental factors are not static but constantly change. For a population of organisms to survive and reproduce under these changing conditions, it must be able to adapt to them. A population's ability to adapt is limited by the genes and alleles of those genes carried by the individuals in the population. Only those individuals that have combinations of genes and alleles that allow adaptations to their particular surroundings are likely to survive and leave offspring. An organism's biological success, or **fitness,** is measured by the number of its own genes present in the next generation. The more offspring an individual contributes to the population, the more biologically successful that individual is.

## New Species Evolve from Existing Species

Each different kind of organism is a different **species.** What exactly, though, is a species? The definition of species has changed over the years as our understanding of evolution and molecular biology has improved.

### Typological Definition of Species

In the past a species was defined as an organism with a definable set of characteristics that is visibly different from other similar organisms. Two fishes the same size and shape but having a different color pattern would be considered different species in the historical sense. This definition of species is called **typological** and is based on **morphology,** or the structure and appearance of the organism. For the purpose of identification, a museum specimen considered to be representative of the species is designated the **type specimen,** and other specimens are then compared to the type. If they appear similar based on certain morphological characteris-

tics, they are considered to be the same species, and if they are different, they are considered different species.

Unfortunately, this is not the best way to define a species. As we noted previously, there is a large amount of variation in natural populations, and not all members of the same species are exactly alike. For instance, in some marine animals, such as fiddler crabs (*Uca pugnax*) and rosy razorfish (*Hemipteronotus martinicensis*), the males (Figure 5-16a) and females (Figure 5-16b) look quite different—a condition known as **sexual dimorphism.** In other species, such as the bluehead wrasse (*Thalassoma bifasciatum;* see Figure 5-16c and d) and the spotted hagfish (*Bodianus pulchellus*), the juveniles look distinctly different from the adults. In many marine snails, such as the communal nerite (*Nerita communis*), which is found in the Philippine Islands, individuals in a population show a variety of colors and patterns and are quite dissimilar (see Figure 5-16e). According to the strict typological concept of species, these animals would be considered different species, when indeed they are really the male and female of the same species, the juvenile and adult forms, or ecological variants, respectively. The typological definition of a species has caused a great deal of confusion over the years, and many of the same types of organism have been erroneously identified as separate species. Today, whenever possible, several specimens are defined as types to demonstrate the range of natural variation.

### Modern Species Definition

By modern definition, a species is one or more populations of potentially interbreeding organisms that are reproductively isolated from other such groups. This definition is not as easy to apply as the typological definition and is often not practical (as in the case of organisms that reproduce asexually), but it is more useful in establishing relationships among organisms. Modern research techniques allow biologists to analyze and compare the genes (DNA) and the proteins these genes produce. Such analysis provides the information necessary for applying the modern genetic definition of a species to organisms being studied.

**Reproductive isolation** means the members of different species are not in the same place at the same time or are physically incapable of breeding. This inability to reproduce with individuals outside the gene pool prevents the genes of one species from mixing with the genes of another. There are several ways that organisms of different species are prevented from reproducing in nature. These preventive mechanisms, called **isolating mechanisms,** can be divided into two categories: those that prevent fertilization from ever occurring and those that prevent successful reproduction following fertilization.

***Isolating Mechanisms That Prevent Fertilization***  Isolating mechanisms that prevent fertilization include differences in habitat, breeding time, anatomy, and behavior. For instance, if one species of snail lives on sandy beaches in the Caribbean, and a similar species lives in mangrove swamps, they will not be able to interbreed because they never encounter each other. This mechanism is referred to as **habitat isolation.**

In invertebrate animals different species frequently have incompatible copulatory organs that prevent similar

(a)

(b)

(c)

(d)

(e)

**Figure 5-16 Morphological Diversity Within a Species.** Some species, such as this rosy razorfish (*Hemipteronotus martinicensis*), exhibit sexual dimorphism, that is, the (a) male and (b) female look distinctively different. (c) The juvenile bluehead wrasse (*Thalassoma bifasciatum*) does not look the same as (d) the adult of the species. (e) Snails such as *Nerita communis* exhibit a large range of variation as adults.

species from reproducing with each other. This is an example of **anatomical isolation.**

Some animals exhibit special behaviors during the breeding season, and only members of the same species recognize the behavior as courtship. This constitutes **behavioral isolation.** The male common cuttlefish (*Sepia officinalis*), for instance, presents a striped color pattern that identifies him to a female *Sepia officinalis* as a prospective mate. Only female *Sepia officinalis* recognize the color pattern as a mating display. Members of different species do not recognize or respond to this courtship display and thus do not attempt to mate with the displaying individual.

Most organisms reproduce only at certain times of the year or under certain conditions (temperature, salinity, hours of light). If the time that members of one species are ready to reproduce does not coincide with the time members of a related species are reproducing, then the two species will be separated by **temporal isolation.**

Members of closely related species that succeed in copulating generally do not produce offspring because biochemical or genetic differences between their gametes prevent successful fertilization. This is usually referred to as **biochemical isolation.**

***Isolating Mechanisms That Prevent Successful Reproduction following Fertilization***   Incompatible genes or biochemical differences can prevent a fertilized egg from developing further. In some instances, fertilization and development may be successful, but the offspring, called **hybrids,** may be infertile (unable to reproduce) or ecologically weak (unable to successfully compete with the well-adapted parental species) and thus quickly die. These mechanisms prevent successful reproduction even if mating and fertilization do take place.

## Process of Speciation

**Speciation** refers to the various mechanisms by which new species arise. Speciation frequently begins when two or more populations of the same species become geographically isolated from each other or when segments of a single large population become geographically isolated from other segments. This type of speciation is called **allopatric** ("different homeland") **speciation.** Although allopatric speciation is not the only type of speciation, it is thought to be responsible for the formation of the greatest number of species. Geological changes such as earthquakes, mountain building, and the formation of islands can separate populations. This geographic isolation because of geological change is often the first step in evolution. Geographic isolation is followed by the appearance of biological isolating mechanisms. This occurs because, once two populations have be-

come physically isolated, there is no longer any gene flow between them. From that point on, natural selection will operate on each population independently. This is especially true when the separated populations are subject to different environmental conditions.

For example, more than 4 million years ago the Caribbean Sea and Pacific Ocean were connected between North and South America. The geological changes that formed Central America between 3 and 4 million years ago separated many species of fish and invertebrates into Pacific and Caribbean populations. These isolated populations have evolved independently as they adapted to the different temperature, salinity, sediments, food availability, predators, and other conditions in the two seas. Although some species from the two sides of Central America are morphologically similar, others have diverged in such traits as color, habitat preference, breeding behavior, and breeding season. These differences act as reproductive isolating mechanisms, and the populations can no longer interbreed, even if the physical barrier between them is removed (as in the building of the Panama Canal) or if they jointly colonize a common area. These populations, formerly the same species, have become separate species (Figure 5-17).

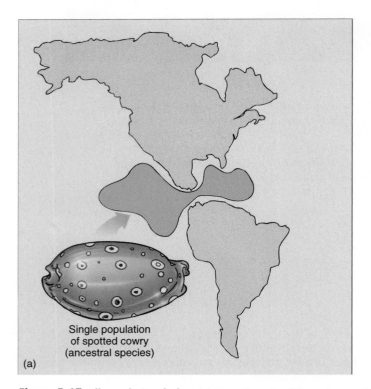

(a) Single population of spotted cowry (ancestral species)

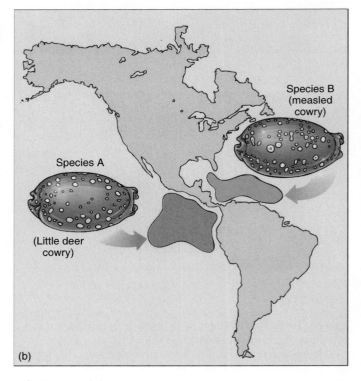

Species B (measled cowry)

Species A

(Little deer cowry)

(b)

**Figure 5-17 Allopatric Speciation.** (a) More than 4 million years ago the Pacific Ocean and the Caribbean Sea were connected and home to many populations, such as the ancestral population of spotted cowries. (b) When the land mass known as Central America was formed between 3 and 4 million years ago, many animal populations, including the spotted cowry population, became geographically isolated from each other, resulting in the evolution of similar yet distinct species. The measled cowry (*Cypraea zebra*) is found in the Gulf of Mexico, while the little deer cowry (*Cypraea cervinetta*) is found along the Pacific coast of Central America. These two species are quite similar and very likely share a common ancestor. When brought together today in an aquarium, they can no longer mate successfully.

## In Summary

Charles Darwin introduced the theory of evolution by natural selection in 1859 in his book *On the Origin of Species by Means of Natural Selection.* Darwin recognized that the amount of resources necessary for survival is limited. His theory suggests that only those organisms possessing adaptations that make them more successful in competing for resources will survive, whereas those that cannot cope will die out. Natural selection functions by favoring the survival of individuals with the combination of traits best suited to their particular place and time. These individuals are also the most likely to reproduce successfully, passing their advantageous traits to their offspring. As environments change, new sets of characteristics are favored. Over long periods, new characteristics accumulate in populations, and new, better-adapted species gradually replace older ones that are not as well adapted.

Although the genetic mechanisms that govern inheritance play a significant role in evolution, they were unknown during Darwin's time. The modern synthetic theory of evolution is basically Darwin's idea refined by modern genetics. The variations among sexually reproducing individuals in natural populations occur because, in general, each individual possesses a unique combination of genes and their alleles. Individuals that possess genes for traits that best adapt them to their environment are the most likely to survive and reproduce, thus contributing more of their genes to the population's gene pool. Biological success is measured in terms of the number of genes an organism contributes to the population.

In the past, a species was defined as a group of organisms with an observable set of characteristics that was different from other, similar organisms. Today a species is defined as one or more populations of potentially interbreeding organisms that are reproductively isolated from other such groups. New species may form when environmental or genetic changes cause portions of a population to become reproductively isolated from the rest of the group. ●

## LINNAEUS AND BIOLOGICAL CLASSIFICATION

We use several schemes to classify the organisms we study in marine biology. The terms *plankton, nekton,* and *benthos* classify organisms based on where and how they live. The terms *herbivore, carnivore,* and *decomposer* classify them according to ecological roles. Placing organisms into domains and their subdivisions is a classification scheme that indicates the evolutionary relationships of the organisms being studied.

While studying living organisms, biologists find it helpful, and indeed necessary, to use a universal system of naming the more than one million known species. Naming organisms enables scientists all over the world to exchange information about them. Most organisms do not have nonscientific names, and those that do, have names that vary from one country to another and even among geographic regions within a country. Biologists need a system of naming things so that every researcher in the field calls the same organism by the same name. To deal with the problems of classifying and naming the large number of organisms that inhabit the earth, biologists have developed the science of **taxonomy,** which deals with describing, classifying, and naming organisms.

## Binomial System of Naming

The system that biologists now use for naming organisms was introduced by Karl von Linné (or, in the Latin form, Carolus Linnaeus), a Swedish botanist, in the mid-1700s. During Linnaeus's time, organisms were classified on the basis of a lengthy description. Linnaeus found this method too cumbersome and developed a standardized method that he refined over the years. In 1750 he introduced the idea of **binomial nomenclature,** a system of naming that uses two words to identify an organism. The first word of a proper scientific name is the **genus** (plural, genera), and Linnaeus placed all organisms that shared certain traits in the same genus. For instance, almost all butterflyfishes belong to the genus *Chaetodon* (Figure 5-18). The second part of the two-part name is the **species epithet,** which identifies a particular kind, or **species** (plural is also species), of organism in the genus. For instance, *Chaetodon longirostris* is the longnose butterflyfish, whereas *Chaetodon ocellata* is the spotfin butterflyfish. The first letter of the genus name is always capitalized and the species epithet is always in lowercase. When the scientific name appears in print, it is in italics or the genus and the species epithet are each underlined.

When Linnaeus developed the system of binomial nomenclature, he placed organisms into genera on the basis of similar morphology. He then grouped organisms into larger categories called classes and orders. These larger groups were made up of organisms that bore a superficial resemblance but were not in all cases related. Today we still use the binomial system of nomenclature, but organisms are now classified on the basis of evolutionary relationships, not solely on the basis of structural similarities. If two species share a genus name, it means they not only appear similar but it is thought that they also share a common ancestor. This is determined by the distribution of shared characteristics and molecular similarities.

## Taxonomic Categories

Several categories are currently used to show the complex evolutionary relationships among related organisms. The major categories, from most inclusive to least inclusive (most specific), are domain, kingdom, phylum, class, order, family, genus, and species. Students should always keep in mind that classification is a human endeavor, and that the important thing is not the memorization of these or other categories but a basic understanding of why each organism is classified as it is.

### Early Schemes of Classification

During Linnaeus's time, and until the 1960s, biologists grouped all living organisms into one of two kingdoms: Plantae and Animalia. The kingdom Animalia contained the organisms that fed on other organisms (heterotrophs) and that

(a)

Jon L. Hawker

(b)

Fred McConnaughey 1985/Photo Researchers, Inc.

**Figure 5-18 Butterflyfishes.** Two different species of butterflyfish, (a) *Chaetodon longirostris* (long-nose butterflyfish), and (b) *Chaetodon ocellata* (spotfin butterflyfish).

were capable of independent movement. All other organisms (autotrophs and saprotrophs, or decomposers) were classified in the kingdom Plantae.

With the invention of the electron microscope and the accumulation of more information from all fields of biology, it became apparent that many prokaryotes were more closely related to each other than they were to eukaryotes and that many unicellular organisms were not closely related to multicellular organisms. There were also major differences among organisms within the groups of multicellular organisms. This led to the creation of more kingdoms.

The kingdom Prokaryotae was introduced and contained all of the prokaryotes. Fungi were placed in their own kingdom, and eukaryotes that did not fit the definition of a plant, animal, or fungus were placed in the kingdom Protista.

## Modern Classification

With the advent of modern molecular techniques, our knowledge of the relationships among organisms has increased dramatically and most biologists agree that the old system of classification does not adequately reflect the actual evolutionary relationships. Continuing studies at the molecular level have led to new categories, and classification schemes continue to change as evolutionary biologists attempt to accurately represent evolutionary relationships among organisms.

*Domains* During the 1980s biologists began to recognize that not all prokaryotes were closely related. Molecular studies resulted in a proposal for a new category higher than kingdom, the **domain.** The domains **Archaea** and **Eubacteria** both contain single-cell, prokaryotic organisms that differ from each other in several ways at the biochemical level. All of the eukaryotic organisms were placed in the domain **Eukarya.** Today most biologists recognize the three-domain system of classification (Figure 5-19).

*Kingdoms* The domain Eukarya contains three well-defined kingdoms and many phyla that are clearly not evolutionarily related. The three kingdoms in the domain Eukarya are Fungi, Plantae (plants), and Animalia (animals). **Fungi** are eukaryotic organisms that are not capable of photosynthesis and whose cell walls contain the polysaccharide chitin. Some fungi are unicellular whereas others are multicellular. They are primarily saprotrophs, although there are some parasitic species. Yeasts, molds, and mushrooms are a few examples of fungi. **Plants** have cells with cell walls containing the polysaccharide cellulose. All plants are multicellular and capable of photosynthesis. **Animals** are all multicellular. Their cells lack cell walls and they are ingestive heterotrophs. This means that they rely on other organisms for food, which they take into their bodies to digest.

*Protists* The domain Eukarya also contains the organisms generically referred to as **protists.** Protists are eukaryotic organisms that do not fit the definition of animal, plant, or fungus. Some examples of protists are algae, protozoans, and slime molds. Although the majority of protists are unicellular, there are also multicellular forms. Their cells may or may not have cell walls, and the composition of the cell walls can be quite variable. Nutritionally they can be autotrophs, heterotrophs, saprotrophs, and some can even be combinations of two nutritional types. Some biologists group all protists into a separate kingdom, Protista, but as more and more molecular studies are done, it has become clear that grouping all of these organisms together is artificial. It makes more sense to deal with these organisms as separate phyla in the domain Eukarya. Their relationships to each other and to other eukaryotes are complex, and biologists are just beginning to unravel them with the aid of modern molecular techniques. One of many possible schemes for classifying eukaryotes is shown in Figure 5-20.

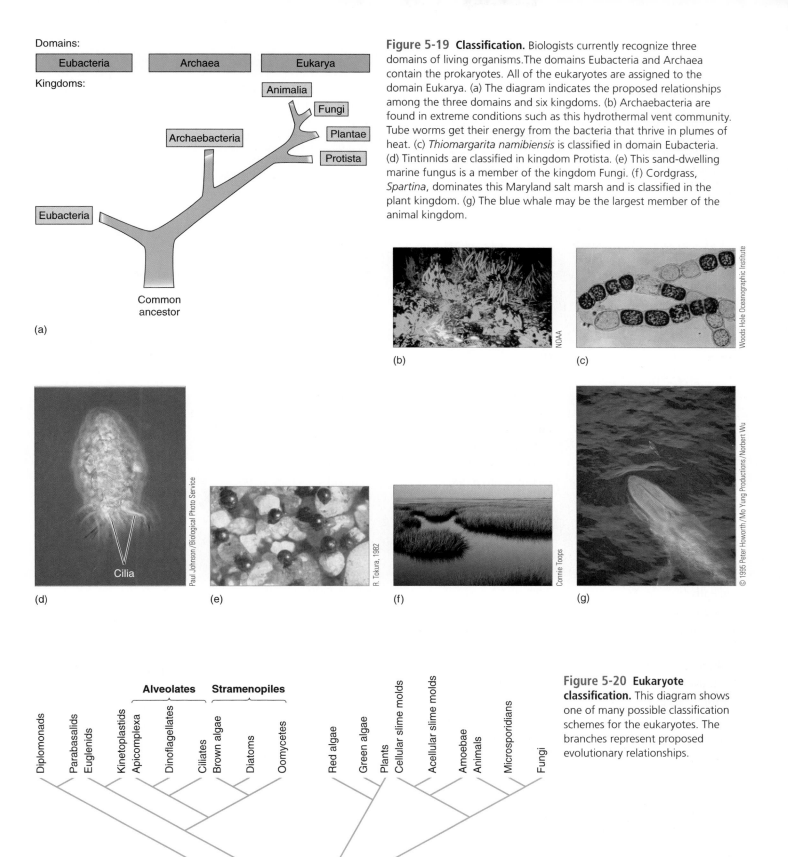

**Figure 5-19 Classification.** Biologists currently recognize three domains of living organisms. The domains Eubacteria and Archaea contain the prokaryotes. All of the eukaryotes are assigned to the domain Eukarya. (a) The diagram indicates the proposed relationships among the three domains and six kingdoms. (b) Archaebacteria are found in extreme conditions such as this hydrothermal vent community. Tube worms get their energy from the bacteria that thrive in plumes of heat. (c) *Thiomargarita namibiensis* is classified in domain Eubacteria. (d) Tintinnids are classified in kingdom Protista. (e) This sand-dwelling marine fungus is a member of the kingdom Fungi. (f) Cordgrass, *Spartina*, dominates this Maryland salt marsh and is classified in the plant kingdom. (g) The blue whale may be the largest member of the animal kingdom.

**Figure 5-20 Eukaryote classification.** This diagram shows one of many possible classification schemes for the eukaryotes. The branches represent proposed evolutionary relationships.

## In Summary

The science of taxonomy deals with naming, describing, and classifying organisms. The categories in the taxonomic scheme are organized to reflect evolutionary relationships. The broadest, most inclusive category is the domain. Organisms are subdivided into one of three domains based on molecular and cellular characteristics. Organisms in the domain Eukarya are placed into kingdoms based on several characteristics, such as cell type, number of cells, and feeding type. Kingdoms are divided into progressively less-inclusive taxonomic categories. From most inclusive to least inclusive, they are phylum, class, order, family, genus, and species. A group of eukaryotes called protists are frequently placed into their own kingdom, but these organisms are not related evolutionarily and are better treated as separate phyla within the domain Eukarya. ●

## SELECTED KEY TERMS

adenosine triphosphate (ATP), *p. 90*

alleles, *p. 97*

allopatric speciation, *p. 102*

amino acids, *p. 86*

anatomical isolation, *p. 101*

animal, *p. 104*

Archaea, *p. 104*

artificial selection, *p. 96*

asexual reproduction, *p. 97*

behavioral isolation, *p. 101*

binary fission, *p. 93*

binomial nomenclature, *p. 103*

biochemical isolation, *p. 101*

carbohydrate, *p. 85*

carbon fixation, *p. 93*

cell wall, *p. 90*

cell, *p. 90*

cellular respiration, *p. 92*

cellulose, *p. 86*

chitin, *p. 86*

chloroplast, *p. 90*

chromosome, *p. 90*

cilia, *p. 90*

crossing over, *p. 99*

cytokinesis, *p. 93*

cytoplasm, *p. 90*

cytosol, *p. 90*

deoxyribonucleic acid (DNA), *p. 88*

diploid number (2*N*), *p. 99*

disaccharide, *p. 85*

domain, *p. 104*

endoplasmic reticulum (ER), *p. 90*

enzyme, *p. 88*

Eubacteria, *p. 104*

Eukarya, *p. 104*

eukaryote, *p. 90*

eukaryotic cells, *p. 90*

evolution, *p. 96*

fertilization, *p. 98*

fitness, *p. 100*

flagellum, *p. 90*

fungus, *p. 104*

gamete, *p. 98*

gene pool, *p. 99*

gene, *p. 88*

genetics, *p. 97*

genus, *p. 103*

glucose, *p. 85*

glycogen, *p. 86*

Golgi apparatus, *p. 90*

grana, *p. 93*

habitat isolation, *p. 100*

haploid number (*N*), *p. 99*

hybrid, *p. 102*

isolating mechanism, *p. 100*

lactose, *p. 85*

lipid, *p. 86*

lysosome, *p. 90*

macromolecule, *p. 85*

maltose, *p. 85*

meiosis, *p. 99*

messenger RNA (mRNA), *p. 88*

mitochondria, *p. 90*

mitosis, *p. 93*

modern synthetic theory of evolution, *p. 97*

molecular biology, *p. 97*

monosaccharide, *p. 85*

morphology, *p. 100*

natural selection, *p. 96*

nucleic acids, *p. 88*

nucleolus, *p. 90*

nucleotide, *p. 88*

nucleus, *p. 90*

organ, *p. 95*

organ system, *p. 95*

organelle, *p. 90*

phospholipid, *p. 86*

plant, *p. 104*

polymer, *p. 85*

polypeptide, *p. 86*

polysaccharide, *p. 85*

prokaryote, *p. 90*

prokaryotic cell, *p. 90*

protein synthesis, *p. 88*

protein, *p. 86*

protist, *p. 104*

recombination, *p. 99*

reduction division, *p. 99*

reproductive isolation, *p. 100*

ribonucleic acid (RNA), *p. 88*

ribosomal RNA (rRNA), *p. 88*

ribosome, *p. 88*

rough endoplasmic reticulum (RER), *p. 90*

selective forces, *p. 97*

sex cell, *p. 98*

sexual dimorphism, *p. 100*

sexual reproduction, *p. 98*

smooth endoplasmic reticulum (SER), *p. 90*

speciation, *p. 102*

species epithet, *p. 103*

species, *p. 103*

starch, *p. 85*

steroid, *p. 86*

stroma, *p. 93*

sucrose, *p. 85*

taxonomy, *p. 103*

temporal isolation, *p. 101*

thylakoid, *p. 92*

tissue, *p. 95*

transfer RNA (tRNA), *p. 88*

triglyceride, *p. 86*

type specimen, *p. 100*

typological definition of species, *p. 100*

vacuole, *p. 90*

## QUESTIONS FOR REVIEW

### Multiple Choice

1. The molecule that carries genetic information is
   a. protein
   b. polysaccharide
   c. lipid
   d. DNA
   e. polypeptide

2. If a cell lacked ribosomes, it would not be able to
   a. synthesize lipids
   b. digest food
   c. synthesize proteins
   d. form membranes
   e. signal to other cells

3. Which of the following organelles is involved in locomotion?
   a. endoplasmic reticulum
   b. flagellum
   c. Golgi apparatus
   d. vacuole
   e. chloroplast

4. The molecule that supplies most of the energy needed for cellular function is
   a. DNA
   b. RNA
   c. starch
   d. ATP

5. Mitochondria
   a. transfer the energy from food molecules to ATP
   b. convert radiant energy into ATP
   c. convert ATP into food molecules
   d. produce oxygen
   e. assist cellular reproduction

6. During the process of meiosis
   a. two new cells are formed that are identical to the parent cell
   b. chromosomes are not duplicated before division occurs
   c. the number of chromosomes in a cell's nucleus is reduced by one half
   d. two gametes fuse to form a new individual
   e. two cells are formed that contain twice as many chromosomes as the parent cell

7. Darwin proposed that evolution occurs as the result of
   a. cosmic forces
   b. human intervention
   c. artificial selection
   d. natural selection
   e. inherent need

8. Biological success is measured by
   a. how long an organism lives
   b. how large an organism grows
   c. how many times an organism mates
   d. how many of an organism's genes are in a population
   e. how successful an organism is at avoiding predation

9. A population of potentially interbreeding organisms that is reproductively isolated from other populations defines
   a. a kingdom
   b. a community
   c. a family
   d. a genus
   e. a species

10. The domain Archaea contains
   a. fungi
   b. protists
   c. single-cell algae
   d. prokaryotes
   e. organisms whose cells have a nucleus

### Short Answer

1. How can you distinguish between flagella and cilia?

2. How can you distinguish prokaryotic cells from eukaryotic cells?

3. What are the four points of Darwin's theory of evolution by natural selection?

4. What are the higher categories in modern classification?

5. Describe how new species are formed.

6. Describe how populations can become reproductively isolated.

7. Describe the characteristics of organisms in each of the three domains.

8. Compare asexual and sexual reproduction.

9. Why does it make more sense not to place protists into one kingdom?

10. What organelles found in eukaryotic cells are involved in energy conversion?

### Thinking Critically

1. While studying plankton samples, a friend asks you how you can distinguish protozoans from small animals like crustaceans. What would you answer?

2. While on a field trip, you collect two snails that look very similar. How would you determine if they were the same or different species using the methods available in 1858? How would you determine it today?

3. While working in the field you collect several similar specimens of marine fishes from populations in two different localities. What experiments would you perform to determine if these fishes were the same or different species? What results might you expect from these experiments and what might the results prove?

Kylie Francis
Mook
E121

## SUGGESTIONS FOR FURTHER READING

Margulis, L., K. Schwartz, and M. Dolan. 1994. *The Illustrated Five Kingdoms: A Guide to the Diversity of Life on Earth.* New York: Harper Collins.

Price, P. W. 1996. *Biological Evolution.* Philadelphia: Saunders College Publishing.

### InfoTrac College Edition Articles

Gould, S. J. 1999. Branching Through a Wormhole, *Natural History* 108(2).

Norris, R. D. 1992. So You Thought Extinction Was Forever? *Oceanus* 35(4).

Scheltema, R. S. 1996. Describing Diversity: Too Many New Species, Too Few Taxonomists, *Oceanus* 39(1).

Stoeckle, M. 2003. Taxonomy, DNA, and the Bar Code of Life, *Bioscience* 53(9).

Vermeij, G. J. 2002. Why Are There No Lobsters on Land or Bats at Sea? The Answer Appears to Be a Biological Version of Beginner's Luck, *Natural History* 111(1).

Winston, J. E., K. L. Metzger. 1998. Trends in Taxonomy Revealed by the Published Literature, *Bioscience* 48(20).

### Websites

**http://www.ucmp.berkeley.edu/history/evolution.html**
Good introduction to evolutionary theory.

**http://www.biosis.org/zrdocs/zoolinfo/syst_tax.htm**
Information about taxonomy and systematics.

# Marine
# Microbes

**M**icrobes are organisms that are invisible to the naked eye. They include viruses, one-celled organisms, and occasionally fungi. Here we will deal with viruses and one-celled organisms and will include the fungi because most marine fungi are one-celled or at least microscopic. These microbes belong to all three domains of life and play many roles in marine ecosystems, such as producer, consumer, decomposer, and parasite. Although unseen by visitors to the ocean, they are the most numerous organisms in the sea.

## MARINE VIRUSES

As in so many biological disciplines, **virology,** the study of viruses, had a late start in the marine sciences. Viral diseases among humans and their domesticated crops and animals were known to ancient civilizations, and Edward Jenner's vaccinations against smallpox were reported more than 300 years ago. It was not until the mid-1900s, shortly after the chemical nature of viruses was becoming understood, that the first marine virus, one that infected bacteria, was discovered. Viruses of marine eukaryotic hosts were reported in the 1970s. Reliable counts of viral abundance were made in the 1980s, and the counts led to a better understanding of the ecological roles of marine viruses in the 1990s. The first decade of the new millennium seems to be a period of molecular studies, with emphasis on genome sequencing and the diversity of marine viruses. We examine marine viruses in this chapter because of their abundance —more than any other organism in the sea—diversity, and significance in marine food webs, population biology, and disease.

NOAA; inset, Dr. Dennis Kunkel/Visuals Unlimited

## Key Concepts

1. Microbial life in the sea is extremely diverse, including members of all three domains of life as well as viruses.

2. Marine virology is an emerging field of study, due to recognition of the critical role that viruses may play in population control of other microbes, in nutrient cycling, and in marine pathology.

3. Photosynthetic and chemosynthetic bacteria and archaeons are important primary producers in marine ecosystems.

4. Heterotrophic bacteria, archaeons, and fungi play essential roles in recycling nutrients in the marine environment.

5. Marine eukaryotic microbes are primary producers, decomposers, and consumers, and some contribute significantly to the accumulation of deep-sea sediments.

6. Populations of several kinds of photosynthetic marine microbes may form harmful blooms that affect other marine and maritime organisms directly and indirectly.

# Viral Characteristics

Most authorities do not consider viruses to be alive because viruses are little more than stores of genetic information, bits of DNA or RNA surrounded by protein. Unlike living organisms, viruses have no metabolism. They rely entirely on host cells for energy, material, and organelles for duplicating themselves, a process called **viral replication,** and can reproduce only inside a host cell. Despite this, **virologists** (scientists who study viruses) use terms such as "life cycle" and "live virus" that sound as if viruses are alive, and the two prevailing hypotheses about the origins of viruses postulate that they come from living cells. One hypothesis states that viruses are highly reduced prokaryotic parasites, whereas the other views them as renegade genes.

Viruses seem to infect all groups of living organisms. Individual viruses are usually host specific, but families of viruses vary widely in the groups they infect. For example, the Papillomaviridae family of viruses, which includes human papillomavirus, is restricted to vertebrates, but the Rhabdoviridae family of viruses infects vertebrates, insects, and plants. Many viral families are specific for bacteria, and they are generally called **bacteriophages** (or simply **phages**), which translates as "eaters of bacteria."

## Viral Structure

Viruses are subcellular particles that vary in size from 10 to 400 nanometers, barely overlapping the lower size range of bacteria. Outside the host cell, a virus particle is called a **virion.** A virion is composed of a nucleic acid core surrounded by a coat of protein called a **capsid** (Figure 6-1). The nucleic acid core can be either DNA or RNA and contains a small number of genes (from 3 to almost 300) that are unable to replicate outside a host cell. The combination of the virus's genetic material and protein is called a **nucleocapsid,** and sometimes it is coated with an **envelope,** a membrane derived from the host's nuclear membrane or cell membrane. The capsid protects the virus from the environment.

Viruses occur commonly in three shapes: icosahedral, helical, and binal (see Figure 6-1). **Icosahedral viruses** have a capsid with 20 triangular faces composed of protein subunits. In **helical viruses,** the protein subunits of the capsid spiral around the central core of nucleic acid. **Binal viruses** are those with icosahedral heads and helical tails. Some more complex virions have filaments and other parts that are used for attachment to and infection of the host cell.

## Viral Life Cycles

Two broad categories of life cycles are recognized in viruses (Figure 6-2). They differ in the length of time the virus remains in its host. The **lytic cycle** is a rapid one of infection, replication of viral nucleic acids and proteins, assembly of virions, and release of virions by rupture (lysis) of the cell. In the **lysogenic cycle,** the viral nucleic acid is inserted into the host genome and may reside there through multiple cell divisions before becoming lytic.

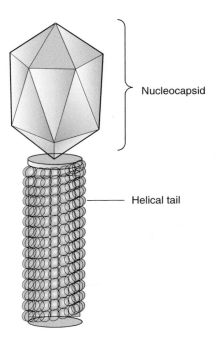

Nucleocapsid

Helical tail

**Figure 6-1  Structure of Marine Viruses.** Many marine viruses as mature virions consist of an icosahedral (20-sided) head and a helical tail. Such viruses, as in the example above, are called binal viruses, having two parts. The head is a nucleocapsid, consisting of a protein coat (capsid) protecting the genetic material (not shown) inside. Icosahedral viruses consist only of the 20-sided head. Helical viruses have a nucleocapsid that looks like a tail without the head.

# Biodiversity and Distribution of Marine Viruses

Marine viruses are ten times more abundant than marine prokaryotes. Densities may reach $10^{10}$ virions per liter (approximately $10^{10}$ per quart) in surface waters (epipelagic zone) and $10^{13}$ per kilogram ($2 \times 10^{13}$ per pound) of sediment. Recent studies have found an incredible diversity of viruses in the oceans. Samples of ocean water and sediments are estimated to contain 100 to 10,000 different viral genotypes, any one of which comprises no more than 2% of the viral community. Many marine viruses are genetically distinct from other known kinds of viruses, but much work is needed before we can characterize them genetically or understand their relationships with other viruses.

We now know that virions in plankton samples are not inactive contaminants but infective agents. Most marine planktonic viruses are icosahedral or binal bacteriophages with lytic life cycles. Members of the Podoviridae, a family of viruses with short tails, seem especially abundant. The bacteriophages include those that infect cyanobacteria (cyano-

**Lytic Cycle**

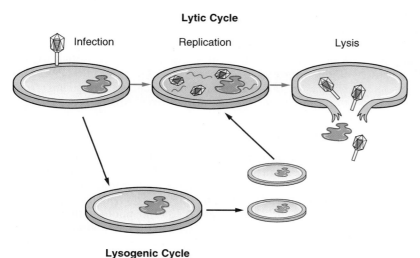

**Figure 6-2 Viral Life Cycles.** Many viruses undergo a rapid cycle of infection, direction of the host genome to replicate viral particles, and release of the mature virions by rupture (lysis) of the host cell after self-assembly of the viral parts. Other viruses add to this life cycle a resting phase in which the viral genome becomes inserted into the host genome, possibly undergoing replication as cells divide through several generations. The virus then becomes lytic and infects other cells. Such a virus is said to be lysogenic.

phages) as well as the abundant heterotrophic bacteria. Eukaryotic phytoplankton also are hosts of marine viruses (phycoviruses), as are marine animals. Some of the animal viruses have lysogenic life cycles (for example, green turtle fibropapilloma retrovirus, seaslug retrovirus). Viruses in marine sediments typically are helical and lysogenic.

## Ecology of Marine Viruses

The ecological roles of marine viruses are largely based on lysis of their hosts. The immediate outcome of lysis is death of the cell, and the longer-term effect is some level of population control of bacteria and other microbes in plankton communities. Plankton diversity may be maintained or restored by the suppression or decline of blooms (for example, red or brown tides) caused by viral infection. This occurs in addition to or instead of limitation by predation and lack of nutrients. Viruses can alter biogeochemical cycles and planktonic food webs by bacterial lysis. This reduces the amount of food available for other microbes and releases nutrients into the seawater, thus increasing community respiration. It also facilitates sedimentation of particles (dead cells) to the seafloor. Because lysis of some photosynthetic microbes may release the greenhouse gas dimethyl sulfide, viruses may be agents of global climate change.

Marine viral populations probably are controlled by several abiotic and biotic factors. Near the surface, wavelengths of visible and ultraviolet light penetrating the ocean can damage or alter viral nucleic acids, causing their elimination. Because they are little more than complex macromolecules, viruses can become adsorbed onto suspended particles (**seston**) and rendered noninfective; in this way, seston are comparable in function to a charcoal filter. Enzymes secreted by bacteria may destroy virions, and eukaryotic microbes may ingest them without harm to themselves. As is the case with many parasites, some virions will attach to nonhost cells and fail to infect them.

## In Summary

Viruses are more abundant than other microbes in the sea. Marine planktonic viruses are icosahedral and lytic and are responsible for the death of many bacteria and phytoplankton in the epipelagic zone. Viruses in marine sediments are mostly helical and lysogenic. •

## MARINE BACTERIA

The domain Eubacteria includes the original inhabitants of the sea, dating back several billion years. Over that time, they have evolved a diversity of metabolic pathways, accomplishing biological tasks in ways not found in members of other domains today. According to the most widely accepted hypothesis on evolution of the Eukarya, bacteria have repeatedly formed symbiotic associations with their own descendents to give rise to much of the microbial diversity covered in the rest of this chapter. Their unique metabolism, widespread presence, and abundance in the sea make bacteria vital components of marine ecosystems. Marine bacteria are primary producers, decomposers, agents in biogeochemical cycles, food for other marine inhabitants, modifiers of marine sediments, and symbionts and pathogens (agents of disease).

## General Characteristics

Bacteria have cells with a simple, prokaryotic organization, a general feature shared with the domain Archaea. Recall that prokaryotic cells lack nuclei and other membrane-bound organelles such as mitochondria and chloroplasts. Bacterial cells have a single chromosome of DNA that is circular and contains relatively few genes, and most are surrounded with a nonliving cell wall that is made from a special combination of sugars and amino acids and that gives support and protection. Bacteria reproduce asexually by **binary fission.** In this process, a cell's genetic material (DNA) is duplicated, a membrane forms between the duplicated DNA molecules (chromosomes), and the cell splits into two daughter cells of roughly equal size. Each daughter cell then grows rapidly and repeats the process. Locomotion in bacteria most often involves a flagellum, with a structure and function quite different from that of eukaryotes. In bacteria the flagellum is a bent filament that turns like a corkscrew, and it is driven by a protein engine fueled by protons.

Bacteria come in a wide range of shapes and sizes. Most marine bacteria are shaped like rods. This shape is called the **bacillus** (Figure 6-3a). Others have a spherical shape, the **coccus** (see Figure 6-3b). They range from one to a few micrometers in length or width. The smallest is about 200 nanometers long, and the largest is 750 micrometers

(about 1/32 of an inch) in diameter. In fact, the two largest bacteria described in the last decade are marine: the bacillus *Epulopiscium fishelsoni* (80 micrometers by 600 micrometers) from the gut of the brown surgeonfish (*Acanthurus nigrofuscus*) from the Red Sea and the coccus *Thiomargarita namibiensis* (750 micrometers in diameter) from nutrient-rich sediments off southwest Africa.

## Nutritional Types

Bacteria acquire nutrients in a variety of ways and play multiple roles in marine ecosystems (Figure 6-4). Some marine bacteria are primary producers (autotrophs), manufacturing organic molecules from inorganic molecules. The source of energy for primary production in photosynthesis is the sun, and in chemosynthesis it is simple and abundant chemicals. Other bacteria are heterotrophs, requiring organic molecules for their nutrition and using them to produce additional bacterial biomass. Heterotrophic marine bacteria obtain external organic matter by absorption across the cell wall and membrane, a style of nutrition called **osmotrophy.** When osmotrophic bacteria encounter organic molecules too large to absorb, they break them down into smaller ones with digestive enzymes they secrete (**exoenzymes**). In this way, heterotrophic bacteria are decomposers, recycling dead organic substances as living bacterial biomass and releasing inorganic molecules that become available to primary producers in biogeochemical cycles. A critical aspect of bacterial metabolism is the ability of some groups to break the strong bond of molecular nitrogen ($N_2$) and to convert the nitrogen into ammonium ion, a form usable by other living organisms. Without such nitrogen fixation, there would be insufficient nitrogen in many habitats to support life in the sea.

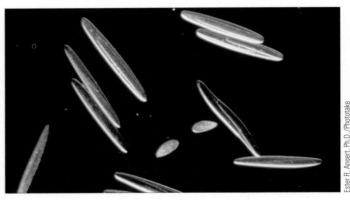

(a)

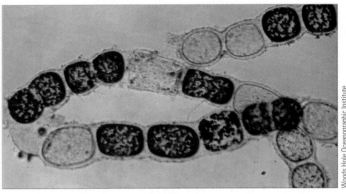

(b)

**Figure 6-3 Bacterial Shapes.** Most marine bacteria come in two shapes: rods (bacilli) and spheres (cocci). (a) *Epulopiscium fishelsoni* is a bacillus that lives in the gut of the brown surgeonfish. Note the enormous size of these prokaryotic cells compared to the much smaller paramecia. (b) *Thiomargarita namibiensis* is a coccus that lives in sediments off the coast of southwest Africa.

### Cyanobacteria

Among the almost two dozen phyla of bacteria, the most familiar are the photosynthetic marine bacteria called **cyanobacteria** (blue-green bacteria, or blue-green algae; phylum Cyanobacteria; Figure 6-5). Cyanobacteria are found in environments that are high in dissolved oxygen, and they produce free oxygen as a product of their photosynthesis (Figure 6-6a). Excess photosynthetic products are stored as **cyanophycean starch** and oils. The photosynthetic activity of blue-green bacteria accounts for a major proportion of the production of organic matter and oxygen in the seas. In particular, species of two genera, *Prochlorococcus* and *Synechococcus,* are among the smallest and most productive bacteria in plankton communities. Of the two genera, *Prochlorococcus* is believed to be the most abundant life form in the sea, and it is well-adapted to the nutrient-poor open ocean of the tropics. Yet it was discovered only recently, in 1988.

The primary photosynthetic pigments in cyanobacteria are **chlorophyll *a*** and, in some, **chlorophyll *b*.** Both pigments are also characteristic of land plants, but chlorophyll *b* is rare among photosynthetic microbes. These pigments absorb wavelengths of light in the red and blue parts of the

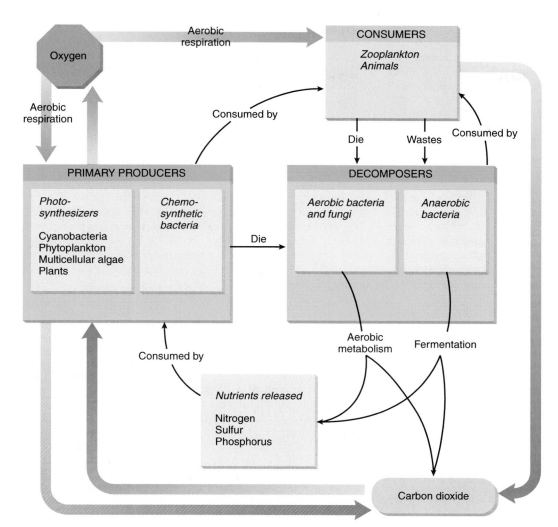

**Figure 6-4 Bacteria and Nutrient Cycling.** Bacteria decompose the bodies of dead organisms as well as animal waste products, releasing nutrients such as carbon, nitrogen, and sulfur back into the environment. Photosynthetic and chemosynthetic bacteria can recycle carbon dioxide and other nutrients back into food molecules for themselves and consumers. Most marine organisms metabolize their food molecules by the process of aerobic respiration, which consumes oxygen and returns carbon dioxide to the environment.

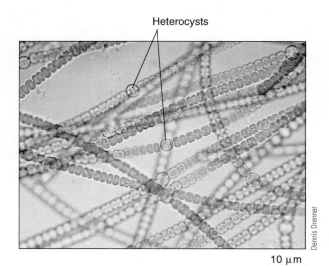

spectrum, and they reflect green wavelengths, which then tend to define the color of the cell. These pigments not only capture light but they also convert it to chemical energy (for more information regarding photosynthesis see Chapter 5). Other pigments found in cyanobacteria, called accessory pigments, capture different wavelengths of sunlight and pass it to the chlorophylls, thereby increasing the efficiency with which the bacteria harvest available light. Among these accessory photosynthetic pigments are the **carotenoids** and **phycobilins**. Carotenoids include the familiar **beta-carotene** (of carrot fame), which imparts a yellow color,

**Figure 6-5 Blue-Green Bacteria.** Phylum Cyanobacteria includes blue-green bacteria, or algae, such as these filaments. The blue-green color seen here is the result of the green pigment chlorophyll *a* and the blue pigment phycocyanin. The enlarged cells are heterocysts, within which nitrogen is fixed for incorporation into amino acids, nucleic acids, and other nitrogen-containing organic compounds.

**Figure 6-6 Bacterial Photosynthesis.** (a) Blue-green bacteria carry out photosynthesis in the same way that green plants do, using water as a source of electrons and hydrogen atoms to make carbohydrates from carbon dioxide and water and releasing oxygen in the process. (b) Purple bacteria and green bacteria use a variety of compounds (such as hydrogen sulfide) rather than water and therefore do not release oxygen. In the reaction shown here, sulfate is released by oxidation of sulfide.

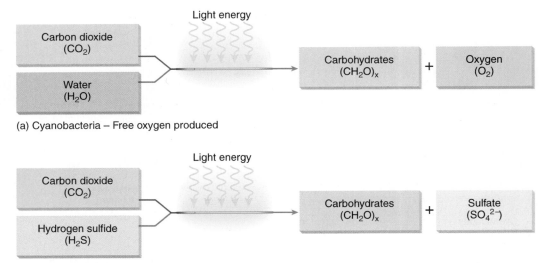

(a) Cyanobacteria – Free oxygen produced

(b) Purple and green bacteria – No free oxygen produced

and the **xanthophylls,** which are brown pigments. The phycobilins are pigments found almost solely in bacteria and other microbes, and they include the reddish pigment **phycoerythrin** and the bluish pigment **phycocyanin.** The concentrations of these accessory pigments in cells change as the quality of light in the sea changes with depth, season, and sea conditions. This response in pigment composition is called **chromatic adaptation** and occurs in many groups of photosynthesizers. In addition to increasing the efficiency of photosynthesis, accessory pigments shield the cell against damaging wavelengths or intensities of light, a protection essential to their survival in clear surface waters at low latitudes. For this reason they are also called **photoprotective pigments.**

Cyanobacteria may exist as single cells or form dense mats or long filaments. Their presence as tufts on rocks may make walking in the intertidal zone treacherous because they secrete **mucilage,** a slippery, biological glue that holds cells of the filaments together. Feltlike mats of cyanobacteria and other microbes on soft sediments help to stabilize sediments, reducing resuspension and thereby increasing water clarity. For millennia cyanobacteria have formed associations called **stromatolites,** a coral-like mound of microbes that trap sediment and precipitate minerals in shallow tropical seas (Figure 6-7). Although they form only a thin living "skin" over the growing mound, the microbes produce a strongly stratified community, some living off the byproducts of others but all relying ultimately on the photosynthesis of cyanobacteria and diatoms (described in a later section). Some microbiologists believe that stromatolites were a dominant type of seafloor community in earlier times, but now they are reduced to patchy occurrences in the Bahamas, Australia, and a few other locations. The decline of stromatolites over the last 500 million years is thought to have resulted from predation by grazing molluscs and crustaceans and from competition with eukaryotic algae, sponges, and other space monopolizers.

## Other Photosynthetic Bacteria

Other groups of photosynthetic bacteria, such as the green and purple sulfur and non-sulfur bacteria, offer a contrast to cyanobacteria because they are anaerobic and do not produce oxygen as a by-product of photosynthesis. These organisms cannot tolerate the high levels of oxygen found in many marine habitats. Sulfur bacteria, for instance, are **obligate anaerobes,** tolerating no oxygen, whereas non-sulfur bacteria are **facultative anaerobes,** respiring when in low oxygen or in the dark and photosynthesizing anaerobically when in the presence of light. The primary photosynthetic pigments in these bacteria are **bacteriochlorophylls.**

**Figure 6-7 Stromatolites.** Stromatolites are formed by dense mats of blue-green bacteria and other microbes. Some fossil stromatolites date back 3.4 billion years.

Sulfur bacteria use the hydrogen sulfide, sulfur, and hydrogen available in anaerobic habitats rather than water and thus do not produce oxygen. For example, the splitting of hydrogen sulfide ($H_2S$) releases elemental sulfur or sulfate rather than oxygen (see Figure 6-6b). Non-sulfur bacteria use small organic molecules more than they use hydrogen or sulfur compounds. Purple sulfur and purple non-sulfur bacteria are found in surface sediments and in locally anoxic waters that get enough sunlight to support photosynthesis. These bacteria, rather than cyanobacteria, are the photosynthetic component of microbial communities that form in mats of drifting seaweeds and seagrasses that accumulate along the shores of quiet lagoons and bays. As objectionable as these putrefying mats are to humans, microbial decay makes them important to the ecology of the seashore. Next we examine bacteria that do not use photosynthesis for energy production: chemosynthetic bacteria.

## Chemosynthetic Bacteria

Light is absent from most ocean waters and sediments, but bacteria still can be autotrophic in these habitats, using inorganic chemicals rather than light as a source of energy. Such bacteria are called **chemosynthetic bacteria**. Chemosynthetic bacteria use energy derived from chemical reactions that involve substances such as ammonium ion ($NH_4^+$), sulfides ($S^{2-}$) and elemental sulfur (S), nitrites ($NO_2^-$), hydrogen ($H_2$), and ferrous ion ($Fe^{+2}$). The energy is used to manufacture organic food molecules, usually with carbon dioxide ($CO_2$) as the carbon source. Chemosynthesis is far less efficient than photosynthesis, and if large quantities of inorganic compounds are not available, rates of cell growth and division are slower. Therefore chemosynthetic bacteria are found in areas where the inorganic substances they require are in abundance, and they usually are anaerobic.

Some chemosynthetic bacteria live around deep-sea hydrothermal vents (Figure 6-8). Many occur in dense concentrations in overlying waters or as microbial mats within or near the warmer (8° to 23°C) vents, where suspension feeders and benthic grazers consume them as food. These bacteria convert the sulfide ions that spew from the vents to sulfur and sulfate and use the energy released to produce food. These chemosynthetic bacteria form the base of a productive food chain that consists of a diverse assembly of organisms, including worms, clams, and crabs, that make up the hydrothermal vent community. Others live symbiotically with benthic invertebrates (for more information regarding vent communities see Chapter 18).

Some chemosynthetic bacteria occur in shallower habitats. The largest bacterium, *Thiomargarita namibiensis* ("sulfur pearl of Namibia"; see Figure 6-3b), lives in the sediments at 100 meters (330 feet) depth off the southwestern coast of Africa. It uses nitrate ions to convert the high concentration of sulfides in its habitat into granules of elemental sulfur within its cell. This bacterium and others like it off Chile and Peru are characteristic of upwelling areas where the high productivity of plankton communities results in nutrient-rich sediments that are high in hydrogen sulfide.

## Heterotrophic Bacteria

Heterotrophic bacteria are decomposers, which use available organic matter in their surroundings to obtain energy and material for synthesis of their own compounds and for general metabolism. Their respiration and fermentation return to the environment many inorganic chemicals that become available to primary producers (see Figure 6-4). These bacteria release exoenzymes with the capacity to digest cellulose, lignin, chitin, keratin, and other natural molecules that are otherwise resistant to decay. In fact, we now know that enzymes evolved by bacteria are able to degrade plastics and other artificial polymers in the sea. Without these heterotrophic bacteria, many kinds of complex substances would accumulate in the oceans and remain unavailable for millennia as nutrients for other organisms. Organic substances in the sea usually occur as particles suspended in the water or accumulated on the ocean floor. Soon after formation, these particles quickly become populated with bacteria. The release of exoenzyme to the outside of the cell presents the risk of dilution in the surrounding environment. To combat this risk, the bacteria secrete mucilage that glues them to the particles. The mucilage keeps the exoenzymes close to the source of food and prevents the digestive byproducts from washing away from the bacterium. In these microbial communities, the different kinds of bacteria on the same food particle provide a variety of enzymes capable of digesting a wide range of nutrients, thus producing a broader "menu" of by-products to absorb.

The association of bacteria with particles in the water column aids in three processes: consolidation, lithification, and sedimentation. Sticky surfaces, resulting from the secretion of mucilage and the alteration of the electric charge of particles, cause adjacent particles to adhere (**consolidation**). The metabolic activity of the bacteria usually alters the acidity or alkalinity (pH) of the area immediately around the particle, often causing minerals to form a cement between particles (**lithification**). Consolidation and lithification increase the aggregation of particles. As aggregate size increases, the ability of the water to hold the particles in suspension declines, and the particles begin to settle (**sedimentation**). Particles might accumulate on the seafloor, at places where layers of water of different density meet (such as thermoclines), or in large cobwebby drifting structures called **marine snow**, formed by mucus secreted by many kinds of plankton. As particles accumulate, they become available to suspension and deposit feeders. These consumers survive on the large microbial communities growing on the particles. In addition to these attached bacteria, free bacteria have a prominent part in marine communities, usually in food webs involving several of the microbial groups that we will discuss later in this chapter.

## Nitrogen Fixation and Nitrification

The major process that adds new usable nitrogen to the sea is **nitrogen fixation**. This process converts the molecular nitrogen dissolved in seawater to ammonium ion (Figure 6-9). Nitrogen is one of the elements that often have

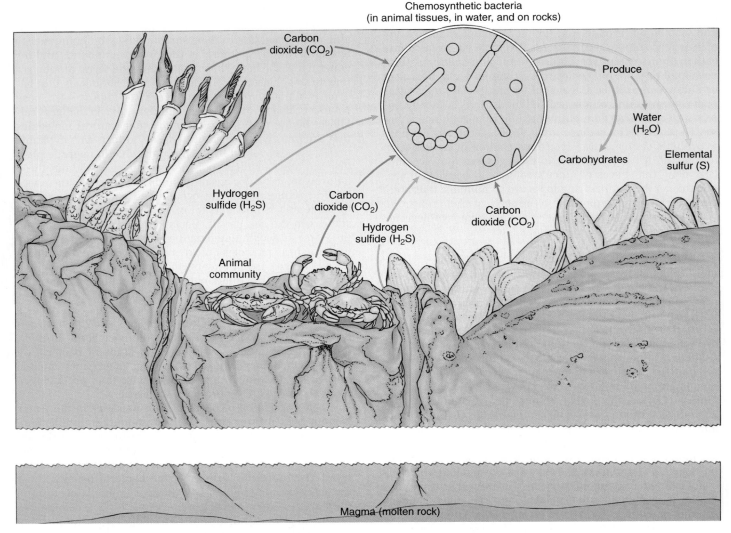

**Figure 6-8 Chemosynthesis.** Bacteria and archaeons that live in deep-sea hydrothermal vent communities can produce food molecules from carbon dioxide and hydrogen sulfide, using the energy derived from chemical reactions rather than from sunlight.

limited availability in marine ecosystems. It is required to form amino acids (the building blocks of proteins), nucleotides (the building blocks of RNA and DNA), and other compounds. Nitrogen is recycled to autotrophic organisms when it is released as metabolic waste from living organisms and when decomposers such as bacteria and fungi break down dead organic matter to ammonium ions. This rapid uptake by primary producers and loss of nitrogen to settling particles may leave a deficit of nitrogen in surface waters. Nitrogen might also be limiting in marine sediments, as in seagrass beds and mangrove swamps.

Only some cyanobacteria (and a few archaeons, described later) are capable of fixing nitrogen. These organisms have an enzyme, **nitrogenase,** that is capable of breaking the strong molecular bond between the two atoms of nitrogen that make up the molecules of nitrogen gas. Be-

cause nitrogenase is sensitive to oxygen, nitrogen fixation must be accomplished in an anaerobic environment. This is, therefore, a difficult task for a group of aerobic autotrophs that produce oxygen as they photosynthesize. In many filamentous cyanobacteria, nitrogen fixation occurs in a **heterocyst,** a thick-walled cell in which photosynthesis is altered to prevent the release of oxygen (see Figure 6-5). What little oxygen diffuses across the cell wall is consumed rapidly by respiration. Other cyanobacteria accomplish nitrogen fixation under environmental conditions that do not require the formation of heterocysts. Any ammonium ion ($NH_4^+$) that diffuses into the surrounding seawater becomes available for other primary producers, but they usually require bacterial conversion of the ammonium to nitrite ($NO_2^-$) and nitrate ($NO_3^-$) ions before they are able to use the nitrogen, a process called **nitrification.**

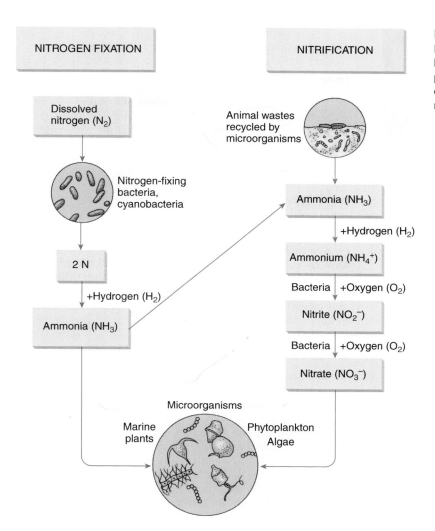

**Figure 6-9** **Nitrogen Fixation and Nitrification.** Nitrogen dissolved in seawater is converted by blue-green bacteria and some archaeons to ammonium ion; this process is called nitrogen fixation. In nitrification, bacteria convert ammonium to nitrites and nitrates, forms of nitrogen that are more readily used by primary producers.

Now that we have examined how some cyanobacteria participate in nitrogen fixation and nitrification, we look at symbiotic associations of bacteria and other marine organisms.

## Symbiotic Bacteria

Symbiosis refers to the biological phenomenon in which two different organisms form a very close association with each other (for more information on symbiosis see Chapter 2). Many bacteria have evolved symbiotic relationships with a variety of marine organisms. The symbiotic relationships that probably have the longest history are those now represented by the mitochondria and chloroplasts of members of the domain Eukarya. Other more recently evolved symbioses involve associations in which the bacterial guest is clearly distinguishable from its eukaryotic host.

In deep-sea vent communities (see Figure 6-8), chemosynthetic bacteria live within the tissues of marine tube worms and clams. These bacteria supply their hosts with organic food molecules through chemosynthesis, whereas the hosts supply the bacteria with carbon dioxide and other essential nutrients, particularly sulfides. Since the discovery of bacterial symbioses in vent communities, similar relationships have been found in shallow-water habitats (for more information regarding bacterial symbiosis see Chapter 18).

Another example of symbiosis is in the phenomenon of bioluminescence. Bioluminescent bacteria are capable of emitting blue-green or yellow light using the chemical energy from metabolic processes similar to those that form adenosine triphosphate (ATP). This biochemical process requires oxygen and might have evolved as a way for bacteria to rid themselves of toxic oxygen in primordial seas. Although many animals produce their own bioluminescence, some have organs called **photophores** with cultures of luminescent bacteria. Hosts include 2 of 19 families of luminescent squid, the colonial planktonic seasquirt (tunicate) *Pyrosoma,* and about a dozen families of fish, most of which live in coastal waters or near the seafloor. The bacterial symbionts include three members of one family of bacteria: *Vibrio fischeri, Photobacterium leiognathi,* and *P. phosphoreum.*

Several squid species have bioluminescent bacteria in glands that are embedded in the ink sac. Bacteria are re-

**Figure 6-10  Bacterial Bioluminescence.** In addition to light production in free bacteria, the bioluminescence of many animals is due to symbiotic associations of bacteria in organs called photophores. In lanterneyes, the photophore occurs below the eye, as shown here in *Photoblepharon palpebratum.* This fish is unusual among luminescent fishes by its habitation of shallow-water reefs. Light emission during its nighttime period of activity can be blocked by raising a black partition like an eyelid. By "blinking," the fish can lure prey, confuse predators, and communicate with other lanterneyes.

John B. Corliss

leased when the squid is disturbed, producing a luminous cloud that might startle or temporarily blind a predator. The bioluminescence allows the squid to escape in the deep sea, just as a cloud of ejected ink provides cover for escape of its shallow-water counterparts. Many species of deepwater fish have also developed symbiotic relationships with bioluminescent bacteria. The bacteria are usually found in pits, sacs, or tubes located in the animal's skin (near the eye or jaw or on a lurelike structure) or in parts of its digestive system. Because bacteria luminesce continuously in the presence of oxygen, emission of light may be controlled by reducing blood flow (which reduces oxygen supply) to the photophore or by shielding the photophore with black pigment. Fishes called lanterneyes in the genera *Photoblepharon* and *Anomalops* contain light glands beneath the eye (Figure 6-10). In *Photoblepharon* the emission of light can be controlled by covering the photophore with a lid of black tissue, much like an eyelid, and in *Anomalops* by rotating the gland so that the light is directed into a pocket of black tissue. These two genera are unusual among bioluminescent fishes because they inhabit shallow-water reefs instead of the dark recesses of the sea, but their photophores are useful during their period of nighttime activity. (For more information regarding bioluminescence in deep-sea organisms see Chapter 18.)

## In Summary

Bacteria have cells with a simple, prokaryotic organization. Chemosynthetic and photosynthetic bacteria extract inorganic nutrients, such as nitrogen, phosphorus, and carbon dioxide, from the environment and incorporate them into organic molecules. Chemosynthetic bacteria use energy derived from chemical reactions (often involving compounds of sulfur) to produce their food molecules, whereas photosynthetic ones use the radiant energy from

the sun. Some producers, such as blue-green bacteria, release oxygen during photosynthesis. Such primary producers as well as heterotrophic bacteria form the base of marine food webs. In addition, marine bacteria play a critical role in nitrogen fixation and nitrification. As decomposers, bacteria return dead organic matter to biogeochemical cycles as inorganic matter that primary producers can incorporate into living biomass. The chemosynthetic bacteria of hydrothermal vent communities and the bioluminescent bacteria found in association with deep-sea organisms are some examples of symbiotic relationships.  ●

## ARCHAEA

The domain Archaea includes microbes formerly considered to be bacteria. Molecular and ultrastructural analyses have revealed the Archaea to be quite distinct from the Eubacteria and Eukarya domains. Also surprisingly, they share a number of features with each of the other two domains and might have a closer evolutionary relationship with the Eukarya.

### General Characteristics

**Archaeons** are small (0.1 to 15 micrometers) and have a much narrower range of sizes than do bacteria. Like bacteria, archaeons are prokaryotes, but they differ from bacteria in other ways. Most have a cell wall, which lacks the special sugar–amino acid compounds of bacterial cell walls. Some archaeons have only a layer of proteins or proteins complexed with carbohydrates covering their cell membrane. The lipids that occur in the cell membranes of archaeons are quite different from those found in the membranes of bacteria. The differences are related in part to the need for stabilizing the membranes under extreme environmental conditions, which characterize the habitats of many of the Archaea. The professional journal *Extremophiles,* first published in 1997, is dedicated to the biology of archaeons and other microbes that thrive in high and low temperatures, high salinities, low pH, and high pressure. Although their adaptations to extreme habitats typify archaeons, we are only recently coming to recognize them also as members of the plankton.

### Nutritional Types

The domain Archaea includes photosynthetic, chemosynthetic, and heterotrophic organisms. Some can fix nitrogen and most are anaerobes. Although many aspects of their nutrition are similar to those of bacteria, other aspects are unusual. Much is yet to be learned about their metabolism.

Most archaeons are **methanogens,** anaerobic organisms that live in environments that are rich in organic matter. The organic matter is metabolized for energy, and the gas methane is a waste product of this metabolism. (Interestingly, some areas of research and engineering attempt to convert what we call "waste" *into* methane as a source of energy!) *Methanococcus jannaschi* and *Methanopyrus kandleri* are two of many marine methanogens, and both live at

# Halobacteria

**Halobacteria** belong to a group of archaeons known as **halophiles,** or "salt lovers." They received their name because they grow best in extremely salty environments such as salt lakes and salt evaporation ponds. In fact, if the salt concentration of their environment drops below five times the concentration of normal seawater, halobacteria will die.

Like many other species, halobacteria obtain their energy by aerobic respiration. Because warm salty water holds very little oxygen, halobacteria frequently experience a shortage of this important gas, and the amount of ATP that these bacteria can synthesize is greatly decreased. When oxygen levels are low and sunlight is plentiful, halobacteria employ a backup system to produce enough ATP to survive. The backup system uses light energy to synthesize ATP.

Light energy is captured by patches of a purple protein called bacteriorhodopsin, a molecule very similar to the visual pigment in vertebrate eyes. The patches of bacteriorhodopsin may occupy as much as 50% of the surface of the cell membrane, and the pigment molecules extend through the membrane such that a portion of the molecule is exposed at both surfaces. When light strikes the bacteriorhodopsin, the molecule undergoes a series of reactions in which a hydrogen ion is pumped from the inside of the cell to the outside.

The flow of hydrogen ions back into the cell provides the energy for ATP synthesis.

Seaweeds and plants use a variety of pigments working together to trap light energy and pass it to a complex system that eventually restores the ATP used in photosynthesis. Halobacteria, on the other hand, use only a single pigment, bacteriorhodopsin. The existence of such a mechanism suggests that primitive prokaryotic cells may have evolved systems to trap sunlight and make ATP long before the evolution of photosynthesis. The ability to use the energy of sunlight certainly would have had a tremendous effect on the evolution of early life. ●

---

deep-sea hydrothermal vents. The first complete genome sequenced for an archaeon was that of *Methanococcus jannaschi,* a study supported in part by the U.S. Department of Energy because of the potential of this microbe to give clues to the industrial production of methane. The scientific world was astounded to learn that 56% of the genes of *Methanococcus jannaschi* were unique, giving further support for the three-domain system of life. Aside from its potential commercial use, methane is a greenhouse gas that may contribute to global warming. Methane production in anoxic habitats in the sea is less likely to affect global warming because much of the methane may be oxidized as it rises into overlying oxygenated waters.

Some methanogens are chemosynthetic, using hydrogen as a source of energy. Indeed, most archaeons that are primary producers are chemosynthetic. On the other hand, one group of archaeons, the Halobacteria, is photosynthetic, but it does not use chlorophyll to capture light for ATP production. Instead, certain planktonic marine halobacteria possess purple proteins, **bacteriorhodopsins,** that they use for trapping light energy. Bacteriorhodopsins are chemically similar to the visual pigment rhodopsin in the eyes of vertebrate animals. The group is called the Halobacteria because they thrive at high salinities (see the accompanying box).

## Hyperthermophiles

Many archaeons tolerate very warm temperatures, but some, categorized as **hyperthermophiles,** are known for their survival at temperatures exceeding 100°C, such as

occur at deep-sea hydrothermal hot vents. Such microbes often have names that refer to heat, fire, and smoke. For example, *Pyrolobus fumarii* translates as "fire lobe of the chimney," a name given to a hyperthermophilic primary producer collected from a black smoker chimney on the Mid-Atlantic Ridge at 3,650 meters (12,045 feet) depth. *Pyrolobus fumarii* cannot use organic matter as food but is chemosynthetic, using carbon dioxide and hydrogen to produce organic compounds. It accomplishes this and also divides at rates of up to once per hour at an optimal temperature of 106°C. It will grow at temperatures between 90° and 113°C and tolerates brief exposure to 121°C. Recently an archaeon called "strain 121" from a hydrothermal vent in the Pacific Ocean better tolerated a temperature of 121°C in the laboratory than did *Pyrolobus fumarii,* and it also survived 2 hours of incubation at 130°C. The heat tolerance of hyperthermophilic archaeons is truly remarkable. This group of microbes holds much potential for exciting revelations of biological adaptation and for biomedical or industrial applications in the future.

## In Summary

Like bacteria, archaeons are prokaryotes. Most archaeons are methanogens, anaerobic organisms that live in environments that are rich in organic matter, and they make significant contributions to methane production. Archaeons have an unsurpassed ability in the natural world to tolerate extreme environmental conditions. ●

# EUKARYA

The domain Eukarya contains all of the organisms with eukaryotic cells. These include plants, animals, and fungi, as well as algae and single-celled animal-like protozoa.

## Fungi

We are familiar with many kinds of terrestrial fungi because they grow in our lawns and on trees (for example, puffballs and bracket fungi), cause diseases in plants (rusts, powdery mildews) and humans (ringworm, athlete's foot), are used in preparation of food and drink (yeasts), produce antibiotics (*Penicillium*), are on our menus (mushrooms, morels, truffles), and spoil our food (bread molds, ergot). Marine fungi are much less obvious to us because of their microscopic size, and they affect our lives much less directly. Less than 1% of known species of fungi are marine, but marine **mycologists** (biologists who study fungi) are finding that, although fungi are not as diverse as some other marine organisms, they are important in marine ecosystems as decomposers, prey, pathogens, and symbionts.

### History of Marine Mycology

The first marine fungus was discovered in 1849, but it was not until the mid-1900s that mycologists recognized that there are communities of fungi unique to the seas. Early progress in marine **mycology** (the study of fungi) was impeded because contaminant spores from land and freshwater habitats could not be distinguished from those that were truly marine. In much of the second half of the 1900s, marine mycology consisted of the cataloguing of new species or focused on descriptive studies of the roles of marine fungi. The difficulties of quantifying fungal biomass have made it hard for mycologists to evaluate critically the contributions of fungi to the ecology of the sea. Application of recently developed techniques is likely to give better estimates of fungal productivity in the future.

### General Features of Fungi

Members of the kingdom Fungi are eukaryotes with cell walls of **chitin,** a carbohydrate that is more commonly associated with crustaceans and other arthropods. Many fungi are unicellular **yeasts** or have a unicellular yeastlike stage in their life cycle. **Filamentous fungi,** on the other hand, grow vegetatively into long filaments called **hyphae** (Figure 6-11a) composed of many cells that may or may not be divided from each other by cell walls. The branching of hyphae produces a tangled mass called the **mycelium.**

Fungi are not photosynthetic. Like most bacteria, fungi are heterotrophic decomposers and recycle organic material. Their exoenzymes are capable of digesting a wide variety of biologically and synthetically produced molecules. Perhaps the most significant of these molecules is lignin. Destruction of the chemical lignin, a major component of wood, allows fungal enzymes called cellulases to reach the fibers of cellulose, which serves as the fungus's food source. Food

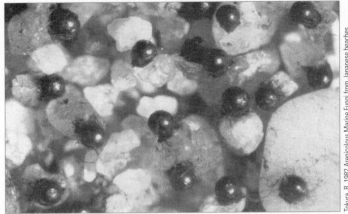

(a)

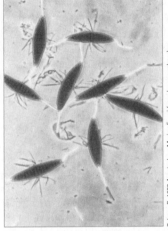

(b)

**Figure 6-11 Arenicolous Marine Fungus.** (a) Threadlike hyphae and large spherical ascocarps comprise the mycelium of this sand-dwelling marine fungus in the genus *Corollospora*. The very fine hyphae give a cobweb appearance over the surface of the grains of sand. Where adjacent hyphae have fused and undergone sexual reproduction, ascospores are aggregated within the large fruiting bodies, called ascocarps, comparable to terrestrial mushroom caps. (b) Within the ascocarps, ascospores mature and then are released. The spines and filaments on the ascospores aid their entrapment in sea foam, which transports them to distant shores.

reserves are stored as **glycogen,** often called "animal starch" because of its role as a storage carbohydrate in the animal kingdom. Unlike many heterotrophic bacteria, most fungi are strict aerobes, requiring a high content of oxygen in their habitats.

The kingdom Fungi is divided into four phyla: Chytridiomycota, with motile cells; Zygomycota, such as black bread mold; Basidiomycota, or club fungi, including the familiar mushrooms; and Ascomycota, or sac fungi. The last three phyla lack motile stages in their life cycles. The ascomycotes

Tokura, R. 1982 Arenicolous Marine Fungi from Japanese beaches. Transaction of the Mycological Society of Japan 23:423–433

Tokura, R. 1982 Arenicolous Marine Fungi from Japanese beaches. Transaction of the Mycological Society of Japan 23:423–433

are the most diverse and abundant phylum of fungi in the sea, constituting 81% of the fewer than 500 known marine species of filamentous fungi. In addition, many of the nearly 200 marine yeasts are ascomycotes.

## Ecology and Physiology of Marine Fungi

Some species of marine fungi are obligately marine, requiring ocean or brackish waters to grow and produce spores. Facultative marine fungi are those that grow and may sporulate (form spores) in marine environments but are primarily of terrestrial or freshwater origin. Most terrestrial and freshwater fungi cannot grow in the sea because of growth-retardant chemicals in the seawater and because of their intolerance of the high salt content. Growth of hyphae requires taking in water across the cell membrane, and only obligate and facultative marine fungi are able to do this in seawater. In addition, the concentration of sodium ion in seawater is toxic to marine fungi, and much of the cellular activity of fungal hyphae is directed toward ridding cells of excess sodium. For these and other reasons, some mycologists argue that marine fungi have evolved from terrestrial and freshwater fungi, probably multiple times.

Marine fungi can also be classified on the basis of the material on which they are found. By far, the most diverse group of marine fungi lives on or in wood of terrestrial origin. They decompose the logs swept by rivers into coastal environments and even into the sea. Logs on the seafloor in anoxic locations are devoid of fungi and are decomposed by bacteria. Second in diversity to marine fungi that decompose wood are those that inhabit coastal salt marshes of temperate zones. Grasses (*Spartina*) that grow in these marshes are intertidal. The fungi that live on them range from terrestrial species growing on the rarely submerged flower stalks to facultative marine species living on the higher parts of stems and leaves to obligate marine fungi growing on the lower plant parts frequently covered by seawater. Decomposition of plant matter in the anoxic sediments of salt marshes is the role of marine bacteria. The next most diverse are those fungi that live on marine algae worldwide and the mangroves of tropical coasts. Again, because mangroves are intertidal, fungi on mangrove trees range from those least adapted to marine life and found on branches, leaves, and upper parts of roots to those best adapted to marine life and found on lower but exposed parts of the root system. Among the fungi with lowest diversity are those that dwell on sand (see Figure 6-11a). Just as beaches are physically and physiologically stressful habitats for plants and animals, so they are for fungi. In the open sea, marine fungi in plankton communities decompose the chitinous remains of dead crustaceans that rain down to the seafloor or become caught in marine snow.

## Reproduction of Marine Fungi

Marine yeasts reproduce asexually by **budding,** a kind of mitosis that produces daughter cells of unequal size. Filamentous marine fungi reproduce asexually, often by the production of spores called **conidiospores** that form at the tips of hyphae. The terminal spores on the end of a string of conidiospores are released and carried away by currents. Upon contacting an appropriate surface, the conidiospore germinates into a mycelium.

Sexual reproduction frequently involves a spore-forming structure called a **fruiting body.** Filamentous marine ascomycotes reproduce sexually by production of a fruiting body called an **ascocarp** (see Figure 6-11a). Whereas terrestrial ascocarps often consist of a cluster of spore-bearing sacs, each called an **ascus** (plural, asci), marine species typically have ascocarps of one ascus. A zygote nucleus is formed when two haploid nuclei from different mycelia fuse. The zygote nucleus then develops by meiosis and mitosis into eight cells that mature into **ascospores** (see Figure 6-11b). Released ascospores will become new adult fungi. A similar process occurs in marine basidiomycotes.

Mushrooms and fruiting bodies of other terrestrial fungi are elevated on long stalks above the ground and release their spores into air currents. Some terrestrial ascomycotes explosively shoot ascospores into the air to aid dispersal. In contrast to their terrestrial counterparts, marine fungi live in a denser medium. Long stalks of terrestrial species would easily break in water currents, and an explosive mechanism of dispersal would not be effective. Marine ascocarps are small (maximally 2 to 4 millimeters [0.08 to 0.16 inches]) and have hard walls and short stalks. Ascocarps of sand-inhabiting fungi, which live in an abrasive environment, have the hardest walls of all marine fungi. Fungi on mangrove prop roots in the intertidal zone produce ascocarps that barely break the surface of the root, reducing exposure of these delicate structures to the erosive effects of tidal currents. The ascocarp walls of these fungi dissolve and passively release ascospores into the surrounding seawater. Ascospores often have spines or threads that are sharp or sticky, improving their adherence to surfaces when they settle. Projections also allow the ascospores of sand-dwelling fungi to become caught in sea foam as a transport mechanism to other beaches (see Figure 6-11b).

## Maritime Lichens

**Lichens** are mutualistic associations between a fungus and an alga. The fungi most often are ascomycotes, and the algae are usually green algae or blue-green bacteria. The fungus provides attachment, the general structure of the lichen, minerals, and moisture. The alga produces organic matter through photosynthesis. The association allows lichens to inhabit environments that are inhospitable to either member alone.

Several of the 3,600 species of lichen in North America are restricted to the seashore, where they most often occur on rocks in the high intertidal zone or above it in the spray zone (Figure 6-12). Zonation of lichens on rocky shores is an outcome of their differing tolerances to submersion by tides, exposure to salt spray, and abrasion by ice. Maritime lichens rarely occur on drift lumber, on coastal trees and shrubs, and in sand dunes. As prevalent as they are on some rocky shores, there seems to be little information on the role that maritime lichens play in ecological communities by the sea.

(a)

(b)                              (c)

**Figure 6-12 Maritime Lichens.** (a) This rocky intertidal zone in British Columbia, Canada, has a broad white band of the volcano lichen *Coccotrema maritimum* (b) in the spray zone and a broad black band of sea tar *Verrucaria maura* in the upper intertidal zone (c).

## In Summary

Marine fungi are microscopic decomposers and pathogens. Most are sac fungi that can degrade the cell walls of terrestrial, maritime, and marine plants. Some fungi digest chitin and other decay-resistant molecules that otherwise would accumulate in seafloor sediments. Marine fungi take advantage of water currents and sea foam for the transport of spores. They form lichen associations with green algae and blue-green bacteria in maritime communities. ●

## Stramenopiles

The **stramenopiles** are a diverse group of eukaryotic organisms. One feature that unifies the group, but is not the sole character, is the nature of the two flagella these cells possess (Figure 6-13). One flagellum is a simple form, usually with a light-sensing body at the base, and the second flagellum bears many hairlike filaments called **mastigonemes,** with a thickened base and a branching tip along the shaft. Whereas one flagellum senses light, the other is an effective "feathery" swimming organelle. Members of some groups have flagellated stages that appear only very briefly in their life cycles, but once observed, their presence reveals them as stramenopiles. Other groups seem to have lost the

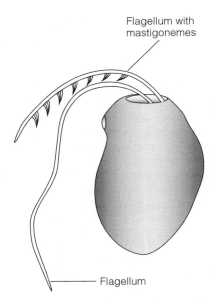

Flagellum with mastigonemes

Flagellum

**Figure 6-13 Stramenopiles.** Stramenopiles have two flagella. One is a simple flagellum and the second has many hairlike mastigonemes.

simpler light-sensitive flagellum and retained the complex one. The term **heterokont** refers to the different (*heteros*) form of the two flagella (*kontos,* a pole). The closest relatives of the stramenopiles seem to be the dinoflagellates, which also have heterokont flagella.

Stramenopiles can be nonphotosynthetic or photosynthetic. Photosynthetic stramenopiles are usually golden brown and are referred to as **ochrophytes** (as in the color ochre). They include diatoms, silicoflagellates, and brown algae (treated as a separate phylum in the next chapter). Most ochrophytes have chlorophyll *a*, chlorophyll *c* (molecularly unrelated to the true chlorophylls), the yellow beta-carotene, and usually the brown pigment fucoxanthin. The end product of their photosynthesis is the complex carbohydrate **laminarin.**

### Diatoms

Diatoms are among the most distinctive organisms in samples of marine phytoplankton, and they also are important members of benthic communities. Diatoms are abundant in the oceanic province at high latitudes, whereas at low latitudes they are predominately found in coastal waters or in areas of upwelling. Benthic habitats include soft sediments (as a part of microbial mats) or associations with living organisms. In addition, dense populations of diatoms may occur in sea ice. Wherever they occur, diatoms often contribute a major portion of primary production. Many kinds of diatom can survive long periods of dim light or darkness if there is sufficient dissolved organic matter for heterotrophy. Diatoms typically store their food reserves as lipids (oils and fatty acids) as well as laminarin. The lipids help to make diatoms more buoyant and provide a food source of high caloric value to herbivores such as copepods.

***Diatom Structure***    A unique feature of diatoms is the glassy cell wall, called a **frustule,** a two-part organic box impregnated with silica (silicon dioxide), which accounts for up to 95% of the diatom's weight (Figure 6-14a). Each half of the frustule is called a **valve,** one larger than the other and fitting over it like the cover of a shoebox. The surface of the frustule is ornamented with pores, knobs, grooves, spines, hooks, and other structures whose functions are only partly known. The intricate geometric patterns, however, are useful in identifying different species. As the diatom cell inside grows, the overlapping valves, which cannot grow, spread to accommodate the greater volume. Diatoms occur in two general shapes: those with radially symmetrical valves and those with bilaterally symmetrical valves. The radially symmetrical diatoms generally are planktonic, and the bilaterally symmetrical are more typical of the benthos. Diatoms may be as small as 2 micrometers in diameter and as long as 4 millimeters.

***Locomotion in Diatoms***    Some benthic diatoms can move by secretion of mucilage from pores and grooves, but the mechanism is still poorly understood. Movement is too slow to serve for avoidance of predators, but the ability to move might lessen the chance of depleting sources of nutrients on a microscopic scale. The secretion of mucilage also can elevate the diatom on a stalk for better exposure to the overlying water column, and it can bind sediment particles, increasing the stability of sediment where a surface layer of diatoms forms a mat over mud and sand.

***Reproduction in Diatoms***    Diatoms reproduce asexually by fission. Unlike the cells of other algae, diatoms never share cell walls, and most often the daughter cells separate after fission and live individually. In some species, however, the frustules of adjacent diatoms might remain connected to form filaments. Connection may be by a siliceous cement, by mucilage, or by interlocking ornamentation. Formation of chains of cells may reduce their rate of sinking in seawater and may reduce herbivory by plankton or small fishes whose mouths or filtration apparatuses cannot handle the larger chains.

When a diatom cell divides, each daughter cell inherits only one of the two valves of the parental frustule (see Figure 6-14b). The cell is able to complete its frustule only by secreting a new, smaller valve within the inherited one. The daughter cell that inherits the larger valve has the potential to grow to the same size as the parent cell. The other daughter cell is limited in its growth by the size of the smaller parental valve it inherited. In each generation, half the cells are smaller than the parents, and the average cell size of the population decreases over successive generations. When the cell size reaches about 50% of the maximal cell size, the diatom forms an **auxospore,** which casts off the small frustule, increases the volume of its cell, and secretes a new frustule of normal dimensions. Diatoms smaller than one third of the parental size cannot form auxospores and they

die. This period of growth is usually preceded by sexual reproduction.

Under poor environmental conditions, marine diatoms will form resting spores. During this process, the frustule becomes thicker by addition of silica, the cytoplasm becomes denser, stored organic reserves increase, and the cell sinks to the bottom sediments or becomes entangled in marine snow. Spores must be resuspended under favorable environmental conditions to germinate.

Diminishment of silica in the water is a major limiting factor in diatom populations, especially seasonally. Without silica, diatoms cannot reproduce or form spores.

***Diatomaceous Sediments***    The silica of diatom frustules cannot be decomposed by bacteria, and the frustules are almost insoluble in seawater, even under great pressure. When diatoms die, their frustules sink and accumulate on the seafloor as siliceous oozes, especially at high latitudes. Eventually, some of these accumulations may form sedimentary rock. Over millions of years, geological events have moved some of these deposits to the surface. The deposits are called **diatomaceous earth,** and they are mined commercially. Because of the small size of the diatoms and the small pores in their frustules, diatomaceous earth makes an excellent material for filtering swimming pool water, beer, and champagne. It is a mild abrasive used in products such as silver polish and toothpaste. It is also used in soundproofing and insulation products. The nutrient reserves that are stored in diatoms as lipids also accumulate in the siliceous oozes. Most of the world's petroleum reserves were formed over the past 150 million years from diatom productivity and death.

## Other Ochrophytes

Most other ochrophytes belong to small and poorly studied groups. One group, the silicoflagellates, may be extremely abundant in cold marine waters. They have basket-shaped external skeletons composed of silica as in diatoms. The cell wraps around the skeleton, which then appears to be internal. Silicoflagellates also possess one or two flagella, at least one being the specialized flagellum of ochrophytes. Standard plankton sampling often destroys the cytoplasm, and only the skeleton remains in the samples (Figure 6-15). A second group, the pelagophyceans, includes the bloom-forming alga *Aureococcus anophagefferens,* a nontoxic coastal species. Its brown tides can devastate beds of seagrasses by blocking sunlight and beds of commercially valuable molluscs by clogging their filter-feeding mechanisms.

## Labyrinthomorphs

**Labyrinthomorph** stramenopiles once were classified with slime molds and are often called "fungoid protists" and "zoosporic fungi." They lack chloroplasts and have an osmotrophic nutrition like that of heterotrophic bacteria. Their cells are spindle shaped, and they secrete a mucous coating, through which they move. They produce free-swimming stages, called zoospores, having heterokont flagella. Members of this group seem to be quite tolerant of the wide

**Figure 6-14  Diatoms.** (a) A sample of representative diatoms shows the amazing diversity of their glassy cell walls called *frustules*. (b) During asexual reproduction, diatoms divide by mitosis, each time producing a smaller cell. When a cell reaches a certain critically small size, it stops dividing, forms gametes, and sexual reproduction occurs. Regrowth of the zygote by auxospore formation restores the normal size of the cell.

100 µm

(a)

Philip Harrington / The Stock Market / Corbis

**Asexual Reproduction**

Mitosis

Mitosis

Mitosis

Mitosis

Mitosis

Cells' division continues until cells become too small to divide

**Sexual Reproduction**

New cell

Frustule formation

Growth of the cell (auxospore)

Zygote

Gamete from another

Gametes formed

Gametes released

(b)

variation in salinity found in waters and sediments of coastal and estuarine habitats, where they are responsible for degradation of large amounts of plant detritus. Some are parasites of marine algae, seagrasses, and marine invertebrates.

***Labyrinthulids***    The **labyrinthulids** include *Labyrinthula zosterae*, now known to have caused the wasting disease of eelgrass (*Zostera marina*) in the North Atlantic in the early 1930s. As much as 90% of eelgrass acreage was lost in some geographic regions because of this disease. Recovery has

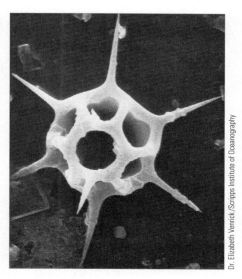

**Figure 6-15 Silicoflagellates.**
Silicoflagellates are tiny phytoplankton related to diatoms. Their external shells are composed of silica.

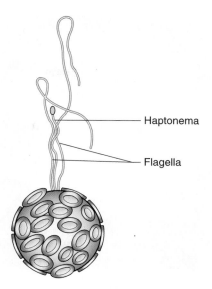

**Figure 6-16 Haptophyta.** Members of the phylum Haptophyta have a unique structure, the haptonema, that is used to collect food.

been full in parts of its range, but eelgrass beds have not returned to other locations. The loss taught marine biologists a lesson about the critical role of seagrasses in the ecology of the seas. A related but unnamed species may have been responsible for a wasting disease of turtlegrass (*Thalassia testudinum*) in Florida Bay during 1987. Species of *Labyrinthula* seem to occur on other seagrasses as well as on eelgrass and turtlegrass without causing disease. They are likely to be part of the normal seagrass flora, opportunistically invading living tissue only when the seagrass is under stress.

***Thraustochytrids***  The **thraustochytrids** (thraw-stow-KY-tids) *Thraustochytrium* and *Schizochytrium* are planktonic and benthic decomposers, and some species are pathogens of marine organisms, including shellfish in both natural populations and aquaculture facilities. In addition to their beneficial role in marine environments and in contrast to their pathogenic impact, thraustochytrids are used commercially as a source of human dietary supplements. Oils extracted from *Schizochytrium* species are high in the polyunsaturated omega-3 fatty acid docosahexaenoic acid (DHA). DHA is known to be necessary in human neurological development and function and may reduce the concentration of undesirable lipids in blood.

## Haptophytes

The phylum Haptophyta is closely related to the stramenopiles and the alveolates (a group discussed later in the chapter). It includes photosynthetic organisms that were formerly grouped with the diatoms and silicoflagellates. The composition of their photosynthetic pigments is similar, but the haptophytes possess two similar simple flagella, both used for locomotion, and a unique associated structure, the **haptonema** (Figure 6-16). The haptonema arises from the cell surface between the two flagella but does not undulate to propel the cell. The haptonema is very long in heterotrophic haptophytes, intermediate in length in ones that mix photosynthesis and heterotrophy, and short or absent in photosynthetic haptophytes. The haptonema captures food on its sticky surface and then bends around to transfer the food to the posterior end of the cell, where ingestion occurs by phagocytosis.

Most haptophytes are called *coccolithophores,* named after their surface coating of disc-shaped scales, or **coccoliths,** made of calcium carbonate (Figure 6-17). Coccoliths are produced internally within membranous vesicles and then are transported to the exterior of the cell. Production of coccoliths accounts for up to 40% of carbonate production in modern seas. Dead coccolithophores contribute heavily to formation of calcareous oozes on the seafloor and, where uplifted by geological processes, to some of the chalk cliffs along the seashore. Although some other haptophytes live in fresh water, coccolithophores are exclusively marine. They are significant members of the phytoplankton, particularly in the nutrient-poor open sea of the tropics, where they are heavily grazed by copepods and a variety of other zooplankton. It is believed that the high reflectance of the chalky coccoliths and their release of the greenhouse gas dimethyl sulfide may have a strong impact on global climate change, especially by bloom-forming species such as *Emiliania huxleyi.*

Because the shells of both diatoms and coccolithophores are relatively inert and do not decompose (although calcium carbonate dissolves at great depths), the amount of these

**Figure 6-17  Coccolithophores.**
Coccolithophores are tiny phytoplankton with cells covered by calcareous plates called *coccoliths*.

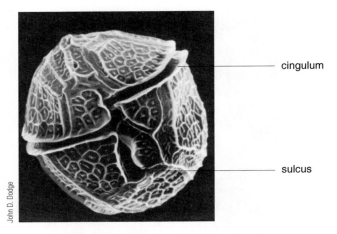

cingulum

sulcus

**Figure 6-18  Dinoflagellates.** This scanning electron micrograph of the armored dinoflagellate *Gonyaulax* shows the intricate plates of cellulose within the pellicle. The two grooves that house the flagella (not shown) are the cingulum around the middle and the sulcus, which branches from the cingulum.

sediments in the seafloor can be used to estimate what surface conditions were like in the past. Large drills are used to take core samples from the seafloor. These samples allow researchers to analyze the diatom and coccolith content and estimate the rate at which they accumulated in the bottom sediments. When surface waters were warm and carbon dioxide was plentiful in seawater for photosynthesis and calcification, there were large numbers of these organisms, and their remains rapidly accumulated on the bottom as they died. When climates were cool or carbon dioxide concentrations were lower, there were fewer of these organisms, and their remains accumulated more slowly.

Heterotrophic haptophytes are among the smallest (2 to 20 micrometers) eukaryotic plankton, giving them the ability to consume large numbers of yet smaller bacteria. In this way they are responsible for the transfer of bacterial biomass to higher levels of marine food chains, largely through predation by marine ciliates, a group described in the next section.

## Alveolates

Scientists have recently regrouped several kinds of microbes into the phylum Alveolata on the basis of molecular biology and structure of their cells. **Alveolates** have membranous sacs (**alveoli**) beneath their cell membranes. If the combination of cell membrane and alveoli is complex, this structure at the cell surface is called a **pellicle.** A pellicle differs from a cell wall because it exists within rather than outside the cell membrane. Marine alveolates include dinoflagellates and ciliates. A third group, the apicomplexans, are strictly parasitic. The groups within Alveolata are treated as separate phyla by many authorities.

## Dinoflagellates

**Dinoflagellates** are globular, single-celled organisms (although a few species are colonial) with two flagella that lie in grooves on the cell surface (Figure 6-18). The pellicle often contains thin plates of cellulose within the alveoli. A unique chemical called **dinosporin** is associated with the cellulose plates, especially in dormant (encysted) stages of the life cycle found in sediments. About 90% of known species of dinoflagellate are marine, and although most species are planktonic, some are benthic, some symbiotic, and others parasitic. Dinoflagellates range in size from 2 micrometers to 2 millimeters.

***Dinoflagellate Structure***   The two flagella of dinoflagellates are heterokont, and both arise from a pore on the side of the cell. A simple flagellum encircles the cell in a horizontal groove, the **cingulum,** and produces a spinning motion. A longer flagellum, with hairlike filaments like those on the specialized flagellum of stramenopiles, trails down a longitudinal groove, the **sulcus,** and imparts most of the forward motion to the cell. The thin cellulose plates in the pellicle are absent or few in **unarmored dinoflagellates,** and they occur in multiple layers in **armored dinoflagellates.** The plates of armored dinoflagellates are so thick that they are easily observed under the microscope, giving the impression of a soccer ball. The number, size, and shapes of plates are used by specialists to tell species apart. The surfaces of many armored dinoflagellates are ornamented with spines and other structures that reduce predation and the rate of sinking or that increase surface area for absorption of nutrients.

***Dinoflagellate Nutrition***   Dinoflagellates obtain food by several means. The chloroplasts of photosynthetic dinoflagellates contain chlorophylls *a* and *c* and the accessory photosynthetic pigments beta-carotene and **peridinin.** The latter pigment, a xanthophyll, gives dinoflagellates their typical golden-brown color. They store food as starch of the same composition as that of green plants. Many photosynthetic dinoflagellates are **mixotrophic,** which means that they supplement photosynthesis by either osmotrophy (absorbing nutrients) or phagotrophy (engulfing nutrients). Only half the species of dinoflagellate are photosynthetic; the other half live mainly by a combination of phagotrophy and osmotrophy.

Because dinoflagellates are among the larger phytoplankton, their lower ratio of surface area to cell volume makes them less efficient in absorbing phosphorus and nitrates from seawater. These inorganic nutrients are largely obtained by eating other plankton (diatoms, other dinoflagellates, and many other kinds of microbes, as well as metazoan plankton, including copepods and fish eggs).

***Reproduction in Dinoflagellates***   Dinoflagellates reproduce asexually by fission of the cell into two, at rates of up to one division per day. Sexual reproduction is known in a number of species and occurs by fusion of similar gametes. In most cases meiosis then occurs, resulting in adult dinoflagellates that are haploid. Life histories often include the production of cysts, dormant stages resistant to decay (by the presence of dinosporin) and to environmental stress. Cyst formation may follow seasonal sexual reproduction and is otherwise associated with the onset of unfavorable environmental conditions, such as reduced light, temperature, and nutrients.

***Ecological Roles of Dinoflagellates***   Dinoflagellates play an important role in marine ecosystems. Together with diatoms and coccolithophores, they are a major component of the phytoplankton that provides food directly or indirectly to many marine animals. Their flagella, mixotrophic nutrition, ability to migrate vertically in the water column, and lack of dependence on silicon give dinoflagellates an ecological advantage over diatoms. For these reasons, dinoflagellates are more abundant than diatoms in tropical waters, particularly in the open sea where inorganic nutrients are in low concentration. On the other hand, aside from predation, to which diatoms are subjected as well, dinoflagellates are less tolerant of stormy seas and other turbulence. Their relatively large size and multiplated structure increase the chance of their cells being torn apart. Some dinoflagellates are parasitic and live in the intestines of marine crustaceans called copepods. Other species, collectively called **zooxanthellae** (ZOH-oh-zan-thel-ee), lack the flagella found in most other dinoflagellate species and are symbionts of jellyfish, corals, and molluscs. The zooxanthellae are photosynthetic and provide food for their host organisms, and the hosts provide carbon dioxide, other essential nutrients, and shelter (for more information regarding zooxanthellae see Chapter 15).

***Harmful Algal Blooms***   Some marine dinoflagellates are responsible for the phenomenon known as **harmful algal blooms** (HABs, formerly called "red tides"). HABs occur when photosynthetic dinoflagellates undergo a population explosion. The specific cause of these population explosions, or blooms, is not known. During a bloom, the number of organisms can be so great that they color the water red, orange, or brown. The species that cause HABs produce potent toxins that affect the nervous system, kidneys, muscles, or other parts of animals, either directly or through consumption of contaminated food. HAB toxins can kill fish, marine mammals, and even humans. As bacteria decompose animals killed by the toxins, the oxygen content of the water may be depleted, leading to the death of other organisms otherwise unaffected by the toxins. Dead fish and other marine life can wash up on beaches and decay, creating problems of sanitation and making the beaches unfit for recreational use (see accompanying box).

Some dinoflagellate species produce toxins that are taken up by molluscs, in particular, clams and mussels. The molluscs themselves are not harmed by the toxin, but the toxin is concentrated in their tissues. Other animals may feed on these contaminated shellfish and be injured or killed as a result. Humans who consume contaminated shellfish exhibit neurological problems, such as loss of balance and coordination, a tingling sensation or numbness in the lips and extremities, slurred speech, and nausea. Death can occur if an amount of toxin sufficient to paralyze the respiratory muscles is ingested. This condition is known as **paralytic shellfish poisoning,** or PSP. Dinoflagellate toxins cannot be destroyed by cooking. Contamination of commercial shellfish by HABs results in a substantial loss of revenue to the shellfishing industry.

## Ciliates

**Ciliates** are protozoans that bear cilia for locomotion and for gathering food. The cilia may densely cover the body (that is, the cell) in highly organized columns and rows, or they may be confined to certain patches on the cell. In many ciliates, adjacent cilia are fused in tufts or long rows called **membranelles** (Figure 6-19). Membranelles often occur in the oral region, where an organelle called the **cytostome** serves as a permanent site for phagocytosis of food. Marine ciliates range from 10 micrometers to 3 millimeters long and are members of the plankton and benthos, where they are major links in food chains and form symbiotic and parasitic relationships with many other marine organisms. Because suspension feeding by cilia is a highly efficient method for removing small particles from seawater, ciliates are significant components of microbial food webs. Marine ciliates reproduce asexually by binary fission and sexually by **conjugation** (the transfer of nuclei between fused cells). Fission may occur as often as once every 2 hours.

***Types of Marine Ciliates***   Marine communities have a high diversity and abundance of ciliates, especially in sediments and coastal waters. Among the many groups of marine ciliates, there are three ecologically significant groups: the scuticociliates, oligotrichs, and tintinnids. The scuticociliates have a dense and uniform distribution of cilia on their bodies. Oligotrichs have few cilia, and tintinnids (Figure 6-20) usually lack body cilia and secrete an organic,

# Harmful Algal Blooms

Many kinds of marine algae undergo rapid population increases under conditions that are still poorly understood. Some are macroscopic algae such as the exotic *Caulerpa taxifolia* that has invaded the Mediterranean Sea and continues to expand its coverage of the seafloor. Microscopic algae often bloom as "red tides" in surface waters (Figure 6-A) and wreak havoc in coastal fisheries by killing commercially valuable species. Toxic species are especially notorious for their impact on populations of commercial fish and shell-

fish and the humans, birds, and marine mammals that consume them. In addition, toxin-bearing cells can become airborne as a result of rough sea-surface conditions and enter the lungs of marine mammals and humans when they inhale the mist, causing them respiratory distress. Blooms of non-toxic phytoplankton shade benthic communities by day and consume much oxygen while respiring at night. These harmful algal blooms (HABs) have become an area of significant interest to marine biologists, and the journal *Harmful*

*Algae,* launched in 2002, serves for scientific communication in this emerging discipline.

Toxic marine HABs are caused mostly by dinoflagellates. The toxins they produce are not part of their normal metabolism but are specially produced for protection. Such chemicals, called **secondary metabolites,** are highly complex end products of elaborate metabolic pathways that function under altered growth conditions. The best-known HAB toxins are the **saxitoxins,** a family of neurotoxins produced by dinoflagellates. The human syndrome caused by consumption of saxitoxins is called *paralytic shellfish poisoning.* Other human syndromes caused by other dinoflagellate toxins include **diarrhetic shellfish poisoning, neurotoxic shellfish poisoning,** and **ciguatera fish poisoning,** the latter caused by benthic dinoflagellates. A fifth syndrome, **amnesic shellfish poisoning,** is caused by a toxin produced by blooms of diatoms.

Studying HABs is difficult for several reasons. HAB organisms are difficult to culture in the laboratory. Re-creating the proper conditions for altered growth characteristics and inducing HAB organisms to produce toxins are also problematic. Solutions to these problems should ultimately help marine biologists better understand the causes of these harmful blooms and help them find ways to control or prevent them. ●

*Figure 6-A  Harmful Algal Blooms.* "Red tides" are caused by population explosions of certain species of dinoflagellates, diatoms, and blue-green bacteria.

Sanford Berry/Visuals Unlimited

---

loosely fitting shell called a **lorica** (see Figure 6-19). The physical support and protection given by the lorica may be increased by cementing foreign particles to the organic membrane. The particles may be grains of silica or carbonate sand, or they may be leftovers of consumed food, such as coccoliths and diatom frustules.

***Ecological Roles of Marine Ciliates***  Although some marine ciliates harbor autotrophic symbionts or chloroplasts, most are heterotrophs. They are links in the transfer of pro-

duction from heterotrophic and autotrophic (blue-green) bacteria to higher levels in the food chain. Ciliates less than 30 micrometers long typically graze on bacteria and heterotrophic nanoflagellates (plankton measuring 2 to 20 micrometers) and are a critical link in the bacterial loop of pelagic food webs (for more information regarding pelagic food webs see Chapter 17). Many are herbivores, smaller ones taking in diatoms, cyanobacteria, dinoflagellates, and autotrophic nanoflagellates. Larger ciliates eat other ciliates and small zooplankton. Ciliates in sediments capture bacte-

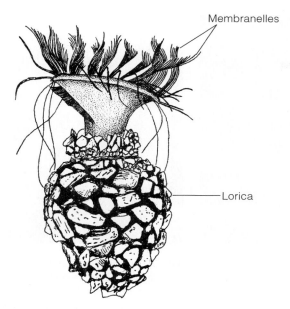

**Figure 6-19 Membranelles.** Membranelles are tufts or long rows of fused cilia that function in locomotion and feeding. This marine ciliate is a tintinnid with its body partly encased in a lorica.

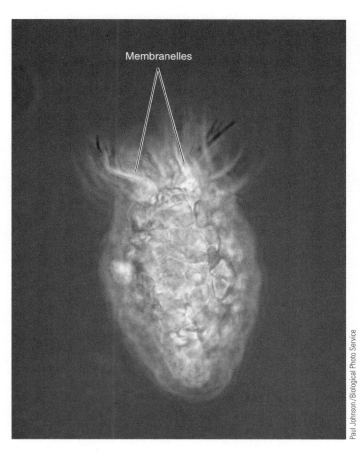

**Figure 6-20 Tintinnid.** Tintinnids are ciliates that are members of the zooplankton.

ria, diatoms, and small animals that live among the grains. Most ciliates capture food by suspension feeding with their membranelles, but some actively chase prey. In turn, small ciliates are prey for larger ciliates, and ciliates in general are consumed by heterotrophic dinoflagellates, amoeboid "protozoans" (described in a later section), and animals, particularly small crustaceans, ciliated animals, and the larvae of many invertebrates. Planktonic ciliates reach their highest densities in nutrient-rich coastal waters, sometimes $5 \times 10^5$ cells per liter, largely consisting of small ciliates that eat bacteria.

## Choanoflagellates

**Choanoflagellates** are a phylum of marine and freshwater flagellated cells that are more closely related to animals than any other group of one-celled microbes, a conclusion based largely on molecular genetics. Their structure is much like that of choanocytes, a type of cell found in sponges (for more information regarding choanocytes see Chapter 8). They may occur as single cells or in colonies, and the colonies may be stalked or embedded in a gelatinous mass (Figure 6-21). The cell often is surrounded by a lorica composed of siliceous rods. A single flagellum extends from the cell surface and is surrounded at its base by a funnel-shaped collar of tentacle-like projections called microvilli. The flagellum bears filaments like the mastigonemes of stramenopiles. Its movement creates a current that draws water and particles through the collar. Choanoflagellates are small organisms, about 10 micrometers or less. They are highly efficient consumers of bacteria, which become trapped between the microvilli and then are taken into food

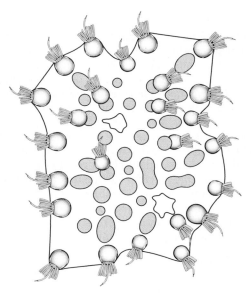

**Figure 6-21 Choanoflagellates.** A colony of the marine colonial choanoflagellate genus *Proterospongia.* The collar is composed of microvilli and surrounds a central flagellum. Cells are united in a gelatinous mass.

vacuoles by phagocytosis at the base of the collar. Although some are attached, free-swimming planktonic choanoflagellates in some locations may outnumber other groups of flagellated plankton, including dinoflagellates.

## Amoeboid Protozoans

Amoeboid protozoans belong to several unrelated "protozoan" groups, once placed within the same phylum. They all have an organelle called a **pseudopod** that is an extension of the cell surface. Unlike the cilium and flagellum, pseudopods can change shape and are used for locomotion by benthic species and for food capture by benthic and pelagic ones. Amoeboid protozoans are heterotrophs, consuming bacteria and other small organisms that they find in the plankton, among grains of sediment, or on various surfaces. Upon contact, the prey becomes attached to the pseudopod. At some point along the pseudopod, the prey is then engulfed by phagocytosis and digested.

Some amoeboid protozoans have an unprotected cell membrane, although most have a shell, or test. The **test** is an externally secreted organic membrane that is often covered with foreign particles or strengthened by mineral secretions. In addition to a test, some amoeboid protozoans have an internal mineral skeleton. Pseudopods extend from a single opening or through multiple pores in the test.

This functionally similar but taxonomically mixed group of marine organisms includes two major phyla: foraminiferans and actinopods. Foraminiferans are abundant and diverse on the seafloor and are perhaps the most ecologically significant planktonic protozoans. The actinopods include the radiolarians, acantharians, and heliozoans, of which the radiolarians predominate.

### Foraminiferans

Foraminiferans, or forams (phylum Foraminifera), are amoeboid protozoans having branched pseudopods that form elaborate, netlike structures called **reticulopods,** used to snare prey. In addition, benthic forams use their reticulopods to crawl through sediments or over surfaces, and planktonic forams may use reticulopods to reduce the rate at which they sink through the water column. Forams consume lots of diatoms and bacteria, and some harbor symbiotic green and red algae and zooxanthellae (dinoflagellates). The symbionts provide nutrients to their hosts and, as in reef-building corals, might play a role in calcification, the process of depositing calcium carbonate in their tests. Some forams can attain sizes of 1 centimeter (3/8 inch) or more, and some fossil species reached 12 centimeters (5 inches) in diameter. Forams reproduce both sexually and asexually, and their life cycles are somewhat complex.

***Foraminiferan Test***  Forams often produce elaborate multichambered tests of calcium carbonate (Figure 6-22a). As forams grow, they add more chambers, and the resulting test frequently resembles a microscopic snail shell. Benthic forams may have chambers in rows or spirals, and planktonic species typically have spherically symmetrical tests. The tests of planktonic species are usually thin and light and bear many spiny processes that help to reduce the sinking rate. At depths greater than 2,000 meters (6,600 feet), ben-

John Clegg/Andrea London

Neal Ericsson

**Figure 6-22  Foraminiferans.** (a) Foraminiferans produce tests of calcium carbonate that contain chambers and resemble the shell of a microscopic snail. They use their pseudopods to capture prey. (b) Over a period of millions of years, foraminiferan tests accumulated in the bottom sediments of the seas around Great Britain. Geological events ultimately moved a large amount to the surface, forming the White Cliffs of Dover.

thic forams attach foreign particles to the organic membrane rather than deposit calcium carbonate because of calcium carbonate's solubility at great pressure.

Although most foraminiferan species are bottom dwellers, a few species with enormous numbers of individuals are members of the zooplankton. The tests of dead planktonic forams are a major constituent of sediments in some areas of the deep ocean floor. These sediments are called **globigerina** (gloh-bij-eh-RYE-nuh) **ooze** because of the large number of tests of the genus *Globigerina,* a planktonic foram of low latitudes. In addition to foram tests, globigerina ooze may consist of as much as 25% coccoliths. Globigerina ooze is not found at depths exceeding 5,000 meters (16,500 feet) because the great pressure dissolves the carbonate tests faster than they can accumulate. Over millions of years, geological change has brought some of this sediment to the surface, where it forms large deposits of chalk. The White Cliffs of Dover on the English Channel, for example, are formed from globigerina ooze (see Figure 6-22b). Forams also contribute to the sand found on the beaches of many islands, including Tonga in the Pacific Ocean and the pink sands of Bermuda and the Bahamas.

***Foraminiferans and Zooxanthellae***  Many forams from nutrient-poor waters maintain ecologically efficient relationships with zooxanthellae that live symbiotically within the foram's cytoplasm. By day the dinoflagellates migrate out along the reticulopods, gaining maximal exposure to sunlight. At night they migrate deep into the central shell. Because it is difficult to keep planktonic foraminiferans alive in the laboratory, marine biologists have not yet been able to study the chemical basis of this relationship thoroughly. Experiments have demonstrated, however, that forams with zooxanthellae, when kept in normal sunlight, grow and reproduce more rapidly than those kept in the dark or those that have had the zooxanthellae chemically eliminated. Although the chemical evidence is lacking, the relationship seems to involve an exchange of materials between the forams and their symbionts similar to that occurring between corals and their zooxanthellae. The forams generate waste products such as carbon dioxide and ammonia, which although toxic to the forams, are nutrients for the zooxanthellae. The intracellular zooxanthellae are able to absorb the nutrients with little effort because they do not have to expend energy to reconcentrate them. This self-contained fertilizing and waste-removal system is extremely efficient for supporting life in nutrient-poor water.

### Radiolarians

Radiolarians are a highly diverse class of zooplankton within the phylum Actinopoda (Figure 6-23), named for the long needlelike pseudopods that they bear. They have a central nuclear region surrounded by an external organic membrane, or **capsule,** perforated with one or many pores. Pseudopods pass through the pores of the capsule and form a region called the **calymma.** The extended pseudopods capture food and reduce sinking rates. In addition to pseudopods, the cytoplasm in the calymma has large vacuoles to control buoyancy.

The cytoplasm of the calymma typically secretes an internal skeleton of silica. The skeleton may consist of numerous spines that radiate outward with the pseudopods, and in some species, the spines may be clustered, resembling children's jacks. Other radiolarian species have an intricate fused skeleton that looks like a spherical glass ornament (*see* Figure 6-23b). Because silica is resistant to dissolving in seawater, even under great pressure, the skeletons of dead radiolarians accumulate on the seafloor as siliceous oozes at depths beyond those at which foram tests dissolve. In contrast to diatomaceous oozes, which form at high latitudes, most **radiolarian oozes** form at low latitudes.

Most radiolarians are planktonic and live in the photic zone, where they use their pseudopods to capture a variety of phytoplankton and zooplankton. The larger radiolarians even capture copepods and other planktonic crustaceans. Zooxanthellae often are found in the calymma, and the radiolarians obtain nutrients from these symbiotic dinoflagellates.

Relatives of the radiolarians have similar construction and function, but they belong to different classes of actinopods. Acantharians secrete a skeleton of radiating spines made of strontium sulfate. Heliozoans, rare in the sea, lack an organic capsule between the nuclear region and calymma. Both groups are mainly planktonic, and many species harbor zooxanthellae.

Pseudopods

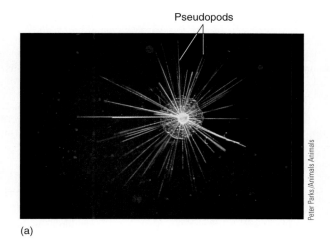

(a)

Peter Parks/Animals Animals

(b)

Dr. Richard Kessel & Dr. Gene Shih/Visuals Unlimited

**Figure 6-23 Radiolarians.** (a) Radiolarians consist of a central nuclear region surrounded by a frothy calymma. Long pseudopods extend from the cell in this live radiolarian to capture food. (b) The cell of the radiolarian is supported by a skeleton of silica, which often bears short radiating spines.

## In Summary

In marine environments, nonfungal eukaryotes make up a large portion of the plankton and benthos. Dinoflagellates, diatoms, coccolithophores, and silicoflagellates are photosynthetic producers and members of the phytoplankton. Dinoflagellates form symbiotic relationships as the zooxanthellae of some eukaryotic microbes and of many invertebrates. Labyrinthomorphs are decomposers and pathogens. Heterotrophic consumers include ciliates (especially the tintinnids), choanoflagellates, foraminiferans, radiolarians, and some dinoflagellates. Among the consumers, a few groups are grazers of bacteria, allowing for the transfer of prokaryotic biomass to higher levels of marine food webs. Some of these producers and consumers incorporate calcium carbonate and silicon dioxide into their cell walls or skeletons, adding to seafloor oozes upon their death and sedimentation.

The eukaryotic groups are distinguished by their cell coverings, the structure of their cell membranes, their possession of cilia, flagella, and pseudopods of various shapes for locomotion and prey capture, the chemistry of pigments and food-storage compounds, life-history characteristics, and many other features. ●

## IN PERSPECTIVE

| Taxonomic Group | Representative Organisms | Form and Function | Reproduction | Biochemistry | Ecological Roles |
| --- | --- | --- | --- | --- | --- |
| Viruses | Bacteriophages, cyanophages, phycoviruses, podoviruses, retroviruses | Icosahedral or helical as free virions, with protein capsid surrounding the genome; some with envelope from host cell membrane | Dependent on host mechanisms to make copies; either lytic or lysogenic | Genome of DNA or RNA, with capsid of protein subunits, and often with a few enzymes | Parasitic, major cause of mortality among plankton, possible control of bloom-forming algae, important in alteration of pelagic food webs |
| Bacteria | Blue-green bacteria, green and purple sulfur and non-sulfur bacteria; *Prochlorococcus*, *Thiomargarita* | Prokaryotic with cell wall containing aminated sugars; some flagellated; shapes mostly include rods and spheres | Binary fission | Widely varied modes of nutrition; photoautotrophs with bacterio-chlorophyll or chlorophylls *a*, *b* | Chemoautotrophs and photoautotrophs, decomposers, nitrogen fixers, symbionts for light production or nutrition; form stromatolites with other microbes |
| Archaea | *Methanococcus*, *Pyrolobus*, Halobacteria | Prokaryotic with cell wall lacking aminated sugars | Binary fission | Widely varied modes of nutrition; some autotrophs with bacterio-rhodopsins; special lipids play role in stabilizing cell membranes under extreme environmental conditions | Chemoautotrophs and photoautotrophs, heterotrophs, methanogens; inhabitants of extreme environments |
| Fungi | Ascomycotes, including yeasts; lichens | Eukaryotic with cell wall of chitin; single-celled as yeasts or forming mycelium of hyphae | Asexual reproduction by budding or conidiospores; sexual reproduction with fruiting bodies and ascospores; passive dispersal of spores | Lignin digestion; glycogen storage; strict aerobes | Heterotrophs, decomposers; symbiotic as the mycobiont in lichens |
| Stramenopiles | Diatoms, silicoflagellates, *Aureococcus*, labyrinthomorphs, brown algae (discussed in Chapter 7) | Eukaryotic; heterokont flagella, one with hairlike filaments; diatoms with frustule of silica | Fission; sexual reproduction; auxospore formation | Chlorophylls *a*, *c*; lipid storage as oils and fatty acids; starch storage as laminarin | Photoautotrophs, some forming harmful blooms; heterotrophic decomposers; producers of seagrass wasting disease; form deep-sea deposits |
| Haptophytes | Coccolithophores | Eukaryotic; two simple flagella with haptonema at base; cell covering of calcium carbonate plates | — | Chlorophylls *a*, *c* | Photoautotrophs and phagocytic heterotrophs in plankton; form deep-sea deposits |
| Alveolates | Dinoflagellates; ciliates, including tintinnids | Eukaryotic; pellicle of membranous sacs; no cell wall but some with loosely fitting lorica; two to many flagella or cilia; cilia often fused as membranelles | Primarily by binary fission; ciliates with sexual reproduction by conjugation | Dinoflagellate pellicle with cellulose plates; autotrophic dinoflagellates with chlorophylls *a*, *c* and storage of simple starch | Photoautotrophs, osmotrophs, phagotrophs; planktonic and benthic; some symbiotic as zooxanthellae or forming harmful blooms |

## IN PERSPECTIVE (continued)

| Taxonomic Group | Representative Organisms | Form and Function | Reproduction | Biochemistry | Ecological Roles |
|---|---|---|---|---|---|
| Choanoflagellates | Choanoflagellates | Eukaryotic; one flagellum with hairlike filaments, collar of microvilli; in gelatinous or branched colonies or as single cells with lorica | — | — | Planktonic and benthic filter feeders on bacteria |
| Foraminiferans | Foraminiferans, *Globigerina* | Eukaryotic; reticulopods; test of calcium carbonate or foreign particles | Asexual fission; some sexual reproduction with complex life cycle | — | Planktonic suspension feeders and benthic grazers; phagocytic; form deep-sea deposits |
| Actinopods | Radiolarians, acantharians, heliozoans | Eukaryotic; perforated organic capsule, needlelike pseudopodia; internal skeleton of silica or strontium sulfate | — | — | Planktonic suspension feeders and benthic grazers; phagocytic; form deep-sea deposits |

## SELECTED KEY TERMS

alveolate, *p. 126*

alveoli, *p. 126*

amnesic shellfish poisoning, *p. 128*

archaeon, *p. 118*

armored dinoflagellate, *p. 126*

ascocarp, *p. 121*

ascospores, *p. 121*

ascus, *p. 121*

auxospore, *p. 123*

bacillus, *p. 112*

bacteriochlorophyll, *p. 114*

bacteriophage, *p. 110*

bacteriorhodopsin, *p. 119*

beta-carotene, *p. 113*

binal virus, *p. 110*

binary fission, *p. 112*

budding, *p. 121*

calymma, *p. 131*

capsid, *p. 110*

capsule, *p. 131*

carotenoids, *p. 113*

chemosynthetic bacteria, *p. 115*

chitin, *p. 120*

chlorophyll *a*, *p. 112*

chlorophyll *b*, *p. 112*

choanoflagellate, *p. 129*

chromatic adaptation, *p. 114*

ciguatera fish poisoning, *p. 128*

ciliate, *p. 127*

cingulum, *p. 126*

coccoliths, *p. 125*

coccus, *p. 112*

conidiospore, *p. 121*

conjugation, *p. 127*

consolidation, *p. 115*

cyanobacteria, *p. 112*

cyanophycean starch, *p. 112*

cytostome, *p. 127*

diarrhetic shellfish poisoning, *p. 128*

diatomaceous earth, *p. 123*

dinoflagellate, *p. 126*

dinosporin, *p. 126*

envelope, *p. 110*

exoenzyme, *p. 112*

facultative anaerobe, *p. 114*

filamentous fungi, *p. 120*

fruiting body, *p. 121*

frustule, *p. 123*

globigerina ooze, *p. 130*

glycogen, *p. 120*

halobacteria, *p. 119*

halophile, *p. 119*

haptonema, *p. 125*

harmful algal bloom (HAB), *p. 127*

helical virus, *p. 110*

heterocyst, *p. 116*

heterokont, *p. 122*

hyperthermophiles, *p. 119*

hyphae, *p. 120*

icosahedral virus, *p. 110*

labyrinthomorph, *p. 123*

labyrinthulid, *p. 124*

laminarin, *p. 122*

lichen, *p. 121*

lithification, *p. 115*

lorica, *p. 128*

lysogenic cycle, *p. 110*

lytic cycle, *p. 110*

marine snow, *p. 115*

mastigonemes, *p. 122*

membranelles, *p. 127*

methanogen, *p. 118*

microbe, *p. 109*

mixotrophic, *p. 127*

mucilage, *p. 114*

mycelium, *p. 120*

mycologist, *p. 120*

mycology, *p. 120*

neurotoxic shellfish poisoning, *p. 128*

nitrification, *p. 116*

nitrogen fixation, *p. 115*

nitrogenase, *p. 116*

nucleocapsid, *p. 110*

obligate anaerobe, *p. 114*

ochrophyte, *p. 122*

osmotrophy, *p. 112*

paralytic shellfish poisoning (PSP), *p. 127*

pellicle, *p. 126*

peridinin, *p. 127*

phage, *p. 110*

photophore, *p. 117*

photoprotective pigment, *p. 114*

phycobilin, *p. 113*

phycocyanin, *p. 114*

phycoerythrin, *p. 114*

pseudopod, *p. 130*

radiolarian ooze, *p. 131*

reticulopod, *p. 130*

saxitoxin, *p. 128*

secondary metabolite, *p. 128*

sedimentation, *p. 115*

seston, *p. 111*

stramenopile, *p. 122*

stromatolite, *p. 114*

sulcus, *p. 126*

test, *p. 130*

thraustochytrid, *p. 125*

unarmored dinoflagellate, *p. 126*

valve, *p. 123*

viral replication, *p. 110*

virion, *p. 110*

virologist, *p. 110*

virology, *p. 109*

xanthophyll, *p. 114*

yeast, *p. 120*

zooxanthellae, *p. 127*

## QUESTIONS FOR REVIEW

### Multiple Choice

1. Most marine viruses are
   a. alive
   b. bacteriophages
   c. human pathogens
   d. lysogenic
   e. noninfective

2. Which of the following is not a method of obtaining organic matter by marine microbes?
   a. chemosynthesis
   b. nitrogen fixation
   c. osmotrophy
   d. phagocytosis
   e. photosynthesis

3. Which of the following characteristics do bacteria and archaeons have in common?
   a. a prokaryotic structure of their cells
   b. chlorophyll *a*
   c. kinds of membrane lipids
   d. production of methane
   e. tolerance of very high temperatures

4. Progress in marine mycology has been slow because
   a. most of them are small yeasts
   b. of the presence of nonmarine contaminants in samples
   c. they do not reproduce sexually
   d. they rarely produce mushrooms
   e. they require anaerobic conditions

5. The auxospore is critical to the life history of diatoms because it
   a. can photosynthesize in the deep dark waters of the ocean
   b. helps disperse the population by flagellar locomotion
   c. is a resting stage that survives environmentally stressful periods
   d. is needed to reconstitute the original cell size
   e. produces many gametes by meiosis

6. Stramenopiles are united by their possession of
   a. different flagella
   b. many flagella
   c. no flagellum
   d. one flagellum
   e. two flagella

7. Which of the following is not an effect of harmful algal blooms?
   a. consumption of oxygen
   b. contamination of shellfish
   c. formation of deep-sea sediments
   d. release of toxins
   e. shading of benthic plants and seaweeds

8. Zooxanthellae are
   a. dinoflagellates
   b. flagellated
   c. foraminiferans
   d. parasites
   e. resting cysts in a life cycle

9. Globigerina ooze is formed by
   a. coccolithophores
   b. diatoms
   c. foraminiferans
   d. radiolarians
   e. tintinnids

10. Select the microbe that is properly paired with its organelle of locomotion.
    a. ascomycote, none
    b. choanoflagellate, collar
    c. ciliate, pellicle
    d. coccolithophore, cilium
    e. phycovirus, flagellum

### Short Answer

1. What biotic and abiotic factors control viral activity in the plankton?

2. Describe how marine bacteria hasten the fall of dead particles from surface waters.

3. Why are archaeons called "extremophiles"?

4. Summarize in outline form the organic and inorganic chemicals that constitute the cell coverings (walls, tests, loricas) of the many groups of marine microbes.

5. Explain why the calcareous skeletons of forams and coccolithophores do not accumulate at the greatest depths of the sea.

6. How do the cells of diatoms and dinoflagellates differ?

7. What significant role do heterotrophic bacteria, fungi, and labyrinthomorphs play in marine habitats?

8. Describe the methods of food capture by tintinnids, foraminiferans, and choanoflagellates.

### Thinking Critically

1. If planktonic viruses lyse bacterial cells at a high rate, what effect might this have on food webs in the pelagic zone?

2. Many planktonic microbes have projections of some sort from their cells, cell walls, tests, and loricas. What are the possible advantages of these structures?

3. Marine microbes exhibit a wide range of sizes. If you wanted to study them, what methods would you use and what problems might you encounter in trying to collect them from seawater?

4. In what ways is *Prochlorococcus* better adapted for life in the open tropical seas than are diatoms?

## SUGGESTIONS FOR FURTHER READING

Capriulo, G. M. 1990. *Ecology of Marine Protozoa.* New York: Oxford University Press.

Fuhrman, J. A. 1999. Marine Viruses and Their Biogeochemical and Ecological Effects, *Nature (London)* 399(6736):541–48.

Hyde, K. D., and S. B. Pointing. 2000. *Marine Mycology: A Practical Approach.* Hong Kong: Fungal Diversity Press.

Kirchman, D. L. 2000. *Microbial Ecology of the Oceans.* New York: Wiley-Liss.

Munn, C. B. 2004. *Marine Microbiology: Ecology and Applications.* New York: BIOS Scientific Publishers.

Plumley, F. G. 1997. Marine Algal Toxins: Biochemistry, Genetics, and Molecular Biology, *Limnology and Oceanography* 42(5, part 2):1252–64.

Porter, J. W. 2001. *The Ecology and Etiology of Newly Emerging Marine Diseases.* Dordrecht, The Netherlands: Kluwer Academic Publishers.

### InfoTrac College Edition Articles

DeLong, E. F. 2003. A Plenitude of Ocean Life, *Natural History* 112(4).

Suttles, C. A. 1999. Do Viruses Control the Oceans? *Natural History* 108(1).

Zimmer, C. 2000. Sea Sickness, *Audubon* 102(3).

### Websites

**http://www.geo.ucalgary.ca/~macrae/palynology/dinoflagellates/dinoflagellates.html** Information on the general biology of dinoflagellates and red tides, with images and links to other sites.

**http://www.indiana.edu/~diatom/diatom.html** Information on diatoms and scientists who study them, with numerous links to other sites.

**http://www.microbeworld.org/home.htm** A website supported by the American Society for Microbiology, with general information and graphics about viruses, bacteria, archaeons, and eukaryotic microbes.

**http://www.radiolaria.org/** Detailed information and graphics about hundreds of radiolarian species and links to and for radiolarian specialists.

**http://www.solaster-mb.org/mb/subindex%20marine%20life.htm** Images and limited information on all forms of marine life, including many microbes.

**http://www.ucl.ac.uk/GeolSci/micropal/radiolaria.html** Detailed information on the biology and study of radiolarians, including many images. The home page provides access to comparable pages on diatoms and foraminiferans.

**http://www.ucmp.berkeley.edu/archaea/archaea.html** Brief descriptions and some images of archaeons. The home page provides access to comparable pages on other microbes.

**http://www.ucmp.berkeley.edu/foram/foramintro.html** Brief descriptions and some images of forams.

**http://www.ucmp.berkeley.edu/protista/radiolaria/radmm.html** Brief descriptions and some images of radiolarians.

# 7

# Multicellular Primary Producers

## Key Concepts

1. Multicellular marine macroalgae, or seaweeds, are mostly benthic organisms that are divided into three major groups according to their photosynthetic pigments.

2. The distribution of seaweeds depends not only on the quantity and quality of light but also on a complex of other ecological factors.

3. Marine algae supply food and shelter for many marine organisms.

4. Flowering plants that have invaded the sea exhibit adaptations for survival in saltwater habitats.

5. Seagrasses are important primary producers and sources of detritus, and they provide habitat for many animal species.

6. Salt marsh plants and mangroves stabilize bottom sediments, filter runoff from the land, provide detritus, and provide habitat for animals.

Most of the primary production in marine ecosystems is done by phytoplankton, but seaweeds and flowering plants also contribute, especially in coastal habitats. Not only do they provide food directly to herbivores but decaying plant parts are a significant source of detritus for detrital food chains. In addition to their role as primary producers, these organisms provide habitats for other marine organisms. They define the structure of coastal marine communities and may be initial colonizers of disturbed areas. In addition, marine flowering plants keep the water clear by trapping sediments, and their root systems stabilize the bottom sediments. In this chapter we will explore the biological adaptations and important ecological roles of these organisms.

## MULTICELLULAR ALGAE

**Seaweeds** (or marine **macroalgae**) are multicellular algae that inhabit the oceans. Some algae, such as filamentous diatoms, dinoflagellates, and cyanobacteria, belong to groups that are predominately unicellular; their biology was covered in Chapter 6. Three other groups of seaweeds—red algae (phylum Rhodophyta), brown algae (phylum Phaeophyta), and green algae (phylum Chlorophyta)—have few unicellular forms, and their biology will be covered in this chapter.

Brown algae are relatives of diatoms and a group of funguslike protists called water molds. Red and green algae appear to be more closely related to each other, and the green algae are even more closely related to land plants. In fact, some authorities place the green algae in the kingdom Plantae. Despite the evolutionary distance, seaweeds have in common many aspects of their biology and of their contributions to the ecology of the seas. Scientists who study

NOAA, inset, Richard N. Mariscal

seaweeds and phytoplankton are called **phycologists** or **algologists**.

## Distribution of Seaweeds

Most species of seaweed are benthic, growing on rock, sand, mud, and coral on the sea bottom, on other organisms, and as part of **fouling communities** (plants and animals that live on pilings, bulkheads, boat hulls, moorings, and other artificial surfaces). Some seaweeds attach to very specific surfaces, whereas other seaweeds are rather nonselective. In general, seaweeds inhabit about 2% of the seafloor. The presence of benthic seaweeds defines the inner continental shelf, where the marine community largely depends on the food and protection that seaweeds provide. Life on the outer continental shelf and in the deep sea is quite different in the absence of seaweeds. (For more information regarding shelf communities see Chapter 16.) The distinction between the inner and outer shelves is based on the **compensation depth** of algae. The compensation depth is the depth at which the daily or seasonal amount of light is sufficient to drive enough photosynthesis to balance the metabolic needs of the alga with nothing left over to support growth.

The environmental factors that are most influential in governing the distribution of seaweeds are light and temperature. Some other abiotic factors critical in governing the distribution of seaweeds are duration of tidal exposure and desiccation (drying out), wave action and surge, salinity, and availability of mineral nutrients. The areas of the world most favorable to seaweed diversity include both sides of the North Pacific Ocean, Australia, southwestern Africa, and the Mediterranean Sea.

### Effects of Light on Seaweed Distribution

The vertical and horizontal distributions of seaweeds are limited in part by the availability of sunlight and, therefore, vary by depth, latitude, sea conditions, and season. It was once thought that the vertical distribution of red, brown, and green algae could be explained by their accessory photosynthetic pigments, the presence of which gives the seaweeds their characteristic colors, a concept known as **chromatic adaptation.** Because green light penetrates deepest in coastal waters and the accessory pigments of red algae absorb mostly green wavelengths, red algae were thought to extend to the greatest depth of primary producers. It followed that green algae, which have pigments absorbing mostly blue and red wavelengths that are diminished rapidly in seawater, should be found at the shallowest depths. Because accessory pigments of brown algae absorb intermediate wavelengths of light, brown algae would be expected to be most abundant at intermediate depths. Indeed, recent evidence seems to support the hypothesis of chromatic adaptation because the depth record (295 meters, or 973 feet) for seaweeds is held by a yet undescribed species of red algae from the Bahamas, but the green alga *Rhipiliopsis profunda* is close behind this record at 268 meters (884 feet).

The concept of chromatic adaptation was proposed in the 1800s, and the hypothesis was accepted for about

**Figure 7-1  Zonation of Seaweeds.** Some of the clearest examples of seaweed zonation can be found in the intertidal zones of rocky shores. In general, the distribution of seaweeds by depth is not explained by their possession of different combinations of photosynthetic pigments (the chromatic adaptation hypothesis). Instead, zonation is explained by the responses of seaweeds to competition, herbivory, and physiological tolerances to many climatic factors.

100 years, until it was realized that such zonation did not occur and that the distribution of seaweeds depended more on herbivory, competition, varying concentration of these specialized pigments, and the ability of seaweeds to alter their forms of growth (Figure 7-1).

### Effects of Temperature on Seaweed Distribution

Temperature affects the distribution of seaweeds. The greatest diversity of algal species is in tropical waters. Farther north or south of the equator, the number of species decreases, and the species themselves are different. Many marine algae in colder latitudes are **perennials,** meaning that they live longer than 2 years. During the colder seasons only part of the alga remains alive, sometimes only a few cells, but most often a mass of stemlike structures. When the temperature warms up in the spring, this body part initiates new growth. Temperature is not usually a limiting factor for algae that live in tropical and subtropical seas, although the temperature in intertidal areas may become too warm and contribute to seasonal mass mortality of many seaweeds and the animals they shelter. At high latitudes, freezing and scouring by ice may eliminate seaweeds from the intertidal and shallow subtidal zones.

## Structure of Seaweeds

The seaweed body is called the **thallus** (Figure 7-2a), and with few exceptions, all cells of the thallus are photosynthetic. Because the thallus lacks the **vascular** (conductive) **tissue** of members of the kingdom Plantae, seaweeds do not have roots, stems, or leaves. The thallus can, however, occur in complex shapes, and some seaweeds are easily mistaken for plants. If most of the thallus is flattened, it may be called a **frond** or **blade.** The structure attaching the thallus to a surface is the **holdfast.** Some seaweeds have a stemlike region, or **stipe,** between the holdfast and blade. Variation in the complexity of these parts of the thallus provides a rich diversity of form (Figure 7-2b–f).

## Biochemistry of Seaweeds

The major distinctions among phyla of seaweeds are in their biochemistry: photosynthetic pigments, composition of cell walls, and the nature of their food reserves.

## Photosynthetic Pigments

The color of the thallus is due to the wavelengths of light that are not absorbed by the combination of pigments in the seaweed. Chloroplasts in all seaweeds have chlorophyll $a$, which along with chlorophylls $b$ (green algae), $c$ (brown algae), and $d$ (red algae) is responsible for photosynthesis. Chlorophylls absorb blue and red wavelengths of light and pass green light. Accessory photosynthetic pigments, such as carotenes, xanthophylls, and phycobilins, absorb different wavelengths of light and pass the energy to chlorophylls for photosynthesis. (For more information concerning photosynthesis see Chapter 5.) In some cases the accessory pigments protect chlorophyll molecules from damage by light (providing a photoprotective function like a sunscreen's).

## Composition of Cell Walls

The cell walls of seaweeds are primarily composed of cellulose. In the case of calcareous algae, the cell wall may be impregnated with calcium carbonate. Many seaweeds secrete a slimy gelatinous mucilage, made of polymers of several sugars, that covers their cells. The mucilage can hold a great deal of water and may act as a protective covering that retards desiccation in intertidal algae exposed at low tide. It can also be sloughed off to remove organisms that may have become attached to the thallus. Some algae have a thick

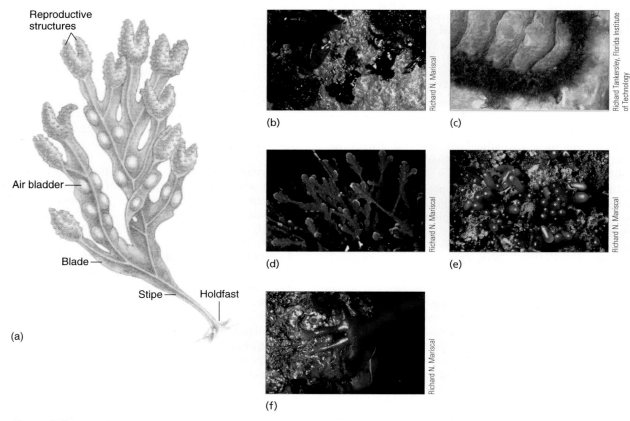

**Figure 7-2 Variable Shapes of Seaweeds.** (a) The thallus of a seaweed may consist of several distinct parts, as in rockweed (*Fucus*). But the shape can vary widely: (b) encrusting, (c) filamentous (epizoic on chiton), (d) branched, (e) and (f) leaflike.

# Seaweeds and Medicine

Seaweeds have been used for centuries to treat a variety of illnesses. Asiatic cultures used seaweeds as long ago as 300 B.C. to treat glandular disorders, such as goiter. A goiter is an enlargement of the thyroid, a gland located in the neck. The thyroid gland requires iodine to produce its secretions, and when an individual's diet is deficient in iodine, the gland frequently enlarges. Because many seaweeds concentrate iodine from seawater in their tissues, consumption of seaweed treats the condition.

The ancient Romans used seaweeds for treating burns and rashes and for healing wounds. The slimy mucilage that coats the blades of many seaweeds effectively blocks air and microorganisms from reaching an affected area. This relieves some of the discomfort, helps prevent infection, and promotes healing. The red alga *Por-*

*phyra* contains vitamin C and was used by English sailors to prevent scurvy, a disease caused by vitamin C deficiency. *Porphyra* was more readily available than citrus fruits during ocean voyages and did not spoil as quickly. Several species of red algae were used to eliminate parasitic worms from the intestines of affected individuals. Kaenic acid, which is extracted from the red alga *Digenia,* is still used for this purpose.

In the past, phycocolloids that were isolated from red algae were used to treat a variety of intestinal ailments. Phycocolloids dissolve slowly and are not digested. They coat the lining of the stomach and intestines so that material in the digestive tract, such as acids, will not irritate it, thus alleviating some of the discomfort related to ulcers and stomachaches. Phycocolloids were used to treat constipation because

they promote retention of fluid in the large intestine, which helps make fecal material easier to move. Today, products from red algae such as agar and carrageenan are used in the treatment of ulcers. The pharmaceutical industry uses a variety of polysaccharides from red algae to coat pills and in the production of time-release capsules.

Probably the most important algal product used in medicine and research today is agar, which is used for culturing microorganisms. Because agar can withstand high temperatures, it can be sterilized. Its porosity allows the movement of nutrients. It is solid at room temperature and resists decomposition by most microorganisms. These characteristics make agar an ideal medium for growing bacteria and fungi for study and research. ●

---

multilayered covering of protein called a **cuticle.** The cuticle is a protective layer and may be so thick that it gives the alga an iridescent sheen. Other cell-wall components have commercial and medicinal value (see accompanying box).

### Nature of Food Reserves

When the amount of photosynthesis exceeds the immediate needs of the seaweed, the excess sugars are converted into polymers and stored in the cells as starches. The chemistry of these starch molecules differs among the groups of algae. Unique sugars and alcohols may also be used as antifreezes by intertidal seaweeds at high latitudes where temperatures can drop below freezing.

### Reproduction in Seaweeds

Seaweeds can reproduce both asexually and sexually. In some, asexual reproduction can occur when the thallus breaks into pieces, and each new piece grows into a new alga. This type of asexual reproduction is called **fragmentation.** In some coastal habitats, huge accumulations of seaweeds, called **drift algae,** form by fragmentation. Communities of drift algae can be physically complex and biotically diverse, but their high demand for oxygen may lead to rapid decline of the community, mass mortality, and decomposition. Some sargassum weeds reproduce entirely by fragmen-

tation and form extensive floating mats covering hundreds of acres of sea surface.

Another type of asexual reproduction involves spore formation, typically by meiosis within a part of the thallus called the **sporangium.** Once created, the haploid spores are dispersed by water currents. Spores (and gametes) of green and brown algae usually have flagella that enable them to move in the water. If the spores settle in a suitable habitat, they germinate and form new thalli (Figure 7-3). The stage of the life cycle that produces spores is the **sporophyte** ("plant that produces spores"), and it is diploid in chromosome number.

Sexual reproduction in seaweeds involves the formation of gametes that fuse to form a diploid, fertilized cell known as a **zygote.** The stage of the life cycle that produces gametes is the **gametophyte** ("plant that produces gametes") and is usually haploid. Gametes are typically produced by mitosis within structures in the gametophyte called **gametangia.** Gametophytes of some seaweeds are differentiated into male (sperm-producing) and female (egg-producing) gametophytes. After dispersal and settlement, the zygote grows by cell division into the sporophyte.

The possession of two or more separate multicellular stages (an asexual sporophyte and sexual gametophyte) in succession is called an **alternation of generations** (see Figure 7-3). Compared with a typical animal life cycle, most

**Figure 7-3** **Life Cycle of the Sea Lettuce *Ulva*.** (a) The sea lettuce *Ulva* is a common marine green alga. (b) Reproduction in sea lettuce: The sporophyte produces two different kinds of motile spores, designated + and −. When the spores germinate, they grow into gametophytes that release two kinds of motile gametes, also designated + and −. The + gamete fuses with a − gamete to produce a zygote. The zygote settles on solid bottom, germinates, and forms a new sporophyte.

(a)

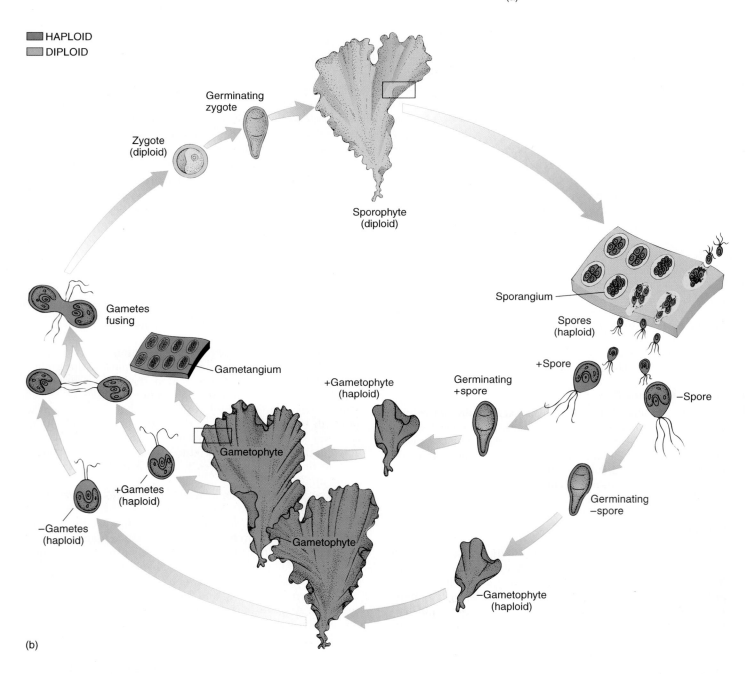

HAPLOID
DIPLOID

Germinating zygote

Zygote (diploid)

Sporophyte (diploid)

Sporangium

Spores (haploid)

+Spore

−Spore

Germinating +spore

Germinating −spore

Gametes fusing

Gametangium

+Gametophyte (haploid)

Gametophyte

−Gametophyte (haploid)

+Gametes (haploid)

−Gametes (haploid)

Gametophyte

(b)

algal life cycles are much more complex, sometimes involving as many as three multicellular generations of diploid sporophytes and haploid gametophytes, which may be of differing size, shape, and longevity.

## Green Algae

**Green algae** (phylum Chlorophyta) are a diverse group of organisms that contain the same kinds of pigments in vascular (land) plants: chlorophylls *a* and *b* and carotenoids. Very few species (about 1,100 species, or 13%) of green algae are marine macroalgae (seaweeds); most are freshwater species and phytoplankton. Green algae are important as seasonal sources of food for marine animals. They also contribute to the formation of coral reefs. Green algae exhibit a rapid growth response to the presence of high levels of nutrients in polluted, near-shore waters and as such are members of fouling communities. Some green algae are introduced exotic species that are of special concern for marine conservation.

### Structure of Green Algae

The majority of green algae are unicellular or are small multicellular filaments, tubes, or sheets. On the other hand, many marine green algae, especially those in the tropics, have a **coenocytic** (see-nuh-SIT-ik) thallus, which consists of a single giant cell or a few large cells containing more than one nucleus and surrounding a large vacuole (Figure 7-4a). Coenocytic algae go through a developmental process in which the cell grows and the nucleus divides, but the cell does not divide. This process is repeated many times, resulting in a large multinucleate cell. Some examples of this body plan are shown in Figure 7-4b–f, and they show that a great diversity of form is possible even with this simplified organization of the cell. Some species appear to have a simple form, as in *Valonia* with a saclike thallus. Although the thallus at first appears quite simple, closer examination shows that it maintains its shape with an elaborate cell wall composed of cellulose. The cellulose is deposited in multiple layers for strength, similar to a multiple-ply automobile tire or plywood. The coenocytic alga *Caulerpa* forms elegant, feather-shaped thalli or structures that look like stems, leaves, and fruit. One of the most beautiful marine algae is *Acetabularia*. Although not technically coenocytic, as its relatives are, the uninucleate thallus is complex. *Acetabularia* has a slender stalk with a delicate, round, umbrella-like cap at the top. The cap functions in photosynthesis. This single-celled alga has been used by cell biologists to study the role of the nucleus in controlling cellular processes. Another interesting seaweed is *Codium*. This alga has a thallus composed of many long filaments wrapped together like a rope. *Codium* is of interest because of its healing response when damaged. A cut end will constrict to prevent loss of the cell sap of its coenocytic thallus. The green alga *Halimeda* has a coenocytic thallus that contains many segments, somewhat like a prickly pear cactus, with cell walls containing deposits of calcium carbonate. Calcareous green algae such as *Halimeda* play an important role, along with coralline red algae and coral animals, in the formation of coral reefs. Deposits of *Halimeda* carbonate may exceed 50 meters (165 feet) depth in the Great Barrier Reef.

### Response of Green Algae to Herbivory

Tolerance, avoidance, and deterrence are three ways that organisms adapt to predation, and the green algae provide good examples of each. Most green algae are small annuals. Rapid growth and the release of huge numbers of spores and zygotes prevent elimination of their populations by herbivores (tolerance). Their small size allows them to occupy crevices on rocky shores and reefs that keep out the larger herbivores (avoidance). In addition to these survival tactics, coenocytic algae illustrate two adaptations to repel herbivores (deterrence). Deposits of calcium carbonate in the cell walls of *Halimeda, Acetabularia,* and other green algae require herbivores to have strong jaws, and they make herbivory expensive by filling the stomachs of grazing invertebrates and fish with minerals that have no nutritional value. Many coenocytic algae produce repulsive toxins that reduce herbivory on their otherwise nutritious thalli. Species of the toxic alga *Caulerpa* that have been introduced into the Mediterranean Sea and to the south Florida and California coasts have become nuisance species for lack of herbivores to control their populations. Although few herbivores tolerate the toxins of coenocytic algae, one group of sea slugs (ascoglossans, molluscs related to nudibranchs; for more information on nudibranchs see Chapter 9) has evolved a piercing-and-sucking feeding style that takes advantage of the coenocytic thallus. A simple radula pierces the cell wall, and large quantities of the cell contents can be withdrawn from a single hole. In addition to getting a good meal, many ascoglossans harvest the chloroplasts and retain the toxins. The chloroplasts, now stolen organelles, are maintained within cells of the slug's body as a source of sugars from photosynthesis, and the algal toxins protect the slug from predation. Some biologists look toward ascoglossan molluscs as a possible means to control nuisance populations of *Caulerpa.*

### Reproduction in Green Algae

A common marine form of green algae is sea lettuce, *Ulva,* shown in Figure 7-3. Its life cycle is representative of the life cycles of green algae. The thallus of this alga resembles a large leaf of lettuce that may exceed 30 centimeters (12 inches) in length. It is found in intertidal and shallow coastal waters, where it attaches to the bottom by means of a tiny holdfast. The life cycle of *Ulva* demonstrates the basic alternation of generations of seaweeds (see Figure 7-3b). The large leafy sporophyte and gametophyte stages are nearly identical. Their flagellated spores (released from the sporophyte) and gametes (released from the gametophyte) are also similar, but spores have four flagella and gametes two. Mating types are designated + and −, and gametes of opposite mating types must fuse for fertilization to occur.

## Red Algae

**Red algae** (phylum Rhodophyta), in contrast to green algae, are primarily marine organisms. About 98% of the 6,000 species are marine, which means that red algae have the highest diversity among the seaweeds. Although they are most diverse in tropical oceans, they can also be found as significant ecological components at higher latitudes. Red algae are mostly benthic in distribution. Some species can

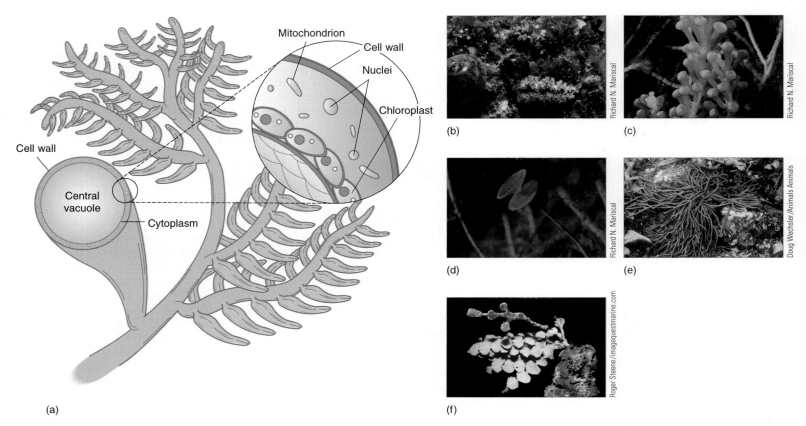

(a)

(b)

(c)

(d)

(e)

(f)

**Figure 7-4 Coenocytic Green Algae.** (a) The thallus of a coenocytic alga consists of a large central vacuole surrounded by a layer of cytoplasm that contains many nuclei. Forms of thalli include (b) the globular thallus of *Valonia* and (c) the feathery thallus of one kind of *Caulerpa*. (d) The umbrella-shaped *Acetabularia* is not technically coenocytic because this enormous cell has only one nucleus. (e) Another form is the branched thallus of *Codium*. (f) The cell wall of *Halimeda* contains calcium carbonate, which contributes to the formation of reef sediments.

survive at depths as great as 200 meters (660 feet) in clear water. In addition to chlorophyll *a* and minor amounts of chlorophyll *d,* red algae contain the accessory pigments phycoerythrins and phycocyanins. (These phycobilins were introduced in Chapter 6.) The normally reddish color of these algae comes from the large amount of phycoerythrins. Phycoerythrin pigments are most effective at absorbing blue and green light and mask the green color of chlorophyll *a*. Red algae, however, are not always red. Their thalli can be a variety of colors, from yellow to black, because phycocyanin pigments absorb more red light than do phycoerythrins and the photoprotective carotenoid pigments, which absorb in several regions of the light spectrum.

### Structure of Red Algae

Almost all red algae are multicellular and are less than 1 meter (3.3 feet) long. Their thalli vary widely in shape and organization (Figure 7-5). Some species develop a large bladelike thallus like that of the green alga *Ulva*. Examples are the species of *Porphyra* that are widely cultivated as **nori** for human consumption in oriental dishes. Many red algal thalli are composed of branching filaments that give the seaweed a bushy appearance, from delicate and fuzzy to robust and crispy. Others that are heavily calcified look like a variety of corals with flat encrusting plates, calcified nodules, erect branches, or leafy blades. Other red algae, along with greens and browns, form low, dense **algal turfs** of filamentous and branched thalli that carpet the seafloor over hard rock or loose sediment.

### Response of Red Algae to Herbivory

The annual red algae are a seasonal food source primarily for sea urchins, fish, molluscs, and crustaceans. Unlike the coenocytic green algae, red algae do not produce many toxins to deter herbivores. Their main defenses against herbivory are making their thalli less edible, changing their growth patterns, and evolving complex life cycles. Some red algae deter herbivores by incorporating calcium carbonate into the cell wall, making the thallus difficult to eat and digest. Others alter their growth forms when subjected to predation, producing growth patterns such as algal turfs that are more difficult for herbivores to graze. As a group, red

**Figure 7-5 Diversity of Red Algae.** (a) The edible sea laver, *Porphyra,* has broad flat blades. (b) The thallus of *Asparagopsis* is filamentous. (c) This *Gracilaria* has cylindrical branches. (d) The encrusting *Porolithon* helps to cement loose calcareous reef sediments together. (e) *Amphiroa* is an example of a coralline alga, which has heavy deposits of calcium carbonate in the cell wall. (f) Many small red algae, such as *Antithamnion,* are epiphytic on larger red algae and other seaweeds and marine plants.

algae are able to tolerate herbivory because of their rapid growth and annual and complex life cycles, which allow them to rapidly replace biomass lost to grazing. Many small red algae avoid large herbivores by growing in crevices.

### Reproduction in Red Algae

Red algae exhibit a variety of life cycles, some of which can be quite complex (Figure 7-6). Two unique features of their life cycle are the absence of flagella, even on spores and gametes, and the occurrence in most species of three multicellular stages: two sporophytes (carposporophyte and tetrasporophyte) in succession and one gametophyte (including male and female, usually as separate thalli). The life cycle can be described as follows: Sperm released by a male gametophyte are carried by water currents to a filament on a part of a female gametophyte that contains an egg cell. After fertilization, the zygote divides while still attached to the female gametophyte, forming the **carposporophyte,** a stage unique to red algae. This "parasitic" stage produces nonmotile diploid spores called **carpospores.** When a carpospore settles on the seafloor or a host plant or animal, it germinates and forms a new adult alga, the **tetrasporophyte.** The tetrasporophyte is comparable to the sporophyte of green algae. The name tetrasporophyte only serves to distinguish it from the carposporophyte and refers to the cluster of four ("tetra") haploid **tetraspores** that result from meiosis in this stage. The tetrasporophyte then releases nonmotile haploid spores, which disperse, settle, germinate, and form a new (haploid) gametophyte generation. The free-living sporophytes and gametophytes are, like *Ulva,* usually hard to tell apart, particularly in tropical red algae.

### Ecological Relationships of Red Algae

A few of the smaller species of red algae are **epiphytes** (*epi* meaning "upon," *phyte* meaning "plant"), that is, they grow on other algae or plants, or are **epizoics** (*zoon* meaning "animal"), which grow on animal hosts. Some red algae grow on a variety of hosts, whereas others grow only on specific host organisms. Some species of smaller red algae are parasitic on larger species of red algae, relying on nutrients from their hosts for their survival.

The cell wall of red algae consists of cellulose, a group of chemicals called phycocolloids (described in the following section) of commercial importance, and often a superficial coating of mucilage or a cuticle composed of protein. As noted previously, the cell wall may be impregnated with calcium carbonate, particularly in a group known as the **coralline red algae** (see Figure 7-5d–e). Coralline algae precipitate calcium from seawater and help to cement bits and pieces of loose coral together. This process, called **consolidation,** is necessary to hold the many parts of a reef together and eventually forms large areas of nearly solid reef. More than half of the mass of some reefs is contributed by coralline algae. Like *Halimeda,* coralline red algae also add to the accumulation of calcareous sediments on the reef. We will examine the role of coralline algae in reef formation in more detail in Chapter 15.

### Commercial Uses of Red Algae

Red algae produce large amounts of **phycocolloids** (a group of polysaccharides) in their cell walls. Several of these are commercially valuable because of their gelling (stiffening) qualities. **Agar,** for example, forms thick gels at very low concentrations and resists degradation by most microbes. Its primary human use is in making solid culture media for growing bacteria in microbiology laboratories. It is also used in foods and pharmaceuticals as a thickening agent. Another phycocolloid from red algae, **carrageenan,** is also used as a thickening and binding agent in ice cream, pudding, and salad dressings and by the cosmetic industry for various creamy preparations.

Some red algae are used by humans as a source of food. Irish moss (*Chondrus crispus*) is widespread on the shores of Ireland and many other parts of the North Atlantic Ocean.

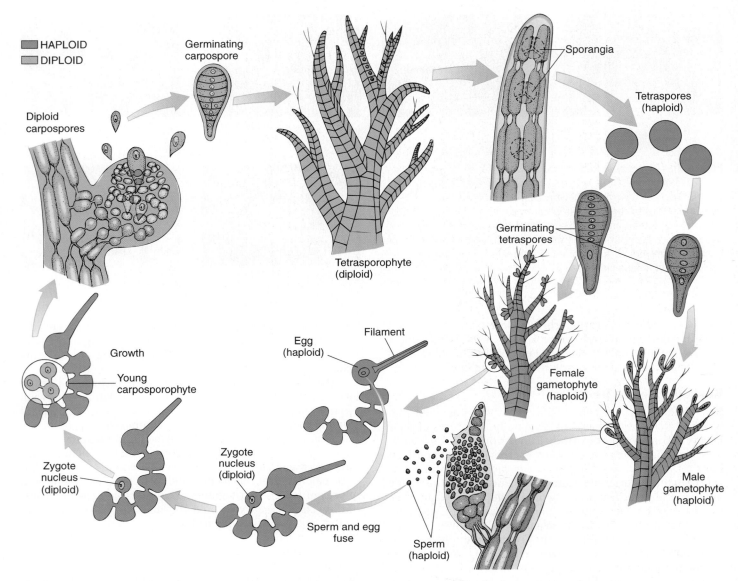

**HAPLOID**
**DIPLOID**

Diploid carpospores

Germinating carpospore

Sporangia

Tetraspores (haploid)

Tetrasporophyte (diploid)

Germinating tetraspores

Female gametophyte (haploid)

Male gametophyte (haploid)

Growth

Young carposporophyte

Egg (haploid)   Filament

Zygote nucleus (diploid)

Zygote nucleus (diploid)

Sperm and egg fuse

Sperm (haploid)

**Figure 7-6  Life Cycle of a Red Alga.** Nonmotile spores released by the tetrasporophyte germinate and grow into either male or female gametophytes. The sperm produced by the male gametophyte are nonmotile. They are carried by water currents to the female gametophyte and fertilize their stationary eggs. The zygote that results develops in place into a new generation, the carposporophyte, which remains parasitic on the female gametophyte. Carpospores released by the carposporophyte disperse, settle, and germinate into the tetrasporophyte.

The Irish make a traditional pudding by boiling small amounts of Irish moss in milk. The dessert was adopted in New England, Canada, Great Britain, and France, where it is called blanc mange. Dulse (*Rhodymenia palmata*) has been used as a foodstuff in England for hundreds of years. In Japan and other parts of the orient, *Porphyra* species (laver, nori) are used in sushi, soups, and seasonings. In addition to being a source of human food, red algae are cultivated to produce animal feed and fertilizer in many parts of Asia.

## Brown Algae

**Brown algae** (phylum Phaeophyta) are represented by such familiar forms as rockweeds, kelps, and sargassum weed. The 1,500 species of brown algae are almost exclusively (99.7%) marine inhabitants. They have a higher species diversity in the sea than green algae and a lower diversity than red algae. With rare exception, such as species of *Sargassum*, they are benthic. Brown algae range in size from

**Figure 7-7  Kelp.** Large kelp, like these off the coast of California, form massive undersea forests.

**Figure 7-8  Sargassum Weed.** Some *Sargassum* species form extensive floating mats in an area of the North Atlantic Ocean known as the Sargasso Sea.

microscopic, filamentous forms to the largest of all algae, the giant kelps (Figure 7-7), which can attain lengths of 100 meters (330 feet). The characteristic olive-brown color of these algae is due to the carotenoid pigment **fucoxanthin,** which is also in diatoms and masks the green of chlorophylls *a* and *c.* Much of the familiar seaweed along the shore and in shallow water is brown algae.

### Distribution of Brown Algae

Brown algae for the most part are more diverse and abundant along the coastlines of high latitudes. In the western North Atlantic Ocean, along the northeastern coast of North America, for example, there are twice as many species of brown algae than in the Caribbean, Bahamas, and Gulf of Mexico. Brown algae are far less diverse than red and green algae in the tropics, but they rival the red algae in the northeast. Most of the large kelp or kelplike brown algae are temperate forms. Tropical sargassum weeds (Figure 7-8) are notable exceptions, some species forming extensive free-floating mats in the subtropical open waters of the North Atlantic Ocean and occurring only sporadically along temperate shores when dispersed in the Gulf Stream.

### Structure of Brown Algae

Most species of brown algae have thalli that are well differentiated into holdfast, stipe, and blade (see Figure 7-2a). Many develop large, flat, leaflike blades, but the shapes of blades vary widely (Figure 7-9). The larger blades frequently

have gas-filled structures, or **bladders,** that help to buoy the blade, allowing it to gain maximal exposure to sunlight. The gas bladders and blades are supported by a large stipe that is attached to the bottom by a branching holdfast. Some kelps, such as the sea palm *Postelsia,* have very thick and flexible stipes that hold the seaweed in heavy surf and surge. As you might suspect from their large size, many brown algae are perennials and offer a year-round source of food and habitat for marine animals.

The cell walls of brown algae are composed of cellulose and a group of phycocolloids called **alginates,** similar to those of red algae and of similar commercial importance. It is primarily the alginates that give the thallus the strength and flexibility to withstand the relentless waves, surge, and currents of the intertidal and subtidal zones of exposed coasts. Although they are not vascular plants, the larger kelps have specialized cells, called **trumpet cells,** that conduct products of photosynthesis to deeper parts of the thallus. The lower stipes and holdfasts of some species cannot produce enough food for themselves by photosynthesis because the blades above absorb and block so much sunlight. In these algae, the trumpet cells carry food (primarily the alcohol **mannitol**) from the blade to the stipe and holdfast to support their metabolic needs.

### Reproduction in Brown Algae

The life cycle of most brown algae consists of an alternation of generations between a sporophyte (often a perennial stage) and a gametophyte (usually an annual). The familiar giant kelp is an example of the diploid sporophyte stage. It reproduces by releasing haploid motile spores into the surrounding water. These spores are produced in extremely

**Figure 7-9 Diversity of Shape in the Brown Algae.** (a) A dense mat of *Fucus* in the intertidal zone. (b) The sea palm *Postelsia* has a tough stipe and blades that resist the heavy ocean surf of exposed Pacific coasts. (c) The flat blade of *Agarum* is perforated with holes, which might reduce resistance to waves and surge. (d) *Padina* looks like a cluster of potato chips or wood shavings. (e) *Dictyota* has a flat branching growth. (f) *Sphacelaria* is an example of a delicate and filamentous brown alga.

large quantities and are a food source for a variety of filter-feeding benthic animals and zooplankton. Spores that settle and germinate form microscopic male and female gametophytes. It is interesting that such a large alga as the kelp has such a small gametophyte stage. The male gametophyte produces flagellated sperm that fertilize an attached egg on the female gametophyte. The zygote germinates in place, as in red algae, and the parasitic sporophyte on the short-lived female gametophyte eventually overgrows and obliterates it.

In a specialized variation on the life cycle of brown algae, *Fucus,* or rockweed, and its relatives produce haploid sperm or egg cells by meiosis at the inflated reproductive tips of the sporophyte in specialized chambers (Figure 7-10), not to be confused with gas bladders. The gametophyte is eliminated from the life cycle. The eggs and sperm are then shed into the water, where fertilization takes place. The resulting zygote germinates as a new sporophyte. Thus only one multicellular stage exists in the life cycle of this group.

In both rockweed and kelp, after the egg is fertilized it attaches to the bottom by mucilage and grows rootlike structures called **rhizoids.** The holdfast eventually develops from the rhizoids and puts forth a shoot. The shoot grows larger and develops one or more growth zones from which further growth proceeds. In rockweed, the growth zone is located at the tips of the thallus. In kelp species, growth zones may be located farther back from the blade, often at the junction of the stipe and blade. As wave action breaks off the tips of kelp or herbivorous animals devour them, the central or basal growth area produces new tissue that is pushed upward and outward to take the place of lost tissue. This pattern of growth offers a renewable resource and is useful in commercial kelp harvest, in which only the canopy of the kelp forest is removed.

### Brown Algae as Habitat

Although the rockweed *Fucus* is abundant on intertidal rocks, the majority of brown algae are found from the low tide line to a depth of about 10 meters (33 feet). The larger forms, the kelps, grow so profusely that they form offshore kelp forests. These forests are very efficient at capturing sunlight and are extremely productive. They are home to a diverse group of marine animals, including sea urchins, fishes, crustaceans, molluscs, sea lions, sea otters, and many more. (For more information on kelp ecosystems see Chapter 16.) An exceptional brown algal group is the sargassum weeds of the Sargasso Sea in the Atlantic Ocean. These free-floating clumps, buoyed by their gas bladders, form a complex three-dimensional habitat that is home to a variety of unique organisms (Figure 7-11).

### Commercial Products from Brown Algae

The alginates (phycolloids) of some brown algae are harvested for commercial use as thickening agents in the textile, dental, cosmetic, and food industries. Brown algae, such as kelps, concentrate iodine from seawater in their tissues. The concentration of iodine in some species of kelp may be 10 times that of the surrounding seawater. Before cheaper methods of obtaining iodine were developed, seaweeds were the main source of this trace nutrient, which is added to table salt to prevent goiter, a thyroid gland disorder. In some areas of the world, especially the Orient, brown algae are used as food. Some coastal countries also use brown algae as cattle feed.

## In Summary

Multicellular marine algae (seaweeds) are mostly benthic organisms. Their division into three major groups (red algae, brown algae, and green algae) is based on the accessory pigments they contain. Algal cells have cell walls in addition to their membranes, and their pigments are located within organelles called chloroplasts. Many algae secrete a slimy covering that protects them from and retards desiccation. Algal life cycles tend to be complex, involving both sexual and asexual generations.

Red algae are most widespread in the tropics. Their red color

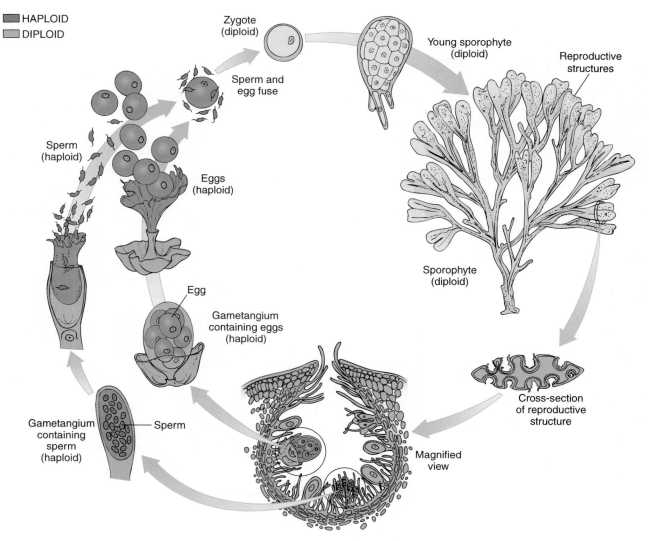

HAPLOID
DIPLOID

Zygote
(diploid)

Sperm and
egg fuse

Young sporophyte
(diploid)

Reproductive
structures

Sperm
(haploid)

Eggs
(haploid)

Sporophyte
(diploid)

Egg

Gametangium
containing eggs
(haploid)

Gametangium
containing
sperm
(haploid)

Sperm

Cross-section
of reproductive
structure

Magnified
view

**Figure 7-10  Life Cycle of *Fucus*.** The life cycle of rockweed eliminates the gametophyte stage that is found in red and green algae as well as in kelps and many other brown algae. The sporophyte is the only multicellular stage in the life cycle and is considered a sporophyte because it is diploid, as are sporophytes of other algae. Gametes form by meiosis in chambers at the tips of reproductive blades. Flagellated sperm and nonmotile eggs are released into the seawater, where fertilization occurs. The zygote then settles to the bottom and develops into a new sporophyte.

**Figure 7-11  Sargassum Weed Habitat.** Free-floating clumps of sargassum weed form a complex three-dimensional habitat that is home to a variety of unique organisms like this sargassum fish.

Larry Lipsky/Tom Stack and Associates

is due to the presence of the pigment phycoerythrin. Brown algae are almost exclusively marine organisms and are most diverse at high latitudes. With the notable exception of sargassum weed, most brown algae can be found along continental shorelines. Many brown algae have flexible bodies, an advantage for avoiding wave shock. Green algae are found in shallow coastal waters. Many marine species of green algae exhibit a coenocytic body plan. Some red and green algae deposit calcium carbonate in their cell walls and are important contributors to the formation of coral reefs. Algae are a food source for many organisms, including humans, and they provide a habitat for many species. They are also a source of many commercial products.  ●

# MARINE FLOWERING PLANTS

Kelp beds are defined by the presence of the kelp, which establish the structure of the bed and which influence interactions among other inhabitants. Similarly, three other marine habitats are defined by their dominant plants: seagrass beds, salt marshes, and mangrove forests. Unlike kelp, which originated in the sea, plants along the shores and shallow ocean waters evolved from terrestrial ancestors. In this section we will study the biology of seagrasses, marsh plants, and mangroves and their adaptations to life in the sea. The ecology of their habitats will be described in later chapters.

## General Characteristics of Flowering Plants

The kingdom Plantae is divided into several major groups. The presence of specialized vessels that carry water, minerals, and nutrients (**phloem**) and give structural support (**xylem**) separates the vascular plants, such as ferns, conifers, and flowering plants, from more primitive types, such as mosses. The presence of phloem and xylem distinguishes the root, stem, and leaf of vascular plants from the holdfast, stipe, and blade of kelp and other seaweeds. The most advanced of the vascular plants reproduce by means of seeds and are called seed plants. The **seed** is a specialized structure containing the dormant embryonic plant (a new sporophyte) and a supply of nutrients surrounded by a protective outer layer produced by the microscopic female gametophyte and parts of the parent sporophyte. Thus the seed represents three generations. There are two groups of seed-bearing plants: conifers, such as pine trees, which bear seeds in structures known as cones, and **flowering plants** (phylum Anthophyta), which bear seeds in structures known as fruits. The **fruit,** produced by the parent sporophyte, is composed of several layers of tissues. It protects the seed and aids in its dispersal. Conifers are exclusively terrestrial plants, but a small number of salt-tolerant flowering plants, known as **halophytes,** are adapted to the marine environment.

## Invasion of the Sea by Plants

Flowering plants have adapted to maritime, estuarine, and fully marine habitats multiple times since they evolved on land about 150 million years ago. It must be kept in perspec-

tive that the ancestors of these plants had become adapted to a terrestrial existence over the preceding 330 million years. Gone were the delicate flagellated stages for dispersal by water, the independent gametophyte stages of the life cycle, the ability to metabolize in a low concentration of oxygen, the reliance on the buoyant effect of water, and the tendency for all parts of the body to absorb minerals from surrounding waters and to photosynthesize under diminished quantity and quality of light. Reinvasion of the sea required some major adaptations in form and function. On the other hand, the presence of flowering plants in marine communities has presented opportunities and challenges to the native species of the sea.

These flowering plants offer a new source of habitat and food, but they compete with the seaweeds for light and with other benthos for space, and their bodies are composed of polymers (cellulose, lignin) that most marine organisms cannot digest. Because of this indigestible cellulose and lignin content, marine plants typically enter microbially based detrital food chains rather than the grazing food chains of herbivores. Marine plants have few competitors, and they tend to form extensive single-species stands that other members of the community have come to depend on. As in any dense population, disease and other disasters may wipe out its members. This potential was realized in the classic case of the wasting disease of eelgrass in the North Atlantic Ocean early in the 1900s (discussed in Chapter 6). The massive die-off of eelgrass had far-reaching effects on both coastal marine and human communities.

## Seagrasses

**Seagrasses** are **hydrophytes,** which means they generally live beneath the water. In contrast to marsh grasses and mangroves, which require periodic exposure by ebb tides to grow and reproduce, seagrasses can grow, flower, go to seed, and germinate while fully submerged. Among flowering plants they are best adapted to marine life and are the only ones that are truly marine. There are about 60 species of seagrasses, which represents about 0.02% of flowering plant species.

### Classification and Distribution of Seagrasses

Seagrasses are not true grasses but are related to lilies and a number of other groups of freshwater plants. It was formerly thought that they invaded the seas from shoreline shrubs, but recent molecular data firmly place seagrasses among freshwater relatives. There are 12 genera of seagrasses in 5 families of 3 clades (groups containing a common ancestor). One clade includes the eelgrasses (*Zostera*) and surfgrasses (*Phyllospadix*); a second clade includes paddlegrasses (*Halophila*), turtlegrasses (*Thalassia*), and the unusual *Enhalus;* a third clade consists of widgeongrass (*Ruppia*), manateegrasses (*Syringodium*), and shoalgrasses (*Halodule*). Half of the species of seagrasses inhabit the temperate zone and higher latitudes, and the other half are tropical and subtropical in distribution. *Ruppia maritima,* widgeongrass, is exceptional in being found in all seas and from the Arctic Circle to Tierra del Fuego at the tip of South America. Seagrass diversity is highest in the Indo–

West Pacific Ocean (45 species), particularly around Australia, where 30 species are found, including 10 that occur nowhere else.

## Structure of Seagrasses

Seagrass beds form primarily from **vegetative growth**. This means they grow in a fashion similar to many lawn grasses by extension and branching of horizontal stems (**rhizomes**) from which vertical stems and leaves arise. Seagrasses have the three basic parts of vascular plants: stems, roots, and leaves (Figure 7-12).

***Stems***    Stems look somewhat like bamboo, with cylindrical sections called **internodes** separated by rings (**nodes**). Rhizomes are horizontal stems with long internodes. They have growth zones at their tips and periodically produce additional growth zones laterally for branching of more rhizomes or for production of vertical stems. Rhizomes generally lie in sand or mud. An exception is the rhizome of surfgrasses. Surfgrasses live on wave-beaten coasts, and their rhizomes are anchored to rocks by a dense tangle of branching roots. Many seagrasses store starch in their rhizomes for overwintering and to provide an early burst of growth in spring. The starchy deposits make seagrass meadows a food source for geese and other migratory birds.

Vertical stems arise from the rhizomes. Vertical stems usually have short internodes and grow upward toward the surface of the sediment. Their slow growth ensures that leaf production can keep up with the accumulation of sediment. The flowering vertical stems of widgeongrass, on the other hand, have long internodes, and the stems grow rapidly to the surface of the water.

***Roots***    Roots, branched or not, arise from the nodes of stems and anchor the seagrass in sediment or to rock. The roots usually bear **root hairs**, cellular extensions from the root surface. Like their terrestrial ancestors, seagrasses depend on roots for absorption of mineral nutrients from the sediment.

Seagrasses and bacteria in the sediments interact with each other by way of the roots. Seagrass roots leak some oxygen into the surrounding sediment, which generally lacks oxygen. Nitrogen-fixing bacteria in the sediments thrive in the anoxic conditions of the seagrass bed, but the thin oxygen-containing region around the roots allows nitrifying bacteria to convert ammonia to nitrites and nitrates. These forms of nitrogen are easily absorbed and are used by seagrasses to make amino acids and other nitrogen-containing organic compounds (for more information regarding the role of microbes in nitrogen cycling see Chapter 6). These physiological relationships between seagrasses and microbes point to the importance of sediment chemistry and to the potential problems from physical disturbance, such as dredging and propeller scarring of seagrass beds.

***Leaves***    Leaves arise from the nodes of rhizomes or vertical stems. Short **scale leaves** of the rhizome protect the delicate growing tips. The long **foliage leaves** arising from vertical shoots have two parts: a lower protective **sheath** that bears no chlorophyll and an upper **blade** that accomplishes all the photosynthesis of the plant. Unlike most land plants,

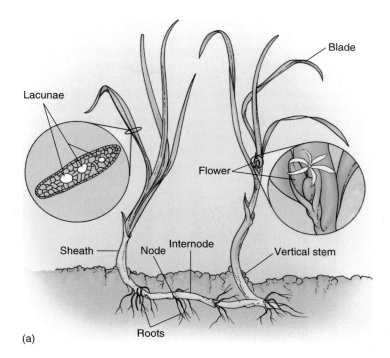

(a)

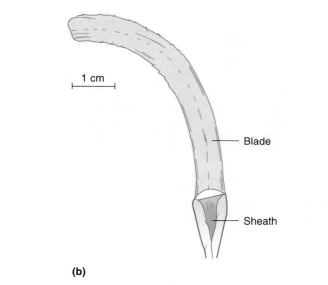

(b)

**Figure 7-12  Seagrass Anatomy.** (a) The horizontal stems of seagrasses are called rhizomes; they grow beneath the bottom sediments. Roots and vertical stems grow from the rhizomes. (b) Leaves, arising in this seagrass from the vertical stems, are divided into an upper green blade for photosynthesis and a lower pale sheath to wrap around the stem and support the blade. Lacunae in the aerenchyme of the blade carry oxygen to other plant parts and make the blade buoyant. The flowers of seagrasses are small and white.

the **epidermis** (surface layer of cells) of seagrass blades contains most of its chloroplasts. In addition to photosynthesis, cells in the leaves, particularly ones in the sheath, control salt balance. Despite a large surface area, seagrass leaves absorb much less mineral nutrient from seawater than roots do from the sediment. Nutrient runoff from human communities into coastal waters, therefore, fertilizes seaweeds and phytoplankton rather than seagrasses. The resulting algal blooms and shading may cause the seagrasses to decline.

Leaves of most species of seagrass are ribbonlike, whether narrow or wide (Figure 7-13a). The long, flat shape is apparently adaptive for life in dense, flowing seawater, providing flexibility and less resistance to the current. By contrast the leaves of manateegrasses are cylindrical (Figure 7-13b), and those of paddlegrasses are oval (Figure 7-13c).

(a)

(b)

(c)

**Figure 7-13 Leaf Shape in Seagrasses.**
(a) A white sea anemone thrives among the broad, flat, straplike blades of a turtlegrass bed. Turtlegrass is the most common species of seagrass in the Caribbean. (b) The cylindrical blades of manateegrass give this seagrass bed the appearance of a field of spaghetti. (c) Paddlegrasses have oval blades.

Leaves undergo periods of growth and aging (**senescence**). The leaves of *Syringodium filiforme,* for example, grow for about 2 weeks. Before growth stops, the tip of the blade begins to fragment, a sign of senescence. Senescence may continue for 2 to 3 more weeks before the blade detaches from the plant. A new blade begins growth 1 or 2 days before growth ceases in the next older blade.

The life cycle of the blades has major significance for epiphytes in the meadow. Seagrasses with small blades having short life cycles (for example, the paddlegrasses) do not support a diverse community of epiphytes, because their animal and plant guests cannot grow and reproduce rapidly enough to keep up with blade production and loss. More robust seagrasses with large, long-lived blades (for example, eelgrasses) not only support a diverse community of epiphytes but also often harbor species that are unique to them as a habitat. If epiphytic growth becomes too great a load in currents, waves, or surge over the seagrass bed, the blade detaches from the sheath at their physically weak junction. This is a safety feature to prevent uprooting the plant. That is why, after severe storms, shorelines of lagoons and bays may be piled high with seagrass blades with rarely a stem to be found. Unless overburdened with incoming sediment during storms, seagrass beds can quickly recover from the damage.

***Aerenchyme***   Internally, a most important gas-filled tissue in seagrasses is **aerenchyme** (AIR-en-kime). All parts of the plant have spaces called **lacunae** between the cells in this tissue (see Figure 7-12a). The lacunae provide a continuous system for gas transport through the plant. Because water carries much less oxygen than air and sediment is devoid of oxygen, the leaves must be a source of oxygen for the rest of the body. Daytime photosynthesis charges the aerenchyme with a high gas pressure. Sometimes during the day seagrass beds can be heard sizzling as oxygen bubbles are released from the surface of the blades. Oxygen diffuses through the lacunae to the stems and roots.

Although the lacunae are spacious, the aerenchyme is reduced to microscopic pores at the nodes and where plant parts join. Oxygen can pass through the pores but water cannot, thus sparing the plant from suffocation by flooding with seawater if one plant part is damaged. This system can be demonstrated by a simple and elegant procedure shown in the accompanying box.

The aerenchyme provides buoyancy to the leaves as gas bladders do for brown algae. This allows the leaves to remain upright in the water column for full exposure to sunlight. Because seawater and aerenchyme help to buoy the leaves of seagrasses, supporting tissues in these plants are much reduced and not very rigid, compared with their terrestrial relatives.

One disadvantage of the aerenchyme is the potential for invasion by pathogenic fungi and labyrinthulids (a group of stramenopiles covered in Chapter 6). As a chemical defense, many seagrasses produce antimicrobials called **tannins,** which are commonly found in plants living in anoxic sediments and which give the waters of cypress swamps their brown tint.

## Reproduction in Seagrasses

Seagrasses vary widely in flowering and setting of seed. Flowers are unknown in some species, and colonization of new sites must occur only by fragmentation, drifting, and rerooting of plants. Those that do flower lack the showy petals of their relatives the lilies, and the flowers are small and inconspicuous (see Figure 7-12), sometimes remaining below the sediment. In most species, flowers are either male or female and are borne on separate plants. Threadlike sperm-bearing **pollen** (actually a male gametophyte) is carried by water currents to female flowers, where a two-part, water-insoluble, species-specific adhesive binds pollen to a **stigma,** the female pollen receptor. This process is called **hydrophilous** ("water-loving") **pollination** and is common among submerged aquatic plants. In the unusual seagrass *Enhalus acoroides,* the whole male flower is released and drifts at the surface until reaching a female flower to fertilize. In surfgrass, rafts of pollen grains either at the surface or barely submerged are painted on the female flowers as the tide and wavelets rise and fall. Seed production is extensive in some seagrasses, so much so that migratory ducks sift the sediment for food in the accumulated seed banks. A few species of seagrass produce seedlings on the mother plant, a process called **viviparity.**

## Ecological Roles of Seagrasses

Seagrasses play several key ecological roles in the marine environment. As primary producers, seagrasses are food for herbivores, long lists of which may be found in the scientific literature. Seagrasses are important in depositing and stabilizing coastal sediments, and seagrass beds provide habitats for many marine species.

***Role of Seagrasses as Primary Producers*** Although the rates of production in seagrass meadows rival those of many terrestrial ecosystems, the direct contribution by seagrasses to the nutrition of animals is much less than for seaweeds for two reasons. First, seagrasses cover less of the seafloor than seaweeds. Except for the paddlegrasses (*Halophila*), seagrasses do not thrive in areas of low light intensity. The maximal depth of occurrence is 90 meters (297 feet). Most live at depths of less than 20 meters (66 feet), and dense beds of high biomass occur at much shallower depths. Second, because of the poor digestibility of seagrass tissue by animals, few animals can use seagrasses as a primary component of their diet. Birds and humans may consume the starchy rhizomes and the seeds. Major herbivores of the leaves are the manatees and dugong, green sea turtle, and some species of fish and sea urchin. Sometimes it is difficult to distinguish the relative contributions of tough seagrass blades and the more easily digested epiphytes to the diets of these consumers.

Seagrasses contribute to marine food webs through fragmentation and loss of leaves. This material is an important source of detritus in microbial food chains in which bacteria and fungi convert the cellulose and lignin of the leaves into microbial biomass. The microbes and detritus are consumed by suspension and deposit feeders of the sea-grass meadow and nearby habitats. Detritus is also exported to offshore communities.

***Role of Seagrasses in Depositing and Stabilizing Sediments*** Seagrasses are important in the deposition and stabilization of sediments. The blades standing in the water column serve as baffles that reduce water velocity. Because slow currents can hold a smaller burden of particles than swift currents, inorganic and organic particles settle in the meadow. Decay of seagrass leaves, stems, and roots adds more organic matter, and the sediment builds up around the slowly growing vertical stems. The rhizomes and roots of seagrasses help to stabilize the bottom, preventing currents and waves from stirring up sediments (resuspension). In these ways, seagrass beds reduce the **turbidity** (cloudiness) of the overlying water. If the water in a seagrass bed becomes too turbid, it can destroy the seagrasses and their dependent organisms. A massive die-off of seagrasses took place in Maryland's Chesapeake Bay estuary during the 1960s, largely due to diminished light associated with excessive sediment runoff from the land surrounding the bay. Other estuaries around North America have experienced gradual decline of seagrass beds due to persistent turbidity in recent decades. The loss of seagrasses makes the turbid conditions worse and can lead to erosion if sediments are resuspended.

***Role of Seagrasses as Habitat*** The considerable acreage and single-species dominance of seagrasses make them a significant habitat for other marine primary producers, consumers, and decomposers. In the presence of seagrasses, an otherwise two-dimensional surface of sand or mud is converted into a three-dimensional space with a greatly increased area on which other organisms can settle, hide, graze, and crawl. Complexity within the sediment increases physically and chemically with the presence of the **rhizosphere,** the system of roots and rhizomes along with the surrounding sediment. Seagrass beds are home to epiphytes such as foraminiferans, diatoms, fungi, delicate seaweeds, and many kinds of invertebrate animals. Microbes, several groups of worms, crustaceans, and molluscs live in the rhizosphere—a greater number and variety than in surrounding bare flats. The water among the blades holds abundant fishes and crustaceans. Species diversity is high, and the young of many commercial species of shellfish and fish thrive in this refuge. It is no wonder that coastal states seek ways to restore seagrass communities that have declined from wasting disease and from the impact of local human populations.

## Salt Marsh Plants

Plants of the salt marsh are much less adapted to marine life than are seagrasses, because they must be exposed to air by the ebbing tide to flourish. This requirement restricts salt marshes to the intertidal zone, where they are a biological filter for terrestrial runoff on its way to the sea. Although these plants grow in waterlogged, anoxic sediments, as do seagrasses, they must tolerate much higher sediment salinities created by evaporation of surface water at low

# Gas Spaces in Marine Plants

Land plants have evolved in an environment that does not support large bodies, because the atmosphere provides little buoyancy, in contrast to the sea. One function of vascular tissue is to support plants growing well above the ground by the tubular design of cells that have thick walls of cellulose and lignin. But height separates the crown of the plant from its source of groundwater. A second role of vascular tissue is to conduct the needed water from the roots to the leaves. An unbroken column of water through the xylem cells reaches heights that are a marvel of physics. If the plant is damaged, the water column may become interrupted with bubbles of air, a condition called an embolism. (Air embolisms also occur in human divers who rise from great depths so rapidly that nitrogen gas dissolved in the blood forms bubbles that clog capillaries and other small vessels, causing severe pain and possibly paralysis or death.) Land plants require adaptations for wound healing that minimize or repair embolisms in the xylem, damage that could cause death to plant parts above the wound.

In seagrasses, a reverse situation applies. Their roots, stems, and leaves have tissues (aerenchyme) with intercellular spaces (lacunae) filled with gas, consisting mostly of oxygen derived from photosynthesis. Aerenchyme allows for diffusion of oxygen to roots and stems that lie in anoxic sediments. (The vascular tissue still carries water, minerals, and organic substances between plant parts.) The buoyancy of seawater reduces the need for vascular tissue to provide physical support for the plant, and the presence of gas-filled aerenchyme helps leaves to stand erect, as gas bladders do for the stipes and fronds of brown algae. Seagrasses face the danger of seawater entering the aerenchyme upon damage and interrupting the flow of oxygen to parts that need it, thereby drowning the plant. The danger is real, for many animals dig up and through marine sediments, a process called bioturbation. The feeding activities of horseshoe crabs, stingrays, migratory geese, manatees, and green turtles break stems and leaves. How is it that local damage from these activities does not result in death of extensive areas of surrounding seagrass that are connected by rhizomes to the damaged plants?

Three seagrass biologists asked two related questions about a quarter of a century ago: Is the gas space of the lacunae really continuous throughout the seagrass? Will seawater also flow through that space? To answer these questions, D. G. Roberts, A. J. McComb, and J. Kuo of the University of Western Australia developed a simple approach. (In science, the simplest approach to observation is often the most elegant and the most instructive.) They connected a segment of the paddlegrass *Halophila ovalis* to a piece of plastic tubing at the cut end of the rhizome (Figure 7-A).

Except for the single cut, the plant was otherwise intact. A syringe filled with air was inserted into the tubing, and the entire experimental setup was submerged in seawater. When the plunger of the syringe was depressed, they observed no effect. Upon cutting a root, a blade, or a distant internode, the investigators observed bubbles of gas rising from the cut ends, thus demonstrating the continuity of the gas spaces in the aerenchyme. When the syringe was filled with water containing a soluble red dye, the biologists could depress the plunger only a short distance before gas bubbles stopped appearing; no red dye escaped the cut ends until a cut was made to the same internode to which the tubing was connected. Fluids could not, therefore, pass from one plant part to the next. Something between plant parts allowed the flow of gas but not the flow of liquid. The seagrass had a mechanism to prevent flooding of the aerenchyme beyond the area of immediate damage. This much of the study by Roberts, McComb, and Kuo can be repeated in any laboratory if fresh seagrass material is available.

An understanding of the healing mechanism required a less simple approach, one using special techniques in preparation of the tissue for examination by electron microscopes. Their continued study revealed that there were many lacunae of relatively large dimension within each part of the plant. Where each part

---

tide. Exposure makes the plants available to terrestrial herbivores, but they otherwise play the same ecological roles as seagrasses.

## Classification and Distribution of Salt Marsh Plants

Salt marshes are well developed along the low slopes of river deltas and shores of lagoons and bays in temperate regions of the world (Figure 7-14). Their distribution is limited at high latitudes by ice scouring and hard bottoms and at low latitudes perhaps by high temperature and weak tides.

Plants of the salt marsh (Figure 7-15) include true grasses (cordgrasses of the genus *Spartina*), needlerushes (*Juncus*), and many kinds of shrubs and herbs such as saltwort (*Batis*) and glasswort (*Salicornia*). These plants generally grow in the middle to upper intertidal zones, where they are protected from the action of waves and tidal currents.

## Structure of Salt Marsh Plants

Smooth cordgrass, *Spartina alterniflora*, is the plant that initiates the formation of salt marshes and dominates the lower marsh. It grows in tufts of vertical stems connected by

joined the next, a barrier, or diaphragm, of small cells interrupted the lacunae. The diaphragms were not solid but instead had very small spaces with diameters of 0.5 to 1.0 micrometers (about 2 to 4 one hundred thousandths of an inch) between the cells; these spaces connected the lacunae of one part with those of the adjacent part. It may be concluded that the surface tension of seawater prevented flow through such narrow openings. Although the diaphragms were finely perforated to allow gas flow, they acted like the bulkheads of a ship hull in restricting the inflow of water to damaged sections. The results of this study were published in 1984 in the journal *Aquatic Botany,* in which much of the world's seagrass literature appears. ●

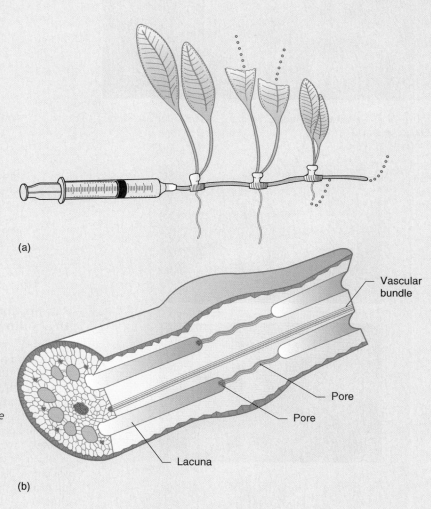

(a)

(b)

**Figure 7-A** *Functions of Seagrass Aerenchyme.* *The upper diagram illustrates the observational procedure of Roberts, McComb, and Kuo in demonstrating the continuity of gas space within the body of a seagrass. Bubbles emerge from the cut ends of plant parts when a submerged plant is charged with air from a syringe. The lower diagram is a three-dimensional interpretation of the structure of two internodes and the intervening node. Microscopic spaces between the small cells of a node connect the lacunae of one internode with those of the adjacent internode, allowing gas to diffuse but preventing water passage.*

rhizomes. Each vertical stem (**culm**) produces additional stems (**tillers**) at its base, giving a tufted or clumped appearance (Figure 7-16). Roots are concentrated at the culms. Long, smooth, flat blades arise from nodes on the culms and tillers, which produce flower heads in warm months. Plants in the well-drained nutrient-rich soils at the seaward edge of the marsh stand as high as 3 meters (10 feet), whereas those in the anoxic, waterlogged, nitrogen-poor soils of most of the lower marsh may grow only one tenth as high. Aerenchyme allows diffusion of oxygen from the blades to the subterranean rhizomes and roots. Flowers are pollinated by the wind, and seeds drop to the sediment or are carried to distant shores by water currents. Other grasses and rushes of the marsh inhabit higher ground. They vary anatomically in the organization of their culms, tillers, and rhizomes and physiologically in their tolerance to salinity, nutrients, and anoxic sediments. Their differences in growth and physiology result in zonation of the marsh (for more information regarding marsh zonation see Chapter 14).

**Figure 7-14 Salt Marsh Plants.** Cordgrass, *Spartina,* dominates this Maryland salt marsh.

**Figure 7-15 Plants of the Salt Marsh.** (a) Smooth cordgrass, *Spartina alterniflora,* is the primary builder of the salt marsh. It forms the leading edge of the marsh and is almost the only plant found in the low marsh. The high marsh may be dominated by dense stands of needlerush (*Juncus*). (b) Succulence is one adaptation of marsh plants such as saltwort (*Batis*), with its water-storing leaves. (c) Glasswort (*Salicornia*), on the other hand, has succulent stems and nearly invisible scalelike leaves.

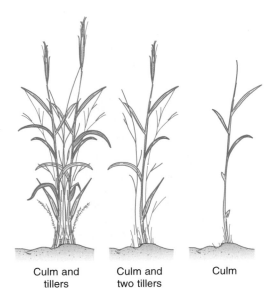

**Figure 7-16 Growth Form of Smooth Cordgrass.** Upright stems (culms) of *Spartina alterniflora* arise from rhizomes and eventually produce additional culms, called tillers, at their bases. This growth pattern gives a cordgrass a clumped appearance.

## Adaptations of Salt Marsh Plants to a Saline Environment

Plants of salt marshes are considered **facultative halophytes,** meaning that they can tolerate salty as well as freshwater conditions. Most seagrasses, in contrast, are intolerant of freshwater. Because they live in sediments with a high salt content, marsh plants tend to lose water to their environment by osmosis and display adaptations similar to those of desert plants to help them retain water. Cordgrasses and rushes have leaves covered by a thick cuticle to help retard water loss, and the stems and leaves contain well-developed vascular tissues for the efficient transport of water within the plant. The leaves of *Spartina alterniflora* have **salt glands** that secrete salt to the outside, whereas vascular tissues maintain a high osmotic pressure not with salts but with organic solutes. Shrubs and herbs of the marsh often have thick, water-retaining, **succulent** parts similar to those of plants that live in hot, arid, terrestrial habitats. In many, the leaves are succulent; in glassworts, which have no leaves, the stems are succulent.

## Ecological Roles of Salt Marsh Plants

Salt marsh plants, particularly the grasses and rushes, contribute heavily to detrital food chains, accumulate and stabilize sediments, and serve as a refuge, feeding ground, and nursery for other marine organisms.

The detritus from salt marsh plants supplies the nutrient needs of a variety of fishes and shellfish. One study suggests that as many as 95% of sport and commercial fish species, many of which spend part or all of their lives in coastal and estuarine waters, are supported by the productivity of detrital food chains in salt marshes.

Like seagrasses, salt marsh plants usually have shallow roots and rhizomes. This arrangement helps to stabilize coastal sediments and to prevent shoreline erosion. The rhizomes of cordgrass also play an important role in recycling the nutrient phosphorus by transferring phosphates produced in the bottom sediments into the leaves and stems of the plant. Marsh plants filter runoff from coastal areas, removing some of the potentially toxic organic pollutants of terrestrial origins and preventing them from entering the sea. They may also help maintain water quality in shallow-water areas by removing excess nutrients entering from terrestrial sources that might contribute to algal blooms.

Unlike seagrasses, marsh plants do not support a heavy community of epiphytes because of their periodic tidal exposure to air, and for the same reason, they are more prone to herbivory by terrestrial animals that are well-adapted to consuming plants. A significant portion of primary production is eaten by grasshoppers, beetles, and other terrestrial consumers of leaves, flowers, and seeds. Leafhoppers and aphids suck nutritive fluid from plant vascular tissue. Marsh plants form at least part of the diet of some crabs, and geese feed seasonally on rhizomes. Although of only local or historical importance, herds of grazing livestock consume large quantities of grasses and rushes in the high marsh.

## Mangroves

Mangroves are in some ways the tropical equivalents of temperate marsh plants, but they more often are trees and shrubs than grasses and herbs. Very little of the mangrove is submerged by the tide. Both similarities and differences in anatomy, physiology, and ecological roles of mangroves and marsh plants are instructive of how terrestrial plants have adapted in reinvading the seas.

### Classification and Distribution of Mangroves

The word **mangrove** comes from the Portuguese *mangue,* which means "tree" and the English "grove," and it is applied to a group of plants that are taxonomically diverse. Defining "mangrove" is subjective, but a conservative account of what constitutes the mangroves of the world lists 54 species of trees, shrubs, palms, and ferns in 16 families. Half of these plants belong to two families: the family of the red mangrove, *Rhizophora mangle* (Figure 7-17a), and the family of the black mangrove, *Avicennia germinans* (Figure 7-17b). Both red and black mangroves are well known for their specialized roots that descend to or rise from the sediment. Five other large mangroves of the tropical Americas include one close relative each of the red and black mangroves, white mangrove (*Laguncularia racemosa*), buttonwood (*Conocarpus erectus*), and *Pelliciera rhizophoreae* of very limited distribution.

Mangroves thrive along protected tropical shores with limited wave action, a low slope, high rates of sedimentation, and soils that are waterlogged, anoxic, and high in salts. Mangroves occur at low latitudes of the Caribbean Sea, Atlantic Ocean, Indian Ocean, and western and eastern Pacific Ocean. They are frequently associated with saline lagoons and are commonly found in tropical estuaries as well as on the protected sides of islands and atolls. Mangroves form the dominant vegetation in communities called mangrove swamps, or **mangals,** which we will study in more detail in

**Figure 7-17 Mangroves.** (a) This young red mangrove consists of a canopy of branches and leaves over a tangled base of prop and drop roots. These above-ground roots not only help to support the large plant in loose mud but also aid in gas exchange for the parts of the root buried in the anaerobic sediments. Can you find the original trunk? (b) The black mangrove looks more like a typical tree, having a well-developed trunk. Spikelike pneumatophores arise from underground cable roots radiating from the base of the trunk. Pneumatophores are the gas-exchange organs of the black mangrove.

(a)    Jon Hawker

(b)    Richard N. Mariscal

Chapter 14. The Indo–West Pacific is the hotspot of mangrove biodiversity as it is also for seagrasses.

## Structure of Mangroves

The representative mangroves featured here are trees with simple leaves and complex root systems. These plant parts must help the tree conserve water in a warm saline environment, supply oxygen to roots in poorly drained, anoxic sediments, and remain stable in shallow soft mud. In addition, the youngest stages in the life cycle of the plants must be adapted to deal with dispersal by tides, navigation among the tangle of above-ground roots, and establishment in their respective zones in the mangal.

***Roots***   Mangrove roots are adapted to grow in loose, shallow, anoxic, saline sediments in regions often visited by hurricanes and typhoons. Root parts are diverse, and different species have their own combinations of parts (Figure 7-18). Tap roots and deep roots generally do not occur in mangroves, and root hairs are lacking. Many are **aerial** (above ground) **roots** and have aerenchyme. The aerial **stilt roots** of the red mangrove arise high on the trunk (**prop roots**) or from the underside of branches (**drop roots**). As the growing tip of the stilt root grows into the mud, aerenchyme is rapidly produced and may constitute as much as 40% of the root volume. The aerenchyme communicates with the atmosphere by way of scarlike **lenticels** on the surface of the stilt

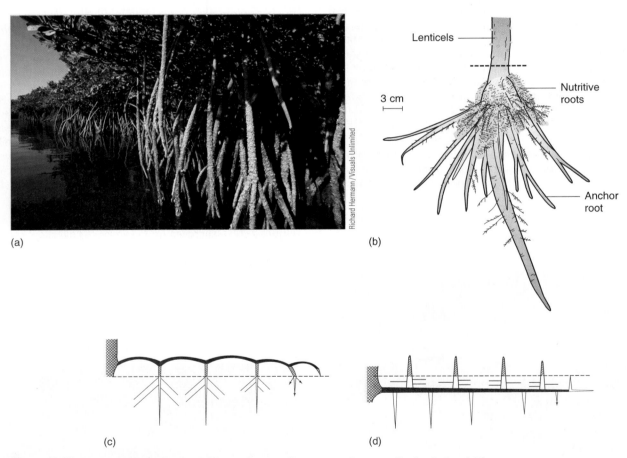

(a)

Richard Hermann / Visuals Unlimited

(b)

Lenticels

3 cm

Nutritive roots

Anchor root

(c)

(d)

**Figure 7-18  Mangrove Root Systems.** (a) Aerial roots of mangroves have scarlike lenticels, which allow air into the aerenchyme of the roots. (b) Within the sediment, mangroves produce anchor roots for stability and nutritive roots for absorption of water and minerals. (c) A red mangrove's looping arches (black) of a prop root give general support, and the part that descends to the substratum (cross-hatch) allows for gas exchange and the production of anchor roots (white) and nutritive roots (black lines). (d) A black mangrove's main structure of the root system is the cable root (black), from which anchor roots (white) descend and pneumatophores arise. The pneumatophores consist of an anchoring part (white), which produces nutritive roots (black lines), and an aerial part (cross-hatch), which serves for gas exchange.

3. Animals cannot produce their own food, so they depend on other organisms for nutrients. In other words, they are heterotrophs.

4. Animals, with the exception of adult sponges, can actively move.

Animals that lack a vertebral column (backbone) are **invertebrates,** whereas animals that have a vertebral column are **vertebrates.** The majority of animals in the sea are invertebrates, and these animals are the subject of this chapter and the next.

# SPONGES

Sponges (phylum Porifera) are simple, asymmetric, **sessile** animals (meaning they are permanently attached to a solid surface, such as bottom sediments or rock). Sponges exhibit a wide variety of sizes and shapes, and their shape is frequently determined by the shape of the bottom sediments or material on which they are growing and by the water currents flowing over them. Although many living sponges are drab, some species are brightly colored. Red, yellow, green, orange, and purple specimens are common.

## Sponge Structure and Function

The structure of a sponge's body is unique in that it is built around a system of water canals. This arrangement is associated with the sponge's sessile lifestyle. A sponge's body is full of tiny holes, or pores, called **ostia** (singular, ostium; Figure 8-1), through which large amounts of water circulate. The water is a source of nutrients and oxygen, and it carries away the animal's wastes. Water enters a sponge's body through the ostia and eventually flows into a spacious cavity called the **spongocoel** (SPUN-joh-seel). Water then exits the spongocoel through a large opening called the osculum. Many species have several spongocoels and oscula.

Sponges lack **tissues,** groups of specialized cells that perform a specific function, and organs. Instead, they have several special cell types that perform specific functions within the animal. These are shown in Figure 8-1. **Collar cells,** or **choanocytes,** have a flagellum. The movement of the collar cells' flagella provides the force that moves water through the sponge's body. A layer of cells called **pinacocytes** provides an outer covering for the sponge. They also line internal chambers that are not lined by collar cells. **Archaeocytes** are cells that resemble amoebas, and like an amoeba they can move through the sponge's body. Archaeocytes can form any of the cell types in the sponge body and

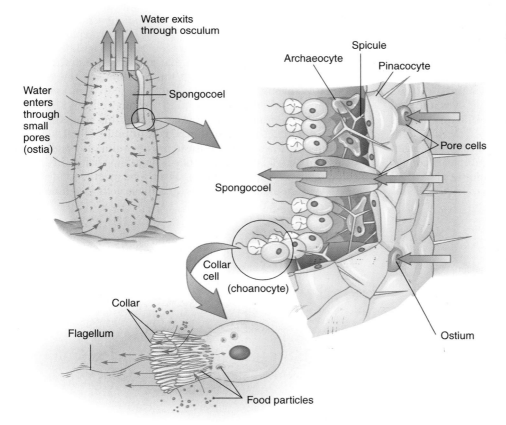

Water exits through osculum

Water enters through small pores (ostia)

Spongocoel

Spongocoel

Collar cell (choanocyte)

Collar

Flagellum

Food particles

Spicule

Archaeocyte

Pinacocyte

Pore cells

Ostium

**Figure 8-1 Anatomy of a Sponge.** A sponge's body has many specialized cells, but no tissues.

play an important role in repair and regeneration. They also transport food and other materials within a sponge's body.

Other specialized cell types produce **spicules** (see Figure 8-1), skeletal elements that give support to a sponge's body. Spicules may be composed of calcium carbonate, silica, or a protein called spongin. **Spongin** forms flexible fibers, allowing the dried sponge skeleton to be flexible as in commercial sponge species (class Demospongia). Sponges known as glass sponges (class Hyalospongia) have spicules made of silica that resemble glass fibers. These are usually fused to form elaborate networks, giving these sponges a delicate appearance. The Sclerospongia is a small group of sponges that live in shaded or dark habitats. They secrete massive skeletons of calcium carbonate, and it is thought that they were the original reef-forming organisms, having evolved before the corals.

## Sponge Size and Body Form

The size of a sponge is limited by its ability to circulate water through its body, and this is determined by the body form (Figure 8-2). The **asconoid** form is the simplest. It is tubular and always small. Sponges with this type of body are usually not found alone but in clusters. As a sponge with an asconoid shape grows, the volume of the spongocoel increases. This increase in spongocoel volume is not, however, accompanied by a proportional increase in the surface area available for collar cells. As a result, there are not enough collar cells to move sufficient amounts of water through the sponge to meet its physiological needs. This is the reason asconoid sponges are always small.

During the evolution of sponges, the problem of water flow and surface area was overcome by folding the body wall and, in many species, reducing the size of the spongocoel. The folding increases the surface area of the collar-cell layer, and the reduction of the spongocoel that occurs because of the folding decreases the volume of water that needs to be circulated. This combination of factors results in increased water flow through the sponge. The more efficient movement of water allows the sponge to grow larger. Sponges display various stages in these changes. Sponges that exhibit the first stages of body-wall folding are called **syconoid** sponges. In syconoid sponges the body wall has

become folded to form internal pockets that are lined with collar cells. The highest degree of folding takes place in **leuconoid** sponges. In these sponges there are many chambers lined with collar cells and the spongocoel is frequently reduced to a series of water canals leading to an osculum. In some leuconoid sponges the number of chambers lined by collar cells may be enormous. For example, the sponge *Microciona prolifera* contains 10,000 chambers per cubic millimeter. Because the leuconoid body plan is the most efficient arrangement, most sponge species are built on this plan. The efficiency of the leuconoid body plan can be seen in *Leuconia,* one of the smaller leuconoid sponges. This sponge is 10 centimeters (4 inches) tall and 1 centimeter (⅓ inch) wide, contains approximately 2,250,000 chambers, and pumps about 22.5 liters (5 gallons) of water per day through its body. It is no wonder that all of the largest sponges have a leuconoid body type.

## Nutrition and Digestion

Because sponges feed on material that is suspended in seawater, they are called **suspension feeders.** They are also referred to as **filter feeders** because they filter their food from the water. Large food particles (1 to 50 micrometers, or 0.00004 to 0.002 inch) are engulfed and digested by pinacocytes and archaeocytes along the sponge's system of canals that carry water from the ostia to spongocoels. Approximately 80% of a sponge's food is trapped by the collar cells and consists of smaller particles (0.1 to 1.0 micrometers, or 0.000004 to 0.00004 inch). Each oval-shaped collar cell that lines the chambers within a sponge has a single flagellum and a collar of fingerlike cellular projections that strain food from the water (see Figure 8-1). The beating of their flagella creates a current that draws water in through the ostia and expels it through the osculum. As the water circulates through the sponge, bacteria, plankton, and detritus are engulfed and digested by various cells.

Very few animals can capture food of this size, and the ability of sponges to utilize this largely untapped food source accounts for the success of these sessile animals. Food particles can be transferred to other cells in the sponge by both collar cells and archaeocytes. Most food particles are digested within the archaeocytes, and these cells also function

**Figure 8-2 Sponge Body Forms.** Sponges exhibit one of three possible body plans: asconoid, syconoid, or leuconoid. Blue arrows represent the direction of water flow. Areas occupied by collar cells are shown in black.

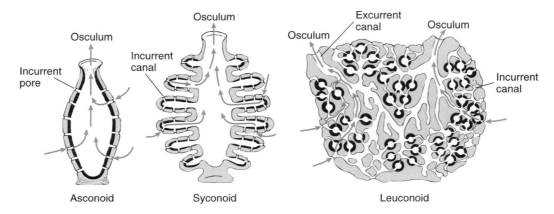

Asconoid          Syconoid          Leuconoid

in the storage of food. Undigested material and nitrogenous wastes leave the sponge with the exiting water currents.

## Reproduction in Sponges

Sponges can reproduce both sexually and asexually (Figure 8-3). Asexual reproduction can involve either budding or fragmentation (essentially the same process that we saw in algae in Chapter 7). In **budding,** a group of cells on the outer surface of the sponge develops and grows into a tiny new sponge. After attaining a certain size, this new sponge drops off and can establish itself near its parent or float with the currents to settle and mature elsewhere. Asexual reproduction by budding is not common. Fragmentation involves the production of a new sponge from pieces that are broken off by physical processes such as waves or storms or by predators. Because the sponge is rather loosely organized anatomically, pieces of sponge can form new sponges if they contain enough of the various cell types. For years, sponge fishers have taken advantage of this characteristic to replenish sponge beds. Large commercial sponges are cut into pieces, tied to concrete blocks or other material that will anchor them, and then thrown back into the water. The pieces of sponge will then eventually grow into new adult sponges. Sponges exhibit marvelous powers of regeneration, and when a piece is broken off or eaten, the sponge can readily replace the missing part.

Most sponges are **hermaphrodites,** which means they can produce both male and female gametes (sex cells, such as sperm and eggs, see Chapter 5) although usually not both at the same time. Sperm cells are formed from modified collar cells. Eggs usually develop from archaeocytes but may also be formed from collar cells. The stimulus to produce gametes is usually a change in the water temperature or **photoperiod** (the relative amount of light and darkness in a 24-hour period). When a sperm cell enters another sponge, it is engulfed by a collar cell. Both cells lose their flagella and the collar cell then transports the sperm to an egg. In most sponges development of the new sponge to the larval stage occurs in the adult's body, although some species release fertilized eggs to develop in the water column. The larval stage of most sponges is an **amphiblastula** that spends time in the water column as plankton before settling and forming a new adult sponge.

## Ecological Roles of Sponges

Although they are simple, sessile animals, sponges interact with other marine organisms in several ways. They compete aggressively for space with other sessile organisms. They are links in some marine food chains. Sponges form many symbiotic relationships, and they provide habitat for other organisms. Sponges also play an important role in the recycling of calcium.

### Competition

The biggest problem that sponges face is finding enough suitable solid material for attachment. Consequently, their primary competitors for space are corals and bryozoans. Some sponge species produce chemicals that either kill corals or inhibit their growth. Other species, such as the boring sponges (family Clionidae), create their own habitat by boring into corals and dead shells. Some species of crab known as sponge crabs (family Dromiidae) attach pieces of sponge to themselves for camouflage and protection and in the process provide a solid surface for the transplanted sponge species (Figure 8-4a).

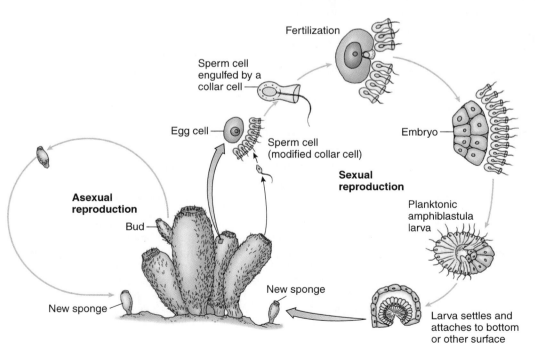

Fertilization

Sperm cell engulfed by a collar cell

Egg cell

Sperm cell (modified collar cell)

Embryo

**Sexual reproduction**

**Asexual reproduction**

Bud

New sponge

New sponge

Planktonic amphiblastula larva

Larva settles and attaches to bottom or other surface

**Figure 8-3 Sponge Reproduction.** Sponges can reproduce asexually (by budding) or sexually.

# Commercial Use of Sponges

Large species of sponges of the class Demospongia are harvested for commercial use (Figure 8-A). Their dried spongin skeletons can hold as much as 35 times their weight in liquid. In the early days of sponge fishing, divers would weight themselves down with heavy rocks while harvesting the sponges. Although today most sponge fishers use diving equipment, sponge divers who fish off the coast of Tunisia still use a lead weight and a rope. For thousands of years, Kalynos, Greece, was a major port for sponge fishers, until a disease killed off the rich sponge beds. Many Greek sponge fishers emigrated to Tarpon Springs, Florida, where their descendants still fish the sponge beds off the Florida coast.

It takes approximately 5 years for a sponge in the wild to reach a marketable size (12.5 centimeters, or 5 inches), and it will retail for about $10.00. Although this is expensive when compared with the price of synthetic sponges, many people are willing to pay the extra money for a sponge that is superior to any synthetic product. For some applications, such as polishing metal, there is no good substitute for a natural sponge.

Sponges may also prove to be sources of novel medications for fighting diseases. A chemical called cytosine arabinoside blocks DNA synthesis in tumors and is used in the treatment of cancer. This substance is isolated from a Caribbean sponge, *Cryptotethya crypta*. Sponges produce antibacterial chemicals and are thus generally free from bacterial infections. These chemicals are being studied to see if they might have uses in the treatment of bacterial diseases of humans and livestock. Other potentially useful drugs from sponges are also being studied. ●

© George Billiris, Sponge Merchant International

**Figure 8-A Commercial Sponge Fishing.** *A sponge fisherman at Tarpon Springs, Florida, is drying some commercial sponges.*

## Predator–Prey Relationships

Very few animals feed on sponges, possibly because ingesting the spicules would be like eating a mouthful of needles. Many sponges produce chemicals that prevent organisms from settling on their surface or that deter grazing. A study of sponges found that 9 of 16 Antarctic sponges and 27 of 36 Caribbean species were toxic to fish. There are a few species of bony fishes and molluscs that will eat sponges, and sea turtles, especially the hawksbill sea turtle (*Eretmochelys imbricata*), feed almost exclusively on them. As much as 95% of the turtle's feces contains glass spicules, indicating the large role of sponges in the turtle's diet. The tough lining of the turtle's mouth and digestive tract prevents the spicules from injuring the animal.

## Symbiotic Relationships

Many sponges are hosts to other organisms. Some of these relationships are mutually beneficial for both the sponge and its symbionts (mutualism), whereas others seem to favor the symbionts without harming the sponge (commensalism). (For more information on symbiosis see Chapter 2.) Some species of sponge host large numbers of mutualistic bacteria that provide food to the sponge while gaining nutrients and protection within the sponge's system of water canals. In fact, in some species, the volume of bacterial cells is twice that of the volume of sponge cells. Cyanobacteria make up 33% of *Verongia,* a sponge that lives in shallow, well-lit habitats. Many species of sponge that live on coral reefs contain symbiotic cyanobacteria in their bodies.

(a)

(c)

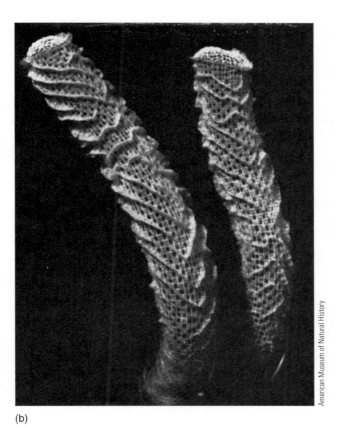

(b)

**Figure 8-4 Types of Sponges.** (a) Sponge crabs camouflage themselves by attaching sponges to their bodies. (b) Venus's flower basket is a deepwater glass sponge from the Philippines. (c) Boring sponges burrow in a mollusc shell.

The spongocoel and internal canals of a sponge can be home to a variety of organisms, such as shrimp and fish that take advantage of the protection and continuous flow of water. For instance, one specimen of loggerhead sponge (*Spheciospongia vesparia*) from Florida was found to contain 16,000 snapping shrimp (*Synalpheus brooksi*)! The glass sponge known as Venus's flower basket (*Euplectella;* see Figure 8-4b) that is found in the deep waters of the tropical Pacific Ocean exhibits an interesting symbiotic relationship with certain species of shrimp (*Spongicola*). A male and a female shrimp enter the sponge's spongocoel when they are young. They feed on plankton in the water that circulates through the sponge's body. Eventually the shrimp grow so large that they cannot escape through the network of spicules that covers the osculum. Their entire life is spent in the sponge. A specimen of this sponge along with the shrimp used to be given in Japanese wedding ceremonies to symbolize lifelong fidelity.

### Sponges and Nutrient Cycling

Members of the family Clionidae (class Demospongia), known as boring sponges (see Figure 8-4c), are important in recycling calcium in the marine environment. These small sponges burrow into coral and mollusc shells. In the process they convert the calcium into a soluble form that is returned to the seawater for use by other organisms.

### In Summary

Sponges depend on their ability to filter large amounts of water through their bodies to survive. Their bodies are asymmetrical and contain several cell types that perform specific functions. Sponges provide habitats for many organisms and play a role in recycling calcium. ●

A sponge provides protection and a sunlit habitat for the cyanobacteria, and the cyanobacteria provide the sponge with nutrients and oxygen. Studies of sponges on the Great Barrier Reef of Australia have found that reef sponges get between 48% and 80% of their food from their symbiotic cyanobacteria. There are also a few species of sponge that contain symbiotic dinoflagellates, a single-cell, photosynthetic organism. (For more information on dinoflagellates see Chapter 6.)

# CNIDARIANS: ANIMALS WITH STINGING CELLS

Even a casual visitor to the seashore can usually identify a jellyfish and knows to avoid contact with one. Many of these creatures can cause injury with their toxic stings. It is not uncommon to find large numbers of these animals close to shore or washed up on a beach after a storm. Jellyfish belong to a group of animals that also includes hydroids, corals, and sea anemones. They are called **cnidarians** (ny-DEHR-ee-uhns; phylum Cnidaria) because of the stinging cell, or **cnidocyte** (NYD-uh-syt), that they all possess. This unique cell is used not only in the capture of prey but also for protection.

## Organization of the Cnidarian Body

Cnidarians have bodies that exhibit **radial symmetry** (Figure 8-5a). This means that many planes can be drawn through the central axis that will divide the animal into two equivalent halves. The body parts of these animals are arranged in a circular pattern around a central axis, much like the spokes of a wheel. Radial symmetry is particularly beneficial to sessile organisms and organisms that are not active swimmers because it allows them to meet and respond to their environment equally well from all sides.

Cnidarians often exhibit two different body plans in their life cycles (see Figure 8-5b and c). The **polyp** (PAHL-uhp) is generally a benthic form characterized by a cylindrical body that has an opening at one end, the mouth, which is usually surrounded by a ring of tentacles. The **medusa** is a free-floating stage that is commonly known as a jellyfish. Many cnidarians exhibit both body plans during their life cycles, although some, such as corals and sea anemones, exist only in the polyp stage. Both the polyp and medusa have an outer layer of cells called the epidermis. Inside the body is a large cavity (gastrovascular cavity) that is lined by a layer of cells called the gastrodermis. Between these two layers is a gelatinous material called mesoglea. Most of the animal's body is the jelly-like mesoglea, which accounts for the common name jellyfish that is given to many cnidarian species.

## Stinging Cells

One of the most important features of cnidarians is their stinging cells (cnidocytes). The stinging cell contains a stinging organelle called the **cnida** (NY-duh). There are more than two dozen types of cnidae (NY-dee, plural form of cnida) found in cnidarians. Some function in locomotion, whereas others function in the capture of prey and in defense. Most of the cnidae are of a spearing type known as **nematocysts** (Figure 8-6) that are located within a capsule in the cell. The capsule has a lid that opens when the nematocyst is discharged. A short, bristlelike structure called the **cnidocil** extends from one end of the stinging cell and acts as a trigger. When the cnidocil comes into contact with prey or some other object, it causes the lid to open and the nematocyst to discharge. The stinging cell can also be triggered through the action of certain chemical substances released by prey organisms. Some nematocysts have a thread

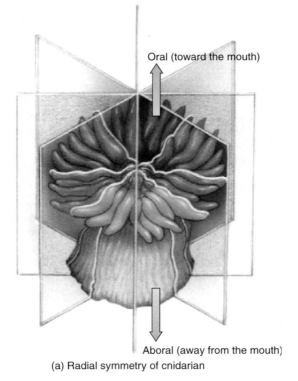

Oral (toward the mouth)

Aboral (away from the mouth)

(a) Radial symmetry of cnidarian

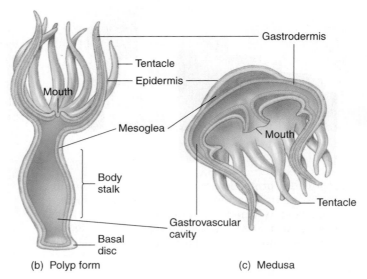

Gastrodermis

Tentacle

Epidermis

Mouth

Mesoglea

Mouth

Body stalk

Tentacle

Gastrovascular cavity

Basal disc

(b) Polyp form

(c) Medusa

**Figure 8-5 Cnidarian Body Plans.** (a) Cnidarians exhibit radial symmetry. Any plane that passes longitudinally through the central axis of the animal will divide it into two equivalent halves. (b) A polyp is one body plan, and (c) a medusa is another.

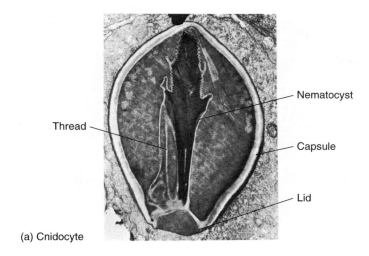

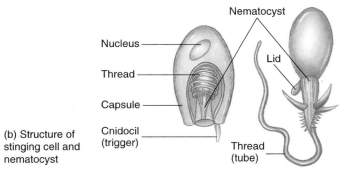

**Figure 8-6 Cnidarian Stinging Cells.** (a) An electron micrograph of a stinging cell (cnidocyte) showes a nematocyst that has not been discharged. (b) The structure of a stinging cell and nematocyst shows its cnidocil, or trigger. *(a, G. B. Chapman, Cornell University Medical College)*

that is used to tangle stung prey. Once the stinging cell has fired, it is reabsorbed by the animal's body, and a new one is formed to take its place. Stinging cells are most common on the tentacles of cnidarians, but they can also be found on the outer body wall and, with the exception of hydrozoans, in the lining of the digestive cavity.

Many species of cnidarian, such as the Portuguese man-of-war (*Physalia*; Figure 8-7a), can cause painful stings. The symptoms include an immediate, intense, burning pain with redness and swelling in the regions of contact. Some individuals may also experience weakness, nausea, headaches, pain, spasms of abdominal and back muscles, dizziness, and increased secretions from the nose and eyes. The sting is generally not fatal to humans unless they are allergic to the toxin; however, the stings of the box jellyfish *Chironex fleckeri* can kill a person within minutes or leave scars that last for life. Death from *Chironex* can happen 3 to 20 minutes after stinging. Nonlethal stings can be severe and slow to heal (see Figure 8-7b). In Australia this animal is commonly known as a stinger and is seasonally abundant around parts of the country, causing the closing of some

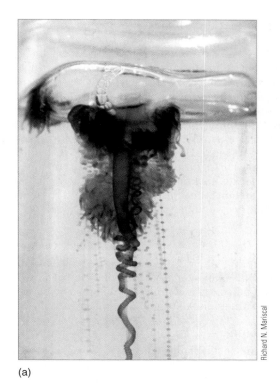

(a)

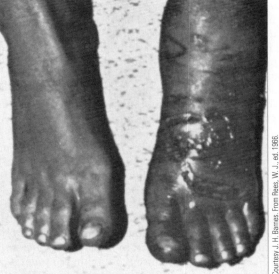

(b)

**Figure 8-7 Jellyfishes.** (a) Although this Portuguese man-of-war appears to be a single jellyfish, it is really a colony of many specialized individuals. (b) This victim's left foot shows 5-day-old injuries produced by a small sting from a jellyfish called a stinger (*Chironex*), or box jellyfish.

swimming beaches. Although it is more common to think of jellyfish stings as being the most harmful, some sea anemone stings can be toxic to humans as well, including the berried sea anemone *Alicia mirabilis* from Europe and *Lebrunia danae* from the Caribbean. The west Australian species *Dofleina armata* is believed to be the most toxic sea anemone.

Not all animals are bothered by the toxic stings of cnidarians. For example, the leatherback sea turtle (*Dermochelys coriacea*) regularly feeds on jellyfish. Some nudibranchs feed on the Portuguese man-of-war and somehow are able to incorporate the nematocysts into their skin for their own protection. Some flatworms can also store nematocysts from their prey and use them in their own defense.

## Types of Cnidarians

Representative cnidarians include hydrozoans, jellyfish, sea anemones, and corals. These marine predators include colonial and solitary benthic organisms as well as some of the largest members of the marine plankton.

### Hydrozoans

**Hydrozoans,** or **hydroids** (class Hydrozoa), are mostly colonial organisms. A **colony** is composed of individual members that are physically connected and adapted to share resources such as food. Most hydrozoan colonies are 5 to 15 centimeters (2 to 6 inches) high, and individual polyps are usually small and inconspicuous. Some exceptions are the deep sea *Branchiocerianthus* with solitary polyps that may reach a length of more than 2 meters (7 feet) and *Corymorpha palmata* from the intertidal mudflats in California that may reach a height of 14 centimeters (5.8 inches). Much of the marine growth that is seen on pilings and rocks along the seashore is composed of hydrozoans (Figure 8-8a).

Colonial forms of hydrozoans usually contain two types of polyp. The **feeding polyp** (gastrozooid) functions in capturing food and feeding the colony. The other type of polyp is the **reproductive polyp** (gonangium), which is specialized for the process of reproduction (see Figure 8-8b). Hydrozoan medusae are usually small. Pulsations of the animal's bell-shaped body drive it upward. This action counteracts the animal's tendency to sink. Although they often turn when swimming, horizontal movement mainly depends on water currents. The medusa of the hydrozoan *Gonionemus* (Figure 8-9) crawls over the bottom, attaching to vegetation with its tentacles. Hydrozoans known as **hydrocorals** (*Millepora* and *Stylaster;* Figure 8-10) secrete a calcareous skeleton around the polyps and resemble hard corals.

Not all hydrozoans form sessile colonies. Members of the hydrozoan orders Siphonophora and Chondrophora produce floating colonies. Perhaps the best known of these is the Portuguese man-of-war (*Physalia;* see Figure 8-7a). The colony is suspended from a large gas sac (up to 30 centimeters, or 12 inches) that acts as a float. It is believed that this structure develops from the original larval polyp, and specialized polyps with gas glands are responsible for keeping the gas sac inflated with a gas similar in composition to air but with a higher concentration of carbon monoxide. Like sessile colonies, these floating colonies contain several types

(a) Colonial hydrozoan

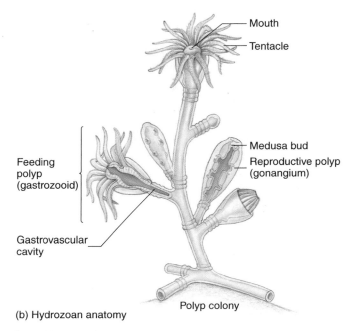

(b) Hydrozoan anatomy

**Figure 8-8 Hydrozoans.** (a) *Tubularia* is an example of a colonial hydrozoan. (b) This drawing of hydrozoan anatomy shows reproductive and feeding polyps.

of specialized polyps. Feeding polyps have a single long tentacle and function in digestion. Fishing polyps have tentacles strewn with stinging cells and are responsible for capturing prey, usually surface fish such as mackerel or flying fish. Other parts of the colony are modified medusae that contain ovaries or testes and function in reproduction.

### Jellyfish and Box Jellyfish

**Scyphozoans,** or true jellyfish (class Scyphozoa; Figure 8-11), can generally exist in both a polyp stage and medusa stage. The predominant stage in the life cycle of these animals is the medusa. Scyphozoan medusae are larger than hydrozoan medusae and are better swimmers. They can swim both vertically and horizontally. Although jellyfish are

**Figure 8-9 Medusa.** Gonionemus is an unusual medusa in that it crawls over the bottom clinging to vegetation rather than floating in the water column.

**Figure 8-12 Jellyfishes.** The stinger from Australia is an example of a box jellyfish.

**Figure 8-10 Hydrocorals.** The fire coral *Millepora* is not a true coral but a colonial hydrozoan that produces a calcerous skeleton.

**Figure 8-11 Scyphozoans.** A common jellyfish in coastal waters is the moon jelly (*Aurelia*). It is not uncommon to see large numbers of these jellyfish on the beach or close to shore after a storm.

capable of swimming by pulsating their bodies, many of them are not strong swimmers, and most float with the currents and are thus part of the plankton. Most medusae are small (less than 10 centimeters, or 3 inches), but some, such as the medusae of the jellyfish *Cyanea,* have a bell diameter of 2 to 3 meters (7 to 10 feet) and tentacles that are 60 to 70 meters (200 to 230 feet) long! Jellyfish have sense organs called **photoreceptors** that allow them to determine if it is dark or light. Many species avoid bright sunlight and come to the surface on cloudy days and at twilight. During bright sunlight and at night they tend to move deeper in the water. Closely related to the true jellyfish are the box jellyfish (class Cubozoa; Figure 8-12). They are voracious predators, feeding primarily on fish. Box jellyfish are tropical animals that have box-shaped bells and are all strong swimmers.

## Anthozoans

Sea anemones, corals, gorgonians, and their relatives are collectively known as **anthozoans** (class Anthozoa). Anthozoans are benthic animals, and all of the adults are sessile. Many species, such as the corals, form large intricately shaped colonies. These animals exhibit only the polyp stage in their life cycle, and many of them bear a resemblance to brightly colored flowers, thus the name anthozoans, which means "flower animal."

***Sea Anemones*** Sea anemones are polyps that are larger, heavier, and more complex than hydrozoan polyps (Figure 8-13a). Unlike hydrozoan polyps, the gastrovascular cavity of sea anemones is divided into compartments that radiate out from the central cavity. Most range from 1.5 to 10 centimeters (3/4 to 3 inches) in diameter but *Tealia columbiana* on the North Pacific coast of the United States and *Stichodactyla* on the Great Barrier Reef may grow to a diameter of 1 meter (3.3 feet) or more. Sea anemones inhabit both deepwater and shallow coastal water worldwide and are most diverse in the tropics. Most species are found attached to a hard surface such as rocks, shells, or submerged

(a)

(b)

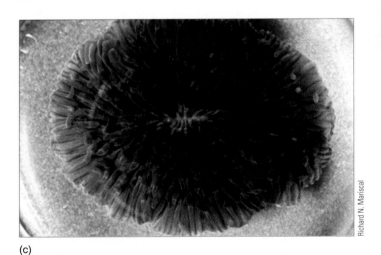
(c)

(d)

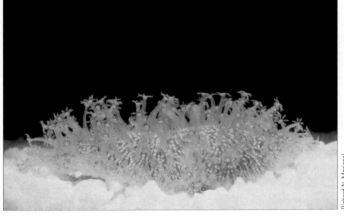
(e)

**Figure 8-13 Anthozoans.** (a) A sea anemone from Puget Sound, Washington; note the young anemones temporarily attached around the column of the parent anemone. (b) Live coral polyps from the Caribbean with their tentacles extended for feeding. (c) Mushroom coral (*Fungia*) is an example of a large solitary coral polyp. (d) An orange sea pen (*Ptilosarcus*) from Puget Sound, Washington. (e) The sea pansy (*Renilla*), which is common along both coasts of the United States, has a body composed of a large primary polyp that gives rise to several small secondary polyps.

wood, but some burrow in sand or mud. Sea anemones are capable of expanding, contracting, and reaching with tentacles to capture their prey. When disturbed, sea anemones will withdraw their tentacles into their oral openings and contract their bodies. Although essentially sessile, many species of sea anemone can change location by gliding on their base, by crawling on the side of their body, or walking on their tentacles. A few species can detach and swim briefly by rapid contractions of their body stalk or tentacles. This behavior is used to escape predators such as sea stars. Some species exhibit aggressive behavior toward other species of sea anemone or clones of themselves. Cnidocytes on specialized searching tentacles fire on contact with foreign anemones. Both animals then move away from each other. This behavior may provide for some spatial separation to avoid crowding.

***Coral Animals***    Coral animals (see Figure 8-13b) are polyps that, unlike sea anemones, secrete a skeleton, which may be hard or soft, around their bodies. The hard or stony corals (**scleractinian** corals) produce a skeleton of calcium carbonate, and the majority of species form large colonies of small polyps. All members of the colony are connected by a horizontal sheet of tissue. Coral reefs are the products of reef-building hard corals, coralline red algae, and calcified green algae. We will discuss corals and coral reef formation in more detail in Chapter 15. Some hard corals such as mushroom and feather corals (*Fungia;* see Figure 8-13c) have solitary polyps as large as 25 centimeters (5 inches) in diameter that are similar in structure to sea anemones.

***Soft Corals***    Soft corals form colonies that look more like plants than animals. Sea fans and sea pens (see Figure 8-13d) are just two examples of this type of coral. Soft corals known as octocorals have 8 tentacles instead of the 12 or more that are typical of other coral species, and the tentacles are feathery. The most common octocorals are gorgonians (order Gorgonacea), such as whip corals, sea feathers, sea fans, and pipe organ coral. Octocorals are common members of reef fauna, especially in the Caribbean. Sea pens and sea pansies (order Pennatulacea) are inhabitants of soft bottoms. They have a large primary polyp with a stemlike base for attachment. The body is fleshy and the upper part gives rise to small secondary polyps. The sea pansy (*Renilla;* see Figure 8-13e) is common along the Atlantic, Gulf, and southern California coasts. They live near shore, where their flattened horizontal surface reduces resistance to turbulent water.

## Nutrition and Digestion

Cnidarians digest their prey in a central cavity called the **gastrovascular cavity.** The gastrovascular cavity functions in both digestion and the movement of materials within the animal. Waste products and indigestible material are forced back out through the mouth after digestion is completed.

Many hydrozoans and anthozoans are suspension feeders, feeding on plankton and organic material floating in the water. Feeding polyps in the hydrozoan colony capture and ingest prey and provide food for the colony. Most colonies feed on any zooplankton small enough to be handled by their feeding polyps. However, the hydrozoan *Rhizophysa eysenhardti,* from the order Siphonophora, which you may recall is a producer of floating colonies, reaches a length of 1 meter (3.3 feet) and has up to 28 feeding polyps. *Rhiophysa eysenhardti* feeds during the day and consumes only fish larvae, approximately 9 fish larvae (5 to 15 millimeters long) per day.

Jellyfish and box jellyfish are carnivorous, feeding mostly on fish and larger invertebrates. The prey is paralyzed by a toxin on the nematocyst and, in some cases, the nematocyst has a long fiber that entangles the prey. The prey is drawn into the mouth and forced into the gastrovascular cavity where it is digested. Some jellyfish, however,

**Figure 8-14  Upside-Down Jellyfish.** The upside-down jellyfish (*Cassiopeia*) feeds on plankton that it traps with its tentacles. It also receives nutrients from symbiotic algae that grow in its tissues.

such as the upside-down jellyfish (*Cassiopeia;* Figure 8-14), feed on plankton that are trapped in mucus produced by modified tentacles. These animals lack a true mouth. The food is passed along fused tentacles by the action of cilia to multiple openings of the gastrovascular cavity, where it is digested. Unlike typical jellyfish, the upside-down jellyfish spends little time swimming. Instead, it usually lies upside down in shallow lagoons, an ideal place for sunlight to reach the symbiotic algal cells that provide nutrition and oxygen to it. The margin of its bell pulsates regularly, drawing a current of water containing plankton across its tentacles. The contractions of the bell also help to hold the animal in place so it is not swept away by strong tidal currents.

Sea anemones generally feed on invertebrates, although some large species can capture fish, and nutrition from symbiotic algae is important to shallow-water species. Many intertidal species feed on crabs and bivalves that are washed down by waves. The prey is captured by the tentacles, paralyzed by nematocysts, and carried to the mouth. Some large species of sea anemone with short tentacles feed on fine organic particles and plankton. Sea anemones contain a groove within their body that is lined by cells with cilia (for more information on cilia see Chapter 5) on their free surface. The beating of the cilia creates a water current through the animal's gastrovascular cavity that brings in oxygen and removes waste. Corals feed like sea anemones and, depending on the size of the polyp, the prey ranges from zooplankton to small fish. In addition to capturing zooplankton, many corals also collect fine particles in mucous films or strands. Cilia move the coral mucus and the trapped food to the animal's mouth for digestion. Although there are exceptions, most Caribbean corals feed at night, whereas Indo-Pacific corals feed both day and night.

## Reproduction

Cnidarians can reproduce both asexually and sexually, and they exhibit a variety of reproductive strategies in their life cycles. Asexual reproduction generally occurs in the polyp

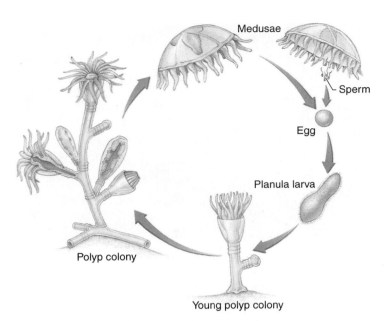

**Figure 8-15 Life Cycle of the Hydroid *Obelia*.** The sexual stage of the life cycle is the medusa. Fertilized eggs develop into planula larva that spend time in the water column before settling and growing into a polyp. The polyp reproduces asexually to form a new colony.

stage and results in the formation of more polyps or tiny medusae. Sexual reproduction usually occurs in the medusa stage.

## Hydrozoans

Hydrozoans generally exhibit both an asexual polyp stage and a sexual medusa stage in their life cycles (Figure 8-15). Reproductive polyps in hydrozoan colonies asexually reproduce by forming tiny medusa-like buds. When mature, the buds are released into the water column where they grow into adult medusae. Medusae can be either hermaphrodites

or the sexes can be separate. Adults release their gametes into the water column, where fertilization and early development occur. The fertilized egg develops into a planktonic larva called a **planula larva** that grows in the water column. The planktonic larva allows these sessile animals to disperse their young to other areas. Eventually the planula settles on a solid surface and begins to form a new colonial hydroid by asexual reproduction.

## Scyphozoans

In adult jellyfish and box jellyfish the sexes are generally separate. The medusae are the sexual stage in the life cycle, and they release gametes into the water column, where fertilization and early development take place (Figure 8-16). The fertilized eggs develop into planula larvae that spend time in the water column before settling on a solid surface to form the polyp stage of the life cycle. The polyps reproduce medusa-like buds by asexual reproduction. The immature buds are released into the water column to grow into mature medusae.

## Anthozoans

Asexual reproduction is common in anthozoans. Sea anemones can reproduce asexually in several ways. **Pedal laceration** involves leaving parts of the base, also known as a **pedal disk,** behind to grow into new individuals. Some species can pinch off parts that will develop into new anemones. In **fission,** the anemone splits into two and each half grows into a new individual. Most sea anemones are hermaphroditic but produce only one type of gamete each reproductive period. Fertilization can occur either in the gastrovascular cavity or in the water column. The larval stage is a planula larva that is a member of the plankton before settling on a solid surface to develop into a new adult.

Large colonies of hard corals are composed of identical individuals that have been produced asexually by budding. Coral species usually have separate male and female forms, although some species are hermaphrodites.

**Figure 8-16 Moon Jellyfish.**
The life cycle and anatomy of the moon jellyfish (*Aurelia*), a typical jellyfish, is shown here.

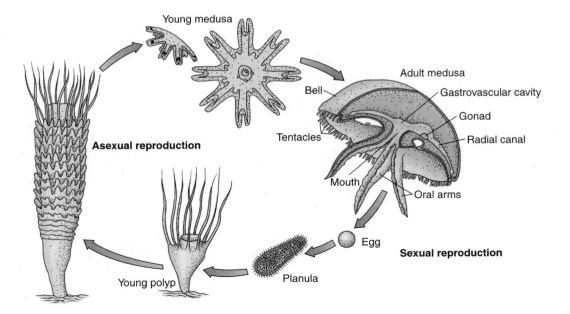

Gametes are shed into the water column, where fertilization and early development occur. The larval stage is the planula larva. After feeding and growing in the water column, the planula settles out on a hard surface and begins to form a new coral colony. (For more information on coral colonies see Chapter 15.)

## Ecological Relationships of Cnidarians

Cnidarians play several key ecological roles in the marine environment. They are predators that feed on a variety of prey, both large and small. Sessile cnidarians, such as corals, provide habitats for numerous marine organisms. Many cnidarians host symbionts that aid in their nutrition and help them grow. Some cnidarians are symbionts of other marine organisms, trading protection for mobility.

### Predator–Prey Relationships

Cnidarians are predators, and because they are well protected by their stinging cells, few predators prey on them. Sea turtles, some fish, and molluscs will feed on hydrozoans and jellyfish. The crown-of-thorns sea star (*Acanthaster*) is one of the chief coral predators on Pacific reefs.

### Habitat Formation

Coral polyps are some of the most ecologically important animals in the sea. They form complex three-dimensional structures that provide habitat for thousands of other organisms. Coral reefs provide a solid surface for sessile marine animals and places for pelagic animals to rest and hide. They act as a buffer to protect coastal organisms from the effects of wave action and storms.

### Symbiotic Relationships

There are many interesting examples of symbiotic relationships involving cnidarians. One is the Portuguese man-of-war (*Physalia*) and the man-of-war fish (*Nomeus*). The man-of-war fish swims among the tentacles of the Portuguese man-of-war where, unharmed by the hydrozoan, it is safe from predators. The fish may serve as bait to lure other fishes into the tentacles of the man-of-war. (For more information on this symbiotic relationship see Chapter 17.)

Reef-forming corals harbor a type of algae called zooxanthellae in their tissues (Figure 8-17a). These symbionts not only provide food to the corals but they also are an important food source for other reef animals. Parrotfishes (family Scaridae), for instance, consume large numbers of polyps to feed on their symbiotic algae. (For more information on corals and zooxanthellae see Chapter 15.)

Symbionts of sea anemones include clownfish, cleaning shrimp, snapping shrimp, arrow crabs, other small crustaceans, and brittle stars. Certain clownfishes (family Pomacentridae) form symbiotic relationships with some of the large species of sea anemone that occur in the tropical Pacific Ocean. These sea anemones secrete fish attractants that differ for each species of anemone and attract only a particular species of clownfish. After a brief period of acclimation, the fish can swim unharmed among the anemone's tentacles, feeding on scraps from the prey captured by the anemone, which the fish themselves may have lured. (For more in-

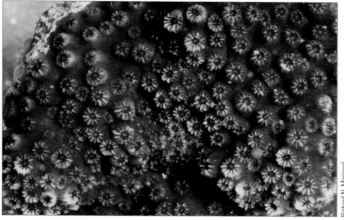

(a)

(b)

**Figure 8-17 Symbiosis.** (a) The dark brown color of this live coral is due to the presence of symbiotic dinoflagellates called zooxanthellae. The zooxanthellae provide nutrition for the coral polyp as well as other reef animals. (b) Sea anemones (*Calliactus tricolor*) attached to the shell of a hermit crab (*Petrochirus diogenes*) from the Gulf of Mexico are another example of symbiosis. Note also the porcelain crab (*Porcellana*) at left, which is commonly found with this species of hermit crab.

formation on this relationship see Chapter 2.) One of the more unusual symbiotic relationships involving anemones is between the hermit crab, *Eupagurus prideauxi,* and the anemone, *Adamsia palliata* (see Figure 8-17b). A young sea anemone will attach itself to the mollusc shell occupied by the young hermit crab. The sea anemone is then carried around by the crab. The anemone's mouth is located just above and behind the crab's, and it feeds on the scraps the crab leaves. Eventually the anemone overgrows the mollusc shell and absorbs it, replacing it with its own body. At this point, the sea anemone and crab both grow at the same rate so that the crab remains covered. Not only does the crab benefit by being camouflaged and protected but it also does not have to find new shells as it grows, as do other species of hermit crab. The sea anemone benefits by being carried from place to place and sharing the crab's food. If the sea anemone is removed from the crab, it will soon die, and the undisguised crab is easy prey. The hydrozoan *Hydractinia echinata* also lives on shells inhabited by hermit crabs. The crabs benefit from the camouflage and the protection of the hydrozoan's venomous stings. The hydrozoan benefits by being carried around from one feeding ground to another, and it does not have to compete for space with other hydrozoan colonies.

## In Summary

Cnidarians exhibit radial symmetry. Cnidarians have evolved a highly specialized stinging cell that they use for capturing prey and for defense. ●

## CTENOPHORES

Although jellyfish are the most familiar of the gelatinous animals that float in the ocean's water, they are not the only ones. Comb jellies are radial animals that closely resemble jellyfish. Their transparent bodies make them difficult to see in the water, and unlike jellyfish, comb jellies lack stinging cells.

### Ctenophore Structure

**Ctenophores** (TEEN-uh-forz), or comb jellies (phylum Ctenophora), are planktonic, nearly transparent, marine animals. They are named for the eight rows of comb plates (ctenes) the animal uses for locomotion (Figure 8-18a). The comb plates are made of very large cilia, and when the cilia of all eight rows beat, the animal travels in a forward direction, mouth first. Ctenophores are weak swimmers and are mostly found in surface waters. Like cnidarians, ctenophores exhibit radial symmetry, but they lack the stinging cells that are the hallmark of cnidarians. The delicate bodies of ctenophores are iridescent during the day. At night, almost all ctenophores give off flashes of luminescence, possibly to attract mates or prey or frighten potential predators. Along with other bioluminescent plankton, they are responsible for the luminescence of many seas.

### Digestion and Nutrition

Ctenophores are carnivorous, feeding on other planktonic animals. Some ctenophores, such as *Pleurobrachia* (see Figure 8-18b), capture prey with branched tentacles that form a large net when fully expanded. Prey, especially small crustaceans, are caught on specialized adhesive cells and then pulled into the mouth, where they are digested in the gastrovascular cavity. Other species, such as *Mnemiopsis* and *Leucothea* use both tentacles and their mucous-covered oral surfaces to capture prey, especially small crustaceans. The ctenophore *Beroe* is cylindrical and has no tentacles. It feeds on other ctenophores. When its mouth comes in contact with another ctenophore, the prey is sucked up whole. The ctenophore *Euchlora rubra* feeds on jellyfish and saves the nematocysts. They move the stolen stingers to their tentacles

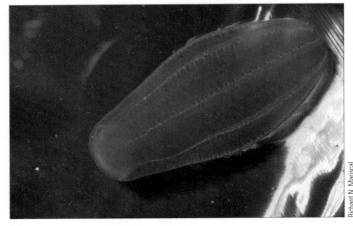

(a)

Richard N. Mariscal

(b)

National Geographic/Getty Images

**Figure 8-18 Ctenophore.** (a) Ctenophores, also known as comb jellies, propel themselves through the water with specialized structures called comb plates. (b) Ctenophores such as *Pleurobrachia* use their tentacles to capture food.

and use them instead of adhesive cells for capturing prey and for defense

### Reproduction

Almost all species of ctenophore are hermaphroditic. Most species appear to shed their eggs and sperm directly into the water column, where fertilization takes place. A few species, though, brood the eggs in their bodies. Fertilized eggs develop into a free-swimming larva called a **cydippid larva** that resembles the adult ctenophore.

### In Summary

Ctenophores exhibit radial symmetry. Ctenophores lack the stinging cell of cnidarians and move by rows of cilia called comb plates. ●

# MARINE WORMS

A large number of marine animals have elongated bodies that for the most part lack any kind of external hard covering. These animals are generally referred to as worms. Most worms gain support for their body from fluid (usually water) that is contained in a body compartment. This condition is known as a hydrostatic skeleton. There is considerable diversity among marine worms. Some are flat, and others are round. Some are small, and others are quite large. Although most are benthic, some have adapted a pelagic lifestyle.

## Flatworms

Flatworms (phylum platyhelminthes) have flattened bodies with a definite head and posterior end. Some flatworms, such as the **turbellarian flatworms** (class Turbellaria), are free-living, whereas others, such as **flukes** (class Trematoda) and **tapeworms** (class Cestoda), are parasites. Flatworms exhibit bilateral symmetry (Figure 8-19a). In **bilateral symmetry,** the body parts are arranged in such a way that only one plane through the midline of the central axis (**midsagittal plane**) will divide the animal into similar right and left halves. Bilateral symmetry allows for a more

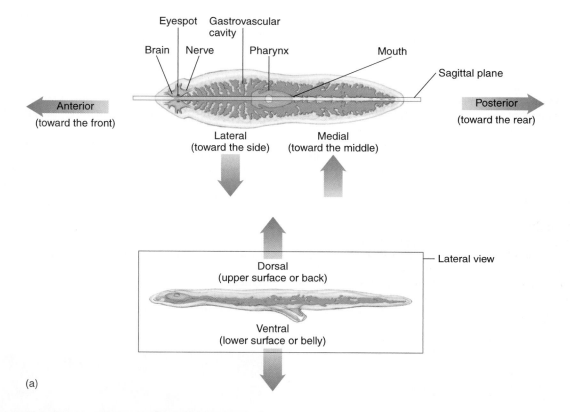

(a)

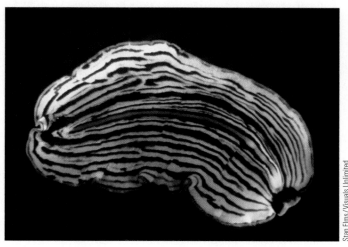

Stan Elms/Visuals Unlimited

(b)

**Figure 8-19 Flatworms and Bilateral Symmetry.** (a) The general anatomy of a turbellarian flatworm. Most animals, like this flatworm, exhibit bilateral symmetry. That is, only one plane (the midsagittal plane) passing through the midline of the long axis will divide the animal into two equivalent halves. Animals that exhibit bilateral symmetry have their sense organs concentrated in one region called the *head*. (b) A marine flatworm from the Pacific Ocean.

streamlined body shape and is more advantageous for animals with active lifestyles than radial symmetry.

The active lifestyle of animals with bilateral symmetry favored the concentration of sense organs at one end of the animal, the end that would first meet the environment. Turbellarians have sensory receptors in their head region that can detect light, chemicals, and movement and that function in maintaining balance. This evolutionary trend toward concentration of sense organs in the head region of an animal is known as **cephalization** (sef-uh-luh-ZAY-shun).

### Types of Flatworm

Turbellarian flatworms range from a few millimeters to 50 centimeters (from less than ⅛ inch to 20 inches) long but only a few centimeters wide and 1 millimeter thick (see Figure 8-19b). A few species are pelagic but most are bottom dwellers that live in sand or mud, under stones and shells, or on seaweed. Flatworms are also common members of the **meiofauna** (MY-oh-fawn-uh), the tiny invertebrates adapted to living in the spaces between sediment particles. A flatworm's body is covered with a layer of cells called the epidermis. The ventral surface of the epidermis is frequently ciliated and contains gland cells that produce mucus. Very small flatworms can swim or crawl by using their cilia. Larger flatworms secrete slime trails of mucus over which they glide, using their cilia to propel them. This form of locomotion is not efficient for larger forms, which also use muscle contractions to propel themselves.

Flukes and tapeworms are parasitic flatworms. Flukes usually have complex life cycles, frequently involving one or more species as an intermediate host. Tapeworms are parasites that live in the digestive tract of their host. The largest tapeworms are those that infest whales. Individuals of the species *Polygonoporous giganteus* found in the intestines of sperm whales have measured as long as 30 meters (99 feet)!

### Nutrition and Digestion

Most turbellarians are carnivorous, feeding on small invertebrates that they locate with chemical-detecting organs called **chemoreceptors.** Small crustaceans, snails, and annelid worms are common prey. They can also feed on detritus or the bodies of dead animals that sink to the bottom. The oyster leech *Stylochus frontalis* feeds on live oysters and *Stylochus triparitus* feeds on barnacles. The commensal flatworm *Bdelloura* lives on the book gill of horseshoe crabs and shares the food of its host. (For a description of commensalism see Chapter 2.) Some species feed on algae, especially diatoms. The turbellarian *Convoluta roscoffensis* relies on zooxanthellae for nutrition and does not ingest food as long as it has the symbiotic algae.

Some species subdue their prey by entangling it in mucus and pinning it against a solid surface until it suffocates. A few are known to stab prey with a penis that terminates in a hard, sharp point, or **stylet,** and projects from the mouth. Once the prey is captured, the animal extends a muscular tube called the **pharynx** out of its mouth, pumps out enzymes onto the prey, and sucks out body fluids or sucks off pieces of food. Turbellarians without a pharynx swallow their prey whole. The food is digested in the gastrovascular cavity. Some species have gastrovascular cavities with lateral branches. This helps increase the surface area for digestion and absorption and compensates for the lack of an internal transport system.

### Reproduction

Flatworms can reproduce asexually and can regenerate missing body parts. Sexually, flatworms are hermaphrodites. When two animals copulate, they usually fertilize each other. This process is known as **reciprocal copulation.** Fertilization is internal and generally development is direct, that is, there is no larval stage. Some species have a free-swimming, planktonic larva. Because of their smallness, turbellarians produce few eggs. By comparison, parasitic flatworm egg production is 10,000 to 100,000 times greater than that of the free-living turbellarian worms.

## Nematodes

**Nematodes** (phylum Nematoda), also known as roundworms, are the most numerous animals on earth. Although the number of species described is only about 12,000, the number of individuals representing those species is staggering, and nematodes can be found in virtually every conceivable habitat. Their bodies are round, slender, elongated, and tapered at both ends (Figure 8-20). Most nematodes are less than 5 centimeters (2 inches) long, although some parasitic species have been reported that are more than 1 meter (3.3 feet). Nematodes play a critical role as scavengers, and many are important parasites of plants and animals. Nonparasitic nematodes are benthic, living in the interstitial spaces of algal mats and sediments. One square meter of bottom mud off the Dutch coast has been reported to contain as many as 4.4 million nematodes. The great numbers of nematodes that are found in estuaries are sometimes

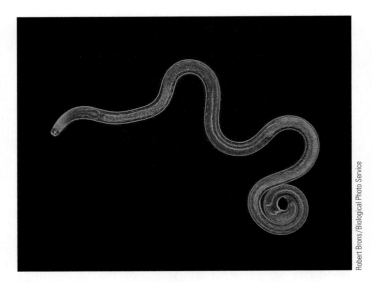

**Figure 8-20 Marine Nematode.** Nematodes are prominent members of the meiofauna. This specimen was taken from bottom sediments in the North Sea.

the result of excess detritus accumulation. This is especially true in small harbors where human waste is abundant. Many free-living nematodes are carnivorous, feeding on small animals, including each other. Others feed on diatoms, algae, fungi, and bacteria. Most are hermaphroditic, but some species have separate sexes.

## Annelids: The Segmented Worms

**Annelids** (phylum Annelida) are worms whose bodies are divided internally and externally into segments that allow them to be more mobile by enhancing leverage (Figure 8-21a). The body wall of annelids contains both longitudinal and circular muscles, allowing them to crawl, swim, and burrow efficiently. The skin of many contains small bristles called **setae** (SEET-ee) that are used for locomotion, digging, anchorage, and protection. Fireworms (family Amphinomidae), which live in coral and beneath stones, have brittle, tu-

bular, calcareous setae containing poison that are used for defense. After penetrating the skin, the setae break off, releasing the poison and causing searing pain. These worms are avoided by fish.

### Polychaetes

The most common annelids in marine environments are **polychaetes** (PAHL-eh-keets; class Polychaeta), but their secretive habits cause them to go unnoticed by most casual visitors to the seashore. Polychaetes are known by common names such as sandworms, clamworms, feather duster worms, tube worms, and lugworms. They range from 1 millimeter to more than 3 meters (10 feet) long, although most are less than 10 centimeters (4 inches). Some are brightly colored red, pink, green, a combination of colors, or iridescent (see Figure 8-21b). They live burrowed in sand and mud, under rocks and corals, in crevices, in shells of other marine organisms, and in tubes they produce themselves.

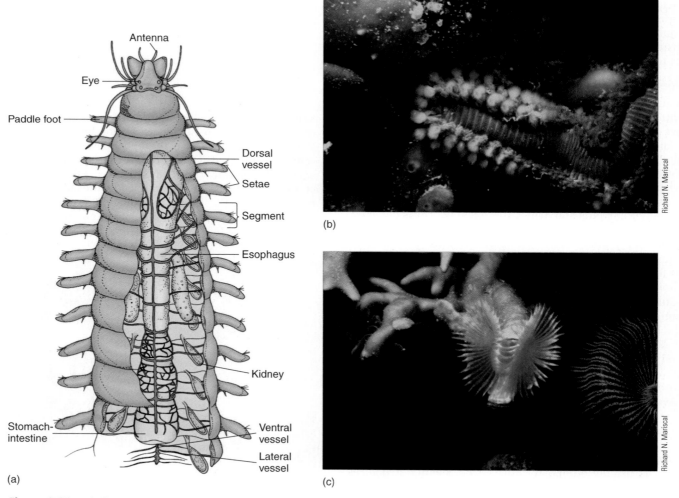

**Figure 8-21 Polychaete Worms.** (a) The general anatomy of a polychaete worm shows its segments. (b) The bearded fireworm (*Hermodice*) is from Bahia Honda in the Florida Keys. (c) The horseshoe worm (*Pomatostegus*), also from the Florida Keys, is a filter-feeder.

Traditionally, polychaetes are divided into two groups: the **errant** (free moving) **polychaetes** that move actively and the **sedentary polychaetes** that are sessile forms. Some errant polychaetes are strictly pelagic, some crawl beneath rocks and shells, some are active burrowers in sand and mud, but some also occupy stationary tubes. Sedentary species construct and live in stabilized burrows or tubes. Some species cannot leave the tube and can project only their head from the opening. Tube worms are sedentary polychaetes that construct their tubes from a variety of materials. Some species sort sand grains from the food particles that they trap and use the sand to enlarge their tubes. Some tubes are made of protein and have a consistency similar to stiff paper (parchment). Others are solid tubes composed of calcium carbonate, whereas still others are built from stones, coral, or sand granules that are held together by mucus. Feather duster, or fan worms (Sabellidae and Serpulidae), live in tubes that they form in or on the bottom sediments or other solid surface. From the opening of the tube emerges a ring of tentacles, each of which is very delicate and frequently brightly colored (see Figure 8-21c). The tentacles are ciliated and used for straining food from seawater as well as providing a surface for gas exchange. They are equipped with light-sensitive organs, and when the worm is disturbed or when a shadow falls on it, it quickly withdraws its tentacles into its tube and out of sight.

## Feeding and Digestion

Some errant polychaetes have mouths that are equipped with jaws or teeth, and they are active predators (Figure 8-22). They tend to be most active at night, when they come out of hiding to search for small invertebrates, including other polychaetes that they capture with their jaws. Tube dwellers may leave the tube partially or completely when feeding, depending on the species. *Diopatra* uses its hood-shaped tube as a lair. Chemoreceptors monitor passing water currents and, when approaching prey is detected, the worm partially emerges from the tube opening and seizes the victim. Species of *Diopatra* may also feed on dead animals, algae, organic debris, and small organisms, such as foraminiferans that are in the vicinity of the tube or become attached to it. Species of *Glycera* live within a system of tubes constructed in muddy bottoms. The system contains numerous loops that open to the surface. Lying in wait at the bottom of a loop, the worm uses its four tiny antennae to detect the movements of prey, such as small crustaceans and other invertebrates. When prey are detected, the worm slowly moves to the burrow opening and seizes the prey with its pharynx. Not all errant polychaetes that possess jaws are carnivores. The jaws may, for example, be used to tear off pieces of algae. These polychaetes generally belong to the same families as the carnivores.

Many of the sedentary burrowers and tube-dwelling polychaetes are filter feeders or suspension feeders. The head is usually equipped with special feeding structures that collect detritus and plankton from the surrounding water. The particles stick to the surface of the feeding structures and are carried to the mouth along ciliated tracts. The crownlike appendages of fan worms (families Serpulidae,

Roger Steene /imagequestmarine.com

**Figure 8-22 Jaws of Errant Polychaetes.** Some errant polychaetes are equipped with jaws that adapt them for a predatory lifestyle.

Sabellidae, and Spirobidae) form a funnel of one or two spirals when expanded outside the end of the tube. The beating of cilia on the surface of the appendages produces a current of water that flows through the feather-like appendages and then flows upward and out. Particles are trapped on the appendages and are carried by the cilia into a groove running the length of each appendage. The polychaete *Chaetopterus* (Figure 8-23) lives in a U-shaped parchment tube

Richard N. Mariscal

**Figure 8-23 Parchment Worm.** The parchment worm (*Chaetopterus*) lives in a U-shaped tube composed of protein. The tube is on the left and the darker colored worm is on the right.

and produces a mucous filter. The mucous film is continuously secreted and is shaped like a bag. Water passes through the mucous bag, which strains suspended detritus and plankton. When large objects are brought in by the water current, the mucous bag is pulled back to let the large objects pass by. The food is continuously rolled into a ball, and when the ball reaches a certain size, the bag is cut loose and rolled up with the food. The ball is then carried to the mouth and the food consumed.

The digestive tract of polychaetes is usually a straight tube extending from the mouth at the anterior end of the worm to the anus located on the last segment. Digested food is absorbed in the animal's intestine. Polychaetes that live in tubes with double openings, such as *Chaetopterus,* use the water current moving through their tube to carry away fecal wastes. Many polychaetes turn around in their tube or burrow and thrust the last segment out of the opening during defecation. Some species produce fecal pellets or strings, which reduce the risks of fouling. Fan worms have a ciliated groove that carries fecal pellets from the anus out of the tube.

Small organic particles from decomposing algae, plants, and animal bodies, as well as planktonic organisms, are continuously settling on the bottom. Feces are also an important source of organic matter that settles to the bottom. This organic material mixes with mineral deposits on the sea bottom and is an important source of food for some animals. Animals that feed on this material are called **deposit feeders.**

**Nonselective deposit feeders** ingest both organic and mineral particles and then digest the organic material, especially the bacteria that grow on the surface of the mineral particles. The remaining organic material and minerals that are not digested are defecated and frequently form piles outside the burrow called **fecal casts,** or **castings.** The small burrowing polychaete *Euzonus mucronatus,* which is about 25 millimeters (1 inch) long and not more than 2 millimeters in diameter inhabits the intertidal zone on the Pacific coast of the United States and forms colonies that occupy large areas of beaches that are protected from waves. In these colonies the number of worms averages 7,500 to 9,000 per square meter. Worms occupying a typical strip of beach 1.6 kilometers (1 mile) long, 3 meters (10 feet) wide, and 30 centimeters (1 foot) thick ingest approximately 14,600 tons of sand each year. Lugworms, members of the family Arenicolidae, are common deposit feeders (Figure 8-24a). They live in an L-shaped burrow whose vertical part opens to the surface. Sand that is rich in organic material is continually ingested. At regular intervals the worm backs up to the surface to defecate castings containing mineral material. **Selective deposit feeders** separate organic material from minerals and ingest only the organic material. Specialized head structures extend over the surface or into the bottom sediments. Deposit material sticks to mucous secretions on the surface of these feeding structures and is then carried to the mouth along ciliated tracts or grooves. Polychaetes such as *Amphitrite* (see Figure 8-24b) and *Terebella* have large clusters of contractile tentacles that stretch over the surface of the bottom sediments. The tentacles of

(a)

(b)

**Figure 8-24 Deposit Feeders.** (a) The lugworm (*Arenicola*) is a deposit feeder. It feeds on the organic material that coats the sand particles it ingests as it digs its burrow. (b) The spaghetti worm (*Amphitrite*) is a selective deposit feeder, feeding on organic matter that sticks to its tentacles.

some large tropical species, such as *Eupolymnia crassicornis,* reach a meter or more in length and move across the surface of the sand like living spaghetti. The deposit-feeding cone worm *Pectinaria* lives buried head down in the sand. It selects clumps of organic material and mineral particles coated with organic matter and, by selecting the organic material specifically, concentrates it from about 32% in the sediment to 42% in its gut. Of the ingested organic matter, only about 30% is used by the worm, and the remainder passes through its gut and ends up in fecal casts.

## Reproduction in Polychaetes

Asexual reproduction is known in some polychaetes. It takes place by budding or division of the body into two parts or into a number of fragments. Polychaetes have relatively great powers of regeneration. Tentacles and even heads ripped off

South Carolina Department of Natural Resources

**Figure 8-25  Red, White, and Blue Worm.** The blue eggs within this worm's body show through the thin body wall.

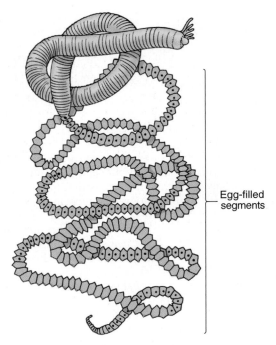

Egg-filled segments

**Figure 8-26  Epitoky.** The epitoke (pelagic reproductive form) of the Samoan palolo worm is shown here. In this species, the anterior part of the worm remains unchanged, while the posterior portion grows to form a chain of egg-filled segments.

by predators are soon replaced. Such replacement is a common occurrence in burrowers and tube dwellers. Some polychaetes, such as *Chaetopterus* and *Dodecaceria,* can regenerate the entire body from a single segment. Cells for regeneration are supplied by the remains of whatever tissues have been lost. Experimental studies indicate that the nervous system is crucial in the regeneration process.

Most polychaetes reproduce only sexually, and the majority of species have separate sexes. Gametes are often shed into the body cavity where they mature. In a mature worm, the body cavity is packed with eggs or sperm, and in species in which the body wall is thin or not heavily pigmented, the presence of the gametes is easily apparent. The blue eggs of the red, white, and blue worm, *Proceraea fasciata* (Figure 8-25), show through the red-banded, whitish body, thus the common name of this species.

There are many different ways for the gametes to be released. A few species have separate, specialized ducts for discharging gametes into the water column. In many species gametes leave through the worm's excretory pores. In some species, such as *Nereis,* gametes are released when the body wall ruptures, although this is probably a specialized adaptation in a few species. Rupturing is more common among polychaetes that become pelagic at sexual maturity. After rupture of the body wall, the adults die.

Some errant polychaetes exhibit an interesting reproductive phenomenon known as epitoky. Epitoky (EP-i-toh-kee) is the formation of a pelagic reproductive individual, or **epitoke,** that is different from the nonreproducing form of the worm. The epitoke is usually most modified in the posterior segments that contain reproductive structures (Figure 8-26). The posterior segments increase in number in the epitoke and are filled with either eggs or sperm. The males and females come to the surface in large numbers at night to shed their sperm and eggs. This behavior, known as **swarming,** brings sexually mature individuals together, increasing the likelihood of fertilization and the species' chances for survival.

Swarming occurs only at specific times of the year and appears to be related to the lunar cycle and tides. The polychaete *Odontosyllis enopla,* which is found in the Caribbean and Bermuda, swarms in the summer about 1 hour after sunset following a full moon. The palolo worm, *Palolo viridis,* an errant polychaete found in the waters of the South Pacific around the islands of Samoa, reproduces in October or November during the last-quarter moon. The palolo worms luminesce when they reach the surface, and when males and females swim around each other releasing gametes, they create small circles of light. The female epitoke releases a substance called **fertilizin** that stimulates the male to shed his sperm. The sperm, in turn, stimulate the females to release their eggs, and fertilization occurs externally. The epitokes provide food for many fish, birds, and some indigenous peoples of tropical islands. In the West Indian palolo worm, *Eunice schemacephala,* swarming takes place in July near the last quarter of the moon. The worm backs out of its burrow in coral, and the posterior epitoke breaks free. The epitoke swims to the surface, and by dawn the ocean surface is covered with them. As the sun rises, the epitokes burst. Fertilization immediately follows rupture and acres of eggs may cover the sea. The fertilized eggs develop into a ciliated

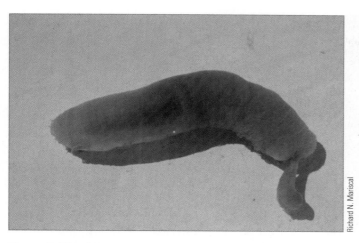

Richard N. Mariscal

**Figure 8-27 Echiuran.** This echiuran worm uses its flat proboscis for feeding.

larval stage by the next day, and in 3 days the larvae sink to the bottom to develop into adult worms.

### Echiurans

**Echiurans,** or spoonworms (class Echiura), are sausage-shaped annelids that resemble sipunculid worms (Figure 8-27). Some occupy burrows in sand and mud. Others live in rock or coral crevices. *Lissomyema mellita* from along the southeast coast of the United States inhabits the remains of dead sand dollars and grooves in discarded mollusc shells. The majority of spoonworms live in shallow water, but some deepwater species are known. They range from 1 centimeter (⅜ inch; *Lissomyema*) to 50 centimeters (20 inches; *Urechis*). Most echiurans are deposit feeders, but at least one, *Urechis caupo* from the California coast, is a filter feeder. *Urechis* forms a mucous net on its burrow wall and then pumps water through at the rate of 1 liter per hour (1.06 quart). Virtually all particles, including typical plankton, are trapped on the net as the water passes through. When loaded with food, the net is detached and swallowed. *Urechis* is known as an innkeeper because of the large number of symbiotic organisms that share its burrow. (For more information on the innkeeper worm see Chapter 14.) Typical deposit feeders such as *Thalassemi* and *Bonnellia* extend their flat, ribbonlike **proboscis** (a tube that extends from the mouth) onto the surface of sediments around their burrows. The proboscis curls, forming a gutter along which deposited particles are carried to the mouth. In echiurans the sexes are separate. When reproducing, they shed their gametes into the water column and fertilization is external. The developing young go through a planktonic larval stage before settling on the bottom and taking up an adult existence.

### Pogonophorans

**Pogonophorans,** or beardworms (class Pogonophora), live in tubes buried in the ocean bottom at depths of 200 meters (656 feet) or more. The body of these animals is cylindrical with a ring of tentacles around the anterior end. Beardworms are unusual in that they lack a mouth and a digestive tract. Evidence suggests that they derive nutrition from dissolved nutrients in the seawater by actively transporting them into their tentacles. Large beardworms (*Riftia*) up to 3 meters (10 feet) long and 5 to 8 centimeters (2 to 3 inches) in diameter are found exclusively in deepwater vent communities. These animals obtain all of their nourishment from large numbers of chemosynthetic bacteria that live in their tissues. (For more information on beardworms and vent communities see Chapter 18.)

### In Summary

Marine worms display bilateral symmetry and cephalization. Flatworms are represented by both free-living and parasitic forms. Turbellarians are free-living flatworms. Flukes and tapeworms are parasitic flatworms. Nematodes are the most numerous animals on earth. They can be parasites or free-living. Annelids are worms whose body is divided internally and externally into segments, allowing them to be more mobile. The most abundant annelid worms in the marine environment are polychaetes. Polychaetes may be either active (errant polychaetes) or sessile (sedentary polychaetes). Some polychaetes are carnivores, whereas others are filter feeders, suspension feeders, or deposit feeders. Polychaetes are important sources of food for other organisms, and burrowing polychaetes play an important role in recycling nutrients. ●

## OTHER MARINE WORMS

Many other wormlike animals inhabit the marine environment, such as ribbon worms, horsehair worms, peanut worms, gastrotrichs, and acorn worms. They differ from those previously discussed in certain anatomical and developmental respects, and as such they are classified into different phyla. Each fills a particular niche in the overall marine ecosystem.

### Ribbon Worms

Ribbon worms (phylum Nemertea) are named because of their ribbonlike bodies (Figure 8-28a). Most are benthic, but some deepwater species are pelagic. At some locations ribbon worms are conspicuous at low tide as they crawl across barnacle and mussel beds. Some species are found coiled under stones at low tide or buried in bottom sediments. Others live in empty mollusc shells, live in seaweed, or swim near the surface. Some form semipermanent burrows in the mud that are lined with mucus or cellophane-like tubes. They resemble flatworms but tend to be longer with thicker bodies. The European species *Lineus longissimus* can reach a length of 30 meters (99 feet). Most are pale, but some are brightly colored with patterns of yellow, orange, red, and green. Some species, including some in the genus *Lineus*, reproduce asexually by fragmentation. The sexes are separate in most species and fertilization is external.

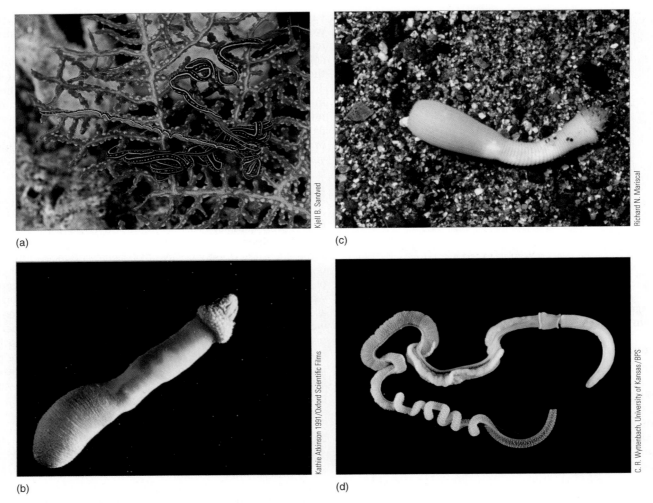

(a)

(c)

Kjell B. Sandved

Richard N. Mariscal

(b)

(d)

Kathie Atkinson 1991/Oxford Scientific Films

C. R. Wyttenbach, University of Kansas/BPS

**Figure 8-28  Other Marine Worms.** (a) A ribbon worm from the Pacific coast of Panama. (b) A peanut worm (sipunculid) from the Great Barrier Reef of Australia. (c) A priapulid worm, which lives buried in sand and mud, from Puget Sound, Washington. (d) An acorn worm from the Atlantic coast of North America.

Ribbon worms are carnivorous and feed primarily on annelids and crustaceans. Prey are captured by a tubelike structure called a proboscis that extends from the mouth. The proboscis coils around the prey and sticky, toxic secretions from the anterior end aid in holding and immobilizing the prey. Some species have a sharp stylet at the tip of the proboscis. Species that have stylets are said to be armed. After the prey is captured, it is either swallowed whole or its tissues are sucked in. The worm *Paranemertes peregrina* of the Pacific coast is an armed intertidal species that feeds on polychaetes. It leaves its burrow and follows the mucous trail of its prey. It must touch the polychaetes for the feeding response to be initiated. Once contact occurs, the proboscis extends and wraps around the prey, stabbing it repeatedly with the stylet. The prey is quickly paralyzed by the injected toxin and then swallowed as the retracting proboscis pulls the prey toward the mouth. After feeding, the worm relocates its own burrow by following its own mucous trail. Other armed ribbon worms feed on small crustaceans. They

kill their prey by piercing the ventral body wall and then forcing their head through the opening. A portion of the digestive tract, the esophagus, is everted through the mouth and the contents of the prey are sucked out and digested. The large, unarmed ribbon worm *Cerebratulus lacteus* of the east coast of the United States enters the burrows of razor clams (*Ensis directus*) from beneath and swallows the anterior end of the clam. Some *Carcinonemerter* species are economically important because they feed on eggs brooded by commercially valuable female crabs.

## Sipunculids

**Sipunculids** (phylum Sipuncula; see Figure 8-28b) are solitary marine worms that live in burrows in mud or sand, in empty mollusc shells, or in coral crevices. When disturbed, these animals contract their bodies so that they resemble a peanut kernel, which gives rise to the common name for some of these species, peanut worms. They range from 2 millimeters to 72 centimeters (28 inches) long, although most are less than 10 cm (4 inches). All are benthic and the majority live in shallow water. Some such as *Sipunculus* live in sand and mud and burrow actively, others inhabit mucous-lined burrows, still others live in coral crevices, empty mollusc shells, annelid tubes, or other protected areas. Several species bore into coralline rock, and at least one bores into wood. Borers direct the anterior end of their body toward the opening to feed. Populations of sipunculids as high as 700 per square meter of coralline rock have been reported from Hawaii. In parts of the tropical Indo-Pacific they are consumed for food.

Sipunculids are either suspension or deposit feeders. They have a proboscis with a ring of tentacles that extends from the mouth. Organic material is trapped in the mucus that coats the tentacles, and the food is then drawn into the mouth by ciliary action. Burrowing species such as *Sipunculus* are direct deposit feeders, ingesting the sand and silt through which they burrow. Some rock borers, such as *Phascolosoma antillarum,* spread their tentacles around the opening while suspension feeding. Other hard-bottom species spread tentacles over rock surfaces and ingest deposited material.

The sexes are separate in sipunculids, and they spawn into seawater. Fertilization is external, and some species develop directly into small worms, whereas others go through a larval stage.

## Priapulids

**Priapulids** (see Figure 8-28c) are benthic worms that bury themselves in sand and mud of shallow and deepwater. They range from 0.5 millimeter to 30 centimeters (1 foot) long. Most of the large body species live in cold water ranging north from Massachusetts on the east coast and California on the west coast. They are also found in the Baltic Sea off Siberia and some are found in the Southern Ocean around Antarctica. Tiny species are widely distributed worldwide, including the tropics, and are important members of the meiofauna. Larger species such as *Priapulus* and *Halicryptus* are thought to be carnivorous. Small species such as *Tubiluchus* and *Meiopriapulus* are probably deposit feeders, feeding mainly on bacteria. A few, such as *Macabeus,* may be suspension-feeding carnivores, feeding on soft-bodied, slow-moving invertebrates, especially polychaetes. The sexes are separate and fertilization is external in large-body species but probably internal in small species. The larvae inhabit bottom mud.

## Hemichordates

**Hemichordates,** or acorn worms (phylum Hemichordata), are sessile bottom dwellers (see Figure 8-28d). They live burrowed in the sediments of intertidal mud or sand flats or under stones. An acorn worm uses its large proboscis to collect food from its surroundings. The food is trapped in mucus and then transported by cilia, located in grooves, to the collar and then into the mouth, where it is swallowed. Some species use their proboscis to dig burrows, which usually have two openings. The head extends from one opening for feeding, and the anus deposits a characteristic mound of fecal material around the second opening.

## Gastrotrichs, Nematomorphs, and Acanthocephalans

Gastrotrichs (phylum Gastrotricha) are small worms that inhabit the spaces between sediment particles, the surface of detritus, and surfaces of submerged plants and animals. The intertidal zone of marine beaches may harbor 40 or more species, and as many as 1,000 individuals have been reported in 20 cubic centimeters of sand. They feed on small dead or living organic particles such as bacteria, diatoms, and small protozoa. The prey are sucked into their digestive tract by the pumping action of a muscular pharynx.

Horsehair worms (phylum Nematomorpha) received their name because of their resemblance to the hairs of a horse's tail. As adults, these animals are free-living, but the juveniles are parasites of arthropods such as deep-sea shrimp. Spiny-headed worms (phylum Acanthocephala) are parasites, mostly of fish, birds, and mammals. They get their name from their cylindrical proboscis that has several rows of spines. The proboscis is used to penetrate the intestine of the host, and the damage can be quite traumatic. Spiny-headed worms are frequently troublesome parasites in seals.

## Ecological Roles of Marine Worms

The many different types of marine worm serve in a variety of ecological roles. The burrowing activity of worms aids in nutrient cycling. Many species of marine worm feed on material too small to be gathered by other marine organisms and channel these nutrients to larger animals in marine food chains. Worm burrows provide habitat for several other species, many of which are commensal symbionts of the worms.

### Nutrient Cycling

Many marine worms are burrowing animals and, like other burrowing organisms, play an important role in the cycling of nutrients. Most oceanic production occurs above the bottom sediments in the photic zone, yet most decomposition occurs on or in the benthos. Over time, many of the nutrients released by the process of decomposition become buried in the bottom sediments. As burrowing organisms tunnel through sediments, many of these nutrients are brought back to the surface, where currents and other natural processes can move them to areas where they can be used by producers to make new organic molecules.

### Predator–Prey Relationships

Marine worms are important links in marine food chains. Many consume organic matter that is unavailable to larger consumers. They in turn provide food for larger marine organisms, channeling nutrients up the food chain.

Marine nematodes are the most abundant members of the meiofauna. They feed on microorganisms and detritus that are too small for larger animals. The nematodes in turn provide food for larger heterotrophs such as fishes, birds, and large invertebrates. In food chains, free-living nematodes are important in channeling nutrients from microscopic producers and consumers to larger consumers.

Echiurans may constitute a significant portion of the diet of some fishes. In one dietary study of leopard sharks caught along the California coast, *Urechis* was found to be the largest of any single food. Apparently the sharks use suction to remove the worms from their burrows. *Urechis* and other echiurans provide shelter for many commensal species in their burrows.

Polychaetes are a major source of food for other marine organisms, both invertebrate and vertebrate. Some errant and many sedentary polychaetes are deposit feeders. These animals feed on the bottom sediments, which include deposit material (organic material that settles to the bottom). The organic material is digested, and the mineral portion is eliminated as castings. These organic nutrients are then channeled to larger organisms when polychaetes are consumed by predators.

## Symbiotic Relationships

Tube-dwelling and burrowing polychaetes that are not carnivorous provide a protected and ventilated retreat for many commensal organisms. These commensals include scale worms, bivalves, and crustaceans, especially small crabs. The 80-centimeter-long (32 inch) Maitre d' worm (*Notomastus lobatus*) lives in clay sediments along the southeast coast of the United States. It occupies a permanent burrow that it shares with at least eight commensals including one scale worm, two other polychaetes, three clams, one crab, and one amphipod. Commensal polychaetes are found in many families, but the scale worms (Polynoidae) contain the largest number. They live in tubes and burrows of other polychaetes and crustaceans, with hermit crabs, on echiuran worms, in the grooves in the arms of sea stars, on sea urchins and sea cucumbers, and with other animals. Many display colors similar to those of the host, making them difficult to see.

## Population Dynamics

As we saw in Chapter 2, populations of marine organisms can be limited by either physical or biological factors or a combination of the two. Burrowing and tube-dwelling polychaetes commonly occur in enormous numbers on the ocean floor and compose a major part of the soft-bottom infauna. A study in Tampa Bay, Florida, reported the average density of polychaetes to be 13,425 individuals per square meter, representing 37 species living in the sediments beneath that area. On the upper continental slope and the deep ocean floor, polychaetes make up 40% to 80% of the infauna. In general, such large populations do not appear to be limited by the resources available, at least not in shallow water. Predation and other pressures usually prevent infaunal populations from ever reaching the carrying capacity of the habitat. When areas in the York River estuary of the Chesapeake Bay were protected from predatory fish and crabs by wire cages, more than half of the species in the polychaete population increased to two or more times the population size in unprotected conditions. Studies such as these support the hypothesis that predation by fishes and crabs plays an important role in limiting the size of infaunal worm populations.

## In Summary

Other wormlike invertebrates include nemerteans, sipunculids, pogonophorans, echiurans, priapulids, and hemichordates. Each of these free-living worms contributes to the flow of energy in marine ecosystems. Horsehair worms and spiny-headed worms have members that are important marine parasites. ●

## IN PERSPECTIVE

| Phylum | Representative Organisms | Form and Function | Reproduction | Type of Feeding | Ecological Roles |
|--------|--------------------------|-------------------|--------------|-----------------|------------------|
| Porifera | Sponges | Asymmetric or radial animals that lack organs and tissues | Asexual by budding or fragmentation. Sexually hermaphrodites. Larval stage is the amphiblastula | Filter feeders and suspension feeders | Provide habitat for other organisms, hosts for symbionts, recycle calcium |
| Cnidaria | Hydrozoans, jellyfish, corals, sea anemones, gorgonians | Radial animals with specialized stinging cells called cnidocytes | Asexual by budding or fission. Sexual reproduction, sexes usually separate. Larval stage is a planula larva. | Mostly carnivorous | Important predators, prey for larger animals, hosts for symbionts, corals form habitat |

## IN PERSPECTIVE (continued)

| Phylum | Representative Organisms | Form and Function | Reproduction | Type of Feeding | Ecological Roles |
|---|---|---|---|---|---|
| Ctenophora | Comb Jellies | Radially symmetric animals but without stinging cells. Move by means of comb plates | Sexually reproducing most hermaphrodites. Larval stage is a cydippid larva | Mostly carnivorous | Members of the zooplankton |
| Platyhelminthes | Turbellarian flatworms, flukes, tapeworms | Bilaterally symmetric body, usually flattened | Mostly hermaphrodites | Turbellarians are carnivores and scavengers. Flukes and tapeworms are parasites | Symbionts, predators, prey, parasites |
| Nematoda | Nematodes | Round, bilaterally symmetrical | Sexes are usually separate | Carnivores, herbivores, scavengers, parasites | Important members of the meiofauna |
| Annelida | Polychaetes, spoonworms, beardworms | Bilateral symmetry, body segmented internally and externally | Some reproduce asexually; the majority sexually. Sexes generally separate | Carnivores, herbivores, filter feeders, suspension feeders, deposit feeders, scavengers | Nutrient cycling, important predators, prey, symbionts, hosts |
| Nemertea | Ribbon worms | Bilateral symmetry. Adults have an eversible proboscis used in feeding | Usually sexual and sexes are separate | Mainly carnivores | Nutrient cycling, predators, prey |
| Sipuncula | Peanut worms | Bilateral symmetry. Tentacles that are used for feeding can be retracted into the body | Most reproduce sexually and sexes are separate | Suspension or deposit feeders | Nutrient cycling, prey organisms |
| Priapulida | Priapulids | Bilateral symmetry | Most reproduce sexually and sexes are separate | Mainly carnivores, some deposit feeders | Nutrient cycling, predators, prey |
| Hemichordata | Acorn worms | Bilateral symmetry | Mostly reproduce sexually and sexes are separate | Mostly deposit feeders | Nutrient cycling, predators, prey |
| Gastrotricha | Gastrotrichs | Bilateral symmetry | Hermaphrodites | Deposit feeders | Nutrient cycling |
| Nematomorpha | Horsehair worms | Bilateral symmetry | Reproduce sexually, sexes are separate. Larvae are parasitic on arthropods | Adults have a degenerate digestive tract and are unable to feed | Parasites, nutrient channeling |
| Acanthocephala | Spiny-headed worms | Bilateral symmetry. Adults have a spiny retractable proboscis | Reproduce sexually and sexes are separate | Feed on nutrients from a host animal | Parasites |

## SELECTED KEY TERMS

amphiblastula, *p. 165*

annelid, *p. 179*

anthozoan, *p. 171*

archaeocyte, *p. 163*

asconoid, *p. 164*

bilateral symmetry, *p. 177*

budding, *p. 165*

casting, *p. 181*

cephalization, *p. 178*

chemoreceptor, *p. 178*

choanocyte, *p. 163*

cnida, *p. 168*

cnidarian, *p. 168*

cnidocil, *p. 168*

cnidocyte, *p. 168*

collar cell, *p. 163*

colony, *p. 170*

ctenophore, *p. 176*

cydippid larva, *p. 176*

deposit feeder, *p. 181*

echiuran, *p. 183*

epitoke, *p. 182*

errant polychaete, *p. 180*

fecal cast, *p. 181*

feeding polyp, *p. 170*

fertilizin, *p. 182*

filter feeder, *p. 164*

fission, *p. 174*

fluke, *p. 177*

gastrovascular cavity, *p. 173*

hemichordate, *p. 185*

hermaphrodite, *p. 165*

hydrocoral, *p. 170*

hydroid, *p. 170*

hydrozoan, *p. 170*

invertebrate, *p. 163*

leuconoid, *p. 164*

medusa, *p. 168*

meiofauna, *p. 178*

midsagittal plane, *p. 177*

nematocyst, *p. 168*

nematode, *p. 178*

nonselective deposit feeder, *p. 181*

ostia, *p. 163*

pedal disk, *p. 174*

pedal laceration, *p. 174*

pharynx, *p. 178*

photoperiod, *p. 165*

photoreceptor, *p. 171*

pinacocytes, *p. 163*

planula larva, *p. 174*

pogonophoran, *p. 183*

polychaete, *p. 179*

polyp, *p. 168*

priapulid, *p. 185*

proboscis, *p. 183*

radial symmetry, *p. 168*

reciprocal copulation, *p. 178*

reproductive polyp, *p. 170*

scleractinian coral, *p. 173*

scyphozoan, *p. 170*

sea anemone, *p. 171*

sedentary polychaete, *p. 180*

selective deposit feeder, *p. 181*

sessile, *p. 163*

setae, *p. 179*

sipunculid, *p. 184*

spicule, *p. 164*

spongin, *p. 164*

spongocoel, *p. 163*

stylet, *p. 178*

suspension feeder, *p. 164*

swarming, *p. 182*

syconoid, *p. 164*

tapeworm, *p. 177*

tissue, *p. 163*

turbellarian flatworm, *p. 177*

vertebrate, *p. 163*

## QUESTIONS FOR REVIEW

### Multiple Choice

1. Each of the following statements concerning sponges is true except one. Identify the exception.
   a. Sponges are living animals.
   b. Sponges have asymmetric bodies.
   c. Sponges reproduce only sexually.
   d. Spicules help to give a sponge support.
   e. The size of a sponge is related to the amount of water it can circulate through its body.

2. If a sponge lacked collar cells it would not be able to
   a. form spicules
   b. produce colored pigments
   c. feed
   d. exchange respiratory gases
   e. protect itself

3. Jellyfish use their nematocysts to
   a. capture prey
   b. digest food
   c. coordinate swimming movements
   d. circulate water through their bodies
   e. eliminate wastes

4. Organs known as *comb plates* are found on animals known as
   a. flatworms
   b. nematodes

   c. cnidarians
   d. ctenophores
   e. annelids

5. Which of the following animals exhibits radial symmetry as an adult?
   a. sponge
   b. flatworm
   c. annelid
   d. nematode
   e. jellyfish

6. Worms with segmented bodies are called
   a. flatworms
   b. nematodes
   c. annelids
   d. acorn worms
   e. ribbon worms

7. An example of a parasitic flatworm would be a
   a. polychaete
   b. nematode
   c. planarian
   d. fluke
   e. ribbon worm

8. Most sedentary polychaetes are
   a. herbivores
   b. carnivores

c. filter feeders

d. parasites

e. symbionts

9. Nematodes play an important role in
   a. recycling calcium in marine environments
   b. removing carbon dioxide from seawater
   c. producing oxygen
   d. recycling organic matter
   e. controlling marine parasites

10. Tiny invertebrates that live in the spaces between sediment particles are collectively called
    a. polyps
    b. plankton
    c. medusae
    d. polychaetes
    e. meiofauna

## SHORT ANSWER

1. How does a sponge's body structure affect its size?

2. What role do boring sponges play in marine environments?

3. What are the advantages of bilateral symmetry?

4. What important ecological contribution do burrowing organisms make to the environment?

5. Describe the ecological role of meiofauna.

6. Describe how sponges feed and reproduce.

7. Describe how the cnidarian stinging cell (cnidocyte) functions.

8. Distinguish between selective and nonselective deposit feeders.

9. Why is radial symmetry advantageous to a sessile organism?

10. Distinguish between hydrozoans and scyphozoans.

## Thinking Critically

1. Development of beachfront property frequently increases the amount of sediment in coastal water. How would you expect this to affect sponges and hydrozoans that are living in the shallow water close to shore?

2. Based on the information in this chapter, construct a food web that includes meiofauna, marine worms, and larger predators such as fishes.

3. Explain with examples how symbiotic relationships can allow marine animals to live in habitats where they normally could not survive.

## SUGGESTIONS FOR FURTHER READING

Brownlee, S. 1987. Jellyfish Aren't Out to Get Us, *Discover* (August):42–54.

Flam, F. 1994. Chemical Prospectors Scour the Seas for Promising Drugs. *Science* 266:1324–25.

Hammer, W. M. 1994. Australia's Box Jellyfish: A Killer Down Under, *National Geographic* 186(2):116–30.

Rudloe, J., and A. Rudloe. 1991. Jellyfishes Do More with Less Than Almost Anything Else, *Smithsonian* 21(11): 100–111.

Ruppert, E. E., R. S. Fox, and R. D. Barnes. 2004. *Invertebrate Zoology,* 7th ed. Belmont, Calif.: Brooks/Cole–Thomson Learning.

### InfoTrac College Edition Articles

Carte, B. K. 1996. Biomedical Potential of Marine Natural Products, *Bioscience* 46(4).

Estabrook, B. 2002. Jellies on a Roll: Jellyfish are "Blooming" around the World and for Many Marine Ecosystems, That Means Trouble, *Audubon* 104(3).

Kunzig, R. 1997. At Home with the Jellies, *Discover* 18(9).

Perkins, S. 2002. Model Predicts Chance of Encountering Jellyfish, *Science News* 162(14).

### Websites

#### Sponges

**www.ucmp.berkeley.edu/porifera/porifera.html** Good overview of sponges with nice pictures.

**www.marinelab.sarasota.fl.us/SPONGE.HTM** A good summary of sponge biology.

#### Cnidarians

**http://water.dnr.state.sc.us/marine/pub/seascience/jellyfi.html** Good overview of jellyfish biology and summaries of some common Atlantic species.

**http://www.ucmp.berkeley.edu/cnidaria/scyphozoa.html** A good overview of jellyfish with links to other cnidarian groups. Good photos of cnidarians.

#### Polychaete Worms

**www.scilib.ucsd.edu/sio/nsf/fguide/annelida.html** Great pictures of a variety of different polychaete worms.

**www.ucmp.berkeley.edu/annelida/polyintro.html** A good overview of polychaetes with good pictures.

# 9 Molluscs, Arthropods, Lophophorates, Echinoderms, and Invertebrate Chordates

## Key Concepts

1. Molluscs have soft bodies that are usually covered by a shell.

2. Molluscs are important herbivores and carnivores in the marine environment.

3. Arthropods have external skeletons, jointed appendages, and sophisticated sense organs.

4. Crustaceans make up a majority of the zooplankton that are a major link between phytoplankton and higher-order consumers in oceanic food webs.

5. Echinoderms exhibit radial symmetry as adults.

6. Echinoderms have internal skeletons and a unique water vascular system that functions in locomotion, food gathering, and circulation.

7. Lophophorates and tunicates are important filter feeders.

8. Arrowworms are carnivorous zooplankton.

The invertebrate animals that will be introduced in this chapter are among the most familiar creatures of sea and shore. If you have ever picked up a seashell or sea star or watched a crab scurry across the beach, you have encountered some of these animals. Not only are these animals very familiar, many are commercially valuable. Scallops, oysters, mussels, crabs, lobsters, and shrimp are only some of the invertebrate animals that provide food for large numbers of people. Many of these animals are also threatened as a result of human activities. In this chapter, you will learn more about these animals and the roles they play in marine environments.

## MOLLUSCS

**Molluscs** (phylum Mollusca) are one of the largest and most successful groups of animals. The term *mollusca* (Latin for "soft") refers to the animals' soft bodies, which in most cases are covered by a shell made of calcium carbonate. Molluscs are represented by familiar animals such as chitons, snails, clams, octopods, and squids. They range in size from microscopic snails to the giant squid, *Architeuthis*. These animals can be found in most marine habitats and exhibit a variety of lifestyles. Some molluscs, such as oysters, scallops, and conchs, are food animals for humans, whereas others, such as shipworms, cause commercial damage.

### Molluscan Body

The generalized molluscan body plan consists of two major parts: the **head-foot** and the **visceral mass** (Figure 9-1). As the name implies, the head-foot region contains the head with its mouth and sensory organs and the foot, which is the animal's organ of locomotion. The dorsal visceral mass contains the other organ systems, including the circulatory

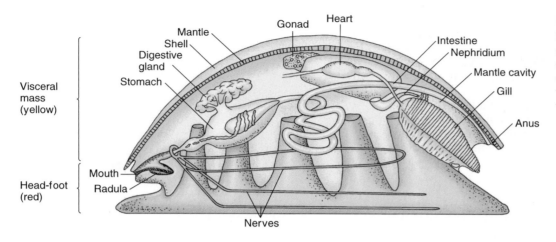

**Figure 9-1 Generalized Molluscan Body Plan.** The generalized molluscan body consists of a head-foot composed of the animal's head and muscular foot and a visceral mass containing the animal's other organs.

(heart and vessels), digestive (stomach, digestive gland, intestine, and anus), respiratory (gill), excretory (nephridium), and reproductive (gonad) systems.

The soft parts of a mollusc are covered by a protective tissue called the **mantle.** The mantle extends from the visceral mass and hangs down on each side of the body, forming a space between the mantle and the body known as the **mantle cavity.** The mantle is responsible for forming the animal's shell in those species that have one. Molluscs such as squids and octopods use the mantle in locomotion, and those molluscan species without gills also use the mantle for gas exchange.

A structure unique to molluscs is the **radula** (Figure 9-2). The radula is a ribbon of tissue that contains teeth and is in all molluscs except bivalves. Depending on the species, the radula is adapted for scraping, piercing, tearing, or cutting pieces of food, such as algae or flesh, which are then moved into the digestive tract. As the anterior portions of the radula are worn out, new teeth are produced at the posterior end. The characteristics of the radula are key in the classification of molluscan species.

## Molluscan Shell

The molluscan shell is secreted by the mantle and normally comprises three layers (Figure 9-3). The outermost layer, the **periostracum,** is composed of a protein called **conchiolin** that protects the shell from dissolution and boring organisms. The middle layer, the **prismatic layer,** is composed of calcium carbonate and protein and makes up the bulk of the shell. The innermost **nacreous layer** is also composed of calcium carbonate, but its crystal structure is different from that found in the prismatic layer, and it is formed in thin sheets.

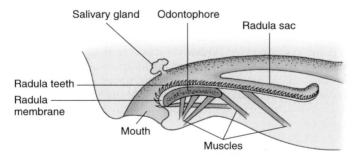

**Figure 9-2 The Radula.** The radula is a ribbon of teeth that is unique to molluscs. It is formed in the radula sac and supported by a mass of muscle and cartilage called the odontophore. Muscles attached to the odontophore move the radula in and out of the animal's mouth.

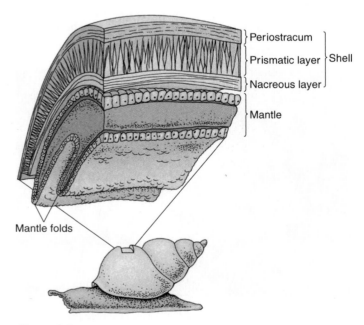

**Figure 9-3 The Mollusc Shell.** The molluscan shell is produced by the animal's mantle and is composed of three layers: the periostracum, prismatic layer, and nacreous layer.

As the animal grows, new periostracum and prismatic layers form at the margin of the mantle. The nacreous layer is secreted continuously and accounts for the increased thickness of the shell in older animals. The iridescent nature of the nacreous layer is due to the arrangement of the calcium carbonate crystals. The thin sheets are a prism, refracting light and producing an iridescent effect. The nacreous layer of oysters is also known as the **mother-of-pearl layer.** Pearls are formed when the nacreous material is layered over sand grains and other irritating particles beneath the animal's mantle.

## Chitons

**Chitons** (KY-tuhnz; class Polyplacophora) have flattened bodies that are most often covered by eight shell plates (Figure 9-4). These plates are held together by a tough girdle that is formed from the mantle, and in some animals, such as *Cryptochiton stelleri* from the Pacific Northwest, the plates are hidden beneath the mantle tissue. Chitons use their large, flat foot to attach tightly to rocks, usually in the intertidal zone. When chitons are removed from a rock, they roll up into a ball for protection.

Most chitons feed on algae and other organisms that they scrape off the surface of rocks with their radulae. An examination of the gut contents of three species of chiton from the coast of Maine revealed the remains of 14 different species of algae and animals, and about 75% of the contents were sediments. The genus *Placiphorella* on the West Coast feeds on small crustaceans and other invertebrates.

## Scaphopods

**Scaphopods** (class Scaphopoda; *scapho* meaning "sheath" and *poda* meaning "foot") are commonly called tusk shells. They derive their name from the shape of their shell, which

**Figure 9-4  Chitons.** Chitons, like this *Tonicella lineata* from California, are usually found on intertidal rocks, where they feed on algae. Their shells are composed of eight separate plates held together by a leathery girdle.

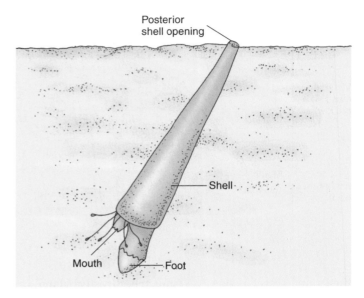

**Figure 9-5  Scaphopods.** Scaphopods, or tusk shells, use their foot to burrow into bottom sediments. They feed primarily on foraminiferans.

resembles an elephant's tusk (Figure 9-5). The tubelike shell is open at both ends, and the animal's foot, which is used for burrowing, protrudes from the larger end. Water enters and exits at the smaller end, bringing in oxygen and removing wastes. They bury themselves in bottom sediments ranging from the intertidal zone to several thousand meters deep, where they feed primarily on foraminiferans. Scaphopods use their foot or special tentacles that emerge from the head to capture their prey.

## Gastropods

**Gastropods** (class Gastropoda; *gastro* meaning "stomach" and *poda* meaning "foot") exhibit a tremendous amount of diversity. Most have shells, but nudibranchs and some others have lost them. Some are so small that they are microscopic, whereas others, like the marine snail, *Syrinx aruanus,* reach lengths approaching 1 meter (3.3 feet).

Gastropods live in a wide variety of habitats, from the littoral zone (shore) to the ocean's bottom, and even pelagic (open ocean) species are known. When they possess a shell, it is always one piece (**univalve**) and it may be coiled (as in snails; Figure 9-6a) or uncoiled (as in limpets; Figure 9-6b). As the animal grows, the **whorls,** or turns, of the shell increase in size around a central axis. Many snails can withdraw into the safety of their shells and close the opening, or **aperture,** with a cover called the **operculum.** The operculum can be horny (stiff protein) or calcareous in composition, and it is not found in all species of snail. Some gastropods, such as limpets, abalones, and slipper snails, are adapted for clinging to solid surfaces such as rocks and shells. All have low, broad shells that can be pulled down tightly and offer less resistance to waves and currents. In these species the animal's large foot is an adhesive as well as movement organ.

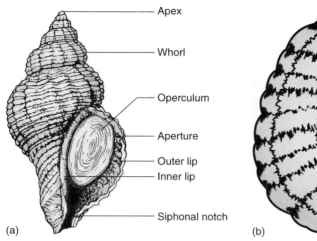

(a)

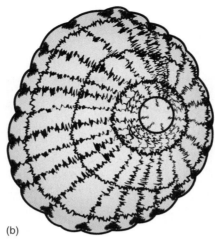

(b)

**Figure 9-6 Gastropod Shells.** Gastropod shells may be coiled (a) as in this oyster drill *Urosalpinx cinerea* or uncoiled (b) as in the limpet *Acmaea.* Each turn of a coiled shell is called a *whorl,* and the whorls wrap around a central axis. Many species have an operculum that is used to close the opening or aperture when the animal withdraws within its shell.

## Feeding and Nutrition

Gastropods exhibit a wide variety of feeding styles. There are herbivores, carnivores, scavengers, deposit feeders, and suspension feeders.

***Herbivores***  Most marine gastropods feed on fine algae that can be scraped from a rock or other solid surface with their radula (Figure 9-7a). A few are adapted to feeding on large algae, such as kelps.

***Carnivores***  Carnivorous gastropods usually locate their prey by following chemical trails in the water. In response to the chemical cues the animal extends its proboscis to search for the food source. The whelk *Buccinus undulatum* can locate a food source as far as 30 meters (99 feet) upstream, but it takes several days for the whelk to reach it.

Gastropods prey on a wide variety of animals, and they have evolved many interesting behaviors to trap and subdue their prey. The flamingo tongue (*Cyphoma*) and its relatives feed on colonial cnidarians, and large snails such as Triton's trumpet (*Charonia*) feed on echinoderms. Helmet snails (*Cassis*) feed on sea urchins. The helmet follows a chemical trail to its prey and pins the urchin down with its heavy shell. With the aid of sulfuric acid secretions, helmets can cut a hole in a sea urchin's shell within 10 minutes. They then insert their mouths into the hole and suck out the contents. Whelks (*Busycon*), tulips (*Fasciolaria*), and murex (family Muricidae) feed on clams. They grip the bivalve with their foot, pulling it open, and then keep it open by wedging the two valves apart with the edge of their shell. They then thrust in their mouth to devour their prey. Oyster drills of the genus *Urosalpinx* can devastate entire oyster beds, boring holes in the shells of their prey and sucking out the contents.

Cone snails (*Conus*) have a radula shaped like a harpoon (see Figure 9-7b). The radula is coated with a toxin that is produced in a poison gland (bulb gland) located near the mouth. These snails feed on worms, fish, and other molluscs, paralyzing their prey with a venomous sting. Six of the approximately 400 species of *Conus* are known to have caused human fatalities.

***Scavengers and Deposit Feeders***  Snails of the genus *Nassarius* are scavengers and deposit feeders. The snail *Ilyanasa obsoleta* can frequently be found in large numbers at low tide on quiet, protected beaches along the Atlantic coast of the eastern United States, where it feeds on organic material deposited by the tide. This same species may also feed on the flesh of freshly dead fish. Other deposit feeders include many common species of horn snail (*Cerithium;* see Figure 9-7c). Conchs (*Strombus*) are deposit feeders as well as grazers on fine algae. They use their large, mobile proboscis to sweep across the bottom like a vacuum cleaner, sucking up food.

***Filter Feeders***  Some gastropods are filter feeders. Slipper limpets (*Crepidula*) feed by ciliary action, using their gills to trap food particles that are then formed into a mucous ball and carried to the mouth for digestion (see Figure 9-7d). Planktonic gastropods, such as sea butterflies (*Gleba* and *Corolla*), form a mucous net that they use to catch the small organisms on which they feed.

## Naked Gastropods

**Nudibranchs** are marine gastropods that lack a shell (Figure 9-8). They are usually brightly colored and sometimes have bizarre shapes because of their gills and appendages. The name *nudibranch* means "naked gill" and refers to the projections from the animal's body called **cerata** that increase the surface area available for gas exchange.

Some nudibranchs feed on cnidarians. Instead of digesting their prey's stinging cells, the nudibranch leaves the cells intact and moves them by means of ciliated tracts in its digestive system. These tracts extend into the cerata. The stinging cells are deposited in the tips of the cerata, where they are used by the nudibranch for its own defense. The bright colors of many species are a warning, advertising that

# Deadly Snails and Medicine

The marine snail *Conus geographus,* or geography cone (Figure 9-A), is one of nature's most sophisticated predators. It has a long, brightly colored proboscis that it uses to attract the attention of fish. A fish apparently mistakes the proboscis for an edible worm. When the fish moves in to bite, the snail fires its poisonous radula into the fish's mouth. The radula penetrates the soft tissues in the mouth and paralyzes the fish. After its prey has succumbed to the venom, the snail begins the leisurely process of digesting its meal.

To date, at least 20 unwary swimmers in the South Pacific have been killed by this animal. Humans are usually stung when they come into contact with the snail while swimming or when they pick it up to admire its colorful shell. Because humans are much larger than the fishes the snail usually preys upon, it takes longer for the venom to move through their body. Death, usually from respiratory paralysis, occurs within 1 hour of the sting.

**Figure 9-A The Geography Cone Snail.** The geography cone (Conus geographus) from the Indian and Pacific Oceans is one of the six species of cone snails known to cause human fatalities. The toxin produced by this animal causes death by paralyzing the victim's respiratory muscles.

In studying how the venom paralyzes its prey, researchers have discovered chemicals that can be used to probe the mysteries of the human nervous system, espe-

cially the relationship between nerves and muscles. Researchers have isolated three different toxins from the snail's venom, one of which is quite similar to the neurotoxin produced by cobras. Investigators have found that some components of the venom interfere with the action of nerve cells and others interfere with the movement of stimuli from nerve cells to muscles. The toxins are highly specific and bind only to certain receptors on cell surfaces. Scientists can use these toxins to locate the receptors in different cells and to study the role receptor molecules play in normal and abnormal function of nerves and muscles. By studying the effects of the toxin on nerves and muscles, researchers hope to gain new insight into how some neuromuscular diseases, such as myasthenia gravis (a progressive disease that causes muscle weakness in the face, neck, and chest) and muscular dystrophy (a group of diseases that cause progressive wasting of the muscles), produce their effects. ●

---

the animals are poisonous or distasteful. Potential predators recognize the bright colors or patterns and avoid them.

## Reproduction and Development

The sexes are separate in most gastropods, although some are hermaphroditic. Most species exhibit internal fertilization. Males of these species use a tentacle-like penis to deposit sperm in or near the female genital opening. The eggs of females that exhibit internal fertilization are usually surrounded by a jellylike substance or covered by protective egg cases. In some species, sperm and eggs are released into the water column and fertilization occurs externally.

In primitive molluscs that shed their eggs directly into the sea, a free-swimming larval stage, called **trochophore** (TROHK-eh-fohr) **larva** (Figure 9-9a), hatches from the egg. More characteristic, however, of marine gastropods is a free-swimming stage called the **veliger larva** (see Figure 9-9b). Veligers can sometimes travel great distances on currents, allowing the otherwise slow-moving gastropods to disperse to other suitable habitats. In some species of marine snail there is no free-swimming larva, and a juvenile that resembles the adult hatches from the egg.

One of the more interesting reproductive strategies occurs in slipper limpets of the genus *Crepidula*. These mostly sessile animals are hermaphroditic and tend to congregate in groups in which several individuals are stacked on top of each other (Figure 9-10). The right margins of the shells are next to each other, allowing the penis of the individual on top to reach the female opening of the individual below. The young of these species are always male. This initial male phase is then followed by a transition period during which the male reproductive tract degenerates, and the animal develops into either a female or another male, depending upon the sex ratio of the group. If males are in short supply, the animals develop into males. If females are limited, the animals develop into females. The sex of the individual is probably controlled by **pheromones**. A pheromone is a hormone that is released into the environment by some members of a population to control the development and behavior of other members. Older males remain males longer if they remain attached to a female. If an older male is removed from the female or isolated, it develops into a female. Once an individual becomes a female, it cannot return to being a male.

(a)

(c)

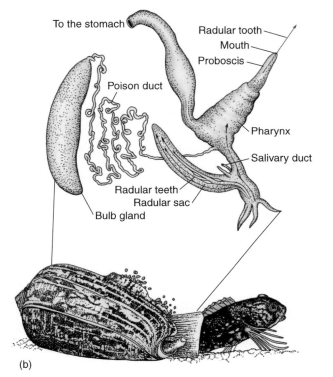

To the stomach

Radular tooth
Mouth
Proboscis

Poison duct

Pharynx

Salivary duct

Radular teeth
Radular sac
Bulb gland

(b)

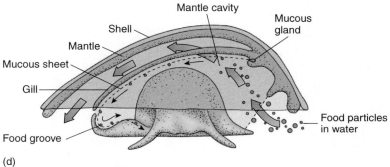

Mantle cavity

Shell

Mantle

Mucous sheet

Gill

Food groove

Mucous gland

Food particles in water

(d)

**Figure 9-7  Gastropod Feeding Strategies.** (a) Cowries are herbivores that graze on algae. (b) A striated cone (*Conus striatus*) and other cone snails use their harpoonlike radula to inject a paralyzing toxin into their prey. After the prey is paralyzed it is slowly eaten by the snail. (c) Horn snails (*Cerithium*) are deposit feeders. At low tide, large numbers can be seen on sand or mud feeding on the organic material left behind by the tides. (d) Slipper limpets use their gills to filter food from the water.

## Bivalves

**Bivalves** (class Bivalvia) are molluscs that have shells divided into two jointed halves called **valves**. This group includes clams, oysters, mussels, scallops, and shipworms. They range from tiny, 1-millimeter clams to the giant clam *Tridacna gigas,* which can grow as large as 1 meter (3.3 feet) and weigh 230 kilograms (507 pounds) or more.

### Bivalve Anatomy

Bivalves have no head and no radula. Their bodies are laterally compressed, and the two halves of the shell that cover it are usually attached dorsally at a hinge by ligaments. The

**Figure 9-8  Nudibranchs.** Nudibranchs, like this one from Washington, are gastropods that lack a shell. The projections from the animal's body are called *cerata* and function in gas exchange.

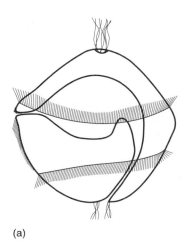

(a)

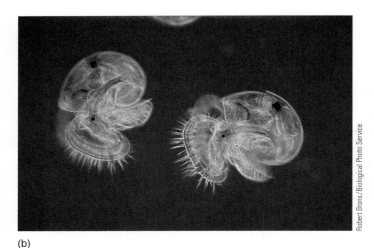

(b)

Robert Brons/Biological Photo Service

**Figure 9-9 Molluscan Larvae.** (a) Primitive molluscs produce trochophore larvae. (b) The veliger larvae of a slipper limpet are more characteristic of many marine molluscs.

oldest part of the shell is the area around the hinge called the **umbo** (Figure 9-11), and growth generally occurs in a concentric fashion around it at the edge of the shell. Large muscles called **adductor muscles** close the two valves. When the muscles relax, the elasticity of the ligaments and the weight of the valves cause them to open. The animal's foot is located ventrally and, in clams, functions in burrowing and locomotion.

For the most part bivalves are filter feeders. In many species, the mantle forms separate **inhalant** and **exhalant openings.** Cilia on the surface of the animal's gills move water through the animal's body, and the openings direct the water flow into and out of the mantle cavity. Water entering the inhalant opening carries food. The food particles, mainly plankton, are filtered from the water by the gills and formed into a mass by a pair of structures called **palps.** The palps move the food mass to the bivalve's mouth where it enters the digestive system.

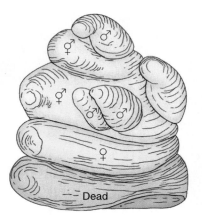

Dead

**Figure 9-10 *Crepidula fornicata.*** Slipper limpets are normally found in groups such as this. The male is on the top, and the next two individuals are in transition from male to female. When the transition is complete, all of the females will be fertilized by the male at the top.

## Bivalve Adaptations to Different Habitats

The evolution of filter feeding has allowed bivalves to successfully colonize several different habitats. Many species burrow into soft bottom sediments and are conspicuous members of the infauna. Others spend their lives attached to solid surfaces as members of the epifauna. Still others are unattached epifauna. A few bivalves have even adapted to burrowing into wood and rock.

***Soft-Bottom Burrowers*** Most species of bivalve are infauna, inhabiting soft bottoms to take advantage of the protection gained from burrowing in sand and mud. They feed on food suspended in the water they draw in from above. A major problem associated with burrowing in soft bottoms is that sediments are brought in with the incoming water currents. To deal with this problem, the mantle in many species is fused around the inhalant and exhalant openings to form tubular structures called **siphons** that can project above the surface of the sediments. With siphons the animals can be completely buried in the mud and only the siphon tips need to project above the bottom.

The geoduck (GOO-ee-duhk; *Panopea generosa*; Figure 9-12a) from the Pacific coast of the United States burrows as deep as 1 meter (3.3 feet) into the mud and can have a siphon more than 1 meter long. Their siphons are so large that they can no longer be retracted between the shell valves. The enormous siphon can account for as much as one half of this animal's 2.75-kilogram (6 pound) body. Bivalves in the genus *Scrobicularia* are deposit feeders. They extend their inhalant siphon above the surface at low tide and, like a vacuum cleaner, suck in deposit material, which is then sorted on the gills. In bivalves such as these, deposit feeding is generally an addition to, rather than a substitute for, filter feeding.

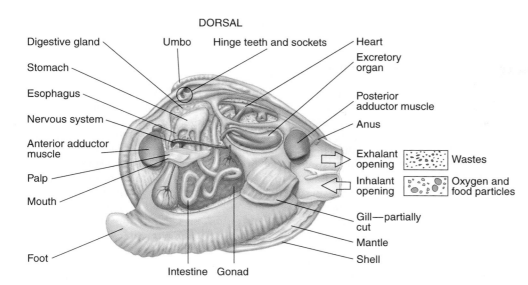

DORSAL

Digestive gland
Umbo
Hinge teeth and sockets
Heart
Excretory organ
Stomach
Esophagus
Posterior adductor muscle
Nervous system
Anus
Anterior adductor muscle
Exhalant opening — Wastes
Palp
Inhalant opening — Oxygen and food particles
Mouth
Gill—partially cut
Foot
Mantle
Shell
Intestine   Gonad

**Figure 9-11  Bivalve Anatomy.** The internal anatomy of a typical marine clam, *Mercenaria mercenaria.* Notice that food and oxygen enter through the inhalant opening while wastes are eliminated by way of the exhalant opening.

***Attached Surface Dwellers***   A number of bivalves are epifauna, attaching to firm surfaces, such as rock, coral, wood, shell, sea walls, or pilings. These bivalves attach either by fusing one of their valves to the hard surface or by a byssus. A **byssus** is a tough protein secreted by a gland in the bivalve's foot and commonly takes the form of threads. Of the bivalves that attach with a byssus, the mussels (family Mytilidae; see Figure 9-12b) are the most familiar. Mussels can be found attached to wharf pilings, sea walls, and rocks or among oysters. Other bivalves that attach by byssal secretions include many of the heavy-bodied arks (family Arcidae), which are very common on tropical corals; mangrove oysters (*Isognomon*), which hang in clusters from mangrove roots; and winged oysters (family Pteridae), which live attached to sea fans and other gorgonians.

Surface-dwelling bivalves that attach by fusion lie on one side and cement one of their valves to a hard surface. Which valve is cemented, the right or the left, depends on the species. Oysters are the most familiar bivalves that attach in this manner. In attached bivalves the foot is reduced to varying degrees, and it is completely absent in those bivalves, such as the oyster, that are attached by one valve.

***Unattached Surface Dwellers***   Bivalves that live free on the bottom or other surface include scallops (family Pectinidae) and file clams (family Limidae). Although some members of both families live anchored by byssal threads, most are unattached and able to move. Scallops and their relatives are able to move by a type of jet propulsion. The animal moves by contracting its adductor muscle, causing

(a)    (b)    (c)

**Figure 9-12  Bivalve Adaptations.** (a) The geoduck *Panopea generosa* is found burrowed in soft sediments along the west coast of the United States. Its siphon can be as long as 1 meter and account for half of this bivalve's body weight. (b) Attached surface-dwelling bivalves like this mussel attach to solid surfaces by way of a byssus formed from threads of tough protein secreted by the animal's foot. (c) Surface dwelling bivalves such as this spiny oyster (*Spondylus*) attach by cementing one of their valves to a hard surface.

the two valves to close forcefully and emit two streams of water backward and to each side. The animal then moves forward in a series of jerky movements (see Figure 9-12c). The swimming ability of scallops and file clams is used primarily to escape predators. Some scallops use the water jets to blow out a depression in the sand surface into which they settle to become less visible. File shells typically nest in crevices beneath stones and swim only when disturbed.

***Boring Bivalves*** Some bivalves are capable of burrowing into wood or stone and can cause a great deal of commercial damage. Members of the genera *Teredo* and *Bankia* are known as **shipworms** and can cause damage to wooden ships and wharves. These bivalves resemble tiny worms with a shell, and the valves bear microscopic teeth that help the animal burrow (Figure 9-13). Pieces of wood that are removed by the burrowing action are swallowed and digested with the aid of enzymes produced by symbiotic bacteria. These bacteria live in a special organ located in the animal's digestive system. The bacteria also fix nitrogen, which helps their hosts compensate for their low-protein diet. Another group of boring clams belongs to the genus *Pholas*. These bivalves can burrow into soft rocks, such as limestone, and play a role in recycling calcium.

### Reproduction in Bivalves

In the majority of bivalves the sexes are separate. During reproduction sperm and eggs are shed into the water column by way of the excurrent water. Fertilization takes place in the water column for most species. The fertilized eggs develop into a free-swimming trochophore larva, and this is followed by a second larval stage known as the veliger larva. The veliger larva develops a shell and settles on the bottom where the animal begins an adult existence. Shipworms and some species of cockle (*Cardium*), scallop, and oyster are hermaphroditic. The oyster *Ostrea edulis* does not release eggs but broods them in a cavity near the gills. In this species sperm are brought in with the incurrent water to fertilize the eggs.

## Cephalopods

Molluscs such as the octopus, squid, and cuttlefish are **cephalopods** (SEF-uh-loh-pahdz; class Cephalopoda). The term *cephalopod* means "head-footed" and refers to the animal's foot, which is modified into a headlike structure. A ring of tentacles projects from the anterior edge of the head, and they are used to capture prey, for defense, in reproduction, and in some species as an aid in locomotion. Cephalopods range in size from 2 centimeters (¾ inch) to more than 18 meters (59 feet). With the exception of the nautiloids, modern cephalopods either have small internal shells, such as some squids, or lack shells entirely, such as the octopods.

### Types of Cephalopod

There are two major groups of living cephalopods. One group, the **nautiloids** (subclass Nautiloidea), is represented by the genus *Nautilus* (Figure 9-14a). These animals are covered by a shell. The other group of living cephalopods is the **coleoids** (subclass Coleoidea). This group includes squids, octopods, and cuttlefish. These animals do not have an external shell.

***Nautiloids*** Nautiloids produce large, coiled shells composed of chambers separated from each other by partitions called **septa.** The chambers are filled with gas and aid in maintaining buoyancy for swimming. The animal inhabits only the last chamber but remains connected to the others by means of a cord of tissue called the **siphuncle** (SY-fuhn-kuhl; see Figure 9-14b). The siphuncle removes the seawater from each new chamber as it forms. The living tissue of the siphuncle removes salts from the seawater in the new chamber by active transport. The water that remains is then more dilute than the animal's body fluids and readily moves

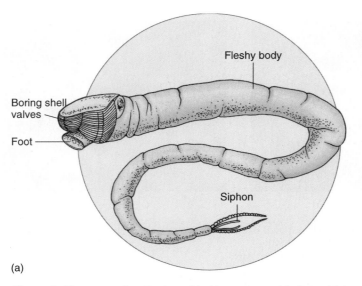

(a)

(b)

**Figure 9-13 Destructive Bivalves.** (a) Shipworms are bivalves with long wormlike bodies and small shells that are used to bore through wood. (b) Shipworms can cause commercial damage.

**Figure 9-14 Shelled Cephalopods.** (a) This chambered nautilus from the tropical Pacific Ocean has changed very little since it first evolved. (b) The nautilus occupies only the last chamber of its shell. The other chambers are filled with gases from the blood, such as carbon dioxide. The animal can regulate the gas content of the chambers with its siphuncle, allowing it to rise toward the surface, sink to the bottom, or maintain neutral buoyancy as necessary.

(a)

W. Gregory Brown 1991/Animals Animals

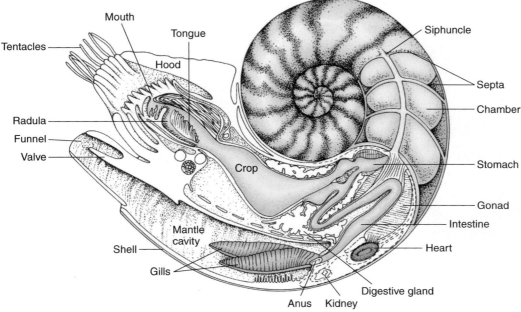

(b)

by osmosis into the animal's body. The excess water is ultimately eliminated by the kidney. The gases that replace the water in the chambers are produced by the respiration of the siphuncle's living tissue. The amount of gas in the chambers can be regulated, allowing the animal to rise or sink in the water column.

The animal's head has 60 to 90 tentacles coated with a sticky substance that helps them to adhere to what they touch. Many of these are chemosensory and tactile (sense of touch) in function but some around the mouth bring food to the mouth. Like other cephalopods, nautiloids move by means of jet propulsion. Water is drawn in through an incurrent siphon and forced out through an excurrent siphon, pushing the animal shell first through the water. A structure known as the funnel directs the jet of water, allowing the nautiloid to steer through the water.

Nautiloids tend to come to the surface at night and dwell on the bottom during the day. Originally it was thought

that they came to the surface at night to feed, but examination of stomach contents indicates that they prey on hermit crabs and scavenge other food on the bottom. After ingesting their food, nautiloids store it temporarily in a large, saclike structure called the **crop** before sending it along to the stomach and intestine for complete digestion.

***Coleoids*** Cuttlefish, squids, and octopods are coleoids. Cuttlefish (*Sepia;* Figure 9-15a) have a bulky body, fins, and ten appendages. The eight short, heavy appendages are called arms, and the two larger appendages are called tentacles. They have small, internal shells, some of which have chambers like those of the nautilus shell. The shell of a cuttlefish, however, is embedded in its mantle.

Squids (see Figure 9-15b) have large cylindrical bodies with a pair of fins derived from mantle tissue. Squids possess 10 appendages called arms that are arranged in 5 pairs around the head. One pair of arms is larger than the rest,

(a)

Biological Photo Service

(b)

Richard N. Mariscal

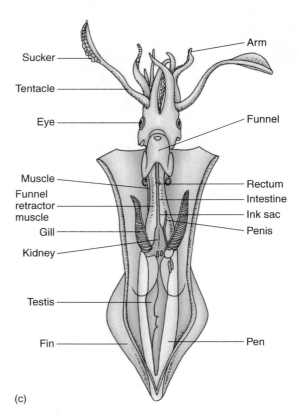

(c)

**Figure 9-15  Cephalopods.** (a) The cuttlefish resembles a squid and has a small internal shell. (b) Squids have streamlined bodies and are active swimmers. The part of the animal's body that looks like a head is actually its foot. (c) The internal anatomy of a male squid.

and these are called **tentacles.** The inner surface of each arm is flattened and covered with cup-shaped suckers that are attached by a short stalk. Toothed structures surround the rim of the sucker. Suckers are present only on the flattened ends of the highly mobile tentacles, which can be projected out with great speed to seize prey. The arms aid in holding the prey after capture. An internal strip of hard protein called the **pen,** which represents a degenerate shell, lends support to the mantle (see Figure 9-15c).

Octopods have eight arms, which are similar to the arms of squids except that the suckers lack stalks and teeth, and octopods do not have tentacles. Their bodies are saclike and lack fins. All of these animals are extremely adept predators, feeding on a variety of fish and invertebrates.

When disturbed, coleoids cloud the water with an inky fluid. An ink gland produces a dark fluid called **sepia** that contains a high concentration of a brown-black pigment called **melanin** (deepwater squid release a white or luminescent fluid). The sepia is stored in an ink sac in the animal's body. When the animal is disturbed or frightened, it can release the ink into the water (Figure 9-16a). The ink cloud resembles the general shape of the animal and distracts predators while the animal escapes.

Squids, octopods, and their relatives can swim, like nautiloids, by jet propulsion produced by forcing water through a ventrally located siphon. The funnel directs the flow of water and thus the direction of movement. Most slow swimming in squid species, however, is achieved by fin undulation. Squids and cuttlefish are very good swimmers, and their bodies are streamlined for efficient movement. Although octopods can swim, they are better adapted to crawling over the bottom.

Coleoids have the most advanced and complex nervous system of any invertebrate. Octopods can be trained to perform simple tasks and are used as model systems to study neural development associated with learning and memory. They have highly developed eyes and can recognize shapes and colors. Octopods seem to have a highly developed tactile sense and can discriminate objects on the basis of touch.

## Color and Shape in Cephalopods

Cephalopods can communicate with each other through movements of their arms and bodies and by color changes. Color changes involve special cells in the skin called **chromatophores** that contain pigment granules. When the granules are dispersed, the color of the animal's skin becomes

(a)                                                    (b)

**Figure 9-16  How Cephalopods Avoid Predation.** (a) Many cephalopods, such as this octopus, escape predators by releasing an inky substance into the water. (b) The blue-ring octopus, *Hapalochlaena maculosa,* from Australia advertises its toxicity with bright blue rings.

darker, and when the granules are concentrated, the color is lighter because of the permanent light background color (Figure 9-17). The animal not only changes general body color but also produces stripes and patterns that communicate information to other members of its species and a warning to other species. For instance, when agitated, the blue-ring octopus (*Hapalochlaena maculosa*) of the Pacific Ocean produces blue rings on its skin to warn potential antagonists (see Figure 9-16b). The bite of this octopus is extremely toxic and can be fatal to humans.

### Feeding and Nutrition

Cephalopods are carnivores. Prey is located with the highly developed eyes and captured by the tentacles or arms. A radula is present in cephalopods but more important is the pair of powerful, beaklike jaws in the oral cavity. The beak can bite and tear off large pieces of tissue, which are then pulled into the oral cavity by a tonguelike action of the radula and swallowed. One of the salivary glands of the cuttlefish and octopods secretes poison. The poison enters the tissues of the prey through the wound inflicted by the jaws.

The diet of a cephalopod depends on the habitat in which it lives. *Nautilus* is a scavenger and preys on benthic organisms, especially hermit crabs. When feeding, it swims forward searching with extended tentacles. Pelagic squids such as *Loligo* and *Alloteuthis* feed on fish, crustaceans, and other squids. *Loligo* will dart into a school of young mackerel, seize a fish with its tentacles, and quickly bite off the head or a chunk behind the head. Cuttlefish swim over the bottom and feed on invertebrates they find on the surface, especially shrimps and crabs.

Octopods live in dens located in crevices and holes. They either forage in search of food or lie in wait near the en-

trances of their dens. They mainly feed on clams, snails, and crustaceans, and the diet of a particular species may include as many as 55 different prey items, although a few items predominate. When feeding, an octopus will leap on its prey, such as a crab, enveloping it in its arms. Sometimes an octopus leaps on clumps of algae or other objects and then feels beneath the object for a possible catch. To remove gastropods from their shells, octopods drill a hole through the shell with their radula and inject poison to kill the occupant. In contrast to cuttlefish and squids, which tear prey with jaws, the feeding habit of octopods is rather like that of spiders. The prey is injected with poison and then flooded with enzymes. The partially digested tissues are taken up into the digestive system and the indigestible remains are discarded.

### Reproduction in Cephalopods

Sexes are separate in cephalopods and mating frequently involves some form of courtship display. Male squid have a modified arm that is used to take a package of sperm, the **spermatophore,** from its own mantle cavity and place it in the mantle cavity of the female near the opening of the **oviduct** (tube that carries eggs to the outside of the body). The eggs are fertilized as they are released from the oviduct. Some species, such as members of the genus *Argonauta,* lay their eggs in delicate shells secreted by modified tentacles of the females. Other species usually attach their eggs to stones or other objects. The females of some octopus species incubate their eggs until they hatch, constantly pumping a stream of water over the eggs to keep them oxygenated and to prevent fungus and other microorganisms from infecting them. Cephalopods usually reproduce only once in their life cycle and then die.

**Figure 9-17  How Cephalopods Change Color.** (a) Tiny muscle cells pull at the edges of the chromatophore, causing the cell to expand and the pigment to disperse, producing a darker color. (b) When the muscle cells relax, the chromatophore returns to its original size, concentrating the pigment; the color of the animal is then lighter.

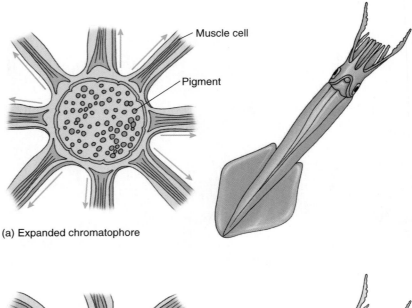

(a) Expanded chromatophore

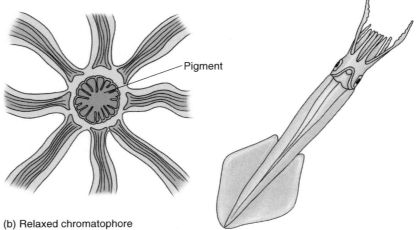

(b) Relaxed chromatophore

## Ecological Role of Molluscs

Molluscs are important to humans as well as to other animals as food, and snail shells are a major source of calcium for some marine birds. According to one study, sperm whales alone consume as large a mass of squid as humans do of all species of fish combined. Other molluscan foods include clams, oysters, mussels, abalone, conch, and scallops. Some snails are intermediate hosts to parasites. Shipworms cause extensive damage to wooden pilings and boat hulls and are becoming increasingly troublesome as improved water quality allows more of them to survive. On the other hand, because few marine organisms are able to break down wood, wood-boring bivalves help to prevent the accumulation of driftwood in the marine environment. A small number of bivalves have evolved commensal relationships. Most commensals are attached by byssus threads but some crawl on their foot like a snail. The hosts are usually burrowing echinoderms, such as heart urchins, brittle stars, sea cucumbers, and shrimplike crustaceans. The only known parasitic bivalve is a species of *Entovolva* that lives in the gut of sea cucumbers.

## In Summary

The phylum Mollusca includes chitons, tusk shells, gastropods, bivalves, and cephalopods. The generalized molluscan body plan consists of two parts: a head-foot and a visceral mass. Molluscs have soft bodies that, in most cases, are covered by a shell of calcium carbonate. The mollusc shell is formed by an organ called the mantle. In cephalopods, the mantle functions in locomotion. Some molluscan species rely on the mantle for gas exchange, though most marine molluscs exchange gases by means of gills. The radula, or ribbon tooth, is a structure unique to molluscs that is used for feeding.  ●

## ARTHROPODS: ANIMALS WITH JOINTED APPENDAGES

**Arthropods** (phylum Arthropoda) represent the most successful group of animals in the animal kingdom, and almost 75% of all identified animal species (most of them insects) belong to this phylum. Several factors contributed to the enormous success of marine arthropods, including the evolution of a hard exterior, jointed appendages, and sophisticated sense organs.

The hard, protective exterior skeleton, or **exoskeleton,** of arthropods is composed of protein and a tough polysaccharide called **chitin** (KY-tin). In many marine species, the exoskeleton is impregnated with calcium salts to give it extra strength. The exoskeleton is flexible enough in the region of the joints to allow movement, and it provides points of attachment for muscles, allowing more efficient movement. The exoskeleton does have its drawbacks, however. Because the exoskeleton is not living, it does not grow with the animal. As the animal grows, it must shed its old exoskeleton and produce a new, larger one. This process is called **molting.** As the exoskeleton increases in size, so does its weight, making it more difficult for the animal to carry around. It is not surprising that the largest arthropods are found in aquatic habitats where the buoyancy of the water helps to counteract the weight of their heavy exoskeletons.

The body of arthropods is divided into segments, and generally, each body segment of the animal has a pair of jointed appendages. Many of these function in locomotion, allowing the animals to move quickly and efficiently. Other appendages have been modified into mouthparts for more efficient feeding, sensory structures for monitoring the environment, and body ornamentation that helps to attract a mate or to make the animal more difficult to see.

Arthropods have highly developed nervous systems. Their sophisticated sense organs allow them to respond quickly to changes in their environment. The high degree of development of the arthropod nervous system has also given rise to a number of behavior patterns that play an important role in their daily activities. Experiments have shown that some species are capable of learning.

There are two major groups of marine arthropods: chelicerates and mandibulates. Chelicerates have a pair of oral appendages called chelicerae and lack mouthparts for chewing food. As a result most predigest their food and suck it up in a semiliquid form. Mandibulates have appendages called mandibles that can be used for chewing food.

## Chelicerates

Chelicerates (keh-LI-suh-rehts; subphylum Chelicerata) are a primitive group of arthropods that includes spiders, ticks, scorpions, horseshoe crabs, and sea spiders. These animals have six pairs of appendages. One pair, the **chelicerae** (keh-LI-suh-ree), is modified for the purpose of feeding and takes the place of mouthparts.

### Horseshoe Crabs

Horseshoe crabs (*Xiphosura;* Figure 9-18) are not true crabs but chelicerates that live in shallow coastal waters. They have not changed much since they first evolved more than 230 million years ago. The body of the animal is in the shape of a horseshoe and comprises three regions: the cephalothorax (sef-uh-loh-THOR-aks), the abdomen, and the telson. The **cephalothorax** is the largest region of the body and contains the more obvious appendages. The **abdomen** is

**Figure 9-18  Horseshoe Crabs.** Horseshoe crabs are not really crabs but are more closely related to spiders. The large anterior portion with the walking legs is the *cephalothorax.* The abdomen contains the gills, and the long taillike structure is the *telson.*

smaller than the cephalothorax and contains the gills. The **telson** is a long spike that is used for steering and defense. The entire body is covered by a hard outer covering called the **carapace.**

Horseshoe crabs move by walking or swimming. They swim by flexing the abdomen. They are scavengers that feed primarily at night on worms, molluscs, and other organisms, including algae. They pick up food with their chelicerae and pass it to the walking legs. The walking legs have structures at their bases that crush the food before passing it to the mouth.

Males are smaller than females, and during the mating season, a male or several males will attach to the carapace of a female. The animals come to shore during high tide to mate, and the female will dig up the sand with the front of her carapace, depositing eggs in the depression. The males riding on her back shed their sperm onto the eggs before they are covered by the female. The eggs are then incubated by the sun and hatch into larvae that return to the sea during another high tide to grow into adults.

### Sea Spiders

Sea spiders (class Pycnogonida; Figure 9-19) are chelicerates that can be found from intertidal waters to depths of more than 6,000 meters (19,700 feet) in all oceans, but especially the polar seas. They have small, thin bodies and usually four pairs of walking legs, although some species may have more. Males in several species have an extra pair of appendages that are used to carry the developing eggs. These are the only marine invertebrates known where the male carries the eggs. Other appendages include chelicerae for capturing prey and sensory structures called **palps.** Sea spiders feed on the juices of cnidarians and other soft-bodied invertebrates. They extract the juices with a long, sucking proboscis.

Doug Allan/Oxford Scientific Films/Animals Animals

**Figure 9-19 Sea Spider.** Sea spiders are marine chelicerates that use their proboscis to suck fluids from their prey, much like terrestrial spiders.

## Mandibulates

Mandibulates have pairs of appendages on the head called mandibles that are modified for feeding. The mandibulates found in the marine environment are **crustaceans** (subphylum Crustacea), a group represented by a variety of animals that range in size from microscopic zooplankton to large lobsters. Some of the more important groups of crustaceans found in the marine environment are decapods, mantis shrimp, krill, copepods, amphipods, and barnacles.

## Crustacean Anatomy

Crustaceans have three main body regions: the head, thorax, and abdomen (Figure 9-20). In some species the head and thorax are fused to form a cephalothorax. Each body segment usually bears a pair of appendages, although in some species abdominal appendages are lacking. Crustacean appendages include sensory antennae (they are the only arthropods that have two pairs) and mandibles and maxillae that are used for feeding. Depending on the species, other appendages may include walking legs and legs modified for swimming (**swimmerets**), reproduction, and defense (chelipeds).

Small crustaceans exchange gases through their body surface, especially through areas at the base of their legs, where the exoskeleton is thin. Larger crustaceans have gills for gas exchange. These are feathery structures that are usually located beneath the carapace or at the base of specialized appendages.

## Molting

Molting is a very important part of the crustacean life cycle. Stages are diagrammed in Figure 9-21. While the animal is molting, it is very vulnerable and usually seeks a hiding place until a new exoskeleton has hardened. Young crustaceans molt frequently, and the time between molts is relatively short. As the animal ages, molting is less frequent, and some species eventually reach an age where they no longer molt. Molting is directly controlled by specific hormones produced by glands in the crustacean's head. The process appears to be initiated by changes in environmen-

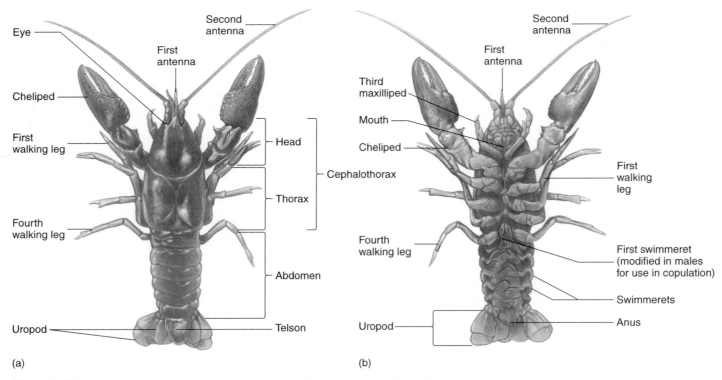

(a)

(b)

**Figure 9-20 Crustacean Anatomy.** The external anatomy of a lobster as seen from (a) a dorsal view and (b) a ventral view.

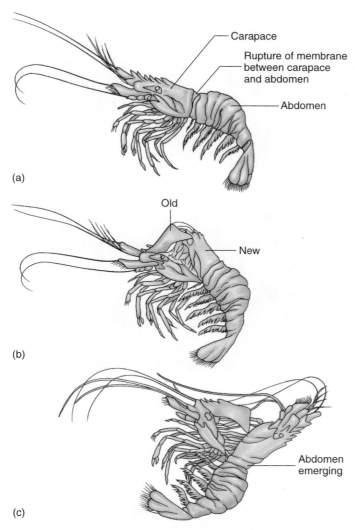

**(a)**

**(b)**

**(c)**

**Figure 9-21** **Molting.** The nonliving exoskeleton of crustaceans must be shed periodically. (a) Old membranes rupture along a joint line. (b) The animal pulls itself free of the anterior portion of the exoskeleton and then (c) crawls out of the posterior part. Once the old exoskeleton has been loosened, it usually takes about 15 minutes for the animal to shed it. After molting, the crustacean's body is soft, and the animal is particularly vulnerable until, in several days, the new exoskeleton is formed.

tal conditions, such as temperature and photoperiod, that alter the level of the hormones.

## Decapods

Crabs, lobsters, and true shrimp are called **decapods** (order Decapoda; *deca,* meaning "ten," and *poda,* meaning "feet") because they have five pairs of walking legs. The first pair of walking legs is modified to form **chelipeds,** pincers that are used for capturing prey and for defense. Although most decapods are relatively small, some can get quite large. The largest of all crustaceans is the giant spider crab (*Macrocheira kaempferi*) from Japan (Figure 9-22a). Specimens are known to have exceeded 4 meters (13 feet) in width and weighed more than 18 kilograms (40 pounds). The North

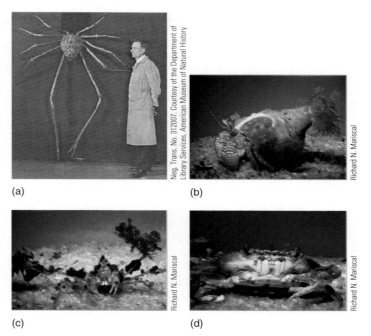

**(a)**　　　　　　　　　　**(b)**

**(c)**　　　　　　　　　　**(d)**

**Figure 9-22** **Decapods.** (a) The Japanese spider crab, *Macrocheira kaempferi,* from the northern Pacific, is the largest living arthropod. (b) Hermit crabs inhabit the empty shells of molluscs. As they grow they must exchange their existing shells for larger ones. This hermit crab inhabits a shell bearing various symbiotic sea anemones. (c) This decorator crab from the Gulf of Mexico attaches a variety of sessile organisms to its shell for camouflage. (d) The blue crab *Callinectes sapidus* (the name means "beautiful swimmer that tastes good") is fished commercially for food along the east coast of the United States. Note that this individual is missing one of its claws, which will begin growing back at the next molt.

Atlantic lobster (*Homarus americanus*) can also get quite large. Two record specimens were captured off the coast of Virginia in 1934. One measured more than 1 meter (3.3 feet) long; one weighed 19 kilograms (42 pounds) and the other 17 kilograms (37 pounds).

***Specialized Behaviors*** Many decapods display interesting adaptations and behaviors that help them to survive in their habitat. Hermit crabs (family Paguridae) have a soft abdomen. They inhabit empty gastropod shells or other suitable enclosures (see Figure 9-22b), and as they grow, they not only molt but also must find a larger shell to accommodate their ever-increasing bodies. When disturbed, these animals will withdraw into their shells and close the opening with a large cheliped. Decorator crabs (family Majidae) attach bits of sponge, anemones, and hydrozoans to their carapace for camouflage (see Figure 9-22c).

Most crabs cannot swim, but members of the family Portunidae, which includes the common edible blue crab (*Callinectes sapidus;* see Figure 9-22d) of the Atlantic coast, are the most powerful and agile swimmers of all crustaceans. The last pair of legs in members of this family terminate in broad, flattened paddles that during swimming move in a figure-eight pattern. The action is essentially like that of a propeller. Portunids can swim sideways, backward, and sometimes forward with great rapidity. They are benthic animals, however, like other crabs, and only swim intermittently.

***Nutrition and Digestion*** Decapods exhibit a wide range of feeding habits and diets. Predators use walking legs, especially chelipeds, to capture their prey. Scavengers use their appendages to pick up pieces of vegetation and dead animals. Most decapods use their mandibles to crush their food, and some species have plates in their stomachs that will further grind and process the food for digestion. The majority of species combine predatory feeding with scavenging. The relative importance of these two habits depends on the individual species and the available food resource.

Large invertebrates are common prey for many crabs. For example, echinoderms and bivalves are the main food of the Alaskan king crab (*Paralithodes camtschatica*) and polychaetes, crustaceans, and bivalves are prey for snow crabs (family Lithodidae). Hermit crabs and many shrimp species are scavengers or feed on detritus. Fiddler crabs (*Uca*) are deposit feeders. They scoop up mud and detritus with their small chelipeds, and the food is filtered and processed in the oral cavity. After organic material has been removed, the mineral residue is spit out in the form of round pellets, which may eventually surround the crab's burrow or cover the surface of the beach. Filter feeders include mole crabs, many burrowing shrimps, some commensal pea crabs, and most porcelain crabs.

***Reproduction*** In decapods the sexes are usually separate and the males have special appendages modified for clasping the female and delivering the sperm. Decapods transmit sperm in packages known as spermatophores, which in most species are delivered to the female by the male's anterior two pairs of abdominal appendages called **copulatory pleopods.** Fertilization is generally internal.

Most decapods brood their eggs and have brood chambers or modified appendages for this purpose. Shrimp shed their eggs into the water column. When most crabs hatch they are in the **zoea larval** stage (Figure 9-23a), which is easily recognized by the very long rostral spine and sometimes lateral spines. The spines appear to reduce predation by small fish. In shrimp species a larva called the **nauplius** (NAW-plee-uhs) **larva** (see Figure 9-23b) hatches from the egg. During subsequent molts the decapod larval stages gradually mature into the adult forms.

## Mantis Shrimp

Mantis shrimp (order Stomatopoda) are highly specialized predators of fishes, crabs, shrimp, and molluscs, and many of their distinctive features are related to their predatory behavior. They range from small species approximately 5 cm (2 inches) long to giant forms more than 36 cm (14.5 inches) long. Most mantis shrimp are tropical, but *Squilla empusa* is common along the North American Atlantic coast. Most mantis shrimp live in rock or coral crevices or in burrows excavated in the bottom sediments.

In mantis shrimp the second pair of thoracic appendages is enlarged and has a movable finger that can be extended rapidly to capture prey and for defense (Figure 9-24). Mantis shrimp can either spear (spearers) or smash (smashers) their prey with these thoracic appendages. Spearers

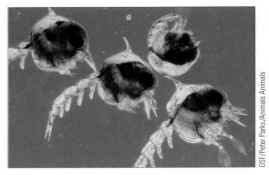

(a)

(b)

**Figure 9-23 Zoea and Nauplius Larvae.** (a) The zoea larva is a common larval stage in crabs and shrimp. The spines may help deter predators such as fish. (b) The nauplius larva is the larval stage of copepods and barnacles.

**Figure 9-24 Mantis Shrimp.** The second pair of thoracic appendages in mantis shrimp are modified into structures that can be used as weapons to capture and kill prey.

feed on soft-bodied invertebrates such as shrimp and fish. Smashers, which live in holes and crevices on rocky or coralline bottoms, stalk their prey, mostly snails, clams, and crabs. Their hard-bodied victims are smashed with the heavy heel of the unfolded appendage. The blows of a mantis shrimp's thoracic appendages are so powerful that captured specimens have cracked the glass walls of aquaria.

Some mantis shrimp pair for life, sharing the same burrow. Others come together only at the time of mating. The female produces an egg mass that can be as large as a walnut and contain as many as 50,000 eggs. A zoea larva hatches from the egg and may stay in the planktonic form for 3 months before settling and taking up an adult existence.

## Krill

Krill (order Euphausiacea) are pelagic, shrimplike creatures that are about 3 to 6 centimeters (1 to 2 inches) long (Figure 9-25). They are filter feeders that feed primarily on zooplankton. Most species of krill are bioluminescent, producing light in a specialized organ called a **photophore.** It is thought that the luminescence is a signal to attract individuals into large masses called **swarms.** It may also function in reproduction to attract mates. Many Antarctic species, such as *Euphausia superba,* live in large swarms that may cover an area of several hundred meters and be as thick as 5 meters (16.5 feet). The density of krill in these swarms can be as great as 60,000 individuals per cubic meter of water. Krill such as *Euphausia superba* constitute the main diet of some whales, seals, penguins, and many fishes. A single blue whale (*Balaenoptera musculus*) can consume a ton of krill in one feeding and may feed four times a day. *Euphausia superba* can molt so quickly that, when alarmed, individuals will literally jump out of their skins. The shed molt may then function as a decoy.

## Amphipods

Amphipods (order Amphipoda) have bodies that tend to resemble shrimp. The posterior three pairs of appendages are directed backward and are used for jumping, burrowing, or swimming, depending on the species. Most amphipods are between 5 and 15 millimeters (0.2 to 0.6 inches) long, although *Alicella gigantea* may reach 14 centimeters (6 inches).

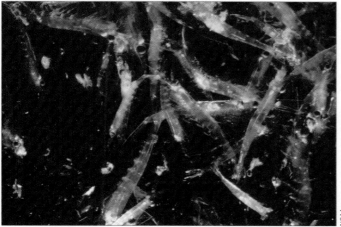

**Figure 9-25 Krill.** Krill are important members of the zooplankton and the favorite food of many marine mammals and penguins. Most species of krill, like this one, are bioluminescent.

Many amphipods are burrowers and some live in tubes that they build. Several tube dwellers, such as species of *Siphonoectes* and *Cerapus,* build tubes of shell fragments and sand grains that they carry with them. A species of *Pseudamphithoides* builds its tube from a species of brown alga. The amphipod uses the bad-tasting chemicals produced by the alga to protect itself from reef fish.

The beach flea *Talitrus* has been shown to use its eyes to obtain astronomical clues for locating the high tide zone where they normally live. If moved above or below the high tide line, they migrate accurately to their normal zone. They use the angle of the sun as a compass along with a map sense of the east–west orientation of the beach they inhabit. Beach fleas have retained gills although they live out of water. As a result beach fleas are restricted to living in moist sand beneath drift or in damp forest leaf litter away from the sea. They feed at night when there is less danger of desiccation.

Some amphipods are herbivores but most are detritus feeders or scavengers. Animal and plant remains are picked up with special appendages called **gnathopods** or detritus is raked from the bottom with antennae. Some species are filter feeders, using their appendages to filter food from the water.

In amphipods males are attracted to females by pheromones. In some species of *Gammarus,* once the male has found a mate, he will carry her beneath him for days. Eggs are fertilized in the female's brood chamber, and when the young hatch they resemble the adults.

## Copepods

Copepods (class Copepoda) are the largest group of small crustaceans. Marine copepods exist in enormous numbers and are usually the most abundant members of the marine zooplankton. Copepods in the genus *Calanus* (Figure 9-26), for instance, constitute a major portion of the zooplankton and are a major food source for several species of commercially valuable fishes as well as some whales, sharks, and birds. Most copepods are 1 to 5 millimeters (less than ¼ inch) but some reach a length of 17 millimeters (⅔ inch). Although most planktonic copepods live in the upper 50 meters (15 feet) of the sea, many species live at greater depths, even in the deep sea. Vertical movement is oriented by light, and many species exhibit daily vertical migrations.

Planktonic copepods are chiefly suspension feeders, feeding on phytoplankton, although some rely heavily on detritus as well. By using radioactive diatom cultures, *Calanus finmarchicus,* a copepod about 5 millimeters (less than ¼ inch) long, was found to collect and digest from 11,000 to 373,000 diatoms every day, depending on their size. A few species of copepod are omnivorous or are strictly predators. Species of *Anomalocera* and *Pareuchaeta,* for example, even capture young fish. Species of *Tisbe* are known to swarm over a small fish and eat its fins, immobilizing it. They then devour the body as it sinks to the bottom. Carnivorous species swim continually as they seek prey.

Male copepods are commonly smaller than females and are usually outnumbered by females. The males are among the few small crustaceans that form spermatophores. The spermatophores are transferred to the female by the tho-

Peter Parks/imagequestmarine.com

**Figure 9-26 Copepod.** Copepods in the genus *Calanus* constitute a major portion of the zooplankton that channels energy from phytoplankton to larger consumers.

racic appendages, and they adhere to the female by means of a special cement. Most copepods shed their eggs into the water column where the young hatch.

## Barnacles

**Barnacles** (class Cirripedia) are the only sessile crustaceans. They can be found attached to rocks, shells, coral, floating timber, and other solid objects. Some species of barnacle attach to ship hulls in large numbers, increasing the ship's drag and slowing it. Other barnacles attach to plants or animals, such as whales and large fishes. The bodies of most barnacles are covered by a shell of calcium carbonate. The shell can be attached directly to a hard surface, as in acorn barnacles (*Balanus;* Figure 9-27a), or be attached by a stalk, as in the goose barnacles (*Lepas;* Figure 9-27b). When the shell is open, feathery appendages known as **cirripeds** extend into the water. The cirripeds filter food such as microorganisms and detritus from the water.

Barnacles are hermaphroditic and usually cross-fertilize using a long, extensible penis to copulate. A nauplius larva hatches from the brooded egg and develops into a **cyprid larva** that has compound eyes and a carapace composed of two shell plates. When the cyprid larva finds a suitable surface, it attaches by its antennae, which contain adhesive glands, and metamorphoses into an adult (see Figure 9-27c).

## Ecological Roles of Arthropods

Arthropods are essential links in food chains and a food source for animals both marine and terrestrial. They are common symbionts of marine organisms in almost every phylum. Some arthropods play an important role in nutrient recycling. Others are fouling organisms that cause commercial damage.

## Arthropods as Food

Many crustaceans are food sources for other marine animals, and some, such as crabs, lobsters, and shrimp, are important food sources for humans. Blue crabs, shrimp, and lobster are fished commercially along many parts of the eastern United States and the Gulf of Mexico. Other species, including the Alaskan king crab and snow crab, are fished commercially along the Pacific coast of the United States. A major part of the diet of many marine animals is copepods. Some planktonic crustaceans feed on phytoplankton, and they are the principal link between phytoplankton and higher trophic levels in many marine food chains. Without them, animal populations in aquatic habitats would collapse. By reproducing efficiently these small crustaceans manage to maintain high populations despite heavy predation.

Krill is harvested for human consumption in the waters around Antarctica. As many as 12 tons have been netted in a single hour, but so far the total annual catch by humans is minuscule compared with consumption by other predators, such as whales. It is assumed that the decline in whale populations, due largely to commercial whaling, has left a surplus of krill available for fishing.

## Arthropods as Symbionts

A number of unrelated groups of shrimps, called cleaning shrimps, remove external parasites (**ectoparasites**) and other unwanted materials from the surfaces of certain reef fish. Some copepods are ectoparasites on fish and attach to the gill filaments, the fins, or the skin in general. Other copepods are commensals or endoparasitic within polychaete worms, the intestines of echinoderms, in tunicates, and bivalves. Cnidarians, especially anthozoans, are hosts for many species of copepod. Commensal crabs called pea crabs (family Pinnotheridae), because of their small size, live in the tubes and burrows of polychaetes. Species also live in the mantle cavities of bivalves and snails, on sand dollars, in tunicates, and in other animals. An Antarctic species of amphipod, *Hyperiella dilatata,* carries a sea butterfly (*Clione antarctica*) attached to its back for protection. (For more information on this symbiotic relationship see Chapter 2.) Barnacles are commensal with a wide range of hosts: sponges, hydrozoans, octocorals, scleractinian corals, crabs, sea snakes, sea turtles, manatees, porpoises, and whales. There are also parasitic species of barnacle.

## Role of Arthropods in Recycling and Fouling

The grass shrimp (*Palaemonetes pugi*) is found in large numbers in tidal marshes along the east coast of the United States. It feeds on small bits of cellulose material it picks from large fragments of detritus and thereby plays an important role in the breakdown of algae and grasses of tidal marsh ecosystems.

Barnacles are among the most serious fouling problems on ship bottoms, buoys, and pilings, and many species have been transported all over the world by shipping. The speed of a badly fouled ship may be reduced by 30%. Much effort and money have been expended toward the development of special paints and other antifouling measures.

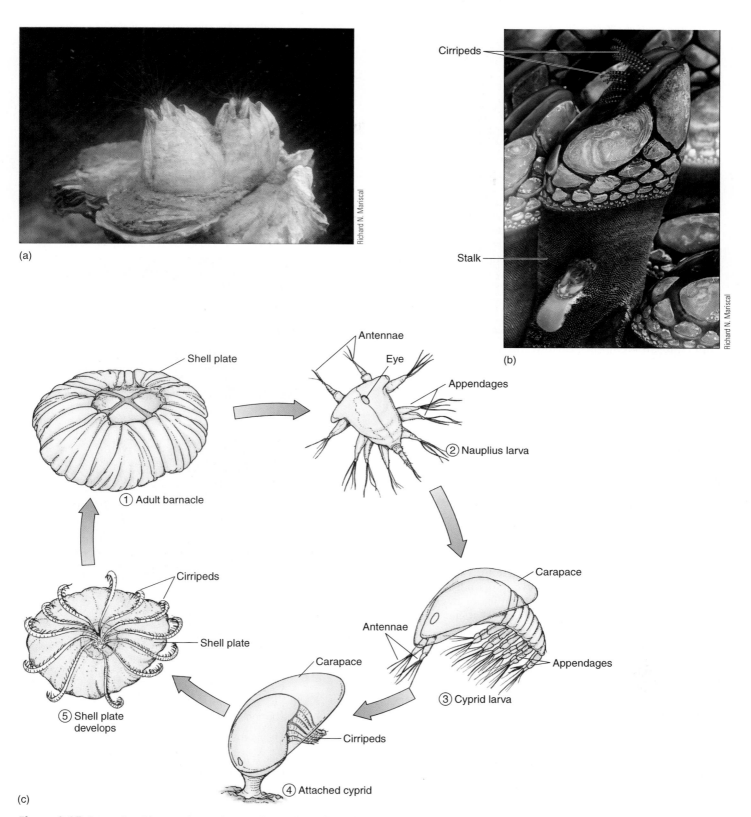

**Figure 9-27 Barnacles.** (a) Acorn barnacles attach to solid surfaces. (b) Goose barnacles attach to solid surfaces by means of a long, fleshy stalk. (c) A generalized life cycle of an acorn barnacle. (1) After fertilization, eggs are brooded in the adult's shell until they develop into nauplius larvae. A single adult may release as many as 13,000 larvae. (2) Like other crustaceans, the nauplius has an eye, antennae, jointed appendages, and an exoskeleton. (3) After several molts, the nauplius develops into a cyprid larva. The cyprid has larger antennae and more body segments and appendages than the nauplius and a thin, but tough, carapace that protects its body. (4) Shortly after becoming a cyprid, the larva settles to the bottom and attaches to a solid surface, using specialized cement glands located on the antennae. Once attached, the cyprid molts and rotates its body so that its appendages, now called *cirripeds,* face upward. (5) The cyprid's carapace acts as a form around which the animal's shell develops. The feathery cirripeds filter food from the surrounding water.

Will Bizily Mook 1B

## In Summary

Arthropods are the most successful group of animals. They have an exoskeleton, jointed appendages, and sophisticated sense organs. The arthropods known as chelicerates lack mouthparts, having instead appendages called chelicerae that are modified for feeding. Chelicerates are represented in the marine environment by sea spiders and horseshoe crabs. Mandibulate arthropods have a pair of appendages called mandibles that are used for feeding. In the marine environment crustaceans are the most abundant mandibulates. Examples of crustaceans include lobsters, crabs, shrimp, barnacles, copepods, amphipods, and numerous other small animals widely distributed in the marine environment. Crustaceans are important members of marine food webs. ●

## LOPHOPHORATES

**Lophophorates** (lohf-uh-FOHR-ayts) are sessile animals that lack a distinct head and possess a feeding device called a **lophophore** (LOHF-uh-fohr). The lophophore is an arrangement of ciliated tentacles that surround the mouth and functions in feeding and gas exchange, and it can be withdrawn when the animal is disturbed. There are three phyla of lophophorate animals: the phoronids, the bryozoans, and the brachiopods.

### Phoronids

**Phoronids** (phylum Phoronida) are small, wormlike animals (Figure 9-28a) that range from a few millimeters to 30 centimeters (12 inches) long. They secrete a tube of leathery protein or chitin that can be attached to rocks, shells, or pilings or buried in bottom sediments. Phoronids feed on plankton and detritus that they catch in the mucus on their tentacles. In addition to reproducing sexually, phoronids can reproduce asexually either by budding or by transverse fission. Phoronids have a planktonic larval stage.

### Bryozoans

**Bryozoans** (phylum Ectoprocta) are small animals (averaging 0.5 millimeters, or 0.02 inches, high) that are extremely abundant (see Figure 9-28b). The majority live in shallow water, forming colonies on a wide variety of solid surfaces, such as rock, shell, algae, mangroves, and ship bottoms. A few species live and move about in the interstitial spaces of marine sand, some bore into coral and other calcareous material, and a few live on soft bottoms. Along with hydroids, bryozoans rank among the most abundant marine epiphytic animals. Large brown algae are colonized by many species of bryozoan, which display distinct preferences for certain types of algae. Evidence indicates that at the time of settling, the larvae of epiphytic species are attracted to the appropriate algal surface, perhaps by some substance produced by the algae.

Bryozoan colonies may appear as white encrustations or as fuzzy growths, or they may resemble hydrozoan colonies. Each colony is composed of thousands of tiny individuals called **zooids**, each of which inhabits a boxlike chamber

(a) Richard N. Mariscal

(b) Richard N. Mariscal

(c) Peter Batson /imagequestmarine.com

**Figure 9-28 Representative Lophophorates.** (a) Phoronids are small, wormlike lophophorates. (b) A colony of *Bugula* (bryozoan) from Newport Harbor, California. (c) The brachiopod *Terebratulina* has a body protected by a shell composed of two asymmetrical valves and fastened to the bottom by a fleshy stalk (pedicle).

that it secretes. When feeding, a zooid extends its lophophore; at other times it is withdrawn into the chamber. Most bryozoans are hermaphrodites with only one sex organ functioning at a time. The vast majority of bryozoans brood their eggs, though a few species shed small eggs directly into the seawater. After a planktonic larval stage, the larvae settle to form new colonies.

### Brachiopods

Some species of **brachiopod** (phylum Brachiopoda), or lamp shells, have changed very little since they first evolved more than 400 million years ago (see Figure 9-28c). Most species

# Induced Defenses

The bryozoan *Membranipora membranacea* forms encrusting colonies on the surface of kelp. Marine biologists have long known that some colonies of this bryozoan have large protective spines and others do not. They also observed that the colonies with spines were frequently being attacked by *Doridella steinberghae,* a nudibranch that specializes in feeding on *Membranipora* (Figure 9-B). These observations prompted Dr. C. Drew Harvell to hypothesize that the protective spines in *Membranipora membranacea* were formed only when the colony was threatened by predators such as *Doridella.* She reasoned that an induced defense, such as the production of spines, might be an effective strategy for surviving attacks by a predator like *Doridella* that feeds slowly and intermittently.

To test her hypothesis, Dr. Harvell set up several aquaria containing *Membranipora membranacea* in her lab. The aquaria were identical in all respects with the ex-

ception that the experimental colonies were exposed to one of two nudibranchs: *Doridella steinberghae,* a *Membranipora* specialist, or *Onchidoris muricata,* a bryozoan generalist. She found that colonies exposed to the predatory nudibranchs produced large spines within 2 days, whereas those in the control tank did not. It did not matter which predator was introduced. The bryozoan colonies responded to each in the same way and formed protective spines.

Dr. Harvell observed that only colonies that suffered direct mechanical damage by nudibranchs produced spines. General mechanical damage alone, however, was not sufficient to induce spine production, because control colonies mechanically damaged during their removal from the kelp did not form spines. This observation suggested that chemical cues from the predator may also be involved in the response. To test this hypothesis, predatory nudibranchs were homogenized, and the homogenate was added to the water in the aquarium with the experimental colonies. The control colonies were maintained under identical conditions except they were given a homogenate made from a marine snail that is also common on kelp. The colonies in the experimental group formed protective spines within 2 days again, whereas none of the colonies in the control group formed spines. The results of this experiment suggest that a chemical substance from the nudibranchs is responsible for the spine-forming response.

In all of the experimental colonies, spines were induced on existing and developing zooids around the entire perimeter of the colony under attack, thus fortifying the edges. Induced spine formation always

followed a consistent pattern, with regular bands of spined zooids forming completely around the colony. This pattern of response seems to indicate that information about a predator attack could be sent to a site removed from the actual attack, triggering a programmed response of spine development throughout the colony. The spines effectively control the pattern and rate of nudibranch feeding. In the field, nudibranch damage was restricted to the central, unspined portions of the colony. The defended zooids along the margin of the colony can usually regenerate new individuals to replace the damaged zooids in the central area, thus maintaining the colony.

In each of the experiments colonies stopped growing spines within a day after the nudibranchs were removed. Spines that were already formed were permanent, and in nature these indicate colonies that have had prior attacks by predatory nudibranchs.

There is an ecological cost associated with the protection. The spine response, which increases with the level of predator attack, is associated with reduced growth of colonies. Spined colonies not only grow more slowly, they also begin reproduction earlier, remain reproductive for shorter periods of time, and die sooner than unspined colonies. Taken together these results confirm that spined colonies, in the absence of predators, have a reduced fitness. This observation supports the hypothesis that the benefits of inducible defenses in the presence of predators are partly balanced by fitness costs. This would explain why not all bryozoan colonies produce protective spines before being attacked by predators. ●

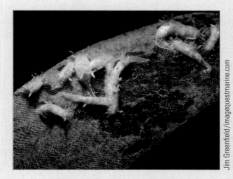

Jim Greenfield /imagequestmarine.com

**Figure 9-B Induced Defense.** *Colonies of the bryozoan* Membranipora membranacea *produce protective spines within 2 days of being attacked by a nudibranch predator.*

are benthic and live in shallow water. They range in size from 5 millimeters to 8 centimeters (0.02 to 3 inches). They have shells with two valves that resemble the shells of molluscs. The two valves of the brachiopod shell are different in size and shape and are dorsal and ventral. Molluscan bi-

valves have right and left valves that are generally similar in size and shape. Another difference between molluscs and brachiopods is the fleshy stalk called the **pedicle** found in many brachiopods. The pedicle attaches the brachiopod shell to a hard surface or is buried in soft sediments. The

*Jeff Patrich*
*Mook 121*

animal's body occupies only a portion of the shell, and an extension of the body wall, the mantle, lines the shell and is responsible for secreting it. Lying in the mantle cavity is a large, horseshoe-shaped lophophore that is used for feeding and gas exchange. Brachiopods feed on detritus and algae that are swept into a groove leading to the mouth by the ciliated lophophore. Sexes in brachiopods are generally separate and eggs and sperm are generally shed into the seawater. A planktonic larval stage lasts for 24–30 hours, after which the larvae settle to develop into adults.

## Ecological Roles of Lophophorates

Lophophorates as a group are filter feeders. In turn they supply food for a number of large invertebrates, especially molluscs and crustaceans. From an economic standpoint marine bryozoans are one of the groups of organisms most responsible for fouling ship's bottoms. About 120 species, of which different species of *Bugula* are among the most abundant, have been taken from ship bottoms.

## In Summary

Lophophorates are sessile animals that possess a feeding device called a lophophore. Lophophorates include phoronids, bryozoans, and brachiopods. ●

# ECHINODERMS: ANIMALS WITH SPINY SKINS

**Echinoderms** (phylum Echinodermata) are represented by well-known animals such as sea stars, sea urchins, and sea cucumbers. The name *echinoderm* means "spiny skinned" and refers to the spiny projections found on many species. It is thought that echinoderms evolved from a bilateral ancestor, because all of the larval forms still exhibit bilateral symmetry. Most of the adults, however, exhibit a modified form of radial symmetry. The move toward radial symmetry may have been an adaptation to a sessile lifestyle similar to that exhibited by some of the living echinoderms, such as sea lilies. As mentioned previously, radial symmetry benefits a sessile animal because it allows the animal to meet its environment equally well on all sides.

Echinoderms are mostly benthic organisms and can be found at virtually all depths. In fact, sea cucumbers and brittle stars are usually the most common form of animal life in deep-sea dredging samples.

## Echinoderm Structure

The spiny covering of echinoderms is the result of the **endoskeleton** (internal skeleton) that lies just beneath the epidermis. The endoskeleton is composed of plates of calcium carbonate (**ossicles**) held together by connective tissue, and the spines and tubercles that produce the spiny surface of the echinoderm project outward from these plates. Around the bases of the spines are tiny, pincerlike structures called **pedicellariae** (ped-uh-suh-LEHR-ee-eh; Fig-

ure 9-29). These structures are found only in some echinoderms. They keep the surface of the body clean and free of parasites and larvae of various fouling species. In some species, they may also aid in obtaining food.

The **water vascular system** is a feature unique to echinoderms. It is a hydraulic system that functions in locomotion, feeding, gas exchange, and excretion. Water enters the water vascular system by way of the **madreporite** and passes through a system of canals that runs throughout the animal's body. Attached to some of these canals are **tube feet.** Each tube foot is hollow and has a saclike structure, the **ampulla,** that lies within the body and a sucker at the end that protrudes from an opening along the arm called the **ambulacral groove.** In some species, however, the terminal suckers are lacking. Later in this section we will examine how different species use their tube feet for locomotion and feeding.

## Sea Stars

A typical sea star (class Asteroidea) is composed of a central disc and five arms, or rays. The mouth is located on the underside, and radiating from the mouth along each ray is an ambulacral groove with tiny tube feet (see Figure 9-29). The **aboral surface** (the side opposite the mouth) is frequently rough or spiny.

Sea stars move when water is pumped into the tube feet from the ampullae, causing them to project from the ambulacral groove. The suckers on the end hold firmly to solid surfaces, and muscles in the tube feet contract, forcing water back into the ampullae and causing the tube feet to shorten. As a result, the animal is pulled over the surface in the desired direction. This type of locomotion is best suited for hard surfaces like rocks. When the animal is moving across sand, the tube feet move in a walking type of motion. Sea star locomotion is generally slow, moving the animal along at the rate of a few centimeters per minute.

### Feeding in Sea Stars

The majority of sea stars are either carnivores or scavengers and feed on all sorts of invertebrates and even fish. Most sea stars locate their prey by sensing substances the prey releases into the water. Some sea stars that inhabit soft-bottom habitats, including species of *Luidia* and *Astropecten,* can even locate prey that is buried in sediments. Sea stars use their tube feet in feeding. When feeding on large bivalves, such as mussels, the sea star wraps around its prey and pries the valves apart (Figure 9-30). The sea star everts a portion of its stomach out of its mouth, and the stomach is inserted into the bivalve, where it digests the prey. Digestive enzymes are supplied by pairs of digestive glands that are located in each of the sea star's rays. When the sea star has finished feeding, it withdraws its stomach back into its mouth and moves away from its prey.

### Reproduction and Regeneration

Sea stars have great powers of regeneration. They can produce new rays if one or more are removed, and some species are able to cast off injured rays and then grow new ones. Several species are capable of regenerating entire individu-

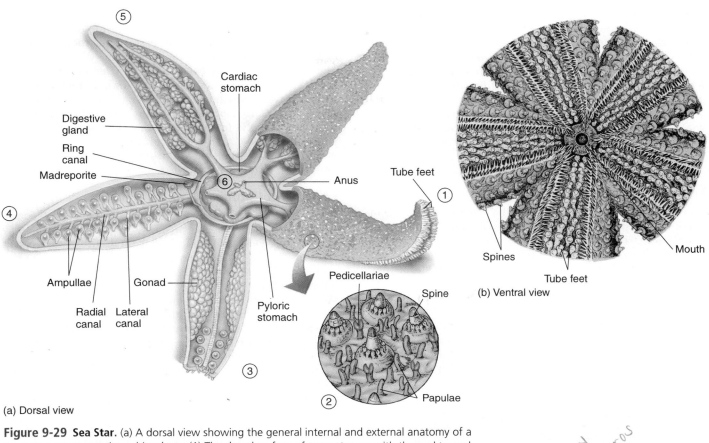

(a) Dorsal view

(b) Ventral view

**Figure 9-29  Sea Star.** (a) A dorsal view showing the general internal and external anatomy of a sea star, a representative echinoderm. (1) The dorsal surface of a sea star ray with the end turned up shows the tube feet on the ventral surface. (2) A close-up view of the dorsal surface of a sea star ray shows the spines are outward projections of the bony plates (ossicles) beneath the skin. The pincerlike pedicellariae keep the surface free of parasites and larvae. The papulae function in gas exchange and excretion. (3) Paired gonads are located in each ray. (4) Radiating from the ring canal, like spokes on a wheel, are the radial canals. The radial canals are attached by short side branches (lateral canals) to the ampullae of the tube feet. (5) Each ray contains digestive glands. (6) The central disk contains the stomachs, anus, madreporite, and ring canal. (b) An enlarged ventral view of a sea star. The mouth is centrally located. Notice the rows of tube feet within the grooves of each ray.

*Brittany*
*Ballesteros*
*Mook*
*121*

**Figure 9-30  Sea Star Feeding.** Some sea stars use their tube feet to pry open the shells of bivalves. When the valves are opened, the sea star everts a portion of its stomach and secretes enzymes to break down the flesh of its prey.

als from a single ray as long as a portion of the central disc is present. A number of sea stars normally reproduce asexually. Commonly, this involves a division of the central disc so that the animal breaks into two parts. Each half then regenerates the missing portion of the disc and arms, although extra arms are commonly produced. Species of *Linckia,* a genus of common sea stars in the Pacific and other parts of the world, are remarkable in being able to cast off arms near the base of the disc. Unlike those of other sea stars, the severed arm regenerates a new disc and rays. Regeneration is a slow process, and it usually requires 1 year before the missing parts are completely reformed.

With few exceptions the sexes in sea stars are separate. In the majority of sea stars the eggs and sperm are shed freely into the seawater, where fertilization takes place. There is usually only one breeding season per year and a single female may shed as many as 2.5 million eggs. In most sea stars the liberated eggs and larval stages are planktonic, although some Arctic and Antarctic species brood their eggs beneath the central disc.

## Ophiuroids

The class Ophiuroidea contains the greatest number of echinoderm species. Brittle stars, basket stars, and serpent stars are all examples of echinoderms known as **ophiuroids.** Ophiuroids are benthic organisms that can be found from shallow water to the ocean's depths. Like sea stars, they have five arms, but the arms are very slender and distinct from the central disc, and they are frequently covered with many spines (Figure 9-31a). Ophiuroids lack pedicellariae, and their ambulacral grooves are closed. The tube feet play a role in feeding and locomotion but do not have suckers. They tend to avoid light, coming out at night to feed and hiding under rocks and in crevices during the day. Many species are burrowers, burying themselves in bottom sediments with only their arms trailing over the seafloor to trap food.

Ophiuroids known as brittle stars get their name because they will detach one or more arms when disturbed, and some species may even shed a portion of their central disc. Once an arm is shed, it undulates wildly, no longer controlled by the central nervous system. The undulating arm distracts potential predators while the brittle star moves away to safety. It later regenerates the missing part. Because the arms of many ophiuroids tend to writhe in a serpentine fashion when they move, they are also called serpent stars.

### Feeding in Ophiuroids

Ophiuroids can be carnivores, scavengers, deposit feeders, suspension feeders, or filter feeders. Brittle stars are mostly suspension feeders, filtering food from the water, or deposit feeders, feeding on organic material that they find on the bottom. When filter feeding, they lift their arms from the bottom and wave them through the water. Strands of mucus strung between the spines on adjacent arms form a net to trap plankton and organic material. The food is then either swept to the mouth by the action of cilia or collected from the spines by tube feet and passed to the mouth. Deposit feeders use their podia in feeding. The podia collect organic particles from bottom sediments, compact them into food balls, and move them toward the mouth. Brittle stars that are predominantly carnivores feed largely on polychaetes, molluscs, and small crustaceans.

Basket stars (see Figure 9-31b) are suspension feeders that can capture relatively large zooplankton (10 to 30 millimeters, or 0.4 to 1.2 inches). They climb up on corals or rocks at night and fan their arms toward the prevailing water current. They capture their prey by coiling the ends of their arm branches around it. Minute hooks on the arm surface prevent the prey from escaping. The basket star removes the collected plankton from the arms by passing them through their comblike oral papillae.

### Reproduction and Regeneration in Ophiuroids

Many ophiuroids can cast off, or **autotomize,** one or more arms if disturbed or seized by a predator. A break can occur at any point beyond the disc and the lost portion is then regenerated. There are some ophiuroids, notably six-armed species of *Ophiactis,* in which asexual reproduction takes place by division of the disc into two pieces, each with three arms.

Although sexes are separate in the majority of ophiuroids, hermaphroditic species are not uncommon. Many species shed their gametes into the water column where fertilization and development take place. Some species, however, brood their eggs in their ovaries or body cavity. The larval stages in all species are planktonic and metamorphosis occurs in the water column before settling.

(a)

Charles Seaborn

(b)

Richard N. Mariscal

**Figure 9-31 Brittle Stars.** (a) Several serpent stars (*Ophiothrix*) on the surface of a sponge. (b) The basket star *Gorgonocephalus arcticus* from Friday Harbor, Washington.

## Sea Urchins and Their Relatives

Sea urchins, heart urchins, sand dollars, and their relatives are known as **echinoids** (class Echinoidea, meaning "like a hedgehog"). Echinoids have a body that is enclosed by a

(a)

(b)

**Figure 9-32 Echinoids.** (a) This purple sea urchin *Strongylocentrus purpuratus* lives off the west coast of the United States. (b) Irregular echinoids such as the heart urchin are adapted to burrowing and have much smaller spines than sea urchins.

hard endoskeleton, or test. These animals are benthic organisms that can be found from shallow water to the ocean's depths. Sea urchins tend to occupy solid surfaces such as rocks, whereas sand dollars and heart urchins prefer to bury themselves in sandy bottoms.

**Regular** (radial) **echinoids** are known as sea urchins. A sea urchin's spheroid body is armed with relatively long, movable spines (Figure 9-32a). Most are 6 to 12 centimeters (2.5 to 5 inches) in diameter, but some Indo-Pacific species may reach nearly 36 cm (14.5 inches).

The bilateral, or **irregular, echinoids** include heart urchins and sand dollars. These animals are adapted for burrowing in sand. Unlike sea urchins, the test in these species is covered with many small spines, which are used in locomotion and keeping sediment off the body surface. Heart ur-

chins (family Spatangidae) are ovoid (see Figure 9-32b). The typical sand dollar (family Mellitidae) has a flattened, circular body. A few species, such as sea biscuits (*Clypeaster*), are shaped somewhat like heart urchins.

### Echinoid Structure

The tube feet of echinoids project from five pairs of ambulacral areas that are derived from the same embryonic structures as the arms of sea stars, and spines project from the test (Figure 9-33). Sea urchins have relatively long spines, and they move mostly by means of their tube feet, with some help from their spines. The spines function in protection and, in some species (*Diadema*), contain a venom. When a person comes into contact with one of these stinging urchins, the spine tip breaks off in the skin and injects the venom,

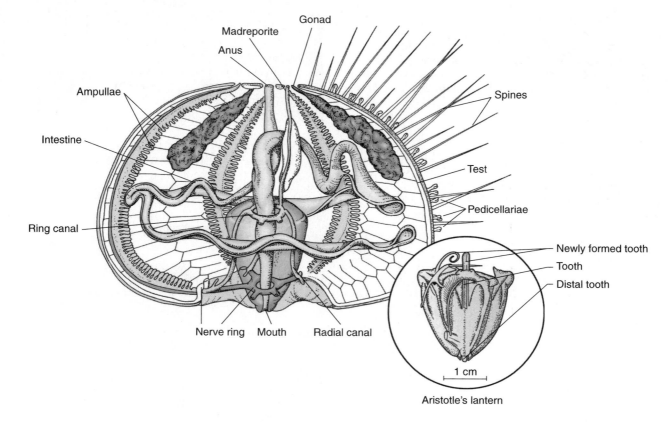

**Figure 9-33 Sea Urchin.** The anatomy of a sea urchin, including Aristotle's lantern.

which causes a severe burning sensation. Some spines have barbed spinelets along their length that make them difficult to remove. The spines allow some species to live along the shore in regions pounded by surf. They help to dissipate the energy of the waves and protect the animal from wave shock. Sand dollars and heart urchins have very short spines that give them a fuzzy appearance.

The sexes are separate in all echinoids. Regular echinoids have five gonads, but most irregular echinoids have only four. Sperm and eggs are shed into the seawater, where fertilization takes place, and the planktonic larvae swim and feed for as long as several months before gradually sinking to the bottom.

## Feeding in Echinoids

Most regular echinoids are herbivores that feed on algae and some marine plants. Irregular echinoids are deposit feeders or suspension feeders.

***Feeding in Regular Echinoids*** The majority of sea urchins are grazers, scraping the surfaces on which they live with their teeth. A sea urchin's mouth contains five teeth that form a chewing structure called **Aristotle's lantern** (see Figure 9-33) because of its resemblance to ancient Greek lanterns and in honor of that early student of echinoderm biology. Although algae is usually the most important food, most sea urchins are generalists and feed on a wide range of plant and animal material. The sea urchin *Lytechinus variegates* inhabits turtlegrass beds and is an herbivore, consuming about 1 gram of grass per day. The ecological role of grazing by some sea urchins was dramatically revealed following the 1983 crash in Caribbean populations of the long-spined urchin *Diadema antillarum*. In some areas where the urchins died out the thickness of the algal mat increased from 1 to 2 millimeters to 20 to 30 millimeters. Boring sea urchins feed on algae that grow on the walls of their burrows as well as fragments of algae and organic matter that are washed into the burrow.

***Feeding in Irregular Urchins*** Irregular urchins are selective deposit feeders. Most species feed on organic material in the sand in which they burrow. Heart urchins use modified podia on their oral surface to collect food. Sand dollars use their podia to pick up particles beneath their oral surface. On the west coast of the United States, the sand dollar *Dendraster excentricus* feeds on suspended particles. This sand dollar lives in quiet water with the posterior half of its body projecting above the sand. Food includes not only particles collected between the spines but also diatoms and algal fragments collected by the tube feet and small crustaceans caught by the pedicellariae.

## Sea Cucumbers

Sea cucumbers (class Holothuroidea; Figure 9-34a) have elongated bodies, and, in most species, the body wall is leathery. They range from 3 centimeters (1.2 inches) long to 1 meter (3.3 feet) long and 24 centimeters (10 inches) in di-

(a)

*Richard N. Mariscal*

**Figure 9-34 Sea Cucumber.** (a) Although it resembles a cucumber, this animal is an echinoderm called a *sea cucumber*. (b) The anatomy of a sea cucumber.

Intestine  Gonad  Ring canal  Stomach  Madreporite  Oral tentacles  Pharynx  Anus  Ambulacral areas  Respiratory tree

(b)

ameter (*Stichopus* from the Philippines). Because of their body shape, sea cucumbers usually lie on one side. They move slowly using their ventral tube feet and the muscular contractions of their body wall. Gas exchange in most sea cucumbers is accomplished by a system of tubules called respiratory trees. The respiratory trees are located in the body cavity on the right and left sides of the digestive tract (Figure 9-34b). The sexes are generally separate. In species that do not brood their eggs, the eggs are incubated on the dorsal body surface. Some species brood their eggs in their body cavity, and the larvae leave the body of the mother through a rupture in the anal region. After hatching, the larvae of all sea cucumbers continue development in the water column as members of the plankton.

### Feeding in Sea Cucumbers

Sea cucumbers are mainly deposit or suspension feeders. Located around the mouth of a sea cucumber are 10 to 30 modified tube feet called **oral tentacles** (see Figure 9-34b). Most sea cucumbers feed on small particles of food that they trap with these tentacles. The tentacles are coated with a sticky mucus, and any small organism that comes into contact with them becomes stuck. Sea cucumbers retract their tentacles into their mouths to remove the food and then extend them to capture more. Other species use their tentacles to shovel sand and bottom sediments into their mouths. They then digest the organic material, leaving conspicuous mounds of sand or fecal-mud castings behind. Many sedentary species that live on hard surfaces or beneath stones are suspension feeders.

### Defensive Behavior

When disturbed, some species of sea cucumber release tubules (**Cuvierian tubules**) from their anus (Figure 9-35). The tangle of tubules resembles spaghetti, and, on contact with seawater, they become sticky. The tubules will stick tenaciously to a predator, which is distracted by the need to clean itself, and the sea cucumber escapes. The tubules are strong enough to ensnare and immobilize crustaceans, and they are distasteful to fishes. This behavior helps slow-moving sea cucumbers deter potential predators.

Other species will **eviscerate,** that is, release some of their internal organs through either the anus or the mouth. In each instance, the animal ultimately regenerates the lost body parts. In some species, evisceration appears to be seasonal and may represent the release of a digestive tract that is full of wastes following a long feeding period. These waste-laden intestines can serve as an important source of food for detritus feeders.

## Crinoids

**Crinoids** (class Crinoidea) are commonly called sea lilies and feather stars, and their bodies look very much like flowers. They are the most primitive of the echinoderms and have a long fossil history. Attached stalked crinoids called sea lilies flourished during the Paleozoic era, and some 80 species still exist today (Figure 9-36a). The stalk of sessile sea lilies may reach a length of 1 meter (3.3 feet). Modern sea lilies,

**Figure 9-35  Sea Cucumber Defense.** This sea cucumber from the Red Sea is releasing tubules of Cuvier. This behavior distracts potential predators and gives the sea cucumber time to escape.

however, live at depths of 100 meters (330 feet) or more and are not commonly encountered. The majority of living crinoids are feather stars (see Figure 9-36b). These are free-living crinoids whose habitat ranges from the intertidal zone to great depths, and some occur in large numbers on coral reefs. Feather stars are free moving. They swim and crawl only for short distances, and swimming is largely an escape response. They cling to the bottom for long periods by means of the grasping **cirri.** Many of the shallow-water species are nocturnal. For instance, three Red Sea species that inhabit coral reefs hide during the day in crevices and deep within branching corals, keeping their arms tightly rolled. Stimulated by the lowered light intensities at sunset, they crawl out of their hiding places to exposed positions and extend their arms to feed.

Crinoids are suspension feeders. They feed on small organisms that are filtered from the water by their tube feet and by mucus nets of the ambulacral grooves. There is still some uncertainty about the precise nature of crinoid food. Gut contents of Red Sea crinoids are largely zooplankton, but *Antedon bifida* from the Indo-Pacific appears to feed largely on detritus.

Crinoids possess considerable powers of regeneration and in this respect are similar to sea stars. Part or all of an arm can be cast off if seized or if the animal is subjected

(a)

(b)

Richard N. Mariscal

**Figure 9-36  Crinoids.** (a) Crinoids such as this sea lily have changed very little since they first evolved. Most modern species live at depths of greater than 100 meters. (b) The majority of living crinoids are feather stars, such as this one from the Pacific Ocean.

to unfavorable environmental conditions. The lost arm is then regenerated. Such regeneration may be important in surviving fish predation. Sexes are separate in all crinoids and eggs and sperm are shed into the seawater. After a free-swimming stage, the larva settles to the bottom and attaches. This is followed by a metamorphosis resulting in the formation of a minute crinoid.

## Ecological Roles of Echinoderms

Because of their spiny skins, echinoderms are not preyed upon by many animals. Some molluscs and sea otters eat sea urchins and sea stars. Many spider crabs feed on echinoderms, which they tear apart or crush with their chelipeds. Humans eat sea urchin gonads and sea cucumbers, and some fishes have mouthparts modified for feeding on echinoderms.

On the other hand, many echinoderms are predators of molluscs, other echinoderms, cnidarians, and crustaceans. The crown-of-thorns sea star (*Acanthaster planci*) is a major predator of corals. A population explosion in the late 1960s caused concern among ecologists worldwide that the sea star would threaten the existence of many Pacific reefs. The corals were able to recover, and the threat is not as grave as once thought. Many theories have been proposed for the population explosion, but the cause is still unknown.

Sea urchins have destroyed several valuable kelp forests of the west coast of the United States. In the Atlantic, excessive populations of sea urchins have become a nuisance by robbing lobster traps. Many former lobster fishers are now sea urchin fishers. They sell the **roe** (ovary with eggs) to Japan for $100 to $150 a pound where it is used in sushi. Throughout the Caribbean the populations of the black sea urchin (*Diadema antillarum*) exploded because of depletion of predatory fishes. The population of *Diadema* then suddenly collapsed because of an infectious disease, leaving the coral reefs without any effective control of algae. The resulting overgrowth of algae contributed to a 90% reduction in coral reef area since 1980. The complexity and unpredictability of these population fluctuations and their effects remind us of how little we understand the role of echinoderms in marine ecology.

In the Pacific near China, sea cucumber beds have been depleted and fishers have begun to threaten beds off South America and the Galápagos Islands. In the future sea cucumbers may be collected as a source of medicine. Pacific islanders have long known that cut up sea cucumbers can be used to poison fish in tide pools. The poison, **holothurin,** has various effects on nerve and muscle and also suppresses the growth of certain tumors.

## In Summary

Echinoderms exhibit radial symmetry as adults, although their larvae exhibit bilateral symmetry, suggesting that they evolved from bilateral ancestors. Echinoderms have an internal skeleton and a unique water vascular system that functions in locomotion and food gathering. Echinoderms are represented by sea stars, brittle stars, sea urchins, sea cucumbers, and feather stars and sea lilies (crinoids). ●

## TUNICATES

**Tunicates** (subphylum Urochordata) are mostly sessile animals that are widely distributed in all seas. These animals are named for their body covering, called the **tunic,** which is largely composed of a substance similar to cellulose.

### Sea Squirts

Tunicates known as sea squirts (class Ascidiacea; Figure 9-37a) get their name because many species, when irritated, will forcefully expel a stream of water from their excurrent siphon. The bodies of sea squirts are generally round or cylindrical and have two tubes projecting from them: an incurrent siphon that brings in water and food, and an excurrent siphon that eliminates water and wastes. Food is trapped on a mucous net that is formed in the animal's pharynx (see Figure 9-37b). The pharynx is also used in gas exchange.

Sea squirts can be solitary, colonial, or compound organisms, which are composed of several individuals called zooids that share a common tunic. A great diversity of species inhabit shallow tropical seas and many minute colonial

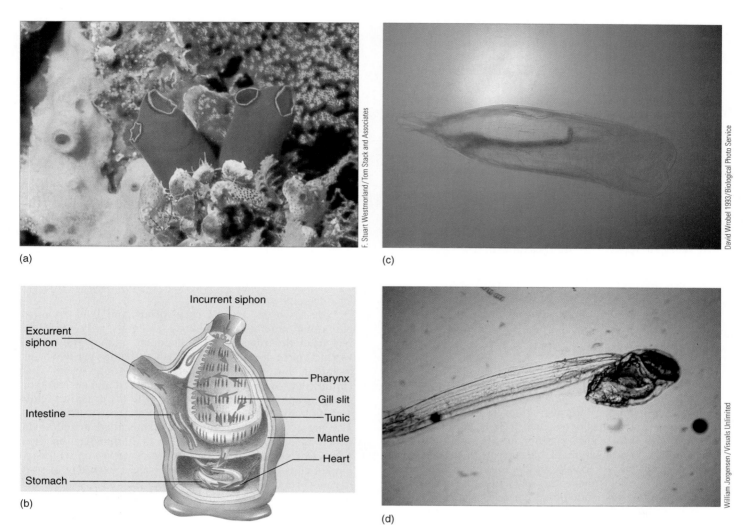

(a)

(b)

(c)

(d)

F. Stuart Westmorland/Tom Stack and Associates

David Wrobel 1993/Biological Photo Service

William Jorgensen / Visuals Unlimited

Incurrent siphon

Excurrent
siphon

Pharynx

Gill slit

Tunic

Mantle

Heart

Intestine

Stomach

**Figure 9-37** **Tunicates.** (a) Two solitary blue and gold sea squirts live on a reef in the Philippine Islands. (b) The blue arrows in this drawing of the anatomy of a sea squirt represent the pattern of water flow through the animal's body, and the red arrows represent the movement of food. (c) Thaliaceans, also known as salps, are planktonic tunicates. (d) A larvacean is a free-swimming tunicate.

forms inhabit crevices in old coral heads and the underside of coralline rock. Others form large conspicuous clusters on gorgonian corals and mangrove roots or massive rubbery lobes on rocks and pilings. Most of the larger sea squirts, such as *Ascidia* and *Molgula,* are solitary.

Sea squirts are filter feeders and remove plankton from the water passing through their pharynx. Some members of the colonial family Didemnidae contain symbiotic algae within the tunic. One of these symbionts, prokaryotes in the genus *Prochloron,* is found only in ascidians. These same ascidians may also have symbiotic cyanobacteria in their tunic. The excess food produced by these photosynthetic symbionts is thought to be used by the tunicate as an auxiliary food source. At least one *Didemnum* species may migrate over the bottom as it grows to optimize the light intensity for photosynthesis by its *Prochloron* symbionts.

All sea squirts can regenerate damaged body parts, but asexual reproduction is characteristic only of colonial tunicates. In these animals, asexual reproduction takes place by means of budding. With few exceptions tunicates are hermaphrodites that release their gametes into the water column, where fertilization takes place. The larval stage resembles a tadpole, and after a free-swimming period of 36 hours or less, the larva settles and metamorphoses to the sessile stage.

## Salps and Larvaceans

**Salps** (class Thaliacea; see Figure 9-37c) are free-swimming tunicates and have their incurrent and excurrent siphons located on opposite ends of their barrel-shaped bodies. As they swim, they pump water through their bodies, extracting food and eliminating wastes. When food supplies are abundant, the populations of salps increase dramatically. Some species are bioluminescent.

**Larvaceans** (class Larvacea) are another group of free-swimming tunicates (see Figure 9-37d). They produce delicate enclosures made of mucus that are used in feeding. When these mucous networks are clogged (approximately every 4 hours) they shed them and produce another in a matter of minutes. Like salps, when food supplies are abundant, larvacean populations explode. Scuba diving at these times among the shed mucous networks has been compared to swimming through a snowstorm.

## In Summary

Tunicates have bodies that are covered with a tunic composed of molecules similar to cellulose. Sea squirts are sedentary tunicates that filter their food from the water. They can be solitary, colonial, or compound animals. Salps and larvaceans are planktonic tunicates. ●

## CEPHALOCHORDATES

**Cephalochordates** (subphylum Cephalochordata) are fish-like chordates that are collectively known as **lancelets.** Lancelets are slender, laterally compressed and eel-like in appearance and behavior (Figure 9-38a). Adult lancelets range in size from 4 to 8 centimeters (0.4 to 3.2 inches). Lancelets are benthic animals that burrow in coarse, shelly, current-swept sands, usually in shallow nearshore areas. Their heads project above the sand into the water from which they filter suspended food particles. Lancelets ingest large quantities of particles from which they extract organic material while eliminating water through gill slits and mineral particles through the anus. The sexes are separate in cephalochordates and fertilization is external. Their life cycles are complex, including a benthic adult and planktonic swimming larva.

Cephalochordates are important as food in parts of Asia. Despite their small size, lancelets are clean muscular animals that lack bones. Fresh animals may be fried for immediate consumption or dried for later use. One amphi-

oxus fishery in southern China recorded an annual catch of 35 tons, or approximately 1 billion, lancelets (*Brachiostoma belcheri*) from a fishing ground 1 mile wide and 6 miles long. Five thousand individuals of *Brachiostoma caribbaeum* per square meter have been reported from Discovery Bay in Jamaica. In parts of Brazil, chickens are herded onto beaches to feed on lancelets.

## ARROWWORMS

**Arrowworms** (phylum Chaetognatha) are common animals found in marine plankton. Most are found in tropical waters in the upper 900 meters (2,970 feet). The body of an arrowworm is torpedo-shaped (see Figure 9-38b) and ranges from 2 to 120 millimeters (0.08 to 2.4 inches) long. Hanging down from each side of the head and flanking the vestibule (the large chamber that leads to the mouth) are 4 to 14 large, curved hooks, called **grasping spines,** that are used to seize prey.

Arrowworms are all carnivorous and feed on other planktonic animals, particularly copepods, which they detect from the vibrations the copepods make. The arrowworm *Sagitta* consumes young fish and other arrowworms, including those of its own species, as large as itself. When capturing prey, arrowworms dart forward and spread their grasping spines. The prey is seized with the grasping spines and its body may be pierced with the arrowworm's teeth. The arrowworm then injects a toxin called tetrodotoxin, which is derived from bacteria, to immobilize its prey. Arrowworms are voracious feeders. *Sagitta nagae,* for example, consumes 37% of its own weight each day in prey. Arrowworms are important ecologically as a trophic link between the primary and higher consumers in marine food chains.

## In Summary

Cephalochordates, also known as lancelets, are small animals that resemble eels. They are found in the bottom sediments along coastal areas where they filter food from the water. Arrowworms are predatory members of the zooplankton that feed on a variety of pelagic animals, including small fish. ●

(a)

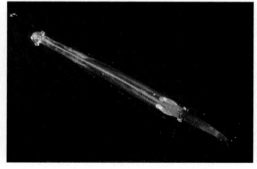

(b)

**Figure 9-38 Lancelets and Arrowworms.** (a) The lancelet *Amphioxus* has a body shaped like an eel. It lives in coastal sediments where it feeds on organic material it removes from particles suspended in the water. (b) The arrowworm is a planktonic carnivore that feeds on copepods, other zooplankton, and small fish.

## IN PERSPECTIVE

| Phylum or Subphylum | Representative Organisms | Form and Function | Reproduction and Development | Type of Feeding | Ecological Roles |
|---|---|---|---|---|---|
| Mollusca | Chitons, snails, nudibranchs, bivalves, tusk shells, nautilus, octopods, squids, cuttlefish | Most are covered with a shell; notable exceptions are nudibranchs and cephalopods. With the exception of bivalves, all have a unique tooth structure called a radula. Gas exchange with gills or mantle cavity. Cephalopods have a highly developed nervous system | Mainly sexual reproduction; separate sexes or hermaphrodites depending on species; larval stages include trochophore and veliger | All types of feeding; herbivores, carnivores, filter feeders, suspension feeders, scavengers, and deposit feeders | Important food organisms; several commensal species; shipworms cause commercial damage |
| Arthropoda | Horseshoe crabs, sea spiders, crabs, lobsters, shrimp, krill, copepods, amphipods, and barnacles | A hard or tough exoskeleton protects the body. Paired, jointed appendages aid in locomotion and feeding. Gas exchange is by gills or through the body surface. Growth by molting | Mainly sexual reproduction; separate sexes or hermaphrodites depending on the species; larval stages include nauplius, zoea, and cyprid, depending on the class | All types: herbivores, carnivores, filter feeders, suspension feeders, scavengers, and deposit feeders | Important food organisms, planktonic crustaceans are main links between phytoplankton and higher-order consumers in marine food chains; many symbiotic species; grass shrimp recycle cellulose, and barnacles are important fouling organisms |
| Phoronida | Phoronids | Wormlike animals that secrete a leathery tube around the body | Can reproduce asexually by budding or transverse fission. About half of the species are hermaphrodites and the other half are separate sexes. Some females brood their eggs, whereas others release them into the water column. The larval stage called an *actinotropha* is planktonic | Use a lophophore to feed on plankton and detritus | Channel food to higher-order consumers |
| Bryozoa | Bryozoans | Form colonies of small individuals called zooids that appear as white encrustations on solid surfaces | Most are hermaphrodites and the cyphonautes larval stage is planktonic | Use a lophophore to filter food from the water column | Channel nutrients from plankton to higher-order consumers |
| Brachiopoda | Lamp shells | Body covered by two asymmetrical valves; many species have a stalk or pedicle that anchors the animal | Sexes are separate, and larval stage resembles a tiny brachiopod and is planktonic | Use a lophophore to feed on detritus and algae | Channel nutrients from plankton to higher-order consumers |

*(continues)*

## IN PERSPECTIVE (continued)

| Phylum or Subphylum | Representative Organisms | Form and Function | Reproduction and Development | Type of Feeding | Ecological Roles |
|---|---|---|---|---|---|
| Echinoderms | Sea stars, brittle stars, basket stars, sea urchins, sand dollars, sea cucumbers, and crinoids | Adults exhibit secondary radial symmetry. All have an endoskeleton and a spiny body surface. A unique water vascular system is used for locomotion, feeding, and circulating internal fluids | Can reproduce asexually and sexually; some species brood their eggs, whereas others release them into the water column. The bilateral larval stages are planktonic. The larval stage of asteroids is the bipinnaria. Ophiuroids and echinoids have a pluteus larva and the larval stage of sea cucumbers and crinoids is the pentacula or vitellaria larva | All types: herbivores, carnivores, filter feeders, deposit feeders, and scavengers | Not preyed upon by many animals because of their spiny exteriors; major predators of bivalve molluscs, cnidarians, crustaceans, and other echinoderms as well as other invertebrates and sometimes fish. Sea urchins can cause extensive damage to kelp and are important grazers that keep algal populations under control |
| Urochordata | Sea squirts, salps, and larvaceans | Body is covered by a tunic composed of a polysaccharide similar to cellulose. Salps and larvaceans are planktonic. Sea squirts are benthic and can be solitary or colonial | Colonial forms can reproduce asexually. Sexual forms are hermaphrodites. Larvae are planktonic and referred to as tadpole larvae | Filter feeders that feed on plankton | Channel nutrients from plankton to higher-order consumers. Some harbor symbiotic photosynthetic bacteria |
| Cephalochordata | Lancelets | Body resembles an eel; adults are benthic | Sexes are separate and fertilization occurs in the water column; larval stage is planktonic | Feed on organic material extracted from particles filtered from the water | Channel nutrients to higher-level consumers |
| Chaetognatha | Arrowworms | Torpedo-shaped body with grasping spines located around the mouth | All are hermaphrodites; when young hatch they resemble adults | Predators that feed on zooplankton, especially copepods, and small fish | Link between the primary and higher consumers in marine food chains |

## SELECTED KEY TERMS

coleoid, *p. 198*

conchiolin, *p. 191*

copepod, *p. 207*

copulatory pleopod, *p. 206*

crinoid, *p. 217*

crop, *p. 199*

crustacean, *p. 204*

Cuvierian tubule, *p. 217*

cyprid larva, *p. 208*

decapod, *p. 205*

echinoderm, *p. 212*

echinoid, *p. 214*

ectoparasite, *p. 208*

endoskeleton, *p. 212*

eviscerate, *p. 217*

exhalant opening, *p. 196*

exoskeleton, *p. 203*

gastropod, *p. 192*

gnathopods, *p. 207*

grasping spine, *p. 220*

head-foot, *p. 190*

holothurin, *p. 218*

inhalant opening, *p. 196*

irregular echinoid, *p. 215*

krill, *p. 207*

lancelet, *p. 220*

larvacean, *p. 220*

lophophorate, *p. 210*

lophophore, *p. 210*

madreporite, *p. 212*

mantle, *p. 191*

mantle cavity, *p. 191*

melanin, *p. 200*

mollusc, *p. 190*

molting, *p. 203*

mother-of-pearl layer, *p. 192*

nacreous layer, *p. 191*

nauplius larva, *p. 206*

nautiloid, *p. 198*

nudibranch, *p. 193*

operculum, *p. 192*

ophiuroid, *p. 214*

oral tentacles, *p. 217*

ossicles, *p. 212*

oviduct, *p. 201*

palp, bivalve, *p. 196*

palp, sea spider, *p. 203*

pedicellariae, *p. 212*

pedicle, *p. 211*

pen, *p. 200*

periostracum, *p. 191*

pheromone, *p. 194*

phoronid, *p. 210*

photophore, *p. 207*

prismatic layer, *p. 191*

radula, *p. 191*

regular echinoid, *p. 215*

roe, *p. 218*

salp, *p. 219*

scaphopod, *p. 192*

sepia, *p. 200*

septa, *p. 198*

shipworm, *p. 198*

siphon, *p. 196*

siphuncle, *p. 198*

spermatophore, *p. 201*

swarm, *p. 207*

swimmeret, *p. 204*

telson, *p. 203*

tentacle, *p. 200*

trochophore larva, *p. 194*

tube feet, *p. 212*

tunic, *p. 218*

tunicate, *p. 218*

umbo, *p. 196*

univalve, *p. 192*

valve, *p. 195*

veliger larva, *p. 194*

visceral mass, *p. 190*

water vascular system, *p. 212*

whorls, *p. 192*

zoea larva, *p. 206*

zooid, *p. 210*

## QUESTIONS FOR REVIEW

### Multiple Choice

1. The molluscan shell is secreted by the
   - a. foot
   - b. skin
   - c. gill
   - d. kidney
   - e. mantle

2. The _____ is a unique, toothlike structure found in many molluscs.
   - a. mantle
   - b. radula
   - c. veliger
   - d. trochophore
   - e. exoskeleton

3. Molluscs that have shells composed of eight plates held together by a fleshy girdle are
   - a. snails
   - b. bivalves
   - c. scaphopods
   - d. cephalopods
   - e. chitons

4. A type of bivalve that can damage wood is the
   - a. scallop
   - b. shipworm
   - c. nudibranch
   - d. oyster
   - e. surf clam

5. Molluscs that have tentacles and a highly developed nervous system are
   - a. gastropods
   - b. bivalves
   - c. scaphopods
   - d. cephalopods
   - e. chitons

6. Arthropod characteristics include
   - a. a water vascular system
   - b. a hard external shell composed of calcium carbonate
   - c. jointed appendages
   - d. soft segmented bodies with hydrostatic skeletons
   - e. a tunic composed of cellulose

7. During molting, arthropods
   - a. grow extra appendages
   - b. shed their old exoskeleton
   - c. are well protected from predators
   - d. shed excess appendages
   - e. reproduce

8. While on a field trip to the seashore, you discover an animal with a spiny skin and a water vascular system. This animal is probably a(n)
   - a. mollusc
   - b. echinoderm
   - c. arthropod
   - d. lophophorate
   - e. tunicate

9. Animals known as *brachiopods*
   a. possess an exoskeleton composed of chitin
   b. exhibit radial symmetry as adults
   c. produce a shell composed of two valves
   d. are all members of the plankton
   e. are also called *tunicates*

10. Which of the following adult animals is likely to be a member of the zooplankton?
    a. crab
    b. squid
    c. sea star
    d. larvacean
    e. sea cucumber

## Short Answer

1. What adaptations allow squids to be successful predators?

2. What is a lophophore and how does it function in feeding?

3. Name four commercially important crustaceans.

4. Explain how the radula is modified in gastropods for different types of feeding.

5. Describe how a sea star uses its water vascular system to move.

6. Explain how slow-moving animals such as sea cucumbers avoid predation.

7. Describe how sea squirts feed.

8. Why are arthropods such a successful group of animals?

9. Distinguish between chelicerates and mandibulates.

10. How are regular and irregular echinoids particularly adapted to their particular lifestyles?

11. Why do bivalves that burrow in soft sediments need siphons?

## Thinking Critically

1. What is the advantage of being able to alter the sex ratio in populations of the slipper limpet, *Crepidula fornicata?*

2. On a field trip to the ocean you discover a gastropod that you have never seen before. Because you are very interested in gastropod feeding, you want to know whether this animal is an herbivore or a carnivore. How could you determine with a fair amount of certainty which type of feeder this animal is?

3. What type of shell characteristics would you expect to observe in a gastropod that spends most of its life burrowing through soft sediments?

4. A sample of a deep-sea dredging contains an animal that is small, flat, and round. It has a ventrally located mouth, a stomach, but no intestines. To which of the groups of invertebrates introduced in this chapter does this animal probably belong? What other characteristics would the animal have to have to confirm your identification?

5. Some of the invertebrates covered in this chapter brood their eggs and other do not. What are the advantages in each of these reproductive strategies?

## SUGGESTIONS FOR FURTHER READING

Abbott, R. T. 1974. *American Seashells,* 2nd ed. New York: Van Nostrand Reinhold.

Bavendam, F. 1996. Flowers of the Coral Seas: Feather Star Crinoids, *National Geographic* 190(6):118–30.

Bavendam, F. 1995. The Giant Cuttlefish: Chameleon of the Reef, *National Geographic* 188(3):94–107.

Bavendam, F. 1991. Eye to Eye with the Giant Octopus, *National Geographic* 179(3):86–97.

Bavendam, F. 1989. Even for Ethereal Phantasms, It's a Dog-Eat-Dog World, *Smithsonian* 20(5):94–101.

Gosline, J. M., and M. E. DeMont. 1985. Jet-Propelled Swimming in Squids, *Scientific American* 252(1): 96–103.

Hadley, N. F. 1986. The Arthropod Cuticle, *Scientific American* 244(7):104–112.

Harvell, C. D. 1992. Inducible Defenses and Allocation Shifts in a Marine Bryozoan, *Ecology* 73(5):1567–76

Harvell, C. D. 1990. The Ecology and Evolution of Inducible Defenses, *The Quarterly Review of Biology* 65(3):232–340.

Ruppert, E. E., R. S. Fox, and R. D. Barnes. 2004. *Invertebrate Zoology,* 7th ed. Belmont, Calif.: Brooks/Cole—Thomson Learning.

### InfoTrac College Edition Articles

Asakura, A. 1995. Sexual Differences in Life History and Resource Utilization by the Hermit Crab, *Ecology* 76(7).

Glausiusz, J. 1997. The Importance of Being Infected: Without Its Bacteria the Hawaiian Bobtail Squid Would Be a Different Animal, *Discover* 18(5).

Jensen, M. N. 1998. Land Hermit Crabs Spurn Leftovers, *Science News* 153(8).

Leonard, G. H., M. D. Bertness, P. O. Yund. 1999. Crab Predation, Waterborne Cues, and Inducible Defenses in the Blue Mussel, *Mytilus edulis, Ecology* 80(1).

McFall-Ngai, M., E. G. Ruby. 1991. Symbiont Recognition and Subsequent Morphogenesis as Early Events in an Animal-Bacterial Mutualism, *Science* 254(5037).

Pechenik, J. A., D. E. Went, J. N. Jarrett. 1998. Metamorphosis Is Not a New Beginning: Larval Experience Influences Juvenile Performance, *Bioscience* 48(11).

Pennings, S. C., T. H. Carefoot, E. L. Siska, M. E. Chase, T. A. Page. 1998. Feeding Preferences of a Generalist Salt-Marsh Crab: Relative Importance of Multiple Plant Traits, *Ecology* 79(6).

Raloff, J. 2001. Wanted: Reef Cleaners, *Science News* 160(8).

Starr, M., J. H. Himmelman, J. Therriault. 1990. Direct

Coupling of Marine Invertebrate Spawning with Phytoplankton Blooms, *Science* 247(4946).

Vecchione, M., R. E. Young, A. Guerra, D. J. Lindsay, D. A. Clague, J. M. Bernhard, W. W. Sager, A. F. Gonzalez, F. J. Rocha, M. Segonzac. 2001. Worldwide Observations of Remarkable Deep-Sea Squids, *Science* 294(5551).

Zimmer, C. 1994. See How They Run, *Discover* 15(9).

## Websites

### Molluscs

**http://coa.acnatsci.org/conchnet/c-101b.html**  Dr. Gary Rosenberg's tutorial on scientific names.

**http://data.acnatsci.org/obis/**  Database of Indo-Pacific marine molluscs.

### Arthropods

**http://www.museum.vic.gov.au/crust/page1.html**  Museum Victoria provides information on crustaceans in Southern Australia.

**http://tolweb.org/tree?group=Crustacea&contgroup= Arthropoda**  Tree of Life's crustacea page.

**http://www.floridasmart.com/subjects/ocean/animals _ocean_crust.htm**  Florida Smart website provides links to information on crustaceans.

### Echinoderms

**http://www.photovault.com/Link/Animals/Aquatic/ oEchinoderms/AAOVolume01.html**  Photo images of echinoderms.

**http://www.ucmp.berkeley.edu/echinodermata/echino dermata.html**  Site provides information and links on echinoderms.

**http://www.calacademy.org/research/izg/echinoderm/**  The California Academy of Sciences echinoderm web page.

**http://www.oceanicresearch.org/echinoderm.html**  Photos and descriptions of echinoderms.

# 10

# Fishes

## Key Concepts

1. Hagfishes and lampreys are jawless fishes.

2. Sharks, skates, and rays have skeletons composed entirely of cartilage.

3. Sharks have streamlined bodies and highly developed senses that help them to be efficient predators.

4. Most marine fishes have skeletons composed primarily of bone.

5. The shape of a fish's body is primarily determined by the characteristics of its environment.

6. Many fishes exhibit coloration and color patterns that help them blend in with their environment.

7. Color in fishes functions in camouflage, species recognition, and communication.

8. Most bony fishes have a swim bladder that helps them maintain neutral buoyancy.

9. Most marine fishes are carnivorous, but herbivores, omnivores, and filter feeders also exist.

10. Most marine fishes are oviparous and produce large numbers of eggs.

11. Fishes such as salmon and eels migrate long distances sometime during their life cycle.

Although hagfish, lampreys, sharks, and bony fishes are traditionally termed *fishes,* these animals are at least as distinct from one another as they are from their chordate relatives, the amphibians, reptiles, birds, and mammals. All are members of the phylum Chordata, animals characterized, at least some time in their development, by the presence of pharyngeal gill slits, a dorsal hollow nerve cord, a notochord (a slender skeletal rod), and a tail lying posterior to the anus.

Marine fishes can be found from the surface waters of the ocean to its deepest trenches. Their species outnumber all other vertebrate species combined. They display an amazing array of adaptations that allow them to exploit virtually every niche. Fish are commercially important for human food, fertilizer, and other products. Indeed, commercial fishing has greatly depleted their stocks in many locations.

## FISHES

The "fishes" along with amphibians, reptiles, birds, and mammals are classified as vertebrates. **Vertebrates** are distinguished from other chordates (tunicates and lancelets) by the presence of **vertebrae,** a series of bones or cartilages that surround the spinal cord and help support the body. The vertebrae also provide attachment sites for muscles, thereby increasing and improving the animal's mobility.

The first fishes to evolve lacked both paired fins and jaws. They probably spent their time scavenging food in the bottom sediments of early seas, a niche they filled for millions of years. About 425 million years ago, some fishes appeared that possessed jaws and paired fins. Fishes with these adaptations could more efficiently obtain food and ultimately replaced all but a few of the early jawless forms. Modern fishes now include most of the highest-level consumers in the sea.

# Surviving in Near-Freezing Water

Of the approximately 25,000 species of fish on earth, only about 275 occur in the Antarctic Sea. Surviving the subfreezing temperatures is a major challenge to all organisms living here. Marine invertebrates that inhabit the bottom of Antarctica's continental shelf regularly encounter temperatures as low as −2.7°C. They don't freeze because their bodies contain the same concentration of salts and minerals as that of the surrounding seawater, as well as a variety of organic compounds, which depresses the freezing point of their body fluids. The blood and other body fluids of marine fishes, however, are more dilute than seawater, and their tissues freeze at approximately −0.2°C. At this temperature, if an ice crystal were even to brush a fish's skin, it would quickly propagate and pierce the skin like a spear. But most Antarctic fishes encounter ice without any difficulty. It was not until the 1960s that physiologists began to study how fishes in this environment can survive under such conditions. They discovered that the fish have developed a relatively simple solution to the problem. The fishes have glycoprotein molecules (molecules

composed of sugar and protein) in their blood and other body fluids that act as antifreeze. These glycoproteins are 200 to 500 times more effective than an equivalent amount of sodium chloride in preventing the formation of ice crystals. Glycoproteins do not actually decrease the temperature at which ice crystals form but lower the temperature at which they enlarge and destroy cells.

Another problem faced by fishes in the Antarctic Sea concerns hemoglobin, a molecule in the red blood cells of vertebrates. Hemoglobin is responsible for the red blood cell's ability to carry oxygen, and at low temperatures it functions very inefficiently. To transport enough oxygen at the low temperatures of the Antarctic waters would require large amounts of hemoglobin and large numbers of red blood cells. Unfortunately, this would make the blood too viscous to flow efficiently. Because of these environmental circumstances, Antarctic fishes gradually became more dependent on oxygen dissolved in their blood plasma, reducing their need for hemoglobin and red blood cells. Indeed, the 17 species of icefish (family Channichthyi-

**Figure 10-A Icefish.** *This icefish from Antarctica lacks the oxygen-carrying protein hemoglobin. The oxygen that its cells need is carried dissolved in the relatively large volume of blood that circulates through its body.*

dae; Figure 10-A) generally lack red blood cells entirely. To compensate for the anemia that results from this condition, these fish have a large heart and wide blood vessels, creating a circulatory system that can handle a large volume of blood under low pressure. This allows the fish to expend minimal amounts of energy to pump the large volumes of blood necessary to carry oxygen in the absence of hemoglobin. ●

## JAWLESS FISHES

Hagfish (class Myxini) and lampreys (class Cephalospidomorphi) lack both jaws and paired appendages. Their skeletons are entirely composed of cartilage and their bodies lack scales. Although superficially similar, hagfish and lampreys are as morphologically distinct from one another as they are from the jawed vertebrates. Because hagfish lack vertebrae, some scientists do not even consider them to be vertebrates. Recent comparisons of DNA sequences among fishes, however, indicate that hagfish are most closely related to the lampreys.

### Hagfishes

Hagfish (Figure 10-1a), sometimes called "slime eels," are bottom-dwelling fish found in ocean waters throughout the world, with the possible exception of Arctic and Antarctic waters. Although they are seldom found at depths shallower

than 600 meters (1,980 feet) in the tropics, some species enter the intertidal zone in the cold coastal waters off South Africa, Chile, and New Zealand. In some parts of the world, hagfish support a thriving commercial fishery, their tanned hide being used for leather goods. The demand for hagfish skin has risen to the point that many populations are depleted. Unfortunately, our knowledge of hagfish biology is insufficient to manage a sustainable fishery.

When feeding, hagfish extend a feeding apparatus composed of two structures, termed **dental plates,** containing horny cusps. They use these dental plates to grasp the flesh of their prey, usually small, soft-bodied invertebrates. As the feeding apparatus is withdrawn into the mouth, the dental plates close tightly, and the cusps tear away the flesh of the prey. A fang above the dental plates keeps live prey from wriggling away between bites. In addition to feeding on live prey, hagfishes act as scavengers on dead or dying whales and other large vertebrates. Because hagfish cannot penetrate the skin of whales or fish, they enter the body cavity

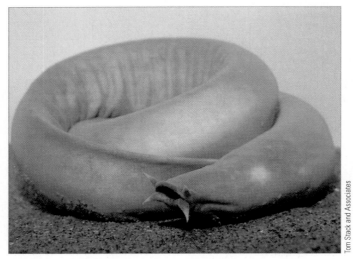

(a)

**Figure 10-1  Hagfish.** (a) This jawless fish is a hagfish from the Pacific Ocean. (b) Hagfish have flexible bodies that they can tie into overhand knots. They tie the knot at their tail, then slide it forward and over their head. This behavior is used to remove excess slime from their body, to escape predators, and to gain leverage when tearing flesh from their prey.

through the mouth, anus, or openings made by other scavengers. They literally eat the carcass from inside out, leaving only the bones and skin.

Hagfishes possess slime glands positioned along their body that can produce large amounts of a milky gelatinous fluid when the animal is disturbed. Although some slime is produced during feeding, its most important function seems to be physical protection. When seized, they produce so much slime that it coats the gills of predatory fish and either suffocates them or encourages them to release their prey. Hagfishes can remove the slime from their bodies by tying their tails in an overhand knot and sliding the knot forward and over the head (see Figure 10-1b).

Little is known about reproduction in the approximately 60 species of hagfish. Although some individuals have both ovaries and testes, the sexes are normally separate. We still do not know where or when hagfish lay their eggs, and relatively few fertilized eggs or juveniles have ever been found. Why females outnumber males by 100 to 1 in some species is also a mystery.

## Lampreys

Unlike hagfish, lamprey species inhabit both freshwater and the ocean. As adults, the approximately 40 species of lamprey possess an oral disk and rasping tongue covered with toothlike plates of keratin. Several species use these plates to grasp prey, rasp a hole in the body, and suck out both tissues and fluids. Marine lamprey species, such as *Petromyzon marinus* (Figure 10-2), spend their adult life in the ocean feeding on other fish but return to freshwater to spawn. In North America, spawning takes place in the spring. Males

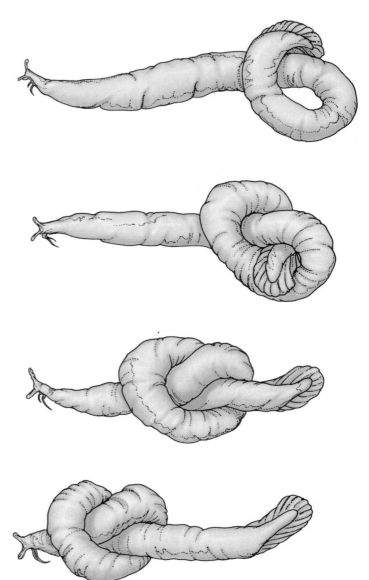

(b)

migrate up rivers and build nests from stones in the shallow riffles of clear streams. The females arrive later and attach to one of the stones of the nest by their oral sucker. The male then attaches to the back of the female, and as the eggs are shed, the male sheds his sperm. The fertilized eggs stick to stones in the nest and are ultimately covered by sand. The adults die shortly after reproducing. Larvae hatch in about 2 weeks and migrate downstream, where they burrow into the river bottom with their mouths directed toward the current. The larvae are filter feeders, and they remain burrowed for 3 to 7 years before finally metamorphosing into adults. Predatory forms travel to the sea where they live for up to 2 years before returning to freshwater to spawn. Many species of lamprey and even some populations of parasitic forms, however, do not feed as adults.

Although the invasion of the Great Lakes by the sea lamprey in the early part of the twentieth century brought

**Figure 10-2 Lamprey.** The marine lamprey *Petromyzon marinus.*

about the near extinction of many lake inhabitants, lampreys, in general, do not seriously deplete their prey populations. Currently, Great Lakes lamprey populations are controlled with the use of lampricides that destroy the larvae of parasitic and nonparasitic species alike. Lampreys have some commercial value as food, and their large nerves make them an excellent subject for neurobiological research.

## In Summary

Hagfishes and lampreys are the only existing representatives of early jawless fishes. Hagfishes are bottom-dwelling predators of soft-bodied invertebrates and scavengers of large vertebrates. Very little is known about their natural history and reproduction. Some lamprey species are parasitic on other fish, rasping a hole in their bodies and sucking out both tissues and fluids. Other species do not feed as adults. The larval form is a filter feeder in freshwater. The invasion of the Great Lakes by the sea lamprey in the early part of the twentieth century brought about the near extinction of many lake inhabitants, but lampreys and their prey are able to co-exist in other areas. ●

## CARTILAGINOUS FISHES

Sharks, skates, rays, and chimaeras are the modern representatives of the cartilaginous fishes, or class Chondrichthyes. Their skeletons are composed entirely of cartilage, although it is often strengthened by the deposition of calcium salts. They possess jaws and paired fins, and their skin is covered with **placoid scales** (Figure 10-3). The teeth of cartilaginous fishes are modified placoid scales. Placoid scales can take several different forms, such as spines or large denticles (toothlike structures) on the back of skates and rays, but they usually form a sandpaper-like covering on sharks.

Cartilaginous fishes can be divided into two major groups, the **holocephalans** (chimaeras, or ratfish) and the **elasmobranchs.** The elasmobranchs have evolved into two general body forms, the typically streamlined bodies of sharks and the dorsoventrally flattened bodies of skates and rays. There are about 760 species of cartilaginous fish including, with the exception of whales, the largest living ver-

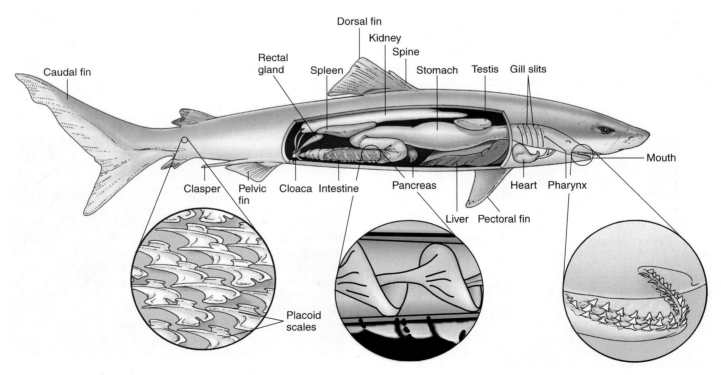

**Figure 10-3 Shark Anatomy.** The external and internal anatomy of a typical shark. Sharks have streamlined bodies and are well adapted for rapid swimming. Their skin is covered with placoid scales, which are similar to the teeth of other vertebrates. Modified placoid scales are found on the jaws and serve as teeth.

**Figure 10-4  Whale Shark.** A whale shark cruises a reef off Western Australia.

tebrate animals. The plankton-feeding whale shark (*Rhincodon;* Figure 10-4), for example, may exceed 15 meters (49 feet) in length and is the largest species of fish.

# Sharks

Sharks generally have streamlined bodies and are excellent swimmers. Using their massive trunk muscles, sharks swim with powerful, sideways sweeps of the tail, or **caudal fin.** The caudal fin of many sharks is termed a **heterocercal tail,** the dorsal lobe being longer than the ventral. Additional fins include one or two dorsal fins as well as paired pectoral and pelvic fins. A sharp spine may be associated with the dorsal fins, as in the spiny dogfish (*Squalus acanthias*). The pelvic fins of male sharks are partly modified to form **claspers** (Figure 10-5). Claspers transfer sperm from the male to the female during reproduction.

Because sharks are slightly denser than water, they sink if they stop swimming. Many sharks are able to compensate for this buoyancy problem with their large livers (in some species the liver may account for 20% of the shark's weight). The shark's liver produces large quantities of an oily material called **squalene.** Squalene has a density less than seawater's (squalene's density is 0.8 $g/cm^3$; the density of seawater is 1.020 to 1.029 $g/cm^3$), and this helps to offset the shark's high density.

## Shark Sensory Systems

A shark's eyes lack eyelids, having instead a clear **nictitating membrane** that covers the eye and protects it. The eyes of many species possess both rods and cones, indicating an ability to perceive color. In contrast to bony fishes, vision seems to be of less importance than **olfaction** (smell) in finding prey.

***Olfaction***    The olfactory receptors are very well developed and located in sacs or pits that are usually located in front of the mouth. Almost two thirds of the cells in a shark's brain are involved in processing olfactory information. Indeed, these animals can detect the presence of a drop of blood diluted in one million parts of water and accurately find the source. It is no wonder that some biologists refer to sharks as "swimming noses." The role that the sense of smell plays in locating prey may explain the shape of the hammerhead shark's head (Figure 10-6). This shark has a nostril at the tip of each end of the "hammer," and as it swims, it moves its head from one side to the other. When the strength of a smell is equal in both nostrils, the shark senses that its prey is straight ahead.

***Lateral Line System***    Another important sensory organ is the **lateral line system,** which consists of canals running the length of the animal's body and over the head (Figure 10-7). At regular intervals, the canals open to the outside and there

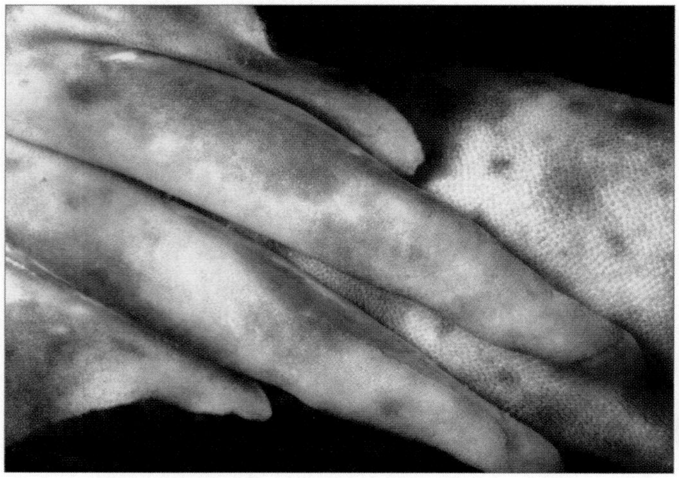

**Figure 10-5  Claspers.** Male sharks have modified pelvic fins termed claspers that are used to transfer sperm from the male to the female during sexual reproduction.

**Figure 10-6  Hammerhead Shark.** The hammerhead shark has a nostril at the tip of each end of its "hammer." As it swims, it moves its head from side to side. When the strength of a smell is equal in both nostrils, the shark senses that its prey is straight ahead.

# Shark Attacks on Humans

Sharks have acquired a rather bad reputation in recent times because of accounts of attacks on humans in various movies, novels, and the popular press. In truth, the annual risk of death from lightning is 30 times greater than that from sharks. Although any large shark may be a potential risk to human beings, most species are actually rather timid and cautious animals. Of the approximately 350 shark species, only 32 have been documented in attacks on humans. The three species that seem to be most often involved in deadly attacks are the white shark (*Carcharodon carcharias*), tiger shark (*Galeocerdo cuvier*), and bull shark (*Carcharhinus leucas*) (Figure 10-B). All these sharks are large and cosmopolitan in distribution, and they feed on large prey such as marine mammals, sea turtles, and large fish. Other shark species that have been implicated in human attacks include the mako (*Isurus oxyrhynchus*), hammerhead (*Sphyrna*), oceanic whitetip (*Carcharhinus longimanus*), Galápagos (*Carcharhinus galapagensis*), and various "reef" sharks.

Since 1958, the International Shark Attack File, administered by the American Elasmobranch Society and the Florida Museum of Natural History at the University of Florida, has compiled data on shark attacks throughout the world. According to their data, most attacks (82%) have occurred in North American waters in recent years. Attacks have also been reported from Australia, Brazil, and South Africa. Surfers are the most common targets,

(a)

Mike Parry/Minden Pictures

(b)

Flip Nicklin/Minden Pictures

(c)

Flip Nicklin/Minden Pictures

**Figure 10-B Dangerous Sharks.** *The three shark species most often involved in shark attacks are the (a) great white shark* (Carcharodon carcharias)*, (b) tiger shark* (Galeocerdo cuvier)*, and (c) bull shark* (Carcharhinus leucas)*.*

followed by swimmers or waders and divers or snorkelers.

## Types of Shark Attacks

Worldwide, there are usually less than 75 shark attacks and fewer than a dozen fatalities reported annually. Three types of attacks are often described, hit-and-run, bump-and-bite, and sneak attacks. **Hit-and-run attacks** occur most often in shallow water under conditions of low visibility on swimmers or surfers who are splashing the water. The shark bites and then releases the swimmer, usually causing relatively minor lacerations on the limbs that are seldom life threatening. **Bump-and-bite attacks,** in contrast, occur on swimmers or divers in deeper water. The shark bumps the victim before attacking. **Sneak attacks** differ in that they occur without warning. Both sneak and bump-and-bite attacks may occur repeatedly and cause deep lacerations that are severe enough to result in the death of the victim.

## Preventing Shark Attacks

Preventing shark attacks involves such commonsense practices as never swimming alone and avoiding areas where people are fishing or where blood and human wastes may be in the water. Don't swim at dusk or at night or where the water is murky or turbid. Also, refrain from splashing. Don't wear shiny jewelry because it may be mistaken for the scales of prey fish. If sharks are sighted, leave the water quickly and calmly. ●

is a free movement of water in and out of them. Within the canals are sensory receptors called **neuromasts** that can detect vibrations in the fluid that fills the canals. Even the slightest movements in the water around the shark stimulate these lateral line sensory receptors. The shark uses its lateral line system to locate prey and potential predators. Some researchers have suggested that the vibrations produced by a swimming human may be similar to those produced by a seal or an injured fish and may account for some sharks being attracted to swimmers.

***Ampullae of Lorenzini*** Sharks and their relatives use the **ampullae of Lorenzini** to sense electrical currents in the water. These organs are scattered over the top and sides of the animal's head. The ampullae of Lorenzini can sense changes of as little as a tenth of a microvolt in the electrical

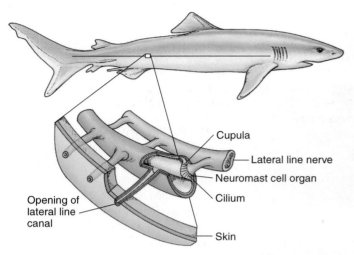

**Figure 10-7 The Lateral Line System.** The lateral line system consists of two canals, one on each side, that run the length of the animal's body. The fluid in the canals freely communicates with the water surrounding the animal. Even tiny vibrations in the surrounding water cause the fluid in the canals to move, moving the cupula of the neuromast cells and sending a signal to the animal's brain.

fields in the water. Marine biologists have speculated that sharks use these organs to sense the tiny electrical fields created by the muscles of their prey.

### Digestion in Sharks

A shark's mouth contains several rows of bladelike, triangular teeth that are used for grasping prey and tearing off large chunks. Shark teeth are continually lost and replaced, some species losing more than 30,000 in their lifetime. Sharks shake their heads when biting their prey, because they cannot move their jaws back and forth to chew. Food chunks are swallowed whole. After the food is swallowed, it passes first to the stomach (see Figure 10-3) and then on to a relatively short intestine that contains a structure called a **spiral valve.** The spiral valve aids in the absorption process by slowing the movement of food and increasing the surface area.

### Osmoregulation in Sharks

Sharks maintain a solute concentration in their internal fluids greater than or equal to that of sea water by retaining large amounts of nitrogenous wastes, mostly **urea** and trimethylamine oxide (TMAO). If they did not balance the salt concentration of the sea, they would lose body water by osmosis. Both the gills and **rectal gland,** a large structure that empties into the intestine, are involved in the excretion of excess sodium chloride, but the kidney excretes other salts. Many species of shark have the ability to enter freshwater by reducing the levels of nitrogenous wastes in their body fluids.

### Reproduction in Sharks

In males, sperm are produced in paired testes (see Figure 10-3) and transferred to the female through a groove in the claspers, the modified pelvic fins described earlier. The

**Figure 10-8 Shark Reproduction.** When nurse sharks copulate, the male holds the pectoral fin of the female in his mouth as he uses his claspers to introduce sperm into the female's genital opening.

female reproductive system consists of paired ovaries and oviducts. The oviducts carry eggs from the ovary to a portion of the oviduct that is modified to function as a uterus. Fertilization is internal in sharks (Figure 10-8). Three reproductive modes are seen in sharks and other fishes, **oviparity, ovoviviparity,** and **viviparity.**

***Oviparity***    In oviparous reproduction, the most primitive mode, eggs are laid outside the body and the embryos develop in a protective case. These egg cases attach to hard surfaces on the seafloor. Shark pups produced in this manner tend to be smaller than those produced by the other reproductive modes because of the limited availability of nutrients within the egg case. Whale sharks, bullhead sharks, and a few other shark species reproduce in this manner.

***Ovoviviparity***    In ovoviviparity, the most common method of shark reproduction, eggs hatch within the mother's uterus but no placental connection is formed. The developing young are nourished by yolk stored in the egg. In the sand tiger and a few other species, the yolk is quickly used and the embryos then feed on unfertilized eggs or other embryos in the uterus. In these species, only a single pup may ultimately be born from each uterus. Sharks exhibiting ovoviviparity include basking sharks, thresher sharks, and saw sharks.

***Viviparity***    In viviparity, the most recent mechanism to evolve, either the young directly attach to the mother's uterine wall or the mother's uterus produces "uterine milk" that is absorbed by the embryo. Viviparous sharks include requiem and hammerhead sharks.

## Skates and Rays

Skates and rays differ from sharks by having flattened bodies, greatly enlarged pectoral fins that attach to the head, reduced dorsal and caudal fins, eyes and **spiracles** (openings for the passage of water) on top of the head, and gill slits on their ventral side (Figure 10-9a and b). The location of the spiracles on the dorsal surface and gill slits on the

# Megamouth Sharks

On November 15, 1976, a group of Navy researchers off the northeastern coast of Oahu in the Hawaiian Islands discovered an unknown species of shark entangled in a nylon drogue (used to position a boat in deep water) they were retrieving. The specimen, now at the Bishop Museum in Honolulu, was subsequently described as a new species (*Megachasma pelagios*) and member of a new genus and family (Figure 10-C). Recent studies indicate that it is the most

**Figure 10-C Megamouth.** *Unknown to science until 1976, this unusual shark feeds on plankton.*

primitive member of the order Lamniformes (containing mako, white, basking, and thresher sharks). These sharks have a huge head and a highly distensible mouth with a silvery lining. Like its relative the basking shark (*Cetorhinus maximus*), the megamouth shark is a plankton feeder. It is thought that this animal feeds on the plankton community of the deep scattering layer (see Chapter 17). They seem to migrate vertically, following their prey from deep water during the daytime to mid-water depths at night. When feeding, it is thought that the large jaws protrude forward and the mouth cavity expands to suck in the plankton. Elongated, denticle-covered structures filter the shrimp and other organisms from the incoming water.

Worldwide, only 18 specimens have been captured or observed since 1976. The largest specimen examined was about 5.5 meters (17 feet) and may have weighed a half ton or more. Their mouth

contains about 50 rows of relatively small teeth, but only 3 rows are functional. In contrast to other deepwater sharks, the cartilaginous skeleton is poorly calcified and soft. Little is known about their reproductive habits because only one mature female has ever been captured.

A large shark like the megamouth probably has few predators, but there is a documented case of an attack by sperm whales (*Physeter macrocephalus*) off Indonesia in 1998. Almost all specimens have scars, perhaps from the bites of cookie-cutter sharks (*Isistius brasiliensis*).

It may be surprising that an animal as large as megamouth could go unnoticed for so long. Part of the reason may be that the animal spends its days in the remote ocean depths. Another reason may be that, because it is a filter feeder, it would not strike at baited hooks like other sharks, and thus would go unnoticed by fishers. ●

---

ventral side is an adaptation for a bottom existence. Water is drawn in through the spiracles and passed out over the gills. This arrangement helps to prevent the delicate gill filaments from becoming clogged with sand or debris. Other characteristics include the lack of an anal fin and the presence of specialized pavementlike teeth that are used for crushing prey, usually invertebrates inhabiting sandy or muddy bottoms. In contrast to other rays, the manta (*Manta birostris*) feeds on plankton. Skates and rays, with the exception of sawfishes (*Pristis*) and guitarfishes (*Rhinobatos*), swim with their pectoral fins. Most of the approximately 500 species are adapted to a bottom existence, but a few species, like the eagle (family Myliobatidae) and manta rays live in open water.

## Differences between Skates and Rays

Skates and rays differ from each other in several ways. Rays swim by moving their fins up and down as a bird moves its wings while flying. Skates, on the other hand, swim by creating a wave that begins at the forward edge of the fin and sweeps down the edge to the back of the fin, allowing the animal to glide easily along the bottom as its fins ripple. The tails of rays are streamlined and contain venomous barbs or spines. The tails of skates, in contrast, have small fins, lack

venomous spines, and are fleshier. Some species of ray grow to a much larger size than skates. The manta ray, for example, can reach a width of 7 meters (22 feet) and weigh more than 1,360 kilograms (3,000 pounds). In contrast, the largest skate (*Raja binoculata*) reaches a length of about 2.4 meters (8 feet).

Most skates are oviparous, releasing their eggs in a leathery, rectangular egg case called a *mermaid's purse* (see Figure 10-9c). Rays are ovoviviparous. At birth, the young rays' tail barbs are flexible and covered with a sheath to prevent injury to the mother during the birth of the pup.

## Defense Mechanisms

Skates and rays have evolved a variety of defenses to protect them from predators. Electric rays (*Torpedo* and *Narcine*) have a pair of electric organs in their head that can deliver up to 220 volts. In addition to defense, electric rays use their electric charge to navigate and to stun prey. Stingrays (*Dasyatis*; see Figure 10-9b) have hollow barbs connected to poison glands. These modified dorsal fin spines may be distributed along the tail or there may be a single barb at the base of the tail. When disturbed, stingrays whip their tails around. If the barb punctures the skin, it injects venom that causes swelling, cramping, and excruciating pain. A

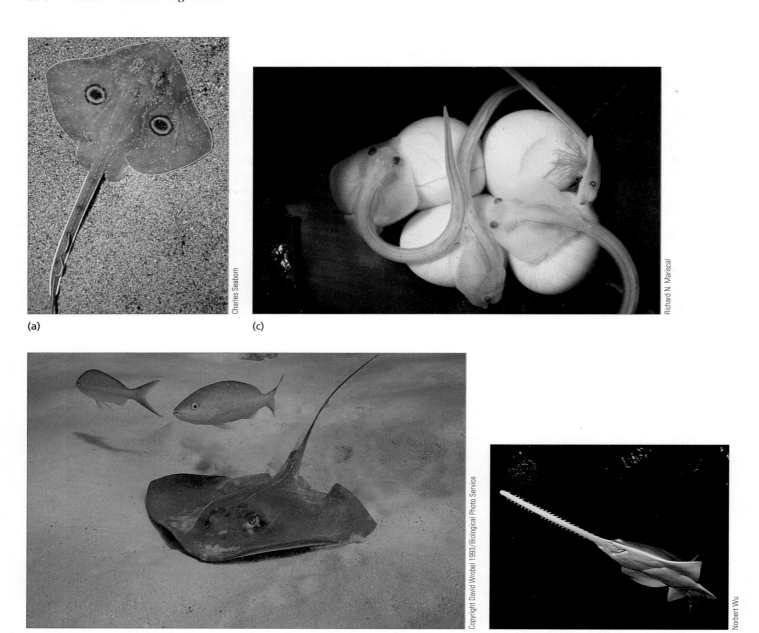

(a)

(c)

(b)

(d)

Charles Seaborn

Richard N. Mariscal

Copyright David Wrobel 1993/Biological Photo Service

Norbert Wu

**Figure 10-9 Skates and Rays.** Both skates and rays have enlarged pectoral fins. (a) Skates lack a stinging spine associated with the tail, whereas (b) rays, like this stingray, have a spine that can inflict a painful injury. Rays are ovoviviparous, whereas skates are oviparous. (c) Skates release their eggs in a leathery egg case called a *mermaid's purse*. This one contains four developing young, each attached to a large yolk sac. (d) Sawfishes are related to skates and rays.

common treatment for stingray injuries is to submerge the injured area in hot water. The heat from the water breaks down the protein toxin. Wounds from stingrays heal very slowly and are prone to bacterial infections.

Sawfishes (see Figure 10-9d) and guitarfishes are very atypical-looking rays. Sawfish have a series of barbs along their pointed rostrums. When disturbed or when feeding, they shake their heads sideways, using the sharp points of the "saw" to inflict injury. Some sawfish can reach a length

of 7.6 meters (24.7 feet) and weigh more than 600 kilograms (1,300 pounds).

## Chimaeras

Chimaeras (subclass Holocephali; Figure 10-10) are given common names such as *ratfish, rabbitfish,* and *spookfish* because of their large pointed heads and long, slender tails. Unlike other cartilaginous fish, the gills are covered with an

**Figure 10-10 Chimaeras.** Relatives of the sharks, chimaeras like this female ratfish from Vancouver Island, Canada, are bottom dwellers that feed on a variety of fishes and invertebrates.

operculum and water is taken in through the nostrils. Males have a clasper on the head as well as claspers on the pelvic fins. Chimaeras are oviparous, producing large eggs in a leathery case. Instead of teeth, chimaeras have flat plates that they use to crush their prey. They feed on a wide variety of foods, including crustaceans, molluscs, echinoderms, and fish. Generally bottom dwellers, they inhabit depths ranging from the shallows to 2,545 meters (8,400 feet) or deeper. Commercially, the 35 known species are of little value. Some are marketed as food in parts of China and New Zealand, and their oils make a fine lubricant.

## In Summary

Sharks, skates, rays, and chimaeras are the modern representatives of the cartilaginous fishes, which comprise the class Chondrichthyes. Their skeletons are composed almost entirely of cartilage, although it is often strengthened by deposition of calcium salts. Sharks have streamlined bodies and are efficient swimmers. They have many highly developed senses and most are predators. Sharks maintain the concentration of solutes in their internal fluids at greater than or equal to that of seawater by retaining large amounts of urea and TMAO. Ovoviviparity is the most common reproductive mode of sharks.

Skates and rays are similar to sharks but have flattened bodies, greatly enlarged pectoral fins that attach to the head, reduced dorsal and caudal fins, and gill slits on their ventral side. They are bottom dwellers that feed primarily on molluscs and crustaceans. Some species of ray are able to generate electricity, which they use for protection and to stun prey. Others have sharp spines on their tails. Some of these spines are hollow, connected to poison glands, and capable of inflicting serious injury. Chimaeras, or ratfish, have large pointed heads and long, slender tails. Unlike other cartilaginous fish, the gills are covered with an operculum and water is taken in through the nostrils. They feed on crustaceans, molluscs, echinoderms, and fish, using flat plates to crush their prey instead of teeth.  ●

## BONY FISHES

The approximately 25,000 species of modern bony fishes (class Osteichthyes) are so diverse that no single characteristic separates them from the cartilaginous fishes. Most forms, however, can be characterized by the presence of a swim bladder (or lung), bone, bony scales, and fin rays. Bony fishes constitute about half of all vertebrates and more than 95% of all fishes. They can be divided into two major lineages, the lobefins (subclass Sarcopterygii), containing the coelacanths (*Latimeria*) and freshwater lungfish, and the ray-finned fishes (subclass Actinopterygii), containing all other species. In many new classifications, these subclasses have been raised to the class level, reflecting the significant differences between the two groups.

### Coelacanths

Coelacanths (Figure 10-11a) are relics of the evolutionary line that gave rise to the tetrapods, or land vertebrates. They are characterized by lobed, paired fins that resemble the limbs of tetrapods. Coelacanths were known from fossils until a living specimen was discovered in 1938. This specimen was given the scientific name, *Latimeria chalumnae*, and it created quite a stir in the scientific community. A second specimen was found in the deep waters of the Indian Ocean around the Comoro Islands in 1952, and several other specimens have since been taken there, in the Mozambique Channel, and as far south as Sodwana Bay, South Africa. In 1998 American and Indonesian scientists discovered a new species of coelacanth, *Latimeria menadoensis*, off Sulawesi, Indonesia, some 10,000 kilometers (6,000 miles) east of where they were originally known.

Most coelacanths live at depths of 150 to 250 meters (495 to 825 feet) in rocky areas with steep subsurface gradients. Their skeletons are made of both bone and cartilage, but the vertebral column is essentially cartilage. Their reduced skeleton along with a fat-filled swim bladder allows the coelacanth to maintain neutral buoyancy. Like sharks, coelacanths maintain high concentrations of urea in the blood to remain nearly isotonic to seawater. They are ovoviviparous.

### Ray-Finned Fishes

The ray-finned fishes are by far the most numerous and dominant group of vertebrates in the ocean. There are so many specialized forms that it is difficult to define common characteristics, but typically, their fins are attached to the body by fin rays rather than fleshy lobes. We can divide them into two major groups. The first group (subclass Chondrostei) contains primitive forms such as the marine sturgeons (Figure 10-11b). Members of this group possess heterocercal tails like those in sharks, a skeleton made primarily of cartilage, and ganoid scales (Figure 10-12a). **Ganoid scales** are very thick and heavy, giving the fish an armored appearance. Members of the second group (subclass Neopterygii) typically have homocercal tails, **cycloid** or **ctenoid scales** (see Figure 10-12b), and more maneuverable

(a)

(b)

**Figure 10-11 Primitive Bony Fishes.** (a) A coelacanth. Notice the muscular bundles at the base of the fins. (b) The scales of this marine sturgeon are modified into bony plates that give the fish an armored appearance.

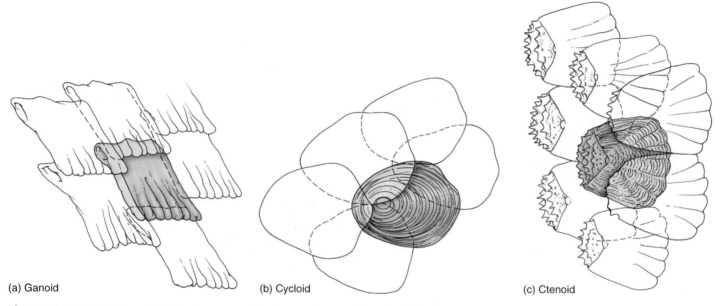

(a) Ganoid

(b) Cycloid

(c) Ctenoid

**Figure 10-12 Fish Scales.** (a) Thick, heavy, ganoid scales are characteristic of primitive bony fishes. More modern bony fishes have lighter, more flexible scales like (b) cycloid scales and (c) ctenoid scales.

fins. **Homocercal tails** have dorsal and ventral flanges that are nearly equal in size, and the vertebral column usually does not continue into the tail. Cycloid and ctenoid scales are thinner and more flexible than ganoid scales and are less cumbersome for active swimmers.

The structure of bony fish fins gives them better control of their movements. Bony fishes possess unpaired median fins and paired fins (Figure 10-13). The median fins consist of one or more **dorsal fins,** a caudal fin, and usually one **anal fin.** Median fins help fishes to maintain stability while swimming. The paired fins consist of **pectoral** and **pelvic**

**fins,** both of which are used in steering. Pectoral fins also help to stabilize the fish.

### Body Shape

The shape of a fish's body is mainly determined by the characteristics of its habitat. Fishes that are very active swimmers, such as the tuna (*Thunnus*) and marlin (*Makaira*), have a **fusiform body shape** (Figure 10-14a) with a very high and narrow tail. This streamlined body form allows these fishes to move through the water with great efficiency. Fishes that live in seagrass or on coral reefs, like butterflyfish

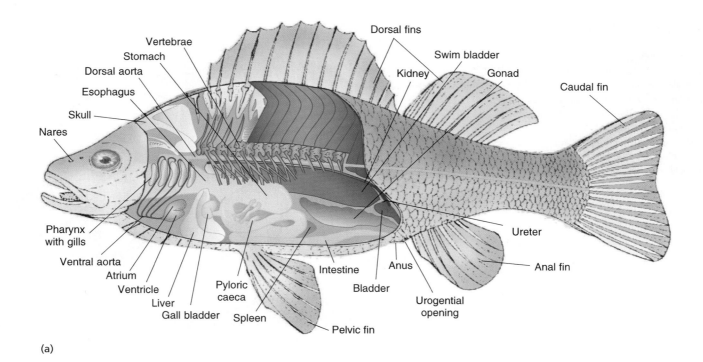

(a)

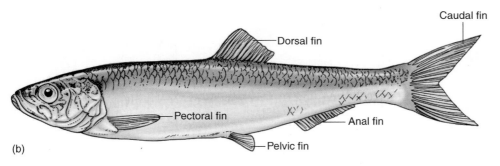

(b)

Caudal fin
Dorsal fin
Pectoral fin
Anal fin
Pelvic fin

**Figure 10-13 Bony Fish Anatomy.**
(a) The general internal and external anatomy of a bony fish. (b) In more primitive species, like the herring (Clupeidae), the pectoral fins are lower on the body and the pelvic fins are more posterior. (c) In more advanced species, like butterflyfishes (*Chaetodon*), the pelvic fins are closer to the throat and the pectoral fins are higher on the body and more vertical.

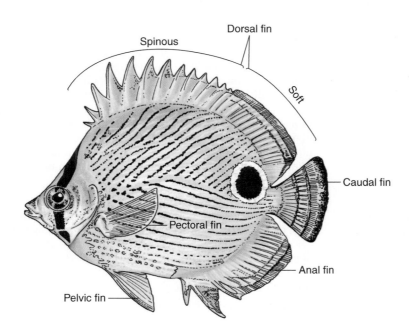

(c)

**Figure 10-14 Fish Shapes.** (a) Fishes that are active swimmers like this marlin have a fusiform body. (b) Reef fishes, such as this butterflyfish, that swim among the corals have laterally compressed or deep bodies. (c) Bottom dwellers like this flounder have horizontally compressed or depressed bodies. (d) Sedentary fishes such as this anglerfish have globular bodies. (e) Burrowing fishes and fishes, like this moray eel, that live in tight crevices have snakelike bodies.

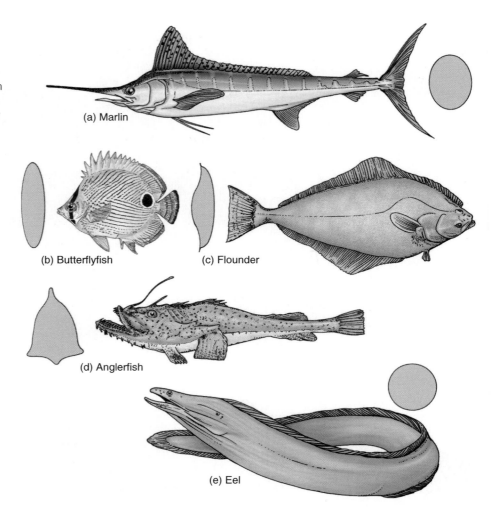

(a) Marlin

(b) Butterflyfish

(c) Flounder

(d) Anglerfish

(e) Eel

(*Chaetodon*) and angelfish (*Pomacanthus*), have a laterally compressed or deep body that helps them to navigate more efficiently through their complex environment (see Figure 10-14b).

Bottom-dwelling fishes, such as the left-eye flounders (family Bothidae), have depressed or flattened bodies (see Figure 10-14c). Flounders begin life looking like normal fish, but early in the juvenile stage, they begin to swim on their side and an eye migrates from what will become the bottom side to the upper side.

Fishes such as the oyster toadfish (*Opsanus tau*), scorpionfish (*Scorpaena*), and anglerfish (*Antennarius*), which exhibit a more sedentary lifestyle, have globular bodies, and their pectoral fins are usually enlarged to help support the body (see Figure 10-14d). Burrowing fishes and fishes that live in tight spaces, such as moray eels (*Gymnothorax*), have long, snakelike bodies, and they lack or have reduced pelvic and pectoral fins (see Figure 10-14e).

## Fish Coloration

Fish colors are of two basic types, **pigments** (biochromes) and **structural colors**. Because most bony fishes use vision as their primary sense in food finding and communication,

color can be important in both concealment and species recognition.

***Pigments*** Pigments are colored compounds found in **chromatophores,** irregularly shaped cells, usually appearing as a central cell body with radiating processes (for more information on chromatophores see Chapter 9). Fish are able to alter their color by moving pigments between the central core and these processes. The flounder, for example, is well known for its ability to alter body color and pattern to match its immediate environment. The control of pigment movement is complex but appears to be under both hormonal and nervous influence. The most common pigments in fishes are the melanins and carotenoids.

***Structural Colors*** Structural colors are produced by light reflecting from crystals located in specialized chromatophores called **iridophores.** Unlike pigments, these crystals are colorless and relatively immobile within the cells. Depending on their orientation, they can produce the mirrorlike silver color of many pelagic fish or the iridescent colors seen in many reef fishes.

***Countershading*** Fishes that live in the open ocean, such as tuna, marlin, and swordfish (*Xiphius gladius*), display a type of coloration known as **obliterative countershading**. A fish with obliterative countershading has a back (dorsum) colored dark green, dark blue, or gray, and the shades graduate on the sides to the belly's pure white. When viewed from above, the dark back blends in with the surrounding dark water. When viewed from below, the white belly blends in with the brightly lit surface. This effectively camouflages the animal even though it is in open water. This same type of coloring is also found in sharks, many marine mammals, and penguins.

***Disruptive Coloration***   Many species of coral reef fish exhibit **disruptive coloration** (Figure 10-15), in which the

**Figure 10-15 Disruptive Coloration.** The vertical lines on this butterflyfish from the Pacific Ocean break up the background color, making the fish more difficult to see. Notice the band that runs through the eye and the eyespot on the opposite end of the body, which make it more difficult for a predator to identify the fish's head.

background color of the body is usually interrupted by vertical lines. This helps to break up the pattern and make it more difficult for predators to see the fish. Often, one of the lines passes through the eye, making it more difficult to be seen, and a dark dot, or eyespot, is present in the area of the tail. Many aquatic predators use the eye to determine which end of their prey is the head. The eyespot on the tail and the line through the eye draw a predator's attention to the wrong end of the fish, making it more likely that the prey will survive an attack.

***Cryptic Coloration***   Some fishes use **cryptic coloration** to blend with their environment, camouflaging themselves to avoid predators or ambush prey. Several pipefish species (*Syngnathus;* Figure 10-16a), for example, avoid predation by mimicking seaweed, both in body pattern and behavior. Many scorpionfish (family Scorpaenidae; see Figure 10-16b) use their irregularly shaped bodies and coloration to blend almost perfectly with the environment. Smaller fishes on which they prey do not see the hidden scorpionfish and on swimming too close become an easy meal.

***Poster Colors***   Many fish associated with coral reefs exhibit **poster colors,** bright showy color patterns that may advertise territorial ownership, aid foraging individuals to keep in contact, or be important in sexual displays. Lionfish (*Pterois volitans,* Figure 10-17) and some other species seem to use bright colors as **warning,** or **aposematic, coloration** to advertise to predators that they are too venomous or spiny to be worth eating.

## Locomotion in Bony Fishes

Bony fishes move about by drifting with the current, burrowing, crawling on the bottom, gliding, and swimming, the latter being the most common method. In swimming, the trunk muscles propel the fish through the water. These

(a)

(b)

**Figure 10-16 Camouflage.** (a) This pipefish from the Solomon Islands in the Pacific Ocean resembles a piece of seaweed. (b) A scorpionfish from the Carribean mimics its environment.

**Figure 10-17 Warning (Aposematic) Coloration.** Some species, like this lionfish, use bright colors to advertise to predators that they are too venomous or spiny to be worth eating.

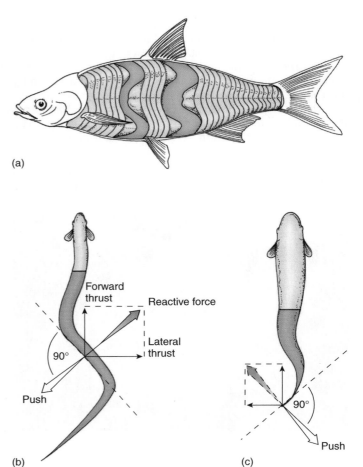

**Figure 10-18 Fish Locomotion.** (a) The trunk muscles of fishes are arranged as a series of W-shaped bands. These muscles contract in sequence from anterior to posterior and alternately from one side of the body to the other. By pushing the body against the water, the fish is propelled forward. (b) Eels undulate the body one full wavelength in swimming. (c) Jacks and other swift swimmers throw the body into a shallow wave of less than one-half wavelength.

muscles are arranged as a series of muscle bands, and each band looks like a letter *W* lying on its side (Figure 10-18a). Movement results when the bands of muscles contract alternately from one side of the body to the other. The muscle contractions originate at the anterior end of the fish and move toward the tail, flexing the body and pushing against the water.

The type of swimming characteristic of elongate fish, such as eels, involves undulating the entire body (see Figure 10-18b). In contrast, swift swimmers such as jacks (family Carangidae), snappers (family Lutjanidae), tuna and mackerels (family Scombridae), and drums (family Sciaenidae) swim by flexing only the posterior portion of the body (see Figure 10-18c). Fishes like cods (family Gadidae) flex their bodies, making a movement somewhere between the full-body undulation of the eel and the posterior body flex of the jacks. Trunkfish (family Ostraciidae) are encased in a dermal skeleton so only the area before the caudal fin can be flexed (Figure 10-19a). Movement is relatively slow, but these fish have toxins and spines to protect them from predators. Many species swim by using their fins alone without body flexure. Triggerfish (family Balistidae, see Figure 10-19b), for example, can move by undulating only their dorsal and anal fins. Wrasses (family Labridae) sometimes move by using the pectoral fins as oars. Flying fish (family Exocoetidae) use their expanded pectoral fins when they glide through the air.

### Respiration and Osmoregulation

Bony fishes use their gills to extract oxygen ($O_2$) from the water, to eliminate carbon dioxide ($CO_2$), and as an aid in maintaining proper salt balance within the body. Gills (Figure 10-20a) are composed of thin, highly vascularized, rod-like structures called **gill filaments.** In these structures, blood flows in the opposite direction from the incoming water (see Figure 10-20b), creating a **countercurrent multiplier system.** The countercurrent flow maintains a stable gradient that favors the diffusion of oxygen in to and carbon

dioxide out of the body. Water passing over the gills, therefore, constantly meets blood coming from the body with a lower oxygen and higher carbon dioxide concentration. This mechanism is very efficient, studies having demonstrated that up to 80% of the oxygen in the incoming water can be extracted through this arrangement versus less than 10% when the flow is concurrent (in parallel).

Water must be continuously moved past the gills to keep the blood properly oxygenated. Most bony fishes ventilate their gills by pumping water across them. First, water enters the open and expanded mouth cavity. Water is then pushed across the gills by contracting the mouth cavity and expanding the chamber surrounding the gills. The gill chamber contracts and water is released from the opercula, moveable flaps of tissue covering the gills. In contrast to this mechanism, very active fishes, such as mackerel, ventilate their gills by ram ventilation, meaning they continuously swim forward at high velocity with their mouth open.

(a)

(b)

**Figure 10-19 Methods of Swimming.** (a) Trunkfish are enclosed in a "box" made of dermal bone so only the caudal peduncle can be flexed. The entire muscle mass on each side is used to flex the short caudal peduncle, producing a sculling motion. (b) Triggerfish swim by undulating only their dorsal and anal fins.

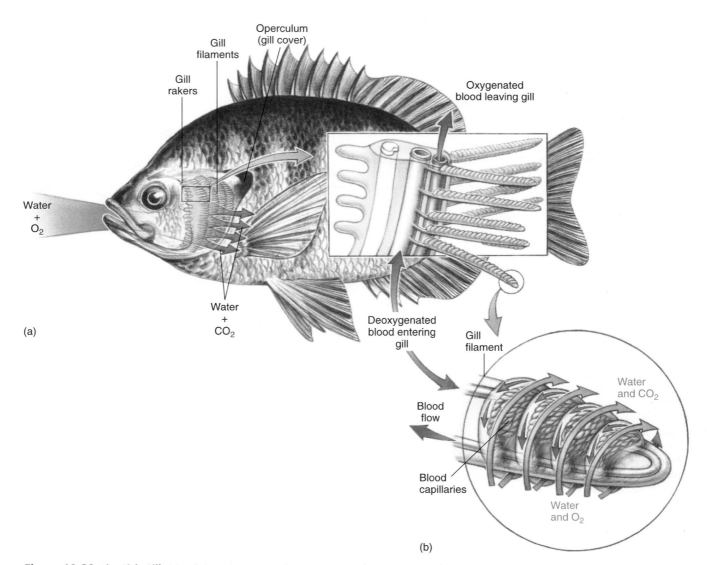

**Figure 10-20 The Fish Gill.** (a) A fish's gills consist of several layers of thin, delicate, fingerlike filaments. (b) The direction of blood flow through the gill filaments is opposite that of the water flow. The countercurrent arrangement allows gas exchange between the blood and the water to occur along the entire length of the blood vessel, resulting in the greatest amount of gas exchange.

**Figure 10-21 Osmoregulation in Bony Marine Fishes.** Marine fishes tend to lose water to their surroundings, especially from the gills. In order to replace the water lost, they drink seawater. For this process to be useful, the fish must retain as much water as possible while eliminating the excess salt. Excess salt is eliminated by the gills, the intestine, and the kidneys.

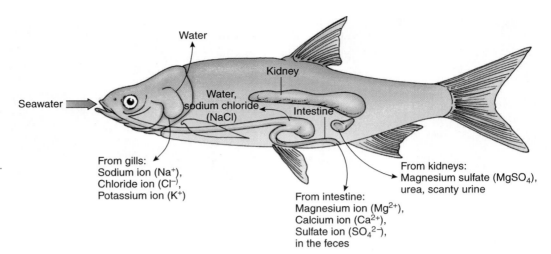

Because the salt concentration of their blood is about one third the concentration of salts in seawater, marine fishes tend to lose much water to their environment. To compensate for this water loss, marine fish drink seawater, remove the excess salt, and retain the water (Figure 10-21). Specialized **chloride cells** on the gills eliminate most of this excess salt. The kidneys and digestive tract remove salts not excreted by this mechanism. Fishes that are adapted to seawater produce negligible amounts of urine, because they need to retain as much water as possible.

## Cardiovascular System

The cardiovascular system of fishes consists of a heart, arteries, veins, and capillaries. Deoxygenated blood, initially collected from the veins by a thin-walled chamber (the **sinus venosus**), is passed to a second chamber (the **atrium**) and then to a muscular **ventricle** (see Figure 10-13a). The ventricle propels the blood forward to the gill capillaries where it is oxygenated. From the gills, the blood is collected by the dorsal aorta and passed to the rest of the body through arteries and the capillaries. Blood pressure is lower in fishes than other vertebrates because blood does not return to the heart after oxygenation. Veins collect blood from the capillaries to complete the circuit.

Many active swimmers, such as tuna, have a countercurrent arrangement of their blood vessels (Figure 10-22) to maintain body-core temperature at 2° to 10°C above that of the surrounding water, thereby increasing the efficiency of their swimming muscles. The veins containing relatively cool blood from the body's surface, pass close to and in the opposite direction from the arteries containing warm blood coming from the body's core. Heat is transferred from the blood in the arteries to blood in the veins. In this way warmed venous blood flows into the core of the body, helping to maintain a higher internal temperature.

## Buoyancy Regulation

Most bony fishes, with the exception of some pelagic species, bottom dwellers, and deep-sea fishes, use a gas-filled sac called a **swim bladder** (see Figure 10-13a) to help them offset the density of their bodies and regulate buoyancy. By

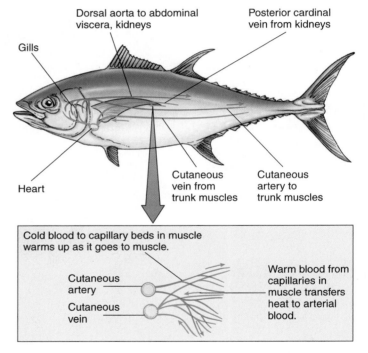

**Figure 10-22 Countercurrent Heat Exchange in Tuna.** Tuna are able to use a countercurrent heat exchange to keep the core of the body several degrees higher than their environment and thereby improve muscle efficiency.

adjusting the amount of gas in the swim bladder, a fish can remain indefinitely at a given depth without any muscular movement and with minimal expenditure of energy. When the fish descends, more gas must be added to the swim bladder, or else the bladder will compress and the fish become denser and sink. On the other hand, as the fish ascends, it must remove gas from the swim bladder, or else the gas will expand and the fish become less dense and rise too rapidly.

Two mechanisms have evolved to allow adjustments in the gas volume of the swim bladder. Some fish, like herrings and eels, adjust the gas volume of their swim bladders by gulping air from the surface or "spitting it out" as needed. Others use a specialized **gas gland** to fill the swim bladder from gases dissolved in the blood. In these fishes, the swim bladder is deflated by diffusion of gases directly into the bloodstream.

Pelagic fishes that are active swimmers, such as mackerels (*Scomber*) and skipjacks (*Katsuwonus pelamis*), do not have swim bladders. Like sharks, these animals must keep swimming or they sink. Bottom dwellers, such as scorpionfishes, lack a swim bladder because they do not need to maintain buoyancy in the water column. Many fishes that live in the deep ocean also lack a swim bladder. We will discuss these fishes and their adaptations in more detail in Chapter 18.

### Nervous System and Senses

Like other vertebrates, the nervous system of bony fishes consists of a brain, spinal cord, associated peripheral nerves, and various sensory receptors. The brain can be divided into several regions, each involved in coordinating an important body function. Areas associated with the senses of smell (olfaction) and vision are particularly well developed.

***Olfaction*** The olfactory receptors of bony fishes are located in **olfactory pits,** blind sacs that open to the external environment. Swimming, movement of cilia within the pit, or constriction of the nasal sacs themselves allows water to move across the olfactory receptors of most fishes. The olfactory sacs and their receptors may be greatly elongated in some fishes, such as eels, that rely heavily on olfaction to find their prey. Some puffers (family Tetraodontidae), in contrast, have greatly reduced olfactory organs, most likely a result of their primary reliance on sight for feeding.

***Taste and Hearing*** Taste receptors of bony fishes may be located on the surface of the head, jaws, tongue, mouth, and **barbels,** whiskerlike processes about the mouth. These receptors are used to detect both food and noxious substances. Like sharks, bony fishes have a lateral line system, which helps them to detect movement in the water. The ears of bony fishes are internal and capable of detecting sounds in the range of 200 to 13,000 hertz. In comparison, humans hear sounds ranging from 20 to 20,000 hertz.

***Vision*** Bony fishes, as a group, rely on vision more than sharks and rays. Their eyes lack eyelids, and they generally do not need to adjust the size of the pupil, because the quantity of light in water is relatively low. If a fish needs to adjust its vision for distance, the entire lens moves backward or forward, much like focusing a camera.

The eyes of most fishes are set on the sides of the head instead of the front. It is believed that a fish sees only a narrow field directly in front of it with each eye. For the most part, fishes have monocular vision, with each eye seeing its own independent field. In addition to possessing black-and-white vision, shallow water species are able to perceive color.

### Feeding Types

The great diversity of bony fishes is reflected in their ability to exploit virtually every food resource available in the marine environment. Included among marine fishes are detritivores, herbivores, carnivores, and omnivores. Various specializations have also evolved that aid in acquiring energy. Herbivorous fishes, for example, typically have longer guts than carnivores, providing them a greater surface area to absorb nutrients from foods containing a high percentage of indigestible matter. Teeth may be highly specialized and positioned on several of the head and face bones.

***Carnivores*** Most bony fishes are carnivores. Prey are usually seized and swallowed whole, because spending time chewing food would block the flow of water past their gills. Many species, such as pufferfish (*Sphoeroides*) and boxfish (*Ostracion*), crush their prey with powerful jaws. Some butterflyfish (*Chaetodon*) use their tiny mouths to feed on individual coral polyps. Groupers (*Epinephelus*) have large mouths with small teeth. They lie in wait in their lairs until their prey comes along. When a mullet (*Mugil*), grunt (*Haemulon*), or large crustacean comes by, the grouper will open its huge mouth, creating suction that draws in the prey. If the prey is too large to fit completely into its mouth, the grouper holds the prey's tail with its small teeth while pharyngeal teeth in the grouper's throat crush, grind, and shear the victim. Flounders lie camouflaged on the bottom, motionless, until a meal in the form of a crustacean, worm, or small fish comes along. The flounder then springs up and grabs the unsuspecting victim.

***Herbivores*** Herbivorous fishes feed on a variety of plants and algae. Surgeonfishes (*Acanthurus*) feed on the algae that grow on rocks and coral. In most species, the teeth are broad and flat with a sharp edge (like a shovel), making them ideal tools for scraping food from these surfaces. In addition, several species have a gizzardlike stomach to grind the vegetable matter. Parrotfish (*Scarus;* Figure 10-23) have teeth that are fused to form a beaklike structure that scrapes algae from the hard surfaces of coral reefs. Some parrotfish species bite off and ingest pieces of coralline algae or hard coral along with the coral's inhabitants. As the material passes through the fish's digestive tract, it is pulverized.

**Figure 10-23 Parrotfish.** Some parrotfish feed on the symbiotic algae of corals. Their beaklike mouthparts allow them to crush the coral to extract the algae.

Richard N. Mariscal

The algae are extracted and digested, and fine white sand is passed as part of the indigestible wastes. Because they extract so little organic material from each mouthful of food, parrotfish feed almost constantly. A large parrotfish may weigh as much as 27 kilograms (59 pounds) and expel as much as 2 or 3 tons of sand per year. Parrotfish have contributed sand to many of the white sand beaches of the world. Obvious scars on the reef surface are an indication of active parrotfish feeding.

***Filter Feeders***   Fish larvae as well as pelagic fishes such as anchovies (*Engraulis*) are filter feeders, feeding on the abundant plankton in the sea. Filter feeders typically use projections from the gill arches called gill rakers (see Figure 10-20a) to filter both phytoplankton and zooplankton from seawater. Most filter-feeding fishes travel in large schools and are an important source of food for larger carnivores.

## Adaptations to Avoid Predation

Just as fishes have evolved a variety of feeding styles, they have also evolved many clever strategies to avoid being eaten. Although many marine fishes exhibit elaborate camouflage that obscures their presence, others have evolved more direct methods of avoiding predation. The pufferfishes and porcupinefish (*Diodon hystrix*), for example, can swallow large amounts of air or water and inflate their bodies to a size that deters potential predators (Figure 10-24a). The rapid change in size frightens some potential predators, whereas others now find the fish too large to fit in their mouths. In the case of the porcupinefish, or spiny boxfish, not only does the fish enlarge itself but also in the process extends spines that normally lie flat against the body, adding an extra measure of protection.

Flying fishes (*Cypselurus*) avoid predation by using enlarged pectoral fins to glide through the air (Figure 10-24b). When frightened or disturbed, these fishes swim forward very quickly and leap out of the water, spreading their large pectoral fins at the same time. This behavior allows the flying fish to glide out of the range of many predators. Pearlfish (*Carapus*) avoid being eaten by slipping into the bodies of such animals as sea cucumbers or bivalve molluscs. At night, some species of parrotfish secrete a mucous cocoon about them, perhaps to discourage nocturnal predators. The surgeonfish is so named because of a pair of razor-sharp spines located on the sides of its tail. When this herbivorous fish is disturbed, the spines snap out like a switchblade knife and cut or stab the intruder. Clingfishes (*Gobiesox*) have a powerful sucker (formed from modified pelvic fins) that it uses to secure its body to rocks in tidal pools, where it feeds on crustaceans that are washed into its jaws. Predators have a difficult time dislodging the clingfish and thus tend to leave it alone. The many species of triggerfish (*Balistes*) have dorsal fins that have the first three spines modified. These spines can be pulled flat against the body or projected up. When a predator tries to swallow a triggerfish, it projects the spines and is rejected as a possible meal. The spines also help the triggerfish to wedge itself into the tight nooks and crannies of coral reefs, making it nearly impossible to be dislodged by a predator. Some fishes have venomous spines that protect them from predators. Scorpionfish and stonefish, for example, have venom glands associated with the dorsal, anal, and pelvic spines. The venom of the Indo-Pacific stonefishes (*Synanceia*) is extremely toxic. There are many recorded instances of people accidentally stepping on them and being severely injured or killed.

## Reproduction in Bony Fishes

Bony fishes exhibit an amazing variety of reproductive strategies. Spawning fish release their eggs into the water column or deposit them on the bottom. Some species hide their eggs, whereas others guard them until they hatch. Still others brood the eggs in their body until the eggs hatch. Most marine bony fishes are oviparous, but a few groups, such as rockfishes (family Scorpaenidae), exhibit ovoviviparity. The surfperches (family Embiotocidae), found along the Pacific coasts of North America and Asia, exhibit viviparity. Their young develop in a uterus-like structure in the ovary and obtain nutrition from the mother through enlarged, highly vascularized fins.

The gonads of bony fishes are paired structures sus-

**Figure 10-24  Avoiding Predation.** (a) A porcupinefish inflates itself to become a large, prickly mouthful. (b) The large pectoral fins of the flyingfish help it to avoid predators by allowing it to leave the water and glide through the air.

Dave B. Fleetham / Tom Stack and Associates

(a)

Norbert Wu / Peter Arnold, Inc.

(b)

# Fish Toxicity

To deter predators, many fish species produce toxic substances. Others accumulate toxic substances from their environment in the course of their normal feeding activities. Scorpionfishes (family Scorpaenidae), for example, use the neurotoxins produced in specialized venom glands associated with grooved spines to discourage potential predators. Likewise, species such as parrotfish (family Scaridae) and surgeonfish (family Acanthuridae) produce toxic secretions from their skin, and puffers (family Tetraodontidae) store toxic substances produced by symbiotic bacteria in their internal organs. Barracuda, groupers, and other reef fish may accumulate toxins while feeding in areas associated with blooms of dinoflagellates such as *Gambierdiscus toxicus.*

Although few humans encounter the venom of scorpionfish, more than 50,000 are sickened annually from eating toxic fish flesh. Most cases of fish poisoning are due to eating reef fish containing various forms of a toxin called *ciguatoxin,* the result being termed *ciguatera poisoning.* Spanish explorers in Cuba first reported symptoms of the disorder more than 500 years ago after eating snails they called *cigua.* The physician on the HMS *Bounty* (of *Mutiny on the Bounty* fame) is said to have died after consuming toxic fish. Most people do not die from ciguatera poisoning, but symptoms may persist for weeks, months, and in some cases, years. Symptoms include joint and muscle pain, weakness, intense itching, numbness in the extremities, and reversal of cold and hot sensations. Abdominal cramps, diarrhea, vomiting, and heart arrhythmia may accompany these symptoms. Most treatments have been unsuccessful, although physicians have recently reported some success with injections of mannitol, a sugar alcohol.

*Gambierdiscus toxicus* grows on dead coral surfaces colonized by filamentous and calcareous algae. Relatively few cases of ciguatera are reported from reefs with living coral. Large groupers, snappers, amberjack, barracuda, and Spanish mackerel are most commonly associated with ciguatera poisoning, but more than 400 species are known to carry the toxin.

The fish acquire toxic levels of the ciguatoxin through food-chain amplification. Grazing fish ingest the toxin and are eaten by carnivorous fish that are, in turn, eaten by humans. Cooking has no effect on the toxin and there are no safe, reliable methods to detect it in fish flesh. Cases are reported from Australia, the tropical Pacific, and the Caribbean. In the United States, most cases occur in Hawaii and Florida.

Approximately 100 people die every year, primarily in Japan and the Philippines, from consuming fugu, or Japanese puffer fish (*Takifugu rubripes*). The liver and other internal organs contain a powerful toxin, tetrodotoxin, which blocks nerve-impulse transmission. It is supposedly 1,200 times more lethal than cyanide. Fugu is considered a delicacy; a meal may cost up to $200. Licensed chefs must prepare the fish, and there is usually a ritual associated with consuming the meal. Most deaths occur outside the main cities, where chefs may not be as well trained in proper fugu preparation. ●

---

pended from the roof of the body cavity by membranes called mesenteries. Sperm from the testes and eggs from the ovaries pass to the outside through special ducts, except in salmon and their relatives, where gametes are shed directly into the body cavity and exit via abdominal pores. In contrast to most land vertebrates, there is typically no association of the reproductive and excretory systems. Development of eggs and sperm is usually seasonal (except in some tropical species), the timing of the reproductive period being influenced by temperature and photoperiod. As in other vertebrates, variation in the level of pituitary and gonadal hormones controls the reproductive process. Courtship varies from none to elaborate ritual displays.

***Pelagic Spawners*** Pelagic spawners include such commercially important species as cod, tuna, and sardines and coral reef species such as parrotfish and wrasses. These species release vast quantities of eggs into the water, the males fertilize them, and the fertilized eggs drift with the currents. There is no parental care. An advantage of this strategy is that the offspring are widely dispersed, but its disadvantage is that mortality is very high. These species compensate by producing large numbers of offspring over a lengthy spawning period.

***Benthic Spawners*** Benthic spawners, such as smelt (family Osmeridae), live closer to shore and produce eggs that are generally larger than those of pelagic spawners and have a large quantity of yolk. These eggs are usually nonbuoyant and are spread over surfaces such as vegetation or rocks. Large numbers of eggs are produced, and there is no parental care. When hatched, the embryos and larvae may be either pelagic or benthic.

***Brood Hiders*** The grunion (*Leuresthes tenuis*) is classified as a **brood hider,** a species that hides its eggs in some way but exhibits no parental care. Grunions swim ashore at high tide during a full moon, burrow partially into the sand, and deposit their eggs. Males curl around the partially buried female and fertilize the eggs as they are laid. The eggs remain in the sand until the next high tide, when they hatch and are washed out to sea.

**Figure 10-25 Males Incubating Eggs.** In some fish species it is the job of the male to incubate and care for the eggs. (a) The male jawfish incubates the eggs in its mouth until they hatch. (b) Male seahorses have a special abdominal pouch in which the eggs are incubated.

(a)                               (b)

***Guarders*** **Guarders** care for their offspring until they hatch and, frequently, through their larval stages. Many species of damselfish, blennies, and gobies exhibit this behavior. Some species of damselfish defend territory, both as spawning sites and sites to cultivate algae. Spawning sites may be prepared nests or simply a bare surface. Females lay eggs at these spawning sites, but only the male guards the offspring. The amount of time spent guarding the eggs varies from a few days to more than 4 months, as in the Antarctic plunderfish (*Harpagifer bispinis*).

***Bearers*** Jawfish (*Ophistognathus macrognathus;* Figure 10-25a) and seahorses (*Hippocampus*) are examples of **bearers.** The female jawfish lays her eggs in the mouth of the male, who incubates them in his mouth until they hatch. When the male must feed, he temporarily deposits the eggs in his protected burrow. The female seahorse lays eggs in a special pouch on the male's abdomen (see Figure 10-25b). After the eggs are deposited, the male fertilizes them and then carries and incubates them until the young hatch.

***Larval Development*** In many species, the currents carry the larvae as members of the plankton. Initially, larval fish are nourished by a yolk sac attached to their abdomen (Figure 10-26). As the larva develops, a mouth and digestive tract form, and ultimately the yolk sac is absorbed. When the larva transforms into a juvenile, it leaves the planktonic community and begins to feed as an adult. Unlike birds and mammals that cease to grow or grow very little after achieving maturity, fishes grow for as long as they live.

***Hermaphroditism*** **Hermaphroditism,** in which individuals have both testes and ovaries at some time in their lives, is known in at least 14 families of bony fishes. Hermaphro-

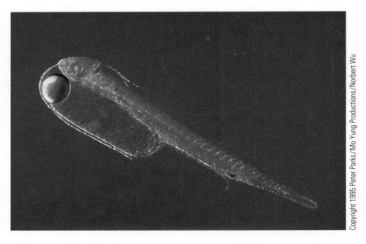

**Figure 10-26 Fish Larva.** After hatching, many fishes still have a yolk sac containing nutrients attached to their abdomen. The size of the yolk sac impairs the larva's ability to swim, and it remains a member of the plankton until the mouth and digestive tract are fully formed and the yolk sac is absorbed.

dites may be **synchronous,** possessing functional gonads of both sexes at one time, or **sequential,** changing from one sex to another.

Hamlets (*Hypoplectrus;* Figure 10-27a), a small group of tropical fish in the sea bass family (family Serranidae), are synchronous hermaphrodites. A partner in a monogamous pair alternately has its eggs fertilized and fertilizes its partner's eggs. Sexual roles may change up to four times during a single mating encounter.

(a)

(b)

**Figure 10-27 Hermaphroditism.**
(a) Hamlets are synchronous hermaphrodites. Members of monogamous pairs alternately offer eggs to be fertilized and, in turn, fertilize their partner's eggs. (b) The striped cleaner wrasse (*Labroides dimidiatus*) is a sequential hermaphrodite practicing protogyny. If a male controlling a harem of females dies, the largest female changes to a male and replaces him.

Sequential hermaphrodites may change from females to males (**protogyny**) or from male to female (**protandry**). The most common pattern is a change from female to male, exhibited in members of at least seven fish families, including the wrasses (family Labridae), parrotfish (family Scaridae), and sea bass (family Serranidae). Typically, a large, dominant male of a species like the striped cleaner wrasse (*Labroides dimidiatus;* see Figure 10-27b) controls a harem of females. If this male dies, the largest female changes to a male and replaces him. Males, therefore, may be primary, beginning as a male, or secondary, resulting from a female. Smaller primary males may occasionally contribute their gametes to the next generation by "sneaking" into group spawning aggregations.

The reverse occurs in the anemone fish (*Amphiprion,* family Pomacentridae), a species that inhabits large sea anemones, where it is protected from predators by the host's stinging cells. Each inhabited anemone contains a single large female, a smaller adult male, and many juvenile males. If the female dies or is removed, the adult male changes to a female and is replaced by the largest of the juvenile males. The large female–smaller male combination seems to result in a higher per capita production of offspring and thus has an evolutionary advantage.

## Fish Migrations

Migratory movements of marine fish are common and may occur daily or seasonally. Daily migrations are usually associated with feeding and predator avoidance. Grunts (family Haemulidae), for example, hover over reefs during the day but at night move to surrounding seagrass beds to feed. Lanternfish (family Myctophidae) are known to perform rapid vertical migrations of 200 meters (660 feet) or more as they follow their planktonic prey. Seasonal migrations of marine fish are usually associated with spawning, changing temperatures, or feeding. In some species migrations occur entirely within saltwater. In other species migrations occur between freshwater and saltwater. The North Pacific population of albacore tuna (*Thunnus alalunga*), for example, winters in midocean but travels to the California-Oregon coast or the coast of Japan for the summer months. Many other species, such as the herring (*Clupea harengus*), annually travel north and south, following sea temperature variation.

Some fish species move between freshwater and saltwater for a purpose other than reproduction. Young mullets (*Mugil cephalus*), for example, spend part of their time in freshwater or estuaries. But as adults, they live most of their life in the ocean and spawn there. Fishes that move from freshwater to seawater to spawn are **catadromous,** whereas those that move from seawater to freshwater to spawn are **anadromous.**

***Freshwater Eels***   The freshwater eels of North America (*Anguilla rostrata*) and Europe (*A. anguilla*) are probably the best-studied catadromous fishes. Adult eels of both species migrate down coastal rivers to the sea during the fall, changing from a dull olive color to silver and acquiring larger eyes. The American species take about 2 months to reach the deepwater spawning grounds in the Sargasso Sea, which lies southeast of Bermuda (Figure 10-28). The European species takes approximately 1 year to make the journey. Here the eels spawn at depths of 300 meters (990 feet) or more, and the adults die after reproducing.

After the young hatch, they develop into leaflike **leptocephalus larvae** that begin their migration back to the rivers of Europe and North America. It takes approximately 3 years for larvae to reach Europe but only a year for the North American species to complete the trip. When they arrive at the coastal rivers, they undergo a metamorphosis, becoming juvenile eels, called "elvers," that migrate into streams and estuaries. Males usually remain in fresh water for 4 to 8 years, whereas females require up to 12 years or more to reach sexual maturity. During the fall season following sexual maturity, the eels migrate downstream and return to the breeding grounds of the Sargasso Sea.

The Japanese freshwater eel (*Anguilla japonica*) has a similar pattern, spawning near the Mariana Islands and drifting back to the Orient via prevailing currents. Unlike the American and European eels, some populations of the Japanese eel never enter freshwater but remain in the sea or estuaries. These populations, along with those inhabiting freshwater, all seem to contribute individuals to the populations that return to the waters near the Mariana Islands to spawn.

***Salmon***   The Atlantic and Pacific species of salmon (Figure 10-29) are examples of anadromous fishes. The six species of Pacific salmon (*Onchorhyncus*) return to their spawning grounds only once and then die after reproducing. In contrast, the single Atlantic species of salmon (*Salmo salar*) may spawn more than once in the streams where it was hatched. Salmon lay their eggs in a shallow depression in

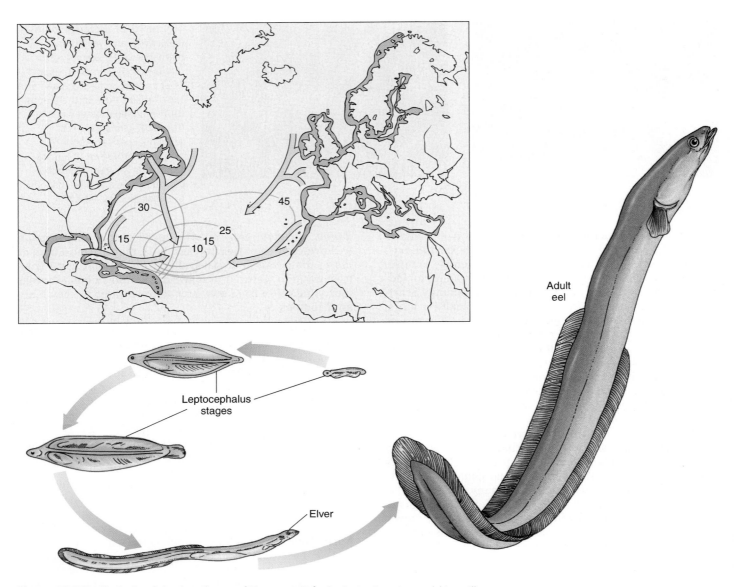

**Figure 10-28  Life Cycle of the American and European Eels.** Both the American eel (*Anguilla rostrata*) and the European eel (*Anguilla anguilla*) breed in the Sargasso Sea. Yellow arrows indicate the migratory paths the eels follow to the Sargasso Sea. Curved lines and numbers in the ocean indicate larval distribution and size of the larva in millimeters. Colored coastal areas show where the elvers enter rivers.

the gravel of a freshwater stream termed a **redd.** The hatchlings of some species return to the sea immediately, but others may spend up to 5 years in freshwater. Populations of some species never return to the sea but remain landlocked. The young salmon mature at sea and eventually return to the headwaters of their native stream to spawn and complete their life cycle.

Populations of salmon have been greatly reduced because of the damming of rivers, pollution, and the transformation for human use of land bordering streams. The Chinook salmon (*O. tshawytscha*), which once ascended the Columbia River for more than 1,600 kilometers (1,000 miles), has been particularly affected by these activities. Hatcheries have had limited success in maintaining stocks, but there is concern about the loss of the genetic diversity of the wild stock (for more information on salmon hatcheries see Chapter 19).

Experiments have shown that salmon are guided upstream by the characteristic odor of their native stream. The odor is thought to be from chemicals in the soil and plants along the stream's edge. There is some controversy as to how the salmon locate the mouth of their river from the open sea. Some suggest that the fish use cues such as the sun and the earth's magnetic field to find the portion of the

Jon L. Hawker

**Figure 10-29 Salmon.** Salmon are anadromous fishes. They reproduce and mature in fresh water but spend their adult lives in the marine environment. This Pacific salmon has just returned to its native river to breed.

coast near their home stream. Others suggest that currents, temperature gradients, and food supplies ultimately bring a salmon to the mouth of its native stream.

## In Summary

Bony fishes move about by drifting with the current, burrowing, crawling on the bottom, gliding, and swimming, the latter being the most common method. Swimming movement results when the bands of muscles contract alternately from one side of the body to the other.

Bony fishes use their gills to extract oxygen from the water and eliminate carbon dioxide. Blood flows in the direction opposite from the incoming water, creating a countercurrent multiplier system that improves the uptake of oxygen from the water. Most bony fishes ventilate their gills by pumping water across them. Gills are also used to maintain salt balance. The kidneys excrete salts that are not removed by other mechanisms.

The cardiovascular system of fishes consists of a heart, arteries, veins, and capillaries. Blood pressure is lower in fishes than other vertebrates because blood is not returned to the heart after

oxygenation. Some fish, such as the tuna, use a countercurrent arrangement of their blood vessels to conserve heat in the core of the body and improve muscle efficiency.

Most bony fishes, with the exception of some pelagic species, bottom dwellers, and deep-sea fishes, use a gas-filled sac called a swim bladder to help them offset the density of their bodies and maintain neutral buoyancy. By adjusting the amount of gas in the swim bladder, a fish can remain indefinitely at a given depth without any muscular movement and with minimal expenditure of energy.

Bony fishes have a keen sense of smell, and most species have very good vision. In general, they rely on vision more than sharks and rays. They also possess sense organs for hearing and a lateral line system for sensing vibrations in the water.

Bony fishes exploit virtually every food resource available in the marine environment. Included among marine fishes are detritivores, herbivores, carnivores, and omnivores. Various adaptations have evolved to aid fishes in capturing and processing their prey. Herbivores, for example, have longer guts, providing greater surface area to absorb nutrients from foods containing a high percentage of indigestible matter. Filter feeders use gill rakers to filter both phytoplankton and zooplankton from seawater.

Marine fishes have evolved a variety of adaptations for avoiding predation. These adaptations include sudden changes in size or color; sharp, sometimes poisonous spines; special behaviors, such as rapidly changing size or darting into crevices; and eluding predators by speed or gliding.

Most marine fishes are oviparous, although both ovoviviparity and viviparity occur in a few groups. Fishes also exhibit an amazing variety of reproductive strategies, from the number of eggs produced to the level of parental care. Hermaphroditism is known in at least 14 families of bony fish. Hermaphrodites may be synchronous or sequential.

Migratory movements of marine fish are common and may occur daily or seasonally. Daily migrations are usually associated with feeding and predator avoidance, whereas seasonal migrations are usually associated with spawning, changing temperatures, or feeding. The movements of freshwater eels and salmon are excellent examples of seasonal migrations. North American and European freshwater eels migrate to the Sargasso Sea in the Atlantic Ocean to reproduce. Many salmon live their adult lives in the sea but return to spawn to the freshwater streams where they hatched. ●

## IN PERSPECTIVE

| Class of Phylum Chordata | Representative Organisms | Form and Function | Reproduction | Type of Feeding | Ecological Roles |
|---|---|---|---|---|---|
| Myxini | Hagfish | Lack vertebrae, jaws, paired fins, dorsal fin, oral disk, and lateral line system | Sexual, with sexes normally separate. No larval stage present. Reproductive behavior is poorly known | Carnivorous, generally feeding on soft-bodied invertebrates | Important as scavengers of carcasses of large vertebrates |
| Cephalospidomorphi | Lampreys | Possess two dorsal fins, cartilaginous vertebrae, oral disk, and lateral line system. Lack jaws and paired fins | Sexual. Reproduce in freshwater and have a larval stage | Larvae are filter feeders. Adults are nonfeeding or parasitic | Predators on large fish. May severely damage prey populations in some lakes |
| Chondrichthyes: subclass Elasmobranchii | Sharks, rays, skates | Possess jaws, paired and median fins, cartilaginous skeletons, placoid scales, and lateral line system | Sexual. Fertilization internal with claspers. Oviparous, ovoviviparous, and viviparous reproductive modes present | Most are carnivorous but a few are planktivorous | Dominant predator in the sea |
| Chondrichthyes: subclass Holocephali | Chimaeras, ratfish, etc. | In addition to the entries for Elasmobranchii, possesses an operculum, crushing plates instead of teeth, and head claspers in males | Sexual. Fertilization internal with claspers. Reproductive mode ovoviviparous | Carnivorous, wide variety of food, crushing their prey | Predators in deep water |
| Sarcopterygii | Coelacanths, freshwater lungfish | Possess jaws, paired and median fins, skeleton composed of cartilage and bone, an operculum, lateral line system, and fins attached with fleshy lobes | Sexual reproductive mode ovoviviparous, but males lack an obvious insertion organ | Carnivores, feeding on large bottom fishes | Predator on other fish |
| Actinopterygii: subclass Chondrostei | Sturgeons | Differs from the entry for Sarcopterygii in skeleton being primarily cartilaginous, scales ganoid, and fins attached with fin rays | Sexual. Reproductive mode oviparous. Fertilization external | Bottom feeders on benthic invertebrates | Predators and scavengers |
| Actinopterygii: subclass Neopterygii | Modern bony fishes | Differs from the entry for Chondrostei in skeleton being primarily bony, scales cycloid, ctenoid, or absent | Sexual reproductive mode generally oviparous but other modes exist. Fertilization generally external | Includes herbivores, omnivores, carnivores, and detritivores | Exploit virtually every feeding niche in the sea |

## SELECTED KEY TERMS

ampullae of Lorenzini, *p. 231*

anadromous, *p. 247*

anal fin, *p. 236*

atrium, *p. 242*

barbel, *p. 243*

bearers, *p. 246*

benthic spawners, *p. 245*

brood hiders, *p. 245*

bump-and-bite attack, *p. 231*

catadromous, *p. 247*

caudal fin, *p. 230*

chloride cells, *p. 242*

chromatophore, *p. 238*

claspers, *p. 230*

countercurrent multiplier system, *p. 240*

cryptic coloration, *p. 239*

ctenoid scales, *p. 235*

cycloid scales, *p. 235*

dental plates, *p. 227*

disruptive coloration, *p. 239*

dorsal fin, *p. 236*

elasmobranch, *p. 229*

fusiform body shape, *p. 236*

ganoid scales, *p. 235*

gas gland, *p. 243*

gill filaments, *p. 240*

guarders, *p. 246*

hermaphroditism, *p. 246*

heterocercal tail, *p. 230*

hit-and-run attack, *p. 231*

holocephalan, *p. 229*

homocercal tail, *p. 236*

iridophore, *p. 238*

lateral line system, *p. 230*

leptocephalus larva, *p. 247*

neuromast, *p. 231*

nictitating membrane, *p. 230*

obliterative countershading, *p. 239*

olfaction, *p. 230*

olfactory pit, *p. 243*

oviparity, *p. 232*

ovoviviparity, *p. 232*

pectoral fins, *p. 236*

pelagic spawners, *p. 245*

pelvic fins, *p. 236*

pigments, *p. 238*

placoid scale, *p. 229*

poster colors, *p. 239*

protandry, *p. 247*

protogyny, *p. 247*

rectal gland, *p. 232*

redd, *p. 248*

sequential hermaphrodite, *p. 246*

sinus venosus, *p. 242*

spiracle, *p. 232*

spiral valve, *p. 232*

squalene, *p. 230*

sneak attack, *p. 231*

structural color, *p. 238*

swim bladder, *p. 242*

synchronous hermaphrodite, *p. 246*

urea, *p. 232*

ventricle, *p. 242*

vertebra, *p. 226*

vertebrate, *p. 226*

viviparity, *p. 232*

warning (aposematic) coloration, *p. 239*

## QUESTIONS FOR REVIEW

### Multiple Choice

1. Lampreys and hagfishes lack
   a. heads
   b. mouths
   c. jaws
   d. tails
   e. fins

2. The skeletons of sharks and rays are composed of
   a. bone
   b. cartilage
   c. soft tissue
   d. fluid
   e. cellulose

3. Shark's teeth are actually modified
   a. cartilage
   b. fins
   c. ctenoid scales
   d. placoid scales
   e. gill supports

4. A sense organ that allows fishes to detect even the slightest movements in the water is the
   a. lateral line
   b. ampullae of Lorenzini
   c. statocyst
   d. olfactory organ
   e. gill arch

5. The presence of _____ in the intestine of the shark slows the passage of food through this organ.

   a. stones
   b. chambers
   c. spiral valves
   d. gill rakers
   e. squalene

6. The type of caudal fin associated with most ray-finned fishes is
   a. homocercal
   b. heterocercal
   c. pelvic
   d. pectoral
   e. dorsal

7. Bony fishes can maintain neutral buoyancy by regulating the gas content of their
   a. blood
   b. gills
   c. intestines
   d. swim bladder
   e. liver

8. Many pelagic fishes have a dark dorsum and a light-colored belly. This type of coloration is known as
   a. cryptic
   b. disruptive
   c. warning
   d. obliterative countershading
   e. mimicry

9. Fishes that live in seagrass or on coral reefs have bodies that are

a. fusiform
b. globular
c. snakelike
d. round
e. laterally compressed

10. An anadromous fish spawns in _____ and lives its adult life in _____.
a. freshwater; saltwater
b. saltwater; saltwater
c. freshwater; brackish water
d. saltwater; freshwater
e. freshwater; freshwater

## Short Answer

1. What are two adaptations that help prevent fishes from sinking because of their relatively high density?

2. Explain what is meant by "disruptive coloration" and give an example.

3. Describe how a fish uses its trunk muscles to swim.

4. Describe how fish gills are ideally suited for gas exchange in water.

5. Explain how the characteristics of the environment influence the shape of a fish's body.

6. Describe how reproduction in an oviparous fish differs from reproduction in an ovoviviparous fish.

7. Explain how you can distinguish a skate from a ray.

8. Why do most carnivorous fishes swallow their prey whole?

## Thinking Critically

1. The jawfish and seahorse depend on the males to take care of the eggs. What is the advantage of this reproductive strategy?

2. What are some of the advantages of hermaphroditism as a reproductive strategy in fish?

3. What characteristics would you expect to observe in a fish that was adapted to a sedentary life hiding among rocks and coral on a coral reef?

4. Several species of marine fish display different colors as juveniles and as adults. What might be the benefit of such an arrangement?

## SUGGESTIONS FOR FURTHER READING

Carruth, Laura L., Richard E. Jones, and David O. Norris. 2002. Cortisol and Pacific Salmon: A New Look at the Role of Stress Hormones in Olfaction and Home-stream Migration, *Integrative and Comparative Biology* 42(3): 574–81.

Conniff, Richard. 1991. The Most Disgusting Fish in the Sea, *Audubon* 93 (March):100–108.

Drollette, Daniel. 1999. World's Biggest Fish, *International Wildlife* 29(1):22–27.

Eisner, Thomas. 2003. Living Fossils: On Lampreys, Baronia, and the Search for Medicinals, *BioScience* 53(3): 265–69.

Erdmann, M. V., R. L. Caldwell, and M. K. Moosa. 1998. Indonesian "King of the Sea" Discovered, *Nature* 395:335.

Groves, Paul. 1998. Leafy Sea Dragons: Masters of Camouflage, Fierce Predators, and Males Become Pregnant, *Scientific American* 279(6):84–89.

Holbrook, Sally J., and Russell J. Schmitt. 2002. Competition for Shelter Space Causes Density-Dependent Predation Mortality in Damselfishes, *Ecology* 83:2855–68.

Klimley, A. Peter. 1995. Hammerhead City: Evidence of Magnetic Navigation in Schools of Hammerhead Sharks, *Natural History* 104:32–39.

Martin, Glen, and Jessica Gorman. 1999. The Great White's Ways, *Discover* 20(6):54–61.

Martini, Frederic H. 1998. Secrets of the Slime Hag, *Scientific American* 279(4):70–75.

Monks, Vicki. 2000. When a Fish Is More Than a Fish: Saving Endangered Salmon, *National Wildlife* 38(2):22–31.

Odgen, J. C., and T. P. Quinn. 1984. Migration in Coral Reef Fishes: Ecological Significance and Orientation Mechanisms. In *Mechanisms of Migration in Fishes,* ed. J. D. McCleave, G. P. Arnold, J. J. Dodson, and W. H. Neill, 293–308. New York: Plenum Publishing Corporation.

Perkins, Sid. 2001. The Latest Pisces of an Evolutionary Puzzle. *Science News* 159(18):282–85.

Ruckelshaus, Mary H., Phil Levin, and Jerald B. Johnson. 2002. The Pacific Salmon Wars: What Science Brings to the Challenge of Recovering Species, *Annual Review of Ecology and Systematics* 33:665–706.

Tennesen, Michael. 2000. A Killer Gets Some Respect: Hawaii's Tiger Sharks. *National Wildlife* 38(5):28–31.

Tsukamoto, K., J. Aoyama, and M. J. Miller. 2002. Migration, Speciation, and the Evolution of Diadromy in Anguillid Eels, *Canadian Journal of Fisheries and Aquatic Sciences* 59(12):1989–98.

### InfoTrac College Edition Articles

Dahlgren, Craig P., and David B. Eggleston. 2000. Ecological Processes Underlying Ontogenetic Habitat Shifts in a Coral Reef Fish, *Ecology* 81(8).

Holden, Constance. 1999. Dispute Over a Legendary Fish, *Science* 284(5411).

Klimley, Peter. 1999. Sharks Beware, *American Scientist* 87(6).

McFarlane, Gordon A., and Jacquelynne R. King. 2003. Migration Patterns of Spiny Dogfish (*Squalus acanthias*) in the North Pacific Ocean, *Fishery Bulletin* 101(2).

Pauly, Daniel, and Reg Watson. 2003. Counting the Last Fish: Overfishing Has Slashed Stocks—Especially of Large Predator Species—To an All-Time Low Worldwide, According to New Data, *Scientific American* 289(1).

Pennisi, Elizabeth. 1989. Much Ado About Eels, *BioScience* 39.

Phleger, Charles F. 1998. Buoyancy in Marine Fishes: Direct and Indirect Role of Lipids, *American Zoologist* 38(2).

Ross, Robert M., George S. Losey, and Milton Diamond. 1983. Sex Change in a Coral-Reef Fish: Dependence of Stimulation and Inhibition on Relative Size, *Science* 221.

Stock, David W., and Gregory S. Whitt. 1992. Evidence from 18S Ribosomal RNA Sequences That Lampreys and Hagfishes Form a Natural Group, *Science* 257(5071).

## Websites

**http://www.flmnh.ufl.edu/fish/** This website of the Florida Museum of Natural History Ichthyology Division links to all sorts of information about fishes. It includes species accounts, pictures, information on current research, and the International Shark Attack File.

**http://www.mbayaq.org/** Website of the Monterey Bay Aquarium includes a webcam of many of their exhibits and information on the many fishes they exhibit.

**http://www.dinofish.com/** Information on the coelacanth.

**http://www.austmus.gov.au/fishes/** The Australian Museum Fish Site includes species accounts and movies of Australian fishes.

**http://www.flmnh.ufl.edu/fish/Sharks/Megamouth/mega.htm** Pictures and information on the megamouth shark.

# 11

# Reptiles and Birds

## Key Concepts

1. The evolution of the amniotic egg gave reptiles a great reproductive advantage.

2. The Asian saltwater crocodile lives in estuaries and is adapted to life in the marine environment.

3. Sea turtles have streamlined bodies and appendages modified into flippers.

4. Sea turtles mate at sea and lay eggs on the same beaches where the females hatched.

5. Sea turtles may migrate long distances between their breeding grounds and their nesting beaches.

6. Sea turtle populations are endangered by a number of human endeavors.

7. The marine iguana of the Galápagos Islands is the only marine lizard.

8. Several species of venomous sea snake live in the marine environment.

9. Shorebirds have long legs for wading and thin, sharp bills for finding food in shallow water and sand.

10. A variety of bird species, including gulls, pelicans, and tubenoses, are adapted to feeding on marine organisms.

11. Penguins are the birds most adapted to life in the sea.

Reptiles and birds generally occupy the higher trophic levels in oceanic food chains. Many are tertiary or higher-order consumers. Their larger size allows them to feed on a wider variety of organisms and reduces the number of predators that can feed on them. They are well-adapted to moving freely about in search of food and have highly developed senses to supply them with a constant flow of information about their environment. A relatively complex nervous system processes this information, and their proportionately larger brains give them a greater capacity for learning. The ancestors of these marine animals were terrestrial, and the move back to the sea led to changes in body form, locomotion, insulation, osmotic balance, and feeding. In this chapter, we will examine how reptiles and birds have successfully met the demands of the marine environment.

## MARINE REPTILES

Reptiles have been successful in both terrestrial and marine environments. The same characteristics that allowed the ancestors of reptiles to conquer land were also useful in allowing their descendants to return to the sea. The ancestors of modern reptiles first began to appear about 100 million years ago. Modern-day reptiles include crocodilians, turtles, lizards, and snakes, all of which are represented in the marine environment.

### Amniotic Egg

A major reason for the success of reptiles was the evolution of an amniotic egg (Figure 11-1). An **amniotic egg** is covered by a protective shell and contains a liquid-filled sac called the **amnion** in which the embryo develops. The egg also contains a supply of food in the form of yolk, stored in a **yolk sac**, and an additional sac, the **allantois** (eh-LAN-toys), for

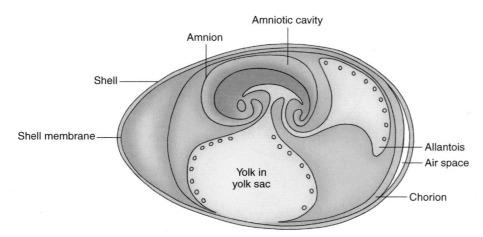

**Figure 11-1 Amniotic Egg.** An amniotic egg is covered by a protective shell and contains a watery interior (amniotic fluid) in which the embryo develops. The yolk sac contains a supply of nutrients for the developing embryo. The allantois serves as a waste receptacle for products of metabolism. The membrane known as the chorion functions in gas exchange.

the disposal of waste. A membrane called the **chorion** lines the inside of the shell and provides a surface for gas exchange during development.

The evolution of an amniotic egg about 340 million years ago gave reptiles a great advantage. It allowed development within the protective egg to continue longer, eliminating the need for a free-swimming larval stage that would be highly vulnerable to predation. It also allowed the eggs to be laid in a dry place out of the reach of aquatic predators.

To produce an embryo within a shelled egg required a means of fertilizing the ovum before the shell was added. The evolution of copulatory organs by reptiles increased the efficiency of internal fertilization before the ovum was encased by a shell and laid by the female.

## Physiological Adaptations

Several other adaptations have helped reptiles survive both on land and in the ocean. The circulatory system of reptiles is more advanced than that of fishes. Circulation through the lungs is nearly completely separate from circulation through the rest of the body (in crocodilians the two circulatory paths are for the most part completely separate). This pattern of circulation results in a more efficient method of supplying oxygen to the animals' tissues and helps to support their active lifestyles. Their kidneys are very efficient in the elimination of wastes and conservation of water, allowing them to inhabit dry regions and the salty environment of the ocean. Reptiles have a skin that is covered with scales and generally lacking glands. This adaptation decreases body water loss in marine environments.

## Marine Crocodiles

Several species of crocodile, including the American crocodile (*Crocodylus acutus*) and the Nile crocodile (*Crocodylus niloticus*), venture into the marine environment to feed. The one best adapted to the marine environment, however, is the Asian saltwater crocodile (*Crocodylus porosus;* Figure 11-2), which inhabits estuaries from India and Southeast Asia to northern Australia. This animal sometimes makes oceanic migrations of several hundred kilometers. One specimen is known to have traveled more than 1,100 kilometers

(683 miles) from Malaysia to the Cocos Islands in the Indian Ocean.

Saltwater crocodiles are large animals that grow to lengths of 6 meters (20 feet). They feed mainly on fishes, and some individuals have been known to attack and kill sharks close to their own size. Saltwater crocodiles are very aggressive and have been known to attack and kill humans in several parts of their range. Like other marine reptiles, saltwater crocodiles drink saltwater, eliminating the excess salt through salt glands on their tongues. As with other crocodiles, this animal lives along the shore, where it makes its nest and lays its eggs.

## Sea Turtles

There are seven species of sea turtle that inhabit the world's oceans (Figure 11-3). They primarily inhabit shallow tropical and subtropical coastal waters, although the leatherback turtle (*Dermochelys coriacea*) can tolerate colder temperatures and has been recorded as far north as Canada and Alaska.

**Figure 11-2 Saltwater Crocodile.** Saltwater crocodiles inhabit estuaries along the Indian Ocean.

(a) Leatherback

(b) Hawksbill

(c) Kemp's Ridley

(d) Green

(e) Loggerhead

(f) Flatback

(g) Olive Ridley

**Figure 11-3  Sea Turtle Species.** The seven species of sea turtle are the (a) leatherback sea turtle, (b) hawksbill sea turtle, (c) Kemp's Ridley sea turtle, (d) green sea turtle, (e) loggerhead sea turtle, (f) flatback sea turtle, and (g) Olive Ridley sea turtle.

## Adaptations to Life at Sea

Marine turtles have changed very little since they first evolved 150 million years ago. Their bodies are covered by protective shells that are fused to the skeleton and fill in the spaces between the vertebrae and ribs (Figure 11-4). The shell is composed of two layers: an outer layer consisting of a protein called keratin, similar to typical reptilian scales, and an inner layer composed of bone. The dorsal surface of the turtle's shell is the **carapace,** and the ventral surface is the **plastron.** The leatherback turtle, the largest of the marine turtles, lacks a shell. Its body is covered by a thick hide that contains small bony plates. The shell of sea turtles is flattened and streamlined for easier movement through the water. It is also reduced in both size and weight compared with that of their terrestrial cousins, an adaptation for buoyancy and swimming. Other adaptations for buoyancy include large fatty deposits beneath the skin and light spongy bones.

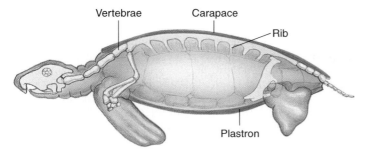

**Figure 11-4 Sea Turtle Anatomy.** The limbs of sea turtles are modified to form flippers. The turtle's shell is composed of an outer layer of hard protein and an inner layer of bone that is directly attached to the animal's backbone.

The front limbs of sea turtles are modified into large flippers, whereas their back limbs are paddle shaped and used for steering and digging nests. On land their movement is awkward, but in the ocean they move with great grace and speed.

## Behavior

Sea turtles are generally solitary animals that remain submerged for most of the time they are at sea. They rarely interact with each other except for courtship and mating. Even when large numbers of turtles congregate on feeding grounds or during migrations, individual animals usually do not interact with each other. During the day sea turtles alternate their time between feeding and resting. Sea turtles sleep on the bottom, wedged under rocks or coral. In deep water, sea turtles can sleep floating at the surface. Turtles breathe air but can stay submerged for as long as 3 hours. Although they generally prefer the safety of open water, it is not uncommon to find some species close to reefs.

## Feeding and Nutrition

Turtles lack teeth but have a beaklike structure that they use to secure food. Six of the seven sea turtle species are carnivorous (Table 11-1). The single herbivorous species is the green sea turtle (*Chelonia mydas*), which is named for its large deposits of greenish fat. The leatherback sea turtle feeds primarily on jellyfish. Their pharynx is lined with sharp spines to hold slippery prey and their digestive system is adapted to withstand the stings of their prey. They may also consume floating plastic bags or plastic trash that they mistake for jellyfish. The turtle cannot digest the plastic, and if enough is consumed, it will block the animal's intestine causing death from starvation.

Sea turtles consume large amounts of salt with their food and the water they drink. To eliminate excess salt, the turtles have salt glands located above their eyes that secrete a concentrated salt solution. The tears produced by these salt glands can have a concentration of salt that is twice that of seawater.

## Reproduction

Marine turtles generally mate at sea, and breeding occurs in cycles that vary from 1 to 5 years. The males remain in the water past the surf line and the females mate with them between trips to land to nest.

***Courtship*** During the mating season, males will court females by nuzzling their heads or nipping at their necks and flippers. If the female does not try to escape, the male will grasp the back of the female's shell with the claws on his front flippers. Once he has successfully mounted the female, he copulates with her (Figure 11-5). Sometimes several males will compete for a single female, and they may even

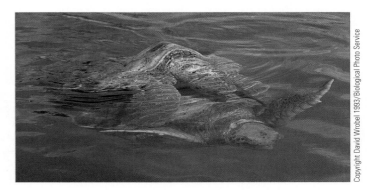

**Figure 11-5 Sea Turtle Reproduction.** Sea turtles, like these green sea turtles, generally mate at sea. The female then comes ashore to lay her eggs in shallow nests in the sand.

| Table 11-1 Food Preferences of Marine Turtles | |
|---|---|
| **Turtle Species** | **Food Preference** |
| Green sea turtle (*Chelonia mydas*) | Turtlegrass and manateegrass |
| Hawksbill sea turtle (*Eretmochelys imbricatus*) | Sponges |
| Kemp's Ridley sea turtle (*Lepidochelys kempii*) | Crabs, shrimp, snails, clams, jellyfish, sea stars, and fish |
| Olive Ridley sea turtle (*Lepidochelys olivacea*) | Jellyfish, snails, shrimp, and crabs |
| Leatherback sea turtle (*Dermochelys coriacea*) | Jellyfish |
| Loggerhead sea turtle (*Caretta caretta*) | Conchs, clams, crabs, horseshoe crabs, shrimps, sea urchins, sponges, fish, squid, and octopuses |
| Flatback sea turtle (*Natator depressus*) | Sea cucumbers, jellyfish, mollusks, prawns, bryozoans, and seaweed |

fight with each other for the chance to mate. Some females mate with several males and lay eggs fertilized by several different males. This behavior may increase the genetic diversity of the population. The sperm that the female receives during mating fertilizes eggs that will be laid on a future trip 2 or 3 years later.

***Nesting***    Nesting occurs most frequently at night. During nesting the female crawls to a dry area of the beach and digs a shallow pit by pushing away loose sand with her flippers and rotating her body (Figure 11-6a). She then digs a hole in the pit for the eggs using her rear flippers. The female lays two to three eggs at a time. During egg laying, the female constantly secretes mucus on the eggs to keep them moist. The clutch size is between 80 and 150 eggs, depending on the species of turtle. When she is finished laying the eggs, the female uses her rear flippers to cover them with sand. She packs the sand down and uses her front flippers to cover the pit and disguise the nest (see Figure 11-6b). Once the nest is concealed, she crawls back to sea to rest before nesting again. A single female will generally lay several clutches of eggs at intervals of 2 to 3 weeks.

***Development and Hatching***    Although the average incubation period is 60 days, the temperature of the nest determines how quickly the turtle embryos will develop. The temperature of the nest also influences the sex ratio. At temperatures above 29.9°C the embryos develop as females. Lower temperatures cause embryos to develop as males.

The eggs usually hatch at night, and digging out of the nest is a group effort. Free of the nest, the hatchlings orient themselves to the brightest horizon, the sea, and then move quickly toward its relative safety (Figure 11-7). As they try to find their way to the water, turtle hatchlings are easy prey for predators such as birds and crabs. If they do not make it to water before daylight, they may die of dehydration when the sun rises. Once in the sea, the hatchlings swim several miles offshore until they are caught in currents that carry them farther out to sea. Even in the sea hatchlings are still not completely safe, because they may become prey for large fishes and sea birds. Some will die as the result of eating tar balls or plastic garbage. Only about 0.1% of hatchlings survive to adulthood.

Young sea turtles spend most of their time feeding and growing in near-shore habitats. After reaching adulthood, they migrate to new feeding areas that will remain their primary feeding grounds for life. Adults leave these primary feeding grounds only to migrate to their nesting beaches to reproduce, after which they return to their feeding grounds.

## Turtle Migrations

Sea turtles migrate hundreds and sometimes thousands of miles from their feeding grounds to their nesting beaches. Females return to the same beaches where they hatched to mate and lay eggs. Each breeding season the females return again and again to the same beaches, sometimes to within a few hundred meters of the spot where they nested previously. Why sea turtles nest on some beaches and not others is not known. The Florida loggerheads, for instance, nest on the central east coast of Florida but not on beaches to the north that appear identical. Their preference for the central

**Figure 11-6 Sea Turtle Nest.** (a) Sea turtles come to land to lay their eggs. After digging a depression in the sand, a pit is hollowed out with the hind flippers and 80 to 150 golf-ball-sized eggs are laid in the nest. (b) Female sea turtles disguise their nests to prevent terrestrial predators from disturbing them. In many nesting areas volunteers go out early in the morning to locate new nests and mark them to prevent accidental intrusions by humans.

(a)

(b)

Scott Tuason / imagequestmarine.com

**Figure 11-7 Turtle Hatchlings.** Hatchlings must move quickly to the sea to avoid bird and terrestrial predators. They orient themselves to the ocean by looking for a bright reflection off the horizon.

coastal area may reflect conditions that existed in the distant past, such as temperature, shape of the shore, and numbers of predators.

The migrations of green sea turtles have been studied extensively. Green sea turtles feed on the turtlegrass and manateegrass found in warm, shallow continental waters but breed on remote islands that can be thousands of kilometers away. One population of turtles feeds on manateegrass along the coast of Brazil, but migrates 2,330 kilometers (1,400 miles) to Ascension Island, a small island between South America and Africa, to breed. Tagging studies showed that not all of these turtles migrate every year; some breed on a 2- or 3-year cycle. The migration begins in December and ends in February to April. Food supplies are scarce around Ascension Island, so after the females have finished laying their clutches of eggs, they return to the Brazilian coast to feed.

There are many hypotheses as to how the green sea turtles, as well as other species, are able to navigate over such long distances. The senses of smell and taste may play an important role, as well as audible signals produced by other animals. Some researchers suggest that the turtles can sense both the angle and intensity of the earth's magnetic field or the forces produced by the earth spinning on its axis. Others believe the turtles may be able to navigate using the sun. Studies to answer these and other questions concerning the natural history of sea turtles are currently being conducted.

### Sea Turtles in Danger

Most species of sea turtle are endangered as a result of several human activities. Beach erosion after commercial development and alteration of beaches for improved boat access has destroyed many nesting sites. Artificial lighting near nesting beaches causes young turtles leaving the nest to crawl toward the lights and away from the water. Many coastal U.S. states, including Florida, and several international nesting sites have strict laws protecting sea turtle eggs and hatchlings. Some Florida coastal communities re-

quire turning off bright lights along the beaches during the seasons when the eggs are laid so that the females and hatchlings will not become confused and lost. Many sea turtles are killed when trapped in fishing nets, especially those used by shrimpers (see box). Turtles are hunted by humans for their meat, eggs, leather, and shells. Humans, however, are not the only predators of turtle eggs; dogs, cats, and wild animals such as raccoons also dig them up to eat.

## Marine Iguana

The only marine lizard is the marine iguana (*Amblyrhynchus subcristatus*) of the Galápagos Islands, off the coast of Ecuador (Figure 11-8). These large lizards are descended from the green, vegetarian iguanas that still inhabit the tropical forests of the mainland. The marine iguana can grow to 1 meter (3.3 feet) long, and most are usually entirely black, although some are mottled red and black, showing a hint of green during the breeding season. It is thought that the dark coloration allows these lizards to absorb more heat energy to raise their body temperatures so they can swim and feed in the cold Pacific waters.

### Feeding and Nutrition

Like their cousins on the mainland, marine iguanas are herbivores. Instead of having a long snout like the forest iguana, however, the marine iguana has a short, heavy snout, similar to that of a bulldog, that is better suited for grazing on the dense mats of seaweed that hug the rocks. Marine iguanas feed by biting with one side of the mouth first and then the other. As the animal tears at the tough algae, it clings to the slippery rocks with powerful clawed toes. To feed under water, the marine iguana swallows small stones to make its body less buoyant.

Whether feeding at low tide or under water, the animal consumes large amounts of saltwater. The excess salt from the water is extracted and excreted by specialized tear and nasal glands. Salt that is extracted by nasal salt glands is periodically expelled by nasal spraying, an unusual but effective behavior.

### Behaviors

Marine iguanas are good swimmers, using lateral undulations of their body and tail to propel them through the water. They avoid heavy surf and rarely venture more than 10 meters (33 feet) from shore. When leaving the water, they tend to ride in with the swell, then swiftly crawl up on the rocks. If they do not find their territory immediately, they touch the rocks with their tongues, which like a snake's carry scent to a receptor in the roof of the mouth. When they locate their own scent, they follow it to their territory, where they rest on the rocks, lying almost motionless above the high tide.

Each male occupies a small territory on the rocks, usually in the company of one or two females. If an intruder wanders by, even by accident, he is immediately attacked and driven off. Most fights between male marine iguanas are the result of deliberate challenges by an intruder. Combat begins with a great deal of posturing, threatening, and bluffing. Each male rises high on its legs and exposes the

# Endangered Sea Turtles

Six of the seven species of marine turtle are endangered by human activities. Pollution, poaching, and destruction of nesting sites are all serious problems for sea turtles. In many areas of the world, one of the greatest threats to turtles is commercial fishing. The Pacific population of leatherback sea turtles, for instance, is in such serious decline that researchers estimate that by 2010 fewer than 50 females will return to the nesting site at Playa Grande, Costa Rica, once one of the most popular nesting sites for Pacific turtles. Marine biologists agree that 50 females is not enough to maintain the population and that, unless there is a change, the Pacific population will become extinct. During the 1988–1989 nesting season, 1,367 leatherbacks nested at Playa Grande. Ten years later, in the 1998–1999 nesting season, the number had fallen to 117 animals. Although the total number of breeding females in the Pacific population is disputed, there seems no doubt that the numbers are declining. Records from other Pacific nesting sites in India, Sri Lanka, and Malaysia show declines similar to those in Costa Rica. Records for a major nesting beach in Mexico indicate that in the 1988–1989 nesting season 70,000 turtles returned to nest, whereas in 1998–1999 fewer than 250 turtles were recorded nesting at the site. Studies cite fishing as the reason for the precipitous decline in the numbers of turtles at Playa Grande and other beaches. Turtles tangle in fishing nets and lines and drown, or they are hauled aboard fishing vessels with the nets and killed. Most fishers are paid by the boatload of fish. This discourages them from spending time saving hooked or tangled turtles.

Pacific leatherbacks are not the only sea turtle in danger. Five species of sea turtles, including the critically endangered Kemp's Ridley, in the Gulf of Mexico and waters off the southeastern United States are also threatened or endangered. Fishing has been cited as contributing to the decline of these turtle populations as well, but over the last 10 years the U.S. government and some shrimp fishers have been trying to change this. The drowning of sea turtles in shrimp trawls was identified as a serious problem in the 1970s. Shrimp nets stay in the water for hours, and if a turtle becomes trapped in the net, it drowns in 40 minutes. During the 1970s and early 1980s it was estimated that as many as 400 turtles per day were dying in shrimp nets. During the 1980s a device called the **turtle exclusion device,** or TED, was invented. This rather simple device when incorporated into the shrimp net reduces turtle mortality by as much as 95%. Shrimp caught in TED nets are designated "turtle safe." The TED is a grating placed in the neck of the net (Figure 11-A). The net is cut just above and in front of the grate. When a turtle is caught in the net, it swims to the grate and is deflected up and out through the opening. The pressure of the water closes the net flap behind the turtle. The much smaller shrimp are pulled through the grate and into the catch pouch of the trawl net. In 1991 the federal government began requiring all U.S. shrimpers to use TED nets. This requirement was met with a great deal of hostility from most shrimpers. They claimed that the TEDs would seriously interfere with their catches. Studies conducted after the introduction of TEDs did not confirm these fears. For instance, shrimp catches in the Gulf of Mexico in 1992, when the TED law was in effect, were higher than in the previous 3 years, when the law was not in effect. In the Atlantic Ocean off South Carolina the catch was the highest in 6 years. Claims for lost and damaged fishing gear declined in the years following enactment of the TED law. Since the introduction of TED nets, strandings and drownings of endangered turtles have been down and turtle nesting activity at some key beaches is up. Although the U.S. government requires that shrimp supplied to U.S. markets be caught in TED nets, there is much resistance in the international community, and not all countries use TEDs, especially in the Pacific. ●

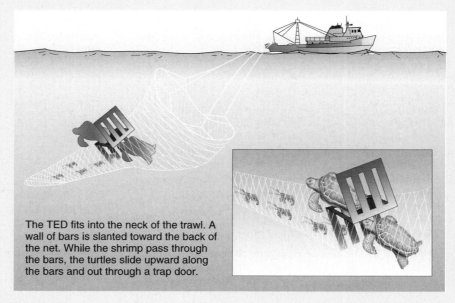

The TED fits into the neck of the trawl. A wall of bars is slanted toward the back of the net. While the shrimp pass through the bars, the turtles slide upward along the bars and out through a trap door.

**Figure 11-A Turtle Exclusion Device.** *Turtle exclusion devices (TEDs) are used in shrimp trawls to save sea turtles from drowning.*

Figure 11-8 **Marine Iguana.** The marine iguana, the only marine lizard, is native to the Galápagos Islands. The scar on this animal's shoulder and side is the result of a shark bite.

Figure 11-9 **Sea Snake.** An olive sea snake patrols the bottom in search of prey.

bright red interior of its mouth. They may also spray each other with a stream of moisture from their nostrils, and as the battle escalates, the combatants butt heads and try to push each other away. In the end, the loser, often the challenger, lies down in a submissive posture, and the winner stands high and nods vigorously. Because such encounters rarely result in serious injury, the survival of the population remains unaffected.

## Sea Snakes

Snakes are descendants of lizards that have lost their limbs as an adaptation to a burrowing lifestyle, a lifestyle that was later abandoned by many species. Although most species of snake are terrestrial or arboreal, there are about 50 species that live in the marine environment (Figure 11-9). Most sea snakes remain close to shore, in the shallow waters of the western Pacific and the Indian Ocean, but the yellow-bellied sea snake (*Pelamis platurus*) has been sighted on several occasions hundreds of miles from land. This species has migrated east and west from the coast of Asia and can be found off the east coast of Africa and the west coast of tropical America. Most of the other species are found in warm coastal waters from the Persian Gulf to Japan and east to Samoa.

### Adaptations to Life in the Sea

Sea snakes have evolved several adaptations that help them survive in a marine environment. As an adaptation for streamlining the body, scales on sea snakes are either absent or greatly reduced. The tail is laterally compressed and used as a paddle for swimming. The nostrils of sea snakes are higher on the head than in terrestrial snakes to aid in breathing while drifting. Specialized valves in the snake's nostrils prevent water from entering when they are submerged. Although all sea snakes breathe air, some species can remain submerged for several hours. The animal's single lung reaches almost to its tail, and its trachea (wind-pipe) has become modified to absorb oxygen, thus acting as an accessory lung. Sea snakes can also exchange gases through their skin when they are under water. These adaptations allow the sea snake to absorb large amounts of oxygen in a very efficient manner. Sea snakes are able to lower their metabolic rate so that they consume less oxygen when submerged, which is another reason they can remain under water for long periods.

### Feeding and Nutrition

Sea snakes feed mainly on fish and eels. Most species have venomous fangs, but a few that have become specialized for feeding on fish eggs have lost them. Because they are slow swimmers they rely more on ambush to capture their prey. Some lie quietly at the surface resembling a piece of drifting wood. Small fish that come near to seek shelter become an easy meal. Others probe crevices and holes on the seafloor for food, sometimes maneuvering their prey into a corner so it will be within easy striking distance. Like their land-dwelling cousins, sea snakes can swallow prey more than twice their diameter. Sometimes two snakes try to feed on the same prey. When this happens, the smaller of the two is usually seen disappearing into the larger one. Some of the fishes that sea snakes feed on have spines, and instead of digesting or voiding the spines, the sea snakes push them out through their body wall. Eyesight is the most highly developed sense in sea snakes. Thus they feed during the day and spend the night on the ocean bottom, rising occasionally to the surface to breathe.

### Reproduction

Sea snakes are less tied to the land than other marine reptiles, with only about three oviparous species, such as the banded sea krait (*Laticauda colubrina*), coming to land to lay their eggs. The rest are viviparous, with the females retaining their eggs within their body until they hatch. The young emerge able to swim and feed immediately. Sea snakes sometimes congregate in enormous numbers, presumably to mate. In 1932 a group of sea snakes that was 3 meters (10 feet) wide and 97 kilometers (60 miles) long was reported near Indonesia. There are many other reports of smaller aggregations.

### Sea Snakes and Humans

Although sea snake toxin is specifically adapted to killing fish, it is also quite toxic to other vertebrate animals, including humans. Sea snakes, however, are timid by nature and rarely bite humans. There are no accounts of sea snakes attacking swimmers or divers, and when sea snakes are caught in fishing nets, they are handled with indifference. In those recorded instances in which sea snakes caused human deaths, the snakes were inadvertently grabbed and were biting in self-defense.

In many parts of their range, sea snakes are eaten by humans. In Japan the consumption of sea snakes is so large that it supports a major fishery. The hunting of sea snakes for their skins has led to their near extinction in some areas.

## In Summary

The evolution of an amniotic egg allowed reptiles to completely sever all ties with their aquatic environment, giving them a great reproductive advantage. Other adaptations, such as more efficient respiratory, circulatory, and excretory systems, as well as a highly developed nervous system, helped reptiles survive on land and, later, in the marine environment.

Several species of crocodile enter the ocean to feed, but the best adapted is the Asian saltwater crocodile. Although usually associated with estuaries, this species sometimes makes long ocean migrations. They feed mainly on fish. In some parts of their range they have been known to attack and kill humans.

Marine turtles have changed very little since they first evolved. Their bodies are streamlined, and their appendages are modified into flippers. Turtles lack teeth, having instead a beaklike structure that is used in securing food. Most sea turtles mate at sea. The female then comes ashore to dig a nest and deposit her eggs. Sea turtles return to the same beaches where they hatched to lay their eggs and sometimes migrate long distances from their feeding grounds to their nesting beaches. Sea turtles are currently endangered because of many human endeavors.

The only marine lizard is the marine iguana of the Galápagos Islands. This relative of terrestrial iguanas feeds on seaweeds and exhibits several adaptations that allow it to feed in the cold marine waters of its habitat. Snakes evolved from the lizard line of reptiles, and several species are found in tropical marine environments. Sea snakes are venomous and feed primarily on fish and eels. Some species that feed on fish eggs have lost their fangs. Only about three of the species come to land to lay their eggs; the others are viviparous. ●

---

## IN PERSPECTIVE

| Type of Reptile | Food | Reproduction | Distribution |
| --- | --- | --- | --- |
| Saltwater crocodile | Fish, birds, and mammals | Oviparous; eggs laid in nest on land | Australia, India, Asia |
| Sea turtles | All but one species carnivorous; green sea turtle is a herbivore | Oviparous; eggs laid in sand above the tidemark | Worldwide in tropics and subtropics |
| Marine iguana | Seaweed | Oviparous; eggs laid in nest on land | Galápagos Islands |
| Sea snakes | Mainly fish and eels | Three species oviparous and lay eggs on land; the remaining species are viviparous and give birth at sea | Tropical and subtropical Pacific and Indian Oceans |

## SEABIRDS

Only about 250 of the approximately 8,500 bird species are adapted for life in and around the ocean. Although less varied than terrestrial and freshwater species, they are quite numerous. Seabirds feed in the sea and sometimes spend months out of sight of land, but they must return to land to breed.

### Adaptations for Flight

Birds are **homeothermic** animals, that is, they maintain a constant body temperature. Their bodies are covered with feathers, and the feathers help insulate the animal. In addition, the light weight and strength of the feather make it the perfect body covering for an airborne animal. Birds have a high rate of metabolism to supply the large amounts of energy needed for active flight and to maintain a rapidly functioning nervous system for flight control. Adaptations such as strong muscles, quick responses, and a great deal of coordination greatly aid birds in flight. An advanced respiratory system and a circulatory system that includes a four-chambered heart provide more oxygen to the active muscles to support the high metabolic rate that flying demands. Their senses, especially sight and hearing, are keener than those of their reptilian ancestors, and they have large brains relative to body weight that allow them to process more sensory information more effectively.

### Adapting to Life in the Sea

Like marine reptiles, seabirds consume large amounts of salt with the food they eat and saltwater they drink. The seawater that these birds consume is about three times saltier than their body tissues. Their kidneys are not able to concentrate such high levels of salt in the urine; therefore, like marine turtles and marine iguanas, seabirds have salt glands (Figure 11-10). These are located one above each eye, and they are capable of producing a solution of sodium chloride that is twice the concentration in seawater. The salt solution that is produced by these glands is released by way of the internal or external nostrils so that seabirds, like gulls and petrels, appear to have constantly runny noses.

### Shorebirds

Shorebirds, or waders, have long legs and thin, sharp bills that they use to feed on the abundance of marine life in the intertidal zone. They range in size from as small as a sparrow to larger than a chicken, and they exhibit countershading with dark dorsums and white ventral surfaces. The group includes oystercatchers, tattlers, curlews, godwits, turnstones, sandpipers, jacanas, surfbirds, phalaropes, herons, and even inland species such as snipes and woodcocks.

#### Oystercatchers, Curlews, and Turnstones

Oystercatchers (family Haematopodidae) have long, blunt, orange bills that are vertically flattened (Figure 11-11). They use their bills to slice through the adductor muscles of partially opened clams, mussels, and oysters so they can feed on the soft flesh. They can also use their bills to pry limpets off rocks, crush crabs, and probe mud and sand for worms and crustaceans.

The long-billed curlew (*Numenius americanus*) uses its 20-centimeter (8 inch) bill like a forceps to extract shellfish from their burrows. Turnstones (*Arenaria*) are heavyset birds with short necks and slightly upturned bills. They use their bills like crowbars for turning over stones, sticks, and beach debris in search of food.

#### Plovers

Plovers (family Charadriidae) are shorebirds with worldwide distribution. They have short, plump bodies, have bills that resemble a pigeon's, and are considerably shorter than other waders. These birds can be found on beaches, mudflats, and grassy fields. It is common to see flocks of hundreds of plovers following the receding waves and plucking small animals from the damp sand. One of the most intrigu-

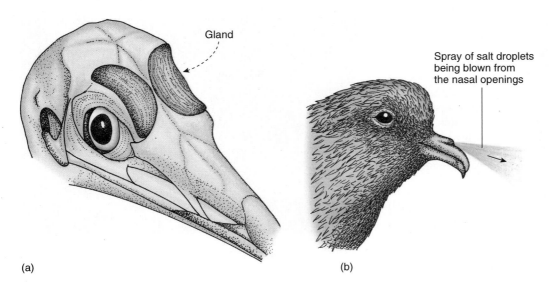

Gland

Spray of salt droplets being blown from the nasal openings

(a)                    (b)

**Figure 11-10 Salt Glands.** (a) Salt glands located above the eyes of seabirds help to excrete the salt the birds ingest when drinking seawater and feeding on marine animals. The glands drain into the bird's nasal passageways, which explains why many marine birds appear to have a constant runny nose. (b) A petrel blowing salt droplets from the nasal openings.

Jon L. Hawker

**Figure 11-11 Oystercatcher.** The oystercatcher uses its sharp bill to slice through the adductor muscle of bivalves.

Steven Holt/Aigrette

**Figure 11-12 Wrybill.** The wrybill is a plover from New Zealand. Notice the unique shape of this bird's bill.

ing plovers is the wrybill (*Anarhynchus frontalis*) of New Zealand (Figure 11-12). It is the only bird in the world with a beak that is bent to the right. The adaptive significance, if any, of this feature is not known.

The nests of plovers are characteristic of waders in general. Most nests are built on the ground in depressions or hollows. The females lay four pear-shaped eggs that fit together like the pieces of an orange, making it easier for a single bird to incubate the clutch. The chicks hatch in about 3 weeks and are covered with a camouflage down. They are active as soon as they hatch and leave the nest within hours to follow their parents and search for their own food.

### Avocets, Stilts, and Sandpipers

Avocets (*Recurvirostra*) and stilts (*Himantopus*) have very long legs, elongated necks, and slender, graceful bodies. Avocets feed by wading through shallow water and moving their partially opened beak from side to side through the water. Stilts have legs that are longer in proportion to their

body size than any other bird except flamingos. They use their straight bills to probe the mud for insects, crustaceans, and other small animals. Both stilts and avocets live in warm climates and breed colonially in marshes or at the edge of lagoons.

The American avocet (*Recurvirostra americana*) is anything but friendly during the breeding season. When another bird comes too close to their nesting sites, the birds angrily charge the intruder. When there are no eggs or young to defend, these birds are quite indifferent to intruders, including humans. This characteristic, combined with destruction of their habitats, has made them rare in the eastern United States.

Sandpipers (family Scolopacidae; Figure 11-13) are relatives of the avocets. As the surf and tide retreat, these birds scurry across the sand to feed on small crustaceans and molluscs. Sandpipers are sociable birds that share the beaches with many other species of shorebird. They are found in most parts of the world and make annual migrations that take some species as far as 10,000 kilometers (6,000 miles).

### Herons

Herons (family Ardeidae), which include egrets and bitterns, are one of the most widespread families of wading bird and are represented on every major continent. Although they may appear rather ungainly, their skinny legs and long necks are aids for hunting. The great blue heron (*Ardea herodias*), for instance, is a stalker that uses its bill as pincers to capture small fish and crustaceans (Figure 11-14). Its appetite for fish is enormous, and it catches all sizes, even those that appear too big for the bird to swallow.

Most herons feed by standing still and waiting for their prey, a small fish or crustacean, to come into range. An exception to this mode of feeding, however, is the little egret (*Egretta garzetta*). This impatient bird stalks its prey along mudflats. Sighting a fish, it will move cautiously until within range, and then with a lightning jab of the beak, seize its prey. It will also vibrate its feet as it moves through shallow water to scare bottom-dwelling prey up to the surface. The

M. Graybill/J. Hodder/Biological Photo Service

**Figure 11-13 Sandpipers.** Sandpipers are a common sight along beaches, where they search for food in the sand.

Figure 11-14 **Great Blue Heron.** This great blue heron is about to land and feed in the shallow water. Herons are wading birds that feed on fishes.

Figure 11-15 **Sea Gulls.** Herring gulls wait on a pier for fishing boats.

snowy egret (*Leucophoyx thula*) also feeds by stirring the bottom with its feet and then rushing about grabbing the small fishes and crabs that it has frightened into motion.

## Gulls and Their Relatives

Gulls (order Charadriiformes) are probably the best-known seabirds because they are found mostly along shores and seaports, with some species venturing far inland. They have a worldwide distribution and are found everywhere the land and ocean meet, including polar regions. Although gulls have webbed feet and oil glands to waterproof their feathers, most species are not true ocean-going birds and so do not stray far from land. They have enormous appetites and are not very selective eaters. Many species will land on the water and feed on floating debris. Relatives that resemble gulls in many respects are the terns. These birds can be found in large numbers along tropical and subtropical shores. Also related to gulls are the hawklike skuas and jaeger birds, skimmers, and alcids.

### Gulls

Herring gulls (*Larus argentatus*) are the most widespread and best known of the gull family (Figure 11-15). These vocal gray and white birds can be found along the shores of both North America and Europe, covering the waterfronts of large seaports and small fishing villages and ranging inland as well. They always travel in large groups. Hordes of these birds can be seen following fishing boats and garbage scows or migrating in groups of tens of thousands of individuals.

***Feeding*** Gulls are noisy, aggressive birds that are efficient predators and scavengers. They are aggressively carnivorous, eating the eggs and young of other birds, stealing prey from other birds, and occasionally, when food is scarce, even eating the young of other gulls. In some areas they are so highly regarded for cleaning garbage from beaches and coastal waterways that they are protected by law.

Gulls exhibit interesting behaviors when it comes to feeding. Some have been observed to pick up clams, mussels, oysters, or sea urchins from shallow water and then fly over rocks, dropping their food repeatedly until the shell breaks. They have also been observed using parking lots as shell-dropping areas. Gulls are so successful at getting food and surviving that in some areas they have reached nuisance proportions, menacing other birds, colliding with airplanes, and fouling buildings and sidewalks with excrement.

***Nesting*** Gulls are highly gregarious and gather in colonies of hundreds of thousands of individuals. They are not picky about nesting sites or nesting materials, using any available material. The female lays two or three eggs, and both sexes help in the incubation. The eggs hatch in 3 or 4 weeks, and the female has the capacity to lay another clutch of eggs immediately if the original clutch is lost. This survival mechanism can be repeated as many as three or four times a year. The newly hatched chicks remain in the nest until they are almost fully grown, camouflaged by their speckled down. Not only predators but also other gulls will feed on the young hatchlings, and it is not uncommon for only one out of every five to survive the 8 weeks after hatching until they are able to fly.

### Terns

Related to gulls are the terns (family Laridae), small, graceful birds with brightly colored and delicately sculpted bills (Figure 11-16). Because they have forked tails, they are sometimes referred to as sea swallows. Although a few species are freshwater birds, most terns live along the shoreline, where they hunt for small fish and invertebrates by plunging into the water but rarely landing on the water. They are generally not the avid hunters gulls are, but they do steal food from other birds. Terns are gregarious nesters; often millions of birds swarm together. Not all terns, however, are so gregarious. The arctic tern (*Sterna paradisaea*) makes solo flights from north of the Arctic Circle to Antarc-

**Figure 11-16 Terns.** Terns nest in large colonies of thousands of birds. Although similar in color to sea gulls, terns have head feathers that project from the back of the head that help to distinguish them from gulls.

**Figure 11-18 Skimmer.** Skimmers fly along the water with their lower mandible just beneath the surface. When the mandible strikes a fish or shrimp, it snaps shut, capturing the prey.

tica. These seasonal migrations allow the bird to spend 10 months of the year in continuous daylight, more than any other living creature.

### Skuas and Jaegers

Some very aggressive relatives of the gulls are the skuas and jaegers (family Stercorariidae). Skuas (*Catharacta;* Figure 11-17) are found in both hemispheres, particularly around the poles, where they breed. One might consider these birds to be the hawks, falcons, and even vultures of the sea. They are omnivorous and keen predators, feeding on garbage, berries, other birds, eggs (especially penguin eggs), lemmings and other small mammals, insects, carrion, and even unprotected chicks of their own species. Jaegers (*Stercorarius*) are predators that steal fish from other birds. They can frequently be seen pursuing terns and other marine birds and robbing them of their prey.

**Figure 11-17 Skuas.** Skuas are related to gulls and fill the same niche in the marine environment that hawks do in a terrestrial ecosystem.

### Skimmers

Skimmers, or scissorbills (family Rynchopidae; Figure 11-18), are perhaps the most interesting of the gull's relatives. They are small birds that grow to a length of 50 centimeters (20 inches) and have two unusual features: pupils that are vertical slits similar to those of cats and a flexible lower jaw that protrudes much farther than the upper bill. These birds fish along coastal rivers and streams, flying along with their lower bill just beneath the surface of the water and creating a thin ripple that attracts small fish to the surface. The skimmer then reverses direction of flight and follows the same line that it previously made in the water. When the mandible strikes a small fish or shrimp, it immediately snaps shut. The birds feed at dusk, dawn, and during the night. They race upstream and then turn to race downstream, occasionally pausing on the bank to rest and eat their catch.

### Alcids

Although auks, puffins, and murres (family Alcidae) look more like penguins, they are actually related to gulls. Alcids are countershaded with black dorsums and white ventral surfaces. Like penguins, these birds are awkward on land. In the water they are remarkably agile, using their stout swimming wings to fly through the water as they pursue their prey of fishes, squid, and shrimp. Alcids and penguins are the result of **convergent evolution,** in which similar selective pressures brought about similar adaptations in unrelated groups of animals (Figure 11-19). They are also ecological equivalents. **Ecological equivalents** are different groups of animal that have evolved independently along the same lines in similar habitats and therefore display similar adaptations.

A major difference between all of the modern alcids and penguins is that alcids can fly, although the largest flightless bird in the Northern Hemisphere was an extinct alcid, the giant auk (*Pinguinus impennis;* Figure 11-20). Standing 750 centimeters (30 inches) high, it was an easy mark for hunters and feather collectors and was hunted to extinction by the middle of the nineteenth century.

**HORNED PUFFIN**

**Colorful head and beak**
(for attracting a mate)

**Salt glands**
(for removing extra salt after
drinking ocean water)

**Stiff, broad beak**
(to catch and hold fish)

**Torpedo-shaped body**
(aerodynamic design
to "fly" through water)

**Black and white color**
(counter-shading
camouflage in ocean)

**Flipper-like wings**
(to help swim)

**Dense, overlapping feathers**
(for insulation and waterproofing)

**Short legs at rear of body**
(for steering underwater)

**Webbed feet**
(for swimming)

**Dense bones**
(for diving)

**GENTOO PENGUIN**

**Figure 11-19 Convergent Evolution.** The bodies of both puffins and penguins have been shaped by convergent evolution, a process in which similar selective pressures brought about similar adaptations in unrelated groups of animals.

**Figure 11-20  Giant Auk.** The giant auk is now extinct, the victim of hunters interested in its meat and feathers.

***Nesting and Reproduction***   Alcids spend their winters in offshore waters and then gather in dense, noisy colonies in the cliffs along the northern Atlantic and Pacific oceans in early spring. Auks (*Alcae*) and murres (*Uria*) nest on ledges and among boulders, whereas puffins (*Fratercula;* Figure 11-21) prefer crevices and burrows on higher bluffs.

The females lay a single egg that is cared for by both parents. Because these nesting sites are rather precarious, eggs of the ledge dwellers are pear shaped, an adaptation that makes them less likely to roll over the edge. As one might expect, both egg and hatchling mortality rates are high. Adults jostling with each other on the ledges cause many eggs and chicks to fall, and others are preyed upon by the ever-present gulls.

***Parental Care of the Young***   Young murres that survive leave the ledges in July and literally plunge into the ocean because they have not yet molted into their flight feathers. Each youngster's parents encourage it to the brink of the ledge where, after a moment of hesitation, it leaps into the

**Figure 11-21  Puffin.** Although this puffin looks like a relative of penguins, it is actually more closely related to the gulls.

ocean. The youngster is quickly joined by one or both parents, and they swim out to sea, where they will spend the winter months.

Like other young alcids, chicks of the common puffin (*Fratercula arctica*) have enormous appetites. Both parents spend most of their time gathering food for their hungry chicks, sometimes carrying as many as 18 fish in a single load. The parents provide the chick with enough fish, mussels, and sea urchins per day to equal its body weight. After 6 weeks of constant care, the adult puffins abruptly leave their chicks and head to the open ocean. The young puffin spends about another week in the burrow or crevice by itself before venturing out to the water. It learns to swim, fly, and dive without the guidance of any adult birds.

## Pelicans and Their Relatives

Pelicans (order Pelecaniformes) are members of a group of birds that includes gannets, boobies, cormorants, darters, frigatebirds, and tropicbirds. Members of this group have webs between all four toes, and in pelicans, cormorants, and frigatebirds, the upper mandible is hooked. Many species are brightly colored. Some, such as the darters (family Anhingidae), have conspicuous head adornments, whereas others, such as the boobies (family Sulidae), have brightly colored faces and feet. Members of this group enjoy worldwide distribution but most live in the tropics and warm temperate regions. Although some species can be seen in the mid-ocean, most are found in coastal areas and some (pelicans, cormorants, and darters) in inland waters.

### Pelicans

Pelicans (family Pelecanidae) prefer warm latitudes and estuary, coastal, and inland waters. They are large animals; some species weigh as much as 11 kilograms (24 pounds). Because they are so large and nest in sizable colonies, the waters where they live must support a large fish population to feed them. Sometimes pelicans have to fly some distance from their nesting sites to find enough food. After breeding, the birds disperse or migrate, sometimes crossing deserts. Their numbers have declined in Europe, Africa, and Asia because of their dependence on inland waters that have been drained or are polluted.

Pelicans feed just under the surface of the water, using their large **gular** (throat) **pouches** as nets. The gular pouch is a sac of skin that hangs between the flexible bones of the bird's lower mandible and is characteristic of pelicans. The brown pelican (*Pelecanus occidentalis*) of the Americas is the most fully marine of the pelicans (Figure 11-22). It patrols about 9 meters (30 feet) above the water, and when it sights a school of small fishes, it folds its wings back and crashes into the water with its gular pouch gaping. The impact stuns the fish, allowing the pelican to scoop up large numbers of them, as well as 8 to 12 liters (2 to 3 gallons) of water, with its pouch. The pelican then returns to the surface with its catch, buoyed by hundreds of air sacs located just beneath the skin (subcutaneous air sacs). It must then let the water drain from its pouch before it can become airborne again. An adult brown pelican will consume 4.5 to 6.5 kilograms (10 to 14 pounds) of fish per day.

**Figure 11-22   Pelican.** A brown pelican with a fish in its gular pouch.

The six other species of pelican hunt in groups in shallow water by forming semicircles and dipping their bills to flush and trap fish. Occasionally they will beat their wings on the water to drive the fish together, making them easier to catch.

### Boobies

Pelicans are not the only birds in this order that exhibit interesting fishing techniques. Boobies and gannets dive into the sea from heights of 18 to 30 meters (59 to 98 feet), and they can catch flying fishes just before they reenter the water.

The red-footed booby (*Sula sula;* Figure 11-23a) is the smallest of the boobies and nests in trees of the Galápagos Islands. The female usually lays only one egg. Blue-footed boobies (see Figure 11-23b) are widely distributed in the Pacific. They nest close to shore on the ground and the female of this species lays two or three eggs. Masked boobies (*Sula dactylatra;* see Figure 11-23c) are the biggest and heaviest species of booby. They live close to cliffs on many of the Galápagos Islands, and the females generally produce two eggs in a clutch. In each of these species, both parents take turns incubating the eggs and feeding the young. Even with such good care, usually only one nestling of the blue-footed and masked boobies will survive to become a fledgling.

The difference in the number of eggs laid by each species is thought to be due to the reliability of their food supply. Because red-footed boobies forage farthest from land and have the most consistent food supply, they lay one egg. The blue-footed boobies feed close to the coastline, which is easier, but the food supply is very unpredictable. The masked booby feeds offshore in relatively deep water in an area that is intermediate between the feeding grounds of the red- and blue-footed boobies. The food resource is generally stable but not as predictable as it is for red-footed boobies, and they spend much more time foraging.

### Cormorants

Cormorants (family Phalacrocoracidae) are some of the most adept avian fishers. Swimming along the surface, they

(a)

(b)

**Figure 11-23 Boobies.** Boobies include the (a) red-footed booby, (b) blue-footed booby, and (c) masked booby.

(c)

scan the water for fish and then plunge to spectacular depths in pursuit of their prey. These birds have been found tangled in fishermen's nets at depths of 21 to 30 meters (69 to 98 feet). Cormorants swallow their prey headfirst so that the fish scales and spines will not catch in their throats. Cormorants lack oil glands to waterproof their feathers, an adaptive advantage because it reduces positive buoyancy, allowing them to expend less energy while diving. Without oil glands, though, the feathers become saturated with water, and so the birds must periodically dry their wings on land before they are light enough to fly again. In Japan and some other Asian countries, cormorants are used by fishers to help catch fish (Figure 11-24). The birds are released with a rope or a steel ring around their throats to prevent them from swallowing the fish that they catch.

Most cormorants are strong fliers, but the flightless cormorant (*Phalacrocorax harrisi*) of the Galápagos Islands has completely lost the ability to fly. This bird has been isolated on two remote islands of the Galápagos group for millions of years. The waters around the islands contain abundant food, and the islands themselves have plenty of nesting material and nesting sites. There are also no natural enemies of the cormorant. A strong swimmer, this cormorant uses its webbed feet for propulsion and keeps its wings folded close to its sides to be more streamlined.

The guano cormorant (*Phalacrocorax bougainvillei*) of the coast of Peru is one of the world's most valuable birds because it produces **guano,** or bird manure, that is particularly phosphate rich and used in making fertilizer. The high phosphate content of this material is the result of the birds' anchovy-rich diet. Commercial harvesters of guano have constructed large wooden platforms in sealed-off peninsulas where the birds can roost safely, isolated from predators.

Copyright 1995 Ted Wood

**Figure 11-24 Cormorant.** In China, cormorants are used for fishing. A rope made of grass is tied around the bird's neck to prevent it from swallowing its catch. This young boy is about to remove a fish from the cormorant's mouth.

## Frigatebirds

Frigatebirds (family Fregatidae; Figure 11-25) have light-weight bodies and wingspans that approach 2 meters (7 feet). Their large wingspan allows them to soar for hours on the lightest of breezes before returning to land to rest. Frigatebirds are another species with no oil glands in their skin to waterproof their feathers. If they are forced to settle on the ocean's surface, there is a good chance that they will drown. Because of this, frigatebirds feed without getting much more than their bills wet. They skim along just above the water, picking up jellyfish, squids, fishes, young sea turtles, and bits of carrion. They are particularly fond of flying fish and can catch them in midflight.

Frigatebirds are also pirates and will frequently steal fish from other birds. They have been observed attacking boobies even after the booby has swallowed its catch. The frigatebird beats and jostles its victim in flight until the booby regurgitates the fish. The frigatebird then catches the meal before it drops back into the water. A frigatebird will sometimes perch on the upper bill of a feeding pelican and steal fish while the pelican is emptying water from its gular pouch.

## Tubenoses

Petrels, albatrosses, and shearwaters (order Procellariiformes) are referred to as tubenoses because of the obvious tubular nostrils on their beaks. The tubes join with large nasal cavities within the head. The significance of these tubes is not known, and they vary in form and length from one species to the next. It has been suggested that these birds have a much better-developed sense of smell than most birds, but this does not explain the peculiar nostrils. These birds fly low over the ocean, and the tube-shaped nostrils may help prevent salt spray from entering the nasal cavity. All tubenoses have large nasal glands that secrete a concentrated salt solution, and the tubes possibly keep the solution away from the eyes. Others hypothesize that the nostrils aid in discerning the strength of air currents.

In addition to the tubular nostrils, members of this family can be distinguished from other birds by the structure of their stomachs. Their stomachs contain a large gland that produces an oil composed of liquefied fat and vitamin A. This yellow oil is used to feed newly hatched young, and because it has a strong unpleasant odor, adult birds use it for defense by regurgitating it through the mouth and nostrils when they are disturbed.

### Albatrosses

Albatrosses are superb gliders with wings nearly 3.5 meters (11 feet) long, the largest of any living bird. The wandering albatross (*Diomedea exulans*) is the largest of the albatrosses and master of the skies. It glides over ocean waves for hours at a time with barely the flick of a wing. With their very large wings, albatrosses are able to take advantage of light sea winds to keep them aloft. If the wind dies, the birds are forced into a labored, flapping flight; if they land on the water, in calm conditions they cannot leave it. For this reason, most species of albatross are restricted in their range to the Southern Hemisphere, the only part of the ocean where winds circle the earth without encountering land. These birds are not likely to move northward because the doldrums, areas of rising air at the equator (for more information on the doldrums see Chapter 4), form an impassable barrier for soaring birds, especially in the Atlantic Ocean. In the Pacific Ocean, however, 3 of the 13 species of albatross have moved northward. Interestingly, these birds still mi-

Richard N. Mariscal

**Figure 11-25 Frigatebird.** A male frigatebird inflates its red throat to attract mates.

**Figure 11-26  Albatross.** A male and female albatross clap bills during a mating ritual.

**Figure 11-27  Storm Petrel.** These Wilson's storm petrels exhibit a characteristic flapping type of flight. Early mariners thought the storm petrel could walk on water and named this bird after St. Peter.

grate to the Southern Hemisphere to mate and breed at the same time that the southern species do, even though it is the middle of winter in the Northern Hemisphere.

As a rule, albatrosses come to land only to breed, and some young birds may spend the first 2 years of their lives at sea before they return to nest. During the mating season, males court the females with elaborate displays that include gyrations, wing flapping, and bill snapping (Figure 11-26). After mating, the female normally lays a single egg on a nest that is shaped like a volcano, and both parents take turns at incubating the egg. The young hatch 60 to 80 days later, usually in March, and are fed first on stomach oil and then regurgitated fish until they are able to feed themselves. Because it takes so long for the parents to raise a single chick, most species mate every other year rather than annually.

### Petrels

Storm petrels (family Hydrobatidae) are small birds with long legs. They exhibit a characteristic erratic type of flight that looks more like a fluttering moth than a flying bird. Wilson's storm petrels (*Oceanites oceanicus*) are the most abundant seabirds. They feed with their legs extended and their feet paddling rapidly just below the surface of the water so that the birds appear to be walking on water (Figure 11-27). For such small birds, storm petrels have an extremely long life span (20 years or more). They usually form long-term pair bonds and return to the same nesting sites every year. They do not breed until they are 4 or 5 years old, which is late for a bird of this size. They breed seasonally and, like other species in this group, lay only a single egg.

Although diving petrels (family Pelecanoididae) have the nostrils, beak, and stomach glands that are characteristic of tubenoses, they look more like the auks of the Northern Hemisphere. Diving petrels are found exclusively in the Southern Hemisphere. They require cold water throughout the year and prefer gray skies, rough seas, and high precipitation and generally do not stray far from their colonies. Diving petrels can spot small crustaceans and fish from the air. Sighting its prey, the petrel performs a headlong dive

and continues the chase by "flying" under water. Grasping its meal in its beak, the bird emerges from the water and reenters the air without missing a wing beat.

### Penguins

The birds that are most highly adapted to life in the sea are penguins (order Sphenisciformes). Although penguins are usually associated with Antarctica, only two species, the emperor penguin (*Aptenodytes forsteri*) and the Adelie penguin (*Pygoscelis adeliae*), live there. Fourteen other species of penguin live on barren, rocky islands in the Antarctic Sea and along the shores of South America. One species, the Galápagos penguin (*Spheniscus mendiculus*), lives as far north as the equator. It is able to survive in this atypical region because of the cold, food-rich waters of the Humboldt Current that bathe the western coast of South America.

On land, penguins are somewhat awkward, moving by hopping or waddling on short legs or tobogganing on their bellies as they push on the ice with their flipperlike wings. In the sea, they are as swift and agile as many fishes, literally flying through the water by flapping their wings. They can swim at speeds of 25 kilometers (15 miles) per hour for long distances and can achieve bursts of speed twice as fast while pursuing prey. Their torpedo-shaped bodies are streamlined to offer less resistance to the water, and their flat, webbed feet, stretched out behind them, are used for steering. While swimming, penguins break the surface to breathe. They arc through the air like a dolphin and disappear into the water again with scarcely a ripple.

Penguins feed on the fishes, squid, and krill that abound in the cold southern seas. In turn, penguins are preyed upon by leopard seals and killer whales. On land, adult penguins have virtually no enemies, but gull-like skuas (discussed previously in this chapter) that live on the outskirts of penguin colonies take a heavy toll on eggs and chicks.

Adelie and emperor penguins breed on the Antarctic continent. Adelies lay their eggs during the summer when the temperatures are above freezing, but the emperor pen-

**Figure 11-28 Penguins.** A male emperor penguin incubating an egg.

*Bruno P. Zehnder/Peter Arnold, Inc.*

guins lay their eggs in the middle of the Antarctic winter, when temperatures drop to −63°C (−80°F) and blizzard winds blow at 130 to 170 kilometers (81 to 106 miles) per hour. This strategy may help to ensure that the young will be mature enough by summer for life in the sea.

The female emperor penguin lays one egg and then leaves for her feeding grounds, a trip that may take her over vast stretches of ice. While she is gone, the egg is incubated by the male (Figure 11-28). For 2 months he stands with the egg on his feet, a fold of skin on his lower abdomen covering the egg to keep the developing embryo warm. During the incubation period, the male fasts, living off his reserves of fat. If the egg hatches before the mother returns, the chick is fed a secretion from the father's **crop,** a digestive organ that stores food before it is processed. When the mother re-

turns, her crop is filled with fish, and she feeds the chick. The male is now free to travel to open water to feed. When the chick is about 6 weeks old, it requires the services of both parents to provide it with enough food, but by the time summer arrives, the young penguin is capable of feeding itself. The young birds do not have to travel as far to reach the sea as the parents did earlier because of the seasonal breakup of the ice.

## In Summary

Instead of a body covering of scales, birds have feathers. Feathers are light, facilitating flight, and provide good insulation. In addition to feathers, the evolution of powerful muscles, homeothermy, and a highly developed nervous system that can coordinate quick responses helped these animals adapt to a flying lifestyle. Seabirds feed in the sea and sometimes spend months out of sight of land. They must return to land, however, to breed.

Shorebirds, or waders, have long legs for wading and thin, sharp bills for finding food in the shallow water and sand. Some well-known shorebirds are oystercatchers, turnstones, plovers, sandpipers, surfbirds, and herons.

Gulls are efficient predators and scavengers. Because gulls are found mostly along shores, they are probably the best-known seabirds. Gulls are highly gregarious and gather in colonies of 100,000 or more individuals. Terns, skuas, skimmers, puffins, and murres are all related to the gulls. Terns are found along the shoreline, where they hunt for small fishes and invertebrates. Skuas are aggressively carnivorous and will even eat the young of their own species. Skimmers are small birds that fly along the surface of the water, scooping up food with their enlarged mandible. Puffins and murres resemble penguins but, unlike penguins, can still fly. Penguins and puffins are examples of ecological equivalents and their similar body form is the result of convergent evolution. Puffins and murres live in the Northern Hemisphere. They breed on bluffs and cliffs and feed on fish.

Pelicans are large-bodied birds that feed by scooping up large amounts of water with their gular pouch. Boobies, cormorants, and frigatebirds are related to pelicans.

Birds known as tubenoses are represented by albatrosses and petrels. Albatrosses are superb gliders, having the largest wings of any living bird. Storm petrels are small birds that exhibit a characteristic erratic flight. As a group, tubenoses feed primarily on fish.

Penguins are the birds most adapted to life at sea. Their streamlined bodies and flippers instead of wings make them excellent swimmers. Most penguin species are found in the Antarctic, where they feed on fish, squid, and krill. In turn, penguins are preyed upon by leopard seals and killer whales. ●

## IN PERSPECTIVE

| Bird Group | Characteristics | Diet | Reproduction |
|---|---|---|---|
| Shorebirds | Plump body; long stiltlike legs, long sharp beaks | Infauna, fish, and epifaunal invertebrates | Eggs laid in nests away from shore |
| Gulls and their relatives | Webbed feet; gulls and terns gregarious; alcids resemble penguins (ecological equivalents) | Fish and shellfish; gulls will also eat garbage and human food | Eggs laid in nests close to shore; alcids nest on ledges of cliffs |
| Pelicans and their relatives | Pelicans have gular pouch for catching food; cormorants dive to great depths to capture prey, skin lacks oil glands | Fish | Eggs laid in nests near shore |
| Tubenoses | Albatrosses and petrels glide on air currents; wingspan long compared with body length | Fish and squid | Eggs laid in protected areas on land |
| Penguins | Most aquatic of marine birds, with streamlined bodies and wings adapted as flippers | Krill and fish | Eggs laid on land; emperor penguin male incubates the egg during the Antarctic winter |

## SELECTED KEY TERMS

allantois, *p. 254*
amnion, *p. 254*
amniotic egg, *p. 254*
carapace, *p. 256*
chorion, *p. 255*

convergent evolution, *p. 266*
crop, *p. 272*
ecological equivalent, *p. 266*

guano, *p. 269*
gular pouch, *p. 268*
homeothermic, *p. 263*
plastron, *p. 256*

turtle exclusion device, *p. 260*
yolk sac, *p. 254*

## QUESTIONS FOR REVIEW

### Multiple Choice

1. The key to the evolutionary success of reptiles is
   a. a skin covered with scales
   b. a four-chambered heart
   c. the amniotic egg
   d. an efficient excretory system
   e. a large brain
2. Most marine reptiles must still return to land to
   a. sleep
   b. reproduce
   c. feed
   d. molt
   e. die
3. Marine iguanas feed on
   a. grass
   b. crustaceans
   c. fishes
   d. seaweed
   e. molluscs
4. Marine crocodiles drink saltwater and eliminate the excess salt by way of
   a. their urine

b. their feces
c. skin glands
d. their gills
e. specialized glands on the tongue

5. Birds that exhibit soaring flight have
   a. small bodies
   b. few feathers
   c. long wings
   d. large heads
   e. small feet

6. _____ use their gular pouch as a net to catch fish.
   a. Petrels
   b. Pelicans
   c. Albatrosses
   d. Gulls
   e. Penguins

7. The seabird that plays an important role in keeping beaches clean is the
   a. albatross
   b. pelican
   c. petrel
   d. tern
   e. gull

8. Two types of seabird that are ecological equivalents are
   a. gulls and skuas
   b. petrels and albatrosses
   c. penguins and puffins
   d. herons and sandpipers
   e. boobies and frigatebirds

9. Skuas
   a. are related to pelicans
   b. feed on krill
   c. are most frequently found along tropical shores
   d. are predators of penguin chicks
   e. nest on ledges along the Arctic shore

10. Shorebirds generally have
    a. short legs and long beaks
    b. short necks and webbed feet
    c. long legs and flat beaks
    d. long legs and long beaks
    e. short legs and webbed feet

## Short Answer

1. What similarities are shared by most species of shorebird?

2. Describe how marine turtles are adapted to life in the sea.

3. Describe the nesting behavior of sea turtles.

4. Compare marine iguanas with their more terrestrial cousins.

5. Explain how sea snakes that do not lay their eggs on land reproduce.

6. Explain how birds are well-adapted for flight.

7. Describe how sea birds maintain their osmotic balance despite drinking seawater and eating marine animals.

8. Why are most soaring birds such as albatrosses primarily confined to the Southern Hemisphere?

9. Why do the birds known as alcids resemble penguins more than the gulls and terns they are more closely related to?

## Thinking Critically

1. Why do seabirds that nest in trees or other protected areas tend to lay fewer eggs than seabirds that nest on the ground?

2. Why is it important for marine reptiles and birds to be able to efficiently eliminate salt from their bodies?

3. Which of the following would you expect to suffer more from predation: albatross adults or chicks? Why?

## SUGGESTIONS FOR FURTHER READING

Alderton, D. 1988. *Turtles and Tortoises of the World.* New York: Facts on File.

Doubilet, D. 1996. Australia's Saltwater Crocodiles, *National Geographic* 189(6):34–47.

Ellis, R. 2003. Terrible Lizards of the Sea, *Natural History* 112(7):36–41.

Jackson, D. D. 1989. The Bad and the Beautiful: Gulls Remind Us of Us, *Smithsonian* 20(7):72–85.

Oeland, G. 1996. Emperors of the Ice, *National Geographic* 189(3):52–71.

Rudloe, J., and A. Rudloe. 1989. Shrimpers and Lawmakers Collide Over a Move to Save the Sea Turtles, *Smithsonian* 20(9):44–55.

Shetty, S., and R. Shine. 2002. Philopatry and Homing Behavior of Sea Snakes (*Laticauda colubrine*) from Two Adjacent Islands in Fiji, *Conservation Biology* 16(5):1422–26.

Taylor, K. 1996. Puffins, *National Geographic* 189(1): 112–31.

Trivelpiece, S. G., and W. Z. Trivelpiece. 1989. Antarctica's Well-Bred Penguins, *Natural History* December, 28–37.

### InfoTrac College Edition Articles

Bouchard, S. S., and K. A. Bjornadal. 2000. Sea Turtles as Biological Transporters of Nutrients and Energy from Marine to Terrestrial Ecosystems, *Ecology* 81(8).

Bowen, B. W. 1995. Tracking Marine Turtles with Genetic Markers, *Bioscience* 45(8).

Cohn, J. P. 1999. Tracking Wildlife: High-Tech Devices Help Biologists Trace the Movements of Animals through Sky and Sea, *Bioscience* 48(1).

Deforest, L. N., and A. J. Gaston. 1996. The Effect of Age on Timing of Breeding and Reproductive Success in the Thick-Billed Murre, *Ecology* 77(5).

Hochscheid, S., F. Bentivegna, and J. R. Speakman. 2002. Regional Blood Flow in Sea Turtles: Implications for Heat Exchange in an Aquatic Ectotherm, *Physiological and Biochemical Zoology* 75(1).

Preiser, R. 1997. The Sheltering Junk: Engineers Wanted to Destroy the Flotsam Around Shooter's Island in New York Harbor Until an Ecologist Found That the Debris Is Vital for the Harbor's Birds, *Discover* 18(2).

### Websites

**www.princeton.edu/~wikelski/GIguana.htm** Current research on the Galápagos marine iguanas from Princeton University.

**www.Starfish.ch/reef/reptiles.html** Links to information and images of sea snakes and marine turtles.

**www.seaturtle.org** Comprehensive information about the sea turtle.

**http://www.nbii.gov/datainfo/syscollect/by_type/vertebrates.php** Links to amphibian and reptile collections from the National Biological Information Infrastructure.

# 12

# Marine Mammals

## Key Concepts

1. Mammals have a body covering of hair, maintain a constant warm body temperature, and nourish their young with milk produced by the mammary glands of the mother.

2. Sea otters have thick coats of fur and feed on marine invertebrates near shore.

3. Pinnipeds have limbs modified to form flippers and are better adapted to life at sea than to life on land.

4. Sirenians are totally aquatic mammals that feed on a variety of aquatic vegetation.

5. Cetaceans have a fishlike body shape and are the mammals most suited to life in the sea.

6. Special physiological adaptations allow cetaceans to dive to great depths and to remain submerged for long periods.

7. Cetaceans are intelligent animals that display a range of behaviors for communication and investigating their environment.

8. Some cetaceans use echolocation to navigate, find prey, and avoid predators.

9. Baleen whales have plates of baleen instead of teeth and feed primarily on plankton, such as krill.

10. Toothed whales include the large sperm whales and the familiar dolphins and porpoises.

11. Dolphins are intelligent animals that are capable of learning and sophisticated intraspecies communication.

Marine mammals live mostly or entirely in the water, depend completely on food taken from the sea, and display anatomical features, such as fins, that adapt them for their aquatic lifestyle. Most marine mammals are quite intelligent compared with other marine animals. This trait, combined with their generally friendly nature, makes them very popular with the public. Unfortunately, marine mammals share another common characteristic: the bodies of many contain materials that are commercially valuable to humans. As a result, they have been hunted in large numbers over the centuries and now many species are endangered. During the last 50 years, international conservation measures have helped to reverse the decline in many populations, and the numbers of some species are now gradually increasing.

## CHARACTERISTICS OF MARINE MAMMALS

Most mammals, whether terrestrial or aquatic, have an insulating body covering of hair, and all maintain a constant, warm body temperature (homeothermic). Mothers feed their young with milk, a secretion produced by special glands in the female called **mammary glands.** It is this characteristic that gives the class Mammalia its name.

Marine mammals are placental mammals. **Placental mammals** retain their young inside their body until they are ready to be born. Within the mother they are sustained by her systems through a remarkable organ, the **placenta,** which is present only during pregnancy. Although mammals produce fewer offspring than many other animals, more of the young survive to adulthood because they receive a great deal of parental care.

The same characteristics that allowed mammals to become so well-adapted to terrestrial environments also allowed them to function well in the sea. The hair (fur) of some and the layers of blubber in others effectively decrease the loss of body heat to the surrounding water. Like other homeothermic animals, mammals expend about 10 times as much energy as similarly sized reptiles, and they need more food than a fish of comparable size to support their high metabolic rate. Being homeothermic allows marine mammals to be active feeders night and day and to adapt to a wide range of habitats. Some, such as baleen whales, feed closer to the base of the food chain, whereas others, such as sea otters and toothed whales, are second order or higher consumers.

## In Summary

Mammals can be distinguished from other animals by their body covering of hair, constant warm body temperature, and mammary glands that produce a secretion called milk that nourishes the young. ●

## SEA OTTERS

Sea otters (*Enhydra lutris;* order Carnivora) are found along the coast of California and as far north as the Aleutian Islands of Alaska. Instead of having a thick layer of blubber like other marine mammals, their skin is covered by a thick fur with an underlying air layer that protects the animal from the cold. They have short, erect ears, five-fingered forelimbs that they use with great dexterity, and well-defined hind limbs with finlike feet.

Sea otters seldom venture more than a mile from shore. They favor the areas around coastal reefs and kelp beds, where they spend most of their time floating lazily on their backs (Figure 12-1). They do not come ashore very often except during storms, although some individuals appear to prefer to sleep on shore, possibly to avoid their primary

**Figure 12-1  Sea Otter.** Sea otters bring their food to the surface and eat it while floating on their backs. This otter is preparing to feed on a sea urchin.

predators, sharks and killer whales. Females normally give birth on shoreline rocks to one pup, and the offspring soon follows its mother into the sea.

Sea otters have a large appetite and consume nearly 25% of their body weight in food per day. Their diet consists almost entirely of sea urchins; molluscs, especially abalone; crustaceans; and a few species of fish. They pick up their prey from the seabed and carry it to the surface, where, floating on their backs, they use their chests as eating surfaces. They have been observed bringing a stone up along with a mollusc and using it as a tool to smash open the shell. They also have extraordinarily powerful jaw muscles and strong cheek teeth (molars) to help them break through the hard shells of their prey.

Sea otters are diurnal and gregarious animals and can often be observed playing. They are also quite vocal. Pups cry when hungry, and females coo affectionately to their offspring and mates. In distress, both sexes scream loudly, and their companionship call is a high-pitched squeal that from a distance sounds like a whistle.

Because of the value of their fur, sea otters were hunted nearly to extinction. At one time sea otters ranged from Baja, California, all along the western coast of North America, and west along the Aleutian Islands to Japan. After 170 years of unrestricted killing, the sea otter was almost wiped out. In 1911 only about 1,000 animals were left in the world. Since that time, international treaties have protected the animal, and currently 130,000 animals now occupy about 20% of their original range. Their survival today is one of the successes of the marine mammal conservation movement.

## In Summary

Sea otters inhabit the northern Pacific Ocean. Instead of thick layers of blubber, this animal has a thick coat of fur to keep it warm. Sea otters usually stay close to shore, favoring areas around coastal reefs and kelp beds. Their diet consists mainly of sea urchins, crustaceans, and molluscs (especially abalone). Hunted nearly to extinction because of their valuable fur, they are now making a comeback after international protective measures were enacted. ●

## PINNIPEDS: SEALS, SEA LIONS, AND WALRUSES

The term **pinniped** means "featherfooted" and is the name applied to the group of marine mammals (suborder Pinnipedia) that includes seals, elephant seals, sea lions, and walruses. Although pinnipeds are more at home in the water than on land, they still retain the four limbs that are characteristic of terrestrial mammals. All pinnipeds come ashore to give birth and to molt. Most species also mate on shore, and some species prefer to sleep on land or ice floes where they are safe from sharks and killer whales. Pinnipeds can be found in all oceans, although most species prefer the colder waters of the Northern and Southern Hemispheres. They feed on fish and larger invertebrates and a few, notably the leopard seal (*Hydrurga leptonyx*), feed on homeothermic

# Where Have the Steller's Sea Lions Gone?

Steller's sea lions were once abundant from the North Pacific coast of California to Japan, with an estimated 70% of them living in Alaskan waters. Since the 1970s the population has dropped to fewer than 70,000 animals, and during the same period fur seal and sea otter populations have also mysteriously declined in the region. Marine scientists have long debated the cause of the declines in these populations. Some suggest overfishing of the marine mammals' prey, whereas others suggest climate change, predators, disease, or poaching may be the culprit.

In 2000, biologists working for the federal government concluded that commercial fishers who fished for bottom fishes, such as pollock, were the main threat to sea lion populations because they removed too much of the animals' food source. The proponents of the overfishing theory suggested that commercial-catch restrictions would assure that the sea lions have sufficient prey. As a result, major catch restrictions were placed on Alaska's $1 billion fishery. Because the Alaskan waters are one of the world's most valuable fisheries, this action gained the attention of members of Congress, who asked a committee of marine scientists to look into the problem. In the committee's preliminary report, issued in December 2002, Dr. Robert Paine of the University of Washington stated that fishing is probably not the major reason for the population decline, but this is not to say that commercial fishing is not a significant problem. To investigate the problem further, Alaskan legislators and other members of Congress ordered a study to be done by the National Academies of Science to determine if overfishing was really the problem. In response,

the National Academies of Science have recommended that the U.S. government embark on a record ecological experiment by running a 10-year test in Alaskan waters to see if commercial fishing is indeed a threat to the Steller's sea lion populations. The experiment's design calls for setting up four experimental zones in Alaskan waters, each containing a breeding colony of Steller's sea lions. Fishing would be banned for up to 50 nautical miles around two of the colonies but permitted near the other two. Biologists would then monitor and compare trends in the sea lion populations.

If, as Dr. Paine suggests, overfishing is not the major villain in this drama, then what is? Several major reviews of sea lions and their habitats indicate that the animals have plenty of food. This observation prompted Dr. Alan Springer of the University of Alaska in Fairbanks to ask if excess predation might be the cause of the population decline. Using data that were collected by the International Whaling Commission, Springer and his associates hypothesized that whalers had effectively eliminated fin, sei, and sperm whales from Alaskan waters by 1970. Interestingly, this coincides with the beginning of the population declines of the smaller marine mammals, including Steller's sea lions. Springer and his associates think that a shift in the diet of killer whales may be to blame not only for the decline in sea lion populations but also the populations of fur seals and sea otters.

Killer whales feed on a wide variety of prey, including the great whales. Although biologists debate how often large whales are eaten by killer whales, they do agree that killer whales are the most significant natural predators of great whales with the

exception of humans. According to data collected over the years, the Alaskan harbor seal population crashed in the 1970s, then populations of fur seals and Steller's sea lions began their precipitous decline in the 1980s. Sea otter populations crashed during the 1990s. Springer thinks these trends indicate predation by killer whales. He suggests that a decline in great whales forced the killer whales to start feeding on the fat harbor seals. As that prey became scarce, they shifted to eating the smaller fur seals and the more aggressive Steller's sea lions. As these populations dwindled, the orcas started feeding on the much smaller sea otters. The decline in the sea otter population has led to an increase in the number of sea urchins, one of the sea otter's favorite foods. This in turn has placed more pressure on the kelp forest habitats, because sea urchins graze on young kelps.

Springer and his colleagues estimated how many seals, sea lions, and sea otters have died in the last 30 years, and by calculating how many small marine mammals it would take to feed a killer whale, they were able to show that the population crashes could have been caused by a shift in the killer whale diet of less than 1%. Dr. Andrew Trites of the University of British Columbia in Vancouver, however, questions these results. He points out that computer models of marine ecosystems show no significant impact on pinniped populations when great whales are removed. Dr. Springer and other researchers in the field admit that solving the mystery of the declining populations will not be easy. Perhaps the new initiative by the National Academies of Science will shed light on the problem and suggest solutions. ●

animals such as penguins and other seals. The natural predators of pinnipeds are sharks, killer whales, and humans. Although hunted by humans for more than 200 years, the world's pinniped populations are estimated to be 30 million, though three species, the Guadalupe fur seal (*Arctocephalus townsendi*), monk seal (*Monachus* ssp.), and Steller's sea lion (*Eumetopias jubatus*), are still endangered.

## Pinniped Characteristics

Pinnipeds are divided into three families within the order carnivora: eared seals (family Otariidae), true seals (family Phocidae), and walruses (family Odobenidae). As their name suggests, eared seals, which include sea lions and fur seals, have visible but small external ears (Figure 12-2a). True seals, or **phocids,** and walruses lack external ears and are thus more streamlined for swimming under water (see Figure 12-12b). Eared seals and phocids also differ in their manner of swimming. The main propulsive force for swimming in eared seals is produced by the forelimbs, and they appear to be flying underwater. Their hind limbs remain nearly motionless during swimming and are used for steering. Phocids, on the other hand, propel themselves through the water with a sculling movement of their hind flippers. Walruses use a combination of the two methods, relying primarily on their forelimbs to move their bulky bodies.

The body of pinnipeds is spindle shaped, and many species have several thick layers of subcutaneous fat. The head is round and is carried on a distinct neck so that it can be moved independently of the rest of the body. They have large brains and well-developed senses. The two sets of pinniped limbs are modified into flippers. The arm bones are shorter than those of whales so that movement occurs only at the shoulder. Their hind limbs, which are set well back, overlap the stumpy, vestigial tail. In phocids, the hind limbs serve only for swimming and are dragged about on land. Eared seals have hind limbs that can be rotated at right angles to the body and act as legs on land.

## Swimming and Diving

Pinnipeds are fast swimmers and can achieve speeds of 25 to 30 kilometers per hour (15 to 18 miles per hour) for short distances. They are also expert divers, remaining under water for as long as 45 minutes without coming up for a breath. Pinnipeds exhale before diving to decrease the amount of air in their lungs and thus make themselves less buoyant. While under water, the animal's metabolism slows by 20% and heart rate decreases to conserve oxygen. During the dive, blood is redistributed so that vital organs, such as the brain and heart, receive sufficient amounts of oxygen to maintain proper levels of function. These physiological adaptations allow pinnipeds to feed on animals in deeper water as well as at the surface.

Some pinnipeds are capable of deep diving. A Weddell seal (*Leptonychotes weddelli*) living in the cold waters of Antarctica has carried a depth recorder as deep as 596 meters (1,968 feet) and remained there for as long as 70 min-

External ear

Rear flippers can rotate forward

Front flippers can rotate backward

(a)

Patti Murray/Animals Animals

No external ears

Front and back flippers cannot rotate

(b)

Marty Snyderman/Visuals Unlimited

**Figure 12-2  Pinnipeds.** Pinnipeds are well adapted to the marine environment, with torpedo-shaped bodies and limbs modified to form flippers. (a) Eared seals, like this fur seal, have visible external ears and long necks. They also have front flippers that rotate backward and rear flippers that can be moved forward to better support the animals' body weight as they move on land. (b) True seals, or phocids, lack external ears, have short necks, and have front and rear flippers that cannot be rotated and, thus, make movement on land awkward.

utes. The record, however, belongs to the northern elephant seal (*Mirounga angustirostris*). In 1988 a female of this species carried a recorder to a depth of more than 1,250 meters (4,125 feet). Other members of this species were found to regularly dive deeper than 697 meters (2,300 feet), making elephant seals the deepest-diving pinnipeds.

## Reproduction in Pinnipeds

Most pinnipeds leave the water during breeding season and congregate on well-established breeding beaches. The bulls arrive first to establish territories and to await the arrival of females. Some species, such as elephant seals (*Mirounga*) and fur seals (*Callorhinus*), are **polygynous** (pah-LIJ-eh-nehs; *poly* meaning "many" and *gyn* meaning "female"), with bulls establishing harems of 15 or more females. The females arrive ready to give birth to pups conceived the previous year. They then mate again almost immediately after giving birth. Although some species mate every 2 years, most mate annually.

Gestation lasts between 9 and 12 months, depending on the species, and typically one or two pups are born. The length of time that the pups nurse (**lactation period**) depends upon the species and the habitat. Pinnipeds that breed in the coldest habitats, mainly phocids, have the shortest lactation periods, from a few days to 2 to 3 weeks, whereas the sea lions of temperate waters will generally nurse their young for as long as 6 months. One reason for the short lactation period of many phocids is that females fast or feed very little during lactation, placing a severe physiological stress on their body. There is also a limit to the amount of weight that the mother can lose before she will be too weak to survive. Another reason that pinnipeds in cold habitats have short lactation periods has to do with the nature of the habitat. Many species that live in cold environments breed on pack ice. The ice is unstable in many cases, and as it breaks up, there is the possibility that shifting ice will crush and injure or kill the pup. The shorter the lactation period, the faster a pup develops insulation, which the pup needs to enter the water and survive the cold climate.

## In Summary

Pinnipeds have four limbs that are modified into flippers. This group includes seals, sea lions, elephant seals, and walruses. Although they are more at home in the water, they can and do come onto land to mate, give birth, and molt. Their bodies are spindle shaped, and many species have several layers of fat under the skin to provide insulation. ●

## Eared Seals

There are two groups of eared seals: sea lions and fur seals. Sea lions have a coarse coat of nothing but hairs, whereas fur seals have thick, dense underfur beneath the stiff, outer guard hairs.

### Sea Lions

The best known of the eared seal pinnipeds is the California sea lion (*Zalophus californianus*; Figure 12-3a). They are the trained seals of zoos and circuses and are intelligent animals that display an eagerness to learn tricks. These sleek pinnipeds are naturally playful and chase each other in the water while vocalizing with a variety of honking and barking sounds. Before becoming popular attractions at animal parks, sea lions were hunted for their blubber.

Sea lions are highly social animals and usually congregate in groups when they come ashore. During the mating season, breeding beaches become battlegrounds as bulls compete with each other for territory and females. Harems consist of as many as 12 females protected by a dominant bull that can be 2.5 meters (8 feet) long and weigh as much as 1 ton. The bull aggressively guards his territory and harem.

### Fur Seals

Fur seals (see Figure 12-3b) can be distinguished from the short-haired sea lions by their thick, woolly undercoats. This thick fur is prized by hunters and is quite valuable on the fur market, a fact that has resulted in massive fur seal slaughters in the past. They are smaller than sea lions, with bulls averaging 2 meters (7 feet) and weighing about 270 kilograms (600 pounds). There are eight species of fur seal in the Southern Hemisphere, but only one in the Northern Hemisphere. Both northern and southern species exhibit similar habits, although the northern species has been studied more extensively.

Of the nine species, the northern fur seal (*Callorhinus ursinus*) is the most abundant. At one time it was hunted relentlessly for its valuable fur, and by 1914 only 200,000 animals remained. This animal is now protected by international treaties, and only a specified number of young males

**Figure 12-3 Eared Seals.** (a) Sea lions, like the California sea lion (*Zalophus californianus*), are eared seals that are insulated with a blubber layer. (b) This northern fur seal (*Callorhinus ursinus*) has a thick coat of fur for insulation.

(a)     (b)

are allowed to be killed each year. As a result of these protective measures, the number of northern fur seals is estimated to be 1.5 million animals. Similar conservation measures have been taken to protect the several species of southern fur seal (*Arctocephalus*), such as those that breed on the islands off the coast of South America.

## Phocids, or True Seals

The forelimbs of true seals, or phocids, are set closer to the head and are smaller than the hind limbs. As a result, they are less well-adapted to life on land than other pinnipeds. On ice floes and land, they move by dragging their bodies and take every opportunity to slide or roll instead of crawl. Most phocids congregate during the breeding season, and males establish their territory on the land. Instead of forming a harem, however, a male mates with a single female, and the couple remains together for the entire breeding season.

The most abundant pinniped is probably the crabeater seal (*Lobodon carcinophagus;* Figure 12-4a), with a population estimated at more than 15 million animals. When these animals were first described in 1842, biologists erroneously thought they fed on crabs, thus the name. The crabeater seal actually feeds on plankton, primarily krill, that it strains from the water with its teeth.

Some of the most familiar species of seal are the harbor seals (*Phoca*) found along cooler coasts of the Northern Hemisphere. Although most species are marine, a few inhabit freshwater lakes in Canada and Finland. All harbor seals spend some time on land, inching along with a caterpillar-like locomotion. In water they use alternating strokes of their flippers to propel themselves. Harbor seals

**Figure 12-4  Seals.** Examples of true seals are the (a) crabeater seal, (b) harp seal with pup, (c) leopard seal, and (d) elephant seal with trunk inflated.

are generally solitary feeders that eat almost any type of fish, crustacean, or mollusc.

A close relative of harbor seals is the harp seal (*Phoca groenlandica*; see Figure 12-4b). They are found in the cold waters of the Atlantic and Arctic oceans. Newborn harp seals have long, silky, yellowish white fur that turns pure white a few days after birth. The white fur of the pups, known as "white coat" to fur merchants, is highly prized and expensive. During the mid-1960s conservationists drew worldwide attention to the cruelty with which harp seal pups were killed and skinned in the major hunting grounds of Canada. Deluged with complaints, the Canadian government implemented stricter hunting laws and stronger control over the killing and skinning of these animals.

The leopard seal (*Hydrurga leptonyx*; see Figure 12-4c) is the only phocid that eats homeothermic prey. Although leopard seals feed on fish and krill like other phocids, penguins, sea birds, and other seals make up the bulk of their diet. They are powerful swimmers and lurk patiently along the edges of ice floes in the Antarctic Sea, waiting for their unwary prey to enter the water. They are less dangerous on land because they must laboriously drag their bodies across the ice floes and cannot move quickly enough to catch their prey. Their primary predators are killer whales.

The giants among the pinnipeds are elephant seals (*Mirounga*; see Figure 12-4d). The common name for these animals comes from not only their large size but also the unique proboscis on the males. This structure is actually an enlarged and inflatable nasal cavity. It is usually limp and deflated, but, during the mating season, it can be inflated to form a trunk 50 centimeters (20 inches) long. The inflated trunk acts as a resonating chamber that amplifies the bull's roar, warning rival bulls to stay away from his territory. The inflated proboscis also plays a role in attracting mates.

There are two species of elephant seal. The largest of the two is the southern elephant seal (*Mirounga leonina*), which grows to a length of 6 meters (20 feet) and weighs as much as 4 tons. It is found on islands at the tip of South America and in the Antarctic. The northern elephant seal (*Mirounga angustirostris*) rarely exceeds 5 meters (16 feet) and is estimated to weigh as much as 3 tons. This closely related species can be found on the Pacific coast of North America, from Vancouver Island in Canada to Baja California. Northern elephant seals breed from December to mid-March. This allows the young pups to first enter the sea at an optimal time, during the spring upwelling. During summer and fall, the adults and pups remain at sea, where they feed primarily on squid.

## In Summary

Eared seals, which include fur seals and sea lions, have visible external ears. True seals and walruses lack external ears. Eared seals use their front limbs primarily to propel themselves through the water. True seals use their hind limbs, and walruses use a combination of both. The hind limbs of eared seals can be rotated at right angles to the body axis and act as legs on land. Fur seals were at one time relentlessly hunted for their coats, but now they are protected by international law. ●

## Walruses

Walruses (*Odobenus rosmarus*) lack external ears and have a distinct neck, as well as hind limbs that can be used for walking on land (Figure 12-5). They are 3 to 5 meters (10 to 16 feet) in length and weigh up to 1,364 kilograms (3,000 pounds). The canine teeth of their upper jaw have developed into tusks in the males, which are used for fighting with other males. Most bulls carry the scars of battles during the mating season. The tusks also help the animal to hoist its massive body onto ice floes. It plants its tusks in the ice like a pick ax and uses them as an anchor. Tusks first appear when the animals are 5 years old and continue to grow for the rest of the animal's life.

The typical family group consists of a large, dominant bull that presides over a harem of as many as three females and half a dozen calves of various ages. One or two calves are born at the end of an 11-month gestation period, and they will stay with their mother until they are between 4 and 5 years old. Initially, young walruses have a yellow-brown fur. As they age, the fur gets paler and eventually disappears. Although females are devoted mothers, old bulls are known to kill the young, and the remains of infant walruses have been found in the stomachs of adult males.

Walruses are found in the Arctic region of the Northern Hemisphere, where they feed on fishes, crustaceans, molluscs, and echinoderms. At one time, walruses were killed by hunters for their ivory tusks and by Eskimos and other northern peoples for their succulent flesh. As a result, the walrus populations declined to such low levels as to make the species endangered. Protective laws have virtually eliminated walrus hunting for sport and for ivory. However, native peoples who subsist on walrus meat are exempt from the restrictions and continue to take the animals for food.

## In Summary

Walruses are restricted to the Arctic seas, where they feed on fishes, crustaceans, echinoderms, and molluscs. At one time these animals were slaughtered for their ivory tusks, but now they are protected by law. ●

**Figure 12-5  Walrus.** A large male on an ice floe.

# SIRENS: MANATEES AND DUGONGS

Manatees and dugongs are collectively referred to as **sirenians** (order Sirenia). They get their name from the mythical sirens of Homer's *Odyssey.* These sweet-voiced creatures, who lured Ulysses and his men with temptations, were probably the forerunners of the mythical mermaids. Mermaids have been a part of the mythology of the sea for more than 2,500 years. Pliny the Elder, a respected Roman naturalist, gave detailed descriptions of them, and Christopher Columbus wrote in his log, "Today I saw three mermaids, but they were not as beautiful as they are painted." No doubt what Columbus and other seafarers saw and mistook for mermaids were manatees and their relatives, although it is difficult to imagine how anyone could mistake these large, gray animals with wrinkled skin and whiskers for the alluring maidens of legend.

At one time sirenians enjoyed a wide distribution. Now they are confined to coastal areas and estuaries of tropical seas. They share many similarities with whales. Both have streamlined, practically hairless bodies, forelimbs that form flippers, a vestigial pelvis with no hind limbs, and tail flukes. Sirenians are completely aquatic animals and are helpless on land, not even able to crawl as do the pinnipeds. Sirenians are by nature gentle animals that become quite tame and trusting in captivity and in their natural habitat. There are several preserves in Florida where divers and snorkelers can swim with manatees and even touch them and rub their bellies. Only in those areas of the world, such as India, where they have been hunted nearly to extinction have sirenians become shy and elusive. There are two families of sirenians, one represented by the manatees (family Trichechidae) of the Atlantic Ocean and Caribbean Sea and the other by the dugongs (family Dugongidae) of the Indian Ocean.

## Dugongs

The primary differences between manatees and dugongs are in anatomy and habitat. Manatees inhabit both the sea and inland rivers and lakes. Dugongs are strictly marine mammals living in coastal areas, where they feed on shallow-water grasses. The head of the dugong is larger and the flippers are shorter than those of the manatee. The dugong's tail is notched, whereas the manatee's tail is rounded. There is only a single species of dugong (*Dugong dugon;* Figure 12-6a). Large numbers of this animal were once found along coastal areas of the Indian Ocean, Red Sea, and South China Sea. It has been hunted extensively for its tasty flesh, and today the species is endangered and found only in isolated areas of its once extensive range.

## Manatees

There are three species of manatee. The northern manatee (*Trichechus manatus;* see Figure 12-6b) is found from the southeastern United States to northern South America. The Brazilian manatee (*Trichechus inunguis*) is a freshwater species endemic to the Amazon and Orinoco rivers of South America. The African manatee (*Trichechus senegalensis*) lives in coastal habitats, rivers, and lakes of western Africa. All three species of manatee have been extensively hunted, and the northern species is now protected by law in Florida.

Manatees mate and give birth under water, and the male remains with his mate even after the breeding season. The female gives birth to usually a single calf after an 11-month gestation period. Manatees are strict vegetarians and consume large amounts of shallow-water plants. A single manatee will consume at least 27 kilograms (60 pounds) of aquatic plants per day. When manatees eat, they guide the water plants to their mouths with their flippers. This observation may account for the association with mermaids. Be-

**Figure 12-6 Sirenians.** (a) Dugongs are found along coastal areas of India and Asia. (b) Northern manatees can be found along the Florida coast where they feed on seagrass and other vegetation.

David B. Fleetham /Oxford Scientific Films /Animals Animals

NOAA

(a)

(b)

cause of their voracious appetites, they provide a service by keeping coastal waterways free of nonnative plant species, such as the water hyacinth, that tend to take over their new habitats. In Guyana, manatees have been introduced to clear weeds from the canals that transport irrigation water to sugarcane fields.

The greatest threat to northern manatees are the propellers of motorboats. Many of these relatively slow-moving animals are mauled or killed by boats. Although there are strict laws governing the speed and use of motorboats in areas where the manatees congregate, there are few regulations governing the canals and waterways that connect these freshwater bodies to the sea. As the manatees migrate from the freshwater areas to the ocean, they must travel these waterways, and it is then that they are in the greatest danger.

## Steller's Sea Cow

Neither the manatee nor dugong are found in Arctic waters. The only sirenian from this area, Steller's sea cow (*Rhytina gigas*), is extinct. During 1741–1742, Georg Wilhelm Steller, a German, was the physician and naturalist for Russia's expedition to the northern Pacific Ocean led by Captain Vitus Bering. Steller discovered the enormous mammal when his ship was wrecked off one of the Commander Islands. He made many notes on the physical characteristics and behavior of this animal and correctly identified it as a previously unknown species of Sirenia. Desperate for food, Steller and his companions killed and ate several of the creatures. When their ship was repaired, they returned to Russia with the furs of sea otters and Arctic foxes that they had also killed for food. Word spread quickly that a fortune in furs was available for the taking on the Commander Islands, and hundreds of hunters descended on them. The hunters found the meat of Steller's sea cow tasty and easy to obtain, and 27 years after its initial discovery, the species was extinct. Fossil evidence suggests that Steller's sea cow had a much wider range than just the Commander Islands and that they were most likely eliminated from most of the range before Steller "discovered" the species.

## In Summary

Sirenians are represented by manatees and dugongs. Although at one time these animals enjoyed a wide distribution, they are currently confined to coastal areas and estuaries of the tropics. Sirenians are vegetarians feeding on shallow-water grasses and a variety of water plants. ●

# CETACEANS: WHALES AND THEIR RELATIVES

Of all marine mammals, **cetaceans** (seh-TAY-shenz; whales, dolphins, and porpoises) are the ones most extensively adapted to a marine environment. Humans have been fascinated and awed by these animals for centuries. Ancient Greeks and Phoenicians worshiped the dolphin or "sacred fish," as did many Polynesian cultures. In Australia more than 5,000 years ago, aborigines painted on stones recognizable pictures of dolphins. The biblical story of Jonah being swallowed by a whale, the appearance of marine mammals in the art of the Middle Ages, and the Herman Melville novel *Moby-Dick* all attest to our fascination with these marine mammals.

## General Characteristics of Cetaceans

The bodies of cetaceans closely resemble those of fishes, and early naturalists even included them both in the same group of animals. Their exact ancestry is not known, but it is generally agreed that they evolved from an ancient group of land-dwelling carnivorous mammals. Recent discoveries in Pakistan, Egypt, and the Arabian desert of fossil whales with fore and hind limbs and teeth similar to carnivores support this hypothesis. Other evidence for the terrestrial origins of cetaceans can be seen in developing fetuses. Whale fetuses are remarkably similar to those of land mammals. Early in development they have four limbs. The rear appendages disappear externally before birth, although there are a vestigial pelvis and leg bones in some species (Figure 12-7). The front appendages persist and develop into flippers (pectoral fins) with the bone structure of a five-fingered hand. Initially, the animal's nostrils are located at the end of its snout, but before birth they migrate to the top of the head to form the characteristic **blowhole.** The location of the blowhole allows these animals to surface and breathe with a minimum of effort.

The cetacean body is streamlined, and beneath the skin is a uniformly thick layer of fat called blubber. The fat provides insulation to conserve heat and is an energy reserve, as well as a source of water when the fat is metabolized. In becoming streamlined, the neck has disappeared, and the head has become continuous with the rest of the body. The seven cervical (neck) vertebrae of cetaceans are very compressed, and in some species they are fused into a single structure. The result of these modifications is that, in many cetaceans, the head cannot be moved relative to the body.

Cetaceans have no external ears or projecting nostrils to hamper their movements through the water. The whale ear consists of a small opening on the side of the head that leads to an eardrum. The opening is plugged with wax to prevent water from entering and damaging the eardrum. The inner ear is contained in a bone that floats freely in the soft tissue surrounding the skull.

The cetacean body is essentially devoid of hair, with the exception of a few hairs on the head. The skin of cetaceans lacks the sweat glands that are characteristic of other mammals. Because sweat glands work via evaporative cooling, they do not work under water. The absence of these glands also helps cetaceans to conserve water in their salty environment. When whales need to lose heat, they shunt blood to their blubber layer.

The forelimbs of cetaceans have become modified into flippers that can move only up and down and twist a bit. They function as stabilizers. The cetacean tail consists of flat **flukes** composed of dense connective tissue and acts as

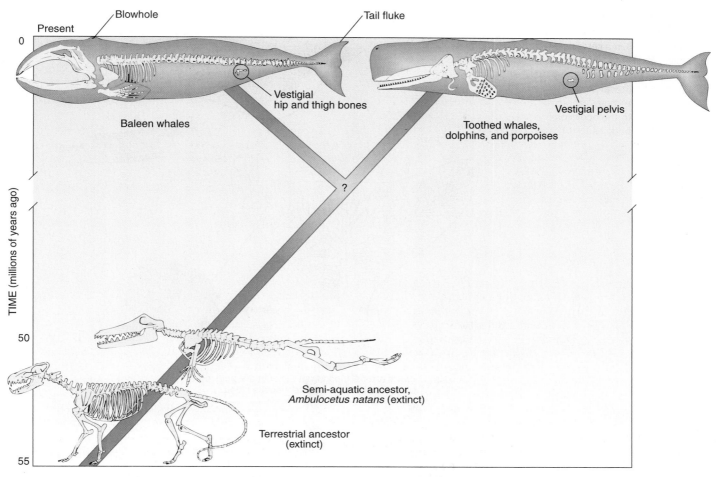

**Figure 12-7 Cetacean Evolution.** Whales evolved from four-legged, terrestrial carnivores. Their ancestors may have moved about on land in addition to swimming in shallow seas. Modern cetaceans are well adapted to life in the sea, with streamlined bodies, appendages modified to form flippers, and a large tail flipper or fluke.

the main propulsion organ. The tail flukes not only supply the force of propulsion but also regulate vertical movement. Most cetaceans also have a dorsal fin that apparently helps them to control roll.

The flippers (as well as the tail flukes) also serve an important role in retaining body heat (Figure 12-8). The arteries that carry blood to the flippers are surrounded by veins that carry blood back to the heart, an example of countercurrent flow (for more information regarding countercurrent flow see Chapter 10). As warm blood in the arteries moves out into the flipper, it transfers heat to the cold venous blood that is returning to the core of the body. Most of the heat is returned to the body, and cold blood is carried to the uninsulated flippers and flukes.

## Adaptations for Diving

Perhaps the most striking adaptations of cetaceans involve those of the circulatory and respiratory systems that allow them to dive to great depths and remain submerged for relatively long times (Figure 12-9). Before diving, a whale

takes an enormous breath and then blows it out. During the inhalation, oxygen is rapidly transferred to the blood. Expelling the air from the lungs then makes the animal less buoyant. Cetacean lungs are proportionately large for the body size, and they contain a large number of small air sacs, or **alveoli.** This feature increases the internal lung surface area that is exposed to blood vessels and allows for a more efficient diffusion of gases into and out of the blood. The lungs and rib cage of whales are structured so that they collapse easily upon descent. As a result, the lungs contain little air during a dive. Because gas volumes change dramatically with changes in pressure, this adaptation helps the animal to avoid the problems of compression and decompression while diving and surfacing.

While diving, the animal's metabolism and heart rate decrease, and blood is preferentially shunted to vital organs and tissues such as the brain and spinal cord. The portion of the brain that controls breathing is the medulla oblongata. In most mammals this region is very sensitive to high levels of carbon dioxide in the blood, which trigger inhalation. The medulla of cetaceans is less sensitive to carbon dioxide, so

**Figure 12-8 Countercurrent Blood Flow in Cetacean Flippers and Tails.**
The blood vessels in a cetacean's (a) flippers and (b) tail fluke are arranged so that (c) some of the heat carried by arterial blood is transferred to the cooler venous blood returning to the body core, thus conserving body heat.

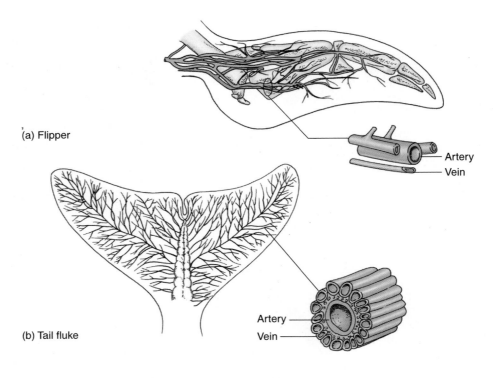

(a) Flipper

Artery

Vein

(b) Tail fluke

Artery

Vein

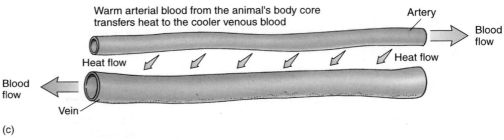

Warm arterial blood from the animal's body core transfers heat to the cooler venous blood

Artery

Blood flow

Heat flow

Heat flow

Blood flow

Vein

(c)

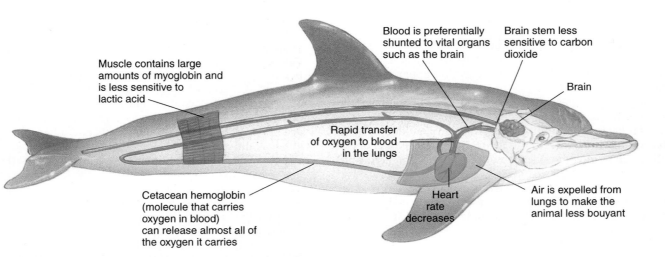

Muscle contains large amounts of myoglobin and is less sensitive to lactic acid

Blood is preferentially shunted to vital organs such as the brain

Brain stem less sensitive to carbon dioxide

Brain

Rapid transfer of oxygen to blood in the lungs

Cetacean hemoglobin (molecule that carries oxygen in blood) can release almost all of the oxygen it carries

Heart rate decreases

Air is expelled from lungs to make the animal less bouyant

**Figure 12-9 Diving Physiology.** Physiological changes in the respiratory system, nervous system, circulatory system, and muscles allow whales to dive to great depths and remain submerged for over 1 hour.

they can remain submerged longer without feeling the urgency to breathe.

Other adaptations for diving involve the transport and storage of oxygen. The molecule **hemoglobin** in red blood cells is responsible for carrying oxygen in the blood and for giving blood its red color. Compared with terrestrial mammals, the blood of whales contains huge amounts of hemoglobin. They also have large blood volumes for their size. These adaptations allow whales to absorb and transport more oxygen, allowing them go longer periods without having to take another breath.

Mammalian muscle tissue contains a molecule called **myoglobin** that is a reservoir of oxygen for muscle activity. Cetacean muscles contain high levels of myoglobin and thus store proportionately more oxygen for muscle activity. Their muscles are also less sensitive to the molecule lactic acid. **Lactic acid** is a waste produced during vigorous or extended muscle activity when there is not enough oxygen. In humans and terrestrial mammals, moderate amounts of lactic acid can impair muscle function, but cetaceans can tolerate much higher accumulations of lactic acid with no ill effects.

When a cetacean surfaces, the top of its head emerges first, thus exposing the blowhole to air. From this opening the animal emits a spout that consists of vapor condensing in the cool air as well as mucous droplets. The mucous droplets may play a role in eliminating nitrogen gas from inhaled air by absorbing it before it reaches the lungs, thus preventing a condition known as **the bends.** The bends occur when nitrogen gas that is dissolved in the blood comes out of solution, and forms gas bubbles as the pressure decreases during an ascent. These bubbles can interfere with the circulation of blood to body tissues and can cause tissue damage or even death. As a cetacean descends, the increased pressure of the water causes the alveoli, or gas sacs, in the lungs to collapse. Any residual air remaining in the lungs is forced into airways, where gas exchange with the blood cannot occur. This prevents nitrogen gas from entering the blood and also helps the animal to avoid the bends.

Cetaceans also exhibit adaptations for preventing water from entering their respiratory passages. The larynx, or voice box, does not open into the back of the throat as it does in some mammals but rather into the nasal chambers. This allows cetaceans to open their mouths under water without the danger of food or water entering their respiratory passageways.

## Cetacean Behaviors

Cetaceans are intelligent, inquisitive animals that exhibit a wide range of interesting behaviors. Some of these behaviors help cetaceans learn more about their surroundings, whereas others may play a role in communication.

### Spy Hopping

Whales sometimes stick their heads straight up out of the water and survey their surroundings. This behavior is known as **spy hopping** (Figure 12-10a). During spy hopping the whale uses its strong flukes to push itself partially out of the water, often exposing its entire rostrum and head. The whale does not appear to maintain its position by swimming but rather relies on buoyancy control and positioning with its pectoral fins and tail flukes. Generally, the whale's eyes will be slightly above or just below the surface of the water, allowing it to see whatever is on the surface nearby. The behavior can last for several minutes depending on what the whale is viewing. Spy hopping often occurs when the whale is interested in a boat or some other object rather than in other whales that may be nearby. This behavior may also help whales establish their bearings while in coastal waters.

### Breaching

**Breaching** refers to a behavior in which a whale completely or almost completely leaves the water (see Figure 12-10b). They begin this behavior under water by swimming quickly, accelerating as they go. Whales are strong swimmers and some species, such as the adult humpback whale, can accelerate to full speed with as few as three or four pumps of their flukes. They hit the surface at full speed in a nearly vertical position and exit the water as they pump their flukes for the last time. At some point the whale usually begins twisting its body in a corkscrew movement. This action may start in the approach and continue as the whale shoots from the water into the air, causing the pectoral fins to fling wide. Some whales can rise out of the water the length of their body, as much as 45 feet for humpback whales. The animal usually assumes a horizontal position as it falls back to the

(a)

(b)

(c)

(d)

(e)

**Figure 12-10 Cetacean Behaviors.** Some cetacean behaviors are (a) spy hopping, (b) breaching, (c) tail slapping, (d) fluke up, and (e) flipper flapping.

water, usually hitting the water on its back. Although many whale watchers think that breaching is a sign of playfulness, it is probably not. It may be a way for a whale to establish dominance. It may also communicate with special emphasis that an individual is arriving or leaving.

Sometimes a whale breaches several times in a row, a behavior called serial breaching. Serial breaching is usually observed in whales that have started to become active after resting, have just left a social group, or appear excited or irritated. Usually the first breach of the series is the largest and the rest become progressively smaller, with less of the body exiting the water.

A related behavior is the **head lunge,** in which a whale breaks the surface and falls forward instead of backward. This behavior includes a quick, horizontal burst through the water with open mouth, and is most often associated with feeding.

### Slapping

Whales like to lift their huge tails high above the water and slap them down on the surface with such force as to make a huge splash and produce a noise that is loud enough to be heard for great distances. This behavior, called **tail slapping** or **tail lobbing** (see Figure 12-10c), may be associated with marking position and is sometimes interpreted as being an aggressive behavior. Tail cocking is another aggressive behavior in which a whale can cock its tail up in the air and bring it down on an opponent. The part of the whale's body closest to the tail fluke is called the **peduncle.** In the **peduncle slap** the whale swings the rear portion of its body, sometimes as far forward as its dorsal fin, out of the water and then drops it down sideways on the water or another whale. Tail slash and tail swish involve moving the tail from side to side across the surface of the water to create turbulence. These are also thought to be aggressive displays.

After arching their body some whales may bring their flukes above the surface. If the tail is brought straight up so that the ventral (bottom) surface is visible, the behavior is called fluke up (see Figure 12-10d). If the fluke clears the water but remains turned down, the behavior is called fluke down.

In flipper flapping a whale rolls over onto its back and flaps its flippers in the air (see Figure 12-10e). Some whales lie on the surface and lift one or both of their pectoral fins up out of the water and wave them about. In pec slapping, a whale rolls on its side and slaps the water with its pectoral fin to create a forceful splash and loud sound. No one knows why they do these behaviors, but some researchers speculate that they are a form of communication.

Pectoral fins can be used to stroke the body of another whale, a behavior known as pectoral stroking. Mothers and calves will stroke each other as a bonding behavior and pectoral stroking in adults is associated with courtship and mating.

## Reproduction and Development

Many cetaceans travel in groups called **pods** that are composed of both adults and young. Normally cetaceans bear only one offspring at a time, rarely two. The young are fed extremely rich milk, containing 40% to 50% fat and 10% to 12% protein. Cow's milk contains 2% to 4% fat and 1% to 3% protein. The high caloric content of whale's milk allows the infant whale to grow at an exceedingly fast rate and to produce enough body heat to resist the cold until it develops its thick layer of blubber. Baby blue whales, for instance, double their birth weight in the first week and then gain some 91 kilograms (200 pounds) a day thereafter.

## Types of Whales

Whales are divided into two suborders (Figure 12-11). Baleen whales (suborder Mysticeti) lack teeth. Instead they use structures called **baleen** to filter their food from the water. The largest whales are baleen whales. Toothed whales (suborder Odontoceti) have teeth and feed on larger prey. The familiar dolphins and killer whales are two examples of toothed whales.

### Baleen Whales

Baleen whales have enormous mouths to accommodate plates of baleen, which take the place of teeth (Figure 12-12). Each plate of baleen has an elongated triangular shape and is anchored at its base to the gum of the upper jaw. The baleen plate is composed of **keratin** (a tough protein in mammalian hair and nails) fibers that are fused, except at the inner edge, where they form a fringe. Hundreds of these plates, each one 1 to 4 meters (3.3 to 13 feet) long, form a tight mesh on the sides of the mouth. The diet of baleen whales consists mainly of plankton, especially krill, or fish. A baleen whale feeds by opening its mouth and swimming into dense groups of krill or schools of fish. When it has filled its mouth, it closes it and expels the water through the baleen plates, straining out the prey in the process. The retained food is then moved by the tongue to the back of the oral cavity and swallowed. The process is repeated until, at the end of a feeding, a large whale has as much as 10 tons of krill or fish in its stomach. Humpback whales (*Megaptera novaeangliae*) sometimes capture their food by blowing a ring of bubbles called a **bubble net** that traps krill near the surface (Figure 12-13). When not actively feeding, the whale covers the baleen with the underlip, which projects above the lower jaw. Baleen whales include the right whales, rorquals, and gray whales.

*Right Whales*   Right whales (family Balaenidae) can be distinguished from other cetaceans by the lack of a dorsal fin and the absence of grooves on the throat and chest. Little is known about most species of right whale, and they are considered rare. Right whales received their name from early whalers. When lookouts sighted a whale, if it was the "correct" whale for hunting, they would yell, "It's the right whale!" thus the origin of the name. Right whales were the favorite catch of early whalers because they were relatively slow and easily harpooned, floated when killed, and could be sliced up with relative ease. As the result of overhunting, three of the four species of right whale were nearly extinct by the end of the nineteenth century, and the fourth species, the Greenland, or bowhead, whale (*Balaena mysticetus*), is the rarest of all whales.

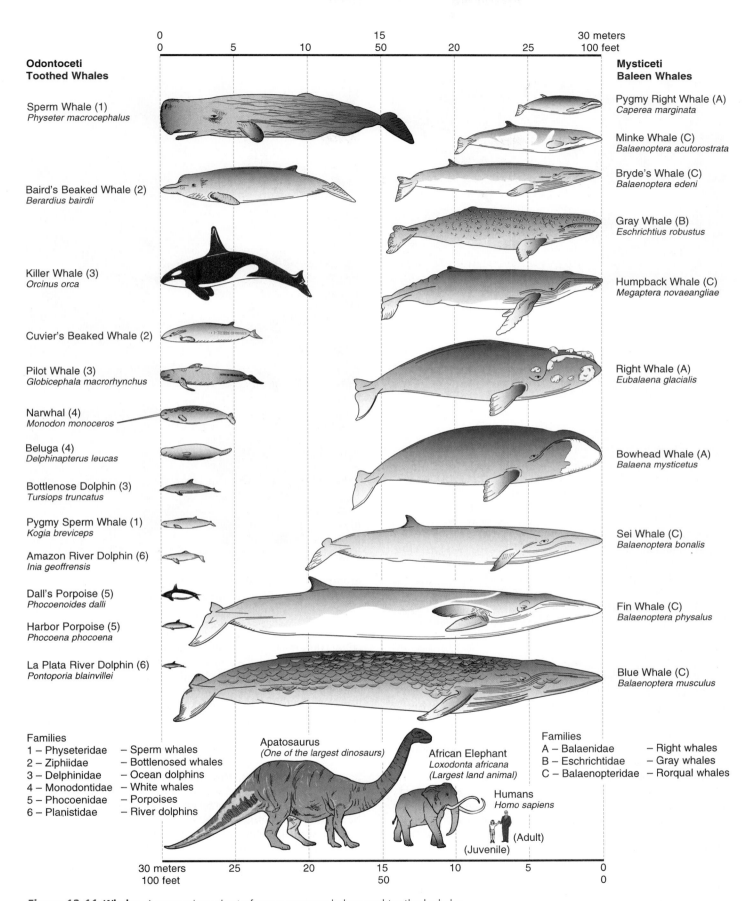

**Figure 12-11 Whales.** A comparison chart of some common baleen and toothed whales.

**Figure 12-12 Baleen.** This piece of baleen is from a humpback whale. Baleen whales use these structures to strain plankton from the water column.

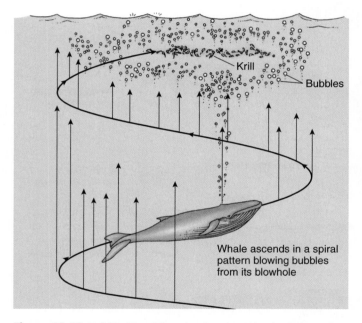

Krill

Bubbles

Whale ascends in a spiral pattern blowing bubbles from its blowhole

**Figure 12-13 Bubble Net.** When feeding, humpback whales will frequently blow rings of bubbles as they rise to the surface. Small prey hesitate to cross the bubble barrier and become easy prey for the whales.

***Rorquals*** Rorquals (family Balaenopteridae) have the dorsal fin and ventral grooves that the right whales lack. The ventral grooves, or pleats, under the mouth allow the throat to expand while the animal is feeding. Rorquals are generally slender and streamlined and are capable of relatively fast swimming. Some of the better-known rorqual species are the blue whale, fin whale, and humpback whale.

The blue whale (*Balaenoptera musculus*) has the distinction of being the largest of the whales and may be the largest animal that has ever lived (Figure 12-14). They range in size from 24 to 30 meters (80 to 100 feet) and weigh more

**Figure 12-14 Blue Whale.** The blue whale, which may be the largest animal that has ever lived, feeds on plankton, especially krill.

than 100 tons (as much as 1,600 humans or 25 elephants!). Because of excessive hunting, however, large individuals are rarely seen anymore. Whaling records indicate that a blue whale consisted of 50 tons of muscle; 8 tons of blood; 1 ton of lungs; and 60 tons of skin, bones, and internal organs (other than the lungs). The heart alone weighed 591 kilograms (1,300 pounds) and measured about 1 meter (3.3 feet) in each direction.

The fin whale (*Balaenoptera physalus*) is the second-largest whale species, growing to between 19 and 22 meters (63 to 73 feet) long and weighing between 45 and 75 tons. It gets its name from the large dorsal fin that often slopes backward. The animal is dark gray to brownish black with a white ventral surface. Fin whales are found in every ocean in the world but rarely come close to shore. Like other whale species, the fin whale migrates to polar waters in the summer, where they feed primarily on krill, and to warmer waters in breeding season.

The humpback whale (*Megaptera novaeangliae*) can be distinguished by the low hump on its back; the large bumps, called **bosses,** on its snout; and the very long pectoral fins (flippers) that may be one third the animal's length, or 5 meters (15 feet). Humpbacks are slow-moving, heavyset creatures that can reach up to 15 meters (50 feet) long and 12 meters (40 feet) around. It is one of the slowest whales, swimming 3 to 8 kilometers (2 to 5 miles) per hour. Humpbacks are mainly coastal inhabitants and will frequently enter harbors and even venture up the mouths of large rivers. These characteristics have made the humpback a very vulnerable species. Although it has been protected by the International Whaling Commission since 1966, it is still considered to be an endangered species. There are three distinct populations of humpback whale: in the North Atlantic, the North Pacific, and the Southern Hemisphere. Of these three, only the North Atlantic population is showing signs of recovery from the impact of whaling.

Humpbacks spend most of their time in polar seas feeding on krill and other plankton that are abundant there, but they come to warmer waters to mate and bear their young. They are a popular sight for tourists during March in the waters off Maui, Hawaii, where the North Pacific population comes to breed. Because humpbacks are coastal species, they are more accessible than other whale species and have been the subjects of many studies. Much research has focused on their vocalizations, or "songs," that may play a role in mating and intraspecies communication.

Until modern whaling techniques were introduced, rorquals like the blue whale were relatively safe from human hunters. They were too large to approach in small boats and fast enough to get out of range of hand-hurled harpoons. On the few occasions when a rorqual was killed, it usually frustrated the whalers by sinking or swamping the whaleboats. All of this changed in the last century with the introduction of harpoon guns, motorized boats, factory ships, and techniques for injecting air into the whale carcass so that it would float and could be towed for great distances (Figure 12-15). Because the average blue whale could furnish 120 barrels of high-grade oil, they became the primary tar-

gets of the whaling industry, and their numbers plummeted. In the 1930–1931 whaling season, the worldwide catch of blue whales was 29,649. By the 1964–1965 season, whalers could find only 372 blue whales to kill. In 1966, the International Whaling Commission gave the blue whale worldwide protection, and their numbers have started to increase again.

***Gray Whale***    At one time, three different populations of gray whales (*Eschrictius gibbosus*) existed in the western and eastern Pacific Ocean and the North Atlantic Ocean. Today, only the eastern Pacific population survives. The other two populations were hunted to extinction by the whaling industry. The eastern Pacific population, which initially contained an estimated 24,000 individuals, was on the brink of extinction in 1946. The number of individuals in the population at that time was estimated to be 250. In that year, strict laws were enacted to protect the remaining gray whales, and today the population consists of more than 15,000 individuals, allowing them to be removed from the list of endangered species in 1994.

Gray whales migrate from their summer feeding grounds in the Bering Sea to the waters off Baja California to mate and give birth. Females aggressively defend their young, a trait that prompted whalers to refer to them as "devilfish." Gray whales carry large accumulations of barnacles on the skin, more than any other cetacean, and for this reason are sometimes referred to as **mossback whales** (Figure 12-16). Although the gray whale has been recovering from the impact of whaling, its popularity with tourists is now causing problems. Large numbers of whale watchers and their motorboats disturb the whales while they attempt to mate and give birth, resulting in fewer offspring.

## Toothed Whales

Toothed whales include sperm whales, dolphins, porpoises, killer whales, and narwhals. As with other aquatic mammals, the teeth of these whales tend to be simplified. Dol-

**Figure 12-15** **Whaling Ship.** The development of modern whaling vessels allowed whalers to kill more whales, pushing many species to the brink of extinction.

**Figure 12-16** **Gray Whale.** Gray whales are sometimes called *mossback whales* because of the large number of barnacles attached to some animals.

phins and porpoises have between 100 and 200 identical teeth that are fused to the jaws. The teeth grasp the fish on which these animals feed. Sperm whales have functional teeth only in the lower jaw, and the teeth fit into sockets in the upper jaw. It is believed that this is an adaptation for feeding on the slimy cephalopods, which are their primary food. Beaked whales have only a few teeth, sometimes only two in each jaw, and the narwhal is peculiar in that one of its teeth forms a tusk that projects straight out from its head.

***Sperm Whales*** The giant of the toothed whales is the sperm whale, or cachalot (*Physeter macrocephalus*), the third-largest animal on earth behind the blue whale and fin whale. Sperm whales (Figure 12-17a) are sometimes confused with right whales and rorquals, but they have teeth instead of baleen and a massive, blunt snout that projects forward beyond the mouth. Sperm whales have no real dorsal fin but instead have a series of humps on the rear third of their body. Unlike most other species of whale, sperm whales are aggressive, attacking squid, fishes, and on occasion, whalers in small boats. They are polygynous, and males are always accompanied by several females. Although mainly

found in tropical and temperate waters, they do range occasionally into the polar seas.

At one time, the sperm whale was abundant in most seas and for almost a century and a half it was the favorite catch of American whalers. Sperm whales were challenging targets for whalers of the eighteenth and nineteenth centuries because of their large size (up to 18 meters, or 60 feet), toothed jaw, and powerful tail. Whalers judged the risks well worth taking because the animal's commercial value was so great. During the twentieth century, modern whaling techniques decreased the risks, whereas profits remained high. As a result, the once large populations have decreased significantly, and although protected by international treaty, the sperm whale is still considered to be a rare and endangered species.

The sperm whale received its name from an oily, wax-like substance in the animal's head. The snout contains a cavity that is developed from one of the nasal chambers. This cavity contains a thin, colorless, transparent oil that forms a waxy material when it comes into contact with air. Early whalers erroneously thought it to be an enormous reserve of semen and named it **spermaceti** (spur-meh-SEH-tee; see Figure 12-17b). At the height of the whaling industry, spermaceti was one of the most prized products of the sperm whale catch because it was found to be a high-grade wax. It was used in the manufacture of ointments, face creams, luxury candles, and lubricants. Now synthetics have taken its place. It is generally believed that the spermaceti plays a role similar to that of the melon in dolphins and thus is part of the whale's echolocation system (see the later discussion of echolocation).

The sperm whales' fondness for squid and cuttlefish generates a digestive by-product called **ambergris,** which in the past was even more valuable than spermaceti. The ambergris is a secretion of the whale's digestive tract and may function in protecting the enormous digestive system from undigested squid beaks and cuttlefish cuttlebone. When this secretion is exposed to air, it forms a solid, gray, waxy ma-

(a)

Doug Perrine/Jeffrey Rotman Photography

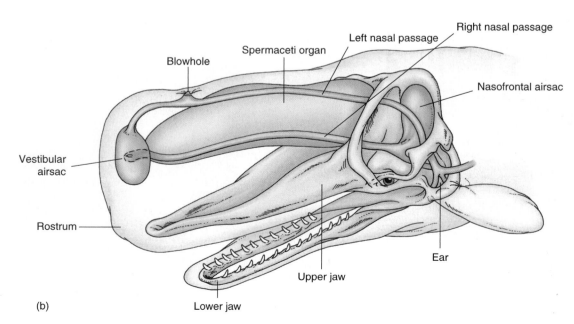

(b)

**Figure 12-17 Sperm Whale.** (a) A sperm whale nears the surface of the water in the North Atlantic Ocean. (b) Spermaceti is believed to function in echolocation and was prized by whalers as a high-quality wax.

terial with a disagreeable odor, but over time it acquires properties that make it highly prized as a base for the most expensive perfumes. Even though synthetic substitutes are available, ambergris is still very much in demand. Lumps of ambergris that are evacuated by the animals sometimes wash up on shore. A 418-kilogram (920 pound) lump of ambergris that washed up on an Australian beach in 1953 sold for $120,000.

***White Whales*** The white whales belong to a small family (family Monodontidae) that contains two genera with one species in each. The beluga whale (*Delphinapterus leucas;* Figure 12-18a) is unique among whales for its white color. In fact, the name is derived from the Russian word for *white.* The beluga lives in the northern polar seas. They usually travel in small family groups and feed on crabs, cuttlefish, flounder, and halibut. On occasion they have been observed to enter and swim up Arctic rivers.

The narwhal (*Monodon monoceris;* see Figure 12-18b) is a close relative of the beluga whale and also lives in Arctic waters. Although the fetal animals have tooth buds, by birth these have all disappeared, with the exception of the

(a)

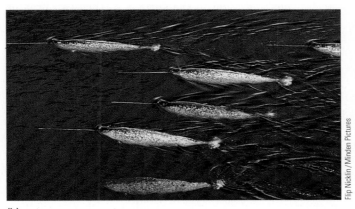

(b)

**Figure 12-18 White Whale.** (a) The beluga whale, *Delphinapterus leucas,* is the only whale species that is totally white. (b) The upper incisor (usually the left) of male narwhals develops into a large tusk that projects straight out from the head.

two upper incisors. These teeth remain undeveloped in the skull of the females, but in males a tusk as long as 3 meters (10 feet) develops from one of the two teeth, almost always the left. In some individuals, the right tooth also develops a tusk, giving an already odd-looking creature an even more bizarre appearance.

During the Middle Ages, the narwhal tusk was passed off as the magical horn of a unicorn. Because the tusk was worth literally more than its weight in gold, the narwhal was hunted relentlessly. In later times, the narwhal was killed incidentally by whalers in the Arctic Ocean who were searching for bowhead whales. Since the demise of the bowhead fishery, the narwhal population has recovered, and the animals survive relatively undisturbed. Their only remaining predator appears to be Eskimos, who kill the animal for its nutritious skin.

***Porpoises*** Porpoises (family Phocaenidae) are related to the familiar dolphins, and both are placed in the same superfamily, Delphinoidea. The word *porpoise* comes from a contraction of the Latin word *porcopiscis* (meaning "pig-fish"), apparently referring to the animal's stocky body. The Greek naturalist Aristotle observed that the porpoise resembled a small dolphin but differed by being broader across the back. Modern biologists point out that the most obvious difference between dolphins and porpoises is that dolphins have a beak, whereas porpoises have a head that is rounded off and no beak.

One of the smallest cetaceans is the harbor porpoise (*Phocoena phocoena;* Figure 12-19a), which measures only 1 to 2 meters (3 to 7 feet) long and weighs nearly 45 kilograms (100 pounds) when fully grown. This animal is widely distributed in the North Atlantic Ocean and elsewhere and is a familiar sight to boaters and beachcombers on both sides of the Atlantic. It is also widely distributed along the Pacific coast of North America. The harbor porpoise is one of the most intelligent cetaceans, and its capacity for learning, as far as is known, is surpassed only by that of the bottlenose dolphins, killer whales, and pilot whales. It feeds on a variety of schooling fish such as herring, sardines, and mackerel, sometimes feeding on fish that have already been caught in nets. A related species, the vaquita, or "little cow" (*Phocoena sinus*), is endemic to the upper Gulf of California and is in danger of extinction as a result of fishers' depleting the stocks of fish and squids on which the animals feed.

For years Europeans considered Dall's porpoise (*Phocoenoides dalli*) a delicacy, and it was quite rare. The rarity was probably more a factor of the porpoise eluding its human predators than few numbers. This animal is perhaps the first to be protected by law. In Normandy as far back as 1098, laws were passed regulating the size and number of the Dall's porpoise harvest. Dall's porpoises feed on fish, preferring herring, whiting, and sole. These animals have been observed swimming up rivers as far as 58 kilometers (36 miles), and one specimen was observed in the Maas River in the Netherlands 323 kilometers (200 miles) from the sea.

***Dolphins*** Because most people think of whales as being giant animals, few are aware that the playful, acrobatic dolphins are also toothed whales. In fact dolphins are the most numerous cetaceans. Dolphins belong to the family Delphinidae and are collectively referred to as **delphinids.** The

(a)

(b)

**Figure 12-19 Comparison of Dolphin and Porpoise.** Dolphins and porpoises are both delphinids. (a) The head of a porpoise is rounded, whereas (b) dolphins can be distinguished by their beaks.

name is derived from ancient Greek mythology. According to legend, Apollo rose from the sea in the shape of a dolphin to lead settlers to Delphi, home of the Delphic oracle.

The common dolphin (*Delphinus delphis*) has a definite beak that is separated from the snout by a groove. This animal is small compared with other cetaceans; adults grow to no more than 3 meters (10 feet). They can be found in all temperate oceans and are among the swiftest cetacean swimmers. They feed on schooling fish such as herrings and sardines that they devour in enormous quantities. The common dolphin is frequently seen escorting ships. When common dolphins find a ship, they encircle it. Some of the dolphins will lead at the bow, while others follow along at the sides. They will leap completely out of the water in a graceful arc that covers several meters and will repeat this behavior for hours and hundreds of kilometers without ever seeming to tire.

The most abundant dolphin species along the eastern coast of the United States and in the Caribbean is the bottlenose dolphin (*Tursiops truncatus;* see Figure 12-19b). These dolphins also inhabit the Mediterranean, Baltic, and Caspian seas, and a subspecies lives in the temperate waters of the Pacific Ocean. These are the dolphins that are most frequently used for scientific studies of cetacean intelligence

(see accompanying box). Bottlenose dolphins are often the species used as performers at aquariums or stars of television and movies. These intelligent animals are quite playful and appear to like the company of others of their species. They have been observed to aid disabled members of their species, pushing them to the surface so they can breathe or helping them to shallow water. Occasionally, dolphins will socialize with other species, including humans. In a coastal region of Brazil, female bottlenose dolphins and their young have assisted fishers for almost 150 years. The fishers cast hand-thrown nets from the shore and wait for the dolphins to chase fish, usually mullet, into them. The dolphins feed on the fish that escape the nets. There are also many stories, some authenticated, about bottlenose dolphins coming to the aid of injured or drowning humans.

Bottlenose dolphins live their lives in carefully structured, complex social groups. Studies of dolphins near Sarasota, Florida, indicate that females associate with other females that are in a similar reproductive state. For instance, mothers with calves swim with other mothers and their calves. Calves remain with their mothers for 3 to 6 years before joining juvenile groups. During this time, a calf learns to identify the other dolphins in the group, its relationship to the other members, how to find food and capture it, and

## Dolphin Behavior

Studies conducted in Hawaii by researcher Lou Herman have greatly increased our knowledge of dolphin learning abilities. A dolphin named Ake (short for Akeakamai) and her mate, Phoenix, learned that they would receive rewards when they mimicked sounds that were generated by a computer. They also were able to make associations. When taught the hand signal to go through the gate in their tank, they also responded correctly to a similar hand signal that meant go through a hoop. They were able to learn that *ball* not only meant the small, red ball that they were first trained with but any other ball, regardless of size, color, or its ability to sink or float.

Though originally taught to respond to hand signals given by their trainers, Ake and Phoenix could also recognize and respond to the same signals when they saw them on a television screen. In a test, dolphins outscored new trainers in interpreting televised gestural commands.

Many animals, including humans, communicate by way of body language. Dolphins are no exception. For instance, a loud opening and closing of the jaws, known as a jaw clap, or the slapping of tail flukes on the water surface can indicate threat or displeasure.

Although dolphins probably communicate with each other in a number of ways, the most important are auditory signals. Bottlenose dolphins produce two types of vocalizations: whistles and pulsed sounds (which include echolocation clicks). It was discovered in the 1960s that each dolphin produced its own unique whistle, called a **signature whistle.** The signature whistle identifies the dolphin that is vocalizing, gives its location, and may relay other information as well. Current research indicates that perhaps as many as 90% of the whistles made by captive dolphins are signature whistles. It appears that a female's whistle is quite different from her mother's, whereas a male's is very similar to his mother's. Because females generally stay with their mothers when they mature, it may be important for them to have distinct signature whistles to avoid confusion. Males usually leave their mothers' group and thus do not have to differ as much in their signature whistle. Calves develop signature whistles between the ages of 2 months and 1 year. The whistles remain unchanged for at least 12 years and possibly for the animal's life.

Although scientists are just beginning to learn the meaning of signature whistles, little is known of the pulsed sounds the dolphins make. One type of pulsed sound is the echolocation click that is used in navigation. Other pulsed sounds may be used for communication. ●

---

where there is danger and where safety. Dolphin feeding behaviors exhibit signs of learned behaviors. Some dolphins in southwestern Florida beat their prey to death with their tail flukes. This appears to be a local behavior that the young learn from their mothers. In the Gulf of Mexico, dolphins have learned to follow shrimp boats for a free meal. This also appears to be a learned behavior that is passed from a mother to her offspring. As they mature, females frequently return to their mother's group, but males travel alone or with one or two other adult males with whom they have grown up.

Bottlenose dolphins appear to be quite clever, sometimes devising ingenious solutions to problems. One example was observed in two captive dolphins trying to extract a moray eel from a rock crevice in their tank. One of the dolphins captured a scorpionfish and, with the fish in its mouth, nudged the moray eel from behind with the spine of the fish. As the eel fled from its crevice, it was captured by the second dolphin.

The largest of the dolphins is the killer whale, or **orca** (*Orcinus orca;* Figure 12-20). Males sometimes reach lengths of 10 meters (33 feet), whereas females are only half as large. These animals received their common name because they are the only cetaceans that feed on homeothermic animals. Killer whales have a high dorsal fin (as much as 1.8 meters, or 6 feet) and broad, rounded flippers. Their

**Figure 12-20 Killer Whale.** Killer whales are the only cetaceans known to feed on homeothermic prey. This whale is exhibiting spy hopping behavior.

coloration is particularly distinctive. They are primarily black with a white patch over the eye and a white ventral stripe. These powerful animals are quite agile and are the fastest cetacean swimmers, reaching speeds of 48 kilometers per hour (30 miles per hour). They can be found in all seas, but are most common in the cold waters of the Arctic and Antarctic Oceans.

Both the upper and lower jaws of killer whales contain large, conical teeth. Although they tend to eat more fish than anything else, they will also prey upon seals, sea lions, baby walruses, penguins and other sea birds, and other cetaceans. The stomach of one captured specimen contained the remains of 13 porpoises and 14 seals. Pods of killer whales have also been known to attack larger whales.

There are no authenticated reports of killer whales killing or injuring humans in the wild, although there was an incident at a Vancouver oceanarium in which a trainer was killed by an orca. There have been a few incidents in which killer whales have rammed boats, usually after the inhabitants of the boat antagonized the whale. There are also reports of killer whales ramming ice floes on which humans are standing, but this appears to be a case of mistaken identity. In captivity, killer whales are quite docile, and their high level of intelligence and ability to learn make them popular performers in aquarium and marine park shows. A relative, the false killer whale (*Pseudorca crassidens*), has a similar appearance but is much smaller and feeds on squid.

Pilot whales (*Globicephala*) are dolphins related to the killer whales. These animals have a globular head, projecting forehead, and a muzzle that forms a small beak. Pilot whales are inoffensive animals that live in pods of several hundred individuals that appear to follow a single leader. They feed mainly on cuttlefish and, when attacked, group closely together. Pilot whales exhibit the odd behavior of following individual members that stray from the pod. Whalers exploit this behavior by harpooning one or two pilot whales, towing them to shallow water, and then waiting for the rest of the pod to follow to mass slaughter. These animals are not protected by any international agreement and are still hunted for dog food and oil. Large numbers of the North Atlantic species are caught each year.

Pilot whales will frequently beach themselves, sometimes in large numbers. There are several theories on these mass strandings. One theory suggests that the cause of the beachings may be a parasite that attacks the inner ear and interferes with the animal's ability to navigate.

## Echolocation

Even in clear water, toothed whales cannot see much farther than 30 meters (100 feet). To compensate for this, they have ears that are modified to receive a wide range of underwater vibrations. This adaptation not only improves the animal's hearing under water but refines it for **echolocation,** which like sonar, allows the animal to distinguish and home in on objects from distances of several hundred meters. Arthur F. McBride, the first curator of Marineland in Florida, was one of the first to discover echolocation in cetaceans in the early 1950s. Every time he tried to capture dolphins for an exhibit by driving them toward a net, the animals would stop short of the net and swim away. It didn't matter whether he tried during day or night, and even in murky water where they could not see the net they would still avoid it. McBride hypothesized that the animals might somehow use sound to sense the presence of the nets in the water.

Later in the 1950s, Winthrop N. Kellogg of Florida State University demonstrated that dolphins use a series of clicking sounds in the same way that humans use sonar. Dolphins can emit a wide spectrum of sounds, lasting from a fraction of a millisecond to as long as 25 milliseconds. To obtain information about their environment, dolphins emit sounds with frequencies that range from less than 2,000 to greater than 100,000 hertz. Only a portion of these sounds can be perceived by the human ear, and these we hear as clicks. The low-frequency clicks are **orientation clicks** that give the animal a general idea of its surroundings. The high-frequency clicks are **discrimination clicks** that give the animal a precise picture of a particular object. The clicks can be produced as a single sound or as a series of sounds strung together.

The dolphin's larynx does not have vocal cords but instead has a ring of muscles that acts as a valve, enabling the animal to control the air flow through the larynx. Air under pressure circulates through the animal's nasal passages, producing the clicks and other characteristic sounds (Figure 12-21). The sounds are directed by being focused in the **melon,** an oval mass of fatty, waxy material that is located between the blowhole and the end of the head. The sounds are directed toward objects in front of the animal but not on objects below a line level with its jaws. This may explain why some toothed whales cannot sense gradually sloping bottoms and sometimes run aground. The clicking sounds travel through the water and bounce off anything solid. The reflected sounds (echoes) are picked up by sensitive areas of the lower jaw that transmit the vibrations to the inner ear. The animal's brain then processes the information from the inner ear to produce a mental image of the target object.

When the sounds bounce off a target, they produce echoes that provide at least four types of information: the direction from which the echo is coming, the change in frequency, the amplitude, and the time elapsed before the emitted sound returns. With this information, dolphins, as well as other toothed whales, can determine not only the range and bearing of an object, but also its size, shape, texture, and density. In tests, blindfolded dolphins have been able to discriminate between two very small objects about 2 inches in diameter and have different shapes at a distance of 3 meters (10 feet).

When a dolphin travels, it usually moves its head from side to side and up and down. This behavior allows the animal to scan a broad path ahead of it. The scanning motions become fast and jerky when the dolphin becomes interested in a small target, such as a fish. As the dolphin scans, it can determine the direction from which echoes are returning and thus determine the bearing of the target. Changes in the frequency of the returning sound give the dolphin information about the size and shape of the object. The amplitude of

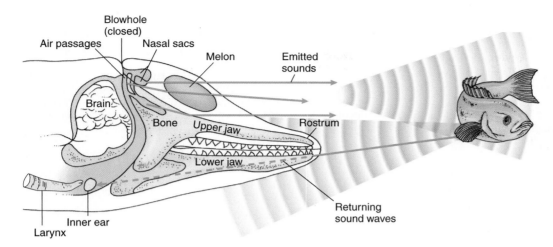

Blowhole
(closed)

Air passages    Nasal sacs

Melon

Emitted
sounds

Brain

Bone    Upper jaw    Rostrum

Lower jaw

Returning
sound waves

Inner ear

Larynx

**Figure 12-21 Echolocation.**
Toothed whales use sound to locate objects, determine shapes and distance, and navigate. Sounds produced by the animal's larynx (voicebox) are focused by a structure in the head called the *melon*. When the sounds strike a target, they are reflected as echoes (returning sound waves). The echoes are picked up by the animal's lower jaw and transmitted to the inner ear. The vibrations received by the inner ear are then converted into signals that can be carried by nerves to the brain for interpretation.

the echo and time elapsed before it returns help the dolphin determine distance.

## In Summary

The mammals that are most suited to life in the sea are the cetaceans. The forelimbs of cetaceans are modified into flippers, and hind limbs are absent. Their tail forms a large horizontal fluke that is used in swimming. Beneath the skin is a thick layer of fat that helps to insulate the body. Some of the most striking cetacean adaptations are those involving the respiratory and circulatory systems that allow them to dive to great depths and remain submerged for more than 1 hour. Cetaceans are intelligent, inquisitive animals that display a variety of interesting behaviors.

Cetaceans known as baleen whales have enormous mouths that contain plates of baleen instead of teeth. The baleen consists of fused keratin fibers, and it strains planktonic food, mainly krill, from the water. Baleen whales include right whales, rorquals, and gray whales. Many species of baleen whale are endangered because of the impact of whaling.

Toothed whales include sperm whales, dolphins, porpoises, killer whales, and narwhals. The largest of the group is the sperm whale, which feeds primarily on squid. Perhaps the best known of the toothed whales are the dolphins that perform in shows at numerous aquariums and oceanariums. Dolphins are strongly social animals. They exhibit problem-solving skills, have long periods to mature with many learning experiences, and are capable of intraspecies and interspecies cooperation.

The harbor porpoise is one of the smallest cetaceans and is a familiar sight to both boaters and beachcombers. The largest of the dolphins is the killer whale, or orca. These are the only cetaceans to feed on homeothermic animals, mainly seals and penguins.

The ears of toothed whales are modified to receive a wide range of water vibrations. This adaptation improves the animal's ability to hear underwater and refines its hearing for the purpose of echolocation. Echolocation is similar to sonar. It allows toothed whales to distinguish and home in on objects from distances of several hundred meters. ●

## IN PERSPECTIVE

| Type of Marine Mammal | Characteristics | Diet | Distribution |
|---|---|---|---|
| Sea otter | Five-fingered forelimbs and well-defined hind limbs with finlike feet; able to move about on land as well as in the water; dense fur covers and insulates the body | Sea urchins, molluscs especially abalone, crustaceans, and some fish | West coast of the United States from California to Alaska and Japan |
| Eared seals (sea lions and fur seals) | Body torpedo-shaped; external ear flaps; limbs modified to form flippers and can be rotated to support body on land; swim using their front flippers and steer with their hind flippers | Fish | Sea lions along west coasts of the Americas; fur seals in the Arctic and Antarctic |
| Phocids (true seals) | Body torpedo-shaped; no external ear flaps; limbs modified to form flippers but cannot be rotated to support the body on land; swim using hind limbs and steer with forelimbs | Fish, krill, squid except for leopard seal, which also feeds on penguins and other seabirds and marine mammals | Worldwide mainly in coastal waters; more species in temperate and polar seas than in the tropics |
| Walruses | Bulky bodies; no external ears; a distinct neck; stiff bristles on muzzle; limbs modified to form flippers; can move awkwardly on land; males have tusks | Bivalves | Arctic regions |
| Sirenians (dugongs and manatees) | Large bodies; forelimbs modified to form flippers; no hind limbs; tail flukes; totally aquatic | Vegetation | All are tropical. Dugongs are found along coastal areas of India and Malaysia. Manatees are found in the Caribbean Sea and along the coast of west Africa |
| Baleen whales | Streamlined bodies that resemble a fish; nostrils located in a blowhole, forelimbs modified to form flippers; no hind limbs; tail flukes; plates of baleen hang down from the upper jaw and take the place of teeth; totally aquatic | Plankton, especially krill, and small fish | Worldwide; generally feed in polar seas and go to warmer tropical waters to give birth |
| Toothed whales | Streamlined bodies that resemble a fish; nostrils located in a blowhole, forelimbs modified to form flippers; no hind limbs; tail flukes; conical teeth sometimes located in only one jaw; totally aquatic | Fish and squid; killer whales also feed on other marine mammals | Worldwide distribution |

## SELECTED KEY TERMS

alveoli, *p. 285*

ambergris, *p. 292*

baleen, *p. 288*

the bends, *p. 287*

blowhole, *p. 284*

bosses, *p. 291*

breaching, *p. 287*

bubble net, *p. 288*

cetacean, *p. 284*

delphinid, *p. 293*

discrimination click, *p. 296*

echolocation, *p. 296*

fluke, *p. 284*

head lunge, *p. 288*

hemoglobin, *p. 287*

keratin, *p. 288*

lactation period, *p. 280*

lactic acid, *p. 287*

mammary gland, *p. 276*

melon, *p. 296*

mossback whale, *p. 291*

myoglobin, *p. 287*

orca, *p. 295*

orientation click, *p. 296*

peduncle slap, *p. 288*

peduncle, *p. 288*

phocid, *p. 279*

pinniped, *p. 277*

placenta, *p. 276*

placental mammal, *p. 276*

pod, *p. 288*

polygynous, *p. 280*

rorqual, *p. 290*

signature whistle, *p. 295*

sirenian, *p. 283*

spermaceti, *p. 292*

spy hopping, *p. 287*

tail lobbing, *p. 288*

tail slapping, *p. 288*

## QUESTIONS FOR REVIEW

### Multiple Choice

1. Mammals feed their young on secretions produced by the mother's
   a. digestive system
   b. oil glands
   c. mammary glands
   d. placenta
   e. salivary glands

2. Phocid seals lack
   a. tails
   b. hind limbs
   c. forelimbs
   d. external ears
   e. nostrils

3. Whales in the suborder Mysticeti have instead of teeth
   a. tusks
   b. bony plates
   c. baleen
   d. strainer nets
   e. suckers

4. Sperm whales feed primarily on
   a. squid
   b. krill
   c. penguins
   d. clams
   e. sea urchins

5. The primary food of baleen whales is
   a. squid
   b. fish
   c. krill
   d. crabs
   e. other whales

6. The spermaceti in the head of a sperm whale is thought to play a role in
   a. digestion
   b. reproduction
   c. excretion of salt
   d. echolocation
   e. swimming

7. The largest dolphin is the
   a. beluga whale
   b. sperm whale
   c. narwhal
   d. killer whale
   e. blue whale

8. The cetacean most often seen in captivity is the
   a. California sea lion
   b. common dolphin
   c. bottlenose dolphin
   d. beluga whale
   e. walrus

9. The only pinniped to feed on homeothermic animals is the
   a. California sea lion
   b. killer whale
   c. elephant seal
   d. walrus
   e. leopard seal

10. Sea otters are protected from the cold by
    a. a thick skin
    b. a thick fur
    c. a layer of blubber
    d. a countercurrent exchange mechanism
    e. high metabolism

### Short Answer

1. What factor contributed to the near extinction of the sea otter?

2. List three characteristics that would help you to distinguish a fur seal from a true seal.

3. Explain the function of the large proboscis in the male elephant seal.

4. Explain how the arrangement of blood vessels in the flippers and tail flukes of cetaceans helps them to retain body heat.

5. Describe the changes that occur in a cetacean's body when it dives for long periods.

6. What is spy hopping?

7. Explain how toothed whales use echolocation to navigate.

8. Describe the effects of whaling over the past several centuries on cetacean populations.

9. Describe how baleen whales feed.

10. How could you distinguish between a manatee and a dugong?

11. Why do mammals produce fewer offspring than most other animal groups?

## Thinking Critically

1. Some marine mammals are polygynous. What is the advantage of this lifestyle?

2. From an ecological standpoint, why aren't more of the large marine mammals predators of birds and other mammals?

3. Toothed whales use echolocation to find their prey, but baleen whales do not. Why?

4. If diving mammals were not able to expel most of the air from their lungs before diving, how would the length and depth of their dive be affected?

## SUGGESTIONS FOR FURTHER READING

Bruemmer, F. 1990. Survival of the Fattest, *Natural History* 99(7):26–33.

Gentry, R. L. 1987. Seals and Their Kin, *National Geographic* 171(4):474–501.

LeBoeuf, B. J. 1989. Incredible Diving Machines, *Natural History* 98(2):34–41.

Marten, K., K. Shariff, S. Psarakos, and D. J. White. 1996. Ring Bubbles of Dolphins, *Scientific American* 275(2): 82–87.

Nicklin, F. 1995. Bowhead Whales: Leviathans of Icy Seas, *National Geographic* 188(2):114–29.

Norris, K. S. 1994. White Whale of the North: Beluga, *National Geographic* 185(6):2–31.

Norris, K. S. 1992. Dolphins in Crisis, *National Geographic* 182(3):2–35.

O'Shea, T. J. 1994. Manatees, *Scientific American* 271(1): 66–72.

Trefil, J. 1991. One of Evolution's Great Missing Links Turns Up in the Egyptian Desert, *Discover* 12(5): 44–48.

Whitehead, H. 1995. The Realm of the Elusive Sperm Whale, *National Geographic* 188(5):56–73.

Whitehead, H. 1985. Why Whales Leap, *Scientific American* 252(3):84–93.

Wiley, J. P. 1987. Manatees, Like Their Siren Namesakes, Lure Us to the Deep, *Smithsonian* 18(6):92–97.

Wolkomir, R. 1995. The Fragile Recovery of California Sea Otters, *National Geographic* 187(6):42–61.

### InfoTrac College Edition Articles

Baker, B. 1995. Marine Mammal Protection Increased, *Bioscience* 45(5).

Baskin, Y. 1993. Blue Behemoth Bounds Back: *Balaenoptera musculus* Seems to Be Inching Away from Extinction, *Bioscience* 43(9).

Boveng, P. L., L. M. Hiruki, M. K. Schwartz, and J. L. Bengtson. 1998. Population Growth of Antarctic Fur Seals: Limitation by a Top Predator, the Leopard Seal? *Ecology* 79(8).

Lantimm, C. A., T. J. O'Shea, R. Pradel, and C. A. Beck. 1998. Estimates of Annual Survival Probabilities for Adult Florida Manatees (*Trichecus manatus latirostris*), *Ecology* 79(3).

Martin, G. 1993. Killer Culture, *Discover* 14(12).

Norris, S. 2002. Creatures of Culture? Making the Case for Cultural Systems in Whales and Dolphins, *Bioscience* 52(1).

Perrin, W. F. 1991. Why Are There So Many Kinds of Whales and Dolphins? *Bioscience* 41(7).

Sritil, K. 1995 Whale Warehouse, *Discover* 16(8).

### Websites

**http://www.tmmc.org/learning/education/teacher_resources/resources.asp** General information on marine mammals as well as a source of classroom materials for teachers.

**http://www.nefsc.noaa.gov/faq/** This site has sections on marine mammals with answers to commonly asked questions.

**http://ourworld.compuserve.com/homepages/jaap/mmmain.htm** General information, pictures, and links to other sites that deal with marine mammals

**http://earthwindow.com/** Great photos of marine mammals.

**http://coreresearch.org/** General information on cetaceans including research opportunities and internships.

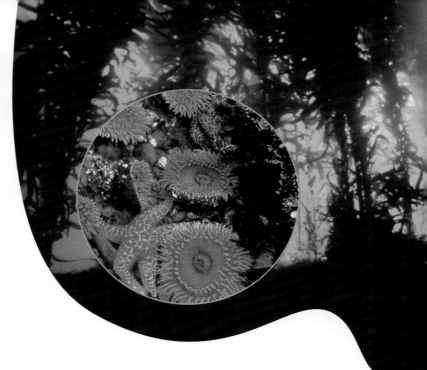

# 13

# Intertidal Communities

B ordering the water where land meets ocean is the realm of the seashore. Sandy beaches and rocky shores offer us places to ponder the wonders of nature or to relax and play in the surf. Although rocky and sandy shores are quite different, they share two features. They are alternately submerged and exposed by the tides and pounded by waves.

The intertidal zone, by definition, is that part of the marine environment that lies between the highest high tide and the lowest low tide. Although it is only a small part of the world's oceans, because of its accessibility it is by far the best studied. Indeed, studies on the intertidal zone have produced many unifying principles that can be applied to other marine systems. This habitat is a stressful environment where organisms must endure searing heat while exposed to air, followed by immersion in sometimes icy water. Despite these conditions, this environment is home to a diversity of organisms that equals or exceeds that found in submerged habitats. To survive in this habitat of extremes, organisms must be particularly hardy and adaptable. In this chapter we will explore how marine organisms survive and make a living in such a seemingly hostile environment.

## CHARACTERISTICS OF THE INTERTIDAL ZONE

The intertidal zone can be sandy beaches, rocky shores, tide pools, sand flats, mud flats, salt marshes, mangrove swamps, or a combination of these. In this chapter we focus on rocky shores and sandy beaches; the other habitats are described in Chapter 14. Regardless of the type of habitat, the inhabitants of the intertidal zone experience daily fluctuations in their environment. The world's coastlines host a variety of

NOAA; inset, B. Kent/Animals Animals

## Key Concepts

1. The intertidal zone is the coastal area alternately exposed and submerged by tides.

2. Organisms that inhabit intertidal zones must be able to tolerate radical changes in temperature, salinity, and moisture and also be able to withstand wave shock.

3. Organisms on rocky shores tend to be found in definite bands, or zones, on the rocks.

4. In contrast to sandy shores, rocky shores provide a relatively stable surface for attachment.

5. Tide pool organisms must be able to adjust to abrupt changes in temperature, salinity, pH, and oxygen levels.

6. Biotic factors are most important in determining the distribution of organisms on rocky shores, but physical factors are most important on sandy shores.

# Connell's Barnacles

As far back as Darwin, biologists have recognized that the organisms that live on rocky coasts are arranged in horizontal layers. These layers are arranged in an orderly and predictable fashion from the supralittoral fringe (splash zone), which receives moisture from waves that crash against the rocks, to the subtidal. A biologist named J. H. Connell from the University of California proposed an explanation for this arrangement of organisms.

Connell performed experiments on the rocky coasts of Scotland, where he noticed that two different genera of barnacles, *Semibalanus* and *Chthamalus,* live next to each other in separate layers on the rocks (Figure 13-A). Members of the genus *Chthamalus* live in the layer of organisms just above the high tide line. In the zone just beneath this, populations of the genus *Semibalanus* replace populations of *Chthamalus.* Barnacles are sedentary animals, and the species that Connell worked with are relatively small and live in large, dense populations. Because of these characteristics, Connell could perform experiments that would have been impossible with larger, more mobile organisms.

On one section of rocky shore, Connell observed the normal growth, reproduction, and daily activities of the two types of barnacles. This section of shore acted as the control for his experiment. On other sections he performed the following experiments. In one experiment, he transplanted barnacles to different zones on the rocks. In another experiment, he completely cleared the barnacles from areas of rock to see which species would colonize.

Connell determined from these experiments that individuals of *Chthamalus* were better adapted to living above the high tide mark. Members of this genus had a much greater tolerance for the higher temperature and desiccation that occurs in this zone compared with the members of the genus *Semibalanus.* The members of the genus *Semibalanus,* however, grow faster than those of *Chthamalus.* When larvae of both species settle on bare rock in the lower zone, the members of the genus *Semibalanus* compete better for the vital resource of space. The *Semibalanus* barnacles would either overgrow the *Chthamalus* barnacles, causing them to starve, or undercut the slower-growing *Chtha-*

*malus* barnacles, literally prying them off the rocks. As a result the *Chthamalus* barnacles did not have the opportunity to become established in this zone. *Chthamalus* barnacles live in the upper zone because they are better adapted to the harsh environment of the supralittoral fringe. *Semibalanus* lives in the lower zone where both species can survive but where *Semibalanus* is a superior competitor.

Experiments such as these show the importance of the interplay between the physical environment and biological factors, such as competition, in determining the distribution of organisms in an ecosystem. ●

---

living organisms. Whether the shoreline is rocky, sandy, or muddy, the interaction of wind, waves, sunlight, and other physical factors creates a complex environment. Organisms that live in this area must be able to tolerate radical changes in temperature, salinity, and moisture, as well as the potentially crushing force of waves, to survive.

During high tide, when the area is submerged by seawater, the inhabitants are most active, foraging for food, finding mates, and reproducing. The water contains food for filter feeders and oxygen for those organisms that use gills to exchange gases. As the tide retreats, the organisms are exposed to air. Those with gills must protect their respiratory structures from drying out and collapsing. Animals that are filter feeders withdraw into protective coverings. The pace

of life slows greatly. In summer the sun bakes the exposed area, causing temperatures to rise and tissues to lose water. During winter months on temperate beaches there is a danger of freezing. Many animals survive during low tide by sealing themselves up in shells or burrows or retreating into cracks and crevices until the high tide returns.

## In Summary

Intertidal organisms must survive in complex habitats that are alternately exposed and submerged by tides. To survive, organisms that inhabit these areas must be able to tolerate radical changes in

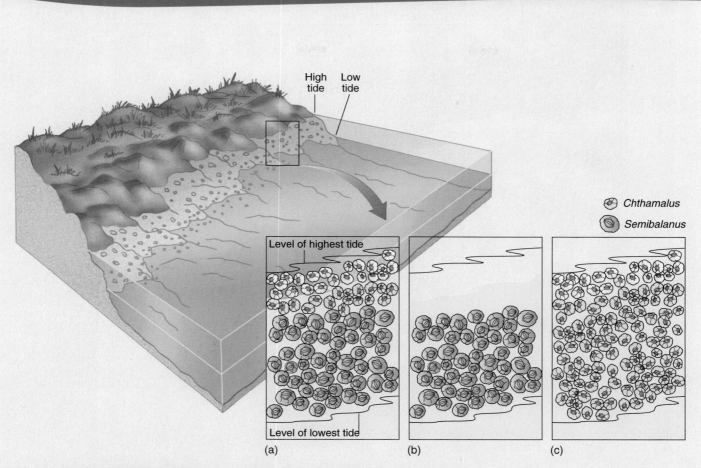

**High tide** **Low tide**

Chthamalus
Semibalanus

Level of highest tide

Level of lowest tide

(a)                (b)                (c)

*Figure 13-A* ***Barnacle Distribution.*** *(a) J. H. Connell noted that two species of barnacles in the genera* Chthamalus *and* Semibalanus *live in separate layers on the rocky shores of Scotland.* Chthamalus stellatus *live just above the high tide line and are replaced by* Semibalanus balanoides *beneath this point. (b) When the population of* Chthamalus *was experimentally removed, the population of* Semibalanus *did not colonize the open space. (c) When the* Semibalanus *population was removed, however,* Chthamalus *quickly colonized the open space.*

temperature, salinity, and moisture, as well as the crushing force of waves. ●

## ROCKY SHORES

Many regions of the world have shores composed of hard materials. In North America, rocky shores are found from California to Alaska on the west coast and from Cape Cod northward on the east coast. In geological terms, these coasts may be recently uplifted, formed from lava flows, or highly eroded areas where sediments have been removed by wind and waves.

## Rocky Shore Zonation

As the tide retreats and the higher regions of the coast emerge from water, prominent horizontal bands defined by color or the distribution of organisms appear. This separation of organisms into definite bands is called **zonation** (Figure 13-1). Unlike shores of sand or mud, rocks provide a relatively stable surface to which organisms can attach, as well as a variety of hiding places. As the tide retreats, organisms in the upper regions are exposed to air, changing temperatures, solar radiation, and desiccation for prolonged periods. The lower regions, on the other hand, are exposed for only a short time before the tide returns to cover them. The fact that zonation patterns in intertidal zones are similar world-

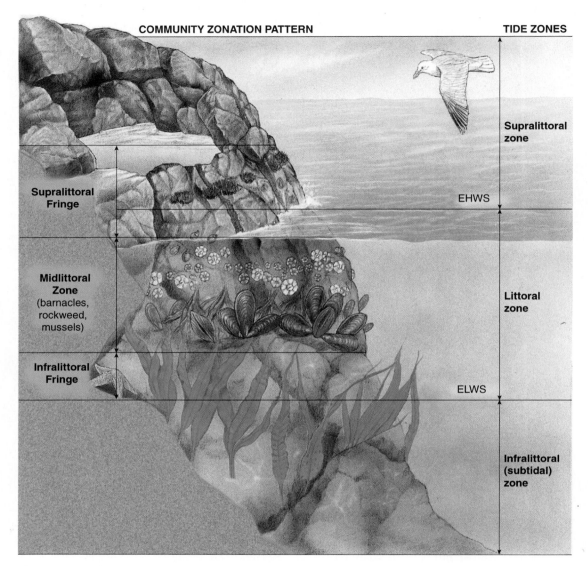

**Figure 13-1  A Rocky Shore.** Since rock provides a relatively stable surface on which organisms can attach, rocky shores exhibit definite bands or zones, each inhabited by organisms adapted to the special conditions of the environment. EHWS: extreme high water of spring tides; ELWS: extreme low water of spring tides.

wide led Alan and Anne Stephenson to propose a universal system to classify this phenomenon, replacing various schemes proposed by earlier authors. Zones were established primarily on the basis of distributional limits of certain common organisms rather than the tides. The width of these zones varies depending on the amount of exposure (Figure 13-2a), slope of the shore (Figure 13-2b), as well as tidal conditions. The particular organisms present may also differ depending on the amount of wave action, tidal cycle, climate, length of exposure, amount of light, shape of the shore, size and shape of the ocean basin, and type of rock. The presence of crevices, overhangs, and caves that retain moisture may also influence the distribution of organisms.

According to the Stephensons, the uppermost area of a rocky shore, which is covered only by the highest (spring) tides and is usually just dampened by the spray of crashing waves, is the **supralittoral fringe,** or **splash zone.** Above this region is the **supralittoral,** or **maritime, zone,** an area above high water that may extend several miles inland. Below the supralittoral fringe is the **midlittoral** (or true intertidal) **zone.** This most extensive part of the intertidal zone is regularly exposed during low tides and covered during high tides. This region may be subdivided into various zones and contains a variety of organisms, barnacles typically being the most prominent in the temperate region. Below this zone lies the **infralittoral fringe,** which extends from the lowest

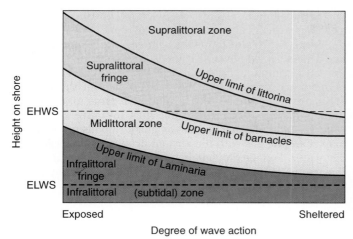

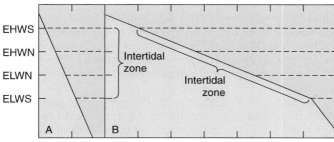

(a)

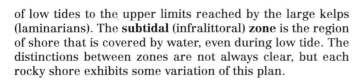

(b)

**Figure 13-2  Vertical Intertidal Zones.** (a) The width of vertical intertidal zones may vary depending on the amount of exposure to wave action. (b) The slope of the shore can also determine the width of vertical intertidal zones. A very steep shore (A) will have narrow intertidal zones, whereas broad bands will be found on shallow sloping shores (B). EHWS: extreme high water of spring tides; EHWN: extreme high water of neap tides; ELWN: extreme low water of neap tides; ELWS: extreme low water of spring tides. (From Lewis, J. R. 1964. *The Ecology of Rocky Shores.* London: English University Press.)

**Figure 13-3  The Supralittoral Fringe.** The supralittoral fringe, or splash zone, receives very little moisture and is inhabited only by a few hardy organisms, such as these yellow lichens and cyanobacteria.

of low tides to the upper limits reached by the large kelps (laminarians). The **subtidal** (infralittoral) **zone** is the region of shore that is covered by water, even during low tide. The distinctions between zones are not always clear, but each rocky shore exhibits some variation of this plan.

## Supralittoral Fringe of Rocky Shores

The supralittoral fringe (or splash zone) receives very little moisture. In northern temperate regions, it is exposed to the drying heat of the sun in the summer and to extreme low temperatures in winter. As a result of these harsh conditions, the supralittoral fringe supports only a few hardy organisms. Gray and orange lichens composed of fungi and algae are common in this zone (Figure 13-3). The fungi in these symbiotic relationships trap moisture for both themselves and their algal partners, and the alga produces nutrients for the pair by photosynthesis. On the North Atlantic

coasts, tarlike patches of cyanobacteria, mostly of the genus *Calothrix*, survive by producing a gelatinous covering that traps and stores moisture.

Sea hair (*Ulothrix*), a filamentous green alga also found on rocks of the North Atlantic coasts, is capable of surviving on the moisture provided by sea spray from waves. During winter months, it grows lower on the intertidal rocks than during the summer. In the winter the reproductive spores the alga produces can survive only where the ocean provides the necessary warmth. The adult algae growing higher on the rocks gradually die out as the air temperature decreases.

The most common animal inhabitants of the supralittoral fringe throughout the world are **periwinkles,** molluscs of the genus *Littorina* and associated genera. The term *littorina zone* is often applied to the supralittoral fringe, although these molluscs also range into the midlittoral zone. On the Atlantic coasts of North America and Europe, rough periwinkles (*Littorina saxatilis*) graze on various types of algae growing at the lower edge of the splash zone (Figure 13-4a). Although basically marine animals, these snails are well-adapted to life out of water. Even though they lack both functional gills and lungs, their mantle cavity is very vascular and is a surface for gas exchange. These snails are so well-adapted to breathing air that they would drown if submerged in water for several hours. To avoid desiccation, rough periwinkles hide in the cracks and crevices of rocks or seal the opening of their shell with mucus, thus trapping moisture in their mantle cavity until temperatures moderate. To prevent her eggs from being damaged by exposure, the female rough periwinkle retains them inside her mantle cavity, where they can be kept moist and oxygenated until they hatch. Other species of periwinkle, such as the common periwinkle (*Littorina littorea*), prevent desiccation by producing planktonic eggs inside jelly coats. Others, such as the

(a)

Andrew Martinez/Photo Researchers, Inc.

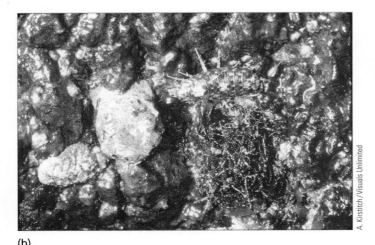

(b)

A. Kirstitch / Visuals Unlimited

**Figure 13-4 Animals of the Supralittoral Fringe.** (a) Rough periwinkles graze on the algae that cover the rocks. (b) These isopods of the genus *Ligia* are scavengers feeding on the available organic material.

northern yellow, or smooth, periwinkle (*Littorina obtusata*), attach gelatinous egg masses to large algae.

Some species of limpets (*Acmaea* and related genera) and crustaceans known as isopods (*Ligia*; see Figure 13-4b) are also common inhabitants of the supralittoral fringe. Limpets are grazers like the periwinkles but the isopods are scavengers and feed on available organic material. Although limpets still retain their gills (or have structures called pseudogills that can act as gills), isopods are adapted to exchanging gas with the air and drown if submerged under water. Female isopods have thoracic pouches in which they carry their eggs to prevent them from drying out.

## Midlittoral Zone of Rocky Shores

When the tide retreats, the midlittoral zone becomes exposed. As the water recedes, intertidal animals must cope with gas exchange, desiccation, temperature extremes, and feeding. Tissues involved in gas exchange must be kept moist, and animals must keep from losing too much body water to the environment as they are exposed to air and the radiant energy of the sun. Filter feeders and organisms that feed on aquatic prey must temporarily suspend feeding and conserve their energy reserves. The solutions to these problems are as varied as the organisms that inhabit the midlittoral zone.

In addition to dealing with the previously mentioned difficulties, organisms that live in the midlittoral zone must withstand the force of the waves as they crash against the rocks during low tide. This force is called **wave shock.** Not only do rock-dwelling organisms have to deal with the crushing force of a wave as it strikes the rocks but they also must deal with the drag that is created as the water moves back out to sea. Animals that live in this zone exhibit compressed or dorsally flattened bodies or shells that dissipate the force of waves. These animals have also acquired mechanisms for adhering tightly to the rocks to prevent them from being dislodged and washed away by the waves (Figure 13-5).

***Upper Midlittoral Zone***   In temperate regions, the most typical organisms of the upper portion of the midlittoral zone are acorn barnacles (*Balanus*) and rock barnacles (*Semibalanus*), which form a line at and below the high tide mark. On some rocky shores, the barnacle population may be as dense as 9,000 individuals per square meter. Barnacles permanently cement themselves to solid surfaces. They are particularly evident on shores pounded by heavy surf where other, less well-adapted organisms would not be able to survive. During low tide, the barnacle closes the opening of its shell with calcareous plates, trapping enough water inside to support the animal until covered again by water at high tide. If the temperature becomes too warm, the barnacle opens its shell just a little, allowing some trapped water to evaporate and cool the animal. When the tide is high, barnacles open their shells and strain food from the surrounding water, using their cirripeds. (For more information regarding barnacles see Chapter 9.)

***Middle and Low Midlittoral Zone***   In the middle and low midlittoral zone, oysters (*Ostrea*), mussels (*Mytilus*), limpets (*Patella* and *Acmaea*), and various periwinkles dominate.

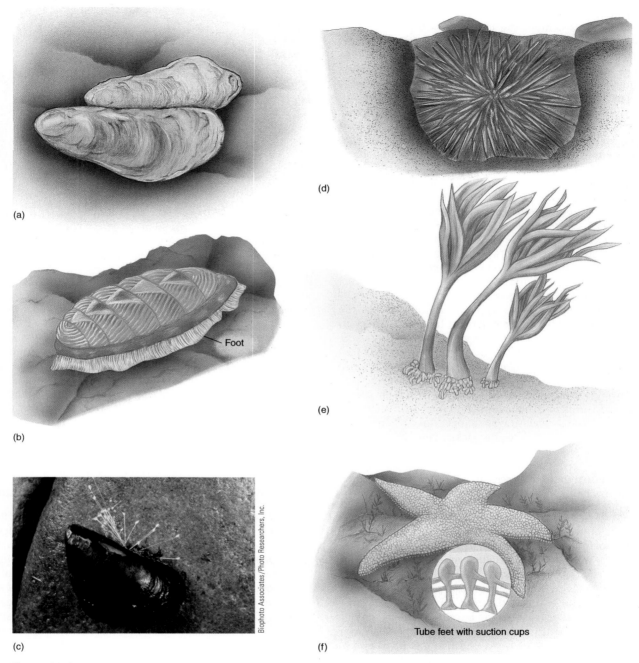

(a)

(b)

Foot

(c)

Biophoto Associates/Photo Researchers, Inc.

(d)

(e)

(f)

Tube feet with suction cups

**Figure 13-5 Adaptations to Wave Shock.** Organisms that live on rocky shores have evolved a number of different strategies for dealing with the crushing force of waves. (a) Some rock dwellers, such as barnacles and oysters, cement themselves to the rock's surface. (b) Chitons and limpets hold tight with their powerful, muscular feet. These animals also present a low body profile, exposing less surface area to the oncoming waves. (c) Mussels attach themselves with tough byssal threads, and (d) rock urchins hollow out cavities in the rocks. (e) Seaweeds have flexible bodies that bend with the flow of water that passes over them. (f) Sea stars use suction cups on the ends of their tube feet to cling to the surface of rocks.

Like barnacles, oysters and mussels fasten themselves to the rocks. Oysters cement the lower valve of their bivalve shell to a solid surface, whereas mussels secrete strong protein fibers, called **byssal threads,** that attach them to the rocks (see Figure 13-5c). At low tide, these bivalves close their shells, trapping enough water inside to provide for their needs until high tide. When the tide is in, they open their shells to filter feed on plankton in the incoming water.

Limpets and chitons also occur in this zone. These molluscs fill a similar niche in that they graze on algae at high tide. At low tide, some limpet species return to an area of rock where they have hollowed out a shallow depression (a *scar*) that perfectly conforms to the shape of their shells. They trap moisture in their mantle cavity to prevent desiccation and cling tightly to the rocks with their muscular foot to prevent waves from dislodging them.

The common periwinkle (*Littorina littorea*) is found on northern Atlantic shores of North America and Europe. Unlike its relative the rough periwinkle, which inhabits the supralittoral fringe, this snail cannot breathe air and must remain moist to exchange gases. During low tide it buries itself in masses of seaweed that trap enough moisture for the snail to survive until the tide returns. At high tide, the common periwinkle feeds on algae that inhabit this area.

In some regions, echinoderms called rock urchins (*Arbacia* on the East Coast and *Echinometra* in the tropics) use their jaws to hollow out a space in the rock. During low tide they wedge themselves firmly in place with their spines (see Figure 13-5d). This behavior, combined with their low profile, helps them to survive wave shock. Rock urchins feed mostly on algae that cling to the rocks. They also feed on drifting bits of seaweeds that they capture among their spines and tube feet. They actively feed during high tide and at low tide return to their depressions, where they are sheltered and less likely to desiccate.

### Seaweeds of the Midlittoral Zone

The most characteristic seaweeds in the midlittoral zone on the northern Atlantic and northern Pacific coasts, especially in colder temperate areas, are brown algae such as rockweeds (*Fucus;* Figure 13-6). Rockweeds are usually small, less than 30 centimeters (1 foot) long, and grow on rocks that do not have full exposure to the sea, because they do not tolerate large waves. In protected areas, such as bays, rockweeds may be as long as 1.8 meters (6 feet) and almost completely cover the rocks. Some species of rockweed use gas-filled chambers called air bladders to buoy them during high tide. (For more information regarding rockweeds see Chapter 7.)

In most rocky intertidal habitats, rockweeds must compete with barnacles for space on the rocks. The large blades of the algae move across the rocky surface in a sweeping action that is powered by waves. This action prevents the cyprid larvae of barnacles from settling down and gaining a foothold, thus providing the algae with more space for growth. Another threat faced by rockweeds is grazing by herbivores. Some species produce chemicals that deter herbivores from feeding on them. *Fucus,* for example, produces a chemical that discourages some herbivores, such as some molluscs, from eating them. Studies have revealed that when herbivorous molluscs begin to graze on *Fucus* the alga

**Figure 13-6 Rockweeds.** Rockweeds such as *Fucus* are common along rocky coasts of the North Atlantic and North Pacific oceans.

produces more of the toxic chemical in the area where the molluscs are feeding. The increased amount of chemical usually succeeds in deterring further grazing.

To prevent desiccation, rockweeds produce a gelatinous covering that retards water loss (see Chapter 7). Their holdfasts anchor them tightly to the rock's surface, and rather than resisting the force of the waves, the blades and stipes of the algae bend gracefully with wave action. As the tide goes out, the seaweeds form large mats that trap water and provide a haven for many species of animal, such as juvenile sea stars, brittle stars, sea urchins, and bryozoans. The shelter provided by large algae, such as rockweeds, may significantly reduce the stress of increased temperature and desiccation on these associated animals during low tide.

### Tide Pools

Not all areas of the rocky intertidal zone are exposed when the tide retreats. Depressions in the rocks retain water, forming areas called **tide pools** (Figure 13-7). Tide pools prevent organisms within them from being exposed to air, but they present their own, unique set of difficult environmental conditions. As the small pool of water heats in the sun, it loses oxygen. This can lead to the suffocation of some inhabitants. During heavy rains, the salinity of tide pools decreases drastically as rainwater accumulates and dilutes the seawater. On the other hand, on a hot sunny day, so much water can evaporate that the salinity increases to dangerous levels. If the tide pool contains a large amount of algae, the oxygen content of the pool will be high during the day while sunlight powers photosynthesis. At night, however, the level of oxygen declines and the level of carbon dioxide rises, resulting in a lower pH. Large tide pools near the low tide mark usually show the least amount of fluctuation, whereas smaller ones and those near the high tide line that are exposed the longest exhibit the widest range of fluctuation.

Regardless of their position on the shore, most tide pools return abruptly to marine conditions as the tide rises and seawater floods the pool. This return to the ocean produces almost instantaneous changes in salinity, tempera-

**Figure 13-7  A Tide Pool.** Organisms that inhabit tide pools must be able to adjust to rapid changes in temperature, pH, salinity, and the oxygen content of the water.

ture, and pH. The organisms that inhabit tide pools have evolved a variety of mechanisms, similar to those found in salt marsh organisms, that allow them to survive in this changeable environment. (For more information regarding salt marsh organisms see Chapter 14.)

***Tide Pool Organisms***  Common tide pool inhabitants include various species of algae, sea stars, anemones, tube worms, hermit crabs, and a variety of mollusc species. Many of the tide pool organisms are filter feeders, feeding on phytoplankton and zooplankton. Anemones and tube worms expose a large amount of surface area while feeding. If disturbed, they retract their tentacles into their bodies or tubes to protect these delicate appendages from damage. This behavior is also an efficient way to avoid desiccation.

The importance of biotic interactions in structuring tide pool communities has been best studied in New England, where one of two algae species dominate depending on the level of grazing (Figure 13-8). *Enteromorpha,* a green alga, outcompetes Irish moss (*Chondrus crispus,* a red alga) in tide pools lacking the common periwinkle (*Littorina littorea*). Where this snail is present, Irish moss is the prevalent alga. *Enteromorpha* persists in certain pools because green crabs (*Carcinus maenas*) eat periwinkles and prevent them from removing this alga. Irish moss persists in other tide pools because gulls eat the crabs and allow the periwinkles to persist. *Enteromorpha* covers the crabs and pro-

tects them from gulls in their pools, whereas gulls can readily capture the crabs in Irish moss pools. Removal and transplantation experiments have demonstrated the importance of these ecological relationships in maintaining both types of pools. Other biotic factors influencing tide pool community structure include recruitment, interspecific competition, and predation. Their importance is unclear because few manipulative studies have been performed.

### Infralittoral Fringe of Rocky Shores

The infralittoral fringe is a transitional area between the midlittoral and the subtidal (infralittoral or sublittoral) zones, the area covered by water even at low tide (Figure 13-9). The infralittoral fringe is submerged except at spring tides. This zone has a rich flora and fauna of organisms that can tolerate limited air exposure. In some areas the rocks are covered with a dense turf of seaweeds, including red, green, and brown algae. Along coasts where the water is cold, large kelps such as *Laminaria* may form a dense cover with smaller algal species and animals living among them. In these cooler waters a variety of molluscs can be found, as well as sea stars and brittle stars. Attached to many of the algae are lacey colonies of bryozoans. Hydrozoans attach to both rocks and the broad blades of algae. On the west coast of the United States, anemones cover the rocks, sea urchins graze on algae and bits of decaying material that they find, and spider crabs (*Libinia*) and Jonah crabs (*Cancer borealis*) scavenge for their food.

## Tropical Rocky Shores

The tropics are generally defined as the area between 20 degrees N and S latitudes, although tropical conditions can extend to higher latitudes as a consequence of warm ocean currents or other conditions. For example, tropical conditions exist in Bermuda (32 degrees N latitude) as a result of the warm water carried by the Gulf Stream. The average incident solar radiation on the sea surface is approximately 30% greater in the tropics than in the temperate zone, and temperatures do not fall much below 18°C for more than a month at a time. The temperature variation is also much less than in the temperate zone, and rainfall is seasonal. These conditions make tropical intertidal systems more stressful in some ways (for example, higher temperatures) but less stressful in others (for example, less variation in temperatures and fewer storms).

### Zonation Patterns on Intertidal Rocks

Zonation patterns in the tropics have many similarities to those seen in the temperate zone but also some significant differences. Tropical rocky intertidal systems have been best studied on the western shore of Panama and in the Caribbean. We will briefly examine zonation patterns in the latter.

### Supralittoral Fringe

In the Caribbean, the supralittoral fringe can be divided into the white, gray, and black zones. The color of each zone is related to the dominant organisms that inhabit the zone.

**Figure 13-8 Tide Pool Communities.** (a) When snails are present, fleshy green algae are eaten, leaving the habitat available for colonization by unpalatable Irish moss and algal crusts. Gulls eat green crabs that are uncovered by the loss of fleshy algae, allowing continuing periwinkle recruitment. (b) In New England tide pools lacking the common periwinkle (*Littorina littorea*), green algae like *Enteromorpha* and *Ulva* outcompete slower-growing forms like Irish moss (*Chondrus crispus*) and algal crusts. The presence of green crabs (*Carcinus maenas*), protected from gulls by an algal canopy, prevents periwinkle recruitment. Both types of tide pools persist because of the activities of green crabs, gulls, and periwinkles.

***White Zone***    The **white zone** is the true border between the land and the sea. No fully marine organisms occur here, but hermit crabs, isopods (*Ligia*), and the knobby periwinkle (*Cenchritis muricatus;* Figure 13-10a) are common. Knobby periwinkles are particularly well-adapted to the harsh conditions experienced at the highest part of the intertidal zone. Their large globular body exposes less surface area to heating by the sun and hot rocks relative to their volume. They also reduce body contact with the hot rock by attaching to it with a mucous thread when wave action is not severe. Their light-colored shell reduces heat gain or loss and the "beads" on their shells act like radiator fins for cooling. They can maintain a body temperature higher than their surroundings at temperatures below 29°C and less than their surroundings at 33°C. They also have a remarkable ability to withstand desiccation using special adaptations in structure and function of the kidney.

***Gray Zone***    Knobby periwinkles also occur in the **gray zone** along with other periwinkle species and nerites (see Figure 13-10b). Nerites are an exclusively tropical group that tends to replace limpets as the dominant form in higher intertidal zones. Their globose shape and ability to hold water in their shell for evaporative cooling make them better

adapted to high-temperature stress and desiccation than the limpets. Nerites may aggregate into large clumps to further reduce heat stress and water loss. The gray zone is also the farthest zone from the low tide line where macroscopic marine algae (*Bostrychia*) grow.

***Black Zone***    The **black zone** is immersed only at the highest spring tides and lacks the knobby periwinkle. Several species of algae (*Bostrychia, Polysiphonia*) and cyanobacteria dominate here. Animal inhabitants include smaller periwinkle snails (*Nodilittorina* spp.), other nerite species, and the fuzzy chiton (*Acanthopleura granulata*).

## Midlittoral Zone

The midlittoral zone is the true intertidal zone. It is inhabited by many exclusively marine species as well as some of the amphibious species described previously. It is usually divided into yellow and pink zones.

***Yellow Zone***    The **yellow zone** is typically yellow or green because of the microscopic boring algae covering its surface (Figure 13-11a). Small populations of barnacles may occur, as well as limpets, the fuzzy chiton, and predatory rock snails (*Thais*). Another inhabitant is the irregular worm

(a)    (b)

**Figure 13-9 The Infralittoral Zone.** Rocky infralittoral (subtidal) zone of the (a) west and (b) east coasts of the United States. The infralittoral zone of rocky shores offers a more stable environment than the intertidal zone and thus exhibits greater diversity.

snail (*Petaloconchus irregularis*), which forms calcareous tubes on the rock surface. They feed by trapping plankton on mucus threads.

***Pink Zone***   In some areas there is a **pink zone** underlying the yellow that is characterized by the widespread encrustation of coralline algae (see Figure 13-11b). Characteristic animals include the irregular worm snail, mats of anemones, keyhole limpets, and various gastropods. In other areas, this zone may be represented by rocks overgrown by short reddish-brown mats of algae. Barnacles, scorched mussels, several chitons, rock snails, and other forms may be found here.

### Infralittoral Fringe

The infralittoral fringe (or surf zone) includes the edge of the lower rocky platform and parts of the reef. *Sargassum* species and other algae may cover the rock surface like turf. Boring urchins (*Echinometra*), anemones, sponges, bryozoa, sea cucumbers, and keyhole limpets are common animal inhabitants.

### Subtidal Zone

Unlike the species-rich subtidal of the temperate zone, the tropical subtidal is relatively barren, with most algae being confined to the infralittoral fringe and lower midlittoral

Courtesy of James Small and Judy Schmalstig

(a)

Michael Giannechini/Photo Researchers, Inc.

(b)

Courtesy of James W. Small

(a)

Courtesy of James W. Small

(b)

**Figure 13-10  Tropical Intertidal Molluscs.** Tropical intertidal molluscs have special adaptations to live in the harsh environment of the rocky intertidal. (a) The knobby periwinkle (*Cenchritis muricatus*) lives on the highest part of the supralittoral fringe. Its globular body exposes less surface area to heating relative to volume, and the light-colored shell reduces heat gain or loss. They attach to the rocks with a mucous thread, reducing body contact with the hot rock, and the "beads" on their shells act like radiator fins for cooling. (b) Nerites are an exclusively tropical group with the ability to hold water extraviscerally, an important adaptation for evaporative cooling.

**Figure 13-11  Yellow and Pink Zones of Tropical Shores.** (a) The yellow zone gets its name from the large number of yellow and yellow-green boring algae. The pink zone is seen below the yellow zone in this photo. (b) The pink zone is named for the pinkish color of the coralline algae that can dominate this area.

platforms. Relatively large brown algae dominate the temperate zone, whereas small, turf-forming red algae dominate in the tropics. A number of explanations have been proposed for these differences, including the effect of seawater temperature, the need for a high surface-to-volume ratio due to low nutrient availability in tropical water, and the relatively constant light intensity that has inhibited the evolution of large algae. The explanation now most favored by researchers suggests that the level of predation (including herbivory) is far greater in the tropics. Important tropical herbivores include surgeonfish, parrotfish, damselfish,

sea urchins, and many species of gastropod. The relative lack of algae in the subtidal zone and its abundance on shallow wave-swept platforms of the intertidal zone, where turbulence interferes with herbivory, support this hypothesis as do experiments that exclude herbivores from the zone. In several herbivore-exclusion experiments, erect algae were able to establish themselves where none were previously found. The ubiquitous presence of such defenses as thallus calcification and noxious chemicals among tropical algae also supports this hypothesis. Higher predation levels would also explain the greatly reduced abundance of barnacles and mussels in tropical intertidal zones, although recent studies in Costa Rica and Panama indicate that recruitment is also greatly influenced by unknown events in the water column. In some parts of the Caribbean, barnacles and mussels are very rare. The fact that competitively inferior crustose algae typically dominate the mid to low intertidal zone can also be attributed to the effects of grazing gastropods and other herbivores. The activity of molluscan herbivores may also inhibit settling by sessile invertebrates.

## Comparison of Temperate and Tropical Rocky Intertidal Systems

High stress and high predation make the tropical intertidal environment rather different from the temperate. Mobile invertebrates are very abundant, but sessile species may be rare and their competition for space less severe, particularly in the upper intertidal zone. Holes and crevices seem to be more important as refuges for these animals than in the temperate zone. Periwinkles may be represented by more species and are more abundant in the tropics than in temperate zones, and exclusively tropical groups like the nerites appear. On the other hand, barnacles and mussels, so abundant in the temperate zone, may be scarce or lacking in the rocky intertidal zone of the tropics. In the temperate zone, large body size or residence in higher regions of the intertidal zone are important means of escaping predators, but these characteristics are less important in the tropics. Because of their greater numbers and diversity of types, consumers have more influence on the availability of refuges in the tropics. In contrast to their role in the temperate zone, macroalgae have significantly less impact on community structure. Although there is some annual variation in the physical environment, these communities change little in space and time.

## Intertidal Fishes

Fishes visiting the intertidal zone can be divided into two categories: residents and temporary inhabitants. Resident species include clingfishes, blennies, gobies, sculpins (*Oligocottus*), and rock eels (*Xiphister*), among others. True residents are reported to make up between 20% and 67% of the inhabitants of tide pools.

True residents typically have special adaptations for surviving the harsh conditions of living in the intertidal zone. Because wave conditions make large body size and lack of mobility disadvantageous, intertidal fish are rarely longer than 20 to 30 cm (8 to 12 inches). Scales are absent (blennies and clingfish), reduced, or very firmly attached (some gobies). Body shape may be compressed and elongate (blennies) or depressed (clingfish). The swim bladder is typically absent or reduced and the body density greater than in pelagic forms. Clingfish (Figure 13-12) and gobies have pelvic fins modified into sucking disks that allow them to maintain position against surging water. The head may be enlarged and, in some clingfish, articulated (able to pivot), an unusual adaptation in fish. The eyes may also be enlarged and set high on the head, an aid to spotting prey above the water level. Most species have a wide tolerance for changes in salinity and temperature. Some can even withstand desiccation to the same extent as amphibians. A few species even leave the water to feed and to avoid the most extreme wave conditions (for example, the Chilean clingfish, *Sicyases sanguineus*). Most fish that leave the water respire through the skin, but some forms (mudskippers) utilize a modified oral-

**Figure 13-12 Clingfish.** Clingfish have pelvic fins that are modified into a sucker. This adaptation allows them to attach firmly to solid surfaces thus reducing the chances of being swept away by strong currents.

pharyngeal cavity. It has also been suggested that intestinal respiration may take place in the Chilean clingfish. Excretion of nitrogenous wastes can be a problem when fish are out of the water. Some forms excrete urea rather than ammonia under these conditions. Intertidal fish can be herbivorous, omnivorous, or carnivorous.

Temporary inhabitants can be divided into tidal visitors, seasonal visitors, and accidental visitors. Tidal visitors move into the intertidal zone at high tide to feed, whereas seasonal visitors move into it for breeding. Accidental visitors are trapped by storms.

## Ecology of the Rocky Shore

Whereas physical factors such as exposure to air, high temperatures, and desiccation determine the upper limits of intertidal organisms, life between the tidemarks is primarily influenced by such biotic factors as level of primary production, recruitment (larval settling), herbivory (grazing), predation, and competition.

### East Coast Rocky Shores

In New England, where wave action is heavy, barnacles (*Semibalanus*) dominate the upper zones primarily because only they have the ability to survive there. Less-resistant species such as periwinkles are washed away by wave action. Below the barnacles, the competitively superior mussels (*Mytilus*) are most abundant. Mussels may completely cover surfaces from which predatory sea stars and dogwinkle snails (*Nucella*) are excluded by wave action. On protected shores, many algal species as well as consumers, including molluscs, crabs, and sea stars, can survive. Algae that are resistant to grazing, such as *Fucus* and *Ascophyllum*, may dominate and form a canopy protecting many types of consumers from desiccation. These consumers in turn may limit the populations of other algae, as well as barnacles and mussels, providing open space for colonization.

## West Coast Rocky Shores

On the Pacific coast, barnacles compete with algae for available space and mussels displace barnacles by growing over them. Ochre sea stars (*Pisaster ochraceus*) maintain a balance by eating enough mussels to prevent them from completely overtaking the barnacles. *Pisaster* has been termed a *keystone predator* because of its role in preventing *Mytilus* from occupying all the space and thus allowing less-competitive forms to colonize the rocks. In many intertidal systems, however, dominant competitors may be controlled by a number of species rather than a single top carnivore. Even the effect of a keystone predator like *Pisaster* may vary over its geographic range.

## Rocky Shores in the Tropics

In tropical intertidal systems, total predation is very strong and the control of competitively dominant species is spread over a number of consumers, hence there are no keystone predators in the usual sense.

## Top-Down and Bottom-Up Factors

Competition, herbivory, and predation are collectively termed **top-down factors** because their effects may flow down the food chain. **Bottom-up factors,** in contrast, are those that affect the basal level of food chains and, thereby, the associated community. Examples of bottom-up factors include nutrient availability and recruitment when it increases prey abundance. Although many studies, including those described above, have demonstrated the importance of top-down effects in structuring intertidal systems worldwide, the relationships are not always simple. In New England, for example, it was found that predation varied along environmental gradients of wave turbulence and thermal or desiccation stress, top-down forces being strongest at benign portions of the environment. Whereas competition and predation are very important in structuring the intertidal community of the Pacific Northwest, community composition in California is determined primarily by recruitment, a bottom-up factor. El Niño nutrient depletion, another bottom-up factor, affects intertidal systems by reducing growth of kelp forests off the coast of southern California and associated populations of rockfish, a major predator on barnacle larvae. Reduction in rockfish numbers, therefore, increases barnacle recruitment. Studies in South Africa have demonstrated the importance of both bird guano as a nutrient source and grazing or predation by limpets and seabirds to the structure of intertidal communities.

It is now apparent that both vertical zonation and the dynamics of intertidal communities are driven by an interplay between biotic factors and local and regional environmental variables. The relative importance of top-down and bottom-up factors in determining community structure varies under different levels of environmental stress and regional resource dynamics. It has been demonstrated, for example, that the relative competitive abilities of the barnacles *Chthamalus fragilis* and *Semibalanus balanoides* depend on the latter's tolerance of physical stress in the high intertidal zone. *Chthamalus* occupies the high intertidal zone because it can withstand higher temperatures than *Semibalanus* but is competitively inferior elsewhere. In the lower intertidal zone, the level of predation on mussels by the sea star *Pisaster ochraceus* depends on small temperature changes due to near-shore upwelling events. In a New England estuary, variation in water flow rates about rocky intertidal sites had major effects on community structure. Sites with higher flow rates demonstrated greater recruitment of planktonic larvae and more particulate food, resulting in increased barnacle and whelk growth (bottom-up effects). In areas of low flow rates, increased predation on mussels, barnacles, periwinkles, and whelks by crabs was seen (top-down effects). Both top-down and bottom-up factors are jointly important determinants of the structure of intertidal systems but their relative contribution varies wherever differences of the physical environment occur along coastlines.

## In Summary

Rocky shores are inhabited by many organisms that tend to form bands, or zones, on the rocks. The infralittoral fringe of a rocky shore receives only the little moisture delivered by sea spray. The midlittoral zone is the area that is alternately exposed and submerged by tides. Organisms that live in this zone display specific adaptations to deal with desiccation and wave shock. Tide pools trap water when the tide recedes, preventing organisms that inhabit them from becoming fully exposed to the air at low tide. However, life in tide pools presents a unique set of difficult environmental conditions. Organisms that inhabit tide pools must be able to tolerate abrupt changes in temperature, salinity, pH, and oxygen content of the water. The subtidal zone is generally covered by water even at low tide. Because this zone is usually more stable, it supports a greater diversity of organisms in the temperate zone but may be barren in the tropics because of the abundance of predators and herbivores. In addition to physical factors, biotic factors such as grazing, predation, and competition all play an important role in determining the distribution of organisms on the rocky shore. ●

## SANDY SHORES

Many temperate and tropical shorelines consist of sandy beaches. On the east coast of North America, sandy beaches extend almost continuously from Cape Cod south to the Gulf Coast. Although the shoreline of western North America is primarily rocky, significant areas of sandy beaches also exist.

## Role of Waves and Sediments

The nature of a sandy beach, the porosity of its sediments, and the ability of animals to burrow into its sediments are all influenced by sediment particle size. Waves play a significant role in determining the types of sediment on a sandy beach. Heavy wave action carries off much of the finer sediment, leaving behind only coarser material. Thus beaches with fine sand are found along coasts with little wave action or in areas protected from heavy surf. The slope of a beach is determined by the interaction of waves, size of sediment particles, and the relationship of swash and backwash water. **Swash** is the water running up a beach after a wave breaks, and **backwash** is the water flowing down the beach. On a **dissipative beach,** wave energy is strong but is dissipated in a surf zone some distance from the beach face. Dissipative beaches are usually flat with fine sediments because they receive less wave action and have gentle swash. In contrast, wave energy is directly dissipated on a **reflective beach.** Heavy wave action sends large swashes up the beach face, producing a steep slope. Coarse sediment is deposited as swash and backwash water collide.

On all sandy beaches, a cushion of water separates the grains of sand below a certain depth. This is especially true of beaches with fine sand where capillary action (for more information regarding capillary action see Chapter 4) is the greatest. Fine sand beaches, therefore, have a greater abundance of organisms due to both greater water retention and the fact that this type of sediment is more suitable for burrowing. Coarse sand beaches drain well, dry out quickly, and therefore support relatively fewer organisms. The variation in slope, particle size, and wave action on sandy shores can be expressed as a continuum of beach types, from highly exposed to very protected. Table 13-1 summarizes the characteristics of four types of sandy beaches along this continuum.

## Comparison of Rocky and Sandy Shores

In sharp contrast to the rocky shore, sandy beaches lack a readily apparent pattern of zonation and superficially appear barren and devoid of life. Where organisms exist, the most important factor determining their distribution is wave action. On exposed beaches, few large organisms can inhabit the surface layer of sediment because it is in constant motion. Organisms that do live here either are mobile or burrow deeply into the sand. Temperature has much less effect on inhabitants of sandy shores than those of rocky shores because of both the insulating properties of sand and the effect of water held in the spaces between sediment particles of its deeper layers. For these same reasons, desicca-

tion and changes in salinity are easier to deal with than they are on rocky shores. Oxygen availability in the sediments, however, may be limited where respiration is high and exchange of water with the sea is reduced because of fine sediments. On sand flats and other beaches with very fine sediments, exchange of water may be so low that anoxic conditions exist below the uppermost layers. Many animals possess siphons or pump water to oxygenate their burrows, whereas others use specialized respiratory pigments to more efficiently remove oxygen from the interstitial water. Anaerobic bacteria are common in such anoxic conditions and often produce foul smelling hydrogen sulfide gas as a byproduct of their breakdown of organic detritus in the sediments.

## Sandy Shore Zonation

Although it is much less defined than on rocky shores, zonation does exist on sandy shores. In addition to a pattern of longitudinal zonation (Figure 13-13) observed as one moves up the beach, vertical zonation exists among organisms buried in the sand. This vertical zonation depends on the amount of water trapped at each level. In a fashion similar to that described for rocky intertidal systems, we can divide a typical sandy beach into three zones: a supralittoral, midlittoral, and subtidal (infralittoral). These zones will be examined in more detail in the following sections.

### Life above the High Tide Line

The supralittoral fringe of a sandy shore stretches from the high tide line to the point where terrestrial vegetation begins. In some areas, sand dunes (Figure 13-14) border the uppermost extent of this zone. Much of this area is unsuitable for habitation because the sun bakes the surface sand, both raising its temperature and drying it. Just below this level is a **zone of drying sand;** moisture reaches this zone only during the highest tides and gradually evaporates over time. Most of the living organisms in this zone are termed **infauna** because they burrow in the sand to survive dry periods and the intense heat of the sun. Infauna typically live in permanent or semipermanent tubes or burrows, or they are able to quickly burrow into the sand. They obtain oxy-

## Table 13-1 Characteristics of Sandy Beaches at Various Levels of Exposure

|  | Exposed (Reflective) | Semi-exposed (Dissipative) | Protected | Very protected |
|---|---|---|---|---|
| Wave action | high | moderate | low | none |
| Wave types | plunging, surging | spilling | — | — |
| Slope | steep | gradual | gradual | flat |
| Width | narrow | wide | various | extensive |
| Oxygen | highest | — | — | lowest |
| Moisture | lowest | — | — | highest |
| Particle size | coarsest | finer | finer | varies |
| Organic matter | lowest | — | — | highest |
| Permanent burrows | no | no | some | frequently |

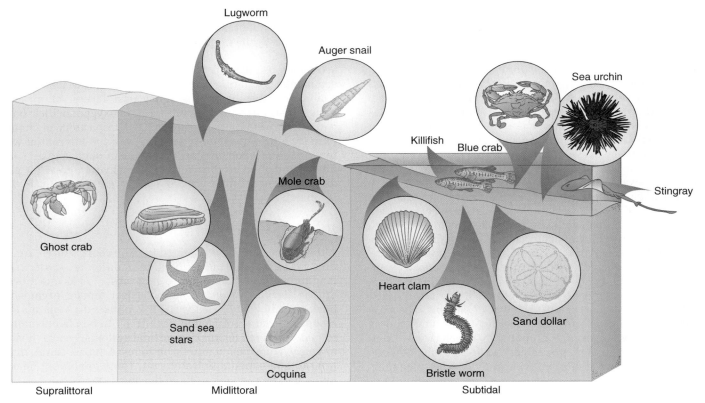

**Figure 13-13  Sandy Shore Zonation.** Although sandy shores do not exhibit the obvious pattern of zonation seen on rocky shores, both horizontal and vertical zonation exist. The vertical zonation of organisms buried in the sand is dependent on the amount of water trapped at each level, time of year, and type of beach.

**Figure 13-14  Sand Dunes.** Dunes form along the coast when sand is blown away from the beach. They are most common along the Atlantic and Gulf coasts, but they also appear on the Pacific coast, as shown here.

**Figure 13-15  Ghost Crab.** Ghost crabs dominate the supratidal zone of sandy shores in temperate and tropical regions. They spend most of the daytime in their burrows, coming out at night to scavenge for food.

gen either through their skin or through gills that are sometimes bathed in supratidal water drawn in by elaborate siphons.

Insects, isopods, and amphipod crustaceans (beach hoppers) are the dominant macroorganisms of the supralit-toral zone of temperate regions. In the tropics and warmer parts of the temperate zone, beach hoppers are replaced by ghost crabs (*Ocypode;* Figure 13-15) and fiddler crabs (*Uca*). The ghost crab gets its name from its light color and its habit of coming out at night to forage for food. During the day, it spends most of its time in its burrow to avoid the heat and predatory birds. The ghost crab is not as aquatic as many of its relatives. Its legs are not adapted for swimming but are useful for scurrying sideways along the beach. When the crab enters the water to escape predators, it runs along the bottom. Female ghost crabs deposit their eggs in the ocean, and the young develop in the water. Young crabs rely on currents to carry them about and eventually to deposit them on a beach, where they can take up an adult existence. Once the crabs arrive on land, they return to the sea only occasionally to moisten their gills or to avoid predators. In either case, their visit is short, and they quickly return to the shore. Like its aquatic relatives, the ghost crab extracts oxygen from water through its gills. A special chamber surrounding the gills traps enough seawater to supply the crab's needs until it can enter the water and replenish its supply.

Adult ghost crabs live in burrows that consist of a deep shaft with a chamber at the end. Sometimes another shaft branches from the chamber and serves as an alternative escape route. The crabs emerge at night to scavenge the beach for food. They spend the early morning hours making necessary repairs to their burrows, and by midday, when the temperatures are reaching a peak, they enter their burrows and seal the opening to escape the high temperatures and desiccation. They reemerge after the sun sets to resume their foraging.

In winter at higher latitudes, ghost crabs move their burrows farther up on the beach, well out of the reach of high tide. They fashion a larger chamber at the end of the shaft that they stock with food for the winter. As the temperatures begin to fall, the crabs retire to their burrows, seal the openings, and live their subterranean existence until the warmth of spring brings them out for another season.

### Life in the Sandy Shore Midlittoral Zone

As in the supralittoral zone, most inhabitants of the midlittoral zone are burrowers. During high tide, they emerge from their burrows to look for food, find mates and reproduce, or extend specialized appendages to filter the water for food and extract oxygen. At low tide, they burrow into the moist sand and retract their appendages to avoid desiccation and exposure to land predators.

***Vertical Zonation***    Like the supralittoral zone, the midlittoral zone exhibits vertical zonation (Figure 13-16). The zones of dry sand and drying sand are not as extensive as in the supralittoral zone. Just below this area is a **zone of retention** that retains moisture at low tide because of the capillary action of water. Worldwide, it is inhabited by several species of isopod crustaceans. Closer to the sea and lower down are the zones of resurgence and saturation. In the **zone of resurgence,** water is retained at low tide and supports a variety of assorted crustaceans and polychaete worms. Farther down is the **zone of saturation.** This zone is constantly moist and supports the greatest diversity of organisms. Polychaetes burrow in this area along with bivalves like coquinas (*Donax*), tellins (family Tellinidae), and Venus clams (family Veneridae). Amphipods and a variety of other small crustaceans also make their homes in this region.

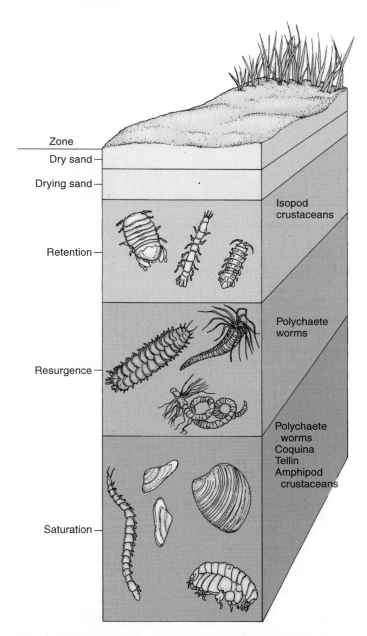

**Figure 13-16** **Vertical Zonation of the Sandy Intertidal Zone.** The sandy intertidal zone exhibits vertical zonation, with each zone characterized by its moisture content. The zones of retention, resurgence, and saturation are home to a variety of burrowing organisms.

***Animals of the Midlittoral Zone***    At low tide, a careful examination of the sandy bottom will usually reveal telltale signs of the presence of burrowing organisms (Figure 13-17). Flat, ribbonlike trails are signs of echinoderms such as sand dollars or sea stars. The sand dollars use tiny spines and their tube feet to pick detritus out of the sand. In Florida and the Caribbean, sand sea stars (*Luidia*) feed mostly on small molluscs and crustaceans that they find burrowed in the sand. These sea stars swallow their prey whole and then regurgitate the shells and hard parts. Unlike the sea stars that inhabit rocky shores, these animals lack suckers at the

G. Karleskint

**Figure 13-17 Animal Trails.** This trail was produced by a gastropod mollusc known as an auger (*Terebra*). Tracks in the sand such as these indicate the presence of burrowing animals, frequently molluscs, crustaceans, or echinoderms.

(a)

J. & M. Bain /Natural History Photographic Agency

(b)

Richard N. Mariscal

**Figure 13-19 Lugworm.** (a) A common burrowing organism in the sandy intertidal zone is the lugworm. (b) The worm ingests sand as it constructs its burrow and deposits it in the form of fecal castings that are stacked in a cone-shaped pile outside the burrow entrance.

ends of their tube feet, so they are not able to cling to rocks. They are, however, very efficient burrowers and can cover themselves with sand in a matter of seconds.

Many snails leave trails that resemble the mounds of elevated earth made by moles. Moon snails (*Polinices*) and olive snails (*Oliva*) are especially common in this habitat and leave hills of sand that are raised by the animal's foot as it crawls along the bottom. Moon snails feed on bivalves by using their radula to drill a neat hole into the bivalve shell (Figure 13-18). Once the shell is pierced, the moon snail inserts its proboscis through the opening and rasps out the flesh from inside the shell. The average moon snail consumes enough clams each week to equal one third of its body weight. Olive snails are also carnivorous. They feed on a variety of organisms but especially prefer small bivalves.

Another prominent inhabitant of the intertidal zone is the lugworm (Figure 13-19a). Lugworms are deposit feeders. They ingest sand as they dig their burrows and digest the organic material that it contains. The sand passes through the worm's digestive system and is deposited dur-

ing defecation at the entrance to the worm's burrow. The coiled, cone-shaped casts are prominent on sandy beaches at low tide (see Figure 13-19b).

***Relationship between Tides and the Activity of Mid-littoral Organisms***    The activity of intertidal organisms is keyed to the movement of the tides. As the tide moves in, the pace of life begins to quicken. During high tide, bivalves, such as cockles (*Cardium*), tellins, and surf clams (*Spisula*), project their siphons from their burrows and begin to filter the water for food and bathe their gills with oxygen. Carnivorous snails, called *whelks* (family Buccinidae), glide along the bottom in search of bivalves, which they locate by sensing the currents produced by the clam's siphons. Sea stars and sand dollars also become more apparent as they come out of the sand in search of food.

Two filter-feeding organisms that move up and down the beach with the movement of the tide are mole crabs (*Emerita*) and coquinas. As the tide moves in, large numbers

H. W. Pratt /Biological Photo Service

**Figure 13-18 Moon Snail.** Moon snails (*Polinices; Lunatia*) are carnivores feeding mostly on bivalves.

(a)

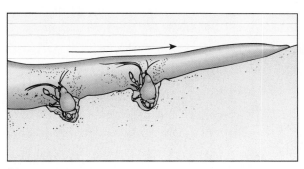

(b)

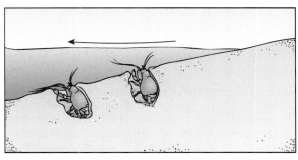

(c)

**Figure 13-20  Feeding Behavior of Mole Crabs.** Mole crabs are filter-feeders that move up and down the beach with the tides. (a) Incoming waves carry the mole crabs up the beach. (b) As the wave dissipates, the crabs burrow into the sand. (c) They then extend their antennae and filter food from the water as the wave retreats.

of mole crabs emerge from the sand (Figure 13-20). Incoming waves carry these creatures up the sand, and as the waves dissipate, the animals force their way backward into the sand and use their large antennae to strain food from the water as it returns to the sea. When the tide retreats, the mole crabs repeat the process in the opposite direction.

**Figure 13-21  Coquinas.** Like mole crabs, coquinas emerge in advance of a wave so the wave will carry them up the beach. As the wave retreats, they burrow into the sand, straining the receding water for food.

Coquinas move along with the tide in a similar fashion (Figure 13-21). As the water covers the sand, thousands of these tiny bivalves come out of their burrows, and the waves carry them up the beach. Before the backwash can carry them back into the water, they burrow into the sand and put out their siphons to feed. As the tide retreats they follow the water back, always staying at the edge of the water.

The incoming tide brings not only food and oxygen but also predators (Figure 13-22). Blue crabs and green crabs (family Portunidae) move in to feed on smaller crustaceans and clams. Small fish feed on crustaceans, worms, and a variety of larval forms. Skates and rays cruise the bottom, preying on crustaceans and molluscs. When the tide recedes again, these predators are replaced by gulls and a variety of shorebirds that scurry across the intertidal zone in search of food.

## Life below the Low Tide Line

The subtidal zone is a truly marine environment exposed during only the lowest spring tides. The variety and distribution of organisms in this zone is primarily influenced by characteristics of the bottom sediments. Where the sediments are bare sand, the inhabitants are predominantly burrowing organisms similar to those inhabiting the midlittoral zone. Several species of bivalves extend their siphons to filter food from the water. Tube worms and other polychaetes either trap their food by filter and deposit feeding or search the sand particles for detritus. Along the southeastern coast of the United States and along Caribbean shores, heart urchins (*Moira*) move slowly along the sandy bottom, feeding on detritus that is mixed with the grains of sand. As they travel just beneath the surface of the sand, they angle their spines backward to avoid resistance, leaving a V-shaped trail as they move along the bottom, half buried in sand. Various molluscan species feed on algae, detritus, or each other.

Along some coasts, fields of seagrass are found in the subtidal zone. In addition to arthropods and molluscs, the seagrass beds host sea urchins, sea stars, brittle stars, sea cucumbers, and anemones. The addition of rocks and bro-

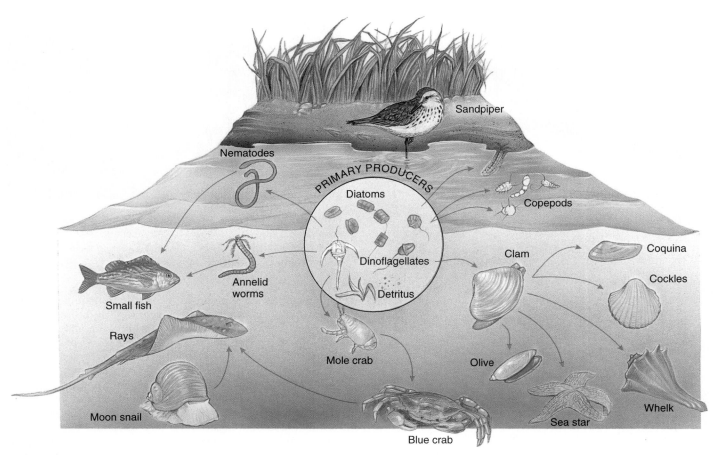

**Figure 13-22 Sandy Shore Food Web.** Detritus and single-celled algae provide food for numerous burrowing animals. These burrowing animals are a source of food for many larger animals including crabs, fishes, and birds.

ken pieces of coral increases the complexity of the bottom and the diversity of life. Because the subtidal zone is rarely exposed, many species of fish are found here, primarily species that move into the midlittoral zone at high tide to feed. The pace of life in the subtidal zone is relatively constant, and during only the lowest ebb tides do these organisms temporarily suspend their activity.

## Meiofauna

In addition to the macrofauna (large animals) described above, many microscopic organisms inhabit the spaces between the sediment particles of the midlittoral and subtidal zones (Figure 13-23). These organisms constitute the **meiofauna** and include those animals that pass through a 0.5-millimeter screen but are retained by a 62-micrometer screen.

### Factors Affecting the Distribution of Meiofauna

The meiofauna are entirely aquatic, so they require water within the interstitial spaces of the sand to survive. Grain size is, therefore, a very important determinant of both the size and types of organisms present there. Coarse-grain sed-

iments have a greater interstitial volume that allows larger organisms to move between the particles, whereas fine-grain sediments have less space and exhibit more burrowing forms (for example, kinorhynchs). Coarse-grain sediments also drain more rapidly, a factor that may affect both the presence and types of organisms. Other factors affecting the distribution and diversity of the meiofauna include water circulation, oxygen availability, temperature, salinity, and wave action. Fine-grain sediments can inhibit water flow and produce anoxic conditions. Anoxic conditions likewise exist at greater sand depth. The lack of oxygen and space severely limits the types of organisms present. Indeed, it has been found that the meiofauna disappears if the median grain diameter is less than 1 millimeter. The upper layers of sand are more variable in temperature but lower layers are well insulated. The fauna of upper layers, therefore, differs from that at lower levels and includes forms with wider tolerance to temperature change. Salinity may also affect the distribution and diversity of organisms, particularly in areas of freshwater runoff or if rain is heavy. Finally, wave action can suspend both the sediments and the organisms living there, making them more susceptible to predation.

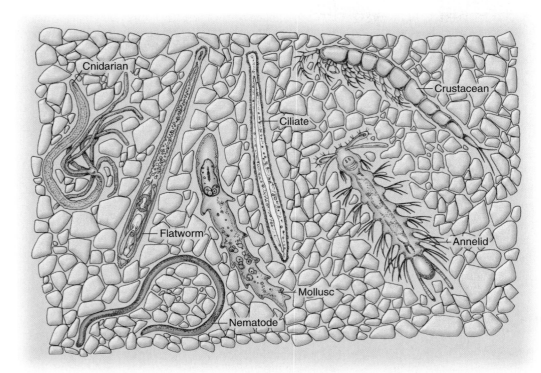

**Figure 13-23 Meiofauna.**
Meiofauna are tiny organisms that inhabit the spaces between sediment particles. Members of the meiofauna that inhabit sandy shores include tiny protozoans, nematodes, cnidarians, flatworms, annelids, crustaceans, and molluscs.

## Characteristics of the Meiofauna

Inhabitants of the interstitial spaces of the sand include invertebrates from a large number of phyla. Ciliates, flatworms, and nematodes are particularly abundant in the meiofauna. Minor phyla represented include the gastrotrichs, kinorhynchs, rotifers, tardigrades, and priapulids, among others. Representatives of the phylum Annelida (both oligochaetes and polychaetes) and several small crustaceans (copepods and ostracods) may also be present. Meiofaunal organisms are generally elongated with few lateral projections, such as setae or spines, that would interfere with their ability to move freely and quickly in the sand. Many are armored to protect them from being crushed by the moving grains of sand. The meiofauna include predators, herbivores, suspension feeders, and detritivores. Because of their small size, the number of offspring produced is very small. Most forms, therefore, exhibit some sort of brood protection. Only a very few forms produce planktonic larvae, and they tend to have relatively larger brood sizes.

## Factors Affecting the Size of Meiofaunal Populations

The sizes of meiofauna populations exhibit seasonal variations, reaching a peak during summer months. Beaches that are protected from wave action have the greatest abundance of meiofauna. Predation on the meiofauna by the macrofauna can be severe in the upper layers of the sediments. For example, the spot (*Leiostomus xanthurus*), a member of the drum family of fishes (family Sciaenidae), has been demonstrated to significantly reduce the meiofauna in the upper 2 centimeters (0.8 inches).

## Ecology of the Sandy Shore

Unlike rocky shores, the fauna of sandy shores is less abundant and does not occupy all available space. The sparseness of the populations, the three-dimensional space, and the abundance of food suggest that competition is not a major factor in determining the distribution pattern of organisms. Likewise, predation may be less important because there are relatively fewer predators among the invertebrate inhabitants. On protected sand flats, however, both predation and disturbance seem very important in determining community structure. The effect of fish and bird predators on the macrofauna is unclear but may be significant. The effect of both competition and predation decreases with greater exposure, hence the physical factors of waves, particle size, and slope may be most important in determining distribution patterns of organisms on sandy shores.

## In Summary

The distribution and types of organisms found on sandy shores are greatly influenced by wave action. Sandy shores can be divided into zones on the basis of the amount of tidal coverage they receive. In addition to longitudinal zonation, sandy shores exhibit vertical zonation, according to the amount of water the sand retains.

In the infralittoral zone of sandy shores, most of the living organisms are buried in the sand. The midlittoral zone is the area of shore that is exposed at low tide and submerged at high tide. Organisms that live in this region are predominantly burrowers. During low tide, they remain hidden in their burrows. At high tide, they emerge and forage for food. Like rocky shores, the subtidal zone of sandy shores is exposed only during spring tides and is truly a marine environment. Where the bottom is bare sand, burrowing organisms predominate. In other areas where the bottom is covered with seagrass, rocks, or coral rubble, a greater diversity of organisms can be found.

The meiofauna are the microscopic organisms that live among the sand grains and include representatives from most invertebrate phyla. Their distribution is affected primarily by physical factors rather than biotic factors. ●

## SELECTED KEY TERMS

backwash, *p. 315*

black zone, *p. 310*

bottom-up factors, *p. 314*

byssal threads, *p. 308*

dissipative beach, *p. 315*

gray zone, *p. 310*

infauna, *p. 315*

infralittoral zone, *p. 304*

maritime zone, *p. 304*

meiofauna, *p. 320*

midlittoral zone, *p. 304*

periwinkle, *p. 305*

pink zone, *p. 311*

reflective beach, *p. 315*

splash zone, *p. 304*

subtidal zone, *p. 305*

supralittoral fringe, *p. 304*

supralittoral zone, *p. 304*

swash, *p. 315*

tide pool, *p. 306*

top-down factors, *p. 314*

wave shock, *p. 306*

white zone, *p. 310*

yellow zone, *p. 310*

zonation, *p. 303*

zone of drying sand, *p. 315*

zone of resurgence, *p. 317*

zone of retention, *p. 317*

zone of saturation, *p. 317*

## QUESTIONS FOR REVIEW

### Multiple Choice

1. A prominent herbivore that can be found grazing on algae at the lower edge of the supralittoral fringe of rocky shores is the
   a. mussel
   b. barnacle
   c. periwinkle
   d. ochre sea star
   e. rock crab

2. A major competitor of barnacles for space on rocks is the
   a. oyster
   b. mussel
   c. periwinkle
   d. rockweed
   e. rock crab

3. The low body profile of animals that live on intertidal rocks is an adaptation that protects against
   a. sunlight
   b. high temperatures
   c. desiccation
   d. predation
   e. wave shock

4. The most common alga in the midlittoral (intertidal) zone of temperate rocky shores is
   a. *Fucus*
   b. *Laminaria*
   c. *Sargassum*
   d. kelp
   e. red alga

5. The most important factor in determining the distribution of life on a sandy beach is
   a. temperature
   b. salinity
   c. pH
   d. wave action
   e. sediment characteristics

6. On warm temperate and tropical sandy beaches, the dominant animal in the supratidal zone is the
   a. periwinkle
   b. sea star
   c. sea urchin
   d. ghost crab
   e. mole crab

7. Most organisms that inhabit the intertidal zone of a sandy beach are
   a. predators
   b. grazers
   c. burrowers
   d. multicellular algae
   e. surface dwellers

8. The vertical zone of the sandy beach that supports the highest number of organisms is the zone of
   a. retention
   b. saturation
   c. resurgence
   d. dry sand
   e. drying sand

### Short Answer

1. What characteristics are exhibited by organisms that inhabit rocky shores?

2. What environmental challenges are encountered by organisms that live in a tide pool?

3. Explain how rough periwinkles avoid desiccation.

4. Describe some of the adaptations exhibited by organisms inhabiting rocky coasts that help them survive wave shock.

5. Describe the vertical zonation of a sandy beach.

## Thinking Critically

1. Do you think pollutants entering from the ocean would have a greater effect on rocky shores or on sandy shores? Explain your answer.

2. What abiotic factor is probably most important in terms of influencing the number of organisms that can inhabit a rocky shore?

3. What kinds of organisms that inhabit sandy shores would be least affected by the recreational use of beaches?

## SUGGESTIONS FOR FURTHER READING

Bertness, M. D. 1999. *The Ecology of Atlantic Shorelines.* Sunderland, Mass.: Sinauer Publishing.

Chadwick, Douglas. 1997. What Good Is a Tidepool? *Audubon* 99 (May–June):50–59.

Connell, J. H. 1961. The Influence of Interspecific Competition and Other Factors on the Distribution of the Barnacle *Chthamalus stellatus, Ecology* 42:710–23.

Fenchel, T. M. 1978. The Ecology of Micro- and Meiobenthos. Annual Review of Ecology and Systematics 9:99–121.

Giménez, Luis, and Beatriz Yannicelli. 1997. Variability of Zonation Patterns in Temperate Microtidal Uruguayan Beaches with Different Morphodynamic Types, Marine Ecology Progress Series 160:197–207.

Horn, M. H., K. L. M. Martin, and M. A. Chotkowski, eds. 1999. *Intertidal Fishes: Life in Two Worlds.* San Diego, Calif.: Academic Press.

Kaplan, E. U. 1988. *A Field Guide to Southeastern and Caribbean Seashores.* Boston, Mass.: Houghton Mifflin.

Lang, R. C., J. C. Britton, and T. Metz. 1998. What To Do When There Is Nothing To Do: The Ecology of Jamaican Intertidal Littorinidae (Gastropoda: Prosobranchia) in Repose, *Hydrobiologia* 378:161–85.

McLachlan, A., and E. Jaramillo. 1995. Zonation on Sandy Beaches, *Oceanography and Marine Biology: An Annual Review, London* 33:305–35.

Menge, B. A. 2000. Top-Down and Bottom-Up Community Regulation in Marine Rocky Intertidal Habitats. *Journal of Experimental Marine Biology and Ecology* 250:257–89.

Menge, B. A., and G. M. Branch, 2001. Rocky Intertidal Communities. In *Marine Community Ecology,* ed. M. D. Bertness, S. D. Gaines, and M. E. Hay, pp. 221–51. Sunderland, Mass.: Sinauer Publishing.

Menge, B. A., and J. Lubchenco. 1981. Community Organization in Temperate and Tropical Rocky Intertidal Habitats: Prey Refuges in Relation to Consumer Pressure Gradients, *Ecological Monographs* 51:429–50.

Winston, J. E. 1990. Intertidal Space Wars, *Sea Frontiers* 36(1):46–51.

### InfoTrac College Edition Articles

Helmuth, Brian S. T., and Gretchen E. Hofmann. 2001. Microhabitats, Thermal Heterogeneity, and Patterns of Physiological Stress in the Rocky Intertidal Zone, *The Biological Bulletin* 201(3).

Menge, Bruce A., Bryon A. Daley, Jane Lubchenco, Eric Sanford, Elizabeth Dahlhoff, Patricia M. Halpin, Gregory Hudson, and Jennifer L. Burnaford. 1999. Top-Down and Bottom-Up Regulation of New Zealand Rocky Intertidal Communities, *Ecological Monographs* 69(3).

Leonard, George H., Mark D. Bertness, and Philip O. Yund. 1999. Crab Predation, Waterborne Cues, and Inducible Defenses in the Blue Mussel, *Mytilus edulis, Ecology* 80(1).

Dial, Roman, and Jonathan Roughgarden. 1998. Theory of Marine Communities: The Intermediate Disturbance Hypothesis, *Ecology* 79(4).

Navarette, Sergio A., and Bruce A. Menge. 1996. Keystone Predation and Interaction Strength: Interactive Effects of Predators on Their Main Prey, *Ecological Monographs* 66(4).

Navarette, Sergio A. 1996. Variable Predation: Effects of Whelks on a Mid-Intertidal Successional Community, *Ecological Monographs* 66(3).

### Websites

**http://www.nps.gov/olym/edtide.htm** Facts about intertidal life.

**http://life.bio.sunysb.edu/marinebio/rockyshore.html** General information and great photos of rocky shores.

**http://bonita.mbnms.nos.noaa.gov/sitechar/rocky.html** Information on rocky intertidal habitats of the Monterey Bay National Marine Sanctuary.

# 14

# Estuaries

## Key Concepts

1. Estuaries form in embayments where freshwater from rivers and streams mixes with seawater.

2. The salinity of water in estuaries varies both vertically and horizontally.

3. Mixing of nutrients from saltwater and freshwater, combined with plentiful sunlight and relatively shallow water, makes estuaries very productive ecosystems.

4. Animals and plants that live in estuaries must be able to adapt to changing salinity.

5. The physical characteristics of estuaries tend to favor benthic organisms.

6. Many commercially valuable fishes and shellfishes spend a portion of their life cycle in estuaries.

7. Estuarine communities include oyster reefs, mud flats, seagrass meadows, salt marshes, and mangrove forests (mangals).

Estuaries form where rivers meet the sea. They are regions of constant environmental change, and although many organisms cannot survive under these conditions, some have adapted quite well and thrive. The nutrients available in freshwater and saltwater complement each other and, when mixed in the relatively shallow, sunlit waters of estuaries, promote high levels of primary production. Animals that can survive the changes in salinity and temperature and the occasional exposure to air that occur in estuaries can take advantage of this high productivity, grow rapidly, and produce large populations.

Estuaries are important to marine ecosystems for a number of reasons. Much of the organic material produced there is exported to, and enriches, adjacent ocean waters. Many fishes and shellfishes of coastal waters spend a part of their lives in estuaries, and the juveniles of many species seek protection from predators in them.

Estuaries also support many commercially important animals that humans rely on for food. Many species of oyster, crab, scallop, and shrimp spend a part or all of their lives in estuaries. Estuaries are nurseries for flounder, fluke, bluefish, tarpon, striped bass, several species of herring, and other fish species. The economic value of estuaries cannot be overstated. About 85% of the fishes and shellfishes that are sold in the commercial markets of the world spend all or part of their lives in estuaries.

Unfortunately, these are also fragile habitats, and the consequences of polluting them are far-reaching. Estuaries are also damaged by dams that block the supply of freshwater. Dams are a major problem along the Gulf of Mexico, where they alter the salinity of estuaries and kill many estuarine organisms.

# PHYSICAL CHARACTERISTICS OF ESTUARIES

Along coastlines there are areas where portions of the ocean are partially cut off from the rest of the sea. These coastal areas are called **embayments** (Figure 14-1). Rivers and streams carry freshwater runoff from the land into some of these embayments. The freshwater mixes with ocean saltwater and forms an **estuary.** All estuaries are partially isolated from the sea by land and diluted by freshwater. These physical characteristics produce a unique habitat inhabited by a variety of hardy organisms that have adapted to the rigors of estuarine life.

## Types of Estuaries

The characteristics of estuaries, such as size, shape, and water flow, can vary greatly depending on the geology of the region where they occur. **Coastal plain,** or **drowned river valley, estuaries** (Figure 14-2a) form between glacial periods, when water from melting glaciers raises the sea level and floods coastal plains and low-lying rivers. The Gulf of Mexico and eastern Atlantic exhibit several examples of coastal plain estuaries. Chesapeake Bay and Long Island Sound are examples of drowned river valleys. Estuaries like the San Francisco Bay were created when earthquakes caused the land to sink, allowing seawater to cover it. This type of estuary is called a **tectonic estuary** (see Figure 14-

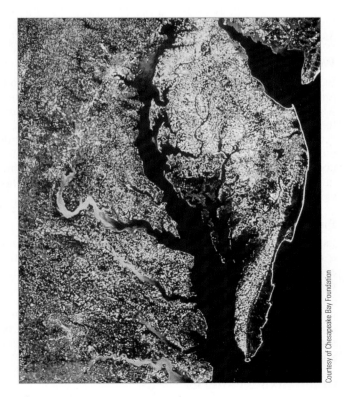

**Figure 14-1 Embayment.** Chesapeake Bay is an example of embayment and an estuary.

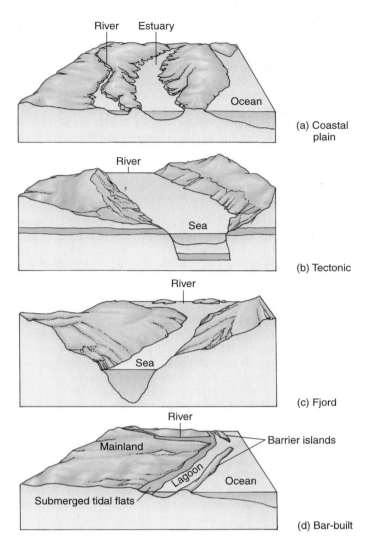

**Figure 14-2 Types of Estuaries.** (a) Coastal plain, or drowned river valley, estuaries form when rising water levels from melting glaciers flood coastal plains. (b) Tectonic estuaries form when geological events, such as an earthquake, cause the land to sink below sea level, allowing seawater to cover it. (c) Fjords form when glaciers carve large valleys in coastal areas. (d) Bar-built estuaries form when geographical barriers, such as an island, form a wall between freshwater and saltwater.

2b). During the last glacial period, glaciers cut deep valleys into some coasts. After the glaciers retreated, these valleys filled with water and formed a type of estuary known as a **fjord** (FYORD; see Figure 14-2c). Spectacular fjords are in Alaska and along the coasts of Scandinavia.

As rivers and streams flow to the sea, they carry along sediments that are ultimately deposited at the mouth of the river. As these sediments accumulate, they form deltas in the upper part of the river mouth, shortening the estuary. **Tidal flats** develop when enough sediment accumulates to be exposed at low tide. The original channel of the estuary is then divided by these tidal flats. At the same time, currents and tides erode the coastal area and deposit sediment on the seaward side of the estuary. When more sediment is deposited than is carried away, barrier islands, beaches, and brackish-water (dilute saltwater) lagoons form. These islands and other geographical barriers form a wall between the freshwater from the rivers and the saltwater from the oceans, forming **bar-built estuaries** (see Figure 14-2d). The Cape Hatteras region of North Carolina, the Texas and Florida Gulf Coasts, the Indian River complex of Florida's east coast, and the coast of northern Europe are all good examples of this type of estuary.

## Salinity and Mixing Patterns

As mentioned in Chapter 4, seawater has an average salinity of approximately 35‰. By comparison, the salinity of freshwater ranges from 0.065‰ to 0.30‰. The concentration of ions in river water also varies from one river drainage to the next and affects the chemistry and salinity of the water in estuaries. Although the quantity of dissolved salt in an estuary is about the same as that in seawater, it is distributed in a gradient that increases from the freshwater to the ocean.

The salinity in estuaries varies both vertically and horizontally. The least-salty waters are located near the mouth of the river where it joins the sea. As one moves farther out to sea, the salinity tends to increase. Salinity can be uniform, or it can be layered. Uniform salinity results when currents are strong enough to thoroughly mix the freshwater and saltwater from top to bottom. In some estuaries, the vertical salinity is uniform during low tide, but at high tide the seawater at the surface moves upstream more quickly than the bottom water. The denser seawater at the surface tends to sink as the lighter freshwater beneath it rises, creating a mixing action from the surface to the bottom, a phenomenon called **tidal overmixing**. Strong winds can also mix freshwater and saltwater. Normally, though, freshwater flows seaward over the seawater moving upstream.

In most estuaries the influx of freshwater from the river more than replaces the amount of water lost to evaporation. As a result the surface water is less dense and flows out to sea, whereas the denser saltwater from the ocean moves into the estuary along the bottom. An estuary with this pattern of circulation is called a **positive estuary**. Most estuaries are positive estuaries. Estuaries in hot, arid regions, however, can lose more water through evaporation than the river is able to replace. This type of estuary is called a **negative estuary**. A negative estuary has a flow opposite that of a positive estuary. The surface water flows toward the river and the water along the bottom moves out to sea. This pattern of circulation occurs because evaporation increases the salinity of the surface water. Some negative estuaries may also experience seasonal or nighttime cooling that increases the density of the water at the surface. The denser surface water sinks and returns to sea along the bottom. The surface water lost to evaporation is replaced by water entering the estuary from the sea. This type of flow pattern traps few nutrients in the estuary, and because the surface water that is drawn in from the ocean is generally nutrient poor, negative estuaries are usually low in productivity. The Laguna Madre estuary in Texas is an example of a negative estuary.

The pattern of water circulation and vertical distribution of salinity are important characteristics of estuaries. On the basis of these mixing patterns, estuaries can be classified into the following types: salt-wedge estuaries, well-mixed estuaries, and partially mixed estuaries.

### Salt-Wedge Estuary

**Salt-wedge estuaries** (Figure 14-3a) occur in the mouths of rivers that are flowing into saltwater. At the surface, freshwater flows rapidly out to sea, whereas at the bottom the denser saltwater flows upstream along the river bottom. The rapid flow of the river prevents saltwater from entering and produces an angled boundary between the freshwater moving downstream and the seawater moving upstream called a **salt wedge**. When the tide rises or the river flow decreases, the salt wedge moves upstream. When the tide falls or the river flow increases, the salt wedge moves downstream. The freshwater normally moves rapidly enough to create a sharp boundary between the freshwater and seawater. The moving freshwater takes saltwater from the face of the salt wedge and mixes it upward with the river water. This action increases the salinity of the surface water that is moving out to sea. Saltwater from the sea is constantly replacing the water removed from the salt wedge, so the salt wedge does not become smaller. Very little of the river water mixes downward with the saltwater; thus the mixing of saltwater with freshwater is essentially a one-way process. In this type of estuary, the circulation and mixing of the water is controlled by the flow rate of the river and tidal currents play only a small role. Some examples of salt-wedge estuaries are at the mouths of the Mississippi, Amazon, and Congo Rivers and at the mouth of the Sacramento River in San Francisco Bay.

### Well-Mixed Estuary

In a **well-mixed estuary** (see Figure 14-3b), river flow is low and tidal currents play a major role in the circulation of the water. The net result is a seaward flow of water and a uniform salinity at all depths. The salinity of the water decreases as it approaches the river. Lines of constant salinity move toward land when the tide rises or when river flow decreases. On the other hand, when the tide falls or river flow increases, the lines of salinity move toward the sea. Delaware Bay is an example of a well-mixed estuary.

### Partially Mixed Estuary

Estuaries that have a strong surface flow of freshwater and a strong influx of seawater are called **partially mixed estuaries** (Figure 14-3c). Tidal currents force the seawater upward, where it mixes with the surface water, producing a seaward flow of surface water. This system of circulation produces a rapid exchange of surface water between the estuary and the ocean. Salinity is increased by the influx of seawater. Chesapeake Bay, San Francisco Bay, and Puget Sound are examples of partially mixed estuaries.

### Other Mixing Patterns

The mixing patterns of estuaries do not always fit neatly into one of the above categories. Some estuaries are inter-

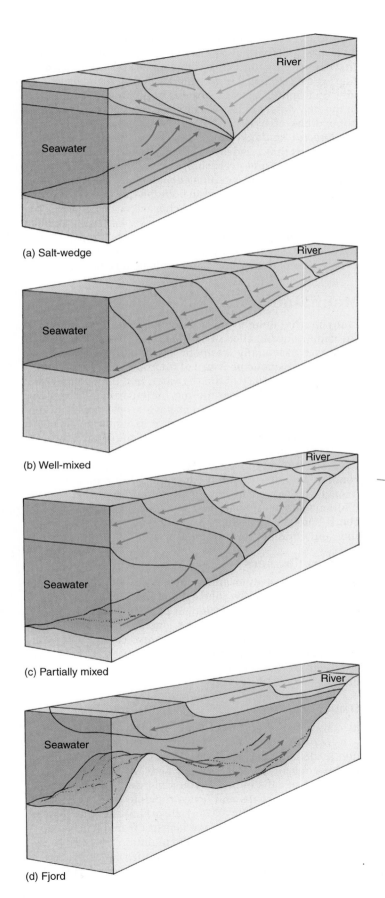

(a) Salt-wedge

(b) Well-mixed

(c) Partially mixed

(d) Fjord

mediate between two of the types we have discussed. Others change from season to season as rainfall, tides, and winds alter the volume of water and strength of currents, as in Galveston Bay, Texas. In **fjords,** river water remains at the surface and moves seaward, mixing little with the saltwater beneath (Figure 14-3d). Salinity increases slowly in the estuary with the slow influx of ocean water.

## Temperature

Salinity is not the only environmental factor that varies in an estuary. Temperature is also important. Because estuaries are relatively shallow, the water temperature changes rapidly with changes in air temperature. Temperatures in estuaries can fluctuate dramatically seasonally and even daily. For instance, in northern temperate regions, water temperature may range from almost 22°C to 30°C, and portions of estuaries may freeze during the winter.

Like other bodies of water, estuaries are heated by the sun. Depending on the season, tidal currents warm or cool an estuary. In some estuaries, the surface water is cooler in the winter and warmer in the summer than deeper water, as a result of the amount of solar energy reaching the surface. This situation produces a winter turnover in which cooler (denser) surface water sinks and is replaced by warmer (less dense) deeper water. Without vertical mixing, nutrients would be swept back out to sea with the next tide. The mixing that occurs because of turnover circulates nutrients vertically between the water and bottom sediments.

## In Summary

Estuaries form in embayments where freshwater from rivers and streams mixes with seawater. The characteristics of estuaries vary with the geology of the regions in which they occur. The mixing of saltwater and freshwater and the distribution of salinity are important characteristics of estuaries. Salt-wedge estuaries have a sharp boundary between freshwater and saltwater. In well-mixed estuaries there is uniform salinity at all depths and a gradient of decreasing salinity as one proceeds toward the source of freshwater. Partially mixed estuaries occur where there is a strong surface flow of freshwater and a strong influx of seawater. Fjords have very little mixing between surface freshwater and deeper saltwater. Many estuaries exhibit a combination of mixing patterns or have mixing patterns that vary with the season.

**Figure 14-3 Mixing Patterns in Estuaries.** Green represents low-salinity freshwater, and blue represents high-salinity saltwater. The darker the blue, the higher the salinity of the water. The arrows indicate the direction of water flow. (a) In salt-wedge estuaries, less-dense freshwater from a river flows rapidly to sea over the denser, slower-moving saltwater flowing upstream. This produces an angled boundary (salt wedge) between the two. (b) In well-mixed estuaries, river flow is low and tidal currents mix freshwater with saltwater, producing uniform salinity at all depths. (c) Partially mixed estuaries have a strong surface flow of freshwater and a strong influx of seawater. (d) In a fjord the river water remains at the surface and mixes very little with the seawater beneath it.

The salinity of estuaries varies both vertically and horizontally. In some, tidal overmixing occurs when denser seawater at the surface sinks and is replaced by less-dense freshwater. Estuaries also exhibit fluctuations in temperature. In some, seasonal temperature change promotes a turnover that circulates nutrients.  ●

# ESTUARINE PRODUCTIVITY

Fishers and those who depend on the sea for food have long known how productive estuaries are. Part of the reason for their high productivity is the mixing of freshwater and saltwater that occurs there. Freshwater runoff from the land often contains nitrogen, phosphorus, and silica. Water at the surface of the open sea often has less nitrogen and silica but more phosphorus and other nutrients. Individually, the two types of water can support only a limited variety of plant and algal growth because of the lack of one or more essential nutrients. When strong mixing currents in the shallow, sunlit waters of an estuary combine the two, however, their nutrients are complemented and high levels of primary production can occur.

Once nutrients enter an estuary, several factors hold them there in forms usable by plants, algae, and animals. Rivers dump loads of silt and clay as they meet the sea. These particles readily absorb any excess nutrients from the surrounding water and release them back to the water when nutrients are in short supply. In this way, the silt and clay are a sort of nutrient buffer, helping to maintain a more or less constant nutrient level in an estuary.

The activities of some estuarine animals also contribute to keeping nutrients in an estuary. When there are large amounts of phytoplankton in an estuary, filter feeders, such as bivalves, remove more phytoplankton from the water than they are able to digest. The excess phytoplankton are eliminated in large, semisolid particles called **pseudofeces.** Pseudofeces contain substantial amounts of nutrients and are relatively large, so they can be easily manipulated by other organisms. Thus the normal feeding behavior of bivalves packages and stores nutrients for use by bottom feeders like gastropod molluscs and polychaete worms.

## In Summary

The mixing of nutrients from freshwater and seawater that occurs in estuaries, along with plentiful sunlight, makes these areas some of the most productive in the marine environment. Nutrients are trapped in estuaries by silt and clay that is brought in by rivers and streams and by the action of some organisms, for instance, the formation of pseudofeces by bivalves.  ●

# LIFE IN AN ESTUARY

Because of the relatively harsh conditions, estuaries contain fewer resident species than the nearby marine or freshwater ecosystems, resulting in less competition for food and space. Because there is less competition, many estuarine species tend to be generalists. That is, they are able to feed on a variety of foods depending on what is available. Species that can tolerate the salinity and temperature changes in estuaries can exploit the area's high productivity, grow rapidly, and multiply into enormous populations. At one time in the rich waters of the Chesapeake Bay, for instance, large populations of oysters (family Ostreidae) and blue crabs (*Callinectes sapidus*) literally carpeted the bottom. Today populations that large are seen infrequently because of pollution, disease, and overfishing. Great South Bay along the coast of Long Island contains enough clams to support New York's largest single fishery. In both the Chesapeake Bay and Great South Bay instances, the characteristics of the populations are typical of those in estuaries. Estuarine populations contain large numbers of individuals belonging to relatively few species, and the dominant animals are those with relatively broad tolerances and ecological requirements.

## Maintaining Osmotic Balance

Many marine animals have body fluids that contain about the same concentration of salts as seawater, and their body fluids are essentially isosmotic to the surrounding water. That is, the osmotic pressure of their body fluids is equal to the osmotic pressure of the seawater, and they neither gain nor lose water. Because the marine environment remains relatively constant, they do not have a problem maintaining water balance. Animals that live in estuaries, however, must have some physiological mechanism for dealing with the varying salinity. Otherwise, their tissues and cells would absorb water and lose salts as they encountered an environment with lower salinity than the sea. Animals that live in estuaries survive by having tissues and cells that tolerate dilution (**osmoconformers**) or maintain an optimal salt concentration in their tissues, regardless of the salt content of their environment (**osmoregulators**).

### Osmoconformers

Animals such as tunicates, jellyfishes, and sea anemones are unable to actively adjust the amount of water in their tissues. When their environment becomes less saline, their body fluid gains water and loses ions until it is isosmotic to the surroundings. These organisms are examples of osmoconformers (Figure 14-4a). The ability of osmoconformers to inhabit estuaries is limited by their tolerance for changes in their body fluid.

### Osmoregulators

In contrast to osmoconformers, osmoregulators employ a variety of strategies to maintain a constant salt concentration in their bodies. Osmoregulators that live in estuarine waters concentrate salts in their body fluids when the concentration of salts in the surrounding water decreases. For instance, some crabs and fishes regulate their salt content in less-saline water by actively absorbing salt ions through the gills to compensate for salt ions lost from their body (see Figure 14-4b). This helps them to maintain a relatively constant body fluid. Some animals can either concentrate salts when their environment is less saline or excrete salts when

(a)

(b)

**Figure 14-4 Osmoconformers and Osmoregulators.** (a) These tunicates are osmoconformers, losing and gaining water and ions until their body fluid is isosmotic to the environment. (b) This hermit crab is an osmoregulator, actively adjusting the concentration of salts in its body fluid when the concentration in the surrounding water changes.

the environment is hypersaline. The latter are generally animals that are semiterrestrial, such as some crustaceans, or that live in areas such as salt marshes and mangrove swamps that occasionally receive large amounts of rain. Other animals, such as the blue crab and polychaete worms in the genus *Nereis,* are osmoregulators at lower environmental salinity and osmoconformers at higher environmental salinity. Many fish species are osmoregulators that can adjust to both high-salt and low-salt environments.

Some estuarine organisms wall themselves off from their external environment to decrease water and salt exchange with their surroundings. Many estuarine animals have a body surface that shows a decreased permeability compared with purely marine forms. This decreased permeability can be the result of increased amounts of calcium in the exoskeleton, as in arthropods, or increased numbers of mucous glands in the skin. Structural adaptations, such as the operculum of a snail, can be used to isolate the body surface from the environment when necessary to prevent salt and water loss or gain.

## Remaining Stationary in a Changing Environment

In addition to changes in salinity, the problem of remaining stationary in a changing environment affects the distribution of organisms in estuaries. The more or less constant movement of water in an estuary makes it difficult for some organisms to remain stationary long enough to feed and carry on other vital functions. Because of this, natural selection favors those organisms that are benthic. Marine plants and algae in estuaries have substantial root systems or holdfasts to prevent moving water from pulling them up and carrying them out to sea. Animals live attached to the bottom, either in the available spaces around other sedentary animals and plants or buried in the small crevices between sediment particles.

Of the nonbenthic animals, crustaceans and fishes are the most dominant, especially their young. These animals usually spawn in the seawater offshore and then spend a portion of their development in the estuary, which is a more protected habitat than the open sea. These animals maintain their position in the estuary by actively swimming or by moving back and forth with the movement of the tides.

## Estuaries as Nurseries

Although estuaries are challenging habitats for many animals, they provide excellent habitats for the juveniles and young of many species to grow and develop. The high level of nutrients and lower number of predators allow juveniles and young to attain a size or stage that gives them a better chance of survival in the open sea.

The striped bass (*Morone saxatilis;* Figure 14-5), for instance, spawns at the border of freshwater and water of low salinity. As the larvae and young mature, they move downstream toward water with higher salinity. The shad (*Alosa*) is an anadromous species that spawns in freshwater but spends its adult life in the marine environment. Young shad spend their first summer in an estuary feeding and growing before moving out to the open ocean. Species such as the croaker (family Sciaenidae) spawn at the mouth of an estuary, and then their young move upstream to feed in plankton-rich, low-salinity water. In the eastern United States, young bluefish (*Pomatomus saltatrix*) come to estuaries to feed but spend the rest of their time in the ocean. Many species of shellfish, such as blue crabs and white shrimp (*Penaeus*), also spend a part of their life cycle developing in the relatively protected waters of estuaries.

## In Summary

Animals that live in estuaries must be able to adapt to changing salinity. Osmoregulators use a variety of physiological mechanisms to maintain optimum salt concentrations in their tissues regardless

## Predation for Regulating Benthic Population Size

Robert Virnstein was interested in the population dynamics of benthic communities. He wanted to know whether physical environment, competition, predation, or some other factor most influenced the size of infaunal populations in estuaries. While doing fieldwork in the Chesapeake Bay estuary, Dr. Virnstein observed that populations of infauna living in shallow-water, subtidal communities contained fewer animals in areas that lacked any vegetation for cover compared with similar areas nearby that were covered by vegetation. He reasoned that in the vegetated areas predators, such as crabs, would have difficulty digging through the rhizome mat of submerged grasses as they searched for food, so the animals in those habitats would be more protected. He also observed that the densities of infaunal populations were lowest in summer and fall when predators that feed on bottom-dwelling prey, such as crabs and fish, have been feeding and highest in winter and spring when the predators were either absent or inactive.

On the basis of these observations, Dr. Virnstein hypothesized that large motile predators played a significant role in controlling the number of infaunal animals in this community. To test his hypothesis, Dr. Virnstein used wire-mesh cages to either exclude predators from or confine predators to a small area of estuary bottom. The specific predators that he studied were two fishes, the hogchoker (*Trinectes maculatus*) and the spot (*Leiostomus xanthurus*), and the blue crab (*Callinectes sapidus*). He selected the two fish species because they feed on the bottom and were abundant in the area he was studying. He selected the blue crab because it is known to feed heavily on infauna. For his study he selected a shallow, sandy, unvegetated area of bottom in the lower York River that he thought was representative of the Chesapeake Bay and its estuaries. He marked off 50 × 50 centimeter plots spaced 3 meters apart. Some plots were covered with wire cages to keep predators out, whereas others were covered with wire cages containing one to four predators.

After 2 1/2 weeks he took samples from each of the plots and counted the number of infaunal animals and the number of different species in each and compared his results to areas that had not been enclosed by a wire cage. He found that the number of species had not changed, but the number of animals in areas protected from predators was significantly greater than areas that were not enclosed by cages and significantly less in the caged areas that contained crabs and spot. The plots were sampled again at the end of 2 months and the same measurements were made. This time the differences were even more striking. As before, the cages that excluded predators showed large increases in the numbers of infauna compared with areas that were not enclosed by a cage. The caged areas containing crabs had the least infauna. The cages containing the hogchoker showed the same results as cages containing no predators, indicating that the hogchokers had little controlling influence on the infauna. The cages containing the spot had results that were intermediate between crabs and hogchokers. The infauna that lived near the surface were most affected by predation. Those species that could bury deeply into the sediments, and thus avoid predation, showed few significant differences.

Based on the results of his experiments, Dr. Virnstein concluded that, although the physical stresses of the estuarine environment may be severe, they are not major factors limiting natural population densities of this community. In the York River community, densities of most species increased when protected from predators and no species decreased in density, suggesting a lack of competitive exclusion. He concluded that competitive pressures are not very important in the regulation of population densities in this community and that predation plays the major role. ●

of the salinity of their surroundings. Osmoconformers have tissues and cells that can tolerate changes in salinity. Remaining stationary is another challenge for organisms in an estuary. The characteristics of estuaries tend to favor benthic organisms. Motile organisms must actively work to maintain position or move in and out with the tides.

Because estuaries are highly productive and relatively protected from wave action, they make good nurseries for the juveniles and young of many species. Many important commercial fishes and shellfishes spend at least a portion of their life cycle in the protected waters of an estuary. ●

## ESTUARINE COMMUNITIES

Estuarine communities contain hardy organisms able to tolerate the changeable physical environment in which they live. Many of these organisms are **euryhaline** species. These are species that can tolerate a broad range of salinity such as that in an estuary. Some of the communities in estuaries include oyster reefs, mud flats, seagrass meadows, salt marshes, and mangrove swamps.

**Figure 14-5  Striped Bass.** Some species, such as the striped bass (*Morone saxatilis*), spawn and complete the early stages of their development in estuaries where they can take advantage of the available food and relative protection from large predators.

## Oyster Reefs

A prominent mollusc in the intertidal zone of temperate estuaries is the oyster (family Ostreidae; Figure 14-6a). Like other bivalves, oysters have a free-swimming larval stage in their life cycle that allows these sessile animals to disperse to other areas (see Figure 14-6b). Oyster larvae attach themselves to any solid surface and form extensive oyster beds,

or large reefs composed of oysters growing on the shells of previous generations. Oyster reefs are usually oriented at right angles to tidal currents and occur generally at the point of lower salinity. The currents bring food to the oysters and carry away their waste. Tidal currents also play a role in clearing sediment from the oysters. If the sediments were allowed to accumulate, they would suffocate the oysters.

Oyster reefs provide habitat for a variety of other organisms, including algae, sponges, hydrozoans, bryozoans, polychaetes, molluscs, echinoderms, and barnacles. Many of these organisms depend on the oysters not only for protection and a surface for attachment but also for food. The oyster drill snail (*Urosalpinx*) preys upon the sedentary oysters, drilling through their shells with its radula and feeding on the contents (Figure 14-7). The veliger larvae of oyster drills are quite sensitive to changes in salinity and are more affected by freshwater than adults. Short periods of low salinity kill predatory oyster drills and disease-causing organisms in the oyster beds. However, rapid changes in salinity as the result of prolonged rainfall or hurricanes kill off large numbers of both oysters and oyster drills, as well as the organisms that grow with them.

## Mud Flats

Mud flats are found in bays and around the mouths of rivers wherever the land is protected from wave action. Mud flats contain rich deposits of organic material mixed with small inorganic sediment grains. Detritus from nearby communities and nutrients carried in from the sea by tides contribute to the rich food reserves. Bacteria and other microorganisms thrive in the mud and produce a variety of sulfur-containing gases that give mud flats a characteristic odor of

(a)

**Figure 14-6  Oysters.** (a) Oysters form extensive beds composed of thousands of individuals, providing a habitat for many other species. (b) Life cycle of the commercial oyster *Crassostrea virginica*. Adult oysters shed their sperm and eggs into the water column, where the eggs are fertilized. The fertilized egg undergoes cell division, forming an embryo that develops into a free-swimming trochophore larva. The trochophore becomes a veliger larva with a small, hinged shell. Within 4 weeks, the veliger develops a foot, settles to the bottom, and cements its left valve to a solid surface. The veliger then develops into an immature oyster called a *spat*. The spat grows and develops into an adult oyster.

(b)

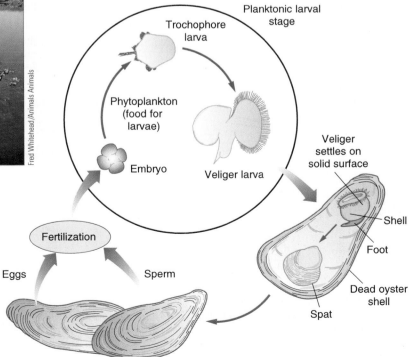

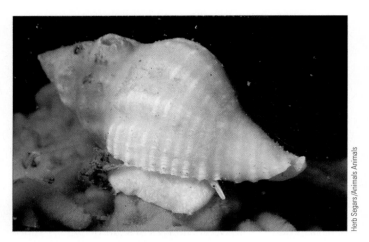

**Figure 14-7  Oyster Drill.** Oyster drill snails can cause extensive damage to oyster beds by feeding on the sedentary oysters.

ter consisting of the decaying remains of local organisms and organic material deposited during high tides. A large amount of this organic matter is channeled to other organisms as the result of bacterial decomposition. The action of bacteria is also important in recycling nutrients such as nitrogen and phosphate back to the sea (for more information on the role of bacteria in recycling see Chapter 6). Bacteria in mudflats are not just producers and decomposers but also primary consumers that serve as a food source for many higher-level consumers. Many deposit-feeding organisms, such as nematodes, polychaetes, some gastropods, and arthropods, ingest organic material to feed on the bacteria that it contains. A variety of small organisms, including filter-feeding animals and the larvae and juveniles of many animal species, also rely on these rich sources of food. The smaller organisms, in turn, are food for larger invertebrates, fishes, birds, and even some terrestrial carnivores such as raccoons.

## Animals of the Mud Flats

Most animal inhabitants of mud flats are burrowing organisms (Figure 14-9). They live just beneath the surface, where they avoid predators and exposure to the drying air during low tides. Unlike sandy beaches, mud flats are not very porous. The silt packs closely together and interferes with the circulation of water that is necessary for carrying oxygen through the sediments. As a result, burrowing animals that exchange gases through their skins generally cannot survive in this habitat unless they circulate water through their burrows or tubes or maintain a "snorkel" connection to the surface.

One common resident of the mud flats of North America is the soft-shelled clam (*Mya arenaria*). Like other bi-

rotten eggs. Mud provides good mechanical support for animals of the flats, many of which have very thin shells or soft bodies. Mud is also cohesive, permitting the construction of a permanent burrow. Sand is frequently mixed with the mud, making it softer and providing a much better bottom material for burrowing organisms.

## Mud Flat Food Webs

The producers on mud flats consist of photosynthetic bacteria, chemosynthetic bacteria, phytoplankton such as diatoms and dinoflagellates, and in some cases, large algae such as the green algae *Ulva* and *Enteromorpha* (Figure 14-8). The main energy base, however, is organic mat-

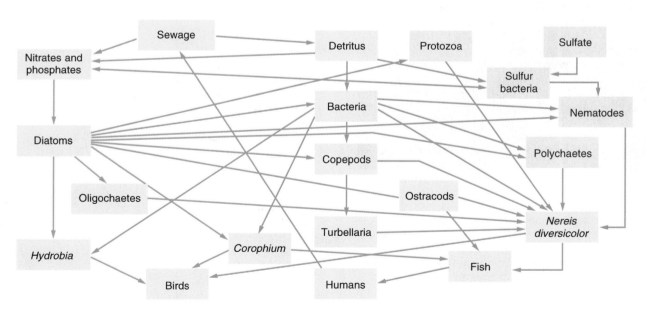

**Figure 14-8  Mud Flat Food Webs.** The energy base of mud flat food webs is usually detritus. A large amount of this organic matter is channeled to other organisms as the result of bacterial decomposition.

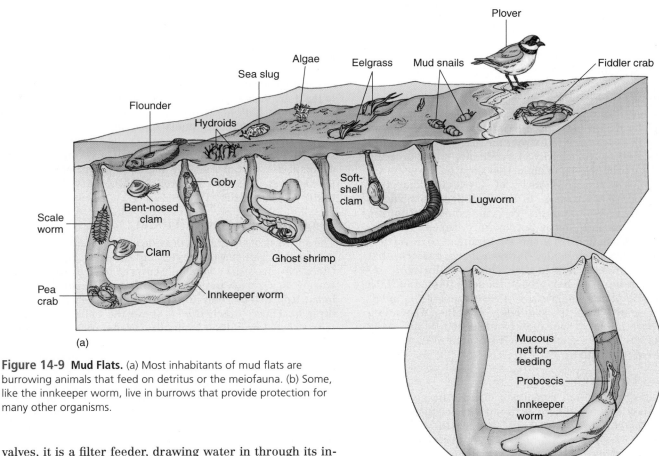

**Figure 14-9 Mud Flats.** (a) Most inhabitants of mud flats are burrowing animals that feed on detritus or the meiofauna. (b) Some, like the innkeeper worm, live in burrows that provide protection for many other organisms.

valves, it is a filter feeder, drawing water in through its incurrent siphon and removing oxygen and planktonic food from it. Wastes are carried away by the water expelled from its excurrent siphon. When the tide retreats, the soft-shelled clam withdraws its siphon and metabolizes anaerobically. In laboratory experiments, soft-shelled clams have survived as long as 8 days without oxygen.

Lugworms (*Arenicola*) are common residents of mud flats worldwide. Concentrations as high as 820,000 individuals per acre have been reported from some areas. Another burrowing worm, found in California mud flats, is the innkeeper worm (*Urechis*). This worm may be 30 centimeters (12 inches) long with a fat, pink, sausage-shaped body. Like lugworms, it lives in a ∪-shaped burrow through which it continuously pumps water to supply oxygen and remove waste. It feeds by producing a mucous net that traps tiny food particles. When the net is full, it is drawn into the animal's mouth and digested and another net is produced to take its place. The innkeeper worm derives its name from the variety of organisms that live with it in its burrow. Some species, such as the small red scaleworm (family Polynoidae) are commensal symbionts in the burrows of innkeeper worms. The scaleworm maintains almost constant contact with the innkeeper worm and feeds on discarded food particles and sometimes on the mucous net. Other common symbionts in the innkeeper's burrow are tiny pea crabs (family Pinnotheridae) that may compete with scaleworms for bits of food and the clam *Cryptomya californica*. Tiny fish

called gobies (family Gobiidae) also inhabit the burrows but do not compete with the other organisms for food. Instead, when pieces of food are too large for their own use, gobies have been observed taking them to the pea crabs.

Innkeeper worms are not the only burrowing organisms on mud flats to take in boarders. The ghost shrimp (*Callianassa*) burrow is composed of a vertical shaft connected to a number of lateral passageways. The lateral tunnels branch repeatedly, sometimes widening to allow the animal to turn around. Within this burrow the shrimp may house small clams, pea crabs, several species of marine worm, and gobies. The ghost shrimp digs its tunnels by scooping up the sediments with its first pair of legs and storing the material in a pouch formed by the fleshy appendages around its mouth. When the pouch is full, the shrimp takes it to the opening of the burrow and dumps it. Ghost shrimp and mud shrimp play an important role in oxygenating the sediments.

## Seagrass Meadows

Stretching seaward from many sand flats and sandy mud flats along the coasts of North and South America, Europe, Asia, and Australia are vast expanses of seagrasses (for more information on seagrasses see Chapter 7). These plants grow in sediments unable to support the holdfasts of large marine algae and thrive in protected waters from the low tide zone to a depth of about 6 meters (20 feet), where they form large underwater meadows.

## Seagrass Productivity

Seagrass meadows can be highly productive communities. Their high level of primary production, however, depends on the ability of seagrasses to extract nutrients from the sediments, as well as the activity of symbiotic, nitrogen-fixing bacteria, the productivity of algae that grow on the seagrass (epiphytes), and the productivity of the green algae that grow among the seagrasses. In the Caribbean, seagrasses such as turtlegrass (*Thalassia*) grow in sediments that are typically high in phosphorus but low in nitrogen. The seagrasses are able to flourish in this environment because they get nitrogen from symbiotic, nitrogen-fixing bacteria that live in the sediments and in their rhizomes (for more information on nitrogen-fixing bacteria see Chapter 6). Like other aquatic plants, seagrasses release some of the nutrients they remove from the sediments into the surrounding water. These nutrients can then be absorbed by the many species of epiphytic algae and used by them to maintain a high level of productivity.

## Seagrass Food Webs

Most seagrasses are not consumed directly by herbivores because they are too tough. In temperate regions where eelgrass (*Zostera*) dominates, waterfowl are the major seagrass herbivores, but they often consume relatively little, less than 5% of the total seagrass production. In the Caribbean, green sea turtles, certain parrotfishes, some sea urchins, and manatees feed extensively on seagrasses. In the case of the green sea turtle and the bucktooth parrotfish (*Sparisoma radians*), about 90% of their diet consists of turtlegrass (Figure 14-10). Seagrasses respond to the graz-

ing by producing more shoots to replace those that are eaten by herbivores. Although few animals feed directly on seagrasses, they provide a substantial food source for many organisms in the form of detritus. The dead leaves and other plant parts provide a food base for bacteria, crabs, brittle sea stars, sea cucumbers, and deposit-feeding worms. Some of the detritus is exported to nearby communities. Many species of snail, both herbivores such as conchs and carnivores such as whelks and tulip snails, thrive in Caribbean and temperate seagrass communities. As snails fall prey to carnivores, their shells become available for a variety of hermit crab species that are mainly scavengers. During high tide, predators such as eels, flounder, and other large fishes come to the seagrass beds to feed.

## Seagrass Meadows as Habitat

The surfaces of seagrasses provide a place of attachment for many tiny organisms (epiphytes and epifauna; Figure 14-11). Older leaves may be completely covered by hydrozoans, tube worms, bryozoans, tunicates, and red algae. Juvenile scallops (*Argopecten*) often attach to seagrass leaves by byssal threads. Those attached well above the bottom sediments tend to avoid predation by benthic predators such as crabs. The epiphytes and epifauna on some turtlegrasses in the Caribbean may block enough radiant energy from reaching the plant to interfere with photosynthesis.

The sand among the blades of seagrass is home to a wide variety of filter feeders, such as adult scallops, jackknife clams (*Ensis*), and some species of sea cucumber. Seagrass rhizoids and root complexes provide sites for more-permanent attachments by tube worms and mussels. The thick growths of seagrass rhizomes make it difficult for small predators to dig out their prey. As a result the number of species is greater in seagrass meadows than in areas where the sediments are exposed.

The high productivity of seagrass communities allows them to support a large and diverse group of organisms, in-

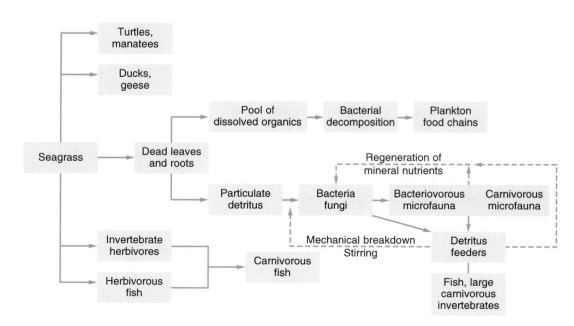

**Figure 14-10 Seagrass Food Webs.** There are not many animals that feed on seagrass directly. Most of the photosynthetic output from seagrass meadows enters the food web as detritus.

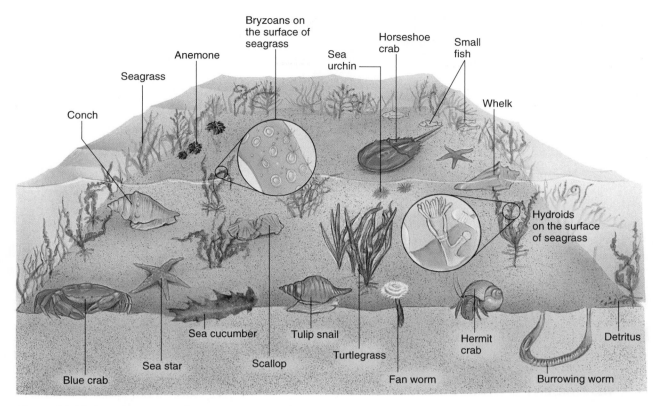

**Figure 14-11** **Seagrass Community.** The seagrass blades provide a suitable surface for attachment for many tiny organisms. The sandy mud provides habitat to a wide variety of animals. Although the seagrass is too tough for most animals to eat, it is an important source of detritus.

cluding the larval and juvenile stages of many animal species. The combination of changing salinity, available hiding places, and shallower water allows small juveniles to be protected from larger predators. In the seagrass nurseries these animals feed and grow without the heavy predation pressure that will face them when they leave their protection and enter the sea.

## Salt Marsh Communities

Salt marsh communities are found on the shoreward side of mud flats in temperate and subarctic regions of the world. The dominant plant life in this community on many North American shores is cordgrass (*Spartina;* see Chapter 7) that has moved from land to the shallow intertidal area. Many commercially important fishes and shellfishes spend a portion of their life cycle in the protection of salt marshes. The Wadden Sea, Europe's largest coastal wetland is estimated to yield $140 million per year in North Sea fish. It is estimated that each acre of U.S. salt marsh has a commercial value of between $50,000 and $80,000 per year, and the value of fish associated with U.S. salt marshes is estimated to be several billion dollars.

### Distribution of Salt Marsh Plants

Salt marshes can be divided into two regions: the **low marsh,** covered by tidal water much of the day and typically flushed twice each day by the tides, and the **high marsh,** covered briefly by saltwater each day and only flushed by the spring tides. In the marshlands of the Atlantic coast and the Gulf Coast of Florida, the low marsh is dominated by the tall **cordgrass** *Spartina alterniflora* that may grow as high as 3 meters (10 feet), whereas the species *Spartina foliosa* is dominant along the coast of California. In the high marsh along both coasts is a thick carpet of short, fine grasses that usually do not grow taller than 60 centimeters (2 feet). These may include salt meadow hay (*Spartina patens*), pickleweed (*Salicornia*), and spike grass (*Distichlis spicata*). Along the Florida Gulf Coast, the most common species in the high marsh is *Juncus.* In areas where tall cordgrass grows thickly, and leaves, stalks, and debris are flushed by tidal currents, the water is very clear. Beds of short cordgrass are not flushed frequently by tidal currents, and dead leaves and debris accumulate on the marsh surface, forming a moist mat that is an ideal habitat for many species. Other primary producers in salt marshes include green algae and benthic diatoms.

### Salt Marsh Productivity

The changing tides bring fresh nutrients into salt marshes where they are mixed and redistributed. This replenishing action of the tides helps to support a high level of production by marsh plants, green algae, and benthic diatoms. Experiments have shown that a Connecticut salt marsh can annually produce 3 to 7 tons of grass and algae per acre, and a typical southern salt marsh may produce as much as 10 tons

**Figure 14-12  Salt Marsh Productivity.** As with other estuarine food webs, most of the productivity of salt marshes enters the marine environment as detritus.

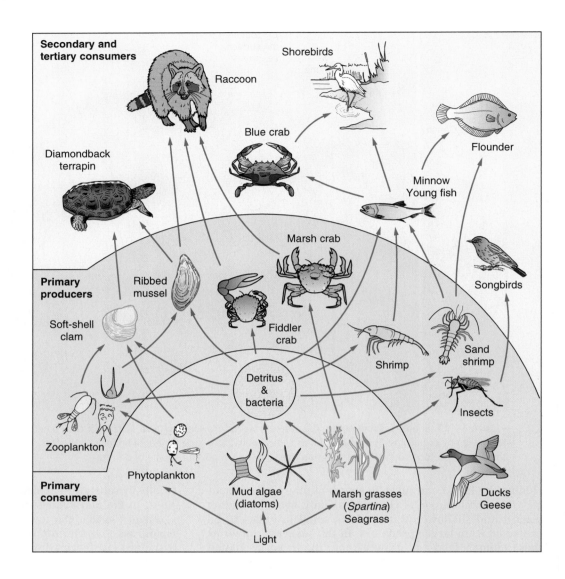

of vegetation per acre. By comparison, a prime wheat field in the Midwest might produce 7 tons of vegetation annually.

About 4% of the plant material in a salt marsh is consumed by insects and another 1% to 2% is consumed by other herbivores (Figure 14-12). Most of the primary production of salt marshes, however, supports detrital food chains. The dead and decaying plant material is a nutrient source for bacteria, which in turn are eaten by a host of deposit feeders. The bacteria convert some of the detritus into bacterial biomass and the rest is decomposed to a mix of organic molecules (dissolved organic matter) and inorganic ions (inorganic nutrients). Some salt marshes export large amounts of nutrients in the form of detritus to nearby communities. The exported detritus primarily supports bacteria and deposit feeders in the neighboring waters. Along the east coast of the United States as much as 40% of the primary production of a salt marsh is exported in the form of dead leaves and dissolved organic matter. Salt marshes farther north export very little, and most of the primary production is consumed by resident organisms in the marsh.

## Animals of the Salt Marsh

Marsh grasses form the basis of a complex and distinct community (Figure 14-13). Some animals that live in this community are permanent residents, whereas others come to the salt marsh only at high tide or low tide to take advantage of the wealth of food.

***Permanent Residents***   On the southern Atlantic coast, marsh periwinkle snails (*Littoraria irrorata*) feed on algae that grow on the surface of marsh grass leaves. The marsh periwinkle, as with most periwinkles, survives out of water because its mantle cavity can function as a type of lung, allowing the snail to absorb oxygen from the air. At low tide it moves to the underside of the leaves, where it is protected from the sun and predators. As the tide moves in, it crawls higher up the plant to avoid being submerged. When high spring tides submerge the entire plant and the snail, it can go without breathing for about an hour, long enough to survive until the tide begins to retreat. The periwinkles exhibit an interesting behavior in that they appear to be able

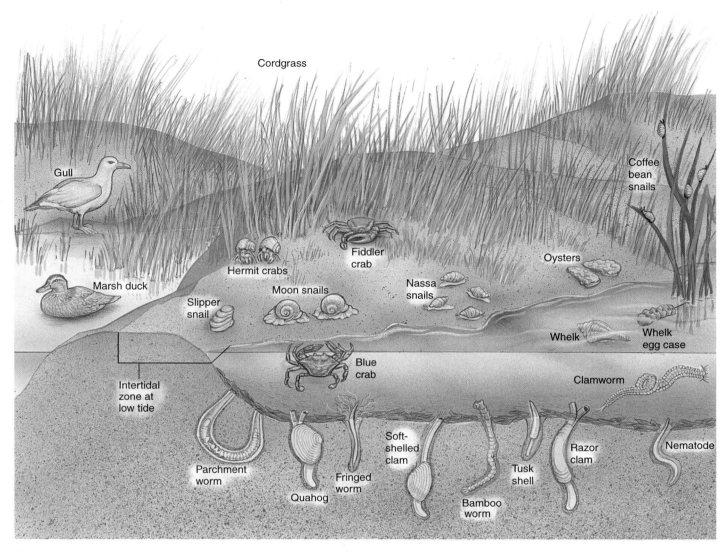

**Figure 14-13 The Salt Marsh Community.** Salt marshes provide habitat for a number of animals. Burrowing animals dominate the low marsh, whereas crabs, periwinkles, and marsh snails are common in the high marsh.

to sense an impending flood tide and crawl to the highest parts of the plant even before the water begins to rise. Periwinkles greatly restrict marsh grass productivity. In experiments in which the snails were removed from the marsh grass and prevented from returning, productivity increased as much as 25%.

The tidal marsh snail (*Melampus bidentatus*) is the dominant gastropod mollusc in many Atlantic coast marshes, frequently attaining densities of more than 1,000 per square meter. This snail is somewhat unusual in that, although it leads a largely terrestrial existence, it possesses an aquatic larval stage. The adults lay eggs in the water, and the larvae hatch and begin to develop in the water column during spring tides when at least part of the high marsh is flooded. The larvae later settle back onto the marsh to take up a more terrestrial lifestyle.

Burrowed in the mud of the low marsh along the Atlantic coast is the ribbed mussel (*Geukensia demissa*). At low tide, the mussel closes its valves to trap moisture and prevent desiccation. During high tide, it opens its valves to filter food particles from the water. Particles that are not acceptable for food are trapped in mucous ribbons and ejected from the mussel as pseudofeces. Another prominent bivalve in some salt marshes along the Atlantic coast is the razor clam (*Solen*). Like ribbed mussels, these clams burrow into the mud and filter feed at high tide. It is not unusual to find large clamworms (*Nereis*), sometimes as long as 19 centimeters (7.5 inches), feeding on the razor clams.

At low tide in more southern salt marshes, purple marsh crabs (*Sesarma reticulatum*) come out of their burrows to feed on the marsh grass. The fiddler crab (*Uca*; Figure 14-14) is one of the most prominent inhabitants of salt

**Figure 14-14  Fiddler Crab.** Fiddler crabs are common inhabitants of salt marshes. The large claw (cheliped) of the male is used to attract mates.

*Courtesy of David Campbell*

marshes, where they dig their burrows at the base of tall cordgrass. As they excavate their burrows, they form the mud being removed into neatly packed pellets that they carry to the opening and stack around it. During high tide, the crab fashions a door from some of the mud pellets and seals the entrance, trapping air in the burrow. At low tide, hundreds to thousands of these animals emerge from their burrows and scurry across the mud in search of food.

Fiddler crabs are omnivores and feed on detritus, algae, and small animals. The fiddler crab is named for the one oversized claw (cheliped) of the male that is used for attracting a mate and defending his territory against other males. Fiddler crabs can exchange gas in both air and water and can survive several weeks without being immersed in water. The space beneath its carapace and just above the legs forms a lung cavity. This space traps air and has a rich blood supply. While the lining of the lung cavity remains moist, the crab can exchange sufficient quantities of gas across it to survive out of water.

Burrowing animals, such as the fiddler crab, play a vital role in the overall marsh ecology. Their burrowing activities constantly bring nutrient-rich mud from deeper down to the surface, where the nutrients have been depleted, and their burrowing allows oxygen to penetrate to deeper sediments.

Amphipods called sandhoppers (*Orchestia*) are also abundant in some salt marshes. These small arthropods are common around the base of cordgrass and under debris. They feed on detritus and are an important source of food for several species of marsh bird. Another prominent arthropod is the grass shrimp (*Palaemonetes*), which is common in the water associated with both salt marshes and seagrass meadows. Seasonally, edible shrimp of the genus *Penaeus* are also very common. These animals feed on detritus and plankton and are important sources of food for fish and birds.

***Tidal Visitors to the Salt Marsh***   Predators that vary with the tides feed in the salt marsh. The most important predators in salt marshes at low tide are birds. Marsh wrens (*Telmatodytes* and *Cistothorus*), clapper rails (*Rallus longi-*

*rostris*), red-winged blackbirds (*Agelaius phoenicius*), and seaside sparrows (*Ammospiza maritima*) all nest in the tall marsh grass. Marsh hawks (*Circus cyaneus*) feed on the many rodents in this area. At low tide along the Atlantic coast, terrestrial predators, such as diamondback turtles (*Malaclemys terrapin*), raccoons (*Procyon*), otters (*Lutra*), and mink (*Mustela vison*), enter salt marshes in search of prey. Many insect and spider species are residents of salt marshes. These animals are not only predators but also food to many species of reptile, bird, and mammal. Herbivorous animals from the land, such as swamp rabbits (*Sylvilagus aquaticus*), white-footed mice (*Peromyscus*), and sometimes deer (*Cervus*), also come at low tide to browse on the marsh plants. At high tide, fishes are the chief higher-order predators. One of the more important fish predators is the killifish (*Fundulus heteroclitis*). The killifish feeds at high tide on small fish, marsh snails, and various amphipods, restricting them to the high marsh where the water is too shallow for the killifish to enter. Blue crabs also come to the flooded marsh at high tide to feed on benthic invertebrates.

### Succession in Salt Marshes

Salt marshes can be the first stage in a succession process that will eventually produce more land. The roots of marsh plants act as a sediment trap, holding sediments carried down by rivers and preventing their removal by tidal currents out to sea. Over time, the area becomes built up with sand and silt, and this in turn becomes mud as it is enriched with decaying material from dead plants and animals. Eventually small islands of mud appear, and as the cordgrass traps more sediment, the size of these islands increases and they begin to merge with each other. As land masses build, high tide covers less and less of them. Tall cordgrass is ultimately replaced by short cordgrass, and that, in turn, is replaced by rushes. At this point, land plants begin to establish themselves, and this is followed by the influx of terrestrial animals.

## Mangrove Communities

Mangrove forests, or mangals, replace salt marshes in tropical regions, where they can cover as much as 75% of the coastline (Figure 14-15; for more information regarding mangroves see Chapter 7). Mangrove forests appear where there is little wave action and where sediments accumulate and muddy sediments lack oxygen. The most highly developed mangals supporting the greatest number of species are along the coasts of Malaysia and Indonesia. Mangals in this region may contain as many as 40 species or more and may exhibit a definite pattern of zonation.

### Distribution of Mangrove Plants

In the Americas, red mangroves are frequently the pioneering species and are usually closest to the edge of the water. These plants grow in the parts of the mangal that experience the greatest degree of tidal flooding. The red mangrove produces seeds that germinate while still attached to the parent plant. When the seeds finally drop off, some take root next to the parent, whereas others are carried away by tidal currents to take root elsewhere (see Chapter 7). Seedlings

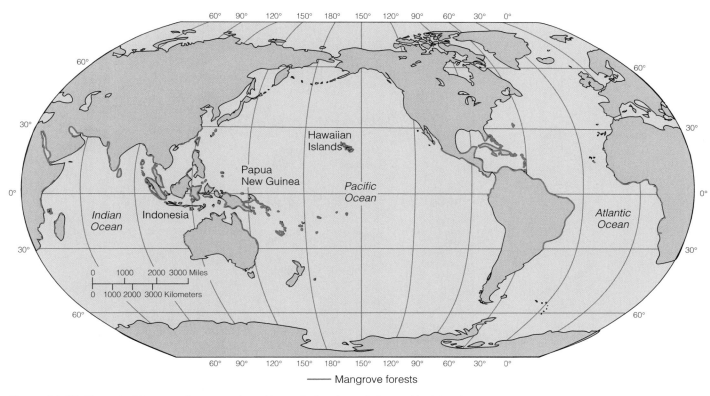

**Figure 14-15 Mangals.** Mangrove forests are found in tropical regions of the world where there is little wave action and where sediments accumulate and muddy sediments lack oxygen.

grow into mature plants that range in size from small shrubs to trees that can reach heights of 30 meters (100 feet).

Shoreward are the black mangroves, occupying areas that experience only shallow flooding during high tide. Closest to land are white mangroves and buttonwoods. Buttonwoods are not true mangroves and represent a transition to terrestrial vegetation. The various types of mangrove usually remain separated by their ability to tolerate flooding by saltwater during high tide and their different tolerances to soil salinity.

## Mangrove Root Systems

Because the mud in which mangroves grow is soft and oxygen poor, their roots are adapted to anchor the trees firmly and to supply sufficient oxygen to those parts of the plant buried in the mud. To help anchor the plants, mangroves have root systems that are shallow and widely spread. Red mangroves gain extra support from prop roots (Figure 14-16), which grow from the trunk and branches. Prop roots also aid in gas exchange for buried roots. Black mangroves have many erect, aerial roots called pneumatophores that branch from horizontal roots beneath the mud. The pneumatophores not only help to support the mangrove but also exchange gases for the buried roots. Without prop roots and pneumatophores, the buried roots of these plants would suffocate, and the plants would die.

Prop roots and pneumatophores form a tangle that slows the movement of water, causing suspended materials

**Figure 14-16 Mangrove Roots.** Prop roots (shown here) give red mangroves extra support and aid in supplying oxygen to the roots underground. Black mangroves have pneumatophores that project upward through the mud from the roots and aid in supplying oxygen to the roots.

to sink to the bottom. The roots collect various sediments and organic material. This, in turn, slows the movement of tidal waters, allowing more sediment to build up. Eventually the area becomes a terrestrial habitat, as the colonizing mangroves continue their growth toward the sea.

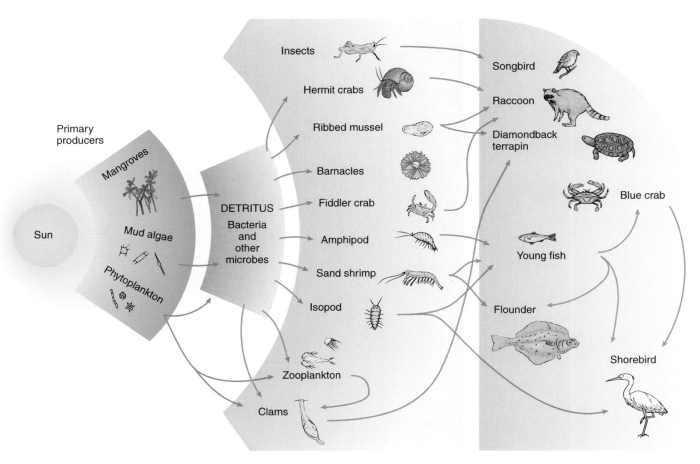

**Figure 14-17 Detritus Food Web.** Detritus from falling leaves and other decaying plant material forms the basis for a productive food web in mangrove communities.

## Mangal Productivity

The primary producers in mangals are mangrove plants, algae, and diatoms. As is the case with salt marsh plants, the tough leaves of mangroves are hard for most organisms to digest. In tropical regions, burrowing and climbing crabs feed on mangrove leaves, but most of the leaves and other detritus are removed by tidal currents to the surrounding water and become the basis of a detritus food web (Figure 14-17). This very productive food web supports a variety of important commercial fishes and shellfishes such as blue crab, shrimp, spiny lobster, mullet (*Mugil*), spotted sea trout (*Cynoscion nebulosus*), and red drum (*Sciaenops ocellatus*).

## Mangroves as Habitat

The prop roots of red mangroves and pneumatophores of black mangroves provide a habitat for a variety of animals. In mangrove forests of the west coast of Florida, prop roots and pneumatophores eventually become encrusted with a purple oyster called the coon oyster (*Lopha frons*; Figure 14-18). This bivalve gets its name from being a favorite food of raccoons. Barnacles and mussels that filter feed when the tide is high compete with coon oysters for space on the roots of red and black mangroves. On the east coast of Florida, mussels and acorn barnacles are dominant. Periwinkle (*Lit-*

**Figure 14-18 Mangroves as Habitat.** Many organisms, such as these oysters, utilize mangrove roots for habitat.

*torina*) snails that are related to the marsh periwinkle graze on algae growing on the stems and prop roots of the mangroves and growing on the shells of the sedentary organisms attached to them. The king's crown conch (*Melongena corona*), a carnivorous snail, feeds on oysters by prying open their valves with its strong muscular foot and then digesting the contents. Mangrove crabs (*Aratus*) go through their larval stages in the water beneath the mangroves. When mature, they crawl up on the mangroves and feed on the leaves.

The roots of mangroves provide a habitat for many of the same organisms found on mud flats and in salt marshes (Figure 14-19). These include fiddler crabs, hermit crabs, marsh crabs, marsh snails, and ghost shrimp. Some animals that cannot tolerate the varying salinity of the salt marsh and mud flats can survive in the more stable environment of

the mangrove forest. These include sea stars, brittle stars, and sea squirts.

In the mangals of Malaysia and Indonesia, fish known as mud skippers (*Periophthalmus chrysospilos*) live burrowed in the mud (Figure 14-20). These fish come out of their burrows at low tide and scoot around on the surface of the moist mud, behaving more like amphibians than fish. The sheltered waters around mangrove roots are also an ideal nursery for crab larvae, shrimp, and fishes. On Florida's

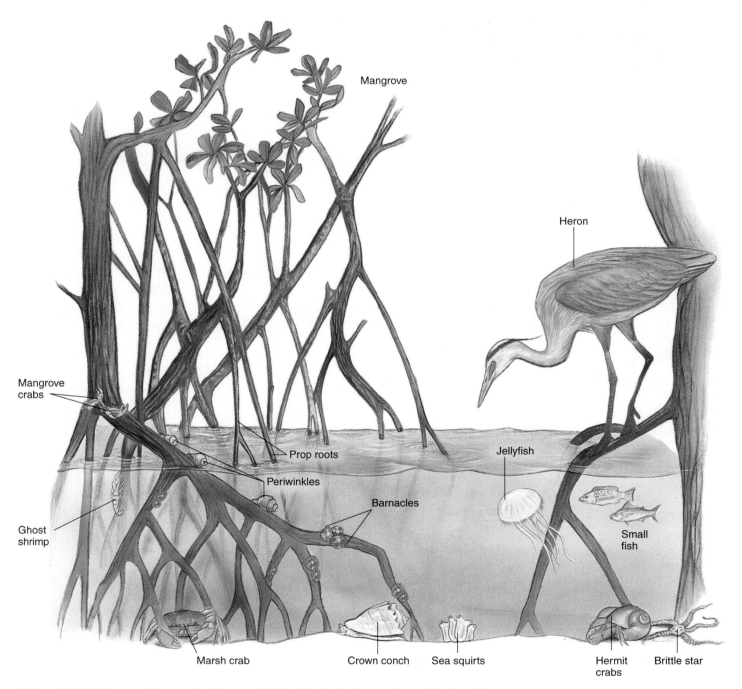

**Figure 14-19  Mangrove Community.** Many animals find shelter in the complex root system of mangrove swamps.

Norbert Wu

**Figure 14-20  Mud Skipper.** Mud skippers, residents of mangrove swamps in Asia, can spend time out of water by exchanging gases at their moist gills.

southern coasts, the role of mangroves as a nursery is equal to that of seagrass communities.

The rich food supplies of mangrove forests along the Gulf Coast of the United States attract a variety of predators, including clapper rails, killifishes (family Cyprinodontidae), diamondback turtles, water moccasins (*Agkistrodon*), and raccoons. Birds such as pelicans, herons, egrets, roseate spoonbills (*Ajaia ajaia*), and wood ibises (*Mycteria americana*) also find the upper branches of mangroves an ideal nesting site. Before much of this habitat was destroyed by drainage projects and pollution, mangals along the west coast of the Everglades provided major nesting sites or rookeries for many bird species.

## In Summary

Estuaries support a variety of distinct communities that include oyster reefs, mud flats, seagrass communities, salt marshes, and mangrove forests. Each community has its unique characteristics and supports distinctive populations of producer and consumer organisms that have become adapted to life in each habitat. •

## SELECTED KEY TERMS

bar-built estuary, *p. 326*

coastal plain estuary, *p. 325*

cordgrass, *p. 335*

drowned river valley estuary, *p. 325*

embayment, *p. 325*

estuary, *p. 325*

euryhaline, *p. 330*

fjord, *p. 325*

high marsh, *p. 335*

low marsh, *p. 335*

negative estuary, *p. 326*

osmoconformer, *p. 328*

osmoregulator, *p. 328*

partially mixed estuary, *p. 326*

positive estuary, *p. 326*

pseudofeces, *p. 328*

salt-wedge estuary, *p. 326*

salt wedge, *p. 326*

tectonic estuary, *p. 325*

tidal flat, *p. 325*

tidal overmixing, *p. 326*

well-mixed estuary, *p. 326*

## QUESTIONS FOR REVIEW

### Multiple Choice

1. Coastal areas where part of the ocean is partially cut off from the rest of the sea is called a(n)
   a. embayment
   b. estuary
   c. inlet
   d. tidal flat
   e. delta

2. The type of estuary formed when earthquakes cause land to sink, allowing seawater to enter, is a
   a. fjord
   b. tectonic estuary
   c. drowned river valley estuary
   d. coastal plain estuary
   e. delta

3. A deep estuary that has high river input but little tidal mixing is a
   a. fjord
   b. tectonic estuary
   c. drowned river valley estuary

   d. coastal plain estuary
   e. delta

4. In a well-mixed estuary, vertical salinity is
   a. stratified
   b. uniform

5. The environmental factor that affects the types of organisms in estuaries more than anything else is
   a. sunlight
   b. temperature
   c. salinity
   d. nutrient density
   e. type of sediment

6. If the concentration of salts in an animal's body tissues varies with the salinity of the environment, the animal would be an
   a. osmoregulator
   b. osmoconformer

7. Most organisms that inhabit mud flats are
   a. active swimmers
   b. terrestrial

c. burrowers
d. predators
e. immobile

8. The mangrove species usually closest to the water in American mangrove forests is the
   a. white mangrove
   b. black mangrove
   c. red mangrove
   d. green mangrove
   e. buttonwood

9. An example of a deposit feeder found on mud flats would be the
   a. jackknife clam
   b. fiddler crab
   c. lugworm
   d. soft-shelled clam
   e. innkeeper worm

10. An animal that feeds directly on seagrasses is the
    a. queen conch
    b. green sea turtle
    c. sea cucumber
    d. fiddler crab
    e. mullet

### Short Answer

1. What are the distinguishing characteristics of an estuary?

2. What is a salt-wedge estuary, and how does it differ from other types of estuaries?

3. What factor(s) contributes to the high productivity of estuaries?

4. What adaptations have evolved in mangroves that help them survive in their habitat?

5. Explain how organisms that are osmoconformers survive in estuaries.

6. Explain how fiddler crabs are well-adapted to life in the salt marsh.

7. Describe the process of succession in a salt marsh.

8. Sketch a chart that traces energy flow in a mud flat.

### Thinking Critically

1. Why is it difficult for burrowing animals that exchange gas through their skin to survive in mud flats?

2. Predict what effect agricultural runoff would have on a neighboring estuary.

3. To control flooding, a series of dams is constructed along a river that feeds a large estuary. What effect will the dams have on the estuary's productivity?

## SUGGESTIONS FOR FURTHER READING

Adam, P. 1990. *Salt Marsh Ecology.* New York: Cambridge University Press.

Bertness, M. D. 1999. *The Ecology of Atlantic Shorelines.* Sunderland, Mass.: Sinauer Publishing.

Boicourt, W. C. 1993. Estuaries: Where the River Meets the Sea, *Oceanus* 36(2):29–37.

Lippson, A. J., and R. L. Lippson. 1984. *Life in the Chesapeake Bay.* Baltimore, Md.: Johns Hopkins University Press.

Rutzler, K., and C. Fuller. 1987. Mangrove Swamp Communities, *Oceanus* 30(4):16–24.

Teal, J. M. 1996. Salt Marshes: They Offer Diversity of Habitat, *Oceanus* 39(1):13–15.

Valiela, I., and J. Teal. 1979. The Nitrogen Budget of a Salt Marsh Ecosystem, *Nature* 280:652–56.

Virnstein, R. 1977. The Importance of Predation by Crabs and Fishes on Benthic Infauna in Chesapeake Bay, *Ecology* 58(6):1199–217.

### InfoTrac College Edition Articles

Boyle, R. H. 1999. Bringing Back the Chesapeake, *Audubon* 101(3).

Dybas, C. L. 2002. Florida's Indian River Lagoon: An Estuary in Transition, *Bioscience* 52(7).

Feinstein, N., S. Yelenik, J. McClelland, and I. Valiela. 1996. Growth Rates of Ribbed Mussels in Six Estuaries Subject to Different Nutrient Loads, *The Biological Bulletin* 191(2).

Harrison, T. R., J. L. Boxhill, L. Z. Santiago Vazquez, K. Foreman, and J. N. Kremer. 1994. Comparison of Phytoplankton and Ecosystem Gross Production in the Quashnet River, an Estuary of Waquoit Bay, Massachusetts, *The Biological Bulletin* 187(2).

Hughes, J. E., L. A. Deegan, B. J. Peterson, R. M. Holmes, and B. Fry. 2000. Nitrogen Flow through the Food Web in the Oligohaline Zone of a New England Estuary, *Ecology* 8(2).

Kinlan, B., E. Duffy, J. Cebrian, J. Hauxwell, and I. Valiela. 1997. Control of Periphyton on *Zostera marina* by the Eastern Mudsnail, *Ilyanassa obsoleta* (Say), in a Shallow Temperature Estuary, *The Biological Bulletin* 193(2).

Leonard, G. H., J. M. Levine, P. R. Schmidt, and M. D. Bertness. 1998. Flow-Driven Variation in Intertidal Community Structure in a Maine Estuary, *Ecology* 79(4).

Tankersley, R. A., M. G. Wieber, M. A. Sigala, and K. A. Kachurak. 1998. Migratory Behavior of Ovigerous Blue Crabs *Callinectes sapidus:* Evidence for Selective Tidal-Stream Transport, *The Biological Bulletin* 195(2).

### Websites

**www.estuaries.org** An overview of estuaries and restoration projects.

**www.epa.gov/owow/estuaries** General description of estuaries, good photos, and information on restoration projects.

**http://water.dnr.state.sc.us/marine/pub/seascience/ dynamic.html** Good description of South Carolina salt marshes and salt marsh ecology.

**http://life.bio.sunysb.edu/marinebio/spartina.html** Description of a New York salt marsh with good photos of salt marsh organisms.

**http://www.epa.gov/owow/wetlands/types/mangrove .html** General description of mangrove forests with photos of organisms.

# 15

# Coral Reef Communities

L ocated in some of the oceans' clear, blue, warm
waters are complex communities teeming with life
known as coral reefs. Like underwater cities, these
communities are crowded with animals representing virtually every major animal phylum. Nowhere else on earth
can you find living organisms with such spectacular colors
or fantastic shapes as on coral reefs. In this chapter you
will learn more about this amazing ecosystem and its
inhabitants.

## WORLD OF CORAL REEFS

Ocean water in the tropics is relatively nutrient poor. It is
this lack of nutrients and the resulting reduced plankton
levels that give the water its blue color. Coral reefs in these
nutrient-poor waters are like oases in a desert. Among
the earth's ecosystems, only tropical rain forests rival them
in the diversity and abundance of organisms present. For
years biologists were puzzled as to how such a diversity
of organisms could be supported in a community that appeared to contain so few primary producers and available
nutrients. The answer to this puzzle was ultimately found in
the symbiotic relationship involving the coral animals that
make up a large part of the reef structure and several species of dinoflagellates known as **zooxanthellae** (for more
information on zooxanthellae see Chapter 6). Together with
algae they form the basis of a community upon which all the
other reef animals depend for food and shelter.

## In Summary

Coral reef communities are species-rich oases in relatively nutrient-
poor tropical and subtropical seas.  ●

## Key Concepts

1. Coral reefs are primarily found in tropical clear water, usually at depths of 60 meters or less.

2. The three major types of coral reefs are fringing reefs, barrier reefs, and atolls.

3. Both physical and biological factors determine the distribution of organisms on a reef.

4. Stony corals are responsible for the large colonial masses that make up the bulk of a coral reef.

5. Reef-forming corals rely on symbiotic dinoflagel-lates called "zooxanthellae" to supply nutrients and to produce an environment suitable for forma-tion of the coral skeleton.

6. Coral reefs are constantly forming and breaking down.

7. The most important primary producers on coral reefs are symbiotic zooxanthellae and turf algae.

8. Coral reefs are oases of high productivity in nutri-ent-poor tropical seas. Nutrients are stored in reef biomass and efficiently recycled.

9. Inhabitants of coral reefs display many adaptations that help them to avoid predation or to be more efficient predators.

10. Coral reefs are huge, interactive complexes full of intricate interdependencies.

# CORAL ANIMALS

**Stony,** or **true, corals** (phylum Cnidaria, class Anthozoa, order Scleractinia) are the primary organisms depositing the massive amounts of calcium carbonate ($CaCO_3$) that make up most of the structure of coral reefs (see Chapter 8). Coralline algae and several other organisms deposit lesser amounts of this mineral. A member of the Cnidarian class Hydrozoa called fire coral (*Millipora;* Figure 15-1), for example, is a significant contributor to the carbonate structure of Caribbean reefs, whereas coralline algae (for more information on coralline algae see Chapter 7) are very important in the structure of Pacific reefs. Coral species that produce reefs are said to be **hermatypic** and are found only in shallow tropical waters. Nonreef building, or **ahermatypic,** corals (Figure 15-2) can grow in deeper water and are found from the tropics to polar seas. Hermatypic corals typically harbor zooxanthellae in their tissues whereas most ahermatypic corals do not.

## Coral Colonies

Much of the structure of a coral reef is composed of large colonies of tiny coral polyps. Colonies begin when a planula larva settles from the plankton and attaches itself to a solid surface, frequently the dead remains of other coral colonies (Figure 15-3). Once attached, the planula develops into a coral polyp and secretes a cup of calcium carbonate, a **corallite,** around its body. The polyp reproduces by budding, and the new polyps formed also form corallites. In turn, these new polyps will form buds, and as asexual reproduction continues, the colony grows. The gastrovascular

**Figure 15-1  Fire Coral.** Fire corals (genus *Millipora*) are not true corals but members of the class Hydrozoa. They are ubiquitous on Caribbean reefs where they are very important in the deposition of carbonate sediments. The name fire coral comes from the fact that contact produces a painful welt.

cavity of each polyp in the colony remains interconnected through tubes, and a thin epidermis overlies the colony surface. The epidermis is usually quite colorful; red, pink, yellow, green, purple, or brown are common. The colony grows both upward and outward and assumes the shape that is genetically determined for the species, with some variation due to environmental stresses.

## Sexual Reproduction in Coral

As a coral colony grows, some polyps develop gonads and are capable of sexual reproduction. Most coral species are hermaphroditic and **broadcast spawners.** They release both sperm and eggs into the surrounding seawater. A smaller number of coral species are **brooders.** Brooders broadcast their sperm into surrounding waters but the eggs are retained within the gastrovascular cavity. When the fertilized eggs reach the planula stage, they are released.

Spawning typically occurs as a brief annual event and is **synchronous** among species on many Pacific reefs. In the Caribbean, however, spawning usually takes place over a longer period and is **nonsynchronous,** different species spawning at different times. The exact time of broadcast spawning and planula release from brooders seems to be determined by the amount of nocturnal illumination.

## Reproduction by Fragmentation

Some branching corals, such as staghorn coral, are fragile, and turbulent water from storms can break branches and topple colonies. If the storms do not kill the fragments, many of them attach and grow, forming new colonies. In one Caribbean study it was found that more than twice the number of new colonies arise from fragments as from larvae. **Fragmentation,** therefore, appears to be a common form of asexual reproduction for branching corals.

## Coral Nutrition

Corals have evolved several strategies for obtaining food. In addition to deriving nutrition from their symbiotic zooxanthellae, they prey on the tiny zooplankton that pass over the reef or other small organisms that venture too close to their stinging tentacles. Some species also use organic material in the surrounding water.

### Symbiotic Zooxanthellae

More than 90% of the nutritional needs of stony corals are supplied by the symbiotic zooxanthellae living within their tissues (Figure 15-4). About 16 species of dinoflagellate are known to form the zooxanthellae. They are acquired either from the environment or directly from the parent. The zooxanthellae provide nutrients to the polyps in the form of glucose, glycerol, and amino acids, and the coral polyp provides a suitable habitat for its symbiont and a variety of nutrients. The zooxanthellae absorb these nutrients directly from the animal's tissues so efficiently that several species of coral release virtually no nitrogen wastes to the surround-

**Figure 15-2 Ahermatypic Corals.** Ahermatypic corals also inhabit the reef community.
(a) Gorgonians (sea fans) and (b) black corals have skeletons made of protein rather than calcium carbonate. Gorgonians are very abundant on shallow Caribbean reefs but black coral is found in deeper water. Black coral is becoming increasingly rare due to its use in making jewelry. (c) Delicate branching soft corals, such as the ones shown here, are more prevalent on Pacific reefs.

ing water. The removal of carbon dioxide and production of oxygen by the zooxanthellae also seems to improve the growth rate of the coral's carbonate skeleton. The presence of zooxanthellae is so important to the life of coral polyps that reef-building corals cannot grow any deeper than sunlight can penetrate to supply the photosynthetic needs of their symbionts.

## Corals as Predators

Like the polyps of other cnidarians, the soft, cylindrical body of a coral polyp has a ring of tentacles surrounding the mouth. The tentacles are strewn with stinging cells called **cnidocytes** (for more information regarding cnidocytes see

Chapter 8). Any small animal that brushes against these cells is immediately paralyzed and passed through the mouth into the digestive cavity. Some coral polyps have short tentacles that produce mucus and are covered with cilia. During feeding, the cilia beat toward the mouth, creating a current that traps small organisms within the mucus and draws it into the mouth. When the polyp is not feeding, the cilia reverse the direction of their beat so that silt and other debris are moved away from the polyp. When they are inactive or disturbed, polyps withdraw into their corallite. Because the polyps of many coral species are nocturnal feeders, day visitors to a reef rarely have the opportunity to view the living animal (Figure 15-5).

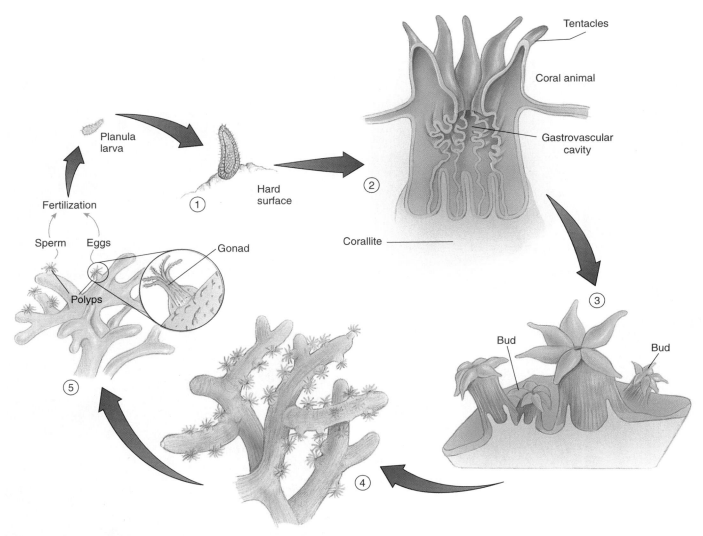

**Figure 15-3 Formation of a Coral Colony.** (1) A coral colony begins when a planula larva attaches to a hard surface. (2) The planula develops into a polyp, and the polyp secretes a cup of calcium carbonate. (3) The polyp reproduces asexually by budding, and each new polyp forms a calcareous cup around its body. (4) The colony continues to grow as the new polyps repeat the budding process. The shape of the colony is genetically determined. (5) As the colony continues to grow, some polyps develop gonads. The sperm and eggs produced by these polyps are shed into the surrounding water, where the eggs are fertilized. The fertilized eggs develop into new planula larvae that begin the process of colony formation again.

**Figure 15-4 Coral Polyp and Symbionts.** Symbiotic dinoflagellates known as *zooxanthellae* live within the tissues of coral polyps and are responsible for the brownish color of many live corals. The zooxanthellae provide the polyps with carbohydrates and remove carbon dioxide. In return, the polyps supply the zooxanthellae with nitrogen compounds and carbon dioxide.

Richard N. Mariscal

**Figure 15-5 Coral Polyp Feeding.** At night coral polyps extend their tentacles and feed on plankton.

### Other Sources of Nutrition

Bacteria living in the tissues of corals can take up nutrients directly from the water in the form of **dissolved organic matter,** the remnants of living organisms. The corals can then feed on these bacteria. In addition, corals are able to absorb organic matter directly across their body surface. Mesenteric filaments, coiled tubes attached to the gut wall, can be extruded from the mouth and used to digest and absorb food outside the body.

## In Summary

Stony (or true) corals are colonial organisms that produce the calcium carbonate deposits that make up most of the structure of a coral reef. Together with the algae they form the foundation of a community upon which all the other reef animals depend for food and shelter. Reef-building, or hermatypic, corals are strictly tropical in distribution and harbor zooxanthellae. Ahermatypic corals, those that do not deposit calcium carbonate, are more widespread and generally lack zooxanthellae. Zooxanthelle are symbiotic dinoflagellates that use coral wastes, produce carbohydrates, and aid in calcium carbonate deposition by corals. Corals obtain about 90% of their energy from zooxanthellae, the remainder coming from plankton and dissolved organic matter. •

## REEF FORMATION

Reef formation involves both constructive and destructive phases. It is not a simple matter of new coral building on top of old existing coral but rather a series of events involving many animal and algal species. Although stony corals form the basic material of the reef, many smaller species provide

**Figure 15-6 Solitary Hard Corals.** Hard corals like this mushroom coral are formed by single polyps that do not form colonies.

a significant amount of additional calcium carbonate to its structure. Not all stony corals form large colonies; some are solitary polyps (Figure 15-6). These, along with the soft corals and many other skeleton-building and shell-building organisms, contribute to the structure of the reef.

The destructive phase of reef formation is called **bioerosion.** It is usually underway well before the death of a coral colony (Figure 15-7). Boring clams or sponges regularly attack the exposed surfaces on the underside of large coral stands. These vulnerable areas are relatively bare and do not receive enough sunlight to support the growth of coral polyps. As the boring organisms riddle the base with tunnels, the coral stand weakens, and a storm or heavy ocean surge will eventually topple the coral, crushing and killing many of the living polyps. As the debris from boring accumulates, it eventually covers and smothers the boring organisms and the boring stops. If this occurs quickly, the chunks of coral left over from the boring will be relatively large. If the process takes a longer time, the large corals will be broken up into much smaller pieces. The breakup of calcareous green algae like *Halimeda* (Figure 15-8) and mollusc shells also adds significantly to the calcium carbonate deposits on the reefs. This sediment covers the reef and settles into the cracks and crevices. Sponges and other organisms may hold this rubble together until encrusting coralline algae (Figure 15-9) deposit more calcium carbonate over it, cementing it. Planula larvae can settle on this hard surface and begin to produce new stands of coral.

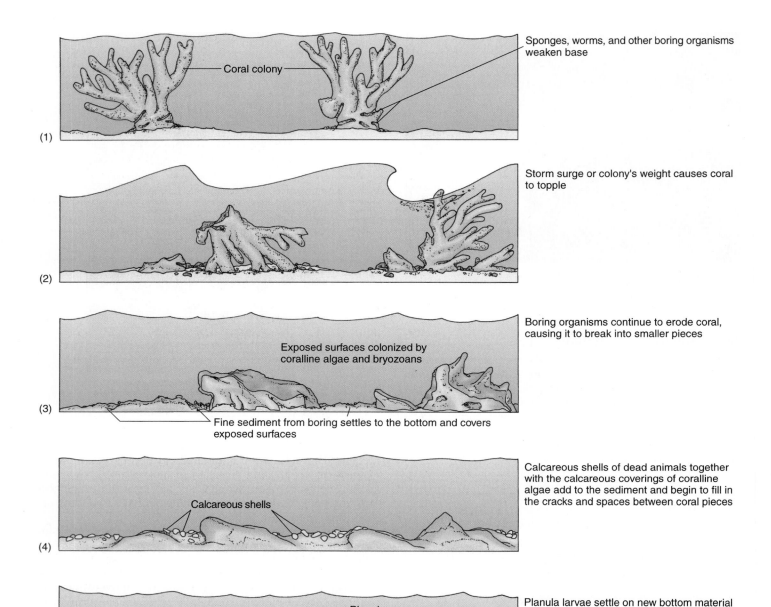

**Figure 15-7  Reef Cycle.** (1) The destructive phase of a reef cycle begins when boring animals burrow through the bases of large corals, weakening them. (2) The weight of the coral or storm surges cause the coral to topple, killing the polyps. (3) The coral is broken into smaller pieces, and bryozoans and coralline algae begin to colonize the exposed surfaces. (4) Shells of dead organisms contribute to filling in cracks and crevices. Coralline algae help to cement the pieces together to form a new, solid surface. (5) Planula larvae begin to settle and form new coral colonies.

**Figure 15-8 Coral Reef Algae.** Calcareous green algae of the genus *Halimeda* are common on coral reefs because of their ability to resist herbivory. The breakup of the thallus of these species contributes to calcium carbonate deposits on coral reefs.

*Courtesy of Matthew Landau*

**Figure 15-9 Encrusting Coralline Algae.** Encrusting coralline algae like *Lithothamnion, Porolithon,* or *Lithophyllum* deposit calcium carbonate over carbonate rubble, cementing it to form a solid surface.

*Courtesy of Matthew Landau*

## In Summary

Coral reefs are constantly forming and breaking down, a process known as bioerosion. Boring organisms weaken the bases of large coral stands, causing them to topple, and tropical storms regularly cause breakdown and destruction of corals. Following the destructive phase of the reef cycle, new solid surfaces form that may ultimately be colonized by new polyps. The shells of other animals and calcareous algae also contribute to the formation of a reef. Encrusting coralline algae cement this calcium carbonate rubble into a solid mass. ●

## CORAL REEF TYPES

Coral reefs are divided into three basic categories on the basis of their structure and relationship to underlying geologic features. The most common type, **Fringing reefs** (Figure 15-10a), border islands or continental landmasses. **Barrier reefs** (see Figure 15-10b) are similar to fringing reefs but are separated from the landmass and fringing reef with which they are associated by lagoons or deepwater channels. The largest barrier reef is the Great Barrier Reef of Australia, which runs about 2,100 kilometers (1,260 miles) along the northeastern coast of Australia to New Guinea (see Figure 15-10d). This reef is so large that it is visible even to astronauts orbiting the earth. The second-largest barrier reef is in the Caribbean Sea off the coast of Belize. **Atolls** (see Figure 15-10c), the third reef type, are usually somewhat elliptical, arise out of deep water, and have a centrally located lagoon. More than 300 atolls are located in parts of the Indian and Pacific oceans, whereas only 10 atolls are in the Atlantic Ocean. The Atlantic Ocean supports fewer reefs in general because the water is too cool and turbid in many areas. The water in the lagoon of an atoll is not isolated but is connected to the open sea by gaps in the reef. Over time, parts of the reef may become exposed and eroded by the action of wind and waves. The sand formed by these physical processes can be the basis for the formation of an island. Sandy islands frequently surround the lagoon. In addition to the main reef types, small **patch reefs** can occur within the lagoons associated with atolls and barrier reefs. Where patch reefs occur, they are usually quite numerous.

Charles Darwin proposed the currently accepted theory for the origin and relationship of these coral reef types, particularly as it applies to atoll formation. According to Darwin's theory (Figure 15-11), as newly formed volcanic islands arise from the sea, corals colonize the shallow areas around them and form a fringing reef. As these volcanic islands slowly sink and erode, the coral growth rate equals the rate of sinking and a barrier reef is formed about the island. At this point, a lagoon separates this barrier reef from what remains of the central island. Finally, as the island completely erodes and sinks below the sea, corals continue to grow to the surface of the barrier reef and an atoll is formed. Water flow is usually inadequate and there is too much sediment where the island once existed for vigorous coral growth to occur, so a lagoon persists. Although Darwin's theory does link fringing reefs, barrier reefs, and atolls, it does not explain how all fringing and barrier reefs develop. Corals reefs probably occur around continents and islands simply because of suitable solid surfaces and conditions for their growth.

## In Summary

There are three basic types of reefs. Fringing reefs surround islands and border continental landmasses. Barrier reefs are separated from the nearest landmass by a lagoon. Atolls form on the tops of

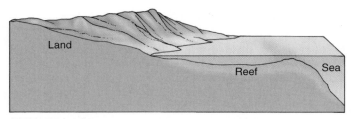

(a) Fringing reef

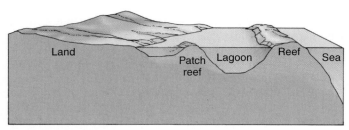

(b) Barrier reef

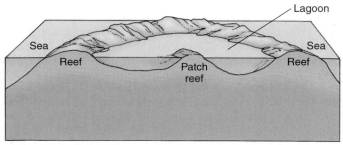

(c) Atoll

(d)

**Figure 15-10  Types of Coral Reefs.** (a) A fringing reef is connected directly to the shore. (b) A barrier reef is separated from the shore by a lagoon. (c) An atoll is a circular reef that encloses a lagoon. Atolls usually form on the cones of extinct volcanos. (d) The largest barrier reef in the world, the Great Barrier Reef, runs along the northeastern coast of Australia.

submerged volcanoes and are generally elliptical with a centrally located lagoon. Charles Darwin proposed the currently accepted theory for the origin of atolls and the relationship of these coral reef types. ●

## REEF STRUCTURE

All three reef types share several common characteristics. On the seaward side, the reef rises from the lower depths of the ocean to a level just at or just below the surface of the water. This portion of the reef is called the **reef front,** or **forereef** (Figure 15-12a). The slope of the reef front can be gentle or quite steep. In some cases, the reef front forms a vertical wall referred to as a **drop-off.** More commonly, fingerlike projections of the reef protrude seaward in a **spur-and-groove formation** (or **buttress zone**). This arrangement disperses wave energy and helps to prevent damage to the reef and its inhabitants (see Figure 15-12b). Between some of these projections (spurs) lie sand-filled pockets (grooves) that allow sediments to be channeled down and away from the living coral surface and provide habitat for many species of burrowing organisms. Coral growth is rich at the reef margin, on the spurs, and along the sides of the grooves but declines with depth because of decreased light availability.

The **reef crest** is the highest point on the reef and the part that receives the full impact of wave energy. Where wave impact is very strong, the reef crest may consist of an algal ridge, composed of encrusting coralline algae and lacking most other organisms. Grooves of the buttress zone penetrate the algal ridge as **surge channels.** Behind the reef crest lies the **reef flat,** or **back reef.** The reef flat may be short or several hundred meters long. It may be shallow or cut through by surge channels several meters deep. The bottom of the flat consists of rock, sand, coral rubble, or some combination of these. Seagrass beds (for more information regarding seagrasses see Chapter 7) are common in the reef flat area. The reef flat of fringing reefs ends at the shoreline. The reef flat of atolls and barrier reefs descends into the lagoon.

Different areas of a reef support different species of coral and other organisms. Coral populations on the reef front are usually found at depths of 10 to 60 meters (33 to 200 feet). Massive, dome-shaped brain corals (*Diploria*) and columnar pillar corals (*Dendrogyra*) are on intermediate slopes. Below this region, species that form platelike formations, such as lettuce leaf and elephant ear coral (*Pectinia, Pavona,* and *Agaricia* species; Figure 15-13a), predominate. Higher up on the reef, where wave stress is greatest, branching species of coral are found. Heavy, spreading branches project toward the sea, where they break the force of incoming waves. In the Caribbean, this upper area is the habitat of elkhorn coral (*Acropora palmata;* see Figure 15-13b). Wave stress is one of the most important factors in determining which species of coral and other organisms can occupy the reef crest.

In more protected areas behind the reef front, the deeper and less-turbulent water supports more delicate

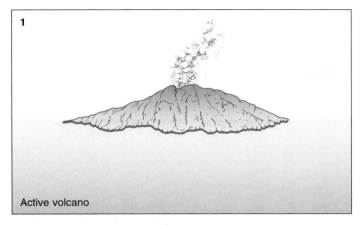

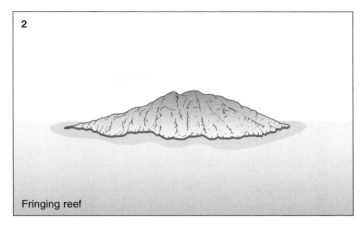

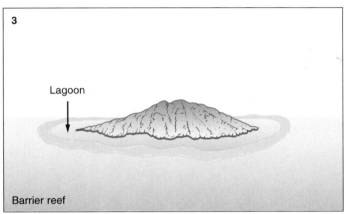

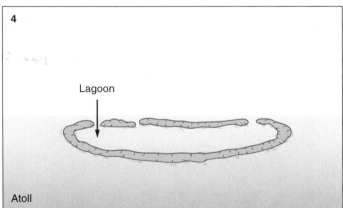

**Figure 15-11  Atoll Formation.** According to Charles Darwin's theory, fringing reefs form in the shallow water around newly formed volcanic islands. Over time the island erodes and becomes separated from the surrounding and continually growing reef. This reef is now termed a barrier reef because it is separated from the island by a lagoon. As the island completely erodes, an atoll is formed as the barrier reef continues to grow making up for the subsidence.

species of coral. In the Caribbean, this is frequently staghorn coral (*Acropora cervicornis*). In the Indo-Pacific region, species such as staghorn, finger (*Stylophora*), cluster (*Pocillopora*), and lace corals (*Pocillopora*) are prominent. Farther from the reef front in shallow, calmer water are small species of coral such as rose (*Meandrina* and *Manicina*), flower (*Mussa* and *Eusmilia*), and star (*Montastraea*).

## In Summary

The three reef types share many similarities in their basic structure. The reef front is typically arranged as a spur-and-groove formation (or buttress zone) that disperses wave energy and prevents damage to the reef and its inhabitants. The highest point of the reef is the reef crest, and the area behind the reef is the reef flat, or back reef. Because of differences in the physical environment, each area supports different combinations of coral and other reef organisms. ●

## CORAL REEF DISTRIBUTION

Temperature, light availability, sediment accumulation, salinity, wave action, and duration of air exposure are the major factors determining the distribution of coral reefs. Coral reefs do not develop in waters where average annual minimum temperature is below 18°C, and the best development occurs where the average annual water temperature is between 23°C and 25°C. Almost all the world's coral reefs are therefore restricted to tropical seas (Figure 15-14). Lack of light explains why coral reefs are rarely found below 60 meters (200 feet). Most grow in depths of 25 meters or less, and they are typically confined to the margins of continents or islands. Sedimentation from land-surface runoff can both reduce light available to zooxanthellae for photosynthesis and clog feeding structures, killing coral polyps. Several examples of the deleterious effect of sedimentation can be seen on South Florida reefs. Salinity level is also important, as demonstrated by the lack of coral reefs in such areas of

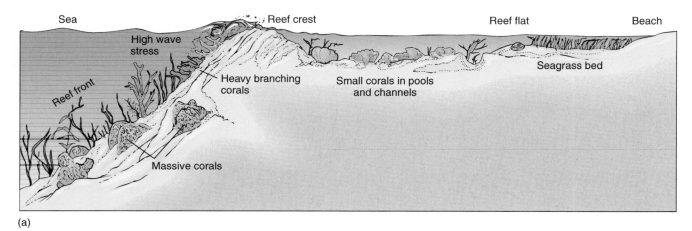

(a)

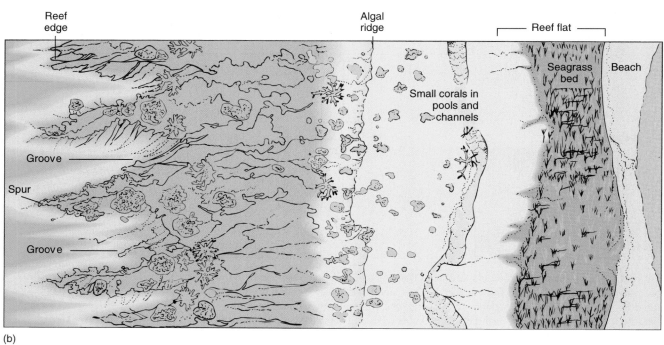

(b)

**Figure 15-12  Reef Characteristics.** (a) The reef front rises from the ocean's depths to the reef crest. The area behind the reef is the reef flat. (b) A view of a reef from above looking down shows the characteristic spur-and-groove formation.

**Figure 15-13  Types of Coral.**
(a) Corals that produce large, platelike formations like this lettuce leaf coral are located farther down the reef front, where wave action is minimal. (b) Areas of a reef that receive heavy wave action contain thick, branching corals such as this elkhorn coral, as well as large, solid corals such as brain coral.

(a)

(b)

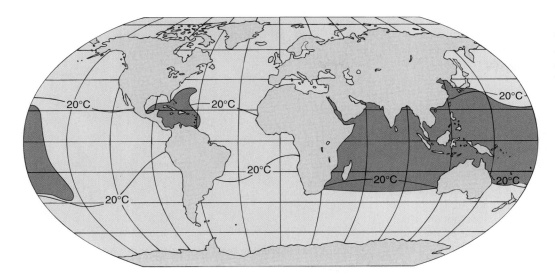

**Figure 15-14  Distribution of Coral Reefs.** Coral reefs (brown areas) are restricted to the tropics and subtropics, the area within the surface 20°C isotherm.

massive freshwater outflow as the mouths of the Amazon and Orinoco Rivers in South America. Moderate wave action is important for a healthy reef because it brings in fresh, oxygenated seawater and removes sediment that may smother coral polyps. But heavy wave action during hurricanes can physically damage reef structure, uprooting corals and destroying habitat. Although some coral species can sustain exposure to air for a few hours, most are killed by it. The lowest tide level, therefore, limits coral upward growth.

### In Summary

Factors limiting reef development include temperature, light availability, salinity, sedimentation, wave action, and duration of exposure to air. Coral reefs are typically confined to shallow water at the margins of continents or islands where the average surface temperature is greater than 18°C. They are rarely found below 60 meters. ●

## COMPARISON OF ATLANTIC AND INDO-PACIFIC REEFS

Atlantic and Indo-Pacific reefs (Figure 15-15) differ in a number of ways. Pacific reefs are of greater antiquity than those in the geologically younger Atlantic Ocean and have a far greater depth of reef carbonates. The buttress zone is deeper on Atlantic reefs and coral growth may extend down to 100 meters. In contrast, coral growth in the Pacific seldom exceeds 60 meters. The proportion of the reef covered by coral may approach 100% on some Pacific reefs but is usually less than 60% on those in the Atlantic. Algal ridges are more common in the Pacific than in the Atlantic because of wind and wave conditions. The hydrozoan *Millipora complanata* (fire coral) dominates on many Atlantic reefs but similar species never dominate in the Pacific. Gorgonians (sea fans and sea whips) are much more abundant in the

(a)

(a)

**Figure 15-15  View of Coral Reefs.** Coral reefs are colorful, complex communities. (a) A Caribbean reef. (b) A Pacific reef.

Atlantic, but Pacific reefs have a greater abundance of soft corals (subclass Alcyonaria). Most Atlantic corals are nocturnal but many Pacific species are diurnal. Atlantic corals often reproduce by fragmentation, whereas Indo-Pacific forms most often propagate by sexual reproduction. Finally, the diversity of coral species is far greater in the Indo-Pacific region than in the Atlantic. The Indo-Pacific has 500 species of stony corals, whereas the Atlantic Ocean has only about 62 species. The dominant coral genus *Acropora* has about 200 species in the Indo-Pacific but only 3 in the Atlantic.

The associated communities of Atlantic and Indo-Pacific reefs also differ. There are more than 5,000 species of molluscs and about 2,200 species of bony fish associated with reefs in the Indo-Pacific versus only about 1,200 mollusc and 550 fish species in the Atlantic. Sponges on Atlantic reefs have far greater biomass than those of the Indo-Pacific and most do not have phototrophic symbionts. Unlike the Pacific, the Atlantic Ocean has no giant clams or sea star species that are major coral predators.

## In Summary

Indo-Pacific reefs are of greater antiquity, have a greater diversity of organisms, have more extensive algal ridges, grow to 60 meters or less in depth, have a far greater depth of reef carbonates, and have coral surface coverage approaching 100%. Atlantic reefs have a deeper buttress zone, have coral growth extending to 100 meters, are commonly dominated by the hydrozoan *Millipora complanata* (fire coral), have a greater biomass of sponges, and have coral coverage rarely exceeding 60%. Atolls are much more common in the Pacific than the Atlantic. Most Atlantic corals are nocturnal but many Pacific species are diurnal. Atlantic corals often recruit by fragmentation, whereas Indo-Pacific forms most often recruit by sexual reproduction. ●

## REEF PRODUCTIVITY

Despite living in nutrient-poor tropical ocean water, coral reefs abound in life and are one of the most productive of all ecosystems. The net annual primary productivity on coral reefs (for more information on primary production see Chapter 2) is estimated to be 2,500 grams of carbon per square meter, a value comparable to tropical rainforests. The annual primary productivity of tropical oceans, in contrast, is estimated to be less than 50 grams of carbon per square meter.

### Source of Nutrients

The high level of productivity of coral reefs requires an abundant source of nutrients, particularly nitrogen and phosphorus. One can assume that reefs close to continents or islands receive nutrient runoff from the land, but the source of the nutrients supporting atolls is unclear. Although some researchers have suggested that upwelling or high flow rates of water across the reef are enough to supply the needed nutrients, many believe that, like rainforests,

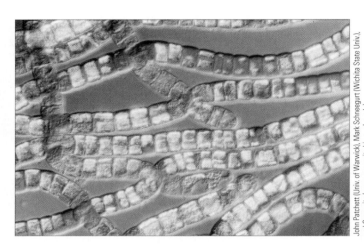

**Figure 15-16 Wind-driven Cyanobacteria.** Cyanobacteria such as *Trichodesmium* are able to fix atmospheric nitrogen. Wind and currents may deposit them on reefs where their subsequent decomposition may supply important nutrients to the community.

reefs have accumulated their nutrients over time. Reefs may be able to maintain their high levels of productivity because they recycle nutrients so efficiently. It has also been suggested that the large amount of nutrients tied up in dissolved and particulate organic matter can be concentrated by reef bacteria or removed by the biofiltering activity of coral and other reef organisms. The accumulation of wind-driven cyanobacteria such as *Trichodesmium* (Figure 15-16) from ocean waters and their subsequent decomposition can likewise supply needed nutrients to the reef. In addition, these cyanobacteria and those associated as symbionts of sponges and other organisms can convert atmospheric nitrogen into forms usable by reef organisms. Many fish species feed away from the reef at night but return to it during the day. Their fecal matter can also be an important source of nutrients to the reef. The reef, therefore, is a nutrient sink for material brought from the outside.

### Reef Photosynthesizers

Reef photosynthetic organisms include the zooxanthellae of the corals and other organisms, benthic algae, turf algae, sand algae, phytoplankton, and seagrasses. Their density is far greater than that in a similar area of open tropical ocean, and their combined biomass is greater than that of the animals inhabiting the reef. The close association of zooxanthellae with the corals and other organisms and a similar association of cyanobacteria with sponges means that nutrients are quickly recycled and not easily lost to the surrounding water. Most other photosynthetic forms are firmly attached to the reef, and most free-living forms contain calcium carbonate, making them too heavy to be easily removed from the reef. Of all the photosynthetic organisms present on the reef, the turf algae seem to be the most important because they process the most organic carbon (Table 15-1).

| Producer community | Productivity ($g\ C\ m^{-2}d^{-1}$) | Approx. cover (%) |
|---|---|---|
| Fleshy benthic algae | 0.1–4.0 | 0.1–5.0 |
| Crustose coralline algae | 0.9–5.0 | 10–50 |
| Turf algae | 1.0–6.0 | 10–50 |
| Zooxanthellae | 0.6 | 10–50 |
| Sand algae | 0.1–0.5 | 10–50 |
| Phytoplankton | 0.1–0.5 | 10–50 |
| Seagrasses | 1.0–7.0 | 0–40 |

**Table 15-1  Primary Productivity of Component Producer Communities on Coral Reefs and Their Distribution**

Source: Larkum, A. W. D. 1983. Primary Productivity of Plant Communities. In *Perspectives on Coral Reefs,* ed. D. J. Barnes, pp. 221–30. Townsville, Australia: The Australian Institute of Marine Science; Chisholm, John R. M. 2003. Primary Productivity of Reef-Building Crustose Coralline Algae, *Limnology and Oceanography* 48(4):1376–87.

## Reef Succession

The ratio of primary production (P, or gross photosynthesis) to community respiration (R), called the **P-R ratio,** is sometimes used as a measure of the state of development (or succession) of a biological community. When the P-R ratio is greater than 1, primary production exceeds the respiratory needs of the community, biomass increases, and the excess biomass is available for growth or harvesting. As systems mature, the P-R ratio approaches 1, little biomass remains available for growth, and the community reaches a steady state or climax. P-R ratios measured for coral reefs are typically close to 1, indicating that the high productivity levels of coral reefs are balanced by high rates of community respiration, efficient nutrient recycling maintains their high productivity, and like tropical rain forests, nutrients are contained in the biomass of the community's inhabitants. This ecosystem is designed to survive in low-nutrient waters by efficiently recycling its nutrients. Increases in productivity are often a result of **eutrophication** (nutrient enrichment), typically manifested as a dramatic proliferation of algae. Unless increased grazing controls the algae, they will grow over the corals and smother them.

## In Summary

Despite existing in nutrient-poor tropical ocean water, coral reefs are one of the most productive of all ecosystems. It is assumed that coral reefs have accumulated their nutrients over time and recycle them so efficiently that they are able to maintain their high levels of productivity. It has also been suggested that the large amount of nutrients tied up in dissolved and particulate organic matter can be trapped by reef bacteria or removed by the biofiltering activity of coral and other reef organisms. Reef photosynthetic organisms include the zooxanthellae of the corals and other organisms, benthic algae, turf algae, sand algae, phytoplankton, and seagrasses. Of photosynthetic organisms present on the reef, the turf algae and zooxanthellae produce the most organic carbon. ●

# CORAL REEF ECOLOGY

Hundreds of species of marine organisms are associated with coral reefs. Not only does coral provide a foundation for the reef food web, their colonies also provide shelter for resident organisms. Close up, the reefs are multicolored, complex structures that support a variety of beautiful and sometimes bizarre creatures. In this complex world of branches, boulders, caves, tunnels, and mazes swarms a multitude of living organisms.

## Coral Reef Community

The coral reef community is made up of an intricate array of producers and consumers. Among the consumers are representatives of many groups of organisms. Some are sessile or sedentary, whereas others actively move through the reef in search of food, shelter, and mates.

### Sponges and Cnidarians

Sponges and many cnidarians are sessile organisms that sift the water currents and sediments in search of food. Large sponges inhabit the deeper portions of the reef front, whereas smaller species inhabit the waters behind the reef. The sea anemone (Figure 15-17) and other cnidarians use stinging cells (cnidocytes) located on their tentacles to capture plankton, small crustaceans, and small fishes that pass by. Sea anemones usually remain fixed in one place, firmly attached to the bottom by means of their muscular, mucus-secreting basal disk. If conditions become unfavorable, however, some species detach and, by undulating their basal disk, crawl snail-like along the bottom at 10 centimeters (about 4 inches) per hour until they find a more suitable area. Other species turn themselves upside down and walk on the tips of their tentacles, whereas others lie on their sides and ripple their bodies, moving along like oversized inchworms.

Richard N. Mariscal

**Figure 15-17  Sea Anemones.** The giant Caribbean anemone, *Condylactis gigantea,* is a common inhabitant of coral reefs in Florida and the Caribbean.

Jon L. Hawker

**Figure 15-18 Reef Filter Feeders.** Reef-dwellers like these annelid Christmas tree worms filter their food from the surrounding water.

## Annelids

Annelids found on coral reefs include featherduster worms (family Sabellidae), Christmas tree worms (family Serpulidae; Figure 15-18), and fireworms (family Amphinomidae). Both featherduster and Christmas tree worms are sessile filter feeders, but fireworms and their relatives are mobile predators. Featherduster worms form tubes of sand and small particles held together by mucus. Christmas tree worms produce calcium carbonate tubes that can contribute to the formation of the reef. Fireworms crawl across the reef surface and feed on coral polyps, colonial anemones, and other sedentary animals. Palolo worms (*Eunice*) burrow through coral and weaken it. They are usually deposit feeders, ingesting organic material found in the sediments.

## Crustaceans

Crustaceans found on reefs include shrimps, crabs, and lobsters. They vary in their feeding habits from active hunters to parasites on other species. Mantis shrimp (order Stomatopoda) have forward appendages that are shaped like the forelimbs of a praying mantis (Figure 15-19). Their fore-

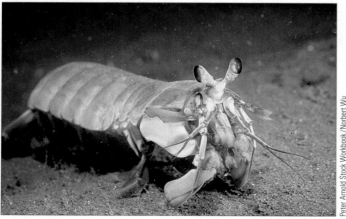

Peter Arnold Stock Workbook / Norbert Wu

**Figure 15-19 Reef-Dwelling Crustaceans.** The anterior appendages of this mantis shrimp are used to capture prey.

limbs are so sharp that they can cut another shrimp in half when they strike. They hide in crevices and under coral, where they wait for their prey to come close enough to attack. Snapping shrimp (*Alpheus*) use a modified cheliped to produce snapping sounds involved in territorial defense. They sometimes use the sound to stun the small fishes on which they feed. Pea crabs (family Pinnotheridae) live on molluscs or sea urchins, either as commensals (feeding on detritus covering their hosts) or parasites.

## Molluscs

Reefs are home to a wide variety of molluscs (for more information on molluscs see Chapter 9). Cowries (family Cypraeaidae; Figure 15-20a) are mainly nocturnal gastropods that come out at night to graze on algae, sponges, and coral polyps. They do not actually feed on the polyps themselves but on the algae growing on their surface. During the day, these animals hide in cracks and crevices and beneath coral formations. Large helmet snails (*Cassis*) partially bury themselves in the sand pockets of the reef. At night they come out to feed on sea urchins and other echinoderms. On Indo-Pacific reefs, the Triton's trumpet snail (*Charonia tritonis*) searches for sea urchins, sea stars, and molluscs that make up its diet (see Figure 15-20b). A similar species, *Charonia variegata,* fills the same niche on Caribbean reefs. Many species of cone snails (family Conidae) are found on reefs. Some live in the cracks and crevices of a reef, and others are found under rocks and pieces of coral or buried in the sand. These venomous snails use their modified radula to sting and paralyze prey. A variety of naked snails (nudibranchs), sporting amazing colors and patterns, feed on sponges, anemones, and jellyfish (see Figure 15-20c).

One of the largest molluscs on Indo-Pacific reefs is the giant clam (*Tridacna gigas;* Figure 15-21). Growing as long as 1 meter (3.3 feet) and sometimes weighing more than 180 kilograms (400 pounds), the shells of these huge clams contribute a great deal of mass to the reef. Like other clams, the giant clam is a filter feeder, but the quantity of plankton available on a coral reef is not great enough to sustain such a large animal. To supplement its energy needs, the clam hosts symbiotic zooxanthellae in its mantle. The clam provides carbon dioxide and nitrogen to the zooxanthellae, which photosynthesize and return oxygen and carbohydrates to the clam. Sometimes the giant clam "harvests" some of its symbionts to supplement its diet. In many areas of the Pacific Ocean, this magnificent clam is overfished and endangered.

Included among the molluscs are such active predators as the octopus (Figure 15-22) and squid. The octopus has keen eyesight, a well-developed nervous system, and the ability to rapidly change its color to blend in with its background or as a social signal (for more information regarding octopods see Chapter 9). Octopods feed on fish, molluscs, and crustaceans, especially crabs. They are nocturnal hunters, returning by day to a den. The den can be a crevice in the reef, a hollowed portion under the coral, or the deserted shell of a large mollusc. The opening to the den is quite small, not much larger than the diameter of one of the octopus's arms. Octopods have an amazing ability to squeeze through small spaces, where potential predators cannot gain access.

(a)

(b)

(c)

**Figure 15-20 Reef-Dwelling Molluscs.** (a) A cowry from the tropical Pacific feeds on the algae that grow on the reef. (b) This Triton's trumpet snail is attacking a crown-of-thorns sea star, which feeds on corals. (c) Many colorful nudibranchs make their home on the reef.

**Figure 15-21 Giant Clam.** Giant clams like this one on the Great Barrier Reef contribute to the mass of coral reefs with their large, heavy shells. They also maintain a symbiotic relationship with zooxanthellae similar to that of corals, contributing to primary production on the reef. Notice how the opening of the clam is facing upward. This allows the animal to drape its mantle over the edges of its shell so as to give the symbiotic zooxanthellae maximum exposure to sunlight.

**Figure 15-22 Reef Octopus.** This octopus is about to feed on the sand crab that it has just captured.

At the entrance of the den, piles of crab carapaces and mollusc shells accumulate, the leftovers of previous meals.

## Echinoderms

Echinoderms found in the reef community include feather stars, sea urchins, brittle stars, sea stars, and sea cucumbers. Feather stars (crinoids) attach themselves to coral with featherlike appendages called cirri. They feed by forming a basket of mucus with their lacy, cilia-covered tentacles and moving them in such a way as to create a current that carries the food to the animal's mouth. Though generally sedentary, the feather star can move by creeping along on

# Destruction of Coral Reefs by Sea Stars

Overpopulation of coral-feeding organisms is another factor causing the loss of coral reefs. Since 1957 about one third of the corals inhabiting Indo-Pacific reefs have been lost to predation by large populations of the crown-of-thorns sea star (*Acanthaster planci;* Figure 15-A). Drastic measures, generally with little success, were taken in the 1960s to deal with destruction on the Great Barrier Reef by this species, but the number of sea stars eventually declined naturally.

Although population explosions of this species seem to be a natural phenomenon, the triggering mechanism is still hotly debated. Some have suggested that the removal of sea star predators like the gastropod *Charonia tritonis* and several fish species by humans have left the crown-of-thorns sea star populations unchecked. Another suggestion was that nutrient runoff in wet

Marilyn and Maris Kazmers/Dembinsky Photo Associates

**Figure 15-A** *Sea Star Reef Destruction. Population explosions of the coral-eating crown-of-thorns sea star (*Acanthaster planci) have devastated many Pacific reefs. These population explosions seem to be a natural phenomenon, but the triggering mechanism is unknown.*

years allowed larger phytoplankton blooms, providing increased food for *Acanthaster* larvae and increasing their numbers. Still another theory suggests that reef destruction by storms causes populations of sea stars to condense and attack the remaining coral reefs en masse. Whatever the cause, explosions of this species cause a dilemma for the managers of coral reef parks in Australia. Currently, sea stars are killed if they threaten particularly important parts of the reef, but otherwise, a population explosion is allowed to run its course. ●

its cirri or using its arms in a swimming movement. Sea urchins graze on the algae that cover some of the rock and coral. Brittle stars scavenge the bottom for food; sea stars feed on molluscs and even the coral polyps themselves (see box). Sea cucumbers eat sediment or strain plankton from the water with anterior tentacles. These echinoderms commonly eviscerate their internal organs when stressed. When a predator sees these organs, it will typically eat them, giving the sea cucumber an opportunity to escape into a crevice or burrow into the sand. Later, the animal regenerates the lost organs. Some species have another defense mechanism. When disturbed they shoot sticky tubules (Cuvierian tubules) from the anus that enmesh small predators.

## Reef Fishes

As a group, the beautifully colored reef fishes are the most prominent and diverse inhabitants of the reef. A single patch of coral may provide habitat for 70 to 80 species. Fish exploit just about every available energy source, from detritus to algae, sponges, coral, various invertebrates, and other fish. How so many species of fish can survive on coral reefs with so much competition is a question that has intrigued scientists for many years. We will explore this question in a later section.

## Species Interactions on Coral Reefs

Space is at a premium on reefs; corals, an algal turf, sponges, or other organisms cover virtually every surface. Because both corals and algae require light to survive, access to light, like space, is also a resource subject to competition.

### Competition among Corals

Fast-growing, branching corals can grow over slower-growing, encrusting, or massive corals and deny them light. In response, the slower-growing forms can extend stinging filaments from their digestive cavity and kill their competitor's polyps. Undamaged polyps on the faster-growing, branching coral, however, may grow very long **sweeper tentacles,** containing powerful nematocysts (stingers), that kill polyps on the slower-growing form. The faster-growing form repairs the damage and continues to overgrow its competitor. In addition to sweeper tentacles and stinging filaments, corals have several other mechanisms available for attack or defense.

In general, slower-growing corals are more aggressive than fast-growing species. In cases where a competitor cannot be overcome, however, corals may survive by taking advantage of differences in local habitats. Massive corals are generally more shade tolerant and able to survive at greater depths. Therefore, on many reefs it is the fast-growing, branching corals that ultimately dominate at the upper, shallower portion of the reef, whereas more massive forms dominate in deeper areas.

## Competition between Corals and Other Reef Organisms

Corals also must compete with other reef organisms, each with its own strategies for survival. Sponges, soft corals, and algae can overgrow stony corals and smother them. Algae are competitively superior to corals in shallow water but less so at depth. Survival of coral in shallow water, therefore, may depend on grazing by herbivorous echinoderms and fishes. In Jamaica, overfishing removed most of the herbivorous fish from coral reefs. Initially, algal growth was kept in check by the grazing sea urchin *Diadema antillarum* (Figure 15-23), but in 1982 a pathogen reduced the population by 99%. As a result, coral cover declined from 52% to 3% and algal cover increased from an average of 4% to 92%. Without grazers, the algae were able to completely overgrow the coral.

### Other Competitive Interactions

Competition may occur among other reef inhabitants. The hydrozoan *Millipora squarrosa* commonly overgrows the gorgonian *Plexaura* or *Briareum,* for example, and fast-growing colonial invertebrates inhabiting coral surfaces can overgrow many other species.

### Effect of Grazing

We have already noted the importance of grazing by urchins and fishes in preventing seaweeds from overgrowing the reef. The dominant algae on a healthy reef are usually

**Figure 15-23  Sea Urchin.** *Diadema antillarum,* the long-spined black sea urchin, has been termed a "keystone species" because of its importance in controlling macroalgae on reefs. In 1982, a pathogen reduced their populations by 99%. Populations have still not recovered, resulting in macroalgal overgrowth of some Caribbean reefs. Their long spines contain toxin that results in a beelike sting.

fast-growing filamentous forms or coralline algae, well protected by calcification and the production of noxious chemicals. These species are inferior competitors to larger, fleshier seaweeds, so grazing allows them and other species to persist. Herbivory is greatest in the shallow forereef areas but decreases with depth, where lower temperatures and light reduce algal growth. The reef is, therefore, a mosaic of microhabitats with different levels of grazing and different algal communities.

An additional complexity arises from the activity of damselfish (family Pomacentridae). Because they are territorial, many damselfish species exclude grazers and other species from certain areas of the reef. Algae grow rapidly in these territories, providing habitat for many small invertebrates but overgrowing the corals. Branching corals tend to dominate in damselfish territories because they are upright and faster growing than the more massive or encrusting forms. Damselfish also control the algal diversity by weeding out undesirable species, making their territories, in many ways, "algal farms." Because damselfish territories can occupy up to 60% of the reef, they are very important in determining its structure.

### Effect of Predation

Although less studied than on rocky shores, predation almost certainly has a significant influence on the community structure of coral reefs. Fish and other predators may preferentially prey on such competitors of corals as sponges, soft corals, and gorgonians, giving competitively inferior reef corals an advantage in securing space. Many species of fish, molluscs, annelids, crustaceans, and echinoderms also feed directly on coral polyps or their mucus. Several surgeonfish (family Acanthuridae) and parrotfish (family Scaridae) may actually pass coral skeletons through their digestive tracts and add sediment to the reef. Both fish and invertebrate corallivores (coral-feeding organisms) seem to attack faster-growing, branching species preferentially, perhaps preventing slower-growing forms from being overgrown. Corallivores, however, rarely ever completely destroy a coral colony except in cases where tropical storms or humans have already done severe damage. The fact that almost all small invertebrates on reefs are so well hidden or highly camouflaged is another indicator of how prevalent predation is on reefs and its importance in determining reef structure.

## Coral Reef Fish Ecology

The diversity and abundance of fishes associated with a coral reef is far greater than that found in any other marine habitat. Although reef structure supplies an abundance of microhabitats for fish, habitat alone is not enough to explain their remarkable diversity. The **competitive exclusion principle** (see Chapter 2) suggests that no two species can occupy the same niche. Yet, 60% to 70% of reef fish can be described as general carnivores, the rest being grazers on coral and algae (about 15%) or omnivorous. Several hypotheses have been proposed to explain the great diversity of these carnivores. One hypothesis, the **competition model,** suggests that factors such as time of day or night, size of prey, position in the water column, and so on provide each species with a unique niche, hence there is no competition. In contrast, the **predation disturbance model** assumes competition among species but suggests that the effect of predation or other causes of mortality keep populations low enough to prevent competitive exclusion. The **lottery model** also assumes competition occurs but suggests that it is unimportant; chance determines which species of larvae settling from the plankton colonize a particular area of reef. As each individual is lost over time, chance again determines which species will occupy the available space. All these models assume that there is no limitation in the availability of fish larvae. The **resource limitation model,** however, suggests that available larvae are limited and that limitation prevents fish populations from ever reaching the carrying capacity of the habitat. Although different researchers have provided data that support one or more of these models, it remains unclear which, if any, best explains how so many species of fish can exist on coral reefs.

## In Summary

Where corals flourish, so do large numbers of other organisms. The corals not only provide a foundation for the food web, but they also provide shelter for resident animals. Competition is very prevalent because space and light are limited. Fast-growing branching corals may grow over slower-growing species, denying them light. The slower-growing forms may extend stinging filaments from their digestive cavity and kill their competitor's polyps. In response, fast-growing corals may produce very long sweeper tentacles containing powerful nematocysts that kill their competitor's polyps. Despite these interactions, fast-growing branching corals ultimately dominate at the upper, shallower portion of the reef, whereas more massive forms, being more shade tolerant, dominate in deeper areas. Corals also compete with other reef organisms, each with its own aggressive and defensive capabilities. Fish and other predators may preferentially prey on such competitors of corals as sponges, soft corals, and gorgonians, giving competitively inferior reef corals an advantage in securing space. Most fish are carnivores. How so many fish species coexist on coral reefs without being excluded by competition is a source of debate.  ●

## THREATS TO CORAL REEFS

The health of coral reefs is affected by both physical and biological phenomena. Storms and climate changes can severely impact coral reefs. Biological factors include disease and predation. Human interaction with coral reefs can also negatively affect the growth and distribution of fragile coral ecosystems.

## Effect of Physical Changes on the Health of Coral Reefs

Hurricanes and typhoons cause massive destruction by toppling coral formations and removing them from the reef. Another physical factor causing reef destruction is the El Niño Southern Oscillation (ENSO), which changes winds, ocean currents, temperatures, rainfall, and atmospheric pressure over large areas of tropical and subtropical ocean (for more information regarding ENSO events see Chapter 4). The 1982–83 ENSO raised water temperatures 2° to 4°C above normal, lowered sea level by 44 centimeters (17 inches), and spawned a number of massive storms, resulting in the destruction of 50% to 98% of the corals on some reefs of the East Pacific and significant destruction elsewhere. The destruction of corals also affected organisms associated with the reef, eliminating many species but encouraging increased populations of others. The corals that survived were subject to increased predation by urchins and sea stars, because natural controls protecting them were lost. On many reefs of the East Pacific, little recovery has been seen. An equally severe ENSO occurred in 1997–98 with devastating effects throughout the Pacific.

## Coral Bleaching

The incidence of **coral bleaching** (Figure 15-24a), a phenomenon by which corals expel their symbiotic zooxanthellae, has most often been associated with the warming of ocean waters by the ENSO or global warming in general, but the actual cause is unclear. Bleaching is a stress response; temperature, amount of radiation, aerial exposure, sedimentation, or a combination of factors could be the trigger. Typically, corals lose 60% to 90% of their zooxanthellae and grow pale in appearance. If the stress is not too severe, many coral species regain their symbionts within a few weeks or months and recover. However, if the stress is prolonged, the corals fail to regain the zooxanthellae and eventually die. Although coral bleaching has been observed for more than 100 years, the incidence has greatly increased throughout the world since 1980. Bleaching associated with the most recent ENSO killed many corals in the Indian Ocean, although those in the Caribbean and on the Great Barrier Reef recovered when the waters cooled.

## Coral Diseases

About 30 diseases associated with significant coral mortality worldwide have been identified since the 1970s. The most common are black band disease, white pox, white plague, white band disease, and yellow blotch disease. Presumably, these diseases were present before the 1970s but remained unknown because of their low incidence.

### Black Band Disease

Black band disease (see Figure 15-24b) is the best-known and studied coral disease. It seems to be caused by a combination of several species of bacteria and a cyanobacterium. In this disease, a distinct dark band of bacteria migrates across the living coral tissue, leaving behind a bare white skeleton. It can spread up to 1 centimeter per day, much faster than the coral can grow. The lack of oxygen and high levels of hydrogen sulfide produced by the microbes presumably kill the coral tissue. The bacteria then use organic compounds from the dying coral. The mode of infection is unclear but black band disease is most prevalent in

(a)

(b)

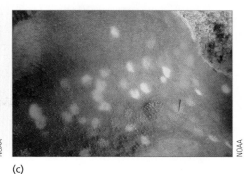
(c)

**Figure 15-24 Coral Diseases.** Coral diseases are having a devastating effect on many reefs. (a) In coral bleaching disease, the zooxanthellae are expelled, possibly as a response to temperature stress. If they are not reacquired within a reasonable time, the coral colony dies. (b) Blackband disease is caused by a consortium of bacteria and cyanobacteria. Because the bacterial species are naturally occurring on reefs, it is unclear what triggers the attack. (c) White pox disease is caused by the fecal bacterium *Serratia marcescens* and has been found in 18 species of corals, killing many within 3 days.

areas where corals experience stress from high temperature or pollution. Although only about 2% of coral is affected worldwide, this disorder is significant because it commonly attacks massive reef-building corals. The disorder was first observed on Atlantic reefs but is now found worldwide.

## White Pox

Since it was first documented in 1996, white pox (see Figure 15-24c) has killed 85% of the elkhorn corals in the Florida Keys and significant numbers of this species throughout the Caribbean. It is highly contagious and grows as rapidly as 2.5 square centimeters per day. The disease is characterized by white lesions and is caused by the bacterium *Serratia marcescens,* found in human and animal feces.

## Other Coral Diseases

Three other coral diseases of importance are white band disease, white plague, and yellow blotch disease. White band disease causes tissue to evenly peel off the skeletons of staghorn and elkhorn corals (*Acropora*). White plague is a similar-appearing disorder that attacks up to 33 species of stony corals. In yellow blotch disease, a spreading yellow patch appears on the coral, the coral dies in the center of the patch, and sediment fills the area. In the Caribbean, this disorder affects only certain members of the genus *Montastraea,* but a similar disorder in the Arabian Gulf affects a number of other genera. The causative agent for one form of white plague disease is the bacterium *Aurantimonas coralicida* but the causes of yellow blotch and white band disease are unknown.

## Human Impact on Coral Reefs

Certainly, human activities have a significant impact on the health of coral reefs. Coastal development can increase runoff and nutrient load, change patterns of water flow, increase sedimentation, and encourage overfishing. We have already discussed the effect of overfishing on Jamaican reefs, noted coral diseases caused by human-sewage bacteria, and remarked that few coral species can withstand heavy sedimentation.

The effect of runoff and nutrient overload on coral reefs has been well documented in long-term studies in Kaneʻohe Bay, Hawaii. Once renowned for its coral gardens, Kaneʻohe Bay was only sparsely populated by humans before 1950. Subsequent increase in development and population brought significant modification of the shoreline, increased runoff, and the release of up to 5 million gallons of sewage directly into the bay each day. The nutrient-rich sewage greatly increased phytoplankton levels and encouraged the growth of seaweeds, particularly green bubble algae (*Dictyosphaeria cavernosa*). Seaweeds overgrew the corals, smothering them, and phytoplankton made the water cloudy. In response to public outcry, sewage water was diverted offshore in the 1980s. Although reefs in the bay have slowly recovered, land runoff and periodic suspension by storms of nutrients stored in sediments continue to cause problems. In the southern part of the bay, bubble algae seems to be making a comeback and excessive growth of the red algal species *Kappaphycus striatum,* introduced from the Philippines in a failed aquaculture experiment, has become a significant problem. Herbivorous fish have failed to control these undesirable species either because of their reduced populations (perhaps a result of overfishing) or the abundance and lack of palatability of these algae. Although the eutrophication of Kaneʻohe Bay's coral reefs is the best-studied example, this problem and others associated with human activities are serious threats to coral reefs worldwide.

## In Summary

The health of coral reefs is affected by both physical and biological phenomena. Physical factors include hurricanes, typhoons, and the El Niño Southern Oscillation (ENSO), which changes winds, ocean currents, temperatures, rainfall, and atmospheric pressure over

# Why Are Florida's Reefs Disappearing?

The Florida Keys are a chain of islands that stretches from Key Biscayne, 5 miles off the coast of Miami, to Key West, 135 miles away (Figure 15-B). The waters that surround these islands contain the only coral reefs that are associated with the continental United States. Florida's reefs contain fewer species of coral and are not as complex as reefs farther south in the Caribbean. Their northern location and colder water during the winter limits the number of tropical species that can exist there. Historically, the southernmost reefs near Key West have been the most diverse and best developed, whereas those located farther north have less diversity and coral growth. In recent years, this pattern has changed dramatically.

Florida's reefs are currently showing signs of significant damage in spite of laws protecting them. Over the past 20 years, scientists have noted major increases in phytoplankton blooms, changes in fish

*Figure 15-B* **The Florida Bay Water Hypothesis.** *(a) Before construction of drainage canals in the 1960s, freshwater flowed south from Lake Okeechobee through the Everglades and into Florida Bay, forming a very productive estuary (light blue color). The less-dense water from the estuary then flowed through the channels in the Florida Keys out and over the coral reefs. (b) The Florida Bay water hypothesis suggests that diversion of freshwater from the Everglades to Biscayne Bay may have caused the water in Florida Bay to become hypersaline (darker blue color). As it flows through the Keys across the reefs, the dense water sinks and damages and kills the corals. It now seems that this hypothesis is incorrect; the true source of the problem is nutrient enrichment from agricultural lands in the Everglades.*

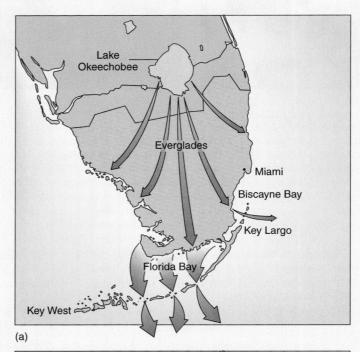

(a)

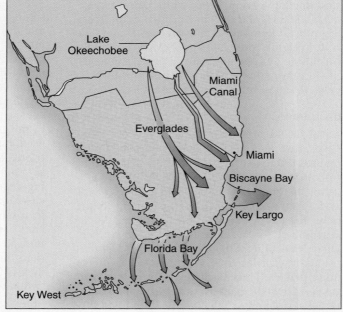

(b)

populations, and losses of seagrass beds, sponges, and coral.

Researchers James Porter and Ouida Meier of the University of Georgia found that, between 1984 and 1989, some reefs lost as much as 29% of their coral species and five out of six reefs surveyed lost as much as 45% of their coverage. Macroalgae have overgrown many coral reefs, and coral diseases appear to be spreading. Surprisingly, of the six study sites examined, the only site that had increases in live coral coverage was the northern site near Key Biscayne. Because of its proximity to metropolitan Miami and coastal pollution, researchers expected this reef to show the greatest amount of damage. When they found that coastal pollution did not seem to be the cause of the problem, researchers were puzzled.

Rather than just concentrating on the immediate area around the reefs, researchers began to look at environmental changes throughout southern Florida. This approach led to the development of the **Florida Bay water hypothesis,** which suggests that the reduced water flow into Florida Bay from the Everglades has led to hypersaline conditions that in turn led to massive losses of the seagrass beds in that area. The decomposition of seagrass and resulting sediment suspension released nutrients that generated algal blooms and other changes.

In the past, much of the rain that fell over central and southern Florida collected in Lake Okeechobee. Lake water drained through the Everglades, the vast wetland that covers the southern end of the Florida peninsula, into Florida Bay, where it mixed with seawater to form a productive estuary (see part a of Figure 15-B). In 1961, to prevent flooding of low-lying wetlands and to make more land available for agricultural use, the U.S. Army Corps of Engineers be-

gan construction of an elaborate system of canals and levees. As a result of this action, the northern portion of the Everglades is now completely dry. The water that once drained into that area is diverted to the east, where it empties into Biscayne Bay. In turn, Florida Bay's water supply was also drastically reduced, which may account for the increased salinity and the resultant damage to the reef and estuary (see part b of Figure 15-B). The Army Corps of Engineers and South Florida Water Management District are now using this hypothesis as a rationale for pumping more freshwater into Florida Bay.

Recently, researchers have begun to question this rationale and whether the pumping of additional freshwater into Florida Bay is such a good idea. Indeed, recent studies have not demonstrated a relationship between high salinity and seagrass die-off in the bay. Much of the seagrass die-off actually occurred when salinity was near normal. Furthermore, if seagrass decomposition fueled the algal blooms, they should be downstream of the seagrass beds. Instead, the blooms are typically upstream of the beds. Local residents have also indicated that massive algal blooms occurred well before seagrass die-off.

An alternate hypothesis is that the algal blooms of Florida Bay are the result of phosphorus-rich waters from western Florida Bay meeting nitrogen-rich waters from the eastern portion. The normal phosphorus source in the west is primarily the erosion of rocks containing large amounts of phosphate along the Peace River. The availability of phosphate to the reefs has not changed in thousands of years, although it may have been somewhat enhanced by phosphate mining over past decades. However, nitrogen availability has greatly increased as farming increased. As noted earlier, much of the Everglades was

drained, converted to agricultural land, and put into sugarcane production. Initially, the nutrient-laden drainage water from sugarcane farming was pumped into Lake Okeechobee. By 1979 the lake had become so nutrient rich (eutrophic) that water was diverted to the Everglades, causing significant changes there as well. The completion of the South Dade Conveyance System at about the same time allowed water from the Everglades canal system to be carried farther away, into Florida Bay. This new system allowed farmers to better drain their lands and further encouraged the development of suburbs west of Miami. The amount of water flow greatly increased again in the early 1990s, resulting in both a 42% increase in nitrate and a 229% increase in ammonia levels in Florida Bay between 1989 and 1994. A later alteration of water flow directing more flow to Taylor Slough in the west and less flow to the east brought nitrogen-rich water even closer to the phosphorus-rich waters in western Florida Bay. Several studies have indicated that the mixing of this water has led to the large algal blooms in north central Florida Bay and altered the entire ecosystem. Some of this nutrient-laden water flows to the middle and lower Florida Keys where it may lead to eutrophic conditions, encouraging algae to overgrow corals and epiphytes to overgrow seagrasses.

Many researchers now believe that the Florida Bay water hypothesis has been disproved and that pumping more freshwater from the Everglades into Florida Bay will only increase algal blooms and the associated ecological problems rather than improve them. A recent proposal to open up more passages along the Florida Keys to allow more Florida Bay water to reach the coral reefs may lower water quality there and cause further degradation. ●

# Is Atmospheric Dust Contributing to the Decline of Coral Reefs?

Scientists continue to debate the reasons for the marked decline in coral reefs throughout the world and their lack of recovery. Although the increasing incidence of coral diseases, rampant overgrowth of coral by macroalgae, reef destruction by human activities, and the mass mortalities of reef-building corals and associated organisms has been widely reported, in relatively few cases has the root cause of reef destruction and lack of recovery been identified. Recently, Virginia Garrison and her colleagues at the U.S. Geological Survey proposed a new hypothesis for the decline of coral reefs. According to their hypothesis, African and Asian dust air masses transport chemical and microbial contaminants to the Americas, the Caribbean, and northern Pacific that adversely affect their coral reef ecosystems. Dust provides nutrients that are normally limited in these systems, encouraging overgrowth by macroalgae, and broadcasts bacterial, viral, or fungal spores that may cause diseases in reef organisms. In addition, industrial contaminants may weaken coral reef organisms and make them more susceptible to pathogens. This dust has been transported for millennia (Charles Darwin

described it in 1846), but the hypothesis suggests that the amount has increased over the last 40 years and the composition changed. Now the dust carries not only plant nutrients but also synthetic organic chemicals, viable microorganisms, trace metals, and other pollutants of human origin.

This hypothesis is far from being verified, but there is some evidence that tends to support it. Caribbean islands formed by coral (for example, Barbados) have red iron and clay-rich soils that are known to be of Saharan dust origin (Figure 15-C). It is also known that the Amazon rain forest in South America derives a significant amount of essential nutrients (mainly phosphate) from Saharan dust. Recent studies in Barbados have shown that the average annual surface concentration of African dust has increased since the 1960s, the period corresponding with rampant reef destruction. Dust levels reached a maximum during the 1983–84 ENSO. This increasing amount of dust may reflect changing land-use practices, a drier climate, and the desertification of parts of Africa. African dust consists primarily of clay soil minerals but also in-

cludes emissions from industry and biomass burning. We have a very limited understanding of the impact of this dust on marine ecosystems, although some researchers have associated it with red tides in the Gulf of Mexico. It has been further suggested that the iron in African dust fuels blooms of the cyanobacterium *Trichodesmium*. *Trichodesmium* can make nitrogen available to other organisms, including dinoflagellates, the organisms causing red tide.

We noted earlier that some species of bacteria are associated with certain coral diseases. Could African dust be a source of these microorganisms? Although we have no data on the oceanic transport of microorganisms in dust, the airborne transport of African locusts to eastern Caribbean islands has been well documented. It has also been demonstrated that the number of fungi and bacteria cultured from air samples is two to three times greater when African dust is in the air. A terrestrial fungus (*Aspergillus sydowii*) identified as the pathogen causing aspergillosis in Caribbean gorgonians (sea fans) was isolated from dust samples taken in the Virgin Islands during a dust event in 1997. The

large areas of tropical and subtropical ocean. Biological factors adversely affecting coral reefs include coral bleaching, black band disease, white pox, white band disease, white plague, and yellow blotch disease. Coastal development has a detrimental effect, increasing runoff and nutrient load, changing patterns of water flow, increasing sedimentation, and encouraging overfishing. •

# EVOLUTIONARY ADAPTATIONS OF REEF DWELLERS

The organisms that inhabit coral reefs display an amazing array of adaptations that help them to survive in this fiercely competitive community. Some have evolved anatomical features such as body armor or color patterns that increase

their chances of surviving and reproducing. Others have evolved adaptive behaviors that increase their success, and still others have formed symbiotic relationships to exploit more specialized niches.

## Protective Body Covering

Some reef animals avoid predation by having tough defensive exteriors. Trunkfishes (family Ostraciidae; Figure 15-25), for example, have a bony skin similar to the shell of a turtle. The armor is not without its drawbacks, because some mobility and growth potential were sacrificed for this protective exterior. Filefishes (family Monacanthidae) are relatives of the trunkfish. Small, prickly scales that give them the texture of sandpaper cover their skin. Triggerfishes (family Balistidae) also have tough, leathery skins that protect them from some predators.

**Figure 15-C Dust Transport.** *Dust is transported in two major global transport systems: (1) from the Sahara and Sahal of Africa to the Americas, Europe, and Near East; and (2) from the Takla Makan and Gobi deserts of China, across China, Korea, Japan, and the northern Pacific to North America. (From Garrison et al. 2003/ Graphic by Betsy Boynton/ USGS.)*

bacterium causing white plague in corals (*Aurantimonas coralicida*) may also be associated with dust, possibly explaining the spread of this disease far from human influence. Although the bacterial consortium causing black band disease may be naturally present on reefs, it has been postulated that the iron in dust may be the trigger that activates these bacteria and allows them to infect the coral. Although many questions remain, African dust is certainly one mechanism by which pathogens are distributed and deposited in both the Atlantic and Pacific. These pathogens along with transported nutrients and contaminants may be another factor causing the deterioration of coral reefs. ●

W. Gregory Brown/Animals Animals

**Figure 15-25 Protective Body Coverings.** This trunkfish from the Caribbean has a tough exterior similar to that of a turtle, making it unappealing to many predators.

## Protective Behaviors

Many behavioral mechanisms have evolved in reef fishes to avoid predation. When attacked, soapfishes (family Grammistidae; Figure 15-26) produce a sudsy coating of mucus that is poisonous and has an unappealing taste. Trunkfishes also produce a sudsy coating of poisonous mucus when disturbed. The pearly razorfish (*Hemipteronotus novacula*), which feeds on small molluscs on the floor of the reef, dives headfirst into the sand and buries itself when disturbed.

**Figure 15-26  Protective Behaviors.** This soapfish produces a toxic mucus when it is disturbed.

**Figure 15-27  Barracuda.** A barracuda patrols the reef off Grand Cayman Island in the Caribbean for fish to feed on.

Pufferfishes (family Tetraodontidae) inflate themselves to look like too much of a mouthful for a potential predator.

A much higher proportion of predators are more active at night than during the daytime. As a result, many fishes that are active in daylight, like parrotfishes (family Scaridae) and wrasses (family Labridae), seek places to hide as twilight approaches. These animals have well-developed eyesight that serves them well during the day, but at night they cannot see well enough to avoid predators. As night approaches, they converge on the reefs and wedge themselves into crevices for the night. Some species of parrotfish take the extra precaution of secreting a mucous cocoon that surrounds their body and masks their scent from nocturnal predators.

## Role of Color in Reef Organisms

Coloration plays a vital role in survival on a coral reef. Reef animals use color for camouflage to be less conspicuous to predators and prey. Color can communicate a warning or a willingness to mate. It also helps animals to advertise their territory or defend it against intruders.

### Color for Concealment and Protection

The coloration of barracuda (Figure 15-27) is a good example of the use of countershading for concealment and protection. Butterflyfishes (*Chaetodon*) exhibit disruptive coloration for this purpose (for more information regarding fish coloration see Chapter 10). Many reef fishes, however, are brilliantly colored. Although in the open water such colorful fishes would stand out, in the reef world, a maze of brilliantly colored corals, algae, and other sedentary organisms, a brightly colored fish simply blends in well. Some fishes, like the Nassau grouper (*Epinephelus striatus*) of the Caribbean, conceal themselves by changing color to match their background. Fishes are not the only animals that have mastered the art of camouflage. Many invertebrates, including shrimp, crabs, molluscs, and worms, also have colors and shapes that allow them to blend with their environment.

Some fishes sport bright colors during the day but take on duller, less conspicuous colors at night. The Spanish grunt (*Haemulon macrostomum*) is silver and yellow by day but becomes dull and blotched at night. The large yellow dorsal fin of the porkfish (*Anisotremus virginicus*) turns black at night. Even in bright moonlight, only a dull outline is visible.

### Other Types of Camouflage

Color is not the only means of camouflage employed by reef organisms; body shape also plays an important role. The thin bodies of trumpetfishes (family Aulostomidae) and pipefishes (family Syngnathidae) help to conceal them as they hover head down among gorgonians that sway gently in the currents. In the Indo-Pacific the extremely venomous stonefish (*Synanceja horrida*) can be found half buried in the sand. Its irregular body resembles rock complete with algal growth. When a small fish swims too close, the stonefish opens its mouth and swallows its unsuspecting prey.

### Warning Coloration

Camouflage is not the only role of coloration in reef organisms. Fishes that have an unpleasant taste or that have specific defensive weapons, such as sharp, venomous spines, frequently display warning coloration. Lionfishes (*Pterois*) with their venomous dorsal spines are an example of a strikingly colored reef fish displaying this type of coloration (for more information regarding warning coloration see Chapter 10).

### Other Roles of Color

Coloration plays a role in defending territories and in mating rituals. Some species, such as the harlequin tusk wrasse (*Lienardella fasciatus*) of Australia (Figure 15-28), display bright colors as they vigorously defend a territory. In essence, the bright coloration is a "no trespassing" sign to others of the same species.

Many species of fish take on special coloration at mating time. The bright colors of adult parrotfishes, for example, indicate the sex of an individual and help attract

Figure 15-28 **Harlequin Tusk Wrasse.** The harlequin tusk wrasse of the Great Barrier Reef displays bright colors while defending its territory.

Norbert Wu

mates. Male gobies also display bright colors at mating time. The brightly colored male attracts the attention of a female, and she is lured to his lair by an elaborate courtship dance.

## Symbiotic Relationships on Coral Reefs

The relationship between coral polyps and zooxanthellae is only one of hundreds of examples of symbiotic relationships between reef organisms. These symbioses range from parasitism through commensalisms to mutualism. Symbiotic relationships help organisms exploit niches and better survive in the fiercely competitive world of the coral reef.

### Cleaning Symbioses

At one time, marine biologists were puzzled by gatherings of many fish species at particular sites on a reef. It was later recognized that these fishes were waiting their turn to be cleaned by one of the reef's many cleaner organisms, such as the cleaner wrasse (*Labroides dimidiatus*). Cleaner wrasses, gobies, and other small fish feed on the parasites of larger fishes (Figure 15-29). These species set up a territory on the coral reef known as a **cleaning station** that "clients" frequently visit to have parasites and dead tissue removed. The cleaners benefit by obtaining a constant food supply, and the clients are kept healthy by the grooming. Wrasses and other species attract potential customers to their stations with their bright colors and a series of movements referred to as a *dance*. Even aggressive predators, such as the moray eel, that usually eat small fish like the wrasse, will submit to cleaning without harming the cleaner. Color changes or ritual postures by clients (also called hosts) may also be used to initiate and terminate the cleaning process. Several species of shrimp, including the peppermint shrimp (*Lysmata wurdemanni*; Figure 15-30), also engage in cleaning symbiosis.

Although the relationship between cleaners and their clients may superficially appear to be a mutualism, many cleaners often feed on their host's mucus, fins, and other healthy tissue in addition to removing parasites and dis-

Figure 15-29 **Symbiosis.** A cleaner wrasse removes parasites from the gills of the mappa pufferfish (*Arothron mappa*).

Chris Newbert / Minden Pictures

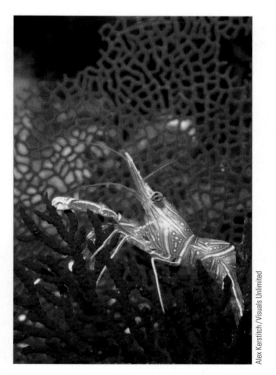

Alex Kerstitch / Visuals Unlimited

Figure 15-30 **Peppermint Shrimp.** This attractive peppermint shrimp (*Lysmata wurdemanni*) is a reef invertebrate that participates in cleaning behavior on other species. Although this species is generally nocturnal, it has recently become popular among aquarium hobbyists.

eased tissue. Indeed, removal of parasites may not be all that important. In most cases, experimental removal of cleaner fish from reefs has little effect on the overall health of the host's population. Hosts, therefore, may tolerate this behavior simply because they enjoy the tactile stimulation.

Some predatory species, such as the sabertooth blenny (*Aspidontus*), take advantage of the cleaning relationship to attack their prey. This blenny resembles a cleaner wrasse both in body shape and color. It goes so far as to mimic the cleaner wrasse's dance that attracts other fish to its cleaning station. When the unsuspecting fish comes near to be cleaned, the blenny attacks, biting pieces of flesh from the unwary victim.

## Other Symbiotic Relationships

Cleaning is not the only example of symbiosis in the coral reef community. The reef is full of examples of organisms that have evolved special relationships with other species. Clownfishes, for example, seek shelter from their enemies in the tentacles of large anemones (see Chapter 2). Pearlfishes (family Carapidae) live in the hindgut of sea cucumbers, the shells of oysters and clams, and in certain species of sea stars. Some species seem to enjoy a commensal relationship with their host, but others are parasites. The tiny conchfish (*Astrapogon stellatus*) is found only in the mantle cavity of the large queen conch (*Strombus gigas*), At night the conchfish emerges to feed on shrimp, sea lice, and other small crustaceans. On Indo-Pacific reefs certain species of gobies are found at the entrance to burrows dug by snapping shrimp. The shrimp digs the burrow and continuously clears it of debris. When predators appear, the gobies hide in the shrimp's burrow. The adult shrimp may benefit from the warning given by the gobies, but on the other hand, the gobies feed on young shrimp.

Hermit crabs rely on the empty shells of dead snails to provide them with a protective covering. As the crabs grow, they exchange smaller shells for larger ones. Some species of hermit crab attach anemones to their shells for added protection. The anemones are carefully transferred to new shells when an exchange is made. The crab massages the base of the anemone until it releases its hold on the old shell. The crab then transfers the anemone to the new shell and holds it in place until the anemone attaches. The anemones benefit by being carried to different feeding grounds. Some species of hermit crabs even feed their anemones by dropping morsels of food into their mouths. The boxer crab (*Dromidia;* Figure 15-31) attaches anemones to its claws, making them even more formidable weapons against potential predators.

Tiny shrimpfishes (family Centriscidae) hide between the spines of sea urchins, where they hover head down with their striped bodies looking very much like the urchin's spines. A species of cardinalfish seeks refuge in the spines of urchins and in return cleans its host's body.

These are only a few examples of the thousands of symbiotic relationships that occur in the coral reef community. In a sense, the entire coral reef community can be thought of as a huge, interactive complex where all the various organisms exist, bound by interdependencies.

## In Summary

Evolution has provided many reef animals with adaptations that help them to avoid predation or to be more efficient predators. Special body coverings protect some animals. Others produce toxic substances to deter predators or exhibit special behaviors that help them avoid predation. Color also plays a vital role in a reef animal's quest for survival. Some color patterns make animals difficult to see, allowing them to avoid predation or to sneak up on their prey. Color also plays a role in defending territories and in mating rituals. Color combines with body shape in some species to disguise and camouflage. The reef is a huge complex full of intricate interdependencies. These intimate relationships are one of the reasons that so many organisms can live and thrive in these communities. ●

**Figure 15-31 Boxer Crab.** The boxer crab uses anemones attached to its chelipeds as defensive weapons.

## SELECTED KEY TERMS

ahermatypic coral, *p. 346*

atoll, *p. 351*

back reef, *p. 352*

barrier reef, *p. 351*

bioerosion, *p. 349*

broadcast spawner, *p. 346*

brooders, *p. 346*

buttress zone, *p. 352*

cleaning station, *p. 369*

cnidocytes, *p. 347*

competition model, *p. 361*

competitive exclusion principle, *p. 361*

coral bleaching, *p. 362*

corallite, *p. 346*

dissolved organic matter, *p. 349*

drop-off, *p. 352*

eutrophication, *p. 357*

Florida Bay water hypothesis, *p. 365*

fragmentation, *p. 346*

fringing reef, *p. 351*

hermatypic coral, *p. 346*

lottery model, *p. 362*

nonsynchronous spawners, *p. 346*

P-R ratio, *p. 357*

patch reef, *p. 351*

predation disturbance model, *p. 362*

reef crest, *p. 352*

reef flat, *p. 352*

reef front or forereef, *p. 352*

resource limitation model, *p. 362*

spur-and-groove formation, *p. 352*

stony (or true) corals, *p. 346*

surge channel, *p. 352*

sweeper tentacles, *p. 360*

synchronous spawners, *p. 346*

zooxanthellae, *p. 345*

## QUESTIONS FOR REVIEW

### Multiple Choice

1. The type of reef that is separated from its associated landmass by a lagoon is called
   a. a barrier reef
   b. an atoll
   c. a fringing reef
   d. a table reef
   e. a patch reef

2. The portion of a reef that rises from the ocean's depths is the
   a. reef crest
   b. reef front
   c. back reef
   d. reef flat
   e. spur-and-groove formation

3. The corals that dominate areas of a reef that receive the most wave energy are
   a. fire corals
   b. elkhorn corals
   c. mushroom corals
   d. cluster corals
   e. flower corals

4. Corals supply their zooxanthellae with
   a. oxygen
   b. lipids
   c. sugars
   d. glycerol
   e. nitrogen wastes

5. Which of the following is the major primary producer on coral reefs?
   a. turf algae
   b. coralline algae
   c. seagrasses
   d. zooxanthellae
   e. phytoplankton

6. An echinoderm that feeds on coral polyps is the
   a. sand dollar
   b. sea cucumber
   c. crown-of-thorns sea star
   d. feather star
   e. pencil urchin

7. Many reef organisms exhibit color patterns and body shapes that help them to
   a. blend in with their background
   b. defend a territory
   c. attract a mate
   d. be more efficient predators
   e. all of the above

8. An example of an animal that exhibits warning coloration is the
   a. stonefish
   b. lionfish
   c. cleaner wrasse
   d. harlequin tusk wrasse
   e. surgeonfish

9. An important nocturnal predator of the coral reef is the
   a. parrotfish
   b. stingray
   c. cowry
   d. octopus
   e. wrasse

10. An important predator of sea stars and sea urchins is the
    a. octopus
    b. squid
    c. shark
    d. Triton's trumpet snail
    e. sea cucumber

## Short Answer

1. What role do coralline and calcareous algae play in reef formation?

2. Explain how the physical characteristics of the reef environment influence the species of corals that inhabit them.

3. Describe how a coral colony is formed.

4. Describe the process of reef formation.

5. If the water surrounding coral reefs is not very productive, how can the coral reef support large numbers of organisms?

6. Describe what is meant by *cleaning symbiosis.*

7. Why don't most coral species grow in the aphotic zone?

8. Why are coral reef fishes so brightly colored?

## Thinking Critically

1. Although illegal, some fishers in the Pacific Ocean region use charges of dynamite to stun fish, causing them to float to the surface, where they can be readily harvested. What effect would this practice have on nearby reef communities? What impact would this practice have on future fish catches?

2. What impact would daily visits by hundreds of sport divers and snorkelers have on a reef community?

3. Why are coral reefs not found along warm coastal areas where large rivers such as the Mississippi, Amazon, and Congo empty into the sea?

## SUGGESTIONS FOR FURTHER READING

Edmunds, Peter J., Robert C. Carpenter. 2001. Recovery of *Diadema antillarum* Reduces Macroalgal Cover and Increases Abundance of Juvenile Corals on a Caribbean Reef. *Proceedings of the National Academy of Sciences of the United States of America* 98(9):5067–71.

Jackson, Jeremy B. C. 1991. Adaptation and Diversity of Reef Corals, *BioScience* 41 (July–August): 475–82.

Jompa, Jamaluddin, and Laurence J. McCook. 2002. The Effects of Nutrients and Herbivory on Competition between a Hard Coral (*Porites cylindrica*) and a Brown Alga (*Lobophora variegata*), *Limnology and Oceanography* 47(2):527–34.

Kaplan, E. H. 1988. A Field Guide to Coral Reefs of the Caribbean and Florida Including Bermuda and the Bahamas. Boston, Mass.: Houghton Mifflin.

Larkam, A. W. D. 1983. The Primary Productivity of Plant Communities on Coral Reefs. In *Perspectives on Coral Reefs,* ed. D. J. Batches, pp. 221–30. Townsville, Australia: Australian Institute of Marine Science.

Levine, Joe. 1999. In Living Colors. *Natural History* 108(7): 40–47.

Porter, J. W., and O. W. Meier. 1992. Quantification of Loss and Change in Floridian Reef Coral Populations, *American Zoologist* 32(6):625–40.

Richter, Claudio, Mark Wunsch, and Mohammed Rasheed. 2001. Endoscopic Exploration of Red Sea Coral Reefs Reveals Dense Populations of Cavity-Dwelling Sponges, *Nature* 413(6857):726–30.

Sale, P. F., ed. 1991. *The Ecology of Fishes on Coral Reefs.* San Diego, Calif.: Academic Press.

Syms, Craig, and Geoffrey P. Jones. 2000. Disturbance, Habitat Structure, and the Dynamics of a Coral-Reef Fish Community, *Ecology* 81(10):2714–29

### InfoTrac College Edition Articles

Coles, S. L., and L. G. Eldredge. 2002. Nonindigenous Species Introductions on Coral Reefs: A Need for Information, *Pacific Science* 56(2).

Garrison, Virginia H., Eugene A. Shinn, William T. Foreman, Dale W. Griffin, Charles W. Holmes, Christina A. Kellogg, Michael S. Majewski, Laurie L. Richardson, Kim B. Ritchie, and Garriet W. Smith. 2003. African and Asian Dust: From Desert Soils to Coral Reefs, *BioScience* 53(5).

Pittock, A. Barrie. 1999. Coral Reefs and Environmental Change: Adaptation to What? *American Zoologist* 39(1).

Wolanski, Eric, Robert Richmond, Laurence McCook, and Hugh Sweatman. 2003. Mud, Marine Snow and Coral Reefs: The Survival of Coral Reefs Requires Integrated Watershed-Based Management Activities and Marine Conservation, *American Scientist* 91(1).

### Websites

**http://coralreef.noaa.gov/** This National Oceanic and Atmospheric Administration (NOAA) website has information on coral reefs and their health.

**http://www.coralreefnetwork.com** The Hawaii Coral Reef Network has links to information on coral reef ecology and reef organisms in Hawaii.

**http://www.motherjones.com/coral_reef/** This Mother Jones website provides information about threats to the health of coral reefs worldwide.

**http://www.uvi.edu/coral.reefer/** This University of the Virgin Islands website has information on corals and their biology.

# 16

# Continental Shelves and Neritic Zone

I n the relatively shallow waters above the continental shelves, plentiful sunlight and abundant nutrients combine to support enormous numbers of primary producers. Grazing on these producers are equally impressive numbers of consumers. In the water column, schools of fish feast on the abundance of food. On the bottom, large numbers of filter feeders, deposit feeders, and scavengers feast on the detritus that rains down from the sunlit waters above. In turn, bottom fishes prey on the abundance of benthic organisms or scavenge for food. In shallower waters with hard bottoms, forests and thickets of sizeable algae provide food and shelter for many animals.

The productive and bountiful coastal waters of the neritic zone are the source of most of the saltwater fish and shellfish that humans consume worldwide. Because of their high productivity, these areas are hotly contested by coastal countries, which vie for the exclusive rights to fish the waters. These areas are not infinitely productive, however, and overfishing in most parts of the world has started to affect the productivity of coastal seas.

## CONTINENTAL SHELVES

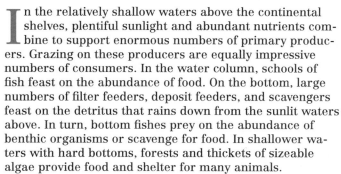

Coastal seas lie over continental shelves (for more information regarding continental shelves see Chapter 3). These shelves extend as little as 1.6 kilometers (1 mile) from the coastline of western North and South America to as far as 1,500 kilometers (900 miles) off the Arctic coast of Siberia (Figure 16-1). Most continental shelves, however, average about 67 kilometers (40 miles) wide. They descend gradually from the shore, reaching depths of 130 meters (430 feet) at their most distant edge. At this point, the gently sloping bottom may become a steep slope or a sheer drop-off. Off the coast of Florida, in the Gulf of Mexico, the continental shelf ends in a drop-off that runs for 800 kilometers (480 miles)

## Key Concepts

1. The number and kinds of benthic organisms on continental shelves are influenced by sediment characteristics.

2. Hard-bottom communities are dominated by epibenthic organisms.

3. In areas north and south of the tropics, kelps (a type of brown algae) dominate the subtidal zone where the water is cold and the sediments are hard.

4. Kelps are important primary producers and provide habitats for many animals.

5. Soft-bottom communities are dominated by suspension feeders and deposit feeders.

6. The distribution of organisms in benthic communities of the continental shelf is patchy.

7. The neritic zone is the water column that lies above the continental shelves.

8. The neritic zone receives high levels of nutrient input from rivers, coastal runoff, and upwellings.

9. The neritic zone supports enormous amounts of phytoplankton.

10. The high productivity of coastal seas supports large numbers of fishes, birds, and marine mammals.

NOAA; inset, Zig Leszczynski /Animals Animals

373

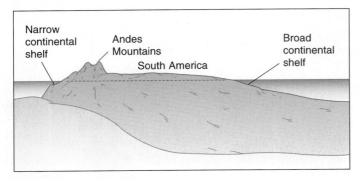

**Figure 16-1 Continental Shelves.** Continental shelves descend gradually from the shoreline out to sea where they end at a steep slope or sheer dropoff. They can be broad, as they are along the Atlantic coasts of North and South America, or narrow, as they are along the Pacific coasts of the Americas.

and at some points has perpendicular drops of 1.6 kilometers (1 mile). By contrast, off the coast of Chile, the continental shelf ends in a steep incline that descends uninterrupted to a depth of more than 6 kilometers (3.6 miles).

Worldwide, rivers annually carry almost 750 million tons of sediment to coastal seas. Some ends up deposited on the bottom of the continental shelves, and the rest is dissolved in the seawater. The sediments contain nutrients such as nitrogen, phosphorus, calcium, and silica that are essential for life in the ocean. Coastal upwellings also supply nutrients to coastal seas. The heavy concentration of nutrients and the large amount of sunlight this area receives combine to make continental shelves and the seas above them highly productive areas.

## In Summary

Coastal seas lie over the continental shelves. Continental shelves vary in their characteristics depending on the geographic and geological characteristics of the coastlines with which they are associated. The water column over the continental shelves is quite productive because of the abundance of sunlight and the high nutrient input from the adjacent coasts.  ●

## BENTHIC COMMUNITIES

A variety of benthic communities can be found on the continental shelf from the shallow subtidal zone to the edge of the continental shelf. In some areas where the bottom is primarily composed of rock, large forests of seaweed form a habitat for a variety of animals that rely on these large algae for food and shelter. Where the bottom is soft, burrowing organisms of all kinds dominate.

On continental shelves, food supply is the major limiting factor. In most areas of a continental shelf, large amounts of food in the form of detritus rain down from the sunlit waters above and fertilize the bottom sediments. The detritus consists of dead plankton, dead and dying larger animals, and organic debris, including human sewage, washed into the

sea from land by rivers and runoff. This continuous food supply supports many filter feeders, suspension feeders, and detritus feeders, which in turn supply food for other animals. Filter feeders and suspension feeders such as sponges, tunicates, and bivalves feed directly on the detritus by filtering large volumes of water containing detritus through their bodies (Figure 16-2). Several species of polychaete worm feed on trapped detritus as they consume bottom ooze in the process of making their burrows. Sea cucumbers use their sticky tentacles to gather food off the bottom, and their relatives the sea urchins move slowly over the bottom, chewing off bits of algae and detritus that have become attached to the firmer bottom sediments.

One of the most important factors in determining the type of organism found in benthic communities is the stability of the environment. The benthic communities of the continental shelves are generally quite stable and not subject to the same changing forces that affect organisms living along the shore.

## Role of Sediments

The number and type of organisms that can live in and on the bottom of continental shelves are greatly influenced by the type and characteristics of the sediments that comprise the bottom. In areas of a continental shelf where currents flow, the bottom is composed of coarse sediments. The moving water tends to carry away the fine, light sediments such as silt, leaving behind larger sand, gravel, and rock particles. The bottom in these areas is constantly shifting and is not a good habitat for **infauna,** animals that burrow in the sediments, and **interstitial animals,** animals that live in the spaces between sediment particles. These animals cannot withstand abrasion or the constant shifting of the sediments. The flow of water, however, does carry with it a large supply of food. **Epifauna** are animals that live on surface sediments. Epifauna composed primarily of sedentary or sessile filter feeders and suspension feeders, such as sponges, anemones, and colonial cnidarians are well-adapted to this type of bottom. The size of the sediment particles provides them with a firm surface for attachments, and the large supply of suspended food provides a constant source of nutrients.

Where bottom currents are weak, the bottom is composed of fine sediments such as silt. The sediments here are more stable and support a variety of infauna that construct permanent burrows, such as polychaete worms, amphipods, and clams. Most of the organisms living in this habitat are deposit feeders, feeding on the organic material that settles from above and becomes trapped in the soft sediments. These bottoms do not support many filter feeders because of the scarcity of suspended food and the fine sediments that interfere with the animal's filtering structures.

## Hard-Bottom Communities

The hard-bottom habitat consists of large sediments that cannot be pushed apart, such as rock and clam shells. These habitats are most often found off rocky coasts. Organisms that live in this type of habitat include sessile organisms that attach to the solid surface by cementing themselves in place

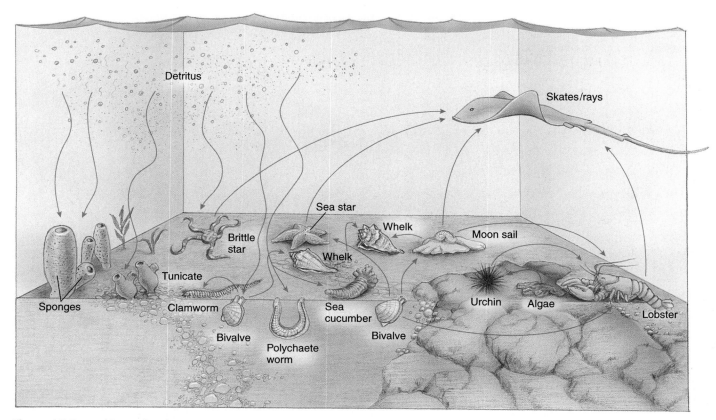

**Figure 16-2 Benthic Food Web.** Detritus forms the basis of the benthic food web in coastal seas.

or by attaching with strong, threadlike structures and those that can freely move over the hard surfaces. Hard-bottom communities tend to be dominated by sessile colonial animals such as bryozoans, hydroids, sponges, and colonial sea squirts. Other common animals in this habitat include anemones, bivalves, snails, crustaceans, and echinoderms. The large number of colonial animals in hard-bottom communities may reflect the inability of these organisms to deal effectively with the changes in temperature and salinity and the problem of desiccation encountered in shallower-water habitats like the intertidal zone.

**Epibenthic organisms,** which live on the surface of the bottom sediments, on hard bottoms are not evenly distributed, and as a result, there is an uneven distribution of benthic organisms. This characteristic distribution of organisms in a population or community is referred to as **patchiness** (Figure 16-3). Groups of organisms are sometimes separated by a few centimeters and at other times by hundreds of meters. Exposure to sunlight is one factor that contributes to the patchy distribution. Vertical and near-vertical rock walls that receive less sunlight are dominated by sessile invertebrates, whereas a variety of algal species dominate the horizontal surfaces in water that is shallow enough to receive sufficient sunlight, except in areas that are heavily grazed by sea urchins.

Disturbance also plays a role in the distribution of benthic organisms. Landslides and shifting sediments have a greater impact on sessile invertebrates than on thick algal turf. On rock walls large patches of diverse organisms dominated by bryozoans, sponges, and tunicates spread for hundreds of meters. Between these patches are patches of less diversity dominated by calcareous algae. Patches of low diversity are maintained by periodic landslides that clear away the more delicate invertebrates and allow the calcareous algae to dominate. The areas of high invertebrate diversity appear to be maintained by larval recruitment. These invertebrates produce larvae that spend a short time in the water column; thus they do not disperse too far from the population and can readily replenish the population. Many of the invertebrate species in these communities also spread asexually. Species that reproduce asexually gain a competitive advantage by being able to either overgrow their neighbors or resist being crowded out by other species.

## Kelp Communities

One of the most productive marine communities on some continental shelves is the kelp bed. Kelp is a type of brown algae (for more information regarding kelp see Chapter 7) that requires a rocky bottom, cold water, and a continuous supply of nutrients to support its high level of photosynthetic activity. Most kelp beds are found in water that is no more than 20 meters (66 feet) deep, but if the water is exceptionally clear, they may occur in water as deep as 30 meters (99 feet). Kelp are strictly cold-water organisms that rarely survive in areas where the average surface water temperature exceeds 20°C (Figure 16-4).

**Figure 16-3 Patchiness.** Patchiness refers to the uneven distribution of benthic organisms that results from the uneven distribution of bottom sediments. Areas of coarse sediments and fine sediments are randomly distributed on the continental shelf, and each supports different types of organisms.

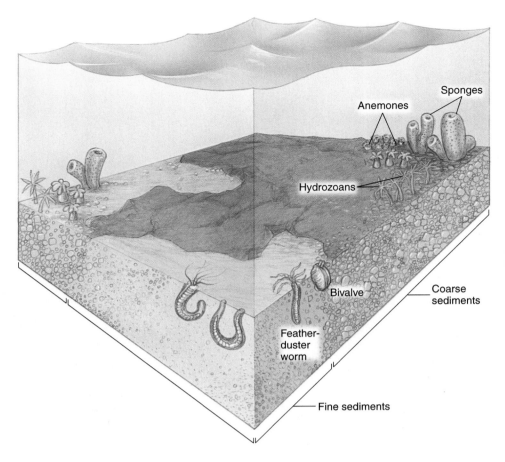

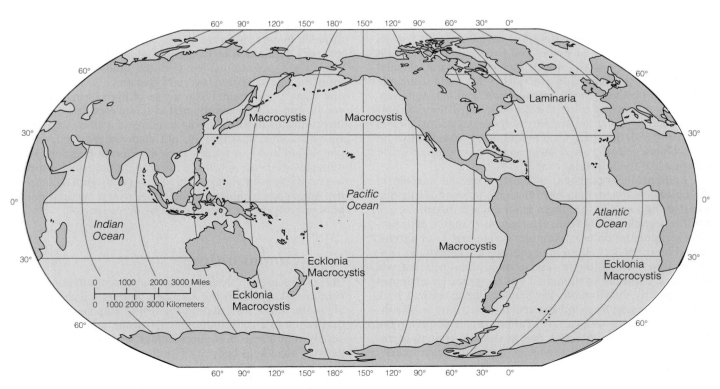

**Figure 16-4 Kelp Distribution.** Kelps are found on continental shelves where the bottom is rocky and the water is cold. The kelp genera listed above are dominant in the indicated areas.

(a)

**Figure 16-5 Kelp Beds.** (a) The giant kelp *Macrocystis*, which forms underwater forests, dominates the kelp zones of the Pacific and southern Atlantic oceans. (b) The kelp *Laminaria*, which forms dense thickets, is shown here growing along the base of a rock outcrop on the Pacific coast.

(b)

## Kelp Beds

Some kelp beds are like underwater forests, forming a **canopy** that shades smaller algal species and an **understory** that is home to many animal species. In the kelp zones of the Pacific and southern Atlantic oceans, *Macrocystis* dominates (Figure 16-5a). The tall, narrow blades of this kelp grow in tight clusters, and there is enough space between them for animals to move about with ease. Beneath the canopy of the giant kelp is a smaller species, the sea palm (*Eisenia*). The stipe of this species is thick and elastic and is able to bend with water currents. The sea palm depends on the minimal light that penetrates the kelp canopy to supply its photosynthesis needs. The rocky bottom is carpeted by several species of red algae that can survive on the wavelengths of light that penetrate the canopy and reach the bottom.

The genus *Laminaria* is dominant in the North Atlantic Ocean. *Laminaria* grows closely packed to form thickets, and the dense growth makes this habitat more suitable for crawling than swimming animals (see Figure 16-5b).

## Kelp Life Cycles

The life cycle of kelp begins when spores settle on rocky bottom that receives enough light to satisfy the organism's needs for energy. The spores germinate and develop into a microscopic form that is preyed upon by herbivores. The density of herbivores in an area is a primary factor in determining whether new algae can establish themselves. Once established, kelps grow quickly. Stipes bearing the flattened blades that carry out photosynthesis grow toward the light at the surface. At the surface, they spread out, forming a canopy similar to that of a terrestrial forest. As they grow, the kelp's appearance changes because each species exhibits a characteristic growth form.

Giant kelps of the genus *Macrocystis* are the world's largest algae, reaching lengths of 20 to 40 meters (66 to 132 feet; Figure 16-6). *Macrocystis* has gas-filled floats at

the base of each blade to help buoy this photosynthetic structure. A mature blade of *Macrocystis* can add as much as 50 centimeters (19.5 inches) of new tissue in a single day when conditions are favorable. An entire individual, consisting of a stipe, floats, and as many as 200 blades, grows

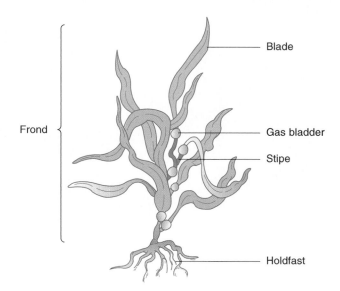

**Figure 16-6 Giant Kelp.** Giant kelps of the genus *Macrocystis* are the world's largest algae. The large blades have floats that buoy them in the water and allow maximum surface exposure to light.

as a single unit and lives for about 6 months. After 6 months, these individuals are replaced by new ones that frequently grow in pairs at the astounding rate of 3 to 5 meters (10 to 16.5 feet) per week under good conditions. *Macrocystis* is a **perennial,** usually living 3 to 7 years before it dies from any number of causes.

   *Laminaria,* on the other hand, grows like a conveyor belt, with new tissue added at the base as older tissue at the tip erodes. Although kelps are constantly growing, they are also constantly eroding, producing an almost steady stream of detritus that plays a significant role in the kelp forest food web.

## Kelp Community

Large numbers of organisms use kelp for food, shelter, or both (Figure 16-7). The dense network of kelp blades slows currents and decreases the force of all but the most energetic storm waves, thus providing convenient shelter for

**Figure 16-7 Kelp Forest Community.** Kelps provide both habitat and food for a large number of marine organisms.

many fishes and mammals such as otters and seals. The large, tree-sized algae greatly increase the amount of useful habitat in the area. In kelp forests of the west coast of the United States, echinoderms, like the omnivorous bat star (*Asterina miniata*), move slowly among the kelp, feeding on both animals and algae. Its relative, the many-rayed sunflower sea star (*Pycnopodia helianthoides*), feeds on urchins, other sea stars, and molluscs that it finds on the kelp forest floor. Fishes such as pipefish, blennies, eels, the Garibaldi (*Hypsypops rubicundus*), and many larger fish species find the kelp an ideal environment in which to live. Sheepshead (*Semicossyphus pulcher*) come to feed on many of the larger invertebrates, and several species of rockfish (*Sebastes*) feed on both invertebrates and other fishes. Without the kelp forests for habitat, these species would either disappear or be present in smaller numbers.

A species of mussel that lives in kelp beds is one of the few multicellular organisms that produces an enzyme that can digest cellulose. This species can feed not only on the bacteria that coat particles of kelp, as do other species, but also on the kelp fragments themselves. Other filter feeders attach directly to the surface of parts of the kelp, some growing over the surface of new blades almost as fast as they are produced. Sponges grow at the base of kelp and, together with the kelp's holdfasts, form a habitat that shelters tiny shrimp, crabs, lobsters, brittle stars, and small fishes. The commercially important molluscs known as abalone (*Haliotis;* Figure 16-8) are common residents of kelp beds of the Pacific coast, and American lobsters are a common resident of North Atlantic kelp beds (Figure 16-9a). Lobsters inhabit holes and crevices in the bottom by day and come out at night to feed on molluscs, crustaceans, and urchins as well as to scavenge. The American lobster (*Homarus americanus*) is fished commercially using pots, or traps. The animals enter these seeking shelter or attracted by bait and are unable to get out (see Figure 16-9b).

Although most of the organisms that live on kelp are filter feeders, some are herbivores that feed directly on the

(a)

(b)

**Figure 16-9 Lobster.** (a) The American lobster (*Homarus americanus*) is a commercially important crustacean that lives in the kelp beds off the New England coast. (b) Lobsters are important commercial shellfish. They are caught in traps like these that are placed on the bottom overnight.

**Figure 16-8 Abalone.** The red abalone (*Haliotis rufescens*) is one of several species of commercially important molluscs found in kelp forests.

kelp itself. Snails crawl along the stipes, using their radula to scrape the outer layer of cells from the alga. Burrowing through the matted holdfasts are hordes of termitelike crustaceans called gribbles (*Phycolimnoria;* Figure 16-10). In some instances, the action of gribbles can be so destructive that the weakened holdfasts can no longer anchor the alga, and storm waves uproot and carry it away. It is thought that in a stable kelp forest gribbles may be the primary cause of mortality among adult kelp.

Courtesy of Pam Brown

**Figure 16-10  Gribbles.** The termitelike gribbles (*Phycolimnoria*) can cause significant damage to kelp.

## Impact of Sea Urchins on Kelp Communities

Kelps are also a favorite food of sea urchins. Wave action and predators that feed on sea urchins usually prevent urchins from doing significant damage to the upper portions of the kelp and the canopy, but there are few predators that prevent these animals from devouring young algae and damaging the important holdfasts. In the United States, a decline in the sea otter population along the west coast and the lobster population along the North Atlantic coast has resulted in population explosions of urchins. Some researchers believe that increased numbers of urchins, along with naturally occurring events such as disease and climate changes, may be primarily responsible for the decline in kelp beds in some areas over the last 50 years. Off the coast of Nova Scotia, the lobster population decreased by as much as 50%, the result of overfishing in the 1970s. During the same time period an estimated 70% of the kelp beds in this area disappeared.

In areas where kelp has been completely removed, enough sea urchins remain to make it nearly impossible for new kelps to reestablish themselves. Although urchins may ultimately destroy the kelp bed, the animals continue to survive. Because they are primarily generalist feeders, sea urchins simply switch to feeding on other species of algae when kelps are not available. In southern California, sea urchins feed on the nutrients from treated sewage. Other animal species that depend on kelp for food or shelter are not as fortunate and soon disappear along with the kelp.

## Soft-Bottom Communities

Soft-bottom communities are found where the sediments are sand or mud or a mixture of both. Where currents are fairly strong, soft-bottom sediments are usually sand. Mud is common where currents are not very strong. Sandy bottoms are usually dominated by suspension feeders, whereas mud bottoms are generally dominated by deposit feeders.

Suspension feeders and deposit feeders do not generally occur in the same area for several reasons. The water above muddy bottoms tends to be quite turbid because of the large number of fine particles suspended in the water. The turbidity of the water is increased by the burrowing and feeding activities of deposit feeders. These animals deposit large amounts of fecal pellets at the surface of the sediments. The fecal pellets are easily broken down by bottom currents and increase the turbidity. The high concentration of suspended sediment particles clogs the feeding organs of suspension- and filter-feeding animals, ultimately causing their death. The activities of deposit feeders also make the bottom sediments more unstable. This makes the habitat unsuitable for suspension feeders because they do not have enough stable surface for attachment.

## Patchiness in Soft-Bottom Communities

Like hard-bottom communities, the distribution of organisms in soft-bottom communities is patchy. The patchiness is the result of several factors such as changes in the bottom sediments, bottom currents, and patterns of larval settlement.

***Changes in Sediment Distribution***   The animals that live in or visit soft sediments can modify what is originally a generally homogeneous area, changing it into an environment characterized by a discontinuous surface. For instance, burrowers, such as the sea cucumber *Molpadia oolitica* (Figure 16-11), can produce fecal mounds that rise several centimeters above the bottom. These mounds are home to suspension feeders such as bivalves and polychaete worms. The areas between mounds are dominated by deposit feeders, whose activity makes the habitat inhospitable to suspension feeders. The feeding activity of predators such as skates, rays, walruses, and whales produces hills, valleys, and furrows in the seafloor changing its contours. This alteration of the seafloor's topography kills some organisms while improving or enhancing the environment for colonization by others.

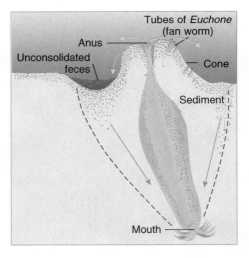

**Figure 16-11  Fecal Mounds.** The burrowing sea cucumber *Molpadia oolitica* builds fecal mounds that provide a more stable surface for other bottom dwellers than the surrounding mud.

***Bottom Currents*** The flow of currents across the seafloor also alters the landscape. If the currents produce more or less constant movement of the sediments, the environment will be unsuitable for organisms that require a stable habitat. Currents may produce ripples on bottoms composed of fine sediments. The troughs between raised areas tend to collect more organic material and are good habitats for deposit feeders. On the other hand, suspension feeders are found on the more exposed crests. In areas where the currents are strong, the pattern of crests and troughs can change so quickly that they cannot be colonized by animals that require a more stable bottom.

***Larval Settlement*** Many invertebrate animals have planktonic larval stages that may stay in the water column for weeks before settling. The pattern of settlement depends on currents, predation, and the type of bottom sediments. The uneven pattern of larval settlement results in a patchy distribution of organisms on the seafloor.

## Soft-Bottom Food Chains

As in hard-bottom communities, the primary source of food for soft-bottom communities is detritus. Suspension feeders feed on detritus and plankton, and deposit feeders feed on detritus and the bacteria that decompose it. Burrowers, sedentary suspension feeders, and slow-moving grazers are preyed upon by more-active animals. Polychaetes known as clamworms (*Nereis*), some as long as 0.5 meter (19.5 inches), feed on small bivalves that burrow in the soft bottom. Carnivorous snails, such as members of the family Buccinidae (whelks), feed on both bivalves and other snails, usually by boring a small hole through the shell so that the contents can be sucked out (Figure 16-12). Sea stars move slowly along the bottom, feeding on mussels, oysters, and scallops. Brittle stars slither across the bottom powered by undulations of their serpentine rays. Some species bur-

row into the bottom, where they feed on decaying material, whereas others feed on detritus and small organisms that are passed along their tube feet toward their mouth. Brittle stars are perhaps the most abundant animals in soft-bottom communities. Some areas of soft bottom off the coast of Great Britain support as many as 80 million brittle stars per square kilometer (0.36 square mile).

Skates, rays, angel sharks (*Squatina*), and batfishes (family Ogcocephalidae) are just a few of the fishes that forage along the bottom, feeding on molluscs and crustaceans that they crush with their powerful jaws (Figure 16-13). These animals are well-adapted to dwelling and feeding on the bottom. They usually have flattened or compressed bodies, retractable fishing lures, and specialized sense organs. They also have well-camouflaged dorsums that allow them to blend with the sea bottom.

A large number of fishes spend their lives on or near the bottom of the continental shelf. Fishes such as haddock (*Melanogrammus*), hake (*Merluccius*), pollock (*Pollachius*), cod (*Gadus*), and rockfish (*Sebastes*) are commercially valuable. They are caught in most of the coastal waters of the Northern Hemisphere. These fishes feed on molluscs, echinoderms, crabs, other fishes, each other, and even their own young. At one time, it was estimated that as many as 10,000 cod lived above each acre of the continental shelf of the Grand Banks, off Nova Scotia, Newfoundland. As the result of overfishing, however, there are so few fishes now that the area can no longer support a major commercial fishery. Another group of commercially valuable fishes, the flatfishes (order Pleuronectiformes), live directly on the bottom. These include flounder (*Platyichthys;* Figure 16-14), halibut (*Hippoglossus*), turbot (*Rhombus*), and various species of sole (*Solea*).

## Succession in Soft-Bottom Communities

When sediments are disturbed, by erosion or landslides, for instance, many animals are removed or killed. The disturbance also exposes deeper layers of sediments that are fre-

**Figure 16-12 Whelks.** Whelks are carnivorous snails that feed on other molluscs.

**Figure 16-13 Cownose Ray.** This cownose ray has powerful jaws to crush the hard shells of the molluscs and crustaceans on which it feeds.

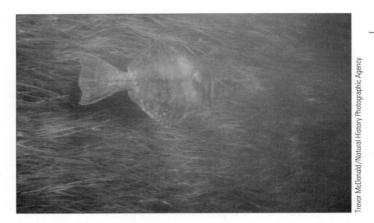

**Figure 16-14 Commercial Fishes.** This flounder represents just one of many important commercial fishes that are caught in coastal seas.

quently anoxic because of the high levels of decomposition that occur there. After the disturbance new organisms arrive by larval settlement to recolonize the area. Early colonizers are usually surface dwellers such as polychaete worms. These animals feed on the accumulating organic material and help to aerate the surface sediments by their burrowing activities. The increased oxygen content in the sediments allows other marine worms along with bivalves and snails to begin colonizing the area. The action of these animals continues to create tunnels in the sediments and improve aeration, making way for more permanent deeper-burrowing organisms.

## In Summary

The number and kinds of organisms that can live on the bottom of continental shelves is greatly influenced by sediment characteristics. Coarse sediments favor sedentary filter feeders and suspension feeders. Fine sediments provide a more stable bottom that favors burrowing organisms that are primarily deposit feeders. Much of the food that supports the primary consumers of bottom communities drifts down from the sunlit waters above and consists mostly of detritus and organic debris.

The seabed of the continental shelf can be either hard bottom or soft bottom. Hard-bottom communities are dominated by epibenthic organisms that live on the surface of the hard sediments. Soft-bottom communities are dominated by burrowing organisms that are generally suspension feeders or deposit feeders. Suspension feeders and deposit feeders do not inhabit the same areas of seafloor because water where deposit feeders are numerous is turbid and the suspended material clogs the food-gathering structures of suspension feeders. The distribution of organisms in benthic communities is uneven, or patchy. The patchiness is due to several factors, including variations in their environment, changes in sediment distribution, bottom currents, and larval settlement. Kelps dominate in areas of the continental shelf where the bottom is hard, the water is cold, and nutrients are readily available. These algae form complex, three-dimensional habitats for large numbers of invertebrate and vertebrate animals. Kelps also serve as an important food source, directly or indirectly, for many of the species that live in and around kelp beds. ●

## NERITIC ZONE

The neritic zone comprises the coastal seas that lie along the edges of the continents and above the continental shelves. These waters contain an enormous number of phytoplankton, the microscopic photosynthetic organisms that form the basis of aquatic food webs. The density of plankton populations in coastal seas is astounding. A 10-milliliter (about 2 teaspoons) sample of surface water may contain thousands of planktonic organisms. Unlike the clear blue waters around coral reefs, coastal seas are green, a sign of high productivity. Grazing on these pastures of phytoplankton are hordes of tiny animals, the zooplankton. The abundant plankton provides food for large schools of fish (Figure 16-15), and these in turn provide food for larger fishes, some marine mammals, seabirds, and humans. Although the neritic zone covers an area equivalent to only about the size of Asia and contains only 10% of the ocean's expanse, it pro-

**Figure 16-15 Commercial Fishes.** The large numbers of plankton in coastal seas support large schools of commercial fishes such as these herring.

# Highly Productive Southern Ocean

At first, one might think that the frigid waters surrounding the continent of Antarctica wouldn't harbor many living organisms. In reality, though, the Southern Ocean, like other coastal seas, is an extremely productive area, supporting not only large numbers of aquatic organisms but also most of the animals that live in Antarctica. The basis of this rich food web is the numerous diatoms that dominate the phytoplankton. Dominant among the zooplankton that feed on these diatoms is krill (*Euphausia superba*). This herbivorous crustacean accounts for almost 50% of the zooplankton in the Southern Ocean. Although krill are found throughout the region, their distribution is not uniform. The greatest concentrations have been recorded in the Weddell Sea, where some swarms of krill have been estimated to be 2 million tons. It has been estimated that between 750 million and 1,300 million tons of krill are produced annually in the Southern Ocean.

During the summer, primary production in the Southern Ocean increases as the pack ice melts and the polar region receives more sunlight. Krill move to the surface to graze on the increased numbers of

diatoms. During the winter, less light, increased pack-ice cover, and increased turbulence lower primary production to near zero. Krill move to deeper water and apparently feed mostly on detritus.

As many as 20 species of squid, some relying heavily on krill as a food source, live in the Southern Ocean. The squids provide food for numerous species of toothed whales, seals, and birds. It is estimated that these animals consume 35 million tons of squid annually from this region. Many fish species are also found in the Southern Ocean. Some species are permanent residents, and others are seasonal visitors, arriving in summer to feed on the large numbers of krill. It is estimated that all of the fish in the Southern Ocean consume 100 million tons of krill annually.

Although Antarctica is home to few species of bird, those that do live there are usually represented by large populations. Their diet consists of crustaceans, mainly krill and copepods, squids, fish, and carrion. Krill accounts for almost 78% of all the food they eat. The total annual direct and indirect consumption of krill by birds is approximately 115 million tons.

Seven species of seal are found in the Southern Ocean. Of these, the crabeater seal's diet consists of 94% krill, the leopard seal's diet consists of 37% krill, and the fur seals of South Georgia island feed almost exclusively on krill. The other seal species feed on fish and squid that in turn rely on krill. Larger baleen whales, such as the blue whale, feed mainly on krill. They consume about 43 million tons of krill annually. Whaling has decreased the number and size of baleen whales in the Southern Ocean. It is estimated that 150 million tons of krill that was formerly consumed by baleen whales is now available to other krill-feeding species. Smaller whales, seals, and penguins appear to be the main beneficiaries of this change, as indicated by increases in the size of their populations, presumably because of the increased availability of food. Krill are also being harvested for human consumption. The krill being netted by fishers is assumed to be available because of the decrease in the size of whale populations. There is still some doubt, however, whether the productivity of krill is sufficient to feed humans without endangering other animal populations in the area. ●

duces nearly 90% of the world's annual harvest of fish and shellfish. By comparison, the open ocean is a barren desert.

## Food Chains in the Neritic Zone

The neritic zone receives freshwater runoff from the neighboring land. The runoff provides enough nutrients to support the growth of relatively large producers called **microphytoplankton** (20 to 200 micrometers) as well as smaller forms, the **nanophytoplankton** (less than 20 micrometers; for more information regarding plankton see Chapter 6). In the colder waters around Antarctica and in the northern Pacific and Atlantic oceans, diatoms dominate the microphytoplankton. Planktonic organisms called coccolithophores are sometimes so numerous in areas of the North Sea that the water turns white. At certain times of the year, they even outnumber the diatoms. In warmer coastal waters, dinoflagellates are more numerous. Unicellular green algae occur in both tropical and polar seas. The composi-

tion of the phytoplankton varies from one region to another, from season to season, and sometimes even within a single season. Grazing by zooplankton and other animals also affects the composition of the phytoplankton. When the number of one type of phytoplankton decreases as the result of heavy grazing, another species will usually proliferate to take its place.

Drifting animals that make up the zooplankton (Figure 16-16) feed on the millions of diatoms, dinoflagellates, and other phytoplankton. The most abundant members of marine zooplankton are crustaceans called copepods. In coastal waters, the concentration of copepods can be as high as 100,000 individuals per cubic meter of water. The primary reasons for such large numbers are the tremendous reproductive capacity of these animals and the rich food supply. After being fertilized, a female can produce as many as 100 eggs, and some species produce a new clutch of eggs every 4 or 5 days. Between 1 and 2 weeks after hatching, the new generation is mature and ready to reproduce. Cope-

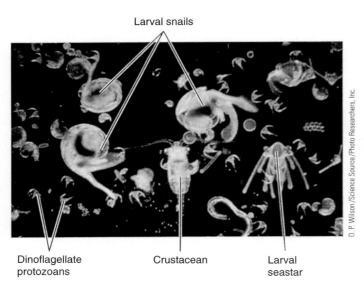

Figure 16-16  **Zooplankton.** Members of the zooplankton include protists, adult crustaceans, jellyfishes, and the larvae of both invertebrate and vertebrate animals.

pods are the primary consumers of diatoms. A single copepod can consume as many as 120,000 diatoms per day.

Because tiny phytoplankton can be either eaten directly by small zooplankton or filtered from the water by benthic filter feeders such as clams and worms, food chains in coastal seas are frequently two or more steps shorter than those of the open sea. Small zooplankton are the preferred food of large fishes like menhaden (*Brevoortia*) and alewives (*Alosa*), whereas large bottom fishes, such as cod and haddock, prey on filter feeders. In both cases there are only three trophic levels between primary producers and consumers of reasonable size (Figure 16-17). Even larger animals such as tuna, sharks, and humans are only four trophic levels from producers. The combination of higher productivity and shorter food chains supports a larger number of higher-level consumers in coastal seas.

## Productivity in the Neritic Zone

The most productive of the planktonic ecosystems are located in upwelling zones, where the combination of winds, ocean currents, and shape of the seafloor interact to bring nutrients into the photic zone from the ocean floor (for more information regarding upwelling zones see Chapter 4). The high productivity of these regions would not be possible without the activity of microorganisms. Bacteria break down the dead remains of plants and animals, releasing the nutrients to be used again. These nutrients are then returned by upwellings to the surface, where they can be recycled by plankton. The shallow banks along the coasts of the North Atlantic and the Pacific coast of Peru are examples of such productive areas. These upwelling areas yield almost half the world's supply of commercial fishes. In these regions, the almost continuous supply of nutrients supports a phytoplankton community dominated by large, chain-forming diatoms (Figure 16-18). These chains, which consist of several

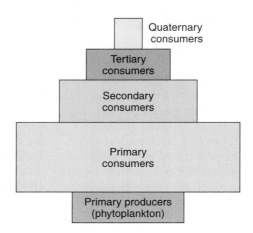

Figure 16-17  **Productivity of Coastal Seas.** In this generalized pyramid, the size of each box indicates the relative amount of biomass at each trophic level. Because of the abundance of sunlight and nutrients, phytoplankton in coastal seas can reproduce at extremely rapid rates and can therefore support about five times their own biomass in primary consumers. These primary consumers, which include enormous amounts of zooplankton, are in turn a rich food supply for large populations of higher-order consumers. (D. P. Wilson/Science Source/Photo Researchers, Inc.)

cells linked together, are large enough to be eaten directly by large zooplankton like the shrimplike krill (*Euphausia*) of the Southern Ocean and small fishes like anchovies (family Engraulidae) of the Pacific Ocean off the coast of South America. Krill and anchovies, in turn, are large enough to be worthwhile prey for large fishes, seabirds, seals, whales, and even humans. The high productivity and short food

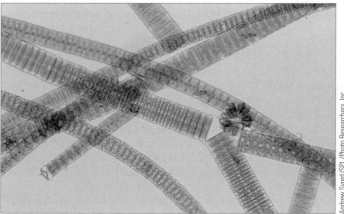

Figure 16-18  **Diatoms.** The phytoplankton of upwelling areas are dominated by large, chain-forming diatoms like these.

chains of upwelling areas support the greatest biomass of any planktonic system.

Biologists estimate that as much as one seventh of the anchovies caught worldwide by humans and seabirds come from Peru's coastal seas in years when upwelling currents are strong. Smaller upwellings along the Pacific coast of the United States and islands like Japan support locally productive fisheries. It is no wonder that so many animals, including humans, come to upwelling areas to feed. The primary production in these regions may be as much as six times higher than that of the open sea, and they produce biomass at a rate more than 36,000 times as fast as that of the open sea. These limited areas represent valuable commercial resources that are protected by laws and require conservation.

## Other Roles of Plankton in Coastal Seas

The ecological importance of plankton goes beyond their role of supplying food for animals. An overwhelming number of animal species spend at least a part of their lives as members of the plankton. Sedentary animals such as barnacles, mussels, and coral, which spend their adult lives fixed in one place, rely on their planktonic larval forms, such as the nauplius, trochophore, veliger, and planula, to colonize new territories. When spawning, these animals release hundreds to thousands of tiny eggs and sperm that unite in the water column and then scatter in the currents. Almost all crabs, shrimps, and lobsters have larval stages that feed and grow as part of the plankton before settling down to lead their adult existence. Even free-swimming fishes such as herring and eels have tiny planktonic larvae and juvenile stages almost as small that join the zooplankton for the first weeks or months of their lives.

Only a small fraction of the larvae in plankton develop into adults. Odds are that only 1 in every 100,000 eggs found in the plankton will survive to adulthood. The health of plankton is important not only because of the role they play in supporting commercial fisheries but also because any disturbance to their community can cause unexpected changes in populations of animals whose adults are anything but planktonic.

## In Summary

The neritic zone comprises the water that lies above continental shelves. This area contains enormous numbers of phytoplankton, which form the basis of an extremely productive food web. The high productivity of these regions is due to a combination of physical and biological factors. Rivers and runoff deposit large amounts of nutrients directly into coastal seas, and the geography of continental shelves favors upwellings that bring nutrients from the bottom to the photic zone, where they can be used by plankton. Phytoplankton support large numbers of zooplankton and some large animals as well. Those large predators that do not prey directly upon the plankton feed on smaller animals that are plankton feeders. ●

## SELECTED KEY TERMS

canopy, *p. 377*

epibenthic organism, *p. 375*

epifauna, *p. 374*

infauna, *p. 374*

interstitial animal, *p. 374*

microphytoplankton, *p. 383*

nanophytoplankton, *p. 383*

patchiness, *p. 375*

perennial, *p. 378*

understory, *p. 377*

## QUESTIONS FOR REVIEW

### Multiple Choice

1. Most of the world's harvest of commercial fish comes from
   a. estuaries
   b. bays
   c. coastal seas
   d. the open ocean
   e. benthic regions

2. The number and types of benthic organisms found on the continental shelf are most influenced by the
   a. amount of sunlight
   b. water temperature
   c. characteristics of the sediments
   d. salinity
   e. water clarity

3. Sedentary and sessile filter feeders and suspension feeders are better adapted to
   a. hard bottoms
   b. soft bottoms

4. The uneven distribution of bottom organisms and sediment type on the continental shelf is referred to as
   a. reticulation
   b. patchiness
   c. diversity
   d. sediment selection
   e. benthic orientation

5. In colder waters, _____ dominate the phytoplankton.
   a. krill
   b. copepods
   c. diatoms

    d. dinoflagellates

    e. kelp

6. The most abundant members of the zooplankton are
    a. krill
    b. copepods
    c. jellyfish
    d. diatoms
    e. dinoflagellates

7. Nutrients produced by bacteria in the bottom sediments are returned to the surface waters by
    a. storms
    b. upwelling
    c. seasonal changes in water density
    d. all of the above

8. In soft-bottom communities suspension feeders and deposit feeders are usually not found in the same areas because
    a. they compete with each other for food
    b. they compete with each other for space
    c. deposit feeders prey on suspension feeders
    d. the suspended silt in regions inhabited by deposit feeders clogs the feeding structures of suspension feeders
    e. the type of sediments inhabited by deposit feeders will not support suspension feeders

9. The world's most productive upwelling area is located off the coast of
    a. California
    b. Peru
    c. China
    d. South Africa
    e. Europe

10. The dominant alga in southern California kelp forests is
    a. *Fucus*
    b. *Laminaria*
    c. *Eisenia*
    d. *Sargassum*
    e. *Macrocystis*

## Short Answer

1. What are the main sources of nutrient input into coastal seas?

2. What factors affect the size of plankton populations?

3. Explain how the type of bottom sediments influences the diversity of life on the floor of the continental shelf.

4. List three explanations for the patchy distribution of organisms in soft-bottom communities.

5. Explain how decreases in the size of American lobster populations have affected North Atlantic kelp beds.

6. Why are kelp beds frequently compared to terrestrial rainforests?

7. Why is the neritic zone such a productive area?

8. Diagram a simple food web for the continental shelf.

9. Describe the process of succession that occurs in soft bottom communities that are disturbed.

10. What are two ecological roles for plankton in the neritic zone?

## Thinking Critically

1. Currently there is concern about the possibility of global warming caused by the greenhouse effect. How might this affect the productivity of coastal seas?

2. Why do so many species of small fish found in coastal seas travel in large schools?

3. Why do disturbances such as landslides have a greater impact on invertebrate populations than on algal populations?

## SUGGESTIONS FOR FURTHER READING

Campbell, D. G. 1992. The Bottom of the Bottom of the World, *Natural History* 101(11):46–52.

Gore, R. 1990. Between Monterey Tides, *National Geographic* 177(2):2–43.

McPeak, R. H., and D. A. Glantz. 1984. Harvesting California's Kelp Forests, *Oceanus* 27(1):19–26.

Smith, W. O., Jr., and D. M. Nelson. 1986. Importance of Ice Edge Phytoplankton Production in the Southern Ocean, *Bioscience* 36:251–57.

Winston, J. E. 1990. Life in Antarctic Depths, *Natural History* 99(4):70–75.

### InfoTrac College Edition Articles

Bustamante, R. H., G. M. Branch, and S. Eekhout. 1995. Maintenance of an Exceptional Intertidal Grazer Biomass in South Africa: Subsidy by Subtidal Kelps, *Ecology* 76(7).

Cohn, J. P. 1998. Understanding Sea Otters, *Bioscience* 48(3).

Reed, D. C., T. W. Anderson, A. W. Ebelin, and M. Anghera. 1997. The Role of Reproductive Synchrony in the Colonization Potential of Kelp, *Ecology* 78(8).

Walsh, R. 2000. The Lobster Pickle, *Natural History* 190(6).

### Websites

**http://www.oceanlight.com/html/kelp.html** Good information on kelp forests with excellent pictures.

**http://www.mbayaq.org/efc/efc_hp/hp_kelp_exhibit
.asp** Information about kelp from the Monterey Bay Aquarium.

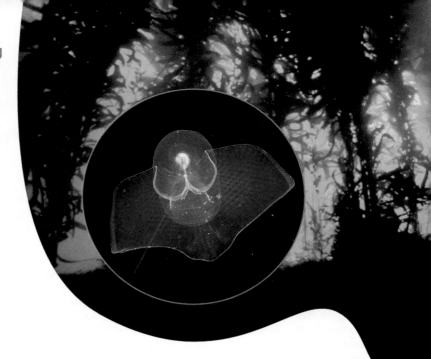

# 17

# The Open Sea

In 1967 a swordfish 3 meters (10 feet) long attacked the research submersible *Alvin* off the coast of Georgia at a depth of about 600 meters (1,980 feet). The entire length of the bill, which measured 1 meter (3 feet), penetrated the vessel's outer fiberglass hull, forcing it to return to the surface. No one was injured in the incident, and the crew ate the fish for dinner after extracting it from the submersible.

The swordfish is just one of several large animals that inhabit the open ocean, an immense area that contains 1.4 billion cubic kilometers (300 million cubic miles) of saltwater and provides several hundred times the living space of all terrestrial habitats combined. But the blue waters of the open ocean are barren compared with the productive coastal seas. Limited amounts of nutrients restrict the numbers of primary producers, and although the open ocean is immense, it supports relatively few large animals such as the swordfish. The combination of low productivity and lack of hiding and resting places makes the oceanic zone a different habitat from neritic seas.

## REGIONS OF THE OPEN SEA

Beyond the shallow coastal seas over the continental shelves (neritic zone) lies the open ocean (oceanic zone). When viewed from above, this open expanse of water appears monotonous. Even though it lacks obvious boundaries, the open sea can still be divided into regions based on the physical characteristics of the water and life forms in them. Vertical zonation depends on the depth to which sufficient light penetrates to support photosynthesis. As you may recall from Chapter 2, the photic zone is the layer that receives enough sunlight for phytoplankton to survive. In clear tropical waters, the photic zone extends down to a maximal depth of 200 meters (660 feet). The photic zone roughly corresponds

NOAA; inset, Richard Hermann

## Key Concepts

1. The open sea is a pelagic ecosystem, in which the living components are plankton and nekton.

2. Plankton range widely in size, taxonomic diversity, and lifestyle.

3. Phytoplankton are the primary producers in open-ocean food webs, and their productivity is limited by the scarcity of nutrients.

4. Bacteria provide a second base to open-ocean food webs, and they allow the scarce nutrients to be efficiently recycled.

5. Limited primary production and food webs with several energy-wasting steps limit the number of large animals the open ocean can support.

6. Gelatinous plankton such as salps and ctenophores play significant roles in open-ocean ecosystems because of their efficient feeding mechanisms, reduction of nutritional quality, and provision as prey for specialist carnivores.

7. Several structural features and behaviors have evolved to keep afloat organisms that are not strong swimmers.

8. Plankton display a number of interesting adaptations that help them avoid predation.

9. Large zooplankton include jellyfishes, gastropod molluscs, and colonial pelagic tunicates.

10. Fishes, squids, and mammals make up most of the nekton in the open sea.

to the **epipelagic zone,** a term used more in reference to the location of pelagic animals in the upper 200 meters of the ocean. Below this is the aphotic zone, which extends to the ocean bottom. In this zone, light rapidly disappears until the environment is totally dark. In this chapter we will be dealing with life in the photic and epipelagic zones. Life in the aphotic zone will be covered in the next chapter.

# LIFE IN THE OPEN SEA

Two groups of organisms inhabit the oceanic zone: plankton and nekton. On the basis of production, biomass, numbers, and biodiversity, life in the oceanic zone consists far more of plankton than nekton. Because nekton are defined by their ability to swim strongly against ocean currents, they are restricted to the large, muscular, streamlined, higher consumers among the animals, such as tuna and other pelagic fishes and whales. Plankton, on the other hand, belong to all three domains of life and many phyla of microbes and animals.

## Classification of Plankton

Many different organisms make up the plankton of the open sea. To accommodate this great diversity, marine scientists use several different schemes to organize plankton into logical groups. Most often the oceanic plankton are classified according to taxonomic group, functional group, size, life history, and spatial distribution in the sea.

### Taxonomic Groups

Particles suspended in the sea are called **seston** and include the nonliving **tripton** (mineral particles, dead organisms, and decaying organic matter or detritus) and the plankton. The two most familiar kinds of plankton are phytoplankton and zooplankton. As you learned in Chapter 2, phytoplankton are primary producers, and they belong to many distantly related groups, most of which were described in Chapter 6. Zooplankton are the heterotrophic eukaryotic microbes and those animals that are so small or are such weak swimmers that their distribution in the sea is determined by ocean currents. **Bacterioplankton** generally include both archaeans and bacteria, many of which are phytoplankton. **Viriplankton** include free viruses, which are the most abundant plankton of all.

### Functional Groups

Some plankton such as viruses, diatoms, and foraminiferans do not move at all and are called **akinetic.** Most plankton, however, are **kinetic** and move by flagella, jet propulsion, undulation of the body, swimming appendages, or other means. Their movements orient them in the water column, help them avoid predators, or change their vertical position. Because of their small size, slow swimming speed, or both, movement of these planktonic organisms does not result in significant changes of location when compared with the effects of water currents and turbulent mixing in moving them.

### Size

Terms originally applied to sizes of planktonic organisms depended on their visibility and methods of collecting them. **Macroplankton** were those organisms that were visible to the naked eye and generally exceeded 1 millimeter. **Microplankton** were small plankton that could be caught with standard plankton nets, whose mesh size became finer with advances in cloth fabrication technology. **Nanoplankton** (also called **centrifuge plankton** to distinguish them from microplankton, or net plankton) passed through the mesh of plankton nets and could be concentrated best by centrifugation.

There has been a trend in recent years to develop a more consistent system for classifying plankton by size. This trend stems from the discovery of viruses and a variety of other kinds of centrifuge plankton of apparently great importance in pelagic ecosystems. One commonly used scheme of classification divides the full range of plankton into seven categories, each containing plankton of a size within one to two orders of magnitude based on the number 2 (Figure 17-1). **Femtoplankton** and **picoplankton** include viruses and the smallest prokaryotes. Most prokaryotes and many kinds of eukaryotic phytoplankton (for example, diatoms and coccolithophores) compose the nanoplankton. The microplankton consist of a mixture of larger diatoms and dinoflagellates, invertebrate larvae, small crustaceans, and other adult invertebrates. Most other animal plankton, including larval fishes, fall into the larger categories of **mesoplankton, macroplankton,** and **megaplankton,** although the latter group also contains sargassum weed and detached rafts of benthic algae.

### Life History

Many kinds of organisms are planktonic throughout their lives and are called **holoplankton.** These include the microbes and many groups of invertebrate animals, especially in the open sea. Arrowworms, salps, siphonophores, comb jellies, copepods, krill, and some specialized groups of gastropod molluscs are holoplanktonic (Figure 17-2a). Many species of scyphozoans, such as the sea wasps, have evolved a life cycle that omits the benthic polyp stage, and these oceanic medusae are, therefore, holoplanktonic.

Not all plankton are planktonic in every phase of their life cycle. In the neritic zone, invertebrates that are benthic as adults often have larval stages in the plankton for feeding and dispersal. These planktonic larvae are called **meroplankton.** Meroplanktonic larvae of benthic invertebrates are less characteristic of the epipelagic zone of the open sea largely because the deep sea floor is too far below the surface waters (4,000 to 6,000 meters, or 13,200 to 19,800 feet, deep) for epipelagic larvae to be an adaptive part of life cycles. On the other hand, the abundant larvae or juveniles of many nektonic fish and squid are small enough to be classified as meroplankton (see Figure 17-2b).

### Spatial Distribution

Communities of plankton can be characterized as neritic or oceanic based on species composition. Neritic plankton communities are often distinguished by the presence of

| Micrometers | Millimeters | Meters | Size category and representatives |
|---|---|---|---|
| | 2000 | 2.0 | |
| | | | Megaplankton Large jellyfishes, colonies of siphonophores and salps, sargassum weed  |
| | 200 | 0.2 | |
| | | | Macroplankton Many gelatinous zooplankton, krill 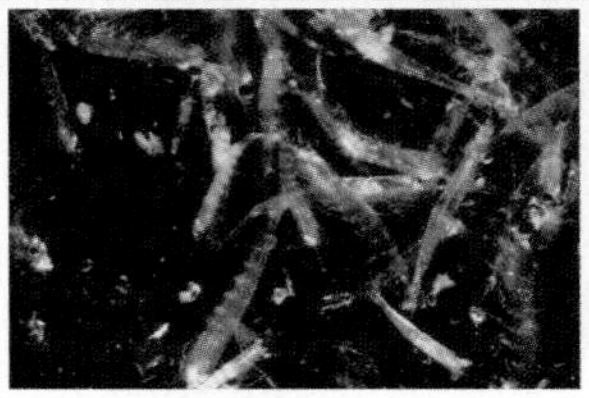 |
| | 20 | 0.02 | |
| | | | Mesoplankton Most adult zooplankton, larval fishes  |
| 2000 | 2.0 | 0.002 | |
| 200 | 0.2 | | |
| | | | Microplankton Many diatoms, dinoflagellates, invertebrate larvae 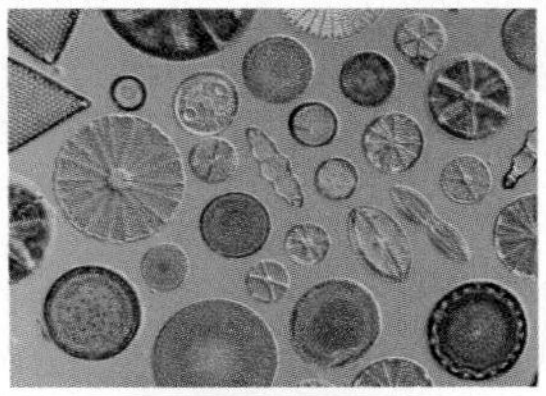 |
| 20 | 0.02 | | |
| | | | Nanoplankton Cyanophytes, coccolithophores, silicoflagellates, green flagellates, ciliates 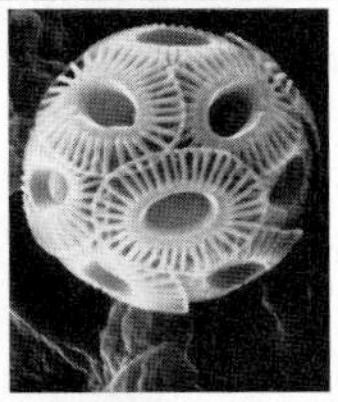 |
| 2.0 | 0.002 | | |
| | | | Picoplankton Many bacteria 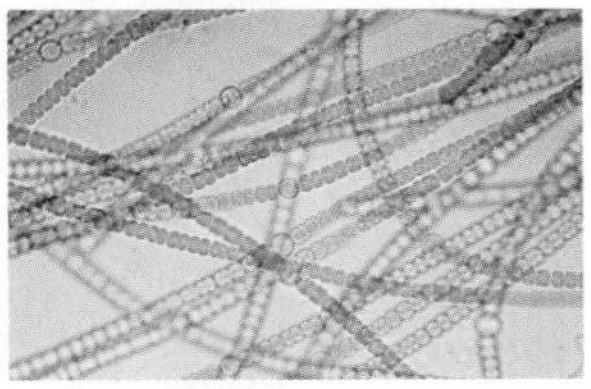 |
| 0.2 | | | |
| | | | Femtoplankton Most viruses |
| 0.02 | | | |

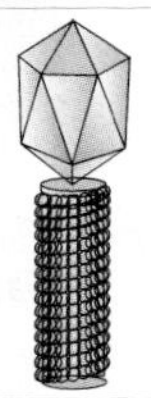

**Figure 17-1 Classification of Plankton by Size.** This chart demonstrates a scheme of classification that divides the full range of plankton into seven categories, each containing plankton of a size within one to two orders of magnitude based on the number 2.

David Wrobel 1994 / Biological Photo Service

1 cm

(a)

NOAA

(b)

**Figure 17-2  Plankton.** (a) Organisms such as this copepod that spend their entire life in the water column as plankton are called holoplankton. (b) Meroplankton, such as this larval fish, spend only a part of their life cycle in a planktonic stage.

meroplankton, such as the nauplius and cyprid larvae of barnacles. At low latitudes, diatoms are an abundant and diverse component of neritic plankton communities. Oceanic plankton communities are less diverse in diatoms and invertebrate meroplankton, but salps, larvaceans, arrowworms, and sea butterflies (a specialized group of snails) are among the more distinctive components of the epipelagic zone.

Plankton that live close to the water's surface are called **neuston** and represent a special case. These organisms often use the surface tension of water to remain at or near the surface. Refer to the photo of the water strider in Chapter 4 (see Figure 4-2). In addition to many microbes that are characteristically neustonic, about a dozen gastropod molluscs, several cnidarians, and a few seaweeds live here. Most of them are buoyed by gas bladders or bubbles and extend above the surface, where wind and water currents direct their paths. Such neuston that break the surface of the water are called **pleuston.**

## Patchiness in the Open Sea

The distribution of plankton in the sea is not uniform. Plankton occur in localized aggregations (that is, they are not evenly spaced) called **patches,** and many factors contribute to the formation of these patches. Areas of upwelling can cause large-scale patchiness in marine plankton and nekton communities by supporting bursts of primary production with the influx of nutrients. Factors such as highly localized variations in sea-surface conditions, vertical mixing of water, downwelling events, and meeting of waters of different density can cause more rapid population growth of phytoplankton in one place and diminished growth elsewhere. Another major contributor to patchiness is grazing by zooplankton. As phytoplankton densities decline with high rates of feeding, zooplankton move about to find adjacent patches for their next meal.

On a smaller scale, plankton biologists have found what they call **micropatchiness** throughout the photic zone. Various marine microbes become attached to particles of organic matter, particularly strands of mucus secreted by zooplankton, and form translucent, cobwebby aggregates called **marine snow** (Figure 17-3). Bits of marine snow and clusters of floating algal threads are home to populations of bacteria up to 10,000 times more concentrated than free bacteria in the open water. Along with these bacteria, which include both primary producers and decomposers, drift hordes of their microscopic predators. Further study of these drifting particles has led to discoveries indicating that the microscopic communities they harbor may in fact be complete ecosystems in miniature. On each of these floating islands of life, bacteria grow, respire, and are eaten. Primary producers reabsorb any nutrients released in this process, and grazing zooplankton feed on the primary producers. The whole system cycles through intense microbial activity with little input from the outside other than sunlight. These floating microenvironments are seen by some biologists as highly evolved, stable associations of primary producers and consumers. By remaining close together, microscopic pro-

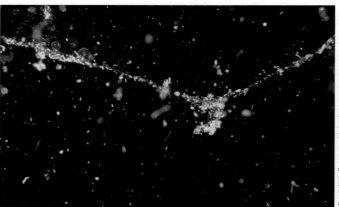

Woods Hole Oceanographic Institute

**Figure 17-3  Marine Snow.** Marine snow is an aggregation of marine microbes attached to particles of organic material such as the mucus secreted by zooplankton. These particles contain both producers and consumers and represent miniature ecosystems.

ducers and consumers can concentrate and store nutrients in their immediate vicinity, allowing them to survive in the nutrient-poor waters that surround them.

## Plankton Migrations

Many open-ocean zooplankton make daily migrations from the surface of the sea to depths of nearly 1.6 kilometers (1 mile). Some marine biologists believe that these movements take advantage of the benefits of feeding on the phytoplankton that live only in the photic zone and reduce to some extent the threat of predation by plankton-eating fishes, which are also more abundant in the epipelagic zone. These migratory zooplankton are often so densely packed that they form what is called a **deep scattering layer,** a mixed group of zooplankton and fishes that gives sonar systems a false image of a nearly solid surface hanging in midwater.

## Megaplankton

Most of the organisms classified as megaplankton are animals. Among these the most prominent are the cnidarians, molluscs, and salps.

### Cnidarian Zooplankton

The largest members of the plankton are jellyfishes. One of the most common is the moon jellyfish (*Aurelia*). This animal's bell-shaped body measures from 15 to 45 centimeters (6 to 18 inches). A fringe of thin, hairlike tentacles surrounds the bottom of the bell, and within the clear body are four oval, pigmented gonads. In addition to the tentacles, the outer surface of the bell is covered by bands of sticky mucus that ensnare any small organism that comes into contact with them. Cilia on the body surface constantly move the

mucus and any trapped food to the edge of the bell. Fleshy projections on the underside of the bell move the large blobs of accumulated mucus and food into the mouth for digestion. The moon jellyfish feeds mainly on copepods and other small zooplankton.

Larger and more dangerous are the jellyfishes known as lion's mane jellyfishes (*Cyanea;* Figure 17-4a). One enormous species in the Arctic Ocean has a bell that averages 2.5 meters (8 feet) in diameter and has tentacles that extend downward for 30 meters (99 feet) or more, giving it the distinction of being the largest member of the zooplankton in the world. Species in temperate regions of the Atlantic and Pacific Oceans tend to be smaller. They feed on surface fishes, and their sting can easily kill a 30-centimeter (12 inch) fish.

One of the jellyfish species best adapted to life in the open ocean is *Pelagia noctiluca* (see Figure 17-4b). This species has a beautiful, pastel body, and at night it is bioluminescent. Unlike most species of jellyfish that have an asexual polyp stage in their life cycle, *Pelagia* larvae mature in the open sea without going through a polyp stage; it is, therefore, holoplanktonic.

### Molluscan Zooplankton

Even some molluscs have become adapted to life in the open sea. Pteropods (from *ptero,* meaning "wing," and *pod,* meaning "foot"), popularly known as sea butterflies (Figure 17-5a), are related to snails. The animal's foot has two large winglike projections, and the shell is much reduced (**thecosome pteropods**) or absent (**gymnosome pteropods**). The projections on the foot propel the animal through the water. Cilia on the surface of these projections on thecosome pteropods create a current that drives small plankton, mainly diatoms, toward the mouth. Some species secrete enormous

(a)

(b)

**Figure 17-4 Jellyfishes of the Open Sea.** (a) The lion's mane jellyfish, *Cyanea capillata,* can achieve bell sizes of 3 meters (10 feet) in diameter and weigh 1 ton. (b) *Pelagia noctiluca* is bioluminescent and lacks an asexual polyp stage in its life cycle.

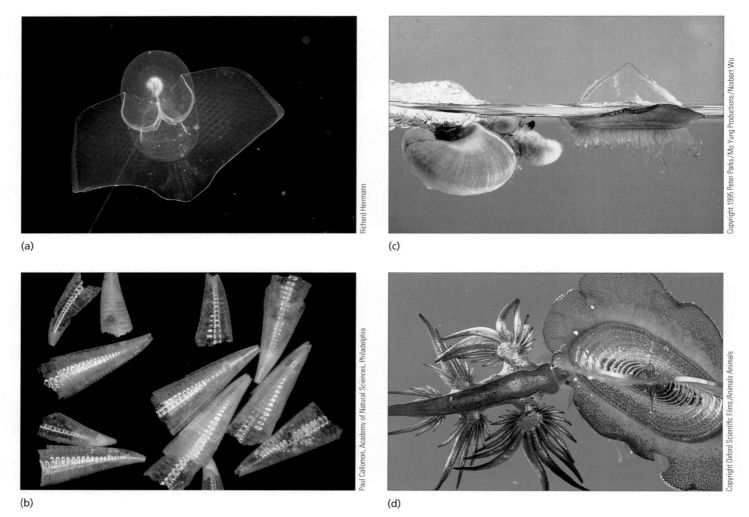

(a)

(b)

(c)

(d)

Richard Herrmann

Paul Callomon, Academy of Natural Sciences, Philadelphia

Copyright 1995 Peter Parks / Mo Yung Productions / Norbert Wu

Copyright Oxford Scientific Films / Animals Animals

**Figure 17-5 Molluscs of the Open Sea.** (a) Pteropods, or sea butterflies, have a foot that is modified to form a pair of winglike structures that animals use to propel themselves through the water column. (b) The thecosome pteropod *Hyalocylis striata* is so abundant in some areas of the South Atlantic Ocean that the ocean floor is covered with the shells of these dead organisms. (c) The purple sea snail keeps itself afloat by clinging to a bubble raft that it produces. This specimen is extending its proboscis toward the nearby siphonophore *Velella* (by-the-wind sailor), one of its favorite foods. (d) The pelagic nudibranch (*Glaucus*) swims upside down just beneath the surface of the water. This specimen is feeding on a by-the-wind sailor.

mucous nets in which phytoplankton are trapped as food. These gastropods are abundant in the open ocean and sometimes form dense groups typically preyed upon by fishes and whales. In some areas of the South Atlantic Ocean, thecosome pteropods (see Figure 17-5b) are so numerous that large areas of the bottom are covered with the shells of these dead animals, forming calcareous sediments called **pteropod ooze.** The less-abundant gymnosome pteropods are carnivores that specialize in preying on the herbivorous thecosomes.

Another prominent mollusc in the open sea is the purple sea snail (*Janthina;* see Figure 17-5c). Unlike pteropods, the purple sea snail has retained a relatively large, although light, shell. It keeps itself from sinking by producing a raft of

bubbles surrounded by mucus. The snail clings upside down to the underside of this bubble raft. This upside-down life-style has led to a reversal of the typical coloration of open-water organisms. What would normally be the animal's dark-colored upper surface is mainly white, whereas its under surface is dark purple rather than light colored. The purple sea snail is widely distributed in all tropical waters, where it feeds on larger members of the zooplankton such as copepods, jellyfishes, and by-the-wind sailors (*Velella*), a siphonophore related to the Portuguese man-of-war.

Some species of nudibranch have also enjoyed success in the open sea. Some of the most numerous are species in the genus *Glaucus* (see Figure 17-5d). These small, blue nudibranchs glide along upside down beneath the surface of the water. Like the purple sea snail, *Glaucus* also displays a reverse color pattern with a dark ventral surface and a lighter-colored upper surface. *Glaucus* feeds on by-the-wind sailors, and like some other species of nudibranch, it retains the stinging cells in its body for its own defense. *Glaucus* has been observed to leave clutches of eggs attached to the floats of by-the-wind sailors after having devoured all of the colony's polyps.

## Urochordates

Collections and observations made by scuba-diving researchers have yielded a wealth of information about pelagic members of the subphylum Urochordata (tunicates) known as salps (thaliaceans) and larvaceans (for more information regarding salps and larvaceans see Chapter 9). Salps have barrel-shaped bodies that are open at both ends. Some species produce and maintain a mucous net inside their oral openings and use it to filter water that is pumped through their bodies. When filled, the net and its contents are digested. Salps eat bacteria too small for most sizable plankton to capture, making that source of organic material available to their own predators. On the other hand, salps produce fecal pellets from planktonic food in abundance, increasing the flow of nutrients downward and out of the photic zone.

Because their bodies are at least 95% water, these adaptable organisms can grow and reproduce very rapidly, creating swarms that can stretch over kilometers and contain up to 500 individuals in each cubic meter of water. **Pyrosomes** (Figure 17-6a), close relatives of salps, produce colonies that measure as long as 14 meters (46 feet), although the usual range is from a few centimeters (1 inch) to 3 meters (10 feet). Each pyrosome comprises hundreds of individual animals that join to form a hollow cylinder that appears to be a single organism. Each member of the colony faces outward and sucks in water and small plankton. The incurrent water enters the center of the colony, flowing through and out the back, moving the colony through the water by means of slow ciliary propulsion. Pyrosomes occur worldwide but most commonly in tropical and subtropical seas.

Sometimes, a salp will be inhabited by an amphipod crustacean of the genus *Phronima* (see Figure 17-6b). This amphipod eats the internal organs of the salp and then uses the salp's exterior as a sort of mobile home in which it lives and broods its young. The salp's body continues to move through the water, propelled by the swimming action of the crustacean.

Larvaceans secrete mucous structures called houses. Water is drawn through the house when the larvacean undulates its tail. Inside the house, tiny plankton are trapped in a wing-shaped feeding filter and digested. Because the houses have no opening for the elimination of feces, larvaceans must abandon their houses several times a day as they become clogged with feces and large plankton and then produce new ones. Interestingly, the discarded houses of larvaceans, which often contain up to 50,000 trapped living phytoplankton cells, immediately become homes for bacteria and end up as particles of marine snow. The types of feeding exhibited by salps, pyrosomes, and larvaceans allow these zooplankton to filter enormous volumes of water very efficiently.

## Nekton

Nekton are the actively swimming organisms whose movements are not governed by currents or tides. Included in this group of animals are some larger invertebrates, fishes, reptiles, birds, and mammals.

### Invertebrates

The invertebrates that reign supreme in the open sea are the squids. Speed, keen eyesight, and intelligence make the squid a formidable predator. The animal's body is streamlined, and it has sucker-laden tentacles for grasping prey. By drawing water into the mantle cavity and forcefully ex-

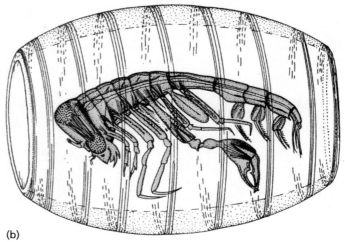

(a)                                                    (b)

**Figure 17-6  Pyrosomes and Salps.** (a) Large colonies of pyrosomes are formed by asexual reproduction from the original tunicate. (b) Amphipods of the genus *Phronima* feed on salps.

pelling it through its siphon, squids can achieve bursts of speed that are faster than the fastest fishes. Although squids normally swim backward, they can reverse their direction by adjusting the position of their siphon. This is especially useful when maneuvering within a school of fish.

## Fish

The nekton include many species of fish, and these species exhibit a variety of adaptations that enable them to survive in this niche. The majority of bony fishes that inhabit the open sea have mouths at the front of the body with lower jaws that protrude farther forward than the upper jaw. This arrangement allows the fish to grab prey from a variety of positions. Some species, like the John Dory (*Zenopsis;* Figure 17-7a), have very extensible jaws. When the fish opens its mouth to seize its prey, the jaws project forward. They then retract quickly as the mouth closes.

(a)

Daniel Heuclin /Natural History Photographic Agency

(b)

Norbert Wu

**Figure 17-7 Fishes of the Open Sea.** (a) The John Dory has jaws that protract for rapid grasping of prey. (b) The jaw of the sailfish is modified to form a bill with which it flails its prey.

***Billfish*** The upper jaw of species like marlin and sailfishes is greatly elongated, forming a bill (see Figure 17-7b). These fishes lack the teeth characteristic of carnivores and use their bills to club their prey. The bill of a swordfish (*Xiphias gladius*) is broad and flat. As they swim into schools of mackerel, menhaden (*Brevoortia*), or squid, they flail their bill left and right, beating their prey and then swallowing their stunned victims whole.

Swordfishes will attack virtually anything, often without any apparent provocation. The broken bills of swordfishes have been found in the timbers of wooden ships, sometimes penetrating two layers of oak planks. Captured blue whales and fin whales have had broken bills embedded in their sides and backs attesting to swordfish attacks.

***Tuna*** The most wide-ranging fishes in the open ocean are tunas, which belong to the mackerel family (Figure 17-8). The largest species of tuna are the bluefin tuna (*Thunnus thynnus*), with individuals achieving lengths of up to 4 meters (13 feet), and the yellowfin tuna (*Thunnus albacares*), which achieves lengths of 2 meters (7 feet) or more. Smaller relatives include the albacore (*Thunnus alalunga,* slightly more than 1 meter [3 feet] long) and the ocean bonito (*Sarda,* not quite 1 meter).

Tunas must swim constantly (some cruising at 27 kilometers, or 16 miles per hour) or they will sink for lack of a swim bladder. This level of activity requires a large amount of energy and a good supply of oxygen. To supply the needed oxygen, these animals must swim fast and move large volumes of water past their gills. One adaptation for fast swimming in some tuna species is a body temperature 8°C to 10°C higher than the surrounding water. The higher temperature allows the fish to metabolize faster. Digestion is more rapid, nerve impulses travel more quickly, and the large skeletal muscles used for swimming contract and relax about three times faster than in other fishes. These adaptations help to account for the tuna's great speed and strength.

Many species of tuna exhibit bursts of speed that can exceed 54 kilometers per hour (more than 33 miles per hour). The main propulsive force comes from the high-keeled, sickle-shaped tail. To decrease resistance in the water, their bodies are streamlined and their body surface is smooth. Their gill covers fit tightly against the sides of their body, and their pectoral fins lie retracted in grooves in their sides. Yellowfin tunas can reach speeds of 75 kilometers per hour (45 miles per hour), and one relative, the wahoo (*Acanthocybium solanderi*), has been clocked at 80 kilometers per hour (48 miles per hour). When they sense food, tunas quickly accelerate from cruising speed to top speed in less than 1 second. Tunas must consume large amounts of food to supply the energy they need for such vigorous swimming. Their diet is varied, consisting mainly of herring, anchovies, mackerel, sardines, flying fishes, and squids.

***Ocean Sunfish*** Quite different from the very active lifestyle of tunas is the more relaxed swimming of the ocean sunfish (*Mola mola;* Figure 17-9). Adults appear to spend most of their time lying on their side at the surface of the water, apparently basking in the sun; this behavior gives them their common name. The ocean sunfish has a highly modified caudal fin and prominent dorsal and anal fins.

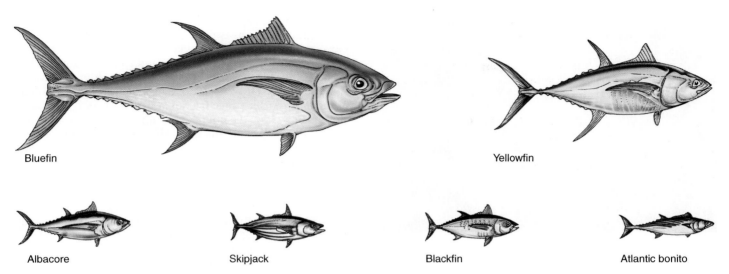

**Figure 17-8 Tuna.** Tunas are fast swimmers that can be found throughout all seas. They range in size from the large bluefin tuna to the small Atlantic bonito.

Bluefin

Yellowfin

Albacore

Skipjack

Blackfin

Atlantic bonito

**Figure 17-9 Ocean Sunfish.** Not much is known of the life history of this large, slow-moving fish.

These fish are large, reaching lengths of more than 3 meters (10 feet) and weights exceeding 1 ton. They feed on larger zooplankton, especially jellyfish, and seem to have few natural predators.

The ocean sunfish is protected by a layer of cartilage 5 to 10 centimeters (2 to 4 inches) thick just beneath the surface of the skin. Although this thick layer offers protection against most predators, it is a haven for parasites. Almost every specimen examined has been full of internal and external parasites, which may cause the death of many individuals. On the other hand, because parasite-infested animals are more easily caught, the data may show a bias.

Analysis of the stomach contents of some specimens reveals the remains of animals found only in deeper water, indicating that these large fish may not be as lazy as once per-ceived but may actually feed well below the surface. Some biologists suggest that the normal habitat for this fish is deep water, and only diseased individuals are found at the surface, where they come to die and are collected. In California, ocean sunfish regularly come near shore in summer to visit cleaning stations near kelp beds.

***Sharks***  Some of the most efficient predators of the open ocean are sharks, whose streamlined bodies are well-adapted to a pelagic lifestyle. Their reproductive behavior also reflects adaptations to life in the open sea. Unlike most bony fishes, which employ external fertilization, sharks copulate (for more information on shark reproduction see Chapter 10), with the male transferring sperm to the female. This increases the likelihood that the maximal number of eggs will be fertilized, and thus fewer eggs need to be produced. Many species are viviparous: the eggs are retained in the female's reproductive tract where they hatch and for a time are nourished on a milky fluid. This strategy helps to ensure better survival of the young when they enter the environment. Hammerhead sharks (*Sphyrna*) and blue sharks (*Prionace glauca*) not only hold their young within the body but also have a connection that supplies nutrients to the young from the mother's blood in a manner similar to that of mammals.

Sharks produce fewer young each breeding season than other fishes because their reproductive strategies are more efficient, and the survival rate of their offspring is better. For instance, whereas most bony fishes produce thousands of offspring each breeding season, the blue shark bears only about 30 young each season. Each is 60 to 70 centimeters (24 to 28 inches) long and fully independent when born, with a full set of teeth for feeding and defense.

***Manta Rays***  Another member of the nekton that is related to the shark is the manta ray, or devilfish (Figure 17-10). Fully adult mantas may measure 6 meters (20 feet) from one fin tip to the other and weigh as much as 1.5 tons. Mariners of old believed that manta rays could grab ships by their an-

**Figure 17-10  Manta Ray.** The manta ray, or devilfish, feeds on small fishes and plankton that it scoops up with its large mouth.

chor chains and drag them to the ocean depths. It was also believed that these creatures would envelop swimmers in their large fins and devour them. Neither of these beliefs is true, of course, but injured or provoked manta rays have been known to reduce small wooden fishing boats to masses of floating splinters. Manta rays feed primarily on small fishes and plankton that they channel into their mouths with the large fleshy extensions called labial flaps.

## Reptiles

Reptiles are not often thought of as oceanic nekton, although they and sharks ruled the seas during the Mesozoic era. In present-day seas, however, the yellow-bellied sea snake (*Pelamis platurus;* Figure 17-11) cruises the tropical waters of the Indian and Pacific Oceans along drift lines in search of small fish to capture as prey. Although the main predators of coastal sea snakes are sea eagles and sharks, the oceanic *Pelamis* seems to have no known enemies. Birds,

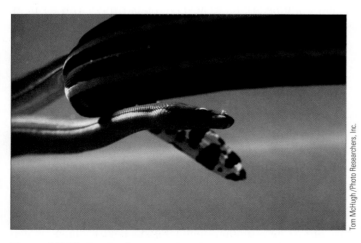

**Figure 17-11  Sea Snake.** The yellow-bellied sea snake, *Pelamis platurus,* is a member of the nekton that cruises the waters of the tropical Pacific in search of small fish on which to feed.

fish, and sharks reject it as food. Laboratory experiments indicate that the meat is distasteful, and large predators apparently learn to avoid it. Unlike its coastal relatives, the yellow-bellied sea snake cannot descend to the sea floor to rub its shedding skin against rocks or coral. Shedding allows the snake to rid itself of accumulated algae and other fouling organisms that increase frictional drag. Instead, this snake slips its body into and out of knots, abrading the skin against itself to remove it and the unwanted guests. Female yellow-bellied sea snakes retain their eggs within the reproductive tract until hatching, giving live birth at sea and relieving themselves of the need to crawl ashore for reproduction.

## Birds and Mammals

The only oceanic birds that can be included in the nekton are the penguins of the southern oceans. The flightless penguins have undoubtedly adapted to life in the open sea except for their need to return to land to reproduce. When foraging at sea, penguins consume large numbers of krill, squid, and fish.

The largest members of the nekton are whales. Baleen whales filter krill, thecosome pteropods, and sometimes fish from the seawater. Toothed whales, such as sperm whales, feed on squid and fish. The biology of marine birds was covered in Chapter 11 and the mammals in Chapter 12; refer to these chapters for further discussion of their lifestyles.

## In Summary

Although the open sea appears uniform and without divisions, it can be divided into regions based on the physical characteristics of the water and the kinds of organisms inhabiting them. It can also be divided into vertical zones according to the amount of sunlight available. The open sea lacks well-defined communities and is termed a pelagic ecosystem. Living components of this ecosystem are the drifting plankton and strongly swimming nekton.

Most members of this ecosystem are plankton. Plankton organisms vary widely in size, nutrition, style of locomotion, life history, and spatial distribution. They belong to many different taxonomic groups of microbes and of multicellular algae and animals. As a result of their diversity, plankton biologists have applied a broad vocabulary to classify the many types of plankton. Plankton are not distributed uniformly in the surface waters; many factors contribute to their patchy distributions.

The nekton consist of animals that are free swimming and whose movements are not governed by currents and tides. Nekton are dominated by squids and a great diversity of fishes. Aside from fishes, few other vertebrates are found in the open sea, but the yellow-bellied sea snake, penguins and other seabirds, and whales are among them. Members of the nekton feed on phytoplankton and zooplankton as well as on each other.  ●

## SURVIVAL IN THE OPEN SEA

Despite lower productivity compared with coastal seas, the epipelagic zone provides an environment that is generally more stable than others we have examined. Sudden fluctuations in salinity and temperature do not occur in the open seas, and even the most violent storms have little effect a few meters below the surface. Organisms that live in the well-lit surface waters, however, face other problems of survival. Both large and small organisms have to work to keep from sinking out of the photic zone, and with no place to hide, animals have to depend on other strategies for avoiding their predators.

## Remaining Afloat

Phytoplankton must remain in the photic zone to carry out photosynthesis, and the distribution of phytoplankton dictates the distribution of zooplankton that rely on these organisms for food. Nekton of the open sea have dense bodies, and remaining on or near the water surface must be accomplished by mechanisms that do not interfere with their pursuit of food. Adaptations for staying afloat include swimming (locomotion) and reducing the sinking rate (Figure 17-12).

### Swimming Methods

Some phytoplankton and many zooplankton and nekton remain afloat by actively swimming. Swimming may involve the use of flagella or cilia, as in many microorganisms, or the use of jet propulsion, as in some invertebrates. Many zooplankton and nekton swim with the aid of their appendages or by undulating their body.

***Flagella, Cilia, and Jet Propulsion***   Dinoflagellates, coccolithophores, and silicoflagellates use their whiplike flagella to swim, as does the blue-green bacterium *Synechococcus*. Tintinnids, other ciliates, and innumerable larval stages of invertebrates use cilia, as do the comb jellies with their ciliated ctenes and *Pyrosoma* with its ciliated pharynx. Jet propulsion is common among jellyfish, siphonophores, salps, and squid.

***Appendages***   Many zooplankton and some nekton use appendages to swim (**appendicular swimmers**). The numerous copepods use their legs and antennae, moving them back and forth as fast as 600 times a second. All other planktonic crustaceans and their larvae are appendicular swimmers as well, but the appendages that are used vary widely among them. The pteropod molluscs, or sea butterflies, flap the winglike extensions of the foot. The legs of water striders scull across the sea surface. Appendicular swimming is very effective for nektonic vertebrates, which may use paired limbs, as do the pinnipeds and diving birds.

***Undulations of the Body***   Finally, many zooplankton and nekton swim by undulation of the body, either side to side (horizontally) or vertically. Notable among the zooplankton are the arrowworms as well as the larvaceans and various specialized members of worm groups. Fish of the open ocean and the whales and dolphins are undulatory swimmers among the nekton.

### Reduction of Sinking Rates

Although swimming slows the rate at which organisms sink to greater depths, the use of other methods to reduce sink-

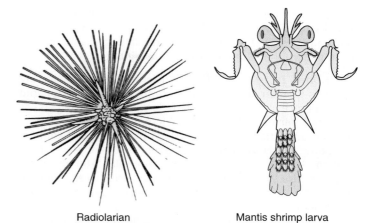

Radiolarian                Mantis shrimp larva

(a) Projections like the needles of the radiolarian and the flattened body of a mantis shrimp larva create friction and slow sinking.

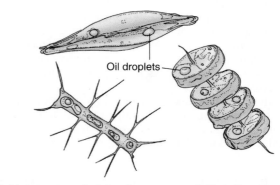

Oil droplets

(b) Diatoms store their food reserves as oil. The oil makes them more buoyant and offsets some of the weight of their shells.

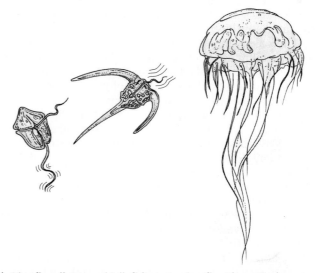

(c) Dinoflagellates and jellyfishes remain afloat by actively swimming.

**Figure 17-12  How Open-Water Organisms Remain Afloat.**
Organisms that inhabit the water column of the open sea slow or prevent sinking by (a) increasing friction, (b) having adaptations that increase buoyancy, or (c) actively swimming.

ing rates allows them to devote more of their swimming activity to orientation, feeding, and avoidance of predation. Sinking rates are reduced in plankton by increasing friction or in plankton and nekton by increasing buoyancy.

***Frictional Drag***    The effect of frictional drag in slowing speed is familiar to anyone who appreciates the shape and performance of fast vehicles (cars, boats, planes, rockets) or has used parachutes. Because friction increases with surface area, friction can be increased by decreasing volume, flattening the body, or increasing the body length. Viruses and small prokaryotes among the femtoplankton and picoplankton are so small that gravity has virtually no effect in overriding the turbulence and viscosity of seawater. The small bodies of nanoplankton have such a large surface area compared with their volume that flagella and stored oils are very effective in keeping these cells afloat. The comparatively lower ratio of surface area to volume of microplankton and larger organisms does not allow generation of much friction, and these plankton often have body projections, which add little weight but increase surface area and thus drag. Spines and other appendages on the surfaces of diatoms, dinoflagellates, radiolarians, and many crustaceans increase their surface area, as do the long pseudopodia of radiolarians and foraminiferans. Other plankton have flattened bodies, which act like a parachute. Elongation of bodies and the formation of long chains of cells among phytoplankton are other ways to increase friction. Although beneficial to plankton, frictional drag is a disadvantage to most nekton.

***Buoyancy***    Adaptations that increase friction do not prevent an organism from sinking; they merely slow it. Most members of the plankton must have other adaptations that allow them to counteract gravity and remain in the sunlit surface waters. In addition to locomotion, buoyancy mechanisms are often used by plankton. In contrast to frictional drag, increasing buoyancy is a most effective adaptation for nekton to maintain their position in the sea. Buoyancy is increased by storage of oils, increasing the water content of the body, exchange of ions, and use of gas spaces.

Because lipids have a density much less than that of seawater, storage of oils, fatty acids, and other lipids is commonly used to compensate for the denser components of bodies. The products of photosynthesis may be stored as oils in the vacuoles of many phytoplankton, particularly diatoms and blue-green bacteria. Copepods, which graze heavily on diatoms, also store food as oil droplets. The use of fats in nekton is well known, because whales, sharks, and codfish have long been harvested for whale oil, squalene, and cod-liver oil. Many fish deposit droplets of oils in their eggs to provide flotation.

The dilution of proteins and minerals with water immediately reduces the density of a body. Although such a mechanism is not suitable for nekton, which have a great need for muscle and skeleton for powerful locomotion, the bony fishes do have somewhat diluted body fluids. Their eggs also contain an elevated content of water, maintained by an impervious egg membrane until hatching. Many zooplankton,

however, use this mechanism to an extreme. See the box on gelatinous zooplankton for more on this phenomenon.

Seawater is composed of many kinds of ions, some of which are heavier than others. Magnesium, calcium, and sulfate ions frequently are replaced by ammonium and chloride ions in the vacuoles and body fluids of phytoplankton, zooplankton, and nekton. Although ammonium ion is toxic, its use in confined parts of the bodies of many squids and crustaceans effectively controls buoyancy. In the body fluids of some squid, as much as four fifths of the positively charged ions are ammonium.

There is no more effective means to reduce density than gas bladders or evacuated chambers. Air and other gases have exceedingly low densities even at great depths in the sea. An organism needs only a small gas space to compensate for the density of well-developed muscle and skeleton. This mechanism is used by both plankton and nekton. Pleuston such as the Portuguese man-of-war and sargassum weed use gas-filled chambers to float at the surface of the ocean. Below the surface, siphonophores have gas bladders, and bubbles of gas occur in the cytoplasm of many blue-green bacteria, radiolarians, and foraminiferans. The rigid chambers of the shells of the nautilus and cuttlefish can be evacuated or filled with fluid to adjust buoyancy (see Chapter 9). Similarly, the flexible swim bladder of bony fishes can be inflated or deflated (see Chapter 10).

## Avoiding Predation

When threatened, animals of the open sea cannot quickly hide in a rock crevice, burrow into bottom sediments, or hide behind a stand of coral. Except for floating mats of sargassum weed, there are no large seaweeds, such as kelps, to provide a quick refuge. Pelagic organisms have evolved a variety of adaptations to make up for the lack of hiding places and increase their chances of survival. Larger members of the plankton, such as jellyfish, have stinging cells to deter potential enemies. Some animals avoid predators with speed, and others, like flying fishes and some species of squid, take to the air. Some find safety in numbers, with many species of nektonic fish swimming in schools and some invertebrates forming colonies. Many animals of the epipelagic zone display adaptations that make them less visible or even invisible in the open water.

Even with protective adaptations, the odds against survival in the open sea are tremendous. Many species compensate for the low survival rate by producing enormous numbers of offspring, thereby increasing the odds that a few will survive to propagate. The female ocean sunfish (*Mola mola*), for instance, produces as many as 30 million eggs in a single breeding season. Only two or three offspring need to survive to perpetuate the species.

### Benefit of Being Less Conspicuous

In an environment that lacks hiding places, camouflage plays a central role in survival. Larger animals from nudibranchs and pelagic snails to sharks and whales exhibit **countershading** (Figure 17-13). These animals have dorsal

# Gelatinous Zooplankton

Because cytoplasm is normally denser than seawater, organisms that cannot compensate for their greater density tend to sink. A common approach to increasing buoyancy in zooplankton is the incorporation of large amounts of water in the body. Proteins and carbohydrates dispersed in a large volume of water contribute much less to body density than when they are concentrated in well-formed muscles, as in the nekton. In addition, zooplankton with watery bodies often replace heavier ions (magnesium, sulfate) with lighter ones (ammonium, chloride). These animals are called **gelatinous zooplankton.**

Gelatinous zooplankton are members of many animal groups. They tend to be suspension feeders but run the gamut of locomotory types. The jellyfishes are typical, as are their relatives the siphonophores and the medusae of other hydrozoans. These cnidarians move by jet propulsion. Exceptions are among the pleustonic cnidarians, such as the Portuguese man-of-war, by-the-wind sailor, and blue buttons, whose movements are driven by wind and water currents. Comb jellies are a common component of the gelatinous zooplankton, moving by their ciliary ctenes. Pteropod molluscs are a group of pelagic snails that swim using winglike extensions of their foot; thecosome pteropods have a reduced shell, and the gymnosome pteropods lack a shell. Some squids have an enlarged body cavity filled with fluid of low density in which magnesium and other ions have been

replaced by ammonium ions; they can "hang" motionless in the water column awaiting unsuspecting prey. A few species of gelatinous sea cucumbers swim by whole-body undulation or by peristaltic operation of an anterior hood or veil. All pelagic urochordates are gelatinous zooplankton, some moving by undulation of the tail (larvaceans), others by peristalsis of the barrel-shaped body (salps), and yet others by cilia on the pharynx (pyrosomes). Undulation is the mode of locomotion in the arrowworms and in the superficially similar leptocephalus larvae of eels and members of the tarpon family.

Gelatinous zooplankton are important in the open sea for three reasons. First, many are highly efficient suspension feeders. Their feeding activities can decimate populations of nanoplankton and microplankton. In this way, they are strong competitors with other zooplankton for food. Second, they have low nutritional value. They convert otherwise rich sources of oils and proteins into the diluted mass of their own bodies. To acquire enough nutrients, other zooplankton and nekton would have to fill their stomachs with large quantities of gelatinous zooplankton and deal with the water and ions consumed. Thus, the possession of gelatinous bodies is an effective predator-deterrent mechanism as well as a buoyancy mechanism. Third, a few animals have become specialized as predators of gelatinous zooplankton. Notable among them are the leatherback sea turtle and the ocean sunfish. In addition, some

of the tubenosed sea birds and the frigatebirds prey heavily on medusae. Some predators of gelatinous zooplankton are found within their own ranks. The comb jellies are sometimes eaten by cnidarian medusae, and some comb jellies eat medusae, salps, and even other comb jellies. There has been speculation recently that leptocephalus larvae, too, eat gelatinous zooplankton.

To learn more about oceanic ecosystems, in some recent studies researchers have donned scuba gear to observe and collect gelatinous zooplankton firsthand. Several important microscopic components of the plankton slip through plankton nets, whereas others, such as the gelatinous ctenophores, siphonophores, and mucous houses of larvaceans are so fragile that collecting nets damage them beyond recognition. Scuba divers can locate ctenophores by the way they reflect light and by the rainbow of colors produced when their rows of beating cilia refract sunlight. Though these animals do not normally survive even the gentlest of standard collecting techniques, a careful diver can scoop them into jars without injuring them. Other species of gelatinous zooplankton, foraminiferans, clusters of floating algal filaments, and marine snow can be collected by hand in a similar fashion. Such labor-intensive techniques are necessary for increasing our understanding of the natural history and ecological significance of gelatinous zooplankton. ●

---

surfaces that are dark blue, gray, or green and ventral surfaces that are silvery or white. This arrangement of colors makes them difficult to see from above and below.

Many planktonic species are inconspicuous because they are nearly transparent. For instance, the most abundant members of the zooplankton, copepods, are often transparent, as are salps, larvaceans, ctenophores, and

jellyfishes. Arrowworms (*Sagitta;* Figure 17-14) are so transparent that they are invisible in the water and were not discovered until 1768, even though they are relatively large (up to 10 centimeters, or 4 inches) and distributed worldwide. Arrowworms are formidable predators with eyes that see in all directions and sensory hairs that detect the slightest movement in the water. They hang nearly

**Figure 17-13  Countershading.**
(a) In a countershaded animal, the body surface that is usually exposed to sunlight is a dark color, while the opposite surface is a paler color or white. In the ocean environment, a uniformly colored fish (b) is highlighted from above and easy to see, while a countershaded fish (c) is effectively camouflaged.

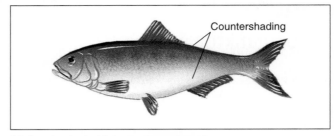

(a) Countershaded fish

(b) Uniformly colored fish in
natural lighting

(c) Countershaded fish in
natural lighting

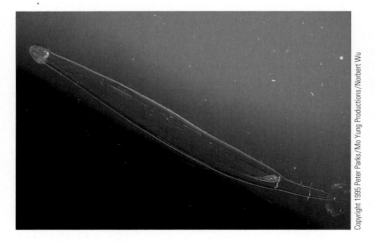

Copyright 1995 Peter Parks/Mo Yung Productions/Norbert Wu

**Figure 17-14  Arrowworms.** Arrowworms are aggressive predators with transparent bodies. Transparency makes the arrowworm difficult to be seen by its own predators.

motionless waiting for prey, such as a copepod or a larval fish, to come within striking distance. When the arrowworm senses prey, it darts toward it with amazing speed, grasping the prey with the hard pincerlike structures around its mouth.

## Safety in Numbers

Some animals increase their chances of survival in the open sea by forming colonies. One of the most successful of these is the cnidarian group Siphonophora. Although siphonophore colonies look like a single individual, they are actually made up of thousands of individuals, none of which can live independently of the colony. Some members of the colony specialize in capturing prey, whereas others digest food, maintain the colony, provide flotation, or reproduce new individuals. Siphonophores are common in all oceans, where they drift on or just beneath the surface of the water, some species propelled by the gentle pulsations of several tiny "swimming bells."

One well-known siphonophore is the Portuguese man-of-war (*Physalia physalis*), which produces a gas-filled float that helps to propel the colony by catching the wind. Sometimes thousands of these colonies passively gather in drift lines caused by currents, forming a mass of siphonophores that stretches for miles across the open sea. The gas sac of a large man-of-war may be as long as 40 centimeters (16 inches), and the colony occasionally submerges the float to keep it moist. If the float dries out, it cracks, the gas escapes, and the colony sinks. The tentacles formed by feeding polyps in the colony can be as long as 8 meters (26 feet) in a large specimen.

Although the stinging cells of the Portuguese man-of-war are capable of killing a large mackerel, one of the colony's main food items, a small fish lives symbiotically with the colony, unharmed by the stings. The man-of-war fish (*Nomeus gronovii*) spends most of its time swimming among the stinging tentacles (Figure 17-15). The little fish averages 8 centimeters (3 inches) long and is deep blue with vertical black stripes, blending quite well with the Portuguese man-of-war's tentacles. How the man-of-war fish is able to avoid being killed by its host is not known. The fish has been observed to feed on some of the polyps, and it is possible that ingestion of the venom helps to immunize it against the stings. In laboratory experiments, the fish survived injections containing 10 times the venom that killed a different fish species of similar size.

## In Summary

One of the challenges to life in the open sea is to remain afloat. Plankton and nekton have adapted to pelagic life in many ways to keep from falling to the deep sea. Whereas many styles of swimming are used by most pelagic organisms to counter the effects of gravity, only the smaller ones can decrease the rate of sinking by increasing frictional drag. Buoyancy is controlled by the incorporation of low-density substances such as lightweight ions, lipids, and gases. Many kinds of zooplankton reduce their density by increasing the content of water in their bodies, giving them the consistency of gelatin. Gelatinous zooplankton are efficient suspension feeders. The combination of their feeding efficiency and low food value greatly reduces the nutritional quality of pelagic communities. They also are prey for food specialists.

A second challenge to life in the open sea is to avoid predation. The general uniformity of surface waters does not provide animals with many places to hide. Instead they have evolved other ways of avoiding predation, including countershading, transparency, schooling, colony formation, toxins, and production of large numbers of offspring. Some plankton have evolved large size (megaplankton); among them are jellyfish, specialized kinds of gastropod molluscs, and colonies of pelagic tunicates. ●

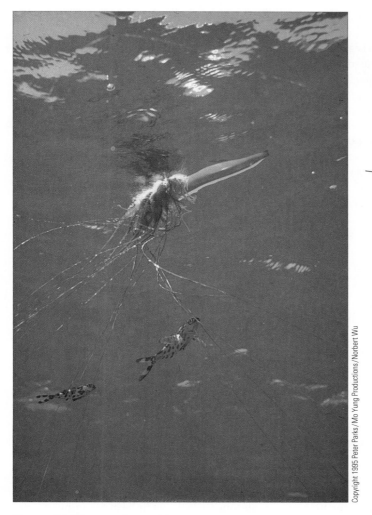

**Figure 17-15  Man-of-War Fish.** The small man-of-war fish gains protection by spending most of its time among the tentacles of the Portuguese man-of-war.

## ECOLOGY OF THE OPEN SEA

The open sea represents a **pelagic ecosystem,** one in which the inhabitants live in the water column. In a pelagic ecosystem, the basis of food chains is the many species of small phytoplankton. Few seaweeds and no vascular plants occur here. There is a good reason for the primary producers in this ecosystem to be so small. These organisms derive the nutrients that they need from the seawater that surrounds them. The smaller the organism, the greater the relative surface area that is exposed for the absorption of nutrients and sunlight and the lesser the amount of nutrients needed to sustain the body. The major herbivores in the open ocean are zooplankton, and these supply food for the nekton.

## Productivity

In the great expanse of the open ocean, there are no vascular plants and few seaweeds that can serve as primary producers to support consumers. The sea floor to which plants and seaweeds might attach is deeply submerged and shrouded in darkness by the light-absorbing properties of the overlying water. In the open sea, all higher forms of life rely on the various species of plankton to supply food.

A dynamic floating ecosystem composed of plankton stretches across the entire surface of the sea and down into the depths for hundreds of meters. Pelagic ecosystems such as these include thousands of microbial and animal species,

# Are Open Oceans Really as Barren as Deserts?

Recent research indicates that the large expanses of the open ocean may not be as devoid of life as previously thought. Data recently collected in the northern Pacific Ocean show two to three times more organic matter produced by photosynthesis than had been reported previously. Some biologists think that the open ocean has not been sampled often enough to catch periods of high productivity, resulting in low productivity assumptions. Others think that the sampling techniques themselves may have been responsible for erroneous results.

Previous sampling methods and contaminated containers may be responsible for low estimates of productivity. Phytoplankton are delicate organisms that can easily be damaged by the collecting techniques. Contaminated or dirty containers may also have caused the death of relatively large numbers of photosynthetic organisms during experiments designed to measure production rates. In coastal and upwelling areas, where phytoplankton levels are very high, some mortality during the experiments would not greatly affect the results or the calculations based on those results. In open-ocean areas, how-

ever, where populations of these organisms are sparse to begin with, the loss of substantial numbers could greatly affect the outcome of the experiments. The experiments that indicated high levels of production in the northern Pacific Ocean were performed under ultraclean conditions to avoid killing any of the phytoplankton. Yet, even with all the care that was taken, microscopic examination of the bottles used in the experiment showed dead and dying organisms.

Data from the Plankton Rate Process in Oligotrophic Oceans Research Project conducted in 1985 showed productivity rates two to four times greater than those previously published. Researchers were careful to avoid contaminating the phytoplankton with trace metals and causing unnecessary mortality. They also used finer filters when gathering samples that could trap even the smallest photosynthesizers such as cyanobacteria and coccolithophores.

Another way of measuring productivity is to use large areas of ocean water separated by natural ocean processes instead of water artificially contained in sample bottles. Researchers at Woods Hole Oceanographic Institution used data that

were collected from the Sargasso Sea near Bermuda over a period of 18 years. Their calculations indicated levels of production so high that there could not be enough nitrogen in the surface water to support it. Two of the researchers, William Jenkins and Joel Goldman, suggest that the nitrogen comes from deeper water brought to the surface at infrequent intervals by storms. They also propose that the portion of ocean studied is a two-layer system. The top layer shows the low productivity associated with the open sea, whereas the level just below that layer and just above the nutrient supply shows much higher productivity. Thus samples taken from slightly different depths may yield quite different results. There is also some evidence for a similar double-layer system in the northern Pacific Ocean.

If indeed open-ocean productivity is greater than assumed, this very large area of the world is more ecologically important on a global scale than previously thought. Additional field studies are being conducted, and previous data are being re-evaluated in an effort to understand better the productivity of this enormous area. ●

each with its own requirements for light, temperature, and nutrients, growth characteristics, and relationships with competitors and predators. Studying such a complex system and its inhabitants is no easy task, and our knowledge about open-water plankton communities continues to grow and change rapidly.

The blue of the open sea contrasts starkly with the green of coastal seas. Although surface waters of the open ocean receive large amounts of sunlight, they do not receive any of the nutrients that wash to sea from the land. This results in low levels of nutrients, such as nitrogen and phosphorus, that are necessary to support the life of phytoplankton. Waters just above the deep-sea floor contain higher concentrations of nutrients but do not receive enough sunlight to make them productive. The problem is compounded by the fact that there is very little mixing of deeper high-nutrient water with the low-nutrient water at

the surface. Only in isolated parts of the open sea are nutrients brought to the surface by upwelling.

The low level of nutrients in the open ocean is most pronounced in tropical waters. These waters have relatively permanent layers that are separated by a thermocline (for more information regarding thermoclines see Chapter 4) with a warmer, less-dense layer of water on top of colder, denser water. This arrangement prevents significant exchange of nutrients from the deep water to the surface. Thus, even though tropical seas are the warmest and offer plankton large quantities of intense sunlight, the low concentrations of nutrients limit the size and composition of phytoplankton populations. At these latitudes, oceanic diatoms are low in abundance and diversity because silicon, needed for making the diatom frustule, is in limited supply. Phytoplankton production in the central South Pacific Ocean is only about half as much as the productivity off the California coast and only about one sixth as much as in some coastal areas off Saudi Arabia or western Africa, where coastal upwelling brings nutrients to the surface. The low numbers of phytoplankton in tropical seas support even fewer numbers of the zooplankton that graze on the phytoplankton.

## Food Webs in the Open Sea

The base of food webs in the open sea is formed by the phytoplankton and heterotrophic bacteria (Figure 17-16). Most of the phytoplankton are nanoplankton, consisting of the smaller diatoms and dinoflagellates as well as coccolitho-

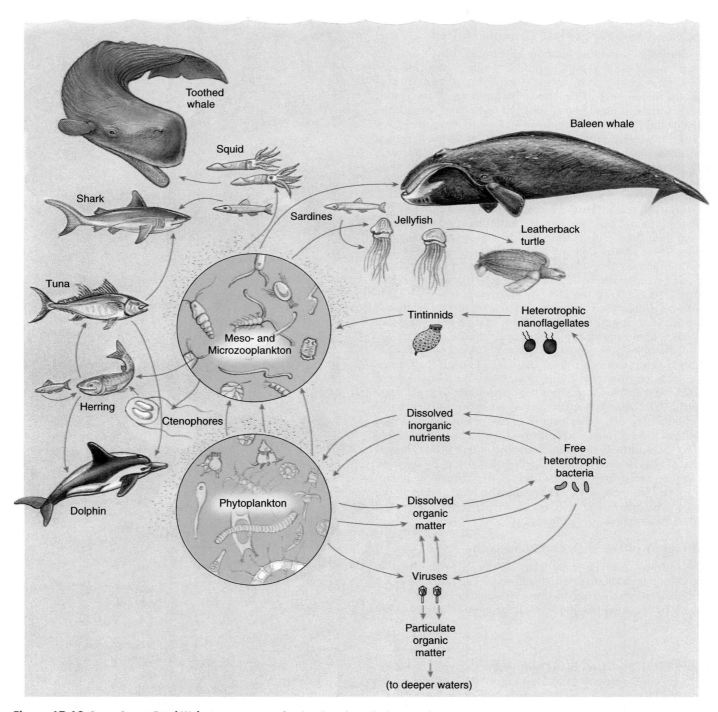

**Figure 17-16  Open-Ocean Food Web.** In open-ocean food webs, phytoplankton and heterotrophic bacteria form two foundations for food chains. The next link in the chains is zooplankton, which are the primary food of many large animals.

phores, silicoflagellates, and naked green flagellates. These photosynthetic nanoplankton dominate the phytoplankton in numbers and maybe in biomass, especially in tropical seas. The larger microphytoplankton might be grazed heavily by larger herbivorous zooplankton such as copepods, whereas nanophytoplankton are important food for invertebrate larvae, tintinnids, foraminiferans, radiolarians, and other smaller zooplankton.

## Dissolved and Particulate Organic Matter

In addition to providing food for herbivores, phytoplankton release some of their photosynthetic products into the surrounding seawater as dissolved organic matter (DOM). It is difficult for one-celled organisms to keep small molecules within their cell membranes and cell walls, and some leakage inevitably occurs. In a nutrient-poor environment, losses of DOM might limit the development of food webs. In recent decades it has become apparent that heterotrophic bacteria play an important role in rapidly recycling DOM in the open sea. The production of bacterial biomass in plankton communities provides a second base to the food chain, but the bacteria are too small to be consumed by the familiar microzooplankton. Instead they are eaten by a poorly known group called heterotrophic nanoflagellates, which in turn are grazed by tintinnids and other ciliates. The ciliates then provide food for larger zooplankton. Bacteria can also metabolize some of the DOM and return it to the water in inorganic form, making the mineral nutrients again available to the phytoplankton. In this way, bacteria form a **bacterial loop** in returning nutrients rapidly to the phytoplankton in waters that are easily depleted of critical nutrients.

Viriplankton, the most abundant plankton in the ocean, also play an important role at the base of ocean food webs. Lysis of bacterioplankton and phytoplankton by viruses releases DOM directly and produces **particulate organic matter (POM)** from the lysed cells. POM may slowly fall to greater depths, where other bacteria degrade it, but the bacterial biomass and recycled nutrients are then no longer available to the community in the photic zone. Although DOM released by viral lysis in surface waters will be absorbed by bacteria in the bacterial loop, the continuous destruction of bacteria by viruses (phages) short-circuits the loop, reducing the rate at which bacterial biomass is converted into biomass of higher consumers.

## Efficiency of Open-Ocean Food Webs

The conversion of biomass from one level to the next in ocean food webs is surprisingly efficient. The phytoplankton or bacterial production of one day may be entirely consumed by the next trophic level, leaving populations to be-

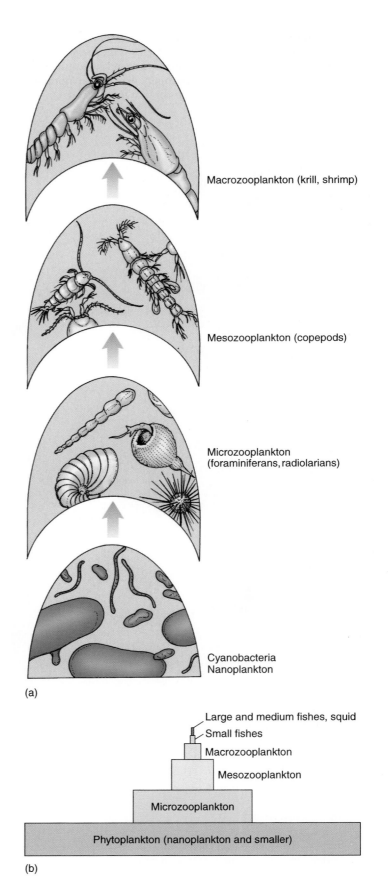

Macrozooplankton (krill, shrimp)

Mesozooplankton (copepods)

Microzooplankton (foraminiferans, radiolarians)

Cyanobacteria Nanoplankton

(a)

Large and medium fishes, squid
Small fishes
Macrozooplankton
Mesozooplankton
Microzooplankton
Phytoplankton (nanoplankton and smaller)

(b)

**Figure 17-17  Planktonic Food Chain.** (a) The nutrient-poor waters of the open ocean support food chains with several links composed of small planktonic organisms. (b) Because open-ocean food chains consist of so many energy-wasting steps, the open seas support relatively few large consumers, as indicated in this generalized pyramid of biomass.

gin the next day at densities of the previous day. Not all of the food consumed can be converted into biomass at the next level because ecological efficiency is never 100% (for more information regarding ecological efficiency see Chapter 2). Conversion rates may, however, be high. For example, bacteria convert DOM into bacterial biomass at 50% to 80% efficiency, and naked microflagellates consuming bacteria may have a conversion efficiency of 40%. Further reduction in conversion efficiency occurs at higher trophic levels, such as the copepods and krill that dominate the larger zooplankton.

The food webs in nutrient-poor open seas may have food chains with five or six links, leading to the nektonic top carnivores among the whales and sharks. Although the food chains may be long, the open sea can support few large animals for two reasons: the limited rate of primary production places a low ceiling on the total biomass that the area can sustain, and declining conversion efficiency along the food chain eventually exacts its toll (Figure 17-17). It is only along the coastlines and at areas of upwelling that huge schools of tunas, flocks of seabirds, herds of seals, and pods of whales can form. The wealth of the open sea resides in its microscopic inhabitants.

## In Summary

The primary producers of the open sea are mainly phytoplankton. Even though the open sea receives large amounts of sunlight, it lacks the supply of nutrients that makes coastal seas so productive. The low level of nutrients is most pronounced in tropical oceans, where the water forms more or less permanent layers that are separated by thermoclines. Bacteria are important in the ability of the community to cycle nutrients efficiently, but viruses that infect bacteria short-circuit this bacterial loop and reduce the efficiency. The open sea does not support many large animals because of limited primary production and food webs involving several energy-wasting steps between primary producers and final consumers. Our understanding of community production and dynamics continues to improve with the application of new techniques for open-ocean studies. ●

## SELECTED KEY TERMS

akinetic, *p. 388*

appendicular swimmer, *p. 397*

bacterial loop, *p. 404*

bacterioplankton, *p. 388*

centrifuge plankton, *p. 388*

countershading, *p. 398*

deep scattering layer, *p. 391*

epipelagic zone, *p. 388*

femtoplankton, *p. 388*

gelatinous zooplankton, *p. 399*

gymnosome pteropod, *p. 391*

holoplankton, *p. 388*

kinetic, *p. 388*

macroplankton, *p. 388*

marine snow, *p. 390*

megaplankton, *p. 388*

meroplankton, *p. 388*

mesoplankton, *p. 388*

micropatchiness, *p. 390*

microplankton, *p. 388*

nanoplankton, *p. 388*

neuston, *p. 390*

particulate organic matter (POM), *p. 404*

patches, *p. 390*

pelagic ecosystem, *p. 401*

picoplankton, *p. 388*

pleuston, *p. 390*

pteropod ooze, *p. 392*

pyrosome, *p. 393*

seston, *p. 388*

thecosome pteropod, *p. 391*

tripton, *p. 388*

viriplankton, *p. 388*

## QUESTIONS FOR REVIEW

### Multiple Choice

1. The layer of ocean that receives enough sunlight to power photosynthesis is the
   a. photosynthetic zone
   b. photic zone
   c. aphotic zone
   d. abyss
   e. benthic zone

2. A collective term for animals whose distribution is not governed by waves or currents is
   a. zooplankton
   b. pelagic
   c. nekton
   d. akinetic
   e. pleuston

3. Productivity in the open ocean is limited in most cases by
   a. sunlight
   b. temperature
   c. space
   d. nutrients
   e. salinity

4. Food webs in the open sea are based on
   a. phytoplankton
   b. seaweeds
   c. dissolved organic matter
   d. viruses
   e. marine snow

5. Life in the open sea is characterized by
   a. uniform distribution
   b. low trophic efficiency

c.  dense populations
d.  food chains of 7 to 10 links
e.  patchiness

6. The rate of sinking can be slowed by anatomical adaptations that increase
   a.  mass
   b.  volume
   c.  friction
   d.  streamlining
   e.  density

7. Which one of the following is not true of gelatinous zooplankton?
   a.  Bodies are high in water content.
   b.  Bodies are high in protein content.
   c.  They reduce density by ion replacement.
   d.  They have reduced skeletons.
   e.  They are consumed by few kinds of predators.

8. The type of coloration exhibited by many large animals of the open sea is called
   a.  countershading
   b.  disruptive
   c.  cryptic
   d.  warning
   e.  reticulated

9. The largest members of the zooplankton are
   a.  copepods
   b.  diatoms
   c.  jellyfishes
   d.  sharks
   e.  whales

10. The greatest diversity of nekton is among the
    a.  squids
    b.  fishes
    c.  reptiles
    d.  birds
    e.  mammals

## Short Answer

1. What are some adaptations in open-ocean plankton that allow them to be productive despite the nutrient-poor environment?

2. What is marine snow and what is its significance?

3. What are some adaptations that help organisms in the open ocean remain afloat?

4. What aspects of reproduction in pelagic sharks indicate adaptations to life in the open sea?

5. Describe some of the ways that animals living in the open sea hide from predators.

6. Describe some of the strategies animals use to increase their survival in the open sea.

7. Describe how the upside-down lifestyle of the purple sea snail and the nudibranch *Glaucus* has influenced their coloration.

8. Describe how a swordfish feeds.

9. Describe how tunas are adapted for fast swimming.

10. Why is the photic zone of the open ocean not as productive as the photic zone of coastal seas?

## Thinking Critically

1. Why does it make good ecological sense for the largest sharks to be plankton feeders rather than predators of other nekton?

2. The water of the North Atlantic Ocean is colder than that of the tropical Atlantic Ocean. Do you think this difference might affect the shapes of phytoplankton living in these two areas? Explain your answer.

3. Why is it difficult to keep large tunas alive in public aquariums?

4. What problems in feeding mechanics, digestion, physiology, or nutrition might be encountered by predators that feed on gelatinous zooplankton? What additional problems might they face in today's seas with drifting bottles and plastic bags?

## SUGGESTIONS FOR FURTHER READING

Hardy, A. 1965. *The Open Sea: Its Natural History.* Boston, Mass.: Houghton Mifflin.

Hardy, J. T. 1991. Where the Sea Meets the Sky, *Natural History* 100(5):58–65.

Klimley, A. P., J. E. Richert, and S. J. Jorgensen. 2005. The Home of Blue Water Fish, *American Scientist* 93(1): 42–49.

Longhurst, A. R., and D. Pauly. 1987. *Ecology of Tropical Oceans.* San Diego, Calif.: Academic Press.

Prager, E. J., and S. A. Earle. 2000. *The Oceans.* New York: McGraw-Hill.

Raymont, J. E. G. 1980. *Plankton and Productivity in the Oceans,* 2nd ed. Vol. 1, *Phytoplankton.* Oxford: Pergamon Press.

Raymont, J. E. G. 1983. *Plankton and Productivity in the Oceans,* 2nd ed. Vol. 2, *Zooplankton.* Oxford: Pergamon Press.

### InfoTrac College Edition Articles

DeLong, E. F. 2003. A Plenitude of Ocean Life, *Natural History* 112(4).

Johnsen, S. 2001. Hidden in Plain Sight: the Ecology and Physiology of Organismal Transparency, *Biological Bulletin* 201(3).

Moore, C. 2003. Trashed: Across the Pacific Ocean, Plastics, Plastics, Everywhere, *Natural History* 112(9).

Phleger, C. F. 1998. Buoyancy in Marine Fishes: Direct and Indirect Role of Lipids, *American Zoologist* 38(2).

### Websites

**http://www.biosci.ohiou.edu/faculty/currie/ocean/**
An academic site with many links to pelagic organisms and the scientists who study them.

**http://jellieszone.com/index.html** Information on jellyfish and other gelatinous zooplankton.

# 18

# Life in the Ocean's Depths

In the early years of marine biology research, scientists believed that no animal life could exist in the dark abyss of the sea. There was no sunlight to power photosynthesis, so there could not be any producers on which consumers would feed. The temperatures were too cold and the pressures too extreme to support life. As we noted in Chapter 1, retrieval of the transatlantic cable in the mid-1800s demonstrated that animal life could exist beneath the photic zone. On July 21, 1951, the Royal Danish research vessel *Galathea* recovered a variety of animals, including sea anemones, sea cucumbers, bivalves, amphipods, and bristle worms from more than 10,000 meters (33,000 feet) deep in the Philippine Trench east of the Philippine Islands. Eight years later, Jacques Piccard and Don Walsh, using the bathyscaphe *Trieste,* discovered animal life at even greater depths at the bottom of the ocean's deepest trench, the Challenger Deep. These and other discoveries proved that animal life could not only survive in the dark recesses of the ocean but also thrive. In this chapter you will be introduced to the forbidding world of the ocean deep and the strange organisms that make it their home.

## SURVIVAL IN THE DEEP SEA

As late as 1843, the British naturalist Edward Forbes concluded that animal life could not exist in the sea below a depth of 55 meters (182 feet). At the time, his hypothesis was completely plausible, because what was known of the conditions of the deep ocean appeared to make it a most inhospitable place for life. As we have noted previously, sunlight rapidly fades with increasing water depth. Even in the clearest ocean water, there is generally not enough light to support photosynthesis below 200 meters (660 feet). Heat energy is also absorbed quickly by surface waters. Temperatures in the deep remain frigid throughout the year.

## Key Concepts

1. Several thousand species have adaptations that allow them to survive in the deep-sea environment.

2. The lack of light has had the most impact in shaping the organisms of the deep sea.

3. Many deep-sea animals exhibit bioluminescence, which helps them find mates and prey in their dark environment.

4. Deep-sea fishes display a variety of adaptations such as sharp teeth, large mouths, and huge stomachs that help them survive in a habitat with limited food.

5. The environmental conditions of the deep sea have been relatively stable for more than 100 million years, and as a result, several organisms have changed very little from when they first evolved.

6. Benthic communities consist of sparse populations that survive on the minimal food available in their environment.

7. Thriving marine communities that depend on chemosynthetic bacteria for primary production exist on the ocean floor around hydrothermal vents.

## Exploring the Ocean's Depths

For centuries humans have dreamed of being able to descend to the ocean's depths to explore them. Some even tried by descending in containers such as wooden barrels equipped with an air hose, but they were limited in the depths they could reach and the observations they could make. In the last century, Dr. William Beebe pioneered underwater exploration. In 1934 he descended to a depth of 918 meters (3,028 feet) in a steel ball called a "diving bell," or "bathysphere." The diving bell had portholes and lights and was supplied with air and power by a mother ship on the surface.

The first self-contained craft that could operate free from a mother ship was the bathyscaphe (BATH-eh-skayf), invented in 1948 by Auguste Piccard. The vessel was basically a sphere that could accommodate two people. It was suspended from a huge float that contained gasoline to power the craft's systems. Auguste's son Jacques and Don Walsh (a lieutenant in the United States Navy) used the bathyscaphe *Trieste* in 1960 (Figure 18-A) to descend to the bottom of the Challenger Deep (11,020 meters, or 36,366 feet).

Today's manned deep-sea submersibles are very sophisticated vessels equipped with the latest in photographic and robotic technology. Of the submersibles used by the United States, the best known is *Alvin* (see photo in Chapter 1), which can carry a crew of three. *Alvin* averages 150 dives per year

and was used by Robert Ballard in 1977 to first view the animals of hydrothermal vent communities. Ballard used *Alvin* again in 1985 to view the sunken passenger ship *Titanic*.

The United States Navy operates a research-and-recovery vessel called the *Sea Cliff*. This vessel can operate at depths

of 6,100 meters (20,000 feet), and in 1986 it was used by the U.S. Geological Survey to examine mineral deposits located off the coasts of California and Oregon. Two other submersibles are the *Johnson Sea Link 1* and the *Johnson Sea Link 2,* both owned by the Harbor Branch Oceanographic Institute in Florida. Each

UPI/Corbis-Bettman

**Figure 18-A** **Early Manned Diving Devices.** *In 1960, the bathyscaphe* Trieste *was used by Jacques Piccard and Don Walsh to descend to the bottom of the Challenger Deep, the deepest point in the ocean.*

Adding to these forbidding conditions is the tremendous pressure of the ocean depths. The pressure at sea level is 1 atmosphere (760 mm Hg [millimeters of mercury] per square centimeter, or about 15 pounds per square inch). About every 10 meters (33 feet) below the surface of the ocean, the pressure increases by another atmosphere. At this rate the pressure exerted on an organism at 1,000 meters (3,300 feet) is 510 kilograms per square centimeter (1,500 pounds per square inch), or about 100 times greater

than the pressure exerted at the surface. At a depth of 10,600 meters (35,000 feet), the pressure is 5,100 kilograms per square centimeter (7.5 tons per square inch).

Even if living organisms could survive in such an environment, they would have to find their food and their mates in total darkness. As adverse as these conditions may seem, they are at least stable, and, over millions of years of evolution, some organisms left the sunlit upper waters and adapted to life in the dark recesses of the sea (Figure 18-1).

can carry a crew of four, and both participated in the efforts to recover pieces of the space shuttle *Challenger* following its disastrous explosion shortly after takeoff in January 1986.

Manned submersibles have been used to photograph sharks, survey coral reefs, and produce motion-picture footage for documentaries on the sea and its creatures. They are also used commercially in offshore petroleum developments and other engineering projects.

The drawbacks of using manned submersibles are that they are expensive to operate and expose the passengers to risk. To deal with these problems, engineers have developed a set of unmanned units called remotely operated vehicles (ROVs; Figure 18-B). These carry photographic equipment, have robotic arms and other types of sampling equipment, and can be controlled from the surface, eliminating the need for an on-board operator. The main disadvantage of ROVs is the need for a sizeable cable through which power is transmitted to the unit and data are transmitted back. However, this one disadvantage is far outweighed by lower construction and operating costs, not to mention the minimal risk to human life.

During the exploration of the *Titanic* in 1986, *Alvin* carried with it a small ROV called *Jason Jr.*, or *JJ. JJ* was attached to *Alvin* and was used to explore the inside of the *Titanic*, something the larger, more cumbersome *Alvin* could not do. The

**Figure 18-B Remote Operating Vehicles, or ROVs.** *Jason, an ROV, is used by marine biologists to explore deep-sea ecosystems.*

Robert Ballard/News Office, Woods Hole Oceanographic Institution, MA

costs of ROVs are decreasing, and they are now frequently used in engineering projects such as offshore drilling work, bridge and dam construction, and laying pipeline. Military and law enforcement agencies are interested in the possible applications of ROVs, and there is a growing demand for them in transportation applications, where they are used for maintenance and routine inspection of large ships. Although ROVs will probably not replace the manned submersibles, they do offer the ability to examine larger areas of the sea for longer times at a greatly reduced cost and risk. ●

## Adaptations to Pressure

Early marine biologists believed that the enormous pressures encountered in the ocean depths would literally crush an organism. Interestingly enough, several thousand species of animal, especially echinoderms, annelids, and molluscs, tolerate the pressures. The fluid pressure in these animals' tissues matches the pressure of the surrounding water. The tissue fluid pressure pushes against the surrounding pressure with an equal but opposite force, preventing the animal's body from being crushed.

## Adaptations to Cold

Nearly all deep-sea animals have body temperatures that are close to the temperature of the surrounding water. Their body temperatures are so low that their metabolism is quite slow. As a result they move more slowly, grow more slowly,

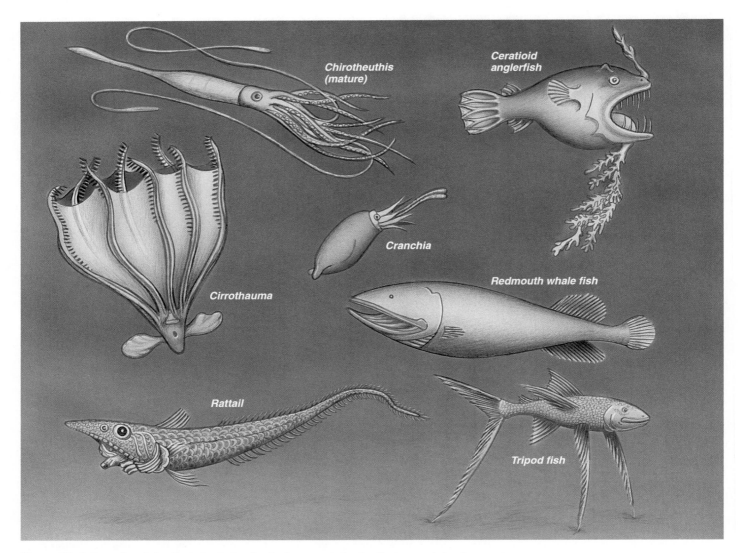

**Figure 18-1 Deep-Sea Animals.** Animals that live in the ocean's depths display a variety of intriguing adaptations that allow them to survive in a dark environment with cold temperatures and extreme pressures.

reproduce less frequently and later in life, and generally live longer than similar species from warmer surface waters. A lower metabolic rate also means that these animals require less food, scarce in the ocean deep, to stay alive.

One advantage of the cold is the increased density of the water. Animals that live in the deep have body densities very close to the density of their environment and do not have to expend energy to keep from sinking, as do animals in the photic zone. (Recall from Chapter 10 that this is the reason deepwater fishes lack a swim bladder.)

## In Summary

Until the mid-1800s, biologists believed that life could not exist in the deepest parts of the ocean. There was no sunlight to support photosynthesis, and the cold temperatures and extreme pressures at great depths were thought to limit, if not exclude, animal life.

Although the conditions of the deep are adverse, they are stable. This has allowed several thousand species to adapt to this habitat, an exciting discovery because the deep sea was expected to be a species-poor region.

Deep-sea animals metabolize more slowly, move more slowly, grow more slowly, reproduce later in life, and generally live longer compared with their counterparts in warmer surface water. Because their metabolic rate is low, deep-sea animals do not have to eat as much or as often to survive. Cold water is denser than warm water, so deep-sea animals do not have the problem of remaining buoyant that their surface relatives have. ●

## LIFE IN THE DARK

Of the three major factors that affect animals living in the ocean deep—light, temperature, and pressure—the lack of light has had the greatest evolutionary impact.

## Color in Deep-Sea Organisms

As we noted in the previous chapter, fishes that inhabit the well-illuminated surface waters exhibit countershading: they have dark dorsal surfaces and lighter ventral surfaces. From 150 to 450 meters (500 to 1,500 feet) below the surface, a region referred to as the **disphotic zone,** or **twilight, zone,** there is still enough light to make countershading a useful means of camouflage. A common resident of these depths is the small hatchet fish (*Argyropelecus;* Figure 18-2). Its body is silvery or iridescent, with a dark dorsal surface and silvery underside. Like many deepwater fishes, hatchet fishes possess rows of light-producing organs called **photophores** along their bodies. These aid in species recognition, and the bioluminescence may make the ventral surface lighter, thus helping in camouflage.

Not all of the animals in this region, however, exhibit countershading. In the twilight zone, as well as the darker zones below, live animals that exhibit a variety of body colors. Black stomiatoid fishes (see Figure 18-10) have an iri-

descent sheen. Fishes known as gulper eels (see Figure 18-9) and swallowers (family Eurypharyngidae) are black or brown. These odd fish prowl the depths at 1,800 meters (6,000 feet), looking for prey. Whalefishes (suborder Cetomimoidei) resemble whales in their general appearance but are only a few centimeters in length. Their fins and jaws are bright orange and red. Deep-sea species of squid are permanently colored deep red, purple, or brown, and many species are bioluminescent. Benthic species of squid are often white. Shrimp that are blood red feed on red arrowworms and scarlet copepods. Interestingly, the bright red and orange colors of these animals appear as shades of black and gray in the depths at which they live. Recent studies by biologists at the Scripps Institution of Oceanography indicate that the red colors of deep-sea crustaceans vary according to the state of the molt cycle and may actually be involved in energy transfer during molting.

## Roles of Bioluminescence

Even though sunlight cannot penetrate to the depths of the ocean, many deep-sea animals produce their own light in the form of bioluminescence. This characteristic is especially common in animals that are found between 300 and 2,400 meters (1,000 and 8,000 feet). Some, such as certain species of squid, crustacean, and fish, have their own luminescent organs, whereas others harbor bioluminescent bacteria in species-specific locations. This symbiosis between bacteria and their host species is an example of mutualism. The host receives light that can be used to locate and recognize a mate or find prey, and the bacteria are given a place to live and supplied with food. Some species of fish are able to control the bioluminescence of their symbiotic bacteria by altering the flow of oxygenated blood to the regions where the bacteria reside. When oxygen levels are high, the bacteria luminesce, and when the oxygen levels are low, they do not.

Bioluminescence occurs when a protein called **luciferin** is combined with oxygen in the presence of an enzyme called **luciferase** and adenosine triphosphate (ATP). During the series of reactions that follows, the chemical energy of ATP is converted into light energy. The process is efficient, producing almost 100% light and little heat. Most bioluminescence associated with marine organisms is blue green, although red and yellow have also been reported.

Luminescent organs can be located at a variety of positions on an animal's body (Figure 18-3). Many deep-sea fishes display rows of photophores along their sides or on their bellies. Other species of fish may have depressions containing bioluminescent bacteria located on their heads or on fleshy growths attached to the head. Deep-sea squids exhibit bioluminescent spots on their tentacles and circling their eyes.

### Camouflage

It would seem that such illuminated animals in a dark world would be easy marks for predators. In the twilight zone, where sunlight is fading into darkness, the presence of bioluminescence on the animal's ventral surface may take the place of countershading in camouflaging the animal. If the

**Figure 18-2 Hatchet Fish.** The silvery underside and dark dorsum of this hatchet fish are characteristic of the coloring of fishes that inhabit the twilight zone.

Norbert Wu

**Figure 18-3 Bioluminescent Organs.** Luminescent organs help species identify each other, locate prey, and avoid predation. They are frequently arranged in rows on the fish's body, as in this Pacific viperfish, or in depressions on the head.

intensity of the bioluminescence matches the intensity of sunlight at that level, the fish will not appear as a shadow or silhouette when seen from below.

## Mating and Species Recognition

Bioluminescence can also play an important role in mating. During breeding times, the pattern of lights identifies an individual as being male or female. In lanternfishes (*Myctophum*), for example, males carry bright lights at the tops of their tails, whereas females have only weak lights on the underside of their tails (Figure 18-4). Like fireflies, some species of deepwater fish signal their readiness to mate by a series of light flashes.

The pattern of lights also serves for species identification, especially among members of closely related species. For instance, one species of lantern fish has three rows of

**Figure 18-5 Anglerfish.** The fleshy projection on the head of this anglerfish contains a bioluminescent "lure" at the end. Smaller fishes are attracted to the light and become an easy meal for the anglerfish.

light spots along its ventral surface, whereas another closely related species displays only two. This would be of particular advantage in finding mates or in forming schools. The marine biologist William Beebe once remarked that the pattern of lights on deepwater fishes was so distinctive that he could recognize species in absolute darkness just by the light pattern that they presented.

## Attracting Prey

Bioluminescence also functions to attract prey. Anglerfish (*Melanocetus johnsonii;* Figure 18-5) and stomiatoids attract prey with bioluminescent lures. Smaller fishes whose normal prey are luminescent are attracted by the light and, when close enough, become a meal for the larger predator. Dr. Beebe observed that a lanternfish apparently uses the light from its ventral surface to find prey. Copepods that were illuminated by the lights from the ventral surface of this fish were quickly eaten. Some species of stomiatoid fish have lights around their eyes that illuminate whatever the fish looks at. They have been observed to fix krill in the beam of light and then rapidly devour them. Deepwater krill also

**Figure 18-4 A lanternfish.** The bright, bioluminescent spots on the tail of this lanternfish identify it as a male. The pattern of bioluminescent spots helps the lanternfish recognize the opposite sex during the mating season.

have illuminated spots around their eyes that may help them find smaller prey.

## Defense

Some animals use bioluminescence for defense. In the dark waters of the deep, the inky cloud produced by surface squids would have no effect in deterring a predator. Some deepwater squid species, however, release a bioluminescent fluid that clouds the water with light, confusing potential predators. This tactic is also used by some species of deepwater shrimp. The scarlet-colored opossum shrimp (family Mysidae; Figure 18-6) gets its name because the female carries her eggs in a pouch under the thorax. When these shrimp are threatened, they release a substance that bursts into a cloud of miniature light particles that look like stars in the sky at night. The sudden burst of light frightens and confuses potential predators, allowing the shrimp to make an escape.

## Seeing in the Dark

Do fishes in the aphotic zone of the sea perceive anything other than the light produced by other animals? To answer this question, the British biologist N. B. Marshall made an extensive study of the brains and eyes of deep-sea fishes, some species living as deep as 900 meters (3,000 feet) below the surface. He concluded that even though there is virtually no light at these depths, these fishes can probably see their food, their own species, and their predators, using what little light is available to stimulate vision.

In contrast to the typical spheriod vertebrate eye, the eyes of many deep-sea fishes are tubular and contain two retinas instead of one (Figure 18-7). One retina is used to view distant objects, while the other sees closer things. Not only do fishes with tubular eyes see better in dim light, but they also have better depth perception. This allows them to judge the distance to their prey quite accurately so that they are less likely to miss their catch. This is a definite advantage in an area where food is so scarce and where speed is not common.

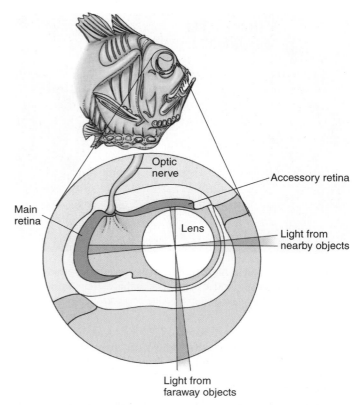

**Figure 18-7  The Eye of Deepwater Fishes.** Fishes that live in the ocean's depths have eyes that are adapted for seeing in dim light. They are tubular and contain two retinas. The shape of the eye gives them better depth perception so they are less likely to miss their prey when they strike.

The biologist G. L. Walls has noted that at depths between 900 and 1,500 meters (3,000 and 5,000 feet) the eyes of fishes become smaller and less functional. The anglerfish has well-developed eyes during its early life, which is spent in well-lit surface waters. As it grows into an adult, it begins to sink and eventually resides at depths of about 1,800 meters (6,000 feet). At this stage in its life, the eyes stop growing and sometimes begin to degenerate.

Deep-sea squids also have strange-looking eyes. They can be barrel-shaped, stalked, or unequal in size. It is thought that these eyes are probably adapted to the amount of light available in their habitat and to their method of locating and capturing prey.

Some animal species, both vertebrate and invertebrate, that occupy the deepest recesses of the sea have tiny eyes that are only slightly functional, or they are totally blind. These organisms rely more on tactile and chemical stimuli to find their prey and mates and to avoid predation.

## Finding Mates in the Dark

Some species of deepwater anglerfish have solved the problem of finding a mate in the dark in an unusual way. Because mates are hard to find, when a male encounters a female, he bites her and remains attached, sometimes for the rest of his life (Figure 18-8). In some species of anglerfish, the male

**Figure 18-6  Bioluminescence as Defense.** For protection the scarlet-colored opossum shrimp releases a substance that produces flashes of light similar to fireworks. This sudden burst of light in an otherwise dark environment frightens potential predators.

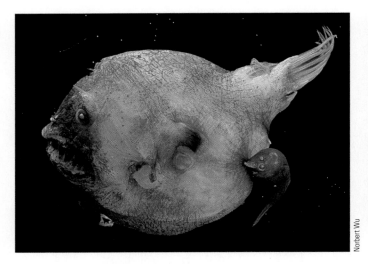

Norbert Wu

**Figure 18-8 Mating in Anglerfishes.** Male anglerfishes attach themselves to a female, and in some species the relationship is permanent, with the male becoming a lifelong parasite of the female.

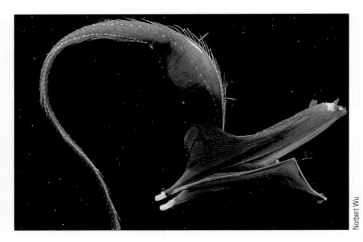

Norbert Wu

**Figure 18-9 Gulper Eel.** Many deepwater fishes such as this gulper eel have huge mouths that can engulf prey as large as the predator.

becomes a lifelong parasite. The skin around the male's mouth and jaws fuses with the female's body, and only a small opening remains on either side of the mouth for gas exchange. The eyes and most of the male's internal organs degenerate, and his circulatory system becomes connected to the female's. The male, in effect, becomes an external, sperm-producing appendage.

It is believed that the female anglerfish releases her eggs in deepwater, and fertilization is external. The fertilized eggs quickly float upward to hatch in surface waters, where the young feed on copepods and other small plankton. As the young grow, females develop fishing lures, and males develop gripping teeth on both their snouts and chins. When the fish mature to adults, they descend and begin their deep-water existence.

## Finding Food in the Dark

In the depths of the sea, food is scarce. There are no photosynthetic organisms to produce food, but organic wastes, scraps of food, and dead and dying organisms drift down slowly from above. Some of this food is consumed before it reaches the bottom, and the rest reaches the floor of the ocean, where it supports a variety of benthic organisms and active deep-sea scavengers. Detritus feeders of the benthic community are, in turn, food for more active predators.

Not all of the deep-sea food chains, however, rely on the rain of detritus from the surface. Many small fishes and invertebrates rise at night to feed in the rich surface waters, returning during the day to the deeper recesses of the sea. These vertical migrations help to bring more food to the lower depths. Larger predators also make daily migrations to feed on smaller organisms.

Even at the best of times, food is scarce, and animals that survive in this environment exhibit some bizarre adaptations for feeding. Many deep-sea fishes appear to be nothing more than a huge, tooth-filled mouth that is attached

to an expandable stomach with a small tail for swimming. Fish species called gulper eels (Figure 18-9) are common at depths below 1,500 meters (5,000 feet). They have jaws that are hinged like a trapdoor and stomachs that can expand to several times their normal size to accommodate prey that is larger than the gulper eel itself. A research vessel captured a 15-centimeter (6 inch) gulper specimen with a 23-centimeter (9 inch) fish coiled in its stomach. Some species of gulper eel grow to 1.8 meters (6 feet) long. Most of this length is in their long, whiplike tail. The tip of the tail is luminescent and may play a role in luring potential prey close to the animal's mouth.

Stomiatoids (Figure 18-10) are also capable of ingesting prey larger than themselves. This gives them the ability to take advantage of the rare large food items. Most species are 15 to 18 centimeters (6 to 7 inches) long. They have large heads with curved, fanglike teeth and elongated bodies that taper into a small tail. They are prominent residents of the deep at depths of 180 to 600 meters (600 to 2,000 feet).

The typical stomiatoid has a fleshy projection called a **barbel** that dangles below its chin or throat. The precise role of these intriguing structures is not known, but it has been speculated that they are lures or used to probe the bottom ooze for food. They may even serve in species recognition during mating.

The characteristics of the barbel vary from species to species. In some the barbel is a short hair, whereas in others it is a whiplike structure that may be 10 times longer than the animal's body. One species that is only 4 centimeters (1.5 inches) long has a barbel that is 40 centimeters (15 inches) long, and a 21-centimeter-long (8.5 inch) species has a barbel nearly 1 meter (3 feet) long. The shape of the barbel is as variable as the length. Some are single strands, whereas others branch repeatedly; still others resemble a strange flower or even a cluster of grapes. Many are also bioluminescent.

Deep-sea anglerfishes found at depths of 1,970 meters (6,500 feet) have a bony projection with a luminescent lure at the tip that is a modification of one of the spines of the

Norbert Wu

**Figure 18-10 Stomiatoid Fishes.** Stomiatoid fishes, like this black sea dragon, usually have a fleshy projection called a *barbel* dangling from their chin or throat. It is believed these structures function as lures to attract prey and as sensors for finding food in the bottom ooze.

fish's dorsal fin. Through evolution, the first spine of the dorsal fin separated from the others, moved forward, and was modified into a fishing pole that, when not in use, lies in a groove on the top of the fish's head. Some species have short, stubby poles, whereas others have long, slender ones. Sets of muscles evolved that can move the pole forward, back toward the mouth, or out of the way when the animal is eating. At the end of the pole is a luminous "lure." The anglerfish's prey may mistake the lure for a worm or shrimp, and as it moves closer to investigate, the anglerfish moves the lure closer to its mouth. When the prey gets close enough, the anglerfish's lower jaw suddenly drops and the gill covers expand, creating suction that sweeps the unwary victim into the anglerfish's mouth, which has become large enough to devour prey as big as the predator. The jaws snap shut, impaling the victim on long, curved teeth on the roof of the mouth. The prey is then swallowed whole, pushed into the stomach by teeth in the predator's throat (pharyngeal teeth).

## In Summary

Of all the factors that have shaped life in the ocean's depths, the lack of light has had the most impact. In the dark, camouflage is not a problem, because it is difficult for one animal to see another. Animals that inhabit the twilight zone use countershading to help camouflage them in this region of minimal light.

Many deep-sea animals exhibit bioluminescence. Some have their own bioluminescent organs, whereas others harbor symbiotic bacteria that produce light. The bioluminescence helps animals to find prey and mates in a world of almost total darkness. It also plays a role in species identification and, in some species, attracts prey. Bioluminescence may also be used as a means of defense.

Food in the ocean's depths is scarce. Some detritus that falls from above is consumed before it reaches the floor, but most supports benthic organisms that are in turn food for more active predators. The nightly migrations of some fish and invertebrates to feed in the surface waters brings more food to the depths. The shortage of food has led to many interesting adaptations in deep-sea fishes. ●

## GIANTS OF THE DEEP

Most deep-sea animals are small compared with those living in surface waters, but a few are giants compared with their shallow-water relatives. One possible reason for the large size of these organisms is that they live longer than their shallow-water relatives and thus have time to grow larger. Sea urchins that live on the ocean floor have bodies 30 centimeters (1 foot) in diameter, compared with shallow-water species only a few centimeters across. In the deep sea off the coast of Japan live hydroids that reach the amazing height of 2.5 meters (8 feet). Sea pens (Figure 18-11a) that are ordinarily 0.5 meter high grow as large as 2.5 meters in the ocean depths; isopod crustaceans (related to terrestrial pill bugs; see Figure 18-11b), as large as 20 centimeters (8 inches). Shrimp species that are bright red or bioluminescent reach lengths of nearly 30 centimeters (1 foot). Some of these species have antennae that are twice their body length, which may be used to capture small prey.

### Giant Squids

Perhaps the most spectacular of the deep-sea giants is a mollusc, the giant squid (*Architeuthis;* Figure 18-12). The giant squid is the largest of all invertebrates, 9 to 16 meters (about 30 to 53 feet) long, including the tentacles. Its arms are as thick as a human thigh and covered with thousands of suckers. The two tentacles can be more than 12 meters (40 feet) long, and their flattened ends bear 100 or more suckers with serrated edges. As in other species of squid, the arms and tentacles capture prey and carry it to the animal's beak, where it is shredded into small pieces. The largest giant squid recorded was washed up on a New Zealand beach in 1888. It measured 18 meters (59 feet) in total length, with the tentacles accounting for 15 meters (49 feet) of the length. Although no giant squid has ever been accurately weighed, estimates run as high as 1 ton for a large specimen.

(a)

(b)

**Figure 18-11 Giant Deep-Sea Animals.** Some animals that live in the ocean depths such as this (a) sea pen and (b) isopod grow much larger than related species that live in the sunlit surface waters.

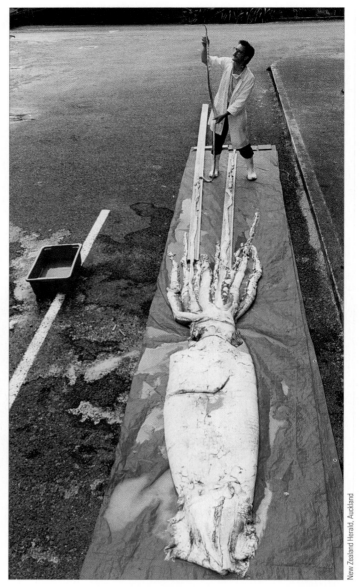

**Figure 18-12 Giant Squid.** The giant squid is the largest of all invertebrates. It feeds on fishes and is preyed upon by sperm whales. This specimen was found in the waters off New Zealand.

Little is known about the natural history of giant squids. They are found in all oceans, spending most of their lives at depths of 180 meters (600 feet) or more. They use the same method of swimming as other squids, a type of jet propulsion. Although there are reports of giant squids passing ships traveling 23 kilometers per hour (14 miles per hour), their anatomy suggests that they are weak swimmers that probably cannot capture active prey. No one knows for sure what these animals eat. It is assumed that they feed on a variety of small invertebrates and fishes. A major predator of giant squids is the sperm whale. Sperm whales have been found with scars from the serrated suckers on their bodies. Analyses of the stomach contents of captured specimens have revealed the remains of giant squids.

## New Species of Deepwater Squid

In 1988 a new and as yet unnamed species of large, deepwater squid was discovered by researchers making observations from a deep-sea submersible. Since the origi-

nal sighting, this animal has been observed at least eight times in water between 1,930 and 4,709 meters (6,367 to 15,580 feet) deep. These squid can reach lengths of 7 meters (23 feet) and have large, billowing fins. Arguably their most unusual feature is their arms, which are significantly longer than those of other known species of squid. The arms project from the squid's body for a short distance and then bend downward at sharp angles, possibly to prevent them from becoming tangled. Researchers report that these squid don't seem to use their arms for grabbing prey but that they may be coated with a sticky substance that allows them to capture small crustaceans for food. Marine biologists speculate that this squid may be quite common in the deep sea because specimens have been observed in the Atlantic, Pacific, and Indian Oceans as well as the Gulf of Mexico, all within a relatively short time. These squid exhibit behaviors different from their shallow-water relatives. For instance, when they release ink they hide in the ink cloud rather than swimming away as squids do in sunlit waters. Pairs have been observed attached to each other, with one squid towing the other through the water, a behavior that is not observed in shallow-water squid. Many questions about this animal, not the least of which is whether it is a new kind of squid, will remain until a specimen can be captured and studied in detail.

## In Summary

Although most deep-sea species are small compared with similar shallow-water species, a few are giants. One possible explanation is that they live longer than shallow-water species and thus have more time to grow. One of the better-known deep-sea giants is the giant squid. Because live specimens have never been captured for study, very little is known about this intriguing animal. ●

## RELICTS FROM THE DEEP

Evolution by natural selection is stimulated by changes in the environment (for more information regarding evolution by natural selection see Chapter 5). Because environmental conditions of the deep sea are believed to have remained nearly stable for more than 100 million years, biologists theorize that the sea's depths hold many organisms that have undergone very little change from their early ancestors. Marine biologists were encouraged to pursue this theory by the discovery in 1864 by Norwegian oceanographers of a large sea lily dredged from 540 meters (1,800 feet). Similar species were known from fossils 120 million years old, but until this discovery no living specimen had ever been discovered. In 1870 another expedition discovered a large, red sea urchin from the depths of the North Atlantic. This genus was previously known only from 100-million-year-old fossils in the white chalk cliffs of Dover, England.

### Spirula

The *Challenger* expedition of 1872 to 1876 netted a living specimen of *Spirula,* a mollusc that is named for its spiral-shaped internal shell (Figure 18-13a). *Spirula* are small molluscs that resemble squids and octopuses. They average 7.5 centimeters (3 inches) in length and have a barrel-shaped body with short, thick arms. The animal swims with its head pointing downward and contains an internal shell divided into gas-filled chambers (see Figure 18-13b). Similar molluscs called belemnites were common in the seas 100 million years ago but disappeared about 50 million years ago, with the exception of the ancestors of the *Spirula.*

### Vampire Squid

In 1903 another strange animal was discovered and named the vampire squid (family Vampyromorphidae; Figure 18-14). The squid received its name because the webbing between the arms and dark color suggested the sinister figure

(a)

Steve O'Shea

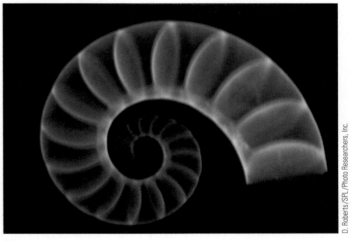

(b)

D. Roberts/SPL/Photo Researchers, Inc.

**Figure 18-13 Spirula.** (a) The spirula (*Spirula spirula*) is a cephalopod mollusc that is similar to extinct molluscs called belemnites that lived in the seas 100 million years ago. (b) The internal shell of spirula is divided into chambers like the shell of the nautilus. The gas content of the chambers helps to offset the animal's density and increases buoyancy.

of a mythical vampire. The vampire squid appears to be the descendant of a group of molluscs that were intermediate between octopuses and squids. These animals, some with webbing between the arms and others with paddle-like fins, disappeared from the fossil record about 100 million years ago. The vampire squid's ancestors may have avoided extinction by retreating to ocean depths of 900 to 2,700 meters (3,000 to 9,000 feet).

The vampire squid's muscles are soft and poorly developed, implying the animal is not a very good swimmer. It probably drifts in the depths or moves feebly with its head down and the arms hanging limp. The arms are connected by a tissue that forms a webbing, and the combination forms a loose bag around the animal's mouth. Originally, the vampire squid was thought to be an octopus, but then researchers found two additional arms, bringing the total to ten,

**Figure 18-14 Vampire Squid.**
This deepwater animal is not really a squid but an intermediate between squids and octopods. It received its name because of the webbing between the arms.

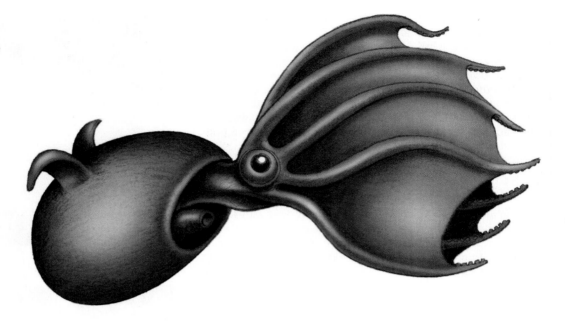

which is characteristic of squids. These two additional arms are long and lack suckers and are quite different from the tentacles of other squids. They are thought to be feelers, and when not in use, they are coiled into special pockets in the animal's web. Because of the animal's differences from squids and octopuses, they are placed in a different order (Vampyromorpha).

Vampire squids apparently spend their entire lives in deepwater. Like other cephalopods, they have good eyesight, and they possess several bioluminescent organs, including two particularly large ones that can be covered by flaps of skin. They are thought to feed on small animals that are not particularly active, possibly using their long feelers to find their prey.

Vampire squids mate in a fashion similar to the octopuses and squids, with the male using one of his arms to insert a packet of sperm into the female genital opening. Spawning and hatching occur at depths of 2,000 to 2,500 meters (6,500 to 8,500 feet). When the young first hatch, they have eight arms. They develop the other two arms, the webbing, and light organs when they are about 2 centimeters (0.66 inch) long.

Vampire squids have been collected in several areas of the world, including off the coasts of South Africa, India, Indonesia, and New Zealand. One was taken from a submarine canyon north of New Zealand at a depth of 3,000 meters (9,850 feet). The largest vampire squid ever taken measured 20 centimeters (8.5 inches) and was a female. No males larger than 12.5 centimeters (5 inches) have ever been recorded.

## Coelocanth

In 1938 fishers caught an unusual fish about 1.8 meters (6 feet) long off the coast of South Africa. The fish had large, thick scales and fleshy bundles between its body and fins. It was identified as the coelacanth (*Latimeria chalumnae;* see

Chapter 10 for photo), a species thought to have been extinct for 70 million years. This first known specimen was taken in relatively shallow water, 73 meters (240 feet). Subsequently, specimens have been caught at depths of 180 to 260 meters (600 to 1,200 feet). The discovery of the coelacanth gave biologists essential insights into the evolution of tetrapods.

## Neopilina

In 1952 researchers aboard the vessel *Galathea* discovered an unusual, limpet-like mollusc called *neopilina* (Figure 18-15) at 3,600 meters (12,000 feet) in the muddy clay off the

William C. Jorgensen/Visuals Unlimited

**Figure 18-15 Neopilina.** This limpet-like mollusc was known only from fossils until the discovery of a live specimen in 1957.

Pacific coast of Costa Rica. This mollusc belongs to the class Monoplacophora and was thought to have been extinct for 350 million years. It measured 4 centimeters (1.5 inches) long and had a lightweight, conical shell. The animal had a large pink and blue foot, probably used to crawl along the bottom mud in search of food that it gathers with the short, fleshy tentacles behind the mouth. Several of the animal's anatomical structures, such as gills, kidneys, and retractor muscles, are repeated several times, a primitive feature in molluscs. Many researchers who study molluscs believe monoplacophorans are an ancestral group that gave rise to modern gastropods, bivalves, and cephalopods. Since the discovery of *Neopilina,* many other species of monoplacophorans have been found in the deep sea.

## In Summary

Because the conditions of the deep have changed very little over the past 100 million years, biologists have speculated that some animals living in the deep may have remained unchanged since they first evolved. Over the years several examples of "living fossils" have been recovered from the ocean's depths, including invertebrates and fishes. ●

## LIFE ON THE SEA BOTTOM

Marine biologists who have visited the sea bottom in submersibles describe a barren landscape with few living organisms compared with communities in sunlit surface waters. Life on the sea bottom is as challenging as life in the abyss, and like the pelagic animals of the deep, benthic fauna have evolved to fill the available niches in this challenging habitat.

## Benthic Communities

Animals that live on the sea bottom must deal with extreme pressure, low temperature, and darkness. The environmental factor that appears to have the greatest impact on benthic organisms, however, is the availability of food.

### Sources of Food for Benthic Organisms

In the dark recesses of the ocean floor there is no light to power photosynthesis, and the cold temperatures slow the growth of bacteria and other microorganisms, so the usual bases of marine food chains are either nonexistent or severely limited. For the most part food to sustain benthic flora comes from the sunlit surface waters. In some areas, potential food in the form of feces, decaying tissue, and other organic matter rains down almost constantly from above. What is not intercepted by pelagic animals accumulates and concentrates on the ocean floor. Although compared with bottom communities along the continental shelf this represents very little food, it does sustain a variety of meiofauna, infauna, suspension feeders, and omnivorous scavengers. Near the margins of continental shelves, turbidity currents (for more information regarding turbidity currents see Chapter 3) deliver organic nutrients to abyssal plains

and trenches. Occasionally, food may arrive on the bottom in the form of a carcass of a whale or some other sizeable organism that sank to the bottom with little interference from pelagic scavengers. The importance of such occasional sinking of large food items was demonstrated by two researchers, John Isaacs and Richard Schwartzlose. In their experiments they lowered cameras mounted to buckets of dead fish to the seafloor. In as little as 30 minutes a large variety of benthic animals, including fishes, snails, echinoderms, and crustaceans arrived to feast on the meal. Shortly after the food was consumed the bottom appeared devoid of life. The few predators found at these depths make a living preying on the small number of sizeable animals that live on top of the sediments.

### Food Chains

In benthic communities that receive sufficient nutrients, bacteria can grow, although slowly because of the high pressures and low temperatures. Meiofauna, such as foraminiferans and nematodes and other small worms, feed on the bacteria, the limited organic matter, and sometimes each other. Larger worms and bivalves make up the infauna that feed on the meiofauna and available organic matter. Unlike bivalves that live in the sunlit waters near the surface, deep-sea bivalves are deposit feeders, using their siphons as vacuums to suck up the available food material that settles on the sediments. Because of the nature of the food source in these habitats, deposit feeders are usually the most abundant of the benthic animals. Deposit-feeding sea cucumbers, brittle stars, and urchins frequently dominate the benthic landscape (Figure 18-16). In areas where sand dominates the bottom sediments, suspension feeders such as giant crinoids and sea pens feed on organic material suspended in the water column. Predators in these communities are primarily fishes, squids, and sea stars.

Farther away from the margins of the continental shelves, the availability of food diminishes even more. Although there is still enough food to support small fauna such

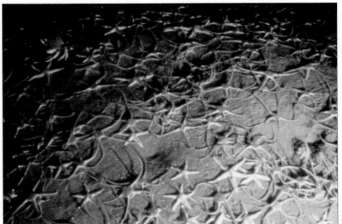

**Figure 18-16 Deposit Feeder.** Because of the nature of the food source in deepwater benthic communities, the most numerous animals are deposit feeders, such these brittle stars.

as bacteria, meiofauna, and infauna, large deposit feeders are generally absent. In the deepest recesses of the ocean, the mid-ocean trenches, food is so scarce that even tiny organisms are rarely found.

## Diversity of Benthic Organisms of the Deep

Although the density of benthic organisms in the deep ocean is only a fraction of that in the benthos of shallower seas, there is a surprising degree of diversity. One hypothesis for the large number of species is that they do not disperse their young as effectively as their relatives in shallower seas. This may contribute to isolation, which in turn is necessary for speciation (for more information regarding speciation see Chapter 5). Another hypothesis is that the environmental conditions of the deep sea are quite stable and the benthic habitat may be much older than benthic habitats on continental shelves or in shallow seas. Because of the stability of the environment, fewer species may go extinct and over time more species may accumulate. Until marine biologists are able to study this ecosystem in more detail the cause of the diversity will remain a mystery.

# Vent Communities

Not all benthic communities in the deep ocean are populated by a relatively few small- to medium-sized animals. In 1977 oceanographers discovered a thriving community off the Galápagos Islands along volcanic ridges in the ocean floor. Mineral-rich water, heated by the earth's core, raises the temperature of the water in these regions (normally 2°C to 8°C) to 16°C. Since this initial discovery, marine scientists have discovered several other vent communities in other parts of the world, including off the coast of Oregon, off the west coast of Florida, in the Gulf of Mexico, and in the central Gulf of California.

Despite the extremes of temperature and pressure, these self-contained **vent communities** are some of the most productive in the sea and demonstrate that communities can exist without solar energy. Not all vent communities surround hydrothermal vents. For instance, the Florida and Oregon vent communities are associated with cold-water seepage areas.

## Formation of Vents

You will recall from our discussion of geology in Chapter 3 that vents form at spreading centers and cold seawater seeps down near them through cracks and fissures in the ocean floor. The water comes into contact with hot, basaltic magma, where it loses some minerals to the magma and picks up others, notably sulfur, iron, copper, and zinc. The superheated water returns to the sea through chimney-like structures formed by minerals that have precipitated from the hot water. Some of these rise as high as 13 meters (43 feet) off the seafloor. The chimneys are divided into two groups: white smokers and black smokers. White smokers produce a stream of milky fluid rich in zinc sulfide. The temperature of this water is normally less than 300°C. Black smokers are narrow chimneys that emit a clear water with

**Figure 18-17 Black Smoker.** The superheated water emitted from these chimneys is rich in sulfides. When the hot, sulfide-laden water mixes with the cold ocean water the sulfides form black precipitates that give the appearance of smoke rising from the chimneys.

temperatures of 300° to 450°C that is rich in copper sulfides. As the clear hot water encounters the cold ocean water, sulfide precipitates, producing the black color (Figure 18-17).

## Vent Communities

Surrounding deep-sea vents for a few meters are rich marine communities (Figure 18-18). The residents of these communities include large clams, mussels, anemones, barnacles, limpets, crabs, worms, and fishes. The clams (*Calyptogena*) are very large and exhibit the fastest known growth rate of any deep-sea animal (4 centimeters, or 1.5 inches per year). Hydrothermal vent communities also contain large worms, called vestimentiferan worms (phylum Annelida) because of the covering that protects their bodies. Some of these worms may be as much as 3 meters (10 feet) long and several centimeters in diameter.

The primary producers in hydrothermal vent communities are chemosynthetic bacteria (see Chapter 6) that oxidize compounds such as hydrogen sulfide ($H_2S$) to provide the

Science Visuals Unlimited / WHOI/D. Foster

**Figure 18-18  Vent Community.** Vestimentiferan worms and crabs are some of the larger animals found in this vent community in the Galápagos Rift.

energy for producing organic compounds from carbon dioxide. Small animals, such as crustaceans, may feed directly on bacteria, whereas soft-bodied animals may be able to absorb organic molecules that are released by the bacteria when they die.

The clams, mussels, and worms are primary consumers, filtering the bacteria from the water or grazing on the bacterial film that covers the rocks. The clams *Calyptogena,* mussels *Bathymodiolus,* and vestimentiferan worms *Riftia* have symbiotic chemosynthetic bacteria in their tissues. *Riftia* has no mouth or digestive system; instead, the soft tissues of its body cavity are filled with symbiotic bacteria. The mussels have only a rudimentary digestive system, and both the clams and mussels harbor large numbers of bacteria in their gills.

The bivalves and worms have red flesh and red blood. The red color is due to the presence of the oxygen-carrying protein hemoglobin. Hemoglobin supplies tissues with the large amounts of oxygen that are necessary for oxidation of molecules such as hydrogen sulfide. It also supplies the oxygen needs of the animal's tissues for growth and maintenance.

Sulfides that the bacteria require for chemosynthesis are provided by the host's circulatory system. *Riftia,* for instance, has a sulfide-binding protein in its blood that allows the animal to concentrate sulfides from the environment and transport them to the bacteria. Most animals would be poisoned by the high concentrations of sulfide circulating in the worm's body, but the sulfide-binding protein has such a high affinity for free sulfide ions that it effectively binds them, thus preventing accumulation in the blood and poisoning of the worm's cells.

## Rise and Fall of Vent Communities

It is generally believed that deepwater vents are not long-lived, forming and then going dormant. Shortly after the vents form, they are colonized by living organisms. These organisms thrive until geological changes inactivate the vent. Without a steady source of nutrients for chemosynthetic bacteria, the vent community dies. At some inactive vent sites, researchers have discovered pieces of clam shells. Because they know how long it takes the clams to grow and that it takes approximately 15 years for the shells to completely dissolve in the conditions around dormant vents, they estimate that active vents last 20 years. Radiometric dating of materials from vent sites supports this estimate.

For the species of vent communities to survive, they must colonize new vents. It has been demonstrated that, with their fast growth rate, large clams can mature in 4 to 6 years. It is believed that their large bodies not only allow them to harbor large numbers of symbiotic bacteria but also to produce large numbers of larvae. If the larvae remain suspended in the water for several weeks to several months, the deepwater currents could carry them over hundreds of kilometers, allowing them to disperse to other vent sites. Some biologists hypothesize that the larvae rise to the surface to be distributed more rapidly by surface currents before sinking back to the seafloor. Another hypothesis suggests that the larvae of vent animals use whale carcasses as an intermediate habitat where they can grow, and in some cases asexually reproduce, before colonizing new vents. More research still needs to be done, however, before any hypotheses are confirmed.

## In Summary

In 1977 oceanographers discovered thriving animal communities on the ocean floor. These communities are located around deep-sea hydrothermal vents that bring superheated water and nutrients from beneath the earth's crust. The basis of the food chain in these communities is chemosynthetic bacteria. These communities are in both the Pacific and Atlantic oceans and are the focus of much current research. ●

# Light from Hydrothermal Vent Communities

Cindy Lee Van Dover, a biologist from the Woods Hole Oceanographic Institution, is interested in energy transfer and predator–prey relationships in hydrothermal vent communities. While studying shrimp populations associated with Atlantic Ocean vents, she made an intriguing discovery. She noticed that, even though these animals lack eyes, they have a reflective spot just behind the head. The spot consists of two lobes, each connected by large nerves to the brain, and it contains a lightsensitive pigment similar to the visual pigment in the eyes of other animals. The evidence suggested that the spots might be modifications of the compound eyes in other species of crustacean. Because the organ lacks a lens, the animal cannot see images, but it can sense the presence of light or similar radiation. Analysis of the pigment suggests that it is sensitive to the longer wavelengths of visible light (red) and some of the infrared spectrum. Because these are the first wavelengths to be absorbed by water, they are the ones least available in the ocean's depths. Why does this deep-sea shrimp have an organ sensitive to this portion of the light spectrum? One hypothesis is that it helps the shrimp locate prey by sensing the source of particular wavelengths of radiation emitted by the prey.

To confirm or deny this hypothesis, it was first necessary to determine if wavelengths of light in this range were present in hydrothermal vent communities. The submersible *Alvin* was used to examine the vent community associated with the Juan de Fuca ridge system off the northwestern coast of the United States. The investigations involved taking digitized pictures of the vent with an electronic still camera, using different types of illumination. The vent was photographed without illumination, illuminated by a halogen lamp mounted on *Alvin,* and finally with illumination from a flashlight shining through *Alvin*'s port. When the photographs were analyzed, the data indicated that radiation coming from the vents was in the longwavelength range of the visible spectrum (800 to 900 nanometers). From these data, researchers concluded that the radiation came from the vent water or substances in it or from the vent chimney's internal wall. The surrounding waters then diffuse the radiation.

The radiation associated with hot vents is an energy source that may be used in some way by bacteria that inhabit the area. The shrimp may use this energy to orient themselves relative to the vent or to locate their food. This discovery of light at the hot vents poses many new questions, and researchers are currently investigating how this energy is related to the vent's biological communities. ●

## SELECTED KEY TERMS

barbel, *p. 414*

disphotic zone, *p. 411*

luciferase, *p. 411*

luciferin, *p. 411*

photophore, *p. 411*

twilight zone, *p. 411*

vent community, *p. 420*

## QUESTIONS FOR REVIEW

### Multiple Choice

1. Compared with animals with higher body temperatures, animals with lower body temperatures tend to
   a. consume more food
   b. move more slowly
   c. grow more quickly
   d. reproduce earlier in their life cycle
   e. all of the above

2. The environmental factor that has had the greatest evolutionary impact on animal life in the deep is
   a. temperature
   b. pressure
   c. light
   d. salinity
   e. oxygen levels

3. The eyes of many deep-sea fishes are
   a. round
   b. spherical
   c. compound
   d. tubular
   e. identical to land vertebrates

4. Many deepwater fishes use _____ to attract prey.
   a. color
   b. bioluminescence
   c. sound
   d. odor
   e. tactile sensors

5. In deepwater animals, bioluminescence functions in
   a. species identification
   b. locating mates

c. locating prey
d. defense
e. all of the above

6. Worms known as *vestimentiferans* lack a _____ system.
   a. respiratory
   b. circulatory
   c. nervous
   d. digestive
   e. reproductive

7. The worm *Riftia* has a sulfur-binding protein in its blood that allows it to
   a. produce its own food
   b. photosynthesize
   c. feed on chemosynthetic bacteria
   d. concentrate high levels of sulfide in its blood
   e. carry oxygen for bacterial metabolism

8. The coelacanth is thought to be related to the ancestor of
   a. molluscs
   b. sea stars
   c. octopuses and squids
   d. tetrapods
   e. vestimentiferan worms

## Short Answer

1. What role does bioluminescence play in life in the deep?

2. What adaptations have evolved in deepwater fishes to help them find and capture prey?

3. What role do vertical migrations play in bringing nutrients into deepwater?

4. Why did early biologists think that animal life could not survive in the ocean's depths?

5. Why are some deepwater animals bright red?

6. Why do some biologists believe that the sea's depths may harbor organisms that have changed very little from their early ancestors?

7. What food sources are available for animals that live on the ocean floor?

## Thinking Critically

1. How might you determine whether an animal in a hydrothermal vent community derived its nutrition from symbionts or from feeding on other vent organisms?

2. The anglerfish spends the early part of its life in the upper regions of the ocean and its adult life in the abyss. What kinds of anatomical changes would you expect to observe as the animal matures?

3. Assuming deepwater fishes could survive the physical environment of surface waters, do you think that they would be able to compete effectively with fishes already adapted to that environment? Explain.

## SUGGESTIONS FOR FURTHER READING

Childress, J., H. Felbeck, and G. Somero. 1987. Symbiosis in the Deep Sea, *Scientific American* 256(5):114–20.

Conniff, R. 1996. Clyde Roper Can't Wait to Be Attacked by the Giant Squid, *Smithsonian* 27(2):126–37.

Grassle, J. F. 1987–88. A Plethora of Unexpected Life, *Oceanus* 31(4):41–46.

Isaacs, L. D., and R. A. Schwartzlose. 1975. Active Animals of the Deep Sea Floor, *Scientific American* 233(10):85–91.

Lutz, R. A., and R. M. Haymon. 1994. Rebirth of a Deep Sea Vent, *National Geographic* 186(5):114–26.

MacDonald, I. R., and C. Fisher. 1996. Life Without Light, *National Geographic* 190(4):86–97.

Van Dover, C. L., 1987–88. Do "Eyeless" Shrimp See the Light of the Glowing Deep Sea Vents? *Oceanus* 31(4):47–52.

Wu, N. 1990. Fangtooth, Viperfish, and Black Swallower: At 3,000 Feet, the Light Goes Out and Life Depends on Strange Adaptations, *Sea Frontiers* 36(5):32–39.

### InfoTrac College Edition Articles

Conley-Early, A. 1995. The Hunt for a Giant Squid, *Sea Frontiers* 41(3).

Menon, S. 1997. Deep Sea Rebirth: For Six Years Richard Lutz Has Been Tracking the Growth of One of Earth's Unique Animal Communities, One that Changes with Every Visit, *Discover* 18(7).

Mullineaux, L., and D. Manahan. 1998. Deep-Sea Diaspora: The LARVE Project Explores How Species Migrate from Vent to Vent, *Oceanus* 41(2).

Shelgrove, P. V., and J. F. Grassle. 1995. The Deep Sea: Desert and Rainforest, *Oceanus* 38(2).

Sonke, J., E. J. Balser, E. C. Fisher, and E. A. Widder. 1999. Bioluminescence in the Deep-Sea Cirrate Octopod *Stauroteuthis syrtensis* Verrill (Mollusca: Cephalopoda), *The Biological Bulletin* 197(1).

### Websites

**http://dsc.discovery.com/convergence/giantsquid/giantsquid.html** Current information about the giant squid.

**http://pubs.usgs.gov/publications/text/exploring.html** Description and photos exploring the ocean floor.

**http://www.whoi.edu** Research and education information from the largest oceanographic institute.

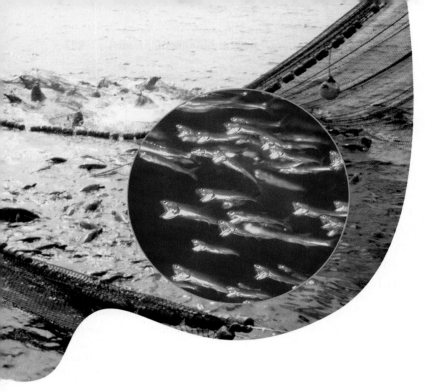

# 19

# Harvesting the Ocean's Resources

## Key Concepts

**1** Fish and shellfish are renewable resources that must be properly managed to produce a sustainable yield.

**2** Increased demand for food from the sea has placed a great deal of pressure on natural fish and shellfish populations.

**3** The advent of mechanized fleets and better fishing techniques, coupled with natural phenomena, has caused a decrease in the size of commercial fish catches.

**4** Overfishing has brought some fisheries to the brink of collapse.

**5** Techniques such as aquaculture have helped relieve fishing pressure on natural populations but not without new impacts on natural environments.

**6** Large numbers of noncommercial animals are killed as a result of current, mechanized fishing techniques.

**7** Our limited knowledge of the basic biology of many commercial species hampers our ability to properly manage and conserve these resources.

**8** The sea is an important source of minerals, including salt (NaCl) and manganese, and the sulfides of valuable metals such as gold and uranium.

**9** Fresh water for drinking and irrigation can be produced from seawater by removing the salt.

**10** The oceans contain energy reserves in the form of fossil fuels and methane hydrate.

The sea has traditionally been a major source of food for human populations, and commercial fishing is an important part of the economy of coastal nations. In the waters of the northeastern United States there are 41 different invertebrate and fish species that each generate more than $1 million a year in fisheries revenue. Along the coast of the four western states, 55 species each yield at least $1 million in fisheries revenue annually. At one time, it was believed that the oceans were an infinite source of food, but as we move into the twenty-first century, we now realize that we must conserve precious fishery resources if they are to last for our generation and future generations.

The sea is also seen as a potential new source of a variety of minerals and materials. The demand for natural resources has increased dramatically in the last 100 years, and the ability of onshore sources to meet the demand is being taxed. As terrestrial mineral resources are depleted, society will look with increasing interest toward the sea as a possible source of these minerals. Whether they will be tapped to the fullest extent depends on a number of factors, and many ecologists are concerned that the development of these resources could have devastating effects on an already stressed ecosystem.

## COMMERCIAL FISHING

Humans have harvested fish and **shellfish** (crustaceans and molluscs) from the ocean for thousands of years, but during the past 50 years there has been a dramatic increase in the amount of fish and shellfish taken from the sea. In 1950 the total world catch of marine fish and shellfish was approximately 21 million metric tons (one metric ton is 1,000 kilograms, or 2,204.6 pounds). By 2000 it reached 86.6 million metric tons (Figure 19-1a). This increase in commercial fish-

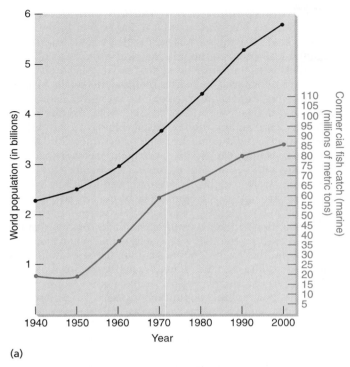

(a)

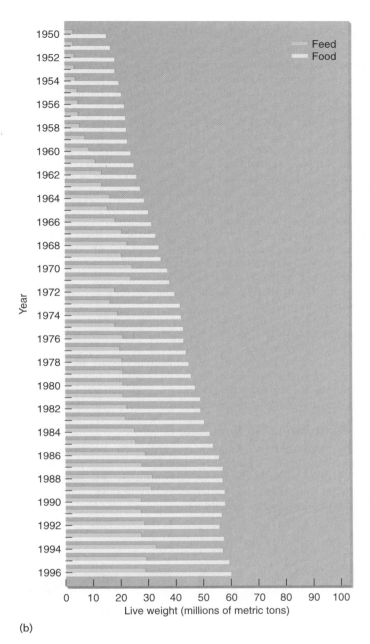

(b)

**Figure 19-1 World Population and Commercial Fishing.** (a) Since 1940, as the population of the world has increased, so has the demand for fishery products. (b) Since 1950, a larger proportion of the world's fish catch has been going to feed livestock rather than humans. (a, Data from the United Nations Organization, the WorldPaper, and FAO; b, FAO)

ing is the result of increased demand brought about by the growth in the number of humans, now more than six billion. Protein from fish and shellfish currently accounts for 16% of all of the animal protein consumed by the human population. Advances in fishing technology, such as the use of giant factory ships to process large catches and the use of sonar and aircraft to locate schools of fish, played a major role in increasing the size of the commercial catch. Recently, however, the world catch has not been increasing proportionately to the increased fishing effort. The total world catch of marine fish and shellfish in 2001 was 83.6 million metric tons, a decrease of 3.45% compared with the previous year. Unfortunately, most populations of ocean species peaked in the 1980s and are now declining.

Over the past 50 years there has also been a change in the use of the commercial catch. In 1950, 90% of the world's catch was used for human consumption, and the remaining 10% was used for fish-meal products, mainly to feed the livestock of more developed countries. By 1980, 60% of the catch went to feeding people and 40% to feeding domestic animals, a trend that continues today (see Figure 19-1b). Recall the ten percent rule of ecology (for more information regarding energy flow through ecosystems see Chapter 2), which states that only about 10% of the energy available at one trophic level is passed on to the next. From this, it becomes apparent that feeding livestock with commercial catch is not a very efficient use of this resource. For each ton of commercial fish fed to animals, only about 0.1 ton of animal biomass would be produced. In addition, most of the commercial fish catch consists of fishes that are in higher trophic levels, such as tunas. This is a greater waste of energy than harvesting fishes at lower trophic levels.

Today such a large variety of food species are taken from the sea that whole ecosystems are being harvested rather than individual species. As the more traditional fisheries become depleted, fishers focus their efforts on species farther down the food chain. Current efforts in the field of fisheries management are attempting to correct some of the problems of recent years and maintain the health of commercial fisheries.

# Fisheries Management

Fish and shellfish are referred to as **renewable resources** because animals that aren't caught can continue to reproduce and replace those that are caught. The goal of fisheries management is to maintain these resources by enacting policies and setting catch limits that will prevent overfishing to the point of extinction and allow enough animals to survive and reproduce so there will be fish and shellfish to catch in the future. To achieve these goals, fisheries biologists need detailed knowledge of a commercial species' life history. Ideally, this would include an understanding of the physical and biological factors that affect an organism's growth and survival, such as its optimal water temperature and salinity, where and on what it feeds, where and when it reproduces, and its migratory patterns. Until biologists better understand the life cycles, behavior, physiology, and ecology of commercial species, they will not be able to develop and implement the plans and conservation measures needed to properly manage these resources now and for future generations.

Managing marine fisheries is difficult because we know so little about the basic biology of commercial species. In addition to traditional experiments, fisheries biologists carefully and systematically monitor populations of commercially valuable species to learn more about them. They then use the information to develop plans to maintain adequate fish catches while not overexploiting the population.

## Monitoring Fish Populations

To make informed decisions regarding the management of species, fisheries biologists need to know how populations are distributed in the ocean environment and their migrational movements. It is also necessary to have some idea of the population size and its age structure.

### Determining Population Distribution and Movement

Some commercial species can be found over a broad geographic range. For the purposes of management, the range is divided into separate populations known as **stocks,** and each stock is assumed to be reproductively isolated from other stocks. Fisheries biologists monitor the stocks by tagging or using molecular markers. In **tagging,** fishes are caught and marked with identification tags, usually made of plastic or metal, and then released (Figure 19-2a). When these fish are caught again by sport or commercial fishers, the tags are removed and sent to the biologists who inserted them along with detailed information about where and how the animals were caught. By mapping the locations where tagged fish are initially caught and released and where they are caught again, biologists are able to track the movement and distribution of members of each stock. Other information, such as how the fish were caught and what bait was used or what they were feeding on, help fisheries biologists better understand the natural history of the species. Because each stock is reproductively isolated from other stocks, individual members of the population possess DNA markers (unique DNA sequences) and proteins with amino acid sequences that are unique to the population. These molecules are markers that identify members of a specific stock. By analyzing these molecular markers, geographic distribution and other biological information can be determined for the species being studied (see Figure 19-2b).

### Determining Population Size and Age Structure
In addition to knowing the geographic range of a stock, it is necessary to estimate its size to develop good management plans. Because fishes and shellfishes of only a certain size have economic value, it is also important to know the number of individuals of various ages and sizes in a given stock. To determine this information, fisheries biologists design

(a)

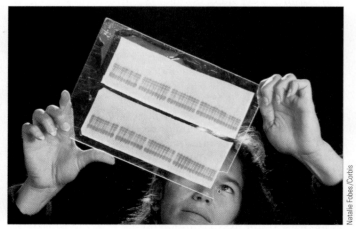
(b)

**Figure 19-2 Monitoring Stocks.** Fisheries biologists can monitor fish stocks in several ways. (a) In tagging, a plastic or metal tag like the one shown here is placed in the fish's fin. When the fish is caught again the catch data and the tag are sent back to the lab that inserted it. (b) These gels show DNA patterns that indicate the two fishes from the same stock.

sampling experiments that involve catching representative samples of the fishes or shellfishes from different areas of their range. This is not easy to do, because some fishes migrate over a large range and the regions sampled may not yield results that are indicative of stock size. Also, most commercial species are not evenly distributed in the ocean. Many occur in patches and some commercial fish species in large groups or schools. It is possible for a sampling procedure to produce erroneous data regarding population size by accidentally sampling a dense patch or school or missing the population altogether. The type of equipment used for sampling, such as net mesh size, may also cause errors by capturing animals of only a certain size and allowing other sizes to escape.

The majority of fisheries data are obtained by monitoring **landings,** the catch made by fishing vessels. These data, however, can be difficult to interpret for some of the reasons mentioned previously. Fishers tend to fish where the stocks are densest, and landings from these patches are not indicative of the population size as a whole. Another complication involves fishing effort. **Fishing effort** takes into account the number of boats fishing, the number of fishers working, and the number of hours that they spend fishing. As a general rule, if the fishing effort increases, so does the catch. This, however, does not mean that there are more fish or shellfish available to catch, it just means that more effort is being expended in catching the available fish and shellfish. On the other hand, if increased fishing effort yields a smaller catch, then it can be assumed that there are probably fewer to catch.

### Fishery Yield

Using the population information, fisheries biologists try to estimate the number of pounds of fish or shellfish that the stock can yield per year without being overexploited. This is known as the **potential yield** of a stock. The information is then used to develop management plans that will maximize the yield over several years while not stressing the population. This is called the **sustainable yield.** Because fishing increases the mortality in a population, too much fishing will reduce the stock and may even drive it to extinction. In some cases, even modest fishing can deplete the stock more rapidly than it renews itself. Such excess fishing can lead to a decrease in the stock and decrease in the yield such that it is no longer profitable to fish it. If the fishing pressure declines, the fishery might recover.

To estimate the sustainable yield, fisheries biologists use a model that takes into account the rates of reproduction and mortality, how quickly the species grows, and what influences recruitment of larvae into the population. The goal of a successful management program then is to set limits that permit a yield that does not eliminate the population. Fishers attempt to maintain a maximum sustainable yield by adjusting the amount of fishing effort to maximize production. A basic assumption is that adult fish stocks are limited by key resources, for example, food or space, and that the growth rate of individual members of a population would slow when the population density is high. If this assumption is true, then reducing the size of the population makes more resources available to the remaining individuals, and this could result in increased reproduction and ultimately more fish to catch. Thus an intermediate-sized population may produce the most fish in the long term.

Fishers also attempt to maintain a sustainable yield by imposing size limits on the individuals caught. If net mesh is too small, many fishes will be caught that are too young or small to reproduce. This would ultimately result in smaller stocks. Because fishes grow more slowly as they age, preferentially catching larger, older fish from a population makes more food available to smaller fish, which can grow more quickly. A disadvantage of nets in general is that none catch only one species of fish, and because each species usually has a different optimal size, one species will be managed properly whereas another is not.

### Problems in Managing Diverse Species

Fisheries diversity is another factor in fisheries management. Traditionally, fisheries biologists attempted to manage only one species at a time, but few single-species plans succeed with multiple species unless they are closely related. While implementing a management plan that meets the needs of one species, it is possible that other species will be negatively affected. For example, to better manage the cod (*Gadus morhua*) fishery in the North Atlantic, large areas in the Gulf of Maine were closed to bottom fishing for cod. Because the regulations were only for cod, dredging (a type of bottom fishing) for scallops (*Placopecten magellanicus*) was still permitted. The dredging operations not only disrupted benthic habitats but also killed and injured commercially valuable fishes, including the cod and yellowtail flounder (*Limanda ferruginei*). The attempt to properly manage one species conflicted with the management of other species. Some fisheries biologists suggest that the way to avoid problems such as this is to take an ecosystem-based or area-based management approach rather than trying to coordinate multiple single-species plans.

## Overfishing

**Overfishing** occurs when fish are caught faster than they reproduce and replace themselves. At least 60% of the world's 200 most valuable fish species are either overfished or fished to the limit. In addition to depleting stocks, overfishing can adversely affect ecosystems by changing genetic and species diversity and damaging or destroying habitat.

### Changes in Genetic Diversity

Overfishing can change the genetic characteristics of a population by selecting for or against certain traits, such as the size at which animals become sexually mature. This can occur when the larger, mature fish in a population are preferentially caught. Constant removal of the larger fish over time tends to favor the survival of smaller fish that mature at an earlier age and smaller size. If heavy fishing removes most fish early in their reproductive life, individuals that mature at a younger age or smaller size have a better chance of surviving and thus an evolutionary advantage. As a result, the genetic variability of the population will shift to a larger proportion of individuals that reproduce at an earlier age or smaller size. In this way, fishing can inadvertently exert a

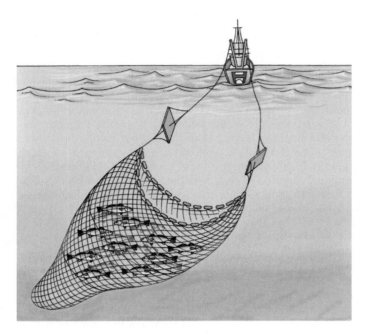

**Figure 19-3 Trawl Fishing.** Trawls are large nets that are dragged along the bottom. The trawl captures virtually everything that enters it.

selective pressure for smaller animals. An example of this is in the Northeast Arctic cod (*Boreogadus saida*) fishery. Between the 1930s and 1950s the cod matured between the ages of 9 and 11 years, and individuals of breeding age had roughly a 40% chance of surviving and reaching their spawning grounds to breed. During the same period, the cod fishery began using trawl nets to increase the size of the catch. **Trawls** are large nets that are dragged along the bottom, capturing virtually everything that enters it (Figure 19-3). The indiscriminate trawling resulted in large catches of fish and increased the mortality of older fish so much that it significantly reduced to 2% the chances of breeding-age cod reaching their spawning grounds. As a result, remaining cod that possessed genes allowing them to grow and mature faster dominated the breeding stock. This resulted in a gradual shift toward fish that mature at about 7 or 8 years of age, earlier and smaller than a half century earlier.

Overfishing can also affect genetic diversity by causing the population to become so small that it loses genetic variability because there are not enough individuals in the gene pool to carry the variety of traits that were once found in the population. An example of this is the orange roughy (*Hoplostethus atlanticus*), a commercial fish from the South Pacific. The orange roughy can live as long as 50 years and does not mature until it is 20 years old. In the 1980s, a large spawning population was discovered off New Zealand and it was fished intensely. By the early 1990s, the heavy fishing of the adults had caused the total biomass of orange roughy to decline by 60% to 70%. Fisheries biologists found that the genetic diversity within the orange roughy population had also decreased significantly during the same period, producing a genetically more homogeneous population that was less able to adapt to environmental changes.

## Changes in Species Diversity

Overfishing can affect the biological diversity of an area by reducing the number of species in an ecosystem. When a population declines to such low levels that the species no longer fulfills its role as prey, predator, or competitor in the ecosystem, it becomes ecologically extinct. This changes the relationships among species that had evolved naturally over time, causing shifts in the competitive and predator–prey relationships in the ecosystem. By removing important species, an entire ecosystem can become restructured, as predators adjust to utilizing new prey species and new predators fill niches vacated by overfished species. This in turn allows new species to become more dominant in the ecosystem. The restructuring of the ecosystem that results can change food chains and impact more fisheries than the one initially overfished.

## Changes in Habitat

Overfishing not only severely depletes the stock of overfished species, but the fishing activities themselves can damage or destroy habitat on which complex communities of marine organisms rely. This in turn disrupts entire ecosystems. For example, some fishing techniques can damage or remove biological structures, such as coral reefs or oyster reefs, resulting in large-scale ecosystem changes due to loss of habitat. Another example is trawling, where heavy nets are dragged along the ocean floor. This type of fishing can damage seagrass beds and rocky habitats and physically move or crush fish and shellfish, while changing the structural characteristics of the habitat on which these organisms rely. Trawling can also disrupt food availability for animals such as shellfish and groundfish, which rely on food in undisturbed sediments.

## Controlling Overfishing

In addition to managing fisheries, many fishing nations are using other methods to take the pressure off heavily exploited stocks and to maintain sustainable yields. These include careful policing of fishing areas, development of new fisheries products from underutilized stocks, and campaigns to inform consumers about the problems of overfishing.

***Coastal Zones***   In 1977 the United States, along with most other coastal nations, increased the area of ocean it controls out to 200 miles off the coast. These zones are called **exclusive economic zones** (EEZs). Other countries can fish in a coastal zone only by specific agreement with the controlling government. What areas can be fished, what fish can be caught, and the limits are all negotiated with the government of the country in whose water the operator is fishing. In the United States, it is the job of the Coast Guard to enforce the agreements, and fisheries observers are placed on fishing vessels to monitor and record the catch. Unfortunately, coastal limits are often not enforced, especially in poor developing countries.

***Developing New Fisheries***   Demand for fish has caused some significant decreases in traditional fisheries such as tuna, while stimulating the development of new fishery-related products. One such product is **surimi,** which is made from Alaskan pollock (*Theragra chalcogramma*), a bottom

**Figure 19-4 Surimi Products.** Many processed seafood products are made from surimi, a refined protein made from a bottom fish, the Alaskan pollock.

fish (Figure 19-4). The flesh is processed to remove the fats and oils that give the fish its characteristic flavor, and a highly refined fish protein, surimi, is produced. The surimi can then be flavored to produce artificial crab, shrimp, lobster, or scallops. Surimi is a major fish product in the Japanese market and is the product with the fastest-growing demand in the U.S. fish market. Unfortunately, the increased demand for this product in recent years is now starting to put pressure on pollock stocks.

***Consumer Education*** Ultimately, consumer demand drives fisheries production. The more a species is purchased by consumers, the more intensely that species will be fished to supply the demand. By educating consumers regarding the problems of overfishing, it is hoped that the demand for endangered species will decline, allowing the stocks to recuperate. Figure 19-5 lists some commercial species of fishes and shellfishes and indicates which are the best to purchase and which should be avoided if you want to support sustainable fisheries.

## Other Factors Affecting Marine Fisheries

Although overfishing is a principal cause, it is not the only activity that causes stocks of marine fish and shellfish to decline. Destruction of coastal habitats, wasteful and destructive fishing practices, and even the practice of aquaculture, or fish farming, have all contributed to the alarming decline in commercial fisheries. These problems, combined with our lack of scientific information about most commercial species, make it difficult to develop adequate management policies and practices to reverse the damage.

### Destruction of Coastal Habitats

Human destruction and development of coastal areas has resulted in a loss of feeding, breeding, and nursery grounds

**Good Fish**
*Fast-growing, abundant, sensibly managed, minimal bycatch and ecological impacts, minimally polluting farming methods*

Anchovies
Bass, bluenosed
Bluefish, Atlantic
Catfish, farmed
Cod, Pacific
Crayfish (crawfish, crawdad)
Crab, Dungeness
Herring and sardines
Halibut, Pacific, Alaskan
Hoki
Mackerel
Mahi-mahi (dorado, dolphinfish)
Mussels, black and green-lipped, farmed
Oysters, farmed
Pollock, Pacific (surimi, krab)
Prawns, white-spotted
Salmon, wild, Alaskan and Californian
Scallops, farmed
Shrimp, pink
Squid (calamari)
Striped bass, farmed
Shrimp, pink
Squid (calamari)
Striped bass, farmed
Sturgeon, farmed
Tilapia, farmed
Trout, farmed
Tuna, Pacific albacore (tombo tuna)
Tuna, yellowfin (ahi)

**Iffy Fish**
*Heavily fished or overfished, capture methods damage habitat and result in excessive bycatch*

Crab, Alaskan King
Crab, Snow
Lobster, clawed, American, Maine
Snapper, tropical (huachinango)
Sole, petrale, English, Dover
Spiny lobsters (crayfish)

**Bad Fish**
*Overfished and unmanaged, ecologically destructive capture destroys habitat and kills massive numbers of non-target animals*

Beluga sturgeon (beluga caviar)
Chilean seabass (Patagonian toothfish)
Clams, dredged
Grouper
Lingcod
Monkfish
Orange roughy (slimehead)
Oysters, dredged
Rockfish (Pacific red snapper, rock cod)
Salmon, Atlantic
Scallops, dredged
Shark (shark cartilage, shark fin)
Shrimp and prawns, farmed, trawled
Swordfish
Tuna, bluefin (maguro)

**Figure 19-5 Choices.** You can help stop overexploitation of marine fish and shellfish by choosing carefully in restaurants and supermarkets. This chart lists species that are abundant and those that should be avoided. *(California Academy of Sciences, San Francisco)*

for marine fish and shellfish. For example, mangrove forests once covered 75% of tropical and subtropical coastlines. Today less than half remain. The destruction of mangrove forests contributes not only to a decline in local fish catches but also to the contamination of coastal waters. In addition, the destruction of mangroves leads to increased salinity (salinization) of coastal soils as seawater evaporates leaving the salt behind. Without the root systems of mangroves to stabilize sediments, coastlines suffer severe erosion, and the loss of mangrove photosynthesis coupled with the burning of mangroves as they are cleared increases greenhouse gases, such as carbon dioxide, in the atmosphere. Draining estuaries for oceanfront development destroys seagrass beds and salt marshes that are spawning and nursery areas for many commercial species. (For more information regarding the impact of shoreline development see Chapter 14.)

### Wasteful and Destructive Fishing Practices

Biologists worldwide are concerned about the number of noncommercial marine animals killed each year during commercial fishing. These animals are referred to as **incidental catch, bycatch,** or in the case of fish, **"trash fish"** and represent a terrible waste of marine resources.

***Drift Nets*** Fishing with drift nets produces especially large numbers of bycatch. **Drift nets** are large nets composed of sections called **tans** (Figure 19-6). Each tan is 40 to 50 meters (132 to 165 feet) long and 7 meters (23 feet) high. A catcher boat sets as much as 60 kilometers (36 miles) of drift net in the evening and retrieves the net in the morning. The commercial catch, usually squid or salmon, is taken to factory ships for processing. The bycatch is dumped overboard. In the North Pacific, drift-net fisheries resulted in the

**Figure 19-6 Drift Net Fishing.**
Drift nets are made up of sections called tans. As much as 60 kilometers of net are set in the evening, and fish and squid get tangled as they swim into the net. The nets are then retrieved in the morning, and the catch is removed and refrigerated. Seabirds and marine mammals may also get trapped in the net as they try to feed on the fish.

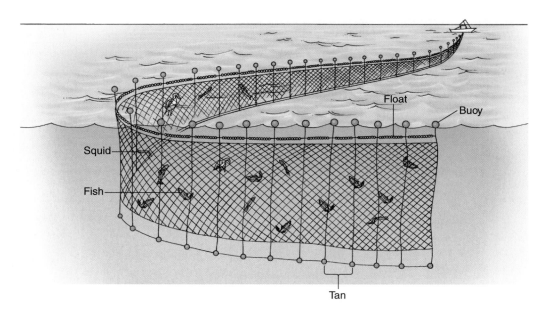

annual death of as many as 15,000 Dall porpoises and 700,000 seabirds, as well as other species such as turtles, fur seals, and sharks. These animals become entangled in the drift nets when they try to feed on the catch. The use of drift nets for commercial fishing was banned by all commercial fishing nations in 1992, largely because of pressure from the United States. Unfortunately, outlaw vessels still use drift nets, and they tend to fish areas where they do the most damage.

***Trawling*** Many fish and shellfish, such as shrimp and scallops, are caught in trawls or dredges that are dragged along the ocean floor. Although trawling is an efficient means of fishing, it damages benthic ecosystems and produces a large bycatch. Of the approximately 1.5 million tons of shrimp and 21 million tons of fish that are caught annually in U.S. waters, 3 to 5 million tons are discarded. The amount of the catch discarded varies from place to place. If the bycatch does not fetch a high enough price or processors are not available for the bycatch, it is discarded. In the past, shrimp fishing in the Gulf of Mexico usually produced 15 pounds of bycatch for each pound of shrimp. The figures are similar for the scallop industry. Almost all of the bycatch is returned, dead, to the ocean or carted off to landfills. Nets now used by shrimpers have a set of deflecting bars that push most of the bycatch below the net and not into it. This change has reduced bycatch to less than 5 pounds per pound of shrimp caught. By comparison, in Southeast Asia the demand for protein is so high that almost all of the incidental catch finds a local market (Figure 19-7) and very little is wasted, but overfishing is still a problem.

It has been estimated that shrimp trawlers catch as many as 45,000 sea turtles annually, with a mortality of 12,000 or more. The introduction of TEDs (turtle exclusion

**Figure 19-7 Bycatch.** In Asia the demand for protein is so high that most of the catch is used, as can be seen by the variety of fish and shellfish in this fish market.

devices) has greatly reduced the turtle mortality, but not every nation requires their shrimp fleet to use TEDs (for more information regarding TEDs see Chapter 11). Trawls for catching tuna can also catch and kill dolphins, although less so in recent years (see the case study of tuna fishing later in the chapter).

***Inefficient Use of the Catch***  Some fisheries use only a portion of the catch and waste the rest. For example, some species of shark are caught only for their fins. After the fins are removed, to make sharkfin soup, a popular dish in some countries, the helpless shark is thrown overboard to die. In the United States, most commercial fishers are independents who sell their catch directly to processors. The biggest demand in the U.S. market is for fish fillets and fish steaks, and commercial fishers preferentially fish for the higher-priced fishes such as swordfish, halibut, and salmon, because the filleting process uses only 20% to 50% of the fish's body weight. Compare this with the government-subsidized commercial fishing of some foreign countries, which processes the catch at sea at a lesser cost and uses substantially more of the animal's flesh. The net result is that it is cheaper to import some marine fisheries' products than to catch them domestically.

## Aquaculture

Many fisheries biologists are looking into the possibility of increasing the commercial yield from the sea through **aquaculture,** the use of agricultural techniques to breed and raise marine organisms. If only one species is raised, the process is called **monoculture;** if several species are raised together, it is called **polyculture.** The benefit of raising several species together in the same pond is that it makes more efficient use of all the available resources. The Chinese were engaged in the aquaculture of oysters as far back as 1000 B.C., and the peoples of Japan and Southeast Asia have used the practice for centuries. In these cultures, the methods have traditionally been labor intensive, and until recently, most operations were small, family-owned farms. Today aquaculture is becoming a big business, especially in low-income countries such as China, which alone accounts for 67.8% of the world's aquaculture production. In 1990 aquaculture of fishes and shellfishes accounted for 13% of the total world production. By 1996 it had risen to 22%, about 18.4 million metric tons (Figure 19-8). As the world fisheries catch continues to decline, aquaculture will become an increasingly important means of supplying commercial species to the world's markets.

***Fish Aquaculture***  A good example of aquaculture today is in Israel, which has a large population concentrated in a small area. The Israelis have acres of fish ponds that produce an annual yield of 6,500 metric tons of carp and mullet in polyculture. In the United States, fish farming produces only 2% of the fisheries' total product, and most of that is catfish and trout, not marine species. Recently, there has been a worldwide increase in the farming of salmon. In 1996 world aquaculture of salmon produced 3 million metric tons, the greatest amounts being produced in Norway, Scotland, and Canada. The Canadian government has been encouraging salmon farming, with increased numbers of

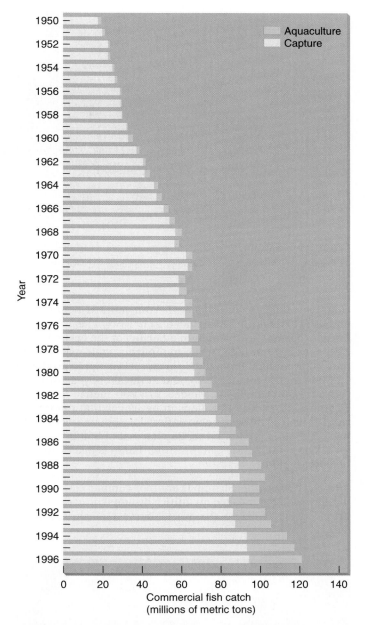

**Figure 19-8 World Production from Capture Fisheries and Aquaculture.** Since 1950, more and more of the commercial catch is being produced by aquaculture. (FAO)

farms opening along the coast of British Columbia (Figure 19-9). Washington, Oregon, California, and Maine allow salmon farming, but the procedure to obtain the proper permits is long and complicated. Texas and Florida have several farms that raise redfish. The initial costs, however, are quite high (approximately $400,000), which limits the number of people engaging in this fishery.

***Raft Culture***  Commercially valuable molluscs are also being raised in aquaculture. Clams, mussels, and oysters are raised worldwide, but especially in Asia. In Europe and Asia a method of aquaculture known as **raft culture** (Figure 19-10a) is popular. In raft culture, juveniles of commercially

**Figure 19-9 Salmon Hatcheries.** Salmon raised in hatcheries such as this one in Newfoundland, Canada, supplement the commercial salmon catch.

valuable molluscs are collected from natural populations and attached to ropes suspended from rafts. This allows the farmer to keep the juveniles in a portion of the ocean where food supplies are high, while minimizing their exposure to natural predators. Raft culture of mussels in Spain produced 132,000 metric tons in 1996, and in the same year raft culture of oysters in Japan yielded 276,000 metric tons. The Japanese also culture scallops in net cages that hang from rafts as well as the more conventional bottom methods. This has resulted in an annual scallop harvest of 250,000 metric tons. The United States, on the other hand, is just beginning to use raft culture.

**Shrimp Farming**   Shrimp is another commercial shellfish that is produced in large quantities by aquaculture (see Figure 19-10b). In 1996 the world harvest of shrimp from

aquaculture was 495,000 metric tons. In Ecuador, shrimp farming is the second largest industry (oil production is the first). Some U.S. companies have shrimp farms in several Latin American countries, where production costs are lower. In Thailand the growing number of shrimp farms along coastal areas is beginning to negatively affect local mangrove communities.

**Eco-Friendly Aquaculture**   At the Woods Hole Oceanographic Institution, biologist John Ryther has developed a model for using sewage waste in oyster aquaculture. Ryther used sewage nutrients from a town of population 50,000 to produce algae to feed oysters. His pilot plant produced an annual yield of 800 metric tons of oyster meat using this method. Because oysters produce large amounts of solid waste, Ryther introduced polychaete worms to feed on the oysters' waste. The worms could then be harvested and sold for fish bait.

In Long Island Sound, warm water from industrial cooling systems is used to increase oyster production. The average water temperature is 30°C (86°F), and it warms the animals enough to increase their metabolism and production. The success of the Long Island project has prompted some biologists to suggest that it would be possible to produce large amounts of oysters by aquaculture in the zones around the equator, where the water is warmer. Nutrients could be supplied from sewage or by pumping nutrient-rich deep water to the surface, producing an artificial upwelling.

**Problems Associated with Aquaculture**   At one time aquaculture was thought to be the solution to global fishing problems, but aquaculture is not always conducted in a sustainable way. Some types of aquaculture may, in fact, be causing some of the same problems it was meant to solve. For instance, shrimp from Ecuador and Asia are produced on coastal farm sites that were once natural mangrove forests. The mangrove ecosystem is destroyed to make room

(a)

(b)

**Figure 19-10 Mariculture.** (a) These oysters are being raised in raft cultures in Ireland. (b) This farm in Borneo raises large shrimp known as prawns. Although the prawns produced by this farm are an important food resource, the operation disrupts the ecology of the mangrove swamps where the farm is located.

for the shrimp farms. To make matters worse, the farms quickly become so polluted from accumulated wastes and the chemicals added to the water to increase production that they are abandoned and more mangroves are cleared to make room for more farms.

Some of the worst problems involve the aquaculture of shrimp and salmon. These animals require large quantities of fish meal and fish oil in their diets. To meet the nutritional demand of these aquacultures, large numbers of fish must be caught, fish that would normally support many fish species in the wild. In addition, so many animals are crowded into a small area in these farms that antibiotics and pesticides must be constantly added to the water to keep them healthy. The excess of these chemicals along with excess nutrients are released into coastal waters, disrupting coastal ecosystems and polluting the ocean. Instead of becoming a substitute for ocean fishing and decreasing pressures on depleted natural stocks, this type of aquaculture adds more pressure and lowers yields in some of the natural fisheries.

## Case Studies

The following are a few examples of commercial fish and shellfish fisheries. Each is a good lesson in how fishing and other human interventions have affected species and what is being done to correct some of the problems.

### Anchovies

Anchovies (family Engraulidae; Figure 19-11a) are small, fast-growing fishes that feed on the bountiful phytoplankton in upwelling areas along the coast of Peru. Because anchovies travel in large, dense schools, it is possible to catch large numbers with nets. Anchovies represent the world's largest fish catch for any single species. Most of the anchovies caught are not used for human consumption but end up as fish meal fed to domestic animals. The commercial fishing of anchovies for fish meal began in 1950, with an annual harvest of 7,000 metric tons. By 1962 world demand for fish meal had so increased that the anchovy fishery yielded 6.5 million metric tons of fish. By 1972 this amount had increased to 12.3 million metric tons. As the tonnage taken increased over the years, the size of the fish being caught decreased, a common effect of overfishing. As a result, more fish needed to be caught to meet commercial demand.

In 1953, 1957 to 1958, 1965, and 1972 to 1973, the Peruvian coast was hit by major El Niños (for more information regarding El Niño see Chapter 4). During an El Niño, the change in wind patterns produces changes in surface currents and upwellings. This leads to a decrease in nutrients in the surface water, a lack of production by phytoplankton, and a massive die-off of zooplankton. The decomposing organic matter robs the surface water of oxygen and adds to the death toll. Seabirds and other animals that depend on anchovies for food die of starvation or migrate.

The intense fishing that produced the record catch of 1972 coupled with an El Niño event later that year caused a dramatic decrease in the 1973 harvest to only 2 million metric tons (see Figure 19-11b). Alarmed by the situation, the Peruvian government set restrictions and placed quotas on the anchovy catch, and the fishery began a slow recovery in 1974 and 1975. Then in 1976 another El Niño caused another major decrease in the catch. From 1977 to 1982 catches again began to show a small increase, reaching 2 million metric tons again by the end of 1982. An El Niño in 1982 and 1983 caused another decline, and between 1983 to 1985, the catch was below 150,000 metric tons. By 1997 the catch was back to 7.6 million metric tons, but a particularly bad El Niño in 1997 to 1998 caused the catch to drop to 1.7 million metric tons. By 2000 the catch had increased again to 11.2 million metric tons. El Niños continue to appear and will continue to cause fluctuations in the anchovy catch. As the anchovy population declines, there is some evidence that sardines may be entering their niche, placing more pressure on the already stressed anchovy population.

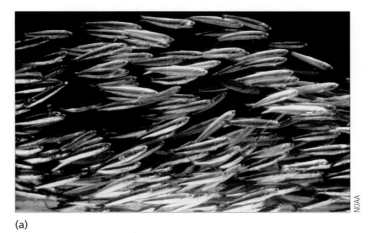

(a)

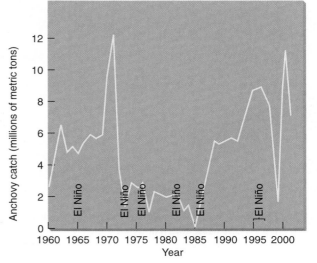

(b)

**Figure 19-11** **Anchovies.** (a) Anchovies swim in dense schools. (b) The Peruvian anchovy catch since 1950 shows the combined impact of overfishing and natural phenomena such as El Niños. (b, Data from the United Nations Food and Agriculture Organization)

There is a general lesson to be learned from these fluctuations in the anchovy fishery. Overfishing, especially when combined with unpredictable natural phenomena, can result in disastrous changes in key marine resources. Overfishing results in fewer reproductive adults in the population, which yields fewer young. Over time, the population continues to get smaller. Without proper safeguards and conservation methods, important fisheries may collapse forever.

## Tuna

Tuna fishing is big business for many countries, including the United States, which has an annual catch that averages 4.6 million metric tons. The fishing of tunas is a sophisticated and expensive operation, with a modern tuna boat costing in the neighborhood of $5 million. To meet the high demand for tuna, commercial fleets set out with state-of-the-art equipment, including **purse seines.** Purse seines are huge nets up to 1,100 meters (3,600 feet) long and 180 meters (595 feet) deep with bottoms that can be closed by pulling on a line, similar to the way a purse or bag of marbles is closed by pulling on the drawstrings (Figure 19-12). These large nets take advantage of the schooling behavior of tuna. When a school of fish is located, a powerful skiff launches from a large vessel called the purse seiner. The skiff hauls the purse seine in a circle around the school and back to the purse seiner. The bottom of the net is then closed off, and the closed seine containing the fish is hauled on board the purse seiner. This method of fishing is preferred because it yields a much larger catch with less effort than other techniques. A single haul can harvest as much as 150 tons of tuna. Although this method of tuna fishing is highly efficient, it has the drawback that dolphins are frequently trapped with the fish and drown. For unknown reasons, dolphins follow schools of tuna on their ocean migrations, swimming in the surface water just above the school. Tuna fishers locate schools of tuna by watching for dolphins. The total number of dolphins killed in tuna fishing is unknown, but it is estimated that the largest kill occurred in 1970, when more than 500,000 dolphins died.

In 1972 the Marine Mammal Protection Act was passed. It was enacted to, among other things, reduce the number of dolphins killed in tuna fishing. The act allows the National Marine Fisheries Service (a division of the National Oceanic and Atmospheric Administration, or NOAA) to set a limit on the number of dolphins killed and to recall the fishing fleet if the limit is exceeded. The limit was set at 20,500 dolphins, and the only time the fleet was recalled for exceeding the limit was in 1977. In addition to setting limits, the law mandates that an observer from the Marine Mammal Protection Agency of NOAA be present on the tuna boats for at least part of the fishing season. These observers not only count the number of dolphins killed but also monitor fishing techniques and methods used to free the dolphins. Based on these observations, the observers attempt to develop new strategies and equipment for protecting the dolphins.

To save dolphins, purse seiners use a technique called **backing down.** In this procedure the skiff draws the purse

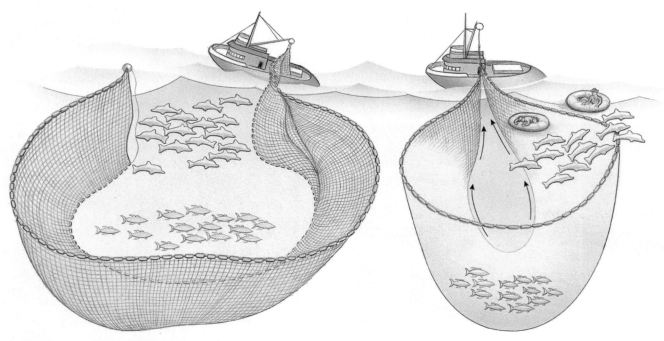

**Figure 19-12 Purse Seine Net.** The purse seine net is set in a large circle around a school of fish. The ends of the net are then drawn together and pulled closed, much like pulling on a drawstring bag. Unfortunately, when this technique is employed in tuna fishing, dolphins are also trapped in the net and drowned.

seine halfway toward the purse seiner. When the tuna are close to the boat and the dolphins are at the edge of the net, the boat backs up, causing the edge of the net to go slack and drop beneath the surface of the water, allowing the dolphins to escape (see Figure 19-12). Some strategies that have been developed have not been used. For instance, not setting the nets at night reduces dolphin kills, but because it also reduces the size of the tuna catch, it is not practiced. Most U.S. tuna boats are now based in foreign countries, where they are free from U.S. fishing restrictions.

By 1990 the National Marine Fisheries Service estimated that the number of dolphins killed in tuna fishing by the U.S. fleet was zero. However, dolphins continue to be killed by foreign fishers who use older fishing methods. In 1990 the major tuna packers in the United States agreed that they would buy tuna only from fishers who did not harm or kill dolphins in the fishing process. To indicate this, their products carry a seal indicating that the tuna is dolphin free.

### Salmon

Another large fishery in which the United States plays a prominent role is the salmon fishery. Pacific salmon (*Oncorhynchus;* Figure 19-13a) are fished in the coastal waters of the Pacific Northwest and Alaska. Salmon breed in freshwater (for more information regarding salmon see Chapter 10) but spend most of their adult lives in the ocean. Most of the commercial and sport salmon catch is made when the adults begin to migrate back to inland streams to reproduce. Because there is a large demand for salmon, the fish bring a high price per pound. As with other highly exploited fishes, salmon are becoming scarce, and their fishing is tightly regulated. Current regulations have shortened the fishing season and decreased the allowable catch, but fishers from several Asian countries still take these fish illegally on the open ocean.

The problem of maintaining salmon populations involves not just overfishing but also protecting their spawning grounds. Salmon require shaded, clear, cool, fast-running, unpolluted streams for reproduction. Damming of rivers for flood control and hydroelectric power has prevented salmon access to many streams. Logging by the lumber industry removes shade and allows runoff of silt that clouds the water. Without shade, the water temperature rises. Pollutants such as human wastes have injured young fish and decreased water quality. Although these problems are being addressed, the measures are not implemented effectively or quickly enough, and the spawning population is quite small.

The price of salmon continues to increase not only because there are fewer fish but also because of the cost of cleaning up their native streams and constructing hatcheries to supplement natural populations. About half of the salmon now caught at sea are from hatchery-reared juveniles (see Figure 19-13b). Hatchery methods, also known as **ocean ranching** or **sea ranching,** raise young fish and return them to the sea, where they can develop into adults and increase the size of the population. In a program in Oregon, the Weyerhauser Company has built hatcheries next to their

paper plant. Warm water from the plant is used to accelerate the early development of the fish. The young are then moved to pens near the sea, imprinted with a chemical substance in the water, and then released. The imprinted adults return to the holding pens later to spawn, and thus far, the project has been successful. Some commercial fishers, however, oppose ocean ranching because they believe it will lead to a few large companies controlling the fishing industry and depressing the price of fish by increasing their availability.

### Shellfish

Although most people think of the commercial catch in terms of fish species, more and more the most important marine fisheries species are shellfish (Figure 19-14). In New England and along the west coast of the United States, invertebrates, not fish, are the most valuable commercial fisheries. Crustaceans, such as crabs, shrimp, prawns, and lobsters, and molluscs, such as clams, mussels, oysters, scallops, octopus, and squid, are the most important. It is estimated that the world's shellfish catch is more than 6 million metric tons of molluscs and 5.5 million metric tons of crustaceans. Although the numbers, compared with fish, seem low, the high demand for shellfish places a high dollar value on each pound of catch. In the United States, oysters are harvested from New England, the Gulf of Mexico, Puget Sound, and Washington coastal estuaries. Lobsters are fished in New England (*Homarus americanus*) and Florida (*Panulirus* and *Scyllarides*). Shrimp are fished in the Gulf of Mexico, the Atlantic Ocean off the coasts of North and South Carolina and Georgia, and along the west coast. Scallops are fished in the Gulf of Mexico and along the southern Atlantic states (*Argopecten*) and the northeastern states (*Placopecten magellanicus*). Crabs and clams are caught along virtually all of the coastal areas. Shellfish fisheries as a whole face the same problems as the rest of the fishing industry, such as overfishing and lack of information regarding the basic biology of species. In addition, they are hard-hit by pollution that contaminates estuaries and near-shore waters and contributes to toxic algal blooms (red tides), which render some shellfish poisonous.

In the late 1970s the demand for king crab (*Paralithodes camtschatica*) increased dramatically (see Figure 19-14c). This animal is fished mainly in the Bering Sea, and as a result of the increased demand, there was a tremendous increase in fishing pressure. Between 1979 and 1980 more than 84,000 metric tons of crab were harvested. By 1982 the catch had declined to 18,000 metric tons, less than one fourth the 1980 catch, and by 1985 the catch had declined to 7,300 metric tons. A combination of overfishing and lack of knowledge of the crab's natural history are blamed for the drop in commercial catch. King crabs migrate across the floor of the Bering Sea, but not enough is known about the migration stage in their life cycle or how many migrate in any given direction to implement appropriate protective measures. In 1986 the fishery came under tight government control, and by 1987 the commercial catch rose to 13,800 tons. After several good years between 1987 and 1993, the catch reached an all-time low of 5,880 metric tons in 1994. With the exception of 1998, when the catch was

**Figure 19-13  Salmon.** (a) Salmon are popular with commercial fishers because the fish are large and bring a high price per pound. (b) Many salmon are now reared in hatcheries and released to the sea. This practice relieves some of the pressure on natural populations.

(a)

Sperm from male

Eggs from female

Salmon hatchery

Eggs develop
in the hatchery

Larval
salmon

Mature salmon
captured from
river

River

Juveniles released into river...

...and migrate downstream to
Pacific Ocean

Salmon grow to
maturity in 1–2 years

Mature salmon enter river and swim
upstream toward spawning area

Pacific Ocean

(b)

(a)

(b)

(c)

**Figure 19-14 Shellfish.** Some of the more commercially important invertebrates include (a) Maine lobster (*Homarus americanus*) (b) Spiny lobster (*Panulirus*) and (c) Alaskan king crab (*Paralithodes camtschatica*).

10,760 metric tons, the catch over the last 10 years has been averaging only 7,000 metric tons. Continued regulation of this fishery should help ensure the future catch, although a major problem, as with other fisheries, will still be the lack of enforcement of protective laws.

## In Summary

In the past 50 years, the demand for food from the sea has increased and so has commercial fishing. Natural populations of fish and shellfish have diminished, and as a result, some fisheries have been brought to the brink of collapse.

Fisheries biologists study the size, distribution, and basic biology of commercial species of fish. They use the information to develop management plans that maximize yields while not overstressing the species. Our limited knowledge of the basic biology of many commercial species, however, hinders our ability to properly regulate and conserve these resources.

Overfishing can change the genetic diversity of a population and the species diversity of ecosystems. Some techniques used in fishing can decrease fish yields by damaging the habitat. Extending coastal fishing zones, developing new fisheries, and educating consumers are some of the ways biologists are trying to conserve marine resources. Biologists are concerned about the number of noncommercial animals that are killed as a result of large-scale mechanized fishing. The incidental catch, or bycatch, represents a large waste of marine resources. Current changes in fishing regulations have resulted in equipment changes that decrease the bycatch and are improving the situation.

Aquaculture represents a way of increasing the commercial catch. Some aquaculture practices, however, contribute to habitat destruction, coastal pollution, and increased fishing of noncommercial catch that supply food for many species.

The cost of commercial fishing is increasing, and because most of the commercial fishers in the United States are small, independent operators, it is hard for them to compete with government-subsidized foreign operators for some catch, and in many instances, it is cheaper for U.S. processors to buy the catch abroad rather than from local fishers. ●

## SALT AND WATER

The most obvious nonfisheries products of the sea are its major constituents: salt and water. About 30% of the world's supply of salt (NaCl) comes from seawater. The remaining 70% comes from terrestrial salt deposits that were formed when water from ancient seas evaporated. In the south of France, Puerto Rico, central California, the Bahamas, Hawaii, and the Netherlands Antilles, sea salt is extracted and refined to produce table salt. To keep the cost of production low, as little industrial energy as possible is used in the extraction process. The process begins by allowing seawater to enter shallow ponds (Figure 19-15). Evaporation of the water produces a concentrated salt solution to which more seawater is added. Finally, the water is allowed to totally evaporate, leaving a thick salt layer behind that can be processed commercially. In colder regions, where evaporation

Copyright 1990 C. Allan Morgan/Peter Arnold, Inc.

**Figure 19-15 Salt Ponds.** A salt pond on the island of Bonaire in the Caribbean. Seawater is trapped in ponds such as this. When the water evaporates, it leaves behind deposits of salt, Bonaire's major export. The high-salinity water teems with brine shrimp and larval brine flies, the favorite foods of West Indian flamingos like this one.

of the water is not feasible, seawater is allowed to enter shallow ponds and freeze. The ice layer that forms on top is nearly pure water, concentrating the salt in the water beneath. This highly concentrated saltwater is then removed and heated to drive off the remaining water.

In addition to salt, the sea provides a vast reservoir of water; however, the salts must be removed—a process known as **desalination**—to make the water useful for irrigation or drinking. The greatest problem with producing freshwater from seawater is the cost. The energy required to power desalination processes is expensive. In general, it costs more to produce freshwater by desalination than to obtain it from groundwater or surface water supplies. Whether desalination is used depends on how badly the water is needed, what other sources of water are available, the uses for the water produced, and the local cost of desalination.

In many areas of the world, the amount of freshwater is the major limiting factor for human populations and industrial expansion. These areas include Israel, Saudi Arabia, Morocco, Malta, Kuwait, and many Caribbean islands. In each of these areas, thousands of cubic meters of freshwater are produced daily by desalination. In Texas and California, desalination plants produce freshwater but at twice the cost

of freshwater from other sources. The water produced by desalination in these states is primarily used for drinking and some light industry. It is too expensive, however, to be used for agriculture, which requires extremely large volumes of water.

## MINERAL RESOURCES

The oceans contain large amounts of minerals, but not all of them can be easily obtained for commercial use. About 60% of the world's supply of magnesium and 70% of the world's supply of bromine come from seawater. Magnesium is combined with other metals to form alloys used in making lightweight, portable tools and business machines and in the aerospace industry. Bromine is used as a disinfecting and bleaching agent and a reagent in numerous commercial chemical applications.

It is estimated that the seas contain as much as 10 million metric tons of dissolved gold and 4 billion metric tons of uranium. The concentrations, however, are on the order of 1 part per billion or less, and no one has yet devised a commercially profitable extraction method.

### Sulfides

During the 1970s and 1980s, deposits of sulfides (mineral compounds containing sulfur) were found at several sites in the oceans. These deposits form when hot molten material beneath the earth's crust rises along rift valleys, heating the rocks and causing them to fracture, forming mineral-rich solutions. When the solutions come into contact with the colder seawater, the minerals precipitate and can form massive deposits on the seafloor more than 10 meters (33 feet) high and hundreds of meters long. Minerals in the form of sulfide ores in various regions of the world include zinc, iron, copper, gold, silver, platinum, molybdenum, lead, and chromium. At this time, however, the technology does not exist for selective sampling of these deposits or for mining them.

During the 1960s, muds containing metallic sulfides of iron, copper, zinc, and small amounts of gold and silver were discovered in the Red Sea. These deposits are 100 meters (330 feet) thick in small basins at depths of 1,900 to 2,200 meters (6,200 to 7,200 feet). The saltwater over these deposits contains hundreds of times the concentration of these minerals compared with normal seawater. Their estimated value is in the billions of dollars.

### Manganese

Manganese is an element used in industry as a component of several alloys. In combination with iron, it strengthens steel and makes it easier to mold. In combination with copper, it forms alloys sensitive to temperature changes and used in temperature-activated switches such as thermostats. Manganese nodules are in many areas on the deep ocean floor. It is estimated that 16 million metric tons of these commercially valuable nodules accumulate on the ocean floor each year. Although first discovered in the nineteenth century by the *Challenger* expedition (see Chapter 1), only during the last 30 years has industry concentrated on ways of mining and extracting them. Since the 1960s, several multinational groups have spent more than $600 million dollars to locate the nodules and develop methods to collect them. Progress has been slow, and by the 1980s many groups had terminated or suspended their activities. The depressed market for metals as well as the question of ownership are two reasons for the decreased activity.

## In Summary

Almost one third of the world's supply of salt (NaCl) is produced from seawater by evaporation. Freshwater for drinking and irrigation can be obtained from desalination, removing the salt from seawater. The oceans contain vast amounts of minerals, but most of them are difficult to reach or expensive to extract. In some areas of the ocean, sulfides of a variety of minerals accumulate in large amounts. Currently, the technology does not exist for sampling or mining these deposits. Manganese in manganese nodules is abundant on the seafloor. ●

## SAND AND GRAVEL

The most widespread seafloor mining operations extract sand and gravel used in cement and concrete and for building artificial beaches. The process of extraction is essentially the same as in land-based operations, and the major portion of the cost is related to transportation distance. Approximately 112 billion metric tons are extracted worldwide annually. The removal of sand and gravel is the only major seabed mining done by the United States at this time. The United States has an estimated 450 billion metric tons of sand reserves off the northeastern coast and large gravel deposits off the coast of New England in the area of the Georges Bank as well as off the coast of New York City. The dredging of sand and gravel, however, severely damages benthic communities and their related ecosystems.

Deposits of calcium carbonate are along the coastal areas of Texas, Louisiana, and Florida. The calcium carbonate from these deposits is used for lime and cement, as a source of calcium oxide for removing magnesium from seawater, and as a gravel substitute in road construction. In the Bahamas, sands are mined for calcium carbonate, and reef sands are mined in Florida, Hawaii, and Fiji.

Coastal sands in some areas of the world contain deposits of iron, tin, uranium, platinum, gold, and even diamonds. For hundreds of years tin has been extracted from sand dredged from the coastal regions of Southeast Asia, from Thailand, along Malaysia, to Indonesia. Today 1% of the world's tin is obtained from coastal sand in this region. In Thailand tin mining has resulted in heavy deposits of silt in the intertidal and subtidal regions—all but destroying the productivity of these coastal habitats. It is estimated that

sands in the shallow coastal waters of Japan hold a reserve of 36 million metric tons of iron.

Since 1972, Russia and the former Soviet Union extracted uranium from bottom sediments of the Black Sea. The United States, Australia, and South Africa extract platinum from some coastal sands. In all of these instances, there is widespread concern about the effects of pollution and habitat destruction that accompany the extraction of minerals from the seabed.

## In Summary

The largest seafloor mining operations extract sand and gravel from the seafloor for use in making cement, concrete, roads, and artificial beaches. ●

## ENERGY SOURCES: COAL, OIL, NATURAL GAS, AND METHANE HYDRATE

Coal, oil, and natural gas are collectively referred to as **fossil fuels** because they are formed from the remains of plants and microorganisms that lived millions of years ago. Another source of fuel, methane hydrate, may be tapped in the future as reserves of fossil fuels are depleted.

### Coal

Coal formed from the remains of plants, such as ferns, that lived in prehistoric swamps. When these plants died, they generally fell into the swamp and were covered with water. The fungi that play a major role in the decomposition of woody plant material could not survive in the anaerobic swamp water, and other decomposers, such as anaerobic bacteria, don't decompose wood very rapidly. As a result, very little decomposition took place, and over time, layers of sediment buried the plant material. The sediment's weight compressed the organic material, and over millions of years, the heat generated from the minimal decomposition and the pressure of the overlying sediment converted the plant material to coal. Some deposits that were eventually submerged under the sea are a source of coal today. In Japan coal is mined from under the sea using shafts that originate on land or that descend from artificial islands.

### Oil and Natural Gas

Ninety percent of the mineral value taken from the sea is in the form of oil and natural gas. Oil and natural gas were formed from the remains of microorganisms, such as diatoms. When these organisms died, their remains settled to the bottom of the sea and were covered with sediments. The pressure and heat of being buried under tons of sediments converted the remains of these organisms to oil and natural gas over millions of years. Major offshore oil deposits are lo-

C. Lockwood / Earth Scenes / Animals Animals

**Figure 19-16 Oil Platform.** Pumping equipment used to remove oil from the seabed is set up on large platforms such as this one in the Gulf of Mexico.

cated in the Persian Gulf, North Sea, Gulf of Mexico, the northern coast of Australia, the southern coast of California, and the coastlines that border the Arctic Ocean. Many areas are still unexplored, such as the continental shelves of Asia, Africa, parts of South America, and Antarctica. Oil has also been discovered off the mouth of the Amazon river in South America and near the Philippine Islands in the Pacific Ocean. To extract the oil and gas, industry has had to develop huge drilling platforms and specialized equipment for drilling and developing wells at great depths (Figure 19-16). This was a particularly challenging task for platforms in the North Sea, where heavy seas and frequent storms are common.

The annual revenue worldwide from offshore oil and gas production is $100 billion. The offshore reserves represent about one third of the world's estimated total reserves. Although the cost of drilling and extracting offshore oil is about three to four times greater than the same development on land, the size of some deposits makes it financially worthwhile. Little is known about the oil reserves in the deep sea, because the deeper the sea, the more expensive the oil recovery. Methods and equipment developed for oceanographic research and deep-sea drilling have provided industry with models for developing the next genera-

tion of oil- and gas-drilling equipment. The development of various offshore sites proceeds slowly because of environmental concerns, legal restraints, and the uncertainty associated with global oil supplies. It appears that petroleum exploration and development will continue to be the main focus of ocean mining for the future.

## Methane Hydrate

Ice crystals that trap methane are known as **methane hydrate,** the ice that burns, and represent the world's largest known fuel reserve. Methane hydrate occurs as layers, nodules, and a space filler in and on ocean sediments of some continental slopes 200 to 500 meters (660 to 1,650 feet) deep worldwide. It forms when methane produced by decomposition of organic matter comes in contact with very cold water under very high pressure such that individual gas molecules are trapped in cages of ice crystals. They appear to be stable for long periods. When sediments containing methane hydrate are brought to the surface where temperatures are warmer and pressures less than the deep water where they are found, the methane gas rapidly escapes, like gas from a carbonated beverage that has been shaken. Because the methane hydrate contains methane, it can be ignited and will burn vigorously. In early 2002 an operation in the Canadian Arctic began to produce a controlled stream of gas from methane hydrate crystals. The results of the experimental attempt indicate that it is at least feasible to produce energy from these sources. Although this is exciting news, geologists point out that only a very small percentage of the known reserves have commercial potential. Most deposits are too thinly spread and it is much too expensive to exploit them at this time.

Because methane is a greenhouse gas, some biologists speculate that the escape of methane from marine sediments may have contributed to ocean warming events that produced climate change in the past. They worry that continued global warming may cause methane to escape from sediments, increasing greenhouse gas levels and aggravating an already bad situation.

## In Summary

The ocean contains large reserves of energy in the form of coal, natural gas, oil, and methane hydrate. Most of what is removed from the sea at this time is natural gas and oil. Offshore reserves of oil and natural gas represent about one third of the world's total reserves.  ●

# Hawaiian Aquaculture

Although Hawaii is better known for its warm coastal water, the deep waters (600 meters, or 2,000 feet) off the islands are cold and nutrient rich, providing a basis for a large and varied aquaculture. Engineers at the Natural Energy Laboratory of Hawaii (NELH), located at Kailua-Kona on the island of Hawaii, have developed ways of using a combination of cold water from the deep and warm surface water to generate electricity, grow health foods, raise seafood and vegetables, and produce drinking water without pollution.

In addition to being cold, the deep-ocean water is rich in nutrients such as nitrates, phosphates, and silicates that are usually scarce in surface waters. NELH uses nine large plastic pipes to bring this water to the surface, creating an artificial upwelling. This technology can potentially enrich areas without natural upwellings. Another benefit of the cold seawater is its purity. It provides an essentially disease-free environment for the species being raised.

Phil and Joe Wilson, who run Aquaculture Enterprises, use the artificial upwelling to raise Maine lobsters in Hawaii without the need for boats or the worry of bad weather. The cold water is actually a little too cold for Maine lobsters, but the Wilsons mix it with warmer surface water to produce a 21°C (70°F) bath that allows the lobsters to thrive. Lobsters in the wild usually require 7 years to grow to a market weight of 0.5 kilogram (a little more than 1 pound). The Wilson brothers can accelerate growth by providing summertime water temperatures year round, thus avoiding the 3-month winter hibernation of animals in the wild. Under these conditions, lobsters reach market size in about 3 years.

Lobster fishers in Maine, however, do not have to worry about the Hawaiian competition. It is still cheaper to catch lobsters than to raise them in aquaculture, and the Wilsons do not have any noticeable impact on the northeastern market. The benefits of the Wilsons' operation is that they can deliver lobsters year round at a fixed price. Recently, the Wilsons have pioneered a lobster aquaculture that utilizes waste products of commercial fishing as a source of food for the lobsters.

Another group, Royal Hawaiian Sea Farms, grows a variety of seaweeds, such as nori (*Porphyra*), ogo (*Gracilaria*), and limu 'ele 'le (*Enteromorpha*). These "sea vegetables" are considered common foods by many Asians and Pacific islanders. Royal Hawaiian's products are highly nutritious and quite popular with supermarkets and health food stores in Hawaii and California. The seaweeds are grown from spores attached to rope nets that sway in large water tanks or from vegetative fragments that grow unattached in large rotating drums (tumble culture). Nourished by the rich seawater and the abundant sunlight, the seaweed can be harvested once a week. Although many of these organisms would grow without the cold nutrient-rich water, the water contributes to more vigorous growth.

Another operation, Cyanotech, grows a microalga called *Spirulina* (a blue-green bacteria). Spirulina is highly prized by health food enthusiasts as a source of B vitamins and beta carotene, and it contains up to 70% protein. Cyanotech produces 9 tons of *Spirulina* per month, which is sold to wholesalers for pills, dips, seasonings, and food additives. Cyanotech also grows algae for the production of phycobiloprotein. The protein is a pigment that fluoresces blue or red and is used in many areas of biomedical research for marking molecules, making them easier to locate and identify. Highly purified forms of this pigment sell for $10,000 per gram. Researchers at Cyanotech are investigating the production of a lower-grade phycobiloprotein that can be sold more cheaply and is used as a coloring agent for cosmetics and food.

Ocean Farms of Hawaii raises thousands of salmon in large ponds, where they are fed on krill. The cold-water zooplankton thrive in the cold water of the artificial upwelling. Ocean Farms also uses water from the artificial upwelling to raise abalone, sea urchins, and oysters. The facility can produce 1.8 million kilograms (4 million pounds) of salmon, 15 million oysters, and 1 million abalone and sea urchins per year. Similar operations are now functioning in St. Croix, U.S. Virgin Islands, and Muroto, Japan. •

## SELECTED KEY TERMS

## QUESTIONS FOR REVIEW

### Multiple Choice

1. The largest seafloor mining operations extract
   a. salt
   b. manganese
   c. sulfides
   d. sand and gravel
   e. calcium

2. The most valuable substance (in dollar amount) that is removed from the sea is
   a. coal
   b. salt
   c. water
   d. oil
   e. sulfide

3. Many _____ are killed as a result of the methods of fishing for tuna.
   a. turtles
   b. seabirds
   c. dolphins
   d. noncommercial fishes
   e. molluscs

4. In the United States, most of the bycatch is
   a. used for feeding livestock
   b. consumed by humans
   c. dumped into landfills
   d. returned to the sea
   e. exported to other countries

5. A major problem associated with preferentially fishing for large fish such as tuna is that
   a. they are harder to catch in nets
   b. they feed at higher trophic levels
   c. fewer can be caught at one time
   d. they yield more waste when processed
   e. they do not swim in schools

6. By 1980 what percentage of the commercial catch was being converted into fish meal?
   a. 10%
   b. 20%
   c. 30%
   d. 40%
   e. 50%

7. Which of the following fishing techniques is most likely to damage habitat?
   a. purse seining
   b. drift netting
   c. gill fishing
   d. trawling
   e. pole fishing

### Short Answer

1. What is the goal of fisheries management?

2. In addition to increased mortality, what are two other problems associated with overfishing?

3. Why is it difficult to accurately determine the size of commercial stocks of fishes and shellfishes?

4. Besides overfishing, what other problems are contributing to the decline in fisheries production?

5. What are some benefits of aquaculture?

6. What is the relationship between the size of the world's population, fishing effort, and the size of the commercial catch of fish and shellfish?

7. How does the cost of fishing influence the kinds of fish caught and the methods used to process them?

8. Considering the vast mineral resources of the sea, why aren't more minerals mined from this rich area?

9. Why is it important for the fisheries industry to be regulated?

10. Why is it not ecologically sound to use anchovies for livestock feed?

11. Why are some commercial fishers opposed to ocean ranching?

### Thinking Critically

1. From an ecological standpoint, would it make more sense for humans to eat anchovies or tuna? Explain.

2. When the size of the commercial catch increases, the size of individual fishes decreases. Why does this occur?

3. Suggest some ways that commercial fishers could decrease the size of their catch without becoming unemployed or going bankrupt.

## SUGGESTIONS FOR FURTHER READING

Bardach, J. 1987. Aquaculture, *Bioscience* 37(5):318–19.

Caddy, J. F., ed. 1988. *Marine Invertebrate Fisheries: Their Assessment and Management.* New York: Wiley Interscience.

Hapgood, F. 1989. The Quest for Oil, *National Geographic* 176(2):226–59.

Parfit, M. 1995. Diminishing Returns: Exploiting the Ocean's Bounty, *National Geographic* 188(5):56–73.

Rudloe, J., and A. Rudloe. 1989. Shrimpers and Sea Turtles: A Conservation Impasse, *Smithsonian* 29(9): 45–55.

Ryther, J. H. 1981. Mariculture, Ocean Ranching, and Other Culture-Based Fisheries, *Bioscience* 30(3): 223–30.

Van Dyk, J. 1990. Long Journey of the Pacific Salmon, *National Geographic* 178(1):2–37.

### InfoTrac College Edition Articles

Baker, B. 1995. Is Overfishing or Habitat Destruction the Key Culprit in Fishery Depletion? *Bioscience* 45(11).

Constanza, R. 1998. Principles for Sustainable Governance of the Oceans, *Science* 281(5374).

Hastings, A., and L. W. Botsford. 1999. Equivalence in Yield from Marine Reserves and Traditional Fisheries Management, *Science* 284(5419).

Jeremy, B., C. Jackson, M. X. Kirby, W. H. Berger, K. A. Bjorndal, L. W. Botsford, B. J. Bourque, R. H. Bradbury, R. Cooke, J. Erlandson, J. A. Estes, T. P. Hughes, S. Kidwell, C. B. Lange, H. S. Lenihan, J. M. Pandolfi, C. H. Peterson, R. S. Steneck, M. J. Tegner, and R. R. Warner. 2001. Historical Overfishing and the Recent Collapse of Coastal Ecosystems, *Science* 293(5530).

Safina, C. 2003. The Continued Danger of Overfishing, *Issues in Science and Technology* 19(4).

Williams, N. 1998. Overfishing Disrupts Entire Ecosystems *Science* 279(5352).

### Websites

**http://www.fao.org/fi/default_all.asp** This United Nations Fisheries site provides information about all aspects of world fisheries and catch statistics.

**http://www.noaa.gov/fisheries.html** This is NOAA's fisheries site providing information on U.S. fisheries and catch statistics for U.S. waters.

**www.montereybayaquarium.org** The Monterey Bay Aquarium site provides information about fisheries and food choices for a healthy ocean.

# 20

# Oceans in Jeopardy

## Key Concepts

1. Dumping wastes into coastal seas decreases their economic and recreational value and creates health hazards.

2. Pollutants enter coastal seas by way of agricultural and urban runoff as well as by direct dumping.

3. Some pollutants accumulate and magnify in food chains, posing serious problems for higher-order consumers.

4. Plastic trash is deadly to large marine animals.

5. Oil spills damage significant amounts of habitat and injure and kill marine life.

6. Development of coastal areas leads to loss of habitat and diminished numbers of marine life.

7. Destruction of wetlands results in decreased ocean productivity.

8. It is not too late to become involved with conserving the oceans and their resources.

In 1968 a group led by Thor Heyerdahl crossed the southern Atlantic Ocean on a papyrus raft. After completing the trip, Heyerdahl reported that the ocean was polluted. His navigator, Norman Baker, noted that even though for days they would see no land, no ships, and no other humans, they would still see garbage and gobs of oil. In the 37 years since this expedition, the condition of the ocean has not improved. If anything, it has gotten worse. For years humans have thought of the ocean as a huge waste receptacle, but now there is concern about what this waste is doing to the environment and what will become of the waste already dumped in the ocean. Once-magnificent recreational beaches are being rendered dangerous. Medical wastes, such as syringes and other biological contaminants, are brought in by the tide. Tar residues clot the sands. Development of oceanfront property has resulted in erosion and damage to offshore habitat that may never be reversed. The United States is not the only country experiencing such problems. Nearly every nation on earth feels the effects of a polluted sea. Without changes in federal and international policy, the situation will only become more serious.

## POLLUTION

More so than at any other time in human history, the ocean is being contaminated by the products of human activity. Trash, sewage, chemicals both toxic and nontoxic, and radioactive materials are just a few of the items polluting the sea. The effects of pollution on marine organisms and ecosystems are as varied as the pollutants themselves. Some injure or kill marine life or interfere with their ability to reproduce. Some so enhance productivity that they trigger algal blooms that can contribute to the death of marine organisms. Others interfere with productivity and produce long-term effects that can impact humans as well as the inhabitants of the sea.

NOAA; inset, AP/Wide World Photos

**Figure 20-1 Ocean Dumping.** Much of the trash on this recreational beach is the result of littering by thoughtless individuals.

## Ocean Dumping

Almost from the beginning of human history, coastal countries around the world have used the sea as a dumping site for trash and garbage (Figure 20-1). Over the years, domestic wastes, industrial wastes, and more recently, radioactive wastes have literally been poured into the ocean. The dumping of human and industrial wastes into coastal waters is an inexpensive but shortsighted solution to eliminating these wastes.

### Trash

In September 2002 the seventeenth annual Coastal Cleanup coordinated by the Center for Marine Conservation cleared 2.8 million pounds of trash and debris from U.S. coastlines in a 3-hour period. The cleanup effort drew 139,746 volunteers nationwide and cleaned 5,148 miles (8,237 kilometers) of the nation's beaches and waterways. The effort in Florida is estimated to have saved the state $500,000 in cleanup costs. The trash collected included 1.3 million cigarette butts and filters (enough to make 82,031 packs of cigarettes), 226,251 glass bottles, and 238,826 metal cans. Medical wastes, including 2,529 syringes, were recovered mostly on beaches of the Northeast and the Gulf Coast (for a more complete list see Table 20-1). Sixty-one percent of the trash collected along the beaches was plastic. This is an increase from previous years and is of special concern because plastics do not degrade quickly and are hazardous to wildlife. A

**Table 20-1 Types of Debris Polluting the Marine Environment**

| Shoreline and Recreational Activities | Ocean and Waterway Activities | Smoking-Related Activities | Dumping-Related Activities | Medical and Personal Hygiene |
|---|---|---|---|---|
| Food wrappers/containers | Rope | Cigarettes/cigarette filters | Building materials | Tampons/tampon applicators |
| Caps, lids | Plastic sheeting/tarps | Cigar tips | Cars/car parts | Condoms |
| Cups, plates, forks, knives, spoons | Fishing line | Tobacco packaging/wrappers | Tires | Diapers |
| Beverage cans | Bait containers/packaging | Cigarette lighters | Batteries | Syringes |
| Beverage bottles (glass) | Buoys/floats | | Appliances (refrigerators, washers, etc.) | |
| Beverage bottles (plastic), 2 liters or less | Strapping bands | | 55-gallon drums | |
| Bags | Bleach/cleaner bottles | | | |
| Straws, stirrers | Oil/lube bottles | | | |
| Balloons | Fishing lures/light sticks | | | |
| Clothing, shoes | Fishing nets | | | |
| Pull tabs | Crab/lobster/fish traps | | | |
| Toys | Light bulbs/tubes | | | |
| Six-pack holders | Crates | | | |
| Shotgun shells/wadding | Pallets | | | |

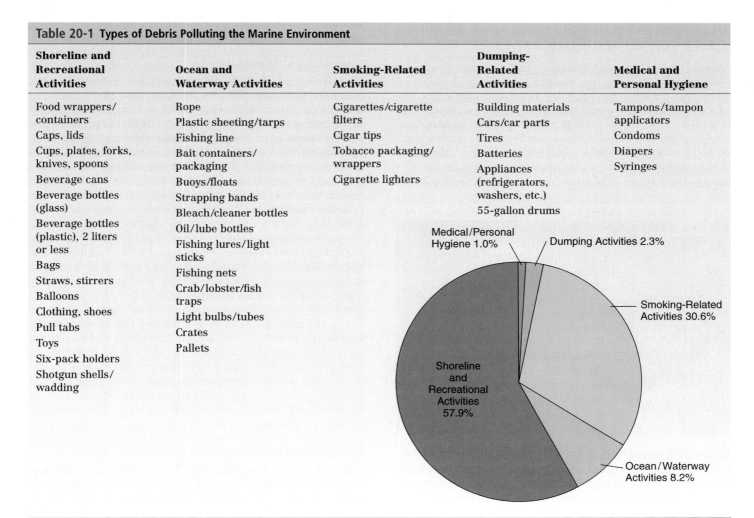

*Source:* The Ocean Conservancy

total of 82 animals were found trapped in the debris. A project report released 10 months after the cleanup concluded that U.S. beaches are still being polluted by a wide variety of materials, especially plastic wastes. If more concerned citizens get involved with projects such as Coastal Cleanup, perhaps the state of our beaches will improve.

## Plastic

It is estimated that naval and merchant ships legally dump 77 tons of plastic into the ocean annually. In addition to this, the National Academy of Sciences estimates that the fishing industry annually discards or loses 149,000 tons of fishing gear (nets, ropes, traps, and buoys, which are mostly plastic) and then dumps another 2,600 tons of plastic packaging material into the sea. To this, add the discarded plastic from pleasure boats, ocean liners, oil platforms, and visitors to the beach. In total, about 1 million tons of plastic waste enters the ocean annually. Carried by the currents, this refuse can appear on even the most remote beaches of the world.

Some of the very characteristics that make plastic such a valuable component in manufacturing—its strength and durability—also make it one of the most hazardous materials dumped into the sea. Many marine biologists consider plastic trash to be as great a killer of marine life as oil spills and toxic chemicals. No one knows for sure how long plastic remains in the marine environment, but it is thought that a plastic six-pack ring could last as long as 450 years.

***Marine Animals and Plastic***   The greatest danger of plastic wastes is to larger marine animals and seabirds. As many as 30,000 fur seals are killed annually when discarded fishing nets and cargo straps ensnare them (Figure 20-2). Discarded netting also traps fish, turtles, and other marine life. Shellfish traps are frequently made in part or entirely of plastic. When these traps are lost, they continue to catch animals. Unable to escape, the animals ultimately die, frequently of starvation. On the western coast of

**Figure 20-2 Plastic Trash.** Plastic trash poses the single greatest threat to large marine animals, especially turtles, birds, and mammals. Animals like this seal get tangled in the plastic and lose appendages or drown. Some consume the plastic, possibly mistaking it for a jellyfish, and die of intestinal blockage.

Florida, more than 100,000 of these traps are set out each year, and approximately 25,000 are lost. Many seabirds die when they become entangled in six-pack rings or fishing line. Whales and dolphins have been found suffocated by plastic bags or sheets of plastic. Seabirds, sea turtles, and mammals all have a tendency to eat plastics. The plastics form an indigestible mass that blocks the digestive tract and causes the animals to die of starvation.

***Controlling Plastic***   The Marine Plastic Pollution Research and Control Act, passed in 1987 and effective as of 1988, prohibits the dumping of plastic in the ocean and requires ports and terminals to provide facilities for the disposal of this waste. The United States Navy, however, can still legally dump trash and plastic into the ocean. Similar laws have been passed by other nations, and an international agreement prohibits dumping plastic waste in the ocean. In the United States, the Coast Guard is mandated to enforce the law, but with all of its other duties and small resources, violations still occur. Other nations are even less successful in enforcing dumping restrictions, and plastic continues to be a problem. Some manufacturers have started to produce plastics that are more biodegradable, with a shorter life span in the environment. Unfortunately, this will probably not do much to help. Only education and responsible action by those who use the sea for commerce and recreation will reduce plastic in the ocean.

## Commercial Dumping

Since 1890, a variety of garbage, sewage, and toxic chemicals have been dumped into the New York Bight, the area lying off the mouth of the Hudson River. Floating debris from the dumping ends up on beaches. By 1934, the problem had reached such large proportions that legislation was passed prohibiting the dumping of floating waste. The legislation did not, however, limit the dumping of building wastes, storm and sewage wastes, or industrial wastes. It has been estimated that 1.4 million cubic meters of solid waste was dumped into the New York Bight between 1890 and 1971—enough to cover all of Manhattan Island to the height of a six-story building. Needless to say, this amount of dumping has greatly decreased water quality and the quality of bottom sediments. The materials dumped were both toxic and oxygen demanding (decomposition of these materials by microorganisms tends to deplete the water of oxygen). Occasionally, the contaminated water would upwell along the coast, killing shallow-water organisms. The offshore dump was finally closed in 1987 by the Environmental Protection Agency. The agency did, however, allow New York City, as well as several New Jersey communities, to continue dumping sewage sludge, a product of sewage treatment, at a site 171 kilometers (106 miles) out to sea at the edge of the continental shelf. More than 8 million tons of sludge is dumped at this site annually.

## Military Refuse

Following the major wars of the last century, the ocean has been the dumping ground for discarded military hardware and munitions (Figure 20-3a). Toxic gases and chemicals removed from Germany after World War II were dumped in the North Atlantic Ocean; in the Pacific Ocean, bays and lagoons were filled with jeeps, tanks, trucks, and munitions. After the Vietnam War, vehicles and munitions were unloaded into the coastal waters of Southeast Asia. Following the first Gulf War in 1991, munitions and military hardware were discarded in the Arabian Gulf. The discarded material can entangle or poison marine birds and mammals. On the positive side, some of this metallic junk has been converted into artificial reefs populated by fishes and invertebrates (see Figure 20-3b).

### Radioactive Wastes

Today's electronic, chemical, and defense industries and nuclear facilities produce large amounts of highly toxic or radioactive wastes that have been stored in landfills and in aboveground repositories. As concern for their effects on human health increases, it has been suggested that they

(a)

(b)

**Figure 20-3 Artificial Reefs.** (a) After military conflicts, such as World War II, wastes are frequently dumped into the ocean. (b) Sunken vessels like this one become artificial reefs, providing habitat for a large variety of marine organisms.

be deposited on the ocean floor, for instance, in the deep-sea trenches far from continental landmasses. Supporters of this proposal point out that because the trenches are subduction zones (for more information regarding subduction zones see Chapter 3) the wastes would ultimately be taken into the mantle of the earth, eliminating the problem of long-term storage. Opponents point out that if the containers holding the waste rupture or decompose before subduction occurs, their contents could cause considerable habitat damage. Considering our general ignorance of the deep-sea benthos and deep-sea food webs, no one really knows the long-range effects or dangers of this disposal method. Subduction-zone disposal is currently restricted by the Ocean Dumping Act of 1972, which requires an environmental impact statement and approval of both houses of Congress before radioactive waste can be dumped in the ocean.

## Pollution via Land and Air

Not all pollutants enter the ocean through indiscriminate dumping. Surface runoff from metropolitan areas contains pesticides, fertilizers from residential areas, chemicals from oil and gasoline, residues from industry, and bacteria from sewage (Figure 20-4). Pesticides and fertilizers used in agriculture can also enter the marine environment as they are rinsed into streams and rivers that flow to the sea.

### Urban Pollution

Throughout the world, major population centers have developed around estuaries and along the rivers that drain into them. In the United States, more than 50% of the population lives within 50 miles of a coastline (including the Great Lakes). Worldwide, the numbers are similar and sometimes higher. This dense population with its needs for energy, industry, and waste treatment has placed a large burden on coastal seas and coastal habitats. Pesticides and fertilizers used on lawns, gardens, and crops wash to the sea during rainstorms and in snow melt. Gasoline, oil, and other petrochemicals that leak on the ground also find their way to the sea. On the east coast, water from storm drains mixes with sewage and overflows sewage treatment facilities after heavy rains and when snow melts. The overflow enters the ocean. This type of pollution damages recreational areas as well as coastal fisheries. Metropolitan pollution in Long Island Sound has killed many species of wildlife, and the stench of the sewage and dead organisms has rendered many of its beaches useless for recreation. In addition to untreated runoff, coastal cities discharge treated sewage that adds its own array of contaminants to ocean water. Chlorine added to drinking water and then later to sewage during treatment forms chlorinated organic compounds in seawater that are toxic to some marine organisms.

Efforts to reduce the discharge of toxic materials into the marine environment are gradually succeeding. Since 1980, the increased use of unleaded fuels has significantly decreased the amount of organic lead entering the marine environment. The concentration of lead in surface waters of the Sargasso Sea dropped by 30% between 1980 and 1984, and in the last 10 years, the amount of lead entering the Gulf of Mexico from the Mississippi River has decreased 40%.

Hal Beral / Visuals Unlimited

Jeffrey Rotman / Peter Arnold, Inc.

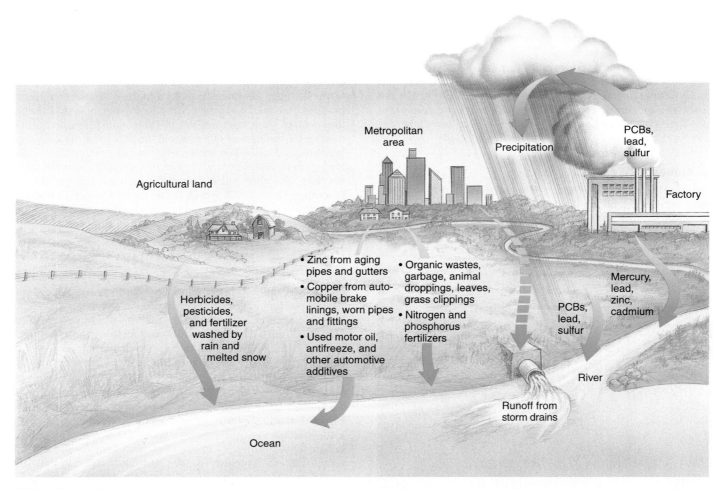

**Figure 20-4  Pollution from Land.** Pollutants, including toxic organic compounds, pesticides, sewage, petroleum products, and heavy metals, are carried into the sea by rivers and washed into the sea by rainwater and melting snow.

## Pesticides and Toxic Materials from Industry

Pesticides, including DDT (dichlorodiphenyltrichloroethane); toxic organic compounds, such as PCBs (polychlorinated biphenyls); and heavy metals such as mercury, lead, zinc, and chromium continue to enter the sea through various routes, even though they are closely monitored and no longer used. The use of DDT was suspended in the United States in the 1960s, and the United States ceased production of PCBs in the late 1970s. Both DDT and PCBs are, however, persistent toxic substances, remaining in the environment and ocean for long periods. DDT is carried to the sea by runoff from the land. The year that DDT was banned, another pesticide made from DDT, dicofol, was approved for use. Hundreds of thousands of tons of this pesticide were used around the nation until 1988, when researchers found that dicofol contained 1% to 7% DDT contaminants.

Seabirds that feed on fish contaminated with DDT lay thin-shelled eggs that break when the parents incubate them. This has led to high mortality of chicks and decreases in the size of some seabird populations, such as brown pel-icans in California, already under pressure from other environmental changes. Even populations of the Bermuda petrel that live far from shores where the pollutant enters have been affected.

PCBs were used in a number of consumer articles, including plastics, solvents, and electrical insulators. They are still present in many electrical devices that were manufactured before 1979. When these devices are burned, the PCBs enter the atmosphere and enter the water supply with precipitation.

***Biological Magnification***  Not all pollutants that enter the marine environment dissolve in the water. Some are absorbed by suspended particles of clay and detritus. The particles clump and form denser masses that sink to the bottom. Toxic materials like DDT, PCBs, and heavy metals always collect in higher quantities in sediments than in the overlying water. They continue seeping into the water for years. The tainted detritus as well as contaminated phytoplankton are used by various marine organisms as a source of food. Because many of these organisms have no means

**Figure 20-5 Biological Magnification.** Toxic substances can accumulate in the tissues of consumers as they feed on organisms that contain low levels of toxic material. Notice how the level of DDT in this example becomes more concentrated in the tissues of organisms as you move up the food chain. Ultimately, the toxic substance will reach a lethal level and cause the death of a higher-order consumer.

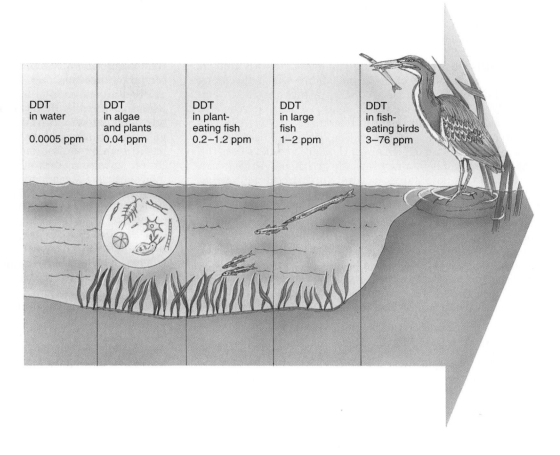

DDT
in water

0.0005 ppm

DDT
in algae
and plants
0.04 ppm

DDT
in plant-
eating fish
0.2–1.2 ppm

DDT
in large
fish
1–2 ppm

DDT
in fish-
eating birds
3–76 ppm

of breaking down the toxins or excreting them, the toxins concentrate in the organisms' body tissues. The toxins are passed along to predators, where they concentrate further, a process known as **biological magnification** (Figure 20-5). At some level, the concentration of toxins can be so high that the next predator to feed on the contaminated flesh will become seriously ill or die. During the 1970s, this is exactly what occurred when mercury from industrial pollution accumulated in the food chain and contaminated tuna. Humans who consumed the contaminated tuna suffered from the effects of mercury toxicity, including neurological problems and death.

Between 1953 and 1960 an industrial plant in Minamata, Japan, dumped industrial wastes containing high levels of mercury into a local embayment. The mercury readily formed organic complexes that entered the food chain and were concentrated in the shellfish and fish that the people of the town used as a major food source. Many cases of severe mercury poisoning and even death resulted. The damage to the nervous system caused by the mercury was particularly noticeable in children whose mothers had consumed contaminated shellfish or fish while pregnant. Sadly, this is not an isolated incident; bottom fish in all of the estuaries of the United States have some amount of toxic materials. Shellfish concentrate certain toxic materials, such as mercury, lead, and cadmium, to several thousand times their concentrations in the surrounding water. Scallops con-

centrate cadmium to 2 million times that of its concentration in the water, and oysters concentrate DDT to more than 90,000 times that of its concentration in water. This contamination of the food supply puts large numbers of people at risk.

***Effects of Toxic Compounds on Plankton*** Phytoplankton are particularly sensitive to chronic pollution by toxic substances, as are many forms of zooplankton and larger crustaceans. Toxic pollutants inhibit photosynthesis, growth, and cell division in marine phytoplankton. They affect the growth and development of filter-feeding zooplankton, as well as the early developmental stages of other organisms, by interfering with cell division. The impact of marine pollution on plankton has barely been studied, but it is known that pollution in the North Atlantic Ocean has resulted in decreased numbers of species of both phytoplankton and zooplankton. This should serve as a warning. In the long term, pollution causes an overall reduction in marine primary productivity, altering interactions at every trophic level throughout the food web.

## Air Pollution

Air pollutants can also have a negative effect on marine ecosystems. The burning of some fossil fuels releases sulfur dioxide, among other pollutants, into the atmosphere. The sulfur dioxide dissolves in water and enters the ocean directly, in precipitation, or indirectly, in river runoff. The water is acidic, which can lower the pH of the seawater.

Greenhouse gases such as carbon dioxide and methane are also a problem. Greenhouse gases trap radiant energy and contribute to global warming. Global warming is thought to be a major cause of coral bleaching, which threatens the extinction of coral reefs. Global warming has also been linked to El Niño events, which change surface currents in the South Pacific and result in decreased productivity and the death of marine organisms. One obvious effect of global warming is melting of glaciers and polar ice. This raises the sea level and adds more freshwater to the sea. A rise in sea level could be devastating to corals that are already being stressed by other environmental changes and disease. On average an upward growth of about 10 millimeters per year is enough to keep the coral within the sunlit waters they require. If water levels rose more quickly, the corals might not be able to keep pace. Without sufficient sunlight the corals' zooxanthellae would decrease productivity or die, resulting in widespread death of coral reefs. Increased ocean levels could also flood barriers that protect estuaries, increasing the input of saltwater and destroying these ecosystems.

## Nutrient Pollution

It was once thought that coastal waters had an infinite capacity to absorb wastes from human populations. But too much waste entered too small an area in too short a time. This has resulted in a decrease in the economic and recreational value of coastal areas and, in some cases, endangered public health and safety. During the 1960s and 1970s, environmental legislation was passed, creating agencies to control and monitor resources, including the ocean. Controlling today's waste, however, is not sufficient, because we have to contend with an accumulation of decades of waste.

### Human Wastes

Cottages built along the shore use septic tanks to process sewage and drain wastes into the sandy soils. Commercial developments intentionally and unintentionally drain a variety of wastes into the ground, and coastal communities dump raw sewage into the shallow water off the coasts. Incoming tides bring sewage-contaminated water onto beaches. This contaminated water, added to the seepage from septic tanks, makes beaches unsafe for humans and contaminates the invertebrates.

***Disease Agents***    Human wastes reduce water quality by adding harmful microorganisms and enriching the water with nutrients. Filter feeders such as clams, mussels, and oysters can concentrate large amounts of harmful microbes in their tissues, such as hepatitis viruses and *Salmonella* bacteria, from sewage, resulting in their meat being unfit for human consumption. In areas where these shellfish are eaten, diseases such as hepatitis and dysentery may occur. In many parts of the United States, bacteria from cities and individual septic systems have forced the closure of clamming beaches and oyster beds. In some cases, the result has been economic hardship for clam and oyster fishers. Local

environmental agencies monitor water quality by counting the number of **coliform bacteria** in the water. Coliform bacteria are found in the intestines of many animals, including humans, and are an indicator of the amount of animal wastes entering the water. Their presence indicates the potential for disease.

***Eutrophication***    Both raw (untreated) and treated sewage add large amounts of nutrients, such as ammonia and urea, to coastal waters. This leads to eutrophication, an increase in the amount of dissolved nutrients in the water. Eutrophication leads to blooms of phytoplankton and other marine microbes. The large increase in phytoplankton populations frequently exceeds the ability of zooplankton to control it by their grazing. The increased amounts of phytoplankton also make the water turbid and interfere with primary production in deeper water. Eventually, the phytoplankton begin to die off, and they are decomposed by aerobic bacteria. As you learned in earlier chapters, excessive decomposition can rob the water of oxygen and result in the death of fish and other marine organisms as well as the aerobic bacteria. Anaerobic bacteria continue the job of decomposition, removing even more oxygen from the water and in many cases producing hydrogen sulfide, a compound toxic to many marine organisms. The lack of oxygen and accumulation of hydrogen sulfide leads to the death of more marine organisms and their decomposition, and the vicious cycle continues.

***Increased Productivity***    In some instances, the addition of sewage and animal wastes can increase the productivity of an area. During the early part of the twentieth century, sewage from London entered the North Sea by way of the river Thames. A small region around the mouth of the river supported a significantly higher catch than other parts of the North Sea, presumably because of the increased nutrient input. Similar increases in coastal productivity have been noticed along the coast of southern California and the northeastern coast of the United States.

### Agricultural Wastes

Another source of marine pollution is surface runoff from agricultural lands. Fertilizers and animal wastes affect marine communities in a manner similar to that of human wastes. Animal feces can add disease-causing microbes to the water like those added by human sewage. It is thought that the outbreak of *Pfiesteria,* a potentially dangerous dinoflagellate, in the Chesapeake Bay region may be related to the contamination of the water by feces from pig and chicken farms.

Surface runoff containing pesticides and nutrients is carried by streams and rivers to the sea, where it can poison and overfertilize the water. Nutrient addition from fertilizers causes eutrophication of coastal waters and supports phytoplankton blooms. The increase in algae outstrips the environment's ability to support it, and the algae die. The decomposition of the dead algae removes excessive amounts of oxygen from the water, and this causes more organisms to die, resulting in even lower levels of oxygen. Agricultural

According to the Pew Oceans Commission report, nitrate runoff pollution from the northern and Midwestern Mississippi River watershed is one of eight major threats to ocean wildlife, polluted beaches and collapse of commercial fishing in U.S. waters. Such pollution has contributed to a growing "dead zone" in the Gulf of Mexico where the Mississippi meets the ocean. Pollution drains oxygen from the ocean ecosystem, causing fish kills and depletion of ocean wildlife.

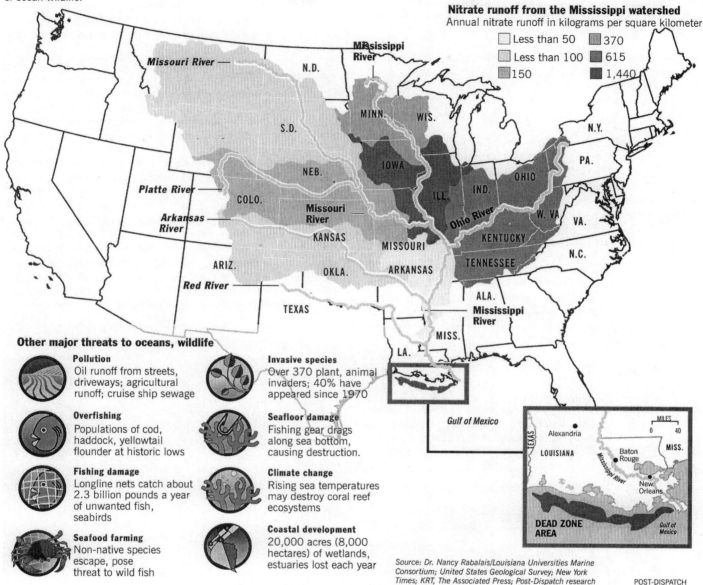

**Nitrate runoff from the Mississippi watershed**
Annual nitrate runoff in kilograms per square kilometer

Less than 50    370
Less than 100    615
150    1,440

**Other major threats to oceans, wildlife**

**Pollution**
Oil runoff from streets, driveways; agricultural runoff; cruise ship sewage

**Overfishing**
Populations of cod, haddock, yellowtail flounder at historic lows

**Fishing damage**
Longline nets catch about 2.3 billion pounds a year of unwanted fish, seabirds

**Seafood farming**
Non-native species escape, pose threat to wild fish

**Invasive species**
Over 370 plant, animal invaders; 40% have appeared since 1970

**Seafloor damage**
Fishing gear drags along sea bottom, causing destruction.

**Climate change**
Rising sea temperatures may destroy coral reef ecosystems

**Coastal development**
20,000 acres (8,000 hectares) of wetlands, estuaries lost each year

*Source: Dr. Nancy Rabalais/Louisiana Universities Marine Consortium; United States Geological Survey; New York Times; KRT, The Associated Press; Post-Dispatch research*

POST-DISPATCH

**Figure 20-6 Hypoxic Zone.** Agricultural nutrients draining into the Gulf of Mexico have contributed to the depletion of oxygen in an area off Louisiana referred to as the "dead zone."
*(Reprinted with permission from the St. Louis Post-Dispatch, copyright 1997.)*

nutrients carried into the Gulf of Mexico by the Mississippi River caused a hypoxic zone (water with abnormally low oxygen content) off the coast of Louisiana (Figure 20-6). This area of the Gulf once supported a major shrimp fishery, but as a result of the eutrophication and resulting decline in oxygen, the fishery has collapsed.

Although there has been increased input of nutrients from fertilizers used in agricultural areas, better sewage treatment has decreased the nutrient content of discharged sewage, so in some areas there has been no net increase in the level of nutrients reaching the sea. Unfortunately, this is not true everywhere.

## Controlling Pollution

In the summer of 1988, many northeastern U.S. beaches had to be closed because of the polluted waters and the amount of dangerous wastes washing up on the beaches. In addition to gummy balls of sewage as much as 2 inches in diameter, medical wastes such as syringes, vials of blood, and medicine bottles were washing up on shore (Figure 20-7). On a New Jersey beach just south of Long Island, New York, more than 100 vials of blood washed up. When tested, 5 vials were found to contain antibodies for HIV (human immunodeficiency virus). The ensuing public scare caused attendance at some beaches to drop by 85%.

In response, Congress passed legislation that prohibited dumping of sewage sludge or industrial wastes in the ocean after January 1, 1992. This legislation does not solve all of the problems, however. Much of the material that contaminated beaches in 1988 continues to contaminate beaches today. It is the product of overflowing storm sewers that empty into the ocean, as well as improperly dumped wastes.

The bigger threat to coastal water quality is not dumping of wastes but increasing coastal populations and improperly controlled residential and commercial development, discussed later in the chapter. In theory we could have both large populations and clean water if we spent the money to treat the sewage. Whether we decide to spend the money depends on ethical, economic, and political considerations.

## Petroleum Pollution

The demand for oil to supply the needs of industry and motorists has exposed the ocean to one of its greatest threats. In addition to the petroleum and petroleum products released into the environment by runoff, industrial discharge, and other processes, each step in production and transport

poses the risk of contamination. When petroleum is pumped from the seafloor by offshore drilling operations, there is the risk of well blow outs and leakage. When it is transported from production sites to processors and industrial areas in tankers, there is the risk of shipwreck. The tanker transport of oil and petroleum products poses the greatest risk of large spills.

### Petroleum Products

Crude oil is a mixture containing nearly 10,000 different polluting chemical compounds. The majority of these chemicals are organic molecules composed primarily of carbon and hydrogen atoms. Crude oil contains two major types of these organic molecules: aromatic hydrocarbons and aliphatic hydrocarbons. **Aromatic hydrocarbons,** such as benzene, naphthalene, and cyclohexane, are made up of carbon atoms in ring structures (Figure 20-8a). **Aliphatic hydrocarbons,** on the other hand, are straight-chain molecules, such as heptane, octane, and nonane (see Figure 20-8b).

When oil is released into water it spreads in a film across the surface. Lighter components of the oil evaporate or are absorbed by particles of clay and other sediments in the water and sink to the bottom. Some inorganic components dissolve in seawater, whereas some of the organic components are degraded by bacteria and fungi. Many bacteria can di-

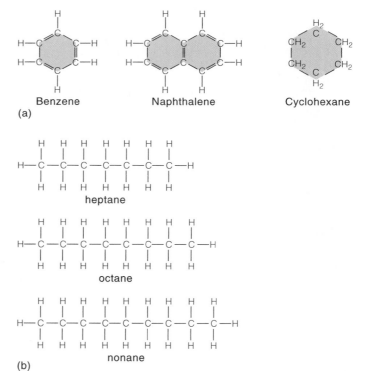

**Figure 20-8 Hydrocarbons.** Petroleum is a complex mixture of (a) aromatic hydrocarbons such as benzene, naphthalene, and cyclohexane, and (b) aliphatic hydrocarbons such as heptane, octane, and nonane.

**Figure 20-7 Medical Wastes.** Toxic and medical wastes, such as this syringe, make recreational beaches unfit for visitors and kill marine organisms.

gest the simpler organic molecules in oil but have difficulty digesting the more complex molecules. These persist in the sea for long times as tarry chunks floating on the surface or lying on the bottom. Little is known about the effect of oil on bottom-dwelling organisms of the deep. We do know that oil can suffocate organisms at the bottom of bays and harbors, and that some of the soluble components of oil are extremely toxic to many forms of marine life, especially eggs and young. Aromatic compounds are generally more toxic than aliphatic compounds, and the smaller the molecules, the more easily they can be taken up by marine organisms. Refined petroleum products such as gasoline and kerosene are even more toxic to marine life than the unrefined components of crude oil.

## Oil Spills

The largest oil spill in the United States occurred in March 1989 when the tanker *Exxon Valdez* ran onto a rocky reef 25 miles out of its home port of Valdez, Alaska. The accident tore five huge holes in the hull, and the ship spilled more than 240,000 barrels (10 million gallons) of oil into Alaska's pristine Prince William Sound (Figure 20-9a).

Several factors intensified the damage and hampered cleanup. Local plans for dealing with oil spills were not designed to handle a spill of this magnitude. Cleanup equipment and personnel were not actively in service, and there was a delay while these were assembled. Rugged terrain and tidal currents of the enclosed area also aggravated the problem. The spill quickly spread over 2,610 square kilometers (900 square miles) of Alaska's most picturesque and biologically productive coastal areas. As the oil moved out of the sound, currents carried it down the Alaskan coast, where it coated rocky shores and damaged marine and terrestrial habitats (see Figure 20-9b). Thousands of birds, fish, and marine mammals died. The cold temperatures of this subarctic region slow the natural decomposition of oil by photochemical reactions and bacteria. The marine life of the area also live longer and reproduce more slowly, so although the region is expected to recover, it will probably take longer than if it occurred in a temperate or tropical zone.

Not all oil spills are the result of shipping accidents. The largest and longest-lasting oil spill in the world occurred in June 1979 when an offshore oil well in the Gulf of Mexico, the *Ixtoc I,* owned and operated by a Mexican company, blew out and caught fire. It required almost a year to cap the well, during which time more than 137 million gallons of petroleum poured into the Gulf of Mexico. Similar events have occurred in the North Sea, Persian Gulf, and off the coasts of the United States. During the Iran–Iraq war, the bombing of oil facilities resulted in large amounts of oil entering the Persian Gulf. One bombed oil rig pumped 50,000 gallons per day into the Gulf for almost 3 months. Miles of beaches on the west side of the Gulf are still covered with a hardened surface of sun-baked oil. Current reports indicate that the oil spills killed 40% of the coral reefs off the coast of Qatar and 30% of the reefs off the Bahrain coast. During the first Gulf War, oil facilities sabotaged when the Iraqi army retreated from Kuwait released thousands of gallons of oil into the Persian Gulf around Kuwait. The damage from this ecological disaster is still being assessed.

Although large oil spills receive the most media attention, more oil and petroleum products enter the sea by way of runoff from urban areas and from spills and leakage that occur in loading and transfer of petroleum products.

## Ecological Effects of Oil Spills

When an oil spill occurs near a coastal area, the damage to the environment and marine life can be substantial. Millions of seabirds, such as cormorants and diving ducks, and thousands of marine mammals, especially seals and sea otters, have fallen victim to oil spills. The effects on invertebrates and algae are just as damaging and have far-reaching eco-

(a)

(b)

*Courtesy of the U.S. Coast Guard*

*Joel Bennett/Peter Arnold, Inc.*

**Figure 20-9  Oil Spills.** (a) Oil booms are used in an attempt to limit the spread of oil leaking from the damaged tanker *Exxon Valdez.* (b) Oil from the *Exxon Valdez* covers some of the rocky coast of Prince William Sound, Alaska, killing intertidal organisms.

logical consequences. The immediate damage from oil spills is quite obvious. The damage to the environment from gasoline, diesel fuel, and other more toxic petroleum products is more insidious. Many of these petroleum products evaporate and disperse quickly, and it is difficult to assess their immediate effects. But because they are constantly released into the sea, they cause chronic problems. Most petroleum transported by sea is offloaded or transferred at major ports and most major ports are located in the world's largest bays and estuaries, putting these delicate, productive, and valuable areas at risk of damage from oil and petroleum products. When oil spills occur in the open sea, the effects are difficult to assess. Because they occur far from land, they are difficult to monitor, and wind, waves, and currents dissipate the oil quickly. Presumably, the damage to marine life is the same as in coastal spills.

***Effects on Birds and Mammals***   A heavy coating of oil impairs a bird's ability to fly and swim. In addition, large outer feathers cannot repel water and down feathers cannot insulate. During cold weather, a spot of oil no bigger than a dime in a vital area can cause death by hypothermia (lowering the body temperature). As birds preen to remove oil from their feathers, they may ingest fatal amounts of oil. In otters, oil destroys the ability of the fur to insulate the animal, causing them to die of hypothermia (Figure 20-10). In otters and other mammals, the oil clogs ears and nostrils

**Figure 20-10 Oil-Related Injury.** This otter was a casualty of the *Exxon Valdez* oil spill. Even a small spot of oil on the feathers of a bird or the fur of a marine mammal can cause the animal to lose body heat and die.

and irritates eyes. Oil has also been shown to cause cancer in sea otters.

***Effects on Invertebrates and Algae***   Less visible, but just as deadly, are the effects on invertebrate, algal, and plant life along the shore. On sandy shores, sand crabs and other organisms that live in spaces between sand particles are killed by the toxic components of oil or are smothered by a coating of oil. Even after tides have washed the beaches clean, a residual layer of oil may persist several meters below the surface. Although some intertidal invertebrates, such as bivalve molluscs, seem to be resistant to oil pollution, their flesh becomes tainted with petroleum chemicals, which are passed along the food chain. On rocky shores, barnacles are fairly resistant to pollution, although in some instances they are smothered by the oil, but grazing molluscs like limpets, periwinkles, and the carnivorous whelks are vulnerable. Toxic components of the oil are a narcotic, causing molluscs to lose their hold on the rocks and be washed away by the tides.

***Community Effects***   The elimination of grazers allows seaweeds to colonize the vacated area. Eventually, the seaweeds become encrusted with oil and are torn away by the waves. The ultimate outcome of oil pollution in the intertidal zone is a decrease in species diversity, a simplification of the food web, and a disproportionate increase in the populations of resistant species (Figure 20-11).

## Oil Spill Cleanup

Current technology for dealing with ocean oil spills includes the use of oil booms and oil skimmers (Figure 20-12) that help to confine the spill to a smaller area and recover some of the oil. These techniques are most efficient when the spill occurs in an area that is easily accessible and where equipment can be rapidly mobilized. In a spill like the one involving the *Exxon Valdez*, where weather, sea conditions, rugged coastline, and distance from cleanup equipment are major factors, little of the oil is actually recovered. In some oil spills along an accessible coast, straw is used to soak up the oil. The oil-soaked straw is then collected by hand and disposed of by burning, which transfers the pollution to the air. Microbiologists have developed a genetically engineered bacterium capable of degrading most of the organic compounds present in crude oil. It is being tested to determine its environmental impact. Current legislation prevents the use of such organisms in nature. Naturally occurring bacteria living near oil seeps on the ocean floor are used with some success to clean up oil spills.

## In Summary

Almost from the beginning of human history, people have dumped wastes into the ocean. These include garbage, sewage, and industrial wastes. Currently there is a proposal to dump radioactive and toxic wastes into oceanic subduction zones. As populations in coastal areas have increased, so has the amount of wastes dumped into coastal seas. This indiscriminate dumping has decreased the

 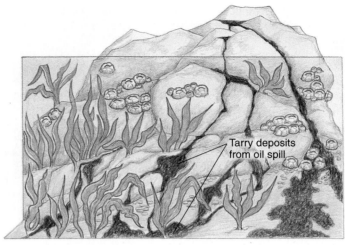

(a) Rocky shore before an oil spill                          (b) Rocky shore after an oil spill

Tarry deposits from oil spill

**Figure 20-11 Generalized Changes in Habitat That Result from Oil Pollution.** (a) A generalized view of a temperate zone rocky shore before an oil spill. (b) The same rocky shore after an oil spill. Oil kills molluscs, allowing seaweeds to overgrow the intertidal rocks. Wave action will eventually remove much of the algae, resulting in a decrease in species diversity.

**Figure 20-12 Oil Spill Cleanup.** Absorbent booms (red) are used to contain and clean up oil spills.

economic and recreational value of some beaches and in some instances has even posed health hazards.

Agricultural and urban runoff contains a variety of pesticides as well as human and industrial wastes. Some substances when incinerated produce gases that dissolve in water and enter the ocean when it rains. Some toxic materials can combine with bottom sediments, increasing the length of time that they remain in the environment and contaminate the water. Legislation and the use of less-toxic materials, such as unleaded fuel, are gradually decreasing the level of pollution in some regions.

Not all pollutants that enter the marine environment dissolve in the water. Some are absorbed by particles, consumed by animals, and channeled through food chains. As one animal feeds upon another, the level of some toxic materials in a food chain in-

creases to levels that can cause disease or death in higher-order consumers.

Plastic trash persists in the environment for long periods. Large marine animals such as birds, reptiles, and mammals are most affected by plastic trash. Plastic traps for catching fish, crabs, and lobsters continue to trap and kill animals even after they are lost or discarded.

Oil spills are another environmental problem. Crude oil contains a mixture of substances, many of which are toxic and can damage marine communities. Although protective measures prohibit spills and mandate cleanup, they are not always effective, and there have been numerous examples of severe environmental damage due to oil spills. ●

## HABITAT DESTRUCTION

The marine environment is also being damaged by the destruction of habitats. As more and more coastal land is developed for housing and recreation, more habitats are lost and so are the species they support. Many of the most vulnerable habitats are those close to shore that are nurseries for commercial species of fish and shellfish. Destruction of these areas not only threatens the survival of many ocean creatures but also aggravates an already bad situation by contributing to the decline in populations of commercially valuable species.

### Wetlands

**Wetlands,** such as salt marshes, provide nutrients, shelter, and spawning areas for a variety of marine organisms, including crabs, shrimp, oysters, and many commercial fishes. In the past, these areas have been drained, filled, and

**Figure 20-13 Wetland Destruction.** This shoreline development at Barnegat Bay, New Jersey, was once a productive wetland that was drained and filled to allow for real estate development.

dredged to provide more ground for industry, channels into ports and harbors for large vessels, and beachfront real estate for the growing number of individuals who desire to live and vacation near the sea (Figure 20-13). It is estimated that in 1776 the United States had between 125 and 215 million acres of wetlands. By 1975 the amount of wetlands had decreased to 99 million acres. It is estimated that 18,000 acres of wetlands were lost annually between 1950 and 1970. The loss of estuarine wetlands has been greatest in Florida, California, Texas, New Jersey, and Louisiana. By the twenty-first century, more than half of the coastal wetlands in the contiguous 48 states has been destroyed, and the other half is threatened as the demand for residential and recreational development along the coasts increases.

We now recognize that wetlands, including estuaries, are among the most productive of marine environments and that they play a major role as a nursery for a number of commercial fish and shellfish (for more information on the importance of estuaries as nurseries see Chapter 14). In the 1970s and 1980s, increased public awareness led to the passage of new laws protecting wetlands in many states and passage of more strict federal regulations. Legislation closely regulates any development or modification of wetlands. Whenever possible, these valuable areas are being carefully managed and restored to their natural state. If estuarine or wetland areas must be altered, the current policies demand that other, damaged areas be restored to replace the lost wetlands. Although this practice sounds good in principle, in the few instances where it has actually been tried it has generally been unsuccessful because of loopholes in the legislation. To further complicate matters, the federal government, in response to pressure from special interest groups, continues to change its definition of what constitutes a protected wetland and what it means to restore an area. Changes such as these severely weaken the laws that were meant to protect these resources. A successful exception is the current restoration of wetlands in the San Francisco Bay area. Environmentalists are cooperating with federal and local government as well as local industries to make their plan for restoring wetlands work. However, wetlands along the west coast are still under heavy pressure for industrial and residential development, and net loss of these valuable areas continues.

## Beaches

Coastal areas are popular vacation and retirement destinations. Travel brochures display pictures of beautiful, immaculate, sandy beaches backed by vigorous shoreline vegetation (Figure 20-14a). Few show the reality—overcrowded beaches backed by hotels, resorts, and shops (see Figure 20-14b). Development of the seashore for recreational and commercial use, combined with intensive seasonal use, has had a severe impact on intertidal areas.

(a)

(b)

**Figure 20-14 Beach Development.** (a) Travel brochures entice vacationers with pictures of beautiful sandy beaches and aquamarine water. (b) In reality, the beach is a small strip of sand crowded with tourists and surrounded by hotels, condominiums, shopping areas, and restaurants.

## Direct Effects of Beach Use and Development on Marine Life

The heavy usage of beaches has had serious effects on intertidal wildlife, especially those of the sandy shores. Beach-nesting birds like the piping plover (*Charadrius melodus*) and the least tern (*Sterna albifrons*) are in danger of extinction along the U.S. Gulf Coast because of disturbances by bathers, pets, and dune buggies. Other shorebirds are subjected to competition for nesting sites and egg predation from the growing populations of several species of large gulls. Gulls are highly tolerant of humans and can thrive on the garbage humans generate (for more information regarding gulls see Chapter 11). Populations of sea turtles and horseshoe crabs are also declining because of a loss of nesting sites on sandy beaches and predation by domestic animals. The construction of beachfront houses, docks, and seawalls disrupts habitats and makes them unavailable for nesting sites, reproduction, and feeding. Some coastal communities are setting aside stretches of beach for bird and turtle nests, levying fines for littering beaches, and participating in projects like the coastal cleanup mentioned at the beginning of this chapter.

## Destruction of Habitat

Development of beaches for recreational use frequently destroys sand dunes and beach vegetation that prevents beach erosion. The loss of vegetation and too much human traffic cause significant beach erosion, and ironically, vacation areas often spend hundreds of thousands of dollars each year to refortify beaches destroyed by the previous season's tourism.

## Interfering with Natural Processes

Human interference with natural processes can have a pronounced effect on beaches. When rivers are dammed to control floods or to produce power, sand and gravel that was once deposited on the coast is instead deposited in the lake behind the dam. The sand and gravel no longer replaces sediments removed by natural erosion processes.

Engineering projects such as breakwaters and jetties in coastal zones also produce changes in beaches. Breakwaters and jetties protect coastal areas from wave action. The areas behind these structures are quiet water. Sediments settle to the bottom in the quiet water rather than being carried by longshore currents to other beaches. **Longshore currents** are generated by waves that break at an angle to the beach. These currents move parallel to the shore in the surf zone carrying sediments, in what is known as the **longshore transport process**.

Historically, coastal engineering projects trigger chain reactions. Facilities constructed in one area create ecological problems in another that must be solved by other engineering projects, which create problems of their own, and so on. According to the U.S. Coastal Survey, 43% of our national shorelines, excluding Alaska, are losing more sediments than they receive. In other words, our beaches are disappearing into the sea. In some areas, eroding beaches are maintained by bringing in sand and gravel from inland areas or dredging it from offshore sandbars. These programs are expensive but necessary if the beaches are to be preserved.

## EPILOGUE

The basic mechanisms of evolution and ecology are inextricably interrelated, and human activity changes those interactions. The scope and pace of the changes caused by humans are very different from those of nature. Nature works by means of small changes over long times. Pollution, oil spills, habitat destruction, and other human changes frequently involve entire biological communities and sometimes occur instantaneously. The result is widespread disruption and damage of marine ecosystems. The environmental damage that results is significant not only because of the damage done this month or this year but because of the long-term changes in ecological relationships and disappearing niches that lead to long-term changes in evolutionary patterns and ultimately species extinction. Although all the various regions of the sea are unique, the knowledge that we gain from studying one area helps us to better understand others. An understanding of the underlying patterns and processes in the sea will help us to judge the impact of our actions and help us determine just how much an area can be modified without jeopardizing its environmental or economic value. Humans will continue to interact with the sea, especially along coastal areas, and problems caused by this interaction will continue to occur. We owe it to ourselves and our children to make the most intelligent and knowledgeable use of these resources. With proper care, these resources can renew themselves and continue to be a source of delight, amazement, and sustenance.

Although the picture of our oceans presented in this chapter is not a pretty one, the situation is not hopeless. Many groups and individuals are working hard to preserve our marine resources. If you would like to know more about current efforts to save the ocean and its creatures or how you as an individual can make a difference, here is a list of some organizations you can contact.

- Cousteau Society, 930 West 21st St., Norfolk, VA 23517. This group encourages protection and preservation of the oceans for future generations. Student volunteers are welcome to work at their Norfolk office. Students might wish to participate in Project Ocean Search, in which people from a variety of backgrounds spend 2 weeks studying an island ecosystem. http://www.cousteau.org/?sPlug=1

- Greenpeace USA, Inc., 1432 U St. NW, Suite 201-A, Washington, DC 20009. This international organization focuses on ocean ecology, hazardous wastes, disarmament, and atmospheric pollution issues. http://www.greenpeace.org/homepage/

- League of Conservation Voters, 1150 Connecticut Ave. NW, Suite 201, Washington, DC 20036. This is a nonpartisan political action group that works to elect

# Effects of Artificial Processes on Beach Formation

An example of the effects of interfering with the natural processes of a coastal zone can be seen in the Santa Barbara harbor project in California. In this section of the California coast, the longshore current transports sediment down the coast from north to south (Figure 20-A). To supply the needs of recreational boaters, a jetty and breakwater were constructed to form a boat harbor. The jetty is on the north side of the harbor and continues out to sea before turning south to form a breakwater. This structure not only shelters the area from waves but also blocks the longshore current. Thus more sand was deposited to the north of the structure than before the jetty was built, and the beach began to grow, whereas beaches south of the structure began to disappear. Eventually, the

beach to the north grew until it reached the seaward limit of the jetty and the longshore currents could again move sand to the south. Instead of carrying sand to more southerly beaches, however, the current deposited sand at the end of the breakwater, forming a sand spit that began to fill in

the harbor, and the southern beaches continued to erode. As a solution, a dredge pumps the sand deposited in the harbor back into the longshore current south of the harbor, which carries it to southern beaches. In this case, interference with a natural process resulted in the expenditure of large amounts of time, money, and energy to re-create a process that nature did for free. ●

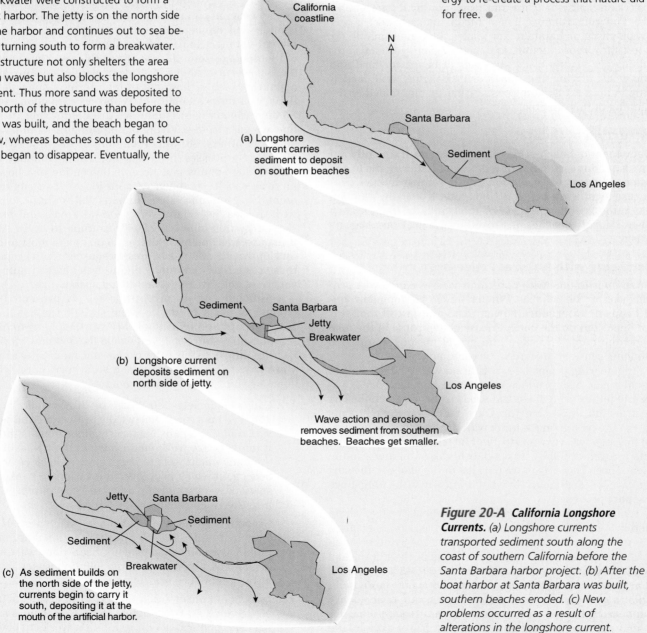

(a) Longshore current carries sediment to deposit on southern beaches

(b) Longshore current deposits sediment on north side of jetty.

Wave action and erosion removes sediment from southern beaches. Beaches get smaller.

(c) As sediment builds on the north side of the jetty, currents begin to carry it south, depositing it at the mouth of the artificial harbor.

*Figure 20-A  California Longshore Currents.* (a) Longshore currents transported sediment south along the coast of southern California before the Santa Barbara harbor project. (b) After the boat harbor at Santa Barbara was built, southern beaches eroded. (c) New problems occurred as a result of alterations in the longshore current.

proenvironment candidates to Congress. They also publish a scorecard that rates current members of Congress on their environmental votes. http://congress.nw.dc.us/lcv/dbq/officials

- National Wildlife Federation, 1400 16th St. NW, Washington, DC 20036. Conservation education is the primary mission of this group. http://www.nwf.org/

- The Nature Conservancy, 1815 North Lynn St., Arlington, VA 22209. This group fosters global preservation of natural diversity by finding, protecting, and maintaining the best examples of communities, ecosystems, and endangered species in the natural world. http://nature.org/

- Resources for the Future, 1616 P St. NW, Washington, DC 20036. This group is concerned with the quality of the environment and the conservation of natural resources. http://www.rff.org/

- World Wildlife Fund, 1250 24th St. NW, Washington, DC 20036. This group strives to save endangered species and acquire wildlife habitats. http://www.panda.org/

## In Summary

The heavy development of coastal regions has severely damaged habitats, especially wetlands and sandy beaches. Development physically disrupts habitat, and it increases siltation and the amount of sewage that enters the ocean. The large numbers of humans, along with their pets, that visit sandy beaches annually destroy habitats, kill marine organisms, and pollute the environment. Destruction of wetlands results in decreases in ocean productivity and in the populations of many important commercial species that rely on these areas as nursery grounds. ●

## SELECTED KEY TERMS

| | | | |
|---|---|---|---|
| aliphatic hydrocarbon, *p. 452* | biological magnification, *p. 449* | longshore current, *p. 457* | wetland, *p. 455* |
| aromatic hydrocarbon, *p. 452* | coliform bacteria, *p. 450* | longshore transport process, *p. 457* | |

## QUESTIONS FOR REVIEW

### Multiple Choice

1. An example of a persistent toxin would be
   a. fertilizer
   b. human waste
   c. DDT
   d. red-tide toxin
   e. lead

2. *Biological magnification* refers to
   a. the increase in size of organisms as trophic levels increase
   b. the increase in populations of algae in response to nutrients
   c. the change in population size from generation to generation
   d. the accumulation of toxins in the flesh of animals at higher trophic levels
   e. the accumulation of pollutants in the marine environment

3. Birds that are tolerant of humans and thrive on human garbage are
   a. pelicans
   b. lesser terns
   c. piper plovers
   d. gulls
   e. sandpipers

4. Sewage from beachfront cottages usually enters
   a. a sewer system
   b. septic tanks
   c. the sea directly
   d. drainage ditches
   e. compost heaps

5. Damage to _____ has the greatest impact on many commercial species of fish and shellfish.
   a. sand beaches
   b. rocky beaches
   c. wetlands
   d. the benthos
   e. the open sea

### Short Answer

1. What environmental problems are associated with plastic trash?

2. What are the major sources of oil pollution?

3. What effect does an oil spill have on the ecology of a rocky shore?

4. What activities are most damaging to wetlands?

5. Describe how toxins and other pollutants can enter ocean food chains.

6. List some of the major problems that are associated with the agricultural runoff that enters the ocean.

7. Describe how an oil spill causes injury to birds and mammals.

8. Describe how recreational and commercial use of beaches affects beach ecology.

9. What are some causes and consequences of eutrophication?

10. How might global warming affect marine ecosystems?

### Thinking Critically

1. Seashell collectors are frequently blamed for decreases in local mollusc populations in Florida. The collectors state that development is more to blame than overcollecting. Do you think that the collectors' argument is valid? Explain.

2. What portion of the east coast of the United States do you think would be more likely to experience algal blooms resulting from agricultural runoff?

3. An industrial waste is being dumped into the ocean off the coast of California, and you are asked to determine if it is accumulating in the aquatic food chains. How might you determine this?

## SUGGESTIONS FOR FURTHER READING

Brower, K. 1989. State of the Reef, *Audubon* 91(2):56–81.

Bruemmer, F. 1995. La Arribada, *Natural History* 104(8): 37–42.

Dolan, R., and H. Lins. 1987. Beaches and Barrier Islands, *Scientific American* 257(1):68–77.

Farrington, J. 1985. Oil Pollution: A Decade of Monitoring, *Oceanus* 29(3):69–77.

Hodgson, B. 1990. Alaska's Big Spill: Can the Wilderness Heal? *National Geographic* 177(1):4–43.

Ragotzkie, R., ed. 1983. *Man and the Marine Environment.* Boca Raton, Fla.: CRC Press.

Ward, F. 1990. Florida's Coral Reefs Are Imperiled, *National Geographic* 178(1):114–32.

Weisskopf, M. 1988. Plastics Reap a Grim Harvest in the Oceans of the World, *Smithsonian* 18(12):58–67.

### InfoTrac College Edition Articles

Eldredge, M. 1993. Stellwagen Bank: New England's First Sanctuary, *Oceanus* 36(3).

O'Reilly, K. 1997. Refuge on the Horizon, *Audubon* 99(5).

Snelgrove, P., and F. Grassle. 1995. What of the Deep Sea's Future Diversity? *Oceanus* 38(2).

Stap, D. 2002. Living on the Edge: With Wetlands Declining and Shorebirds in Trouble, the Question Is: How Do We Reclaim the Habitat the Birds Need to Survive? *Audubon* 104(2).

### Websites

**http://www.coastalcleanup.org/index.cfm** This site has information and statistics about the coastal cleanup effort.

**http://www.oceanconservancy.org** This site deals with ocean conservation issues.

**http://www.coastalstudies.org/** A nonprofit organization that deals with improving marine habitats.

**http://www.audubon.org/** This site lists the National Audubon Society's efforts in the area of marine conservation.

**http://www.alsnyc.org/** The American Littoral Society is involved with conservation of coastlines.

**http://www.cbf.org** The Chesapeake Bay Foundation is involved with conservation in the Maryland region.

**http://www.builderonline.com/article-builder.asp? channelID=59&ArticleType=1&ArticleID= 1000029347** This site is dedicated to conservation of wetlands.

**http://www.theurbandivers.com/elizabethriverproject .htm** This site deals with efforts to restore the Elizabeth River estuary in Virginia.

# Answers to Multiple Choice Questions

**Chapter 1**
1. d    2. b    3. b    4. c    5. c

**Chapter 2**
1. b    2. c    3. c    4. a    5. a    6. c    7. b
8. c    9. c    10. d

**Chapter 3**
1. d    2. d    3. d    4. c    5. b    6. a    7. e
8. d    9. e    10. a

**Chapter 4**
1. b    2. a    3. c    4. d    5. c    6. b    7. e
8. d    9. d    10. c

**Chapter 5**
1. d    2. c    3. b    4. d    5. a    6. c    7. d
8. d    9. e    10. d

**Chapter 6**
1. b    2. b    3. a    4. b    5. d    6. a    7. c
8. a    9. c    10. a

**Chapter 7**
1. a    2. c    3. a    4. d    5. c    6. e    7. b
8. c    9. d    10. e

**Chapter 8**
1. c    2. c    3. a    4. d    5. e    6. c    7. d
8. c    9. d    10. e

**Chapter 9**
1. e    2. b    3. e    4. b    5. d    6. c    7. b
8. b    9. c    10. d

**Chapter 10**
1. c    2. b    3. d    4. a    5. c    6. a    7. d
8. d    9. e    10. d

**Chapter 11**
1. c    2. b    3. d    4. e    5. c    6. b    7. e
8. c    9. d    10. d

**Chapter 12**
1. c    2. d    3. c    4. a    5. c    6. d    7. d
8. c    9. e    10. b

**Chapter 13**
1. c    2. b    3. e    4. a    5. d    6. d    7. c
8. b

**Chapter 14**
1. a    2. b    3. a    4. b    5. c    6. b    7. c
8. c    9. c    10. b

**Chapter 15**
1. a    2. b    3. b    4. e    5. a    6. c    7. e
8. b    9. d    10. d

**Chapter 16**
1. c    2. c    3. a    4. b    5. e    6. b    7. b
8. d    9. b    10. e

**Chapter 17**
1. b    2. c    3. d    4. a    5. d    6. c    7. e
8. a    9. c    10. b

**Chapter 18**
1. b    2. c    3. d    4. b    5. e    6. d    7. d
8. d

**Chapter 19**
1. d    2. d    3. c    4. c    5. b    6. d    7. d

**Chapter 20**
1. c    2. d    3. d    4. b    5. c

# Glossary

**abdomen**  A body region of an animal that corresponds to the belly. In arthropods this region is usually muscular and contains gills.

**abiotic environment (ay-by-AH-tik)**  The physical, or non-living, environment in which an organism lives.

**aboral surface**  The surface opposite the mouth of, for example, a sea star.

**abyssal hill (uh-BIS-suhl)**  Hills on the ocean floor formed by volcanic activity.

**abyssal plain**  A flat expanse at the bottom of an ocean basin.

**abyssal zone**  The ocean bottom that extends from a depth of 4,000 to 6,000 meters.

**acid**  Compound that releases hydrogen ions when added to water.

**adductor muscle (a-DUHK-tir)**  A large muscle that closes a bivalve's shell.

**adenosine triphosphate (ATP)**  The major energy-carrying molecule in cells.

**adhesion**  A property of a liquid by which the liquid is attracted to the surface of objects that carry an electrical charge.

**adsorption**  The process by which ions adhere to the surface of an object.

**aerenchyme (AIR-en-kime)**  A tissue in vascular plants that consists of spaces between the cell walls for carrying gases, usually to roots and stems in sediments that lack oxygen; also used to confer buoyancy.

**aerial root**  A root that is exposed to the atmosphere, not subterranean, such as drop roots, prop roots, and pneumatophores of mangroves.

**aerobic organism**  An organism that requires oxygen to survive.

**agar (AH-gar)**  An algal polysaccharide that forms thick gels at very low concentrations. It is resistant to degradation by most microorganisms and is used to culture bacteria in microbiology laboratories.

**ahermatypic coral**  A coral that lacks zooxanthellae in its tissues and does not form reefs.

**akinetic**  Descriptive of planktonic organisms that have no recognizable method of movement.

**algal bloom**  A dramatic increase in the population size of certain photosynthetic plankton usually associated with the water being enriched with nutrients.

**algal turf**  A dense carpet of low-growing seaweeds covering rock or sediment, similar to a closely cropped lawn.

**alginate (AL-gen-ate)**  A class of polysaccharides found in the cell walls of brown algae; confers flexibility and strength to the thallus; harvested for commercial use as a gelling agent.

**algologist (al-GAHL-o-jist)**  A scientist who studies seaweeds and phytoplankton; synonym, phycologist.

**aliphatic hydrocarbon**  Organic compound in which the carbon atoms form straight chains, such as octane and heptane.

**alkaline**  An adjective meaning basic, that is, the opposite of acidic.

**allantois (eh-LAN-toys)**  An embryonic support membrane found in some vertebrates that functions in the elimination of wastes.

**alleles**  Two or more forms of a single gene.

**allopatric speciation (al-oh-PAT-rik)**  The formation of new species that occurs when two or more populations or parts of a large population become geographically isolated from each other.

**alternation of generations**  The succession of two or more multicellular stages, usually sexual and asexual, in the life cycles of seaweeds, plants, and some animals.

**alveolate**  A group of microbes with membranous sacs (alveoli) beneath the cell membrane.

**alveoli (in animals)**  Tiny air sacs in the lungs of animals.

**alveoli (in alveolates)**  Membranous sacs beneath the cell membrane of a microbe belonging to the group Alveolata.

**ambergris (AM-ber-gris)**  A digestive by-product of sperm whales that is used in making perfume.

**ambulacral groove (am-byu-LAK-ruhl)**  A groove through which the tube feet of an echinoderm extend.

**amino acids**  Molecules that are the building blocks of proteins.

**amnesic shellfish poisoning**  A human syndrome caused by consumption of shellfish that concentrate the toxic chemicals produced by diatom blooms.

**amnion (AM-nee-uhn)**  A liquid-filled sac that contains the developing embryo of some vertebrate animals.

**amniotic egg (am-nee-AH-tik)**  An egg covered by a protective shell and containing a liquid-filled sac called the amnion.

**amphiblastula**  The planktonic larval stage of sponges.

**amphipod**  A type of crustacean with a body similar to a shrimp belonging to the order Amphipoda.

**ampulla**  An expanded area of a tubular structure. In

echinoderms a saclike structure attached to the tube foot.

**ampullae of Lorenzini (am-POOL-ee of loh-ren-ZEE-nee)** Organs in sharks and other cartilaginous fishes that can detect electrical signals in the water.

**anadromous (uh-NAD-ruh-muhs)** A term applied to fishes that reproduce in freshwater and spend their adult lives in the marine environment.

**anaerobic organism** An organism that cannot survive in an oxygen environment.

**anal fin** The ventral, unpaired fin of fish that lies behind the anus.

**anatomical isolation** The prevention of interbreeding between different species in nature due to incompatible copulatory organs.

**anchor root** A root that branches from a major root type to secure the plant in the sediment.

**animal** Multicellular organism that has cells lacking cell walls, that is an ingestive heterotroph, and that can usually move when adult.

**annelid (AN-eh-lid)** A segmented worm belonging to the phylum Annelida.

**Antarctic Circle** The latitude 66.5 degrees S of the equator, which marks the southernmost limit of sunlight at the June solstice.

**anthozoan (an-thuh-ZOH-uhn)** An animal belonging to the cnidarian class Anthozoa, which includes corals and sea anemones.

**aperture** An opening, such as the opening into a gastropod's shell.

**aphotic zone (ay-FOH-tik)** The portion of the pelagic division where sunlight is absent.

**appendicular swimmer** An organism that swims with legs, fins, wings, or other extensions from the body.

**aquaculture** The farming of freshwater or marine organisms.

**Archaea** One of the three domains of life. It contains prokaryotes that lack molecules known as peptidoglycans in their cell walls and can tolerate extreme environmental conditions.

**archaeocyte** Cells in a sponge's body that resemble amoebas and that can form any cell type in the sponge.

**archaeon** Member of the domain Archaea, distinguished from bacteria by its biochemical and genetic makeup and especially noted for its production of methane and tolerance of extreme environmental conditions.

**Arctic Circle** The latitude 66.5 degrees N of the equator, which marks the northernmost limit of sunlight at the December solstice.

**Aristotle's lantern** A chewing structure composed of five teeth found in the mouths of sea urchins.

**armored dinoflagellate** Dinoflagellate with thick layers of cellulose in its alveoli, giving the appearance of having a protective cell covering.

**aromatic hydrocarbon** Organic compound in which the carbon atoms form ring structures, such as benzene and naphthalene.

**arrowworm** A planktonic animal belonging to the phylum Chaetognatha.

**arthropod (AR-thruh-pahd)** An animal belonging to the phylum Arthropoda. Arthropods are characterized by jointed appendages and a hard exterior covering (exoskeleton).

**artificial selection** The process by which farmers and animal breeders select only animals and plants with certain desirable traits for breeding in an effort to produce more organisms with the same desirable traits.

**ascocarp** The fruiting (spore-producing) body of a member of the fungal group Ascomycota.

**asconoid (AS-kuh-noyd)** A type of sponge whose body has only a single spongocoel that does not contain invaginations.

**ascospores** Haploid cells produced within an ascocarp that disperse and become new adult fungi.

**ascus** One component of a fungal ascocarp producing four or eight spores.

**asexual reproduction** The process by which offspring are produced from a single parent without the fusion of sex cells.

**asthenosphere (as-THEN-uh-sfeer)** The region of mantle that lies below the earth's crust.

**atoll (a-TOHL)** A coral reef somewhat circular in shape with a centrally located lagoon.

**ATP** See *adenosine triphosphate*.

**atrium** One of the chambers of the vertebrate heart.

**autotomize** To purposely cast off a body part.

**autotroph (AW-toh-trohf)** An organism that is capable of producing its own food; also known as a *producer*.

**auxospore** A growth stage in the life cycle of a diatom that serves to restore the maximal size of the cell.

**bacillus** A bacterium with a cell shaped like a rod.

**back reef** The area opposite the reef front. Also known as the *reef flat*.

**backing down** A procedure used in tuna fishing to save dolphins in which a skiff draws a purse seine halfway toward the purse seiner and then backs up when the tuna are close to the boat and the dolphins are at the edge of the net, causing the edge of the net to go slack and drop beneath the surface of the water and allowing the dolphins to escape.

**backwash** Water flowing down the beach after a wave breaks.

**bacterial loop** A subcomponent of a food web in which bacteria convert dissolved and particulate organic matter into bacterial biomass, available for consumption by suspension and deposit feeders; particularly applied to bacteria that use DOM released by phytoplankton.

**bacteriochlorophyll** A class of primary photosynthetic pigments in purple and green photosynthetic bacteria.

**bacteriophage** A virus, often called simply *phage*, that infects a bacterium.

**bacterioplankton** Prokaryotic members of the plankton.

**bacteriorhodopsin** A class of light-capturing proteins that serve in ATP production in the Halobacteria.

**baleen (buh-leen)** A proteinaceous structure used to strain food from the water; it takes the place of teeth in baleen whales.

**barbel (BAHR-behl)** A fleshy projection on the head of

some fishes that may fulfill a sensory role or function as a lure.

**bar-built estuary** An estuary formed when islands or geographical barriers form a wall between freshwater from rivers and saltwater from the ocean.

**barnacle** A sessile crustacean whose body is usually covered by plates composed of calcium carbonate.

**barrier reef** A reef separated by a lagoon from the land mass with which it is associated.

**base** Compound that can remove hydrogen ions from solution.

**bathyal zone (BATH-ee-uhl)** The portion of the ocean bottom that extends from the edge of the continental shelf to a depth of 4,000 meters.

**bathygraphic features (bath-eh-GRAF-ik)** The physical features of the ocean bottom.

**bathymetric chart (bath-eh-MET-rik)** Chart of the ocean that shows lines connecting points of similar depth.

**bearers** Fish that carry their offspring with them, either internally or externally.

**behavioral isolation** The prevention of interbreeding between different species in nature due to differences in mating behaviors.

**the bends** A condition that occurs when nitrogen gas dissolved in the blood comes out of solution and forms gas bubbles as the pressure decreases during an ascent from deep water. The bubbles can interfere with the circulation of blood to body tissues causing tissue damage or even death.

**benthic division (BEN-thik)** The division of the ocean environment composed of the ocean bottom.

**benthic spawners** Fish that lay eggs that attach to some type of surface. There is no parental care after spawning and the larvae may become pelagic or remain benthic.

**beta-carotene** A carotenoid pigment that confers a yellow or orange hue to a cell.

**bilateral symmetry** A type of symmetry in which body parts are arranged in such a way that only one plane through the midline of the central axis (the midsagittal plane) divides the organism into similar right and left halves.

**binal virus** A virus composed of an icosahedral head and a helical tail.

**binary fission** A type of asexual reproduction in which one cell splits into two after the original cell has duplicated its genetic material.

**binomial nomenclature (by-NOH-mee-uhl NOH-men-klay-chur)** The system of using two words (that is, a genus-and-species epithet) to name an organism.

**biochemical isolation** The prevention of interbreeding between different species in nature due to molecular differences at the cellular level.

**bioerosion** The gradual wearing away of substrates such as coral reefs through the actions of living organisms.

**biogenous sediments (by-AH-gen-is)** Sediments formed from the remains of living organisms.

**biogeochemical cycles** A combination of biological, chemical, and physical processes that recycle nutrients within the biosphere.

**biological magnification** The concentration of pollutants or toxins in higher trophic levels of a food chain.

**biosphere (BY-OH-sfeer)** The collection of all of the earth's ecosystems together.

**biotic environment (by-AH-tik)** The living portion of an organism's environment.

**bivalve** A type of mollusc whose body is covered by a two-pieced, hinged shell.

**black zone** The lowest zone of the supralittoral fringe of tropical rocky shores. The characteristic color of the rocks is due to the presence of certain species of algae.

**bladder** In seaweeds, an expanded part of an algal thallus that contains gas and is used for buoyancy.

**blade** In seaweeds, the large, flat, leaflike part of the thallus. In vascular plants such as grasses, the outer photosynthetic part of a leaf, usually flattened and free of the stem (compare with *sheath*).

**bloom** A large increase in a population of phytoplankton.

**blowhole** A hole on the top of the head of a cetacean that serves as the opening to the animal's respiratory system.

**bosses** The large bumps on the snout of a humpback whale.

**bottom-up factors** Factors, such as the availability of nutrients, that affect the basal level of food chains and, thereby, the associated community.

**brachiopod (BRAK-ee-uh-pawd)** A type of generally benthic lophophorate with a bivalve shell belonging to the phylum Brachiopoda; it is commonly known as a lamp shell.

**breaching** A behavior thought to be a mating ritual of male humpback whales in which the animal jumps out of the water and comes crashing back down, creating a loud noise.

**breaker** A type of wave, the lower part of which is slowed by friction but whose crest continues to move toward the shore at a speed greater than that of the rest of the wave.

**broadcast spawner** Organism that directly releases eggs and sperm into open water and provides no parental care.

**brood hiders** Fish that hide their eggs but show no parental care after spawning.

**brooders** Organisms that broadcast their sperm into surrounding waters but the eggs are retained within the body cavity.

**brown alga** A seaweed of the phylum Phaeophyta, characterized by possession of chlorophylls *a* and *c* and the pigment fucoxanthin, which gives them their olive-brown color.

**bryozoan (bry-oh-ZOH-uhn)** Tiny lophophorates belonging to the phylum Bryozoa. They are commonly called moss animals and form colonies on a wide variety of solid surfaces.

**bubble net** A ring of bubbles produced by feeding humpback whales that is used to capture food.

**budding** A form of asexual reproduction in microbes and multicellular organisms in which unequal division of the adult produces two individuals.

**buffer** A substance that can maintain the pH of a solution at a relatively constant point.

**bump-and-bite attack** Shark attack mode in which the victim is bumped before being bitten.

**buttress zone** The seaward-sloping portion of a coral reef that consists of alternating ridges and furrows. Also known as a *spur-and-groove formation.*

**bycatch** The noncommercial animals killed during fishing for commercial species. Also known as *incidental catch.*

**byssal threads (BIS-suhl)** Strong protein fibers secreted by mussels that fasten the animal to rocks or another solid surface.

**byssus** A tough protein secreted by a gland in the foot of some bivalves that commonly takes the form of threads. It is used to attach the bivalve to a solid surface such as rock or coral.

**cable root** A subterranean root that arises from a stem or trunk, as in some mangroves, and extends horizontally through the sediment.

**calcareous ooze** Sediment composed of 30% or more of the remains of organisms that produce shells made of calcium carbonate.

**calymma** The vacuolated outermost cytoplasm of a radiolarian.

**canopy** The uppermost layer of a kelp forest.

**capillary action** The ability of water to rise in narrow spaces.

**capillary wave** A wave for which the restoring force is the surface tension of the water.

**capsid** The outer coating of proteins of a mature virus (or virion).

**capsule** In radiolarians, an external organic layer that separates the inner nuclear region from the calymma.

**carapace** The hard dorsal covering of some animals' bodies, such as arthropods or turtles.

**carbohydrate** An organic molecule composed of the elements carbon, hydrogen, and oxygen, frequently in a ratio of $1:2:1$.

**carbon fixation** The process of converting inorganic carbon into an organic form.

**carnivore** An animal that feeds on other animals.

**carotenoids (kuh-RAHT-in-noydz)** Yellow and orange accessory pigments that absorb green light and function in photosynthesis.

**carpospore** A nonmotile, diploid, dispersal stage released by the carposporophyte of a red alga.

**carposporophyte** A multicellular stage in the life cycle of red algae having a diploid thallus arising from the zygote, living parasitically on the female gametophyte, and producing carpospores for dispersal.

**carrageenan (kar-uh-GEE-nuhn)** An algal polysaccharide from red algae that is used commercially as a thickening and binding agent.

**casting** A pile of organic material and minerals defecated by deposit feeders. Also known as *fecal cast.*

**catadromous (kuh-TAD-ruh-muhs)** A term applied to fishes that reproduce in the marine environment and spend their adult lives in freshwater.

**caudal fin (KAW-duhl)** The tail fin of a fish.

**cell** The basic unit of all living organisms.

**cell wall** A tough, rigid structure that surrounds the cell membrane of most cells.

**cellular respiration** The process of producing ATP by reacting organic food molecules with oxygen.

**cellulose** A polysaccharide found in the structural components of plants and algae and in the cell walls of many organisms.

**centrifuge plankton** Components of the plankton that are so small that they are not retained by standard plankton nets; best sampled by centrifugation of water samples; composed mainly of nanoplankton, picoplankton, and femtoplankton.

**cephalization (sef-uh-luh-ZAY-shuhn)** The concentration of sensory organs in the head region of an animal.

**cephalochordate** An animal belonging to the subphylum Cephalochordata. Also known as a *lancelet.*

**cephalopod (SEF-uh-loh-pahd)** An animal belonging to the molluscan class Cephalopoda, which includes squids, octopods, cuttlefishes, and nautiloids.

**cephalothorax (sef-uh-loh-THOR-aks)** The combination of two body regions: the head and the thorax.

**cerata (sir-AH-tuh)** Projections on the body of a nudibranch that function in gas exchange.

**cetacean (seh-TAY-shen)** An animal belonging to the mammalian order Cetacea, which includes whales, dolphins, and porpoises.

**chart** A representation of the oceans and their features.

**chelicera (keh-LI-suh-ruh)** An appendage in chelicerates that is modified for feeding and takes the place of mouthparts.

**chelipeds (KEE-leh-pehdz)** The first pair of legs of many decapods; these two are modified to form claws used for capturing prey and for defense.

**chemoreceptor (kee-moh-ree-SEP-tuhr)** A sense organ capable of detecting changes in the chemical composition of an organism's environment.

**chemosynthetic (kee-moh-sin-THEH-tik)** Having the ability to use the energy from chemical reactions to construct organic food molecules.

**chemosynthetic bacteria** Bacteria that are able to form organic molecules from inorganic molecules using other chemicals rather than sunlight as a source of energy.

**chitin (KY-tin)** A polysaccharide found in the cell walls of fungi and in the exoskeletons of many arthropods.

**chiton (KY-tuhn)** A mollusc belonging to the class Polyplacophora. Chitons are characterized by a shell composed of eight separate plates bound together by a leathery girdle.

**chloride cells** Specialized cells located on marine fish gills that remove excess electrolytes from the body.

**chlorophyll** A green pigment that functions in photosynthesis by absorbing light energy.

**chlorophyll *a*** A primary photosynthetic pigment of most autotrophic microbes, algae, and plants. It absorbs primarily violet and orange light.

**chlorophyll *b*** A primary photosynthetic pigment in few microbes, in green algae, and in all plants. It absorbs primarily blue and orange light.

**chlorophyll *c*** A primary photosynthetic pigment found

in ochrophytes that is molecularly unrelated to true chlorophylls.

**chloroplast**  Organelle found in eukaryotic photosynthetic organisms that contains the pigments necessary for photosynthesis.

**choanocyte (koh-AN-oh-syt)**  A flagellated cell that lines cavities within the body of a sponge. Choanocytes are responsible for moving water through the body of a sponge. Also known as a *collar cell*.

**choanoflagellate**  A group of microbes, similar to the choanocytes of sponges, that filter suspended particles through a collar of microvilli.

**chorion (KOR-ee-uhn)**  An embryonic support membrane found in some vertebrates that functions in gas exchange.

**chromatic adaptation**  The ability of photosynthetic organisms to alter the kind and quantity of photosynthetic pigments in response to changes in the wavelengths and intensity of sunlight. The term *chromatic adaptation* was also used to refer to the now-rejected concept that the apparent vertical zonation of seaweeds can be explained by the way that seawater absorbs wavelengths of light with depth and by the variation in their complement of photosynthetic pigments, which differ in the optimal wavelengths of light used for photosynthesis.

**chromatophore (kroh-MAT-uh-fohr)**  A special cell in an animal's skin that contains pigment molecules.

**chromosome (KROH-muh-sohm)**  Structure consisting of DNA and protein located in the nucleus of eukaryotic cells.

**chronometer (kroh-NAHM-i-ter)**  A clock that keeps accurate time and can be used to determine longitude.

**ciguatera fish poisoning**  A human syndrome caused by consumption of fish that concentrate toxic chemicals produced by zooxanthellae.

**cilia (SIL-ee-uh)**  Hairlike structures found on some cells that function in movement and feeding.

**ciliate**  A group of alveolate microbes characterized by the presence of cilia for locomotion and feeding.

**cingulum**  The groove around the middle of a dinoflagellate that holds one of its two flagella.

**cirri**  Grasping structures found on some species of sea lily.

**cirriped (SER-uh-ped)**  Feathery appendage used by barnacles to filter food from the surrounding water.

**claspers**  Modified pelvic fins of male sharks that transfer sperm to the genital opening of a female.

**cleaning station**  A territory established by one of several cleaner organisms that fishes visit at regular intervals to have parasites and dead tissue removed.

**cnida**  The stinging organelle in the cnidocyte.

**cnidarian**  An animal that belongs to the phylum Cnidaria, which includes hydroids, jellyfish, and corals.

**cnidocil (NYD-uh-sil)**  A bristlelike structure that extends from one end of a cnidocyte and functions as a trigger.

**cnidocyte (NYD-uh-syt)**  The stinging cell found in members of the phylum Cnidaria.

**coastal plain estuary**  An estuary formed between glacial periods, when water from melting glaciers raises the sea level and floods coastal plains and low-lying rivers. Also known as a *drowned river valley estuary*.

**coccolithophore (kahk-oh-LITH-oh-for)**  Photosynthetic protist that belongs to the phylum Chrysophyta. Coccolithophores are covered by calcareous plates.

**coccoliths**  Calcareous plates that cover the surface of phytoplankton called coccolithophores.

**coccus**  A bacterium with a cell shaped like a sphere.

**coenocytic (see-nuh-SIT-ik)**  The condition of having a body consisting of one giant cell or a few large cells containing more than one nucleus.

**coleoid**  A cephalopod belonging to the subclass Coleoidea that does not have an external shell.

**coliform bacteria**  Bacteria in the intestines of animals. Their number in a body of water indicates the amount of animal wastes entering it.

**collar cell**  A flagellated cell, many of which line the cavities within sponges. They are responsible for producing the water current that flows through a sponge's body. Also known as a *choanocyte*.

**colony**  A group of individuals that are physically connected and adapted to share resources such as food.

**commensalism**  A symbiotic relationship in which one organism benefits while the other is neither harmed nor benefited.

**community**  An assembly of populations of different species that occupy the same habitat at the same time.

**compensation depth**  The depth in the sea at which a primary producer can photosynthesize only enough to balance metabolism and not be able to grow; usually dependent on latitude, season, sea surface conditions, water clarity, and kind of primary producer.

**competition model**  A hypothesis that each fish species inhabiting a coral reef has a unique niche and there is no competition between them.

**competitive exclusion**  The process by which the less successful competitor for a limited resource is driven to extinction.

**competitive exclusion principle**  The scientific principle that holds that the less successful competitor for a limited resource is driven to extinction.

**conchiolin (kahn-KY-uh-lin; kahn-CHEE-uh-lin)**  The protein that makes up the periostracum of a molluscan shell.

**conidiospore**  Asexually produced dispersal stage in the life cycle of a fungus.

**conjugation**  A kind of sexual reproduction that involves the exchange of nuclei between two fused cells.

**consolidation**  The formation of large aggregates by chemical or electrostatic attachment of particles to one another, as in the binding of loose sediments on a reef by seaweeds or the aggregation of suspended particles in seawater by bacteria.

**consumer**  An organism that relies on another organism for food. Also known as a *heterotroph*.

**continental drift**  The movement of continental masses as the result of seafloor spreading.

**continental margin**  The region of ocean that lies beneath the neritic zone. It is composed of the continental shelf and the continental slope.

**continental rise**  A gentle slope at the base of a steep continental slope.

**continental shelf**  The edge of a continental landmass.

**continental slope** The transition between the continental shelf and the deep floor of the ocean.

**control set** The trial set in an experiment that does not contain the experimental variable.

**convergent evolution** An evolutionary process in which similar selective pressures bring about similar adaptations in unrelated groups of animals.

**convergent plate boundary** Regions of the ocean floor where the boundaries of lithospheric plates move toward each other and old lithosphere is destroyed.

**copepod** A crustacean belonging to the class Copepoda, usually the most abundant member of the marine zooplankton.

**copulatory pleopod** Specialized appendage on the abdomens of male decapods used for transferring sperm to females.

**coral bleaching** The whitening of coral colonies due to the loss of zooxanthellae from their polyps. Coral bleaching may be caused by such environmental stresses as increased temperature, sedimentation, and pollution.

**coralline red alga** A member of a group of red algae that deposit calcium carbonate in their cell walls; named for their resemblance to coral animals.

**corallite** The calcareous skeleton of a coral polyp.

**cordgrass** A plant in the genus *Spartina*. It is the dominant vegetation in North American salt marsh communities.

**Coriolis effect (kohr-ee-OH-lis)** The apparent deflection of the path of winds and ocean currents that results from the rotation of the earth.

**cosmogenous sediment** Sediments formed from the dust and debris from outer space.

**countercurrent multiplier system** A biological mechanism involving fluids, with a concentration gradient between them, flowing in opposite directions in parallel channels. This arrangement maintains a stable gradient that allows maximum exchange of heat or dissolved substances between the two fluids.

**countershading** A camouflage pattern in which the upper surface of pelagic animals is dark and the lower surface light to blend in with the ocean depths or the surface, respectively, when observed by a potential predator. Also known as *obliterative countershading*.

**crinoid (KRY-noyd)** An animal belonging to the echinoderm class Crinoidea. These generally sessile animals have bodies that resemble flowers and are commonly referred to as sea lilies.

**crop** A digestive organ found in some animals that stores food before it is processed.

**crossing over** The exchange of genetic material between adjacent chromosomes that occurs during meiosis.

**crustacean** An animal belonging to the arthropod class Crustacea, which includes crabs, lobsters, shrimp, and barnacles.

**cryptic coloration** Color patterns that camouflage animals, allowing them to blend with their environment to avoid predators or ambush prey.

**ctenoid scales (TEE-noyd)** Thin, overlapping scales on the more advanced bony fishes. Their exposed posterior edges have small spines.

**ctenophore (TEEN-uh-fohr)** An animal belonging to the phylum Ctenophora. Also known as a comb jelly.

**culm** The specialized kind of vertical stem typical of grasses, sedges, and rushes.

**cuticle** One of several terms used for a layer of an organism outside the most superficial layer of cells, or the outermost layer of a cell wall or exoskeleton of some animals, plants, and seaweeds.

**Cuvierian tubule** Small, sticky tube expelled from the anus of some sea cucumbers to distract predators.

**cyanobacteria (sy-AN-oh-bak-TEER-ee-uh)** A group of photosynthetic prokaryotes that contain chlorophyll *a* and that release oxygen as a by-product of their photosynthesis. Cyanobacteria are sometimes referred to as blue-green bacteria.

**cyanophycean starch** The molecular form in which carbohydrates are stored in cyanobacteria.

**cycloid scales (SY-kloyd)** Thin, overlapping scales with a smooth posterior edge that are found in some of the more primitive bony fishes.

**cydippid larva** The planktonic larva of ctenophores.

**cyprid larva** A planktonic larval stage of barnacles that develops from a nauplius larva and has compound eyes and a carapace composed of two shell plates.

**cytokinesis** The process of dividing a cell into two daughter cells.

**cytoplasm** The combination of cytosol and organelles.

**cytosol** The fluid contents of a cell.

**cytostome** An organelle in a eukaryotic microbe that serves as the site for ingestion of food.

**decapod (DEK-uh-pahd)** An animal with five pairs of walking legs that belongs to the arthropod order Decapoda, which includes crabs, lobsters, and shrimp.

**decomposer** An organism that breaks down the tissue of dead plants and animals, as well as animal wastes.

**deductive reasoning** A process of reasoning in which observations suggest a general principle from which a specific statement can be derived.

**deep scattering layer** A mixed group of zooplankton and fishes that causes sonar systems to generate a false image of a nearly solid surface hanging in midwater.

**deep-sea vent community** A community of marine organisms that depend on the specialized environment found at divergence zones in the ocean floor. Also known as *rift community*.

**deepwater wave** A wave that occurs in water that is deeper than one half the wave's wavelength.

**delphinid (del-FI-nid)** A collective term referring to dolphins and porpoises.

**density** The mass of a substance in a given volume of that substance.

**dental plates** Feeding apparatus of hagfish consisting of plates containing curved rows of sharp, horny cusps. The plates are hinged along the midline, opening and closing like a book.

**deoxyribonucleic acid (DNA)** A double-stranded molecule composed of nucleotides with the shape of a helix that contains an organism's genetic information.

**deposit feeder** An animal that feeds on bottom sediments.

**desalination (dee-sal-uh-NAY-shun)**  The process of removing salt from seawater.

**desiccation (dehs-ik-KAY-shun)**  The process of drying out or losing moisture.

**detritivore (deh-TRY-ti-vor)**  An organism that feeds on detritus.

**detritus (deh-TRY-tuhs)**  Organic matter such as animal wastes and bits of decaying tissue.

**diarrhetic shellfish poisoning**  A human syndrome caused by consumption of shellfish that concentrate toxic chemicals produced by dinoflagellate blooms.

**diatomaceous earth**  An industrial material harvested from exposed deposits of dead diatom frustules and used for filtration and other applications.

**dinoflagellate**  An alveolate microbe with cellulose in its alveoli and with two heterokont flagella for locomotion.

**dinosporin**  A chemical in the cellulose plates of dinoflagellates that increases their resistance to decay.

**diploid number (2N)**  The number of paired chromosomes in a cell's nucleus.

**disaccharide**  Carbohydrate that contains two bonded monosaccharides.

**discrimination click**  High-frequency click used by toothed whales, such as dolphins, to get a precise picture of an object.

**disphotic zone**  The region of ocean between 150 and 450 meters deep where there is not enough light to power photosynthesis. Also known as the *twilight zone*.

**disruptive coloration**  A type of coloration found in some animals in which the background color of the body is broken up by lines that frequently run in the vertical direction.

**dissipative beach**  Type of beach on which wave energy is strong but is dissipated in a surf zone some distance from the beach face.

**dissolved organic matter (DOM)**  Organic molecules dissolved in seawater that were lost or released from living organisms.

**diurnal tide (dy-YUR-nuhl)**  The condition of having only one high tide and one low tide each day.

**divergent plate boundary**  Region of the ocean floor where the boundaries of lithospheric plates move apart as new lithosphere is formed.

**DNA**  See *deoxyribonucleic acid*.

**doldrums**  The area of rising air at the equator.

**DOM**  See *dissolved organic matter*.

**domain**  The taxonomic category that is higher than a kingdom.

**dorsal fin**  A fin on the dorsal surface of a fish.

**downwelling**  The vertical movement of water in a downward direction.

**drift alga**  A seaweed that breaks free from its holdfast and is carried by current or wind, sometimes accumulating in vast beds in quiet waters of bays and lagoons.

**drift net**  Large net composed of sections called tans that may stretch as far as 60 kilometers. Drift nets entangle fish, squid, and other marine animals that swim into them.

**drop root**  An aerial root of a mangrove that arises from a branch and descends to the sediment.

**drop-off**  The vertical wall that is sometimes formed on a reef front.

**drowned river valley estuary**  An estuary formed between glacial periods, when water from melting glaciers raises the sea level and floods coastal plains and low-lying rivers. Also known as a *coastal plain estuary*.

**eastern-boundary current**  Current that flows along the eastern edge of an ocean basin carrying cold water toward the equator.

**ebb tide**  A falling tide.

**echinoderm (eh-KY-noh-derm)**  An animal belonging to the phylum Echinodermata, which includes sea stars, brittle stars, sea urchins, sea cucumbers, and crinoids.

**echinoid**  An echinoderm belonging to the class Echinoidea, which includes sea urchins, heart urchins, sand dollars, and sea biscuits.

**echiuran**  An animal that belongs to the class Echiura in the phylum Annelida. Also known as a spoonworm.

**echolocation**  A process that allows some cetaceans to use sound waves to distinguish and home in on objects from distances of several meters.

**ecological efficiency**  The percentage of energy that is taken in as food by one trophic level and then passed on as food to the next highest trophic level.

**ecological equivalent**  One of two different groups of animal that have evolved independently along the same lines in similar habitats and that therefore display similar adaptations.

**ecosystem**  A system that is composed of living organisms and their nonliving environment.

**ectoparasite**  A parasite that attaches to the outer covering of its host.

**ectotherm**  An animal that obtains most of its body heat from its surroundings.

**Ekman spiral**  The spiral flow of water that results from the movement of water at the surface moving deeper layers of water due to friction. The movement of the successive deeper layers of water are deflected by the Coriolis effect, producing a downward spiral of water to a depth of about 100 meters.

**Ekman transport**  The net movement of water to the 100-meter depth by the Ekman spiral.

**El Niño Southern Oscillation (ENSO)**  Changes in oceanic and atmospheric current patterns that move warm Pacific Ocean surface water farther east than normal, resulting in worldwide changes in weather.

**elasmobranch**  Group of cartilaginous fishes (class Chondrichthyes) containing the sharks, rays, and skates.

**embayment**  Area of coastline where portions of the ocean are cut off from the rest of the sea.

**endoplasmic reticulum (ER)**  A series of membranes that wind through the cytoplasm of eukaryotic cells.

**endoskeleton (EN-doh-SKEL-eh-tuhn)**  An internal skeleton.

**endotherm (EN-doh-therm)**  An animal that maintains a constant body temperature by generating heat internally.

**energy pyramid**  A diagram that pictures the energy flow from one trophic level to the next. The diagram is shaped like a pyramid to indicate that there is successively less energy at each higher level.

**envelope**  An outer covering of a virus (or virion) derived usually from the cell or nuclear membrane of the host.

**enzyme** A biological catalyst that increases the rate of the chemical reactions of metabolism allowing them to be more efficient.

**epibenthic organism** An organism that lives on the surface of hard sediments.

**epidermis (ehp-i-DER-mis)** An outer layer of cells or the cellular covering of an organism.

**epifauna (EP-i-faw-nuh)** Benthic organisms that live on the ocean's bottom.

**epipelagic zone** The waters over the continental shelf or to 200 meters deep in the open sea occupied by zooplankton and nekton.

**epiphyte** Any organism that grows on a multicellular primary producer (that is, on a plant in the broadest sense).

**epitoke** The pelagic, reproductive individual formed by some errant polychaetes.

**epizoic** Any organism that grows on an animal.

**equator** A circle drawn around the center of the earth that is perpendicular to its axis of rotation.

**equatorial upwelling** The upwelling of water along the equator caused by water on either side of the equator being deflected toward the poles by the Coriolis effect. This pulls surface water away from the equator that is replaced by deeper water, producing an upwelling.

**errant polychaete (ER-ent PAHL-eh-keet)** An actively mobile polychaete that has a mouth equipped with jaws or teeth.

**escarpment** A nearly continuous line of cliffs with sharp vertical drops formed by the movement of lithospheric plates along faults.

**estuary** A region where freshwater is mixed with saltwater.

**Eubacteria** The domain of life that contains prokaryotes with peptidoglycans in their cell walls.

**Eukarya** The domain of life that contains all of the eukaryotes.

**eukaryote (yoo-KAR-ee-oht)** An organism whose cells contain a nucleus and membrane-bound organelles.

**eukaryotic cells** Cells that contain a nucleus and membrane-bound organelles.

**euryhaline** Able to tolerate a broad range of salinities.

**eutrophication** Nutrient enrichment of an aquatic system that results in rapid algal growth.

**evaporites** Salt deposits produced when saltwater evaporates.

**eviscerate** The release of internal organs from the mouth or anus. This behavior is used by sea cucumbers to deter predation.

**evolution** The process by which populations of organisms change over time.

**exclusive economic zone (EEZ)** Area of ocean exclusively controlled by a coastal nation.

**exhalant opening** The opening of a bivalve's or tunicate's siphon through which water is expelled from the body.

**exoenzyme** Enzyme released from osmotrophic microbes for external digestion.

**exoskeleton (EK-soh-SKEL-eh-tuhn)** A hard protective exterior skeleton such as that found in arthropods.

**experimental set** The trial set in an experiment that contains the experimental variable.

**experimental variable** In an experiment, the factor that is altered in the experimental group but not in the control group.

**exponential growth** A pattern of population growth characterized by initial slow growth followed by increasingly rapid growth; when graphed the curve is typically a J-shape. Also known as *logarithmic growth*.

**facultative anaerobe** Organism that thrives in the presence or absence of oxygen.

**facultative halophyte** A plant that will thrive in the presence or absence of salt.

**fault** An area where the earth's plates move past each other.

**fecal cast** A pile of organic material and minerals defecated by deposit feeders. Also known as *casting*.

**feeding polyp** A type of polyp found in hydrozoan colonies that functions in capturing food and feeding the colony. Also known as a *gastrozooid*.

**femtoplankton** Plankton that range in size from 0.02 to 0.2 micrometers, composed primarily of viruses.

**fertilization** The process in which a male sex cell or gamete fuses with a female sex cell or gamete to form a zygote.

**fertilizin** A chemical released by some female epitokes that stimulate males to shed their sperm.

**filamentous fungi** Fungi that form mycelia of long thread-like masses.

**filter-feeder** An organism that filters its food from the water.

**first-order consumer** A consumer that feeds directly on a producer. Also known as a *primary consumer*.

**fishing effort** A measure of the number of boats fishing, the number of fishers working, and the number of hours that they spend fishing.

**fission** A type of asexual reproduction in which an adult organism literally tears itself into two pieces that eventually form two individuals.

**fitness** An organism's biological success as measured by the number of its own genes that are present in the next generation of a population.

**fjord (fyord)** A deep valley cut into the coastline by glaciers and filled with a mixture of freshwater and saltwater.

**flagellum (plural, flagella)** Long hairlike structure that functions in cellular locomotion.

**flood tide** A rising tide.

**Florida Bay water hypothesis** A hypothesis that the reduction of water flow into Florida Bay from the Everglades due to human activities has led to hypersaline conditions and caused massive losses of seagrass beds.

**flowering plant** A vascular plant that produces seeds in a fruit.

**fluke** A type of parasitic flatworm belonging to the class Trematoda.

**foliage leaf** A leaf that produces a photosynthetic blade.

**food chain** A sequence of feeding relationships among a group of organisms that begins with producers and proceeds in a linear fashion to higher-level consumers.

**food web** A representation of the complex feeding networks that exist in an ecosystem.

**forced wave** A wave produced by storms at sea.

**forereef**  The outer, seaward margin of a coral reef. Also known as the *reef front*.

**fossil fuel**  Fuel—including coal, oil, and natural gas— formed from the remains of plants and microorganisms that lived millions of years ago.

**fouling community**  An assembly of populations of many kinds of organisms growing on an intertidal or submerged artificial structure.

**fracture zone**  Linear region of unusually irregular ocean bottom that runs perpendicular to ocean ridges.

**fragmentation**  A type of asexual reproduction in which a part of the parent organism breaks off and forms a new individual.

**free wave**  A wave that is no longer affected by the energy from a generating force.

**frictional drag**  Resistance to movement that is generated by the passage of the medium (for example, seawater) over the surface of an object.

**fringing reef**  A reef close to and surrounding newer volcanic islands or that borders continental landmasses.

**frond**  The blade of a large seaweed. Also known as *blade*.

**fruit**  A flowering plant structure formed after pollination from various parts of the flower and the embryo composed of several layers of tissue that surround the seed protecting it and aiding in its dispersal.

**fruiting body**  The general term for a sexually reproductive structure of a fungus, including the stalk and cap of a mushroom.

**frustule (FRUHS-tyool)**  The glassy structure composed of silica that covers diatom cells.

**fucoxanthin**  An accessory photosynthetic pigment in ochrophytes that is brown to golden brown in color.

**fungus (plural, fungi)**  An organism that belongs to the kingdom Fungi. Fungi are eukaryotic, have cell walls containing chitin, and are not photosynthetic.

**fusiform body shape**  Spindle-shaped body of many fishes and other animals characterized by a rounded middle portion and tapering toward each end.

**gametangium**  The part of an algal thallus that produces gametes.

**gamete (GAM-eet)**  A sex cell, such as a sperm, egg, or pollen grain.

**gametophyte (ga-MEE-toh-fyt)**  The stage in the life cycle of an alga or plant during which gametes are produced.

**ganoid scales (GAN-oyd)**  Thick bony scales that do not overlap. Ganoid scales are found in some primitive bony fishes.

**gas gland**  A gland that secretes gas into the swim bladder of fishes or into the float of some large pelagic cnidarians.

**gastropod (GAS-troh-pahd)**  A member of the molluscan class Gastropoda, which includes snails, limpets, abalones, and nudibranchs.

**gastrovascular cavity (gas-troh-VAS-kyoo-ler)**  A central cavity in the body of cnidarians and flatworms that functions in digestion and in the movement of materials within the animal.

**gelatinous zooplankton**  Components of the plankton that have a high content of body water.

**gene pool**  All of the genes in a given population that exist at a given time.

**gene**  A unit of hereditary information located on an organism's DNA.

**generating force**  A force that disturbs the surface of water, producing a wave.

**genetics**  The study of heredity.

**genus (JEE-nuhs)**  The first name in the two-part Latin name for an organism. It is the taxonomic group that is one step above the species level.

**gill filaments**  Thin, rodlike structures that make up the gills found in some animals.

**global positioning system**  A system of satellites that can be used to find an exact position anywhere on earth.

**global warming**  An increase in the earth's average temperature caused by the accumulation of greenhouse gases in the atmosphere.

**globigerina ooze**  A seafloor sediment consisting of the accumulated calcareous tests of foraminiferans.

**glucose**  A monosaccharide that is the basic fuel for all living cells.

**glycogen**  A polysaccharide found in some organisms that serves to store glucose reserves.

**gnathopods**  Specialized appendages found on amphipods that are used for collecting food.

**Golgi apparatus**  Organelle that functions in modifying molecules and forming membrane-bound sacs around them.

**Gondwanaland**  The southern portion of the supercontinent Pangaea composed of what became India, Africa, South America, Australia, and Antarctica.

**grana (singular, granum)**  A stack of thylakoids within a chloroplast.

**grasping spine**  A structure on the head of an arrowworm used to capture prey.

**gravity wave**  A wave for which the restoring force is gravity.

**gray zone**  The middle zone of the supralittoral fringe of tropical rocky shores. This zone is the farthest zone from the low tide line where macroscopic marine algae grow.

**green alga**  A seaweed of the phylum Chlorophyta, characterized by possession of chlorophylls *a* and *b*, which give them their green color.

**greenhouse gas**  A gas, such as carbon dioxide, methane, or a chlorofluorocarbon, that collects in the atmosphere and prevents heat energy from radiating back to space.

**guano (GWAH-noh)**  A phosphate-rich manure produced by birds.

**guarders**  Fish that guard their offspring until they hatch and, frequently, through the larval stage.

**gular pouch (GYOO-ler; GUH-ler)**  A sac of skin that hangs between the flexible bones of a pelican's lower mandible.

**gulf**  A small body of water that is mostly cut off from an ocean or sea by land formations.

**Gulf Stream**  The largest of the western-boundary currents that runs along the western edge of the North Atlantic gyre.

**gymnosome pteropod**  A carnivorous oceanic gastropod that has no shell and swims by flapping lateral extensions of its foot.

**gyre**  The circular flow pattern of water around the edge of an ocean basin.

**habitat**   The specific place in the environment where an organism lives.

**habitat isolation**   The prevention of interbreeding between different species in nature due to their occupying different habitats.

**hadal zone (HAYD-uhl)**   The portion of the ocean bottom that lies at depths greater than 6,000 meters.

**halobacteria**   A group of archaeons that require high concentrations of salt where they live.

**halocline (HAL-oh-klyn)**   A zone in the ocean characterized by a rapid change in salinity with increasing depth.

**halophile**   Organism that grows and reproduces best in the presence of salt.

**halophyte (HAL-oh-fyt)**   A salt-tolerant flowering plant.

**haploid number (N)**   The number of unpaired chromosomes in a cell's nucleus.

**haptonema**   A rodlike organelle that projects between the two flagella of haptophytes and is used to capture prey.

**harmful algal bloom (HAB)**   The formation of a dense population of photosynthetic microbes and macroalgae that presents an environmental threat in natural and managed communities.

**head lunge**   A cetacean behavior in which a whale breaks the surface of the water and falls forward.

**head-foot**   The region of the gastropod body that contains the head, with its mouth and sensory organs, and the foot, which is the animal's organ of locomotion.

**helical virus**   A virus with a capsid of spirally arranged proteins.

**hemichordate**   An animal in the phylum Hemichordata. Also known as an acorn worm.

**hemoglobin**   A pigmented protein in red blood cells that functions in transporting oxygen from the lungs to the tissues.

**herbivore**   An animal that eats only plants and algae.

**hermaphrodite**   An animal that possesses both male and female sex organs.

**hermatypic coral**   A colonial coral that maintains zooxanthellae in its tissues and forms reefs.

**heterocercal tail (het-uh-roh-SIR-kuhl)**   The type of tail in some fishes in which the upper lobe is larger than the lower lobe and the vertebral column bends slightly upward into the upper lobe.

**heterocyst**   A specialized cell of cyanobacteria in which conditions are maintained favorable for nitrogen fixation.

**heterokont**   The condition of a cell that bears two different flagella, one simple and one with mastigonemes.

**heterotroph (HET-uh-roh-trohf)**   An organism that relies on other organisms for food. Also known as a *consumer*.

**high marsh**   The region of a salt marsh closest to shore that is covered briefly by saltwater each day.

**high water**   The greatest height to which a high tide rises.

**hit-and-run attack**   Shark attack mode in which the victim is bitten and then released. The bites usually cause relatively minor lacerations that are seldom life threatening.

**holdfast**   A branching system of fibers at the base of an algal thallus securing the seaweed to the sea floor.

**holocephalan**   A chimaera, a type of cartilaginous fish (class Chondrichthyes) characterized by a large pointed head, a long slender tail, gills covered with an operculum, and males with head claspers.

**holoplankton**   Organisms that are planktonic throughout their life cycles.

**holothurin**   A poison from the flesh of sea cucumbers that protects the animal from predation by fishes and that is used by Pacific islanders to poison fishes.

**homeostasis (HOH-mee-oh-STAY-sis)**   The internal balance that living organisms must maintain to survive.

**homeothermic**   Able to maintain a constant body temperature.

**homocercal tail (hoh-moh-SIR-kuhl)**   A tail in which the upper and lower lobes are generally equal and into which the vertebral column does not extend.

**horse latitudes**   The areas of descending air at 30 degrees N and S latitudes.

**host**   In a parasitic relationship, the organism that supports the parasite.

**hybrid**   The offspring produced by breeding between two different species.

**hydrocoral**   A type of hydroid that secretes a calcareous skeleton around the polyps and resembles hard corals.

**hydrogen bond**   Weak attractive force that occurs between slightly positive hydrogen atoms of one molecule and slightly negatively charged atoms, such as oxygen or nitrogen, on another molecule.

**hydrogenous sediment**   Sediments formed by the precipitation of dissolved minerals from seawater.

**hydroid**   A colonial organism belonging to the cnidarian class Hydrozoa. Also known as a *hydrozoan*.

**hydrophilous pollination**   The mechanism of dispersal of a pollen grain by water currents from the male flower to the female flower.

**hydrophyte (HYD-roh-fyt)**   A flowering plant that generally lives submerged under water.

**hydrozoan (hyd-roh-ZOH-en)**   A generally colonial organism that belongs to the cnidarian class Hydrozoa. Also known as a *hydroid*.

**hyperthermophiles**   Microbes, usually archaeons, that grow and reproduce best at temperatures exceeding 100°C.

**hyphae**   The threadlike filaments that make up the mycelium of a nonyeast fungus.

**hypocotyl**   The initial stem of a young or embryonic plant that arises below the embryonic leaves.

**hypothesis**   An explanation for observed events that can be tested by experiments.

**icosahedral virus**   A virus composed of a DNA or RNA core surrounded by a capsid with 20 sides.

**incidental catch**   Noncommercial animals that are killed each year during fishing for commercial species. Also known as *bycatch*.

**inductive reasoning**   A process of reasoning whereby a general explanation is derived from a series of observations.

**infauna (IN-faw-nuh)**   Benthic organisms that live in the bottom sediments.

**infralittoral zone**   Zone of rocky shores that lies below the extreme low water of spring tides.

**inhalant opening**   The opening of a bivalve's or tunicate's siphon through which water is drawn into the body.

**international date line**   The line that lies at approximately

180 degrees from the prime meridian that marks the beginning of each calendar date.

**internode**  The part of a plant stem that lies between successive nodes, leaves, lateral stems, or roots.

**interspecific competition**  Competition between similar species for a limited resource.

**interstitial animal**  Animal that lives in the spaces between sediment particles.

**intertidal zone**  The region of a shore that is covered by high tide and exposed at low tide.

**intraspecific competition**  Competition between members of a single species for a limited resource.

**invertebrate**  An animal that lacks a vertebral column (backbone).

**ion**  A particle that carries an electrical charge.

**iridophore**  Type of chromatophore that produces iridescence from light striking crystals contained within the cell.

**irregular echinoid**  An echinoid such as a heart urchin or sand dollar that is adapted to a burrowing lifestyle.

**island arc**  A chain of volcanic islands usually found along deep-sea trenches.

**isolating mechanism**  Mechanism that prevents members of different species from reproducing in nature.

**isopycnal (eye-soh-PIK-nuhl)**  Having the same density at the top as at the bottom.

**keratin**  A tough protein found in mammalian hair and nails and in the baleen of cetaceans.

**keystone predator**  An animal whose presence in a community makes it possible for many other species to live there. Also known as a keystone species.

**kinetic**  Descriptive of planktonic organisms that are able to move by flagella, limbs, jet propulsion, or other devices.

**krill**  Pelagic, shrimplike creatures that belong to the arthropod order Euphausiacea.

**labyrinthomorph**  A group of stramenopiles with heterotrophic nutrition that are decomposers or pathogens.

**labyrinthulid**  A labyrinthomorph that includes the pathogen responsible for wasting disease in eelgrass.

**lactation period**  The period of time during which a female mammal nurses her young.

**lactic acid**  A waste product produced in muscle tissue when there is not enough oxygen to support aerobic metabolism.

**lactose**  A disaccharide composed of glucose and galactose.

**lacuna**  Large open gas space in the aerenchyme tissue of plants.

**laminarin**  The molecular form in which carbohydrates are stored in ochrophytes.

**lancelet**  An animal belonging to the subphylum Cephalochordata. Also known as a cephalochordate.

**landing**  The catch made by a fishing vessel.

**larvacean**  A free-swimming tunicate that resembles a tadpole.

**larval settlement**  The process by which planktonic larvae leave the water column and settle to the bottom.

**lateral line system**  A system of canals running the length of a fish's body and over its head that functions in detecting movement in the water.

**latitude**  The angular distance north and south of the equator measured in degrees. Also known as a parallel.

**Laurasia**  The northern portion of the supercontinent Pangaea composed of what became Europe, Asia, and North America.

**lenticel**  A scarlike structure on the surface of a stem serving for passage of atmospheric gas through the cuticle and epidermis into the aerenchyme beneath.

**leptocephalus larva**  Thin, leaflike larval form of eels, tarpon, bonefish, and their relatives.

**leuconoid (LOO-keh-noyd)**  A sponge with a complex body containing many spongocoels and chambers leading to them.

**lichen**  Mutualistic association between a fungus and an alga, often found in dry habitats.

**light–dark-bottle method**  A procedure for measuring productivity that involves using two bottles: one that is clear and allows light to penetrate and one that is opaque and does not allow light to penetrate.

**limiting nutrient**  Nutrient that limits the number or distribution of marine organisms.

**lipid**  Macromolecule composed primarily of carbon and hydrogen. Also known as fat, oil, or wax.

**lithification**  The conversion of an aggregate of particles into a solid mass with a mineral cement.

**lithosphere (LITH-oh-sfeer)**  The part of the earth comprising the crust and upper mantle.

**logarithmic growth**  A pattern of population growth characterized by initial slow growth followed by increasingly rapid growth; when graphed the curve is typically a J-shape. Also known as exponential growth.

**logistic growth**  A pattern of population growth characterized by an initial exponential-growth phase that eventually levels off at zero growth as the population reaches an equilibrium point.

**longitude**  The angular distance east and west of the prime (or Greenwich) meridian measured in degrees. Also known as a meridian.

**longshore current**  Current moving parallel to the shore in the surf zone that is generated by waves breaking at an angle to the beach.

**longshore transport process**  A process driven by longshore currents that carries sediments along the shoreline.

**lophophorate (lohf-uh-FOHR-ayt)**  A sessile animal that lacks a distinct head and possesses a feeding device called a lophophore.

**lophophore (LOHF-uh-fohr)**  An arrangement of ciliated tentacles that surrounds the mouth of animals known as lophophorates. It functions in feeding and in gas exchange.

**lorica**  A loosely fitting external covering of a microbe.

**lottery model**  A hypothesis that competition among fishes on coral reefs is unimportant; chance determines which species of larvae settling from the plankton colonize a particular area of reef. As each individual is lost over time, chance again determines which species will occupy the available space and competitive exclusion does not occur.

**low marsh**  The portion of a salt marsh that occupies the lower intertidal zone and that is covered by tidal water much of the day.

**low water**   The lowest point to which a low tide falls.

**luciferase**   An enzyme that catalyzes the reactions that produce bioluminescence.

**luciferin**   A protein that produces bioluminescence when combined with oxygen in the presence of the enzyme luciferase and ATP.

**lysogenic cycle**   The life history of a virus that remains dormant within the host genome awhile before initiating viral replication.

**lysosome**   Package of digestive enzymes found in eukaryotic cells.

**lytic cycle**   The life history of a virus that has no dormant phase in the host before initiating viral replication.

**macroalga**   Multicellular alga, such as a seaweed, visible to the naked eye.

**macromolecule**   A large organic molecule such as a protein, carbohydrate, lipid, or nucleic acid.

**macroplankton**   Plankton that range in size from 2 to 20 centimeters, composed primarily of animals.

**madreporite (mad-ruh-POHR-ryt)**   The structure at which water enters the water vascular system of an echinoderm.

**magma**   Molten material located deep in the earth's mantle.

**maltose**   A disaccharide composed of two glucose molecules.

**mammary gland**   Special gland in female mammals that produces milk.

**mangal (MAN-guhl)**   A community dominated by plants called mangroves.

**mangrove**   Any of a variety of salt-tolerant trees and shrubs restricted to humid tropical coasts.

**mannitol**   An alcohol and product of photosynthesis in brown algae.

**mantle (earth)**   The thickest layer of the earth and the one that contains the greatest mass of material.

**mantle (mollusc)**   The part of a mollusc's body that secretes its shell.

**mantle cavity**   The space between a mollusc's mantle and its body.

**map**   A representation of the land features of the earth.

**marine biology**   The study of the living organisms that inhabit the seas and of their interactions with each other and their environment.

**marine snow**   A loose network of live and dead particles in the plankton arising from webs of mucus released by many kinds of microbes and zooplankton.

**maritime zone**   Zone of rocky shores that lies above the extreme high water of spring tides. Also known as the *supralittoral zone.*

**mastigonemes**   Hairlike filaments that extend from the shaft of some flagella and that increase the effectiveness of locomotion.

**medusa**   The free-floating form of cnidarian that resembles an umbrella or a bell.

**megaplankton**   Plankton that range in size from 20 to 200 centimeters, composed primarily of the larger jellyfishes, siphonophores, salps, and floating mats of sargassum weed.

**meiofauna (MY-oh-fawn-uh)**   The tiny organisms that are adapted to living in the spaces between sediment particles.

**meiosis**   The process of nuclear division in which the number of chromosomes in the cell's nucleus is reduced to one half in the formation of gametes. Also known as *reduction division.*

**melanin**   A brown or brown-black pigment found in many animals.

**melon**   An oval mass of fatty, waxy material located between the blowhole and the end of the head in cetaceans capable of echolocation. The melon directs and focuses the sound waves produced by an animal.

**membranelles**   Sheetlike or triangular arrangements of cilia that increase the effectiveness of locomotion and feeding.

**Mercator projection**   A type of modified cylindrical projection used to produce maps and charts that distorts the areas near the poles but offers the advantage that a straight line is a line of true direction.

**meridian**   A line of longitude.

**meroplankton**   Planktonic stages of an organism that is benthic or nektonic at other stages in its life history.

**mesoplankton**   Plankton that range in size from 0.2 to 20 millimeters, composed primarily of larger phytoplankton and larvae of zooplankton as well as many kinds of adult crustaceans and other zooplankton groups.

**messenger RNA (mRNA)**   A type of RNA molecule that contains the instructions for synthesizing a protein.

**metabolism**   The sum of all of the chemical reactions that occur within living cells.

**methane hydrate**   Ice crystals containing trapped methane.

**methanogen**   Archaeon with the ability to produce methane gas in its metabolism.

**microbe**   Living organism too small to examine without the aid of a microscope.

**microhabitat**   The smaller subdivisions of a habitat.

**micropatchiness**   The small-scale clumped distribution of organisms within an ecosystem, often explained by the similarly small-scale variation in the physical and chemical environment or disturbance effects of other organisms.

**microphytoplankton (my-kroh-FYT-oh-plank-tuhn)**   Phytoplankton ranging from 20 to 200 micrometers.

**microplankton**   Plankton ranging from 20 to 200 micrometers, composed of many kinds of phytoplankton, marine invertebrate larvae, and smaller adult zooplankton.

**midlittoral zone**   Zone of rocky shores that lies between the supralittoral fringe and the infralittoral fringe. The midlittoral zone is the true intertidal, being inhabited by both marine and amphibious organisms.

**midocean ridge**   A long mountain range that forms along cracks on the ocean floor where erupting magma breaks through the earth's crust. Also known as *ridge system.*

**midsagittal plane**   A plane through the center of the long axis of a bilaterally symmetrical animal that divides the animal into more or less identical right and left halves.

**mitochondria (singular, mitochondrion)**   The membrane-bound organelles in which the energy of food molecules is used to produce ATP.

**mitosis**  The process of nuclear division that occurs in eukaryotic cells and produces two identical nuclei.

**mixed semidiurnal tide**  A tide in which the high tide and the low tide are of different levels.

**mixotrophic**  Nutrition of an organism by combining autotrophy and heterotrophy.

**modern synthetic theory of evolution**  Darwin's theory of evolution by natural selection as modified by modern genetics.

**molecular biology**  The study of the structure and function of macromolecules such as nucleic acids.

**mollusc**  A soft-bodied animal that is a member of the phylum Mollusca. The bodies of most molluscs are covered by a shell.

**molting**  In arthropods, the process in which an old exoskeleton is shed and a new one is formed.

**monoculture**  The process of raising only one species in mariculture.

**monosaccharide**  Simple sugar that usually contains five or six carbon atoms.

**morphology (mohr-FAHL-uh-jee)**  The structure or appearance of an organism.

**mossback whale**  Another name for the gray whale, *Eschrictius gibbosus.*

**mother-of-pearl layer**  Another term for the nacreous layer of pearl oysters.

**mucilage (MYOO-suh-lij)**  The slimy, gelatinous secretion covering algal cells for attachment of cells and for protection.

**mud**  A type of sediment composed of clay and silt.

**mutualism**  A symbiotic relationship in which both organisms benefit.

**mycelium**  The term for the body of a fungus, whether a single cell or a large mass of filaments.

**mycologist**  A scientist who studies fungi.

**myoglobin**  A pigmented protein in vertebrate muscle that stores oxygen.

**nacreous layer (NAY-kree-uhs)**  The innermost layer of a molluscan shell. In oysters, it also is known as the *mother-of-pearl layer.*

**nanophytoplankton**  Plankton smaller than 20 micrometers.

**nanoplankton**  Plankton ranging from 2 to 20 micrometers, composed primarily of small phytoplankton and larger bacteria. Also known as *centrifuge plankton.*

**natural selection**  The mechanism that explains why organisms that possess variations best suited to their particular environments exhibit a better survival rate and reproductive capacity than do less well-suited organisms.

**nauplius larva (NAW-plee-uhs)**  A larval stage in the life cycle of many crustaceans. Nauplius larvae are characterized by three pairs of appendages and a median eye.

**nautical mile**  A unit of distance equal to 1 minute of latitude, 1.85 kilometers, and 1.15 land miles.

**nautiloid**  A cephalopod that has an external shell belonging to the subclass Nautiloidea.

**neap tide**  Tide that exhibits the smallest change between the high and low tide marks. Neap tides occur when the sun and the moon are at right angles to each other.

**negative estuary**  An estuary in which the surface water flows toward the river and the water along the bottom moves out to sea.

**nekton**  All actively swimming organisms whose movements are not governed by currents or tides.

**nematocyst (neh-MAT-uh-sist)**  The stinging organelle found within the stinging cell of cnidarians.

**nematode (NEM-uh-tohd)**  A round, wormlike animal that belongs to the phylum Nematoda.

**neritic zone (neh-RIT-ik)**  The zone of water that lies over the continental shelves.

**net plankton**  Components of the plankton that are large enough to be sampled with standard plankton nets; mainly composed of microplankton and larger organisms.

**neuromast (NOO-roh-mast)**  A sense organ that can detect vibrations in the fluid that fills the canals of a fish's lateral line system.

**neurotoxic shellfish poisoning**  A human syndrome caused by consumption of shellfish that concentrate toxic chemicals produced by dinoflagellate blooms.

**neuston**  Plankton that live at or near the surface of the ocean.

**niche (nish; neesh)**  An organism's role in its environment.

**nictitating membrane (NICK-ti-tay-ting)**  A clear membrane that covers and protects the eye of some vertebrates.

**nitrification (ny-truh-fi-KAY-shun)**  The process by which ammonia from animal wastes and dead tissue is converted into nitrate ions.

**nitrogen fixation**  The process by which some microorganisms are able to convert atmospheric nitrogen into a form that is usable by producer organisms.

**nitrogenase**  The enzyme of nitrogen-fixing bacteria that is capable of breaking the strong bond of nitrogen gas for production of ammonia.

**node**  The jointlike structure on a plant stem at which leaves and sometimes lateral stems and roots arise.

**nonselective deposit feeder**  An animal that ingests both organic and mineral particles and then digests the organic material, especially the bacteria that grow on the surface of the mineral particles.

**nonsynchronous spawners**  Organisms that spawn at different times of the year.

**nori**  An oriental food made from seaweed, often deriving from red algae of the genus *Porphyra.*

**northeast trade winds**  The surface winds that occur between the equator and 30 degrees N latitude.

**nucleic acids**  Polymers composed of basic units called nucleotides.

**nucleocapsid**  The combined core of nucleic acids and surrounding protein coat of a mature virus (or virion).

**nucleolus**  The area of the nucleus where ribosomes are assembled.

**nucleotide**  A molecule consisting of a five-carbon sugar, an organic base, and a phosphate group. Nucleotides are the building blocks of nucleic acids and are found in other important molecules as well.

**nucleus**  A structure in eukaryotic cells that contains the cells' chromosomes and acts as the cellular control center.

**nudibranch (NOO-di-brangk)**  A gastropod mollusc that

does not have a shell and has many projections from its body called cerata.

**nutrient** Any organic or inorganic material that an organism needs to metabolize, grow, and reproduce.

**nutritive root** A final subdivision of a plant root system that absorbs water and minerals.

**obligate anaerobe** Organism that thrives only in the absence of oxygen.

**obliterative countershading** A camouflage pattern in which the upper surface of pelagic animals is dark and the lower surface light to blend in with the ocean depths or the surface, respectively, when observed by a potential predator. Also known as *countershading*.

**observational science** Science in which hypotheses are supported or denied on the basis of observations alone rather than controlled experiments.

**ocean basin** The floor of the ocean.

**ocean productivity** The production of organic matter by photosynthetic and chemosynthetic marine organisms.

**ocean ranching** The process of raising young fish in hatcheries and returning them to the sea when they have reached adulthood. Also known as *sea ranching*.

**oceanic basaltic crust** New earth crust formed when magma erupting from ridge systems cools.

**oceanic zone** The zone of ocean composed of the water that covers the deep ocean basins.

**oceanography** The study of the oceans and their phenomena, such as waves, currents, and tides.

**ochrophyte** A photosynthetic stramenopile characterized by presence of chlorophylls *a* and *c*, the pigment fucoxanthin, and the starch laminarin; ochrophytes include diatoms, silicoflagellates, and brown algae.

**olfaction** The sense of smell.

**olfactory pit** In fish a blind sac that opens to the external environment and contains olfactory organs.

**omnivore** An animal that feeds on both producers and consumers.

**ooze** Sediment composed of fine biogenous particles.

**operculum (oh-PER-kyoo-luhm)** A hard or tough covering found in some molluscan gastropods that closes the opening to the shell when the animal retracts inside. In fishes, the protective covering of the animal's gills.

**ophiuroid** Echinoderm commonly referred to as brittle star or serpent star belonging to the class Ophiuroidea.

**optimal range** The range of environmental factors to which an organism is best adapted.

**oral tentacles** Modified tube feet that are located around the mouth of a sea cucumber and used in acquiring food.

**orca** Another name for the killer whale, *Orcus orcina*.

**organ system** A group of organs that function together for a specific purpose, such as processing of food or gas exchange.

**organelle** Specialized structure in the cytoplasm of eukaryotic cells that performs a specific cellular function.

**organ** A specialized structure made up of more than one tissue.

**orientation click** A low-frequency click produced by toothed whales, such as dolphins, that gives the animal a general idea of its surroundings.

**osmoconformer (ahz-moh-kuhn-FOHR-mer)** An animal whose tissues and cells can tolerate dilution.

**osmoregulator (ahz-moh-REG-yoo-lay-tir)** An animal that can maintain an optimal salt concentration in its tissues regardless of the salt content of its environment.

**osmosis (ahz-MOH-sis)** The movement of water across a semipermeable barrier in response to differences in solute concentration on either side of the barrier.

**osmotrophy** Nutrition of an organism by absorption of small organic molecules from the external medium across the cell membrane.

**ossicles** Plates of calcium carbonate that make up the endoskeleton of echinoderms.

**ostia** The openings into the body of a sponge.

**overfishing** The capture of fish faster than they can reproduce and replace themselves.

**oviduct** A tube that carries eggs to the outside of a female's body.

**oviparity** Reproductive mode characterized by eggs that hatch after leaving the body of the female.

**ovoviviparity** Reproductive mode characterized by females producing large, yolky eggs that hatch in the oviduct but are not nourished by the parent.

**oxide** Chemical compound that contains the element oxygen and one other element.

**oxygen-minimum zone** A region just below the sunlit surface waters in which oxygen is depleted by the resident animal life but not replaced by photosynthesis.

**palp (bivalve)** A pair of structures located near the mouth that forms filtered food into a mass and then moves it to the bivalve's mouth.

**palp (sea spider)** A pair of sensory structures.

**Pangaea** The single continental landmass or supercontinent that existed about 400 million years ago. It was composed of a northern portion, Laurasia, and a southern portion, Gondwanaland.

**parallel** The angular distance north or south of the equator measured in degrees. Also known as *latitude*.

**paralytic shellfish poisoning (PSP)** A toxic condition that occurs following the ingestion of shellfish contaminated by a dinoflagellate toxin.

**parasite** In a parasitic relationship, the organism that lives off its partner.

**parasitism** A symbiotic relationship, in which one organism benefits and the other is harmed.

**partially mixed estuary** An estuary that has a strong surface flow of freshwater and a strong influx of seawater.

**particulate organic matter (POM)** Particles of organic matter suspended in seawater; largely synonymous with *detritus* and *tripton*.

**parts per thousand (ppt)** A measure of the concentration of a solution expressed as the number of grams of solute per kilogram of solvent.

**patch reef** Small patch of reef located in a lagoon associated with an atoll or a barrier reef.

**patches** Unevenly distributed, localized aggregations of plankton in the sea or benthic organisms on the seafloor.

**patchiness** The uneven distribution of organisms in a population or a community.

**pectoral fins** The anterior paired fins of a fish.

**pedal disk**   The base of a sea anemone.

**pedal laceration**   A type of asexual reproduction in which a portion of a sea anemone's basal disk is separated and left to form a new individual.

**pedicellariae (ped-uh-suh-LEHR-ee-eh)**   Tiny pincerlike structures found in some echinoderms that keep the surface of the body clean and free of parasites and the settling larvae of fouling species. In some echinoderm species, they may also aid in obtaining food.

**pedicle (PED-i-kuhl)**   A fleshy stalk found in some lamp shells that fastens the animal to a solid surface.

**peduncle**   The part of a whale's body closest to the tail fluke.

**peduncle slap**   A cetacean behavior in which the whale swings the rear portion of its body, sometimes as far forward as its dorsal fin, out of the water and then drops it down sideways on the water or another whale.

**pelagic division (pe-LAJ-ik)**   The division of the marine environment composed of the ocean's water.

**pelagic ecosystem**   The animals, plants, and microbes of the water column and their realm of the open sea.

**pelagic spawners**   Fish that release vast quantities of eggs into the water. The eggs are fertilized by the males, they drift with the currents, and there is no parental care.

**pellicle**   The complex of alveoli and the cell membrane in alveolates.

**pelvic fins**   The posterior pair of fins of a fish.

**pen**   An internal strip of hard protein that helps support the mantle of a squid.

**perennial**   A plant or alga that completes its life cycle in more than two growing seasons.

**peridinin**   A kind of xanthophyll in dinoflagellates that gives them their golden-brown color.

**period**   The time required for one wavelength to pass a fixed point.

**periostracum (per-ee-AHS-treh-kuhm)**   The outermost layer of a molluscan shell. The periostracum is composed of the protein conchiolin.

**periwinkle**   A mollusc in the family Littorinidae that inhabits the upper part of the intertidal zone.

**pH scale**   A scale that indicates the acidity or basicity of a solution. The pH scale runs from 0 to 14, with 7 being the neutral point. The pH number reflects the concentration of hydrogen ions in a solution. Acids have a pH between 0 and 7, whereas bases have a pH between 7 and 14.

**phage**   A virus that infects a bacterium; shortened form of *bacteriophage*.

**pharynx (FA-ringks)**   A muscular tube that forms part of the digestive tract of an animal.

**pheromone (FAYR-eh-mohn)**   A hormone released into the environment by one individual that controls the behavior or development of other individuals of the same species.

**phloem**   The kind of vascular tissue in plants that carries photosynthetic products from leaves to parts of the plant that need them as food.

**phocid**   A seal that belongs to the family Phocidae. Also known as true seals, these animals lack external ears and are unable to rotate their flippers beneath their bodies.

**phoronid (FOHR-oh-nid)**   A small, wormlike lophophorate that secretes a tube of leathery protein or chitin.

**phospholipid**   A type of lipid that is made up of glycerol, two fatty acids, and a phosphate-containing molecule.

**photic zone (FOH-tik)**   The portion of the pelagic division of the ocean that receives enough sunlight to support photosynthesis.

**photoperiod**   The amount of light and darkness in a 24-hour period.

**photophore (FOH-toh-fohr)**   A specialized organ in some organisms that produces bioluminescence.

**photoprotective pigment**   Complex chemicals in primary producers that reduce the potential harm to chlorophylls by light.

**photoreceptor**   Sensory organ capable of responding to light.

**photosynthesis**   The process by which some organisms use the energy of sunlight to produce organic molecules, usually from carbon dioxide and water.

**phycobilin**   A class of accessory photosynthetic pigments in cyanobacteria and red algae.

**phycocolloid (fy-koh-KAHL-oyd)**   Any polysaccharide that has a gelling effect in water and that can be extracted from the cell walls of seaweeds; includes the commercially important agar, alginates, and carrageenan.

**phycocyanin (fy-koh-SY-uh-nin)**   A blue pigment that absorbs green light. In some organisms, phycocyanin functions as an accessory pigment in photosynthesis.

**phycoerythrin (fy-koh-e-RITH-rin)**   A red pigment that absorbs blue and green light. In some organisms, it functions as an accessory pigment in photosynthesis.

**phycologist (fy-KAHL-ah-jist)**   A scientist who studies seaweeds and phytoplankton; synonym, algologist.

**physiographic chart (fiz-ee-oh-GRAF-ik)**   A chart of the ocean that uses perspective drawing, coloring, or shading to show the various depths.

**phytoplankton (FY-toh-plank-tuhn)**   Tiny photosynthetic organisms that float in ocean currents.

**picoplankton**   Plankton that range in size from 0.2 to 2.0 micrometers, composed primarily of larger viruses and many kinds of bacteria.

**pigments (animals)**   Minute granules typically found in chromatophores that exhibit colors such as white, black, brown, red, or yellow.

**pinacocytes**   Flattened cells that form the outer body surface of a sponge.

**pink zone**   The lower part of the midlittoral zone of tropical rocky shores characterized by the widespread encrustation of coralline algae.

**pinniped (PIN-i-ped)**   An animal belonging to the mammalian order Pinnipedia, which includes seals, sea lions, elephant seals, and walruses.

**placenta (plah-SEN-tuh)**   An organ present only during pregnancy in some female mammals that functions in maintaining the fetus until birth.

**placental mammal**   Mammal that retains its young inside the body until birth.

**placoid scale (PLAK-oyd)**   The type of scale in cartilaginous fishes that has a structure resembling the teeth of other vertebrates.

**plankton**   Organisms that float or drift in the sea's currents.

**plant**   Multicellular, photosynthetic organism whose cells are surrounded by a cell wall composed of cellulose.

**planula larva**   The planktonic larval stage of cnidarians.

**plastron**   The ventral surface of a turtle's shell.

**pleuston**   Components of the neuston that breach the surface of the ocean and are carried by wind as well as water currents.

**plunger**   A type of wave in which the crest curls and curves over and outruns the rest of the wave.

**pneumatophore**   An aerial root of a mangrove that arises from a cable root and grows out of the sediment.

**pod**   A group of related cetaceans.

**pogonophoran**   An animal that belongs to the class Pogonophora of the phylum Annelida. Also known as beardworms.

**polar**   Having a positively charged end and a negatively charged end.

**polar easterlies**   Winds that occur in the Northern Hemisphere, between 60 degrees N and the North Pole and blow from the north and east, and in the Southern Hemisphere, between 60 degrees S and the South Pole and blow from the south and east.

**pollen**   The male gametophyte of a seed-bearing plant; usually carried by wind, water, or pollinating animals to the female gametophyte within the cone or flower.

**polychaete (PAHL-eh-keet)**   A type of annelid worm belonging to the class Polychaeta.

**polyculture (PAHL-ee-kuhl-chur)**   The raising of more than one species in mariculture.

**polygynous (pah-LIJ-eh-nehs)**   Relating to a male having more than one female mate.

**polymer**   A large molecule that consists of multiple identical units linked together.

**polyp (PAHL-uhp)**   A generally benthic form of cnidarian characterized by a cylindrical body that has an opening at one end that is usually surrounded by tentacles.

**polypeptide**   Polymer of amino acids.

**polysaccharide**   A type of carbohydrate that is a polymer of monosaccharides.

**POM**   See *particulate organic matter*.

**population**   A group of individuals of the same species that occupies a specified area.

**positive estuary**   An estuary in which the surface water from the river flows out to sea and saltwater from the ocean moves into the estuary along the bottom.

**poster colors**   Bright showy color patterns in fish and other animals that may advertise territorial ownership, aid foraging individuals to keep in contact, or be important in sexual displays.

**potential yield**   The number of pounds of fish or shellfish that a stock can yield per year without being overexploited.

**P-R ratio**   The ratio of primary production (gross photosynthesis) to community respiration, sometimes used as a measure of the state of development (or succession) of a biological community.

**precipitation nuclei**   Airborne particles that attract water droplets.

**predation disturbance model**   A hypothesis that predation or other causes of mortality keep populations of coral reef fishes low enough to prevent competitive exclusion.

**priapulid**   A benthic worm belonging to the phylum Priapulida.

**primary consumer**   A producer that feeds directly on a producer. Also known as a *first-order consumer*.

**primary productivity**   The process of producing energy-rich organic compounds from inorganic materials.

**prime, or Greenwich, meridian**   The primary line of longitude that passes through the Royal Naval Observatory in Greenwich, England, and is designated 0 degrees longitude.

**prismatic layer (priz-MAT-ik)**   The middle layer of a molluscan shell and the layer that contains most of the mass of the shell.

**proboscis**   A tube that extends from the mouth of some animals.

**producer**   An organism that can produce its own food. Also known as an *autotroph*.

**progressive wave**   A wave generated by wind and restored by gravity that moves in a particular direction.

**prokaryote (proh-KAR-ee-oht)**   Unicellular organism without a nucleus.

**prokaryotic cell**   A cell that lacks a nucleus and membrane-bound organelles.

**prop root**   An aerial root of a mangrove that arises from the trunk and descends to the sediment.

**propagule**   A dispersal phase representing the next generation in a life cycle; specifically applied to viviparous mangroves that do not produce dormant seeds.

**protandry**   Condition in sequential hermaphroditic animals whereby sperm are produced in the life cycle before eggs.

**protein**   A polymer of amino acids.

**protein synthesis**   The process by which ribosomes link amino acids into chains using the instructions provided by messenger RNA.

**protist**   A generic term used for eukaryotic organisms that are not classified as fungi, plants, or animals.

**protogyny**   Condition in sequential hermaphroditic animals whereby eggs are produced in the life cycle before sperm.

**pseudofeces (SOO-doh-fee-sees)**   Large, semisolid particles produced by bivalves and consisting of phytoplankton and detritus that they filter but do not consume.

**pseudopod (SOO-doh-pahd)**   Fingerlike projections of cytoplasm and membrane that function in both locomotion and feeding in amoebae and their relatives.

**pteropod ooze**   Calcareous sediments of the seafloor formed by the accumulation of shells of thecosome pteropods.

**purse seine**   Huge net with a bottom that can be closed off by pulling on a line, similar to pulling the drawstring of a purse.

**pycnocline (PIK-noh-klyn)**   A zone in the ocean that is characterized by a rapid change in density with depth.

**pyramid of biomass**   A pyramid-shaped diagram that indicates the amount of biomass available at successive trophic levels.

**pyramid of numbers**   A pyramid-shaped diagram that indicates the relative number of organisms a series of trophic levels can support.

**pyrosome**   Large colony of pelagic tunicates (salps) that move by cilia of the pharynx. A colony may contain as many as 500 individuals per cubic meter of water and stretch over several kilometers.

**radial symmetry**    The symmetrical organization of body parts around a central axis, similar to spokes on a wheel.

**radiolarian ooze**    A seafloor sediment consisting of the accumulated siliceous skeletons of radiolarians.

**radula (RAJ-oo-luh; RAD-yoo-luh)**    A ribbon of tissue that contains teeth. This structure is unique to molluscs.

**raft culture**    A process in which juvenile commercial molluscs are attached to ropes that are suspended from rafts floating in regions of the ocean where food is plentiful and exposure to natural predators is minimized.

**reciprocal copulation**    The process in which two hermaphroditic animals fertilize each other.

**recombination**    The formation of new combinations of alleles that results from crossing over.

**recruitment**    The addition of new members to a population through reproduction or immigration.

**rectal gland**    A gland associated with the rectal area of cartilaginous fishes that functions in salt secretion.

**red alga**    A seaweed of the phylum Rhodophyta, characterized by possession of chlorophylls *a* and *d*, phycoerythrins, and the phycocyanins, which give many of them their red, purple, or black color.

**redd**    The spawning area of salmon and trout, usually a cleared circular depression on the stream bottom.

**reduction division**    The process of nuclear division in which the number of chromosomes in the cell's nucleus is reduced to one half in the formation of gametes. Also known as *meiosis*.

**reef crest**    The highest point on a coral reef.

**reef flat**    The area opposite the reef front. Also known as the *back reef*.

**reef front**    The outer, seaward margin of a coral reef. Also known as the *forereef*.

**reflective beach**    Type of beach with a steep slope on which wave energy is directly dissipated.

**regular echinoid**    An echinoid with a spherical body. Also known as a sea urchin.

**renewable resource**    A resource, such as fisheries, that can replenish itself.

**reproductive isolation**    The absence of interbreeding between two populations.

**reproductive polyp**    A type of polyp found in hydrozoan colonies that is specialized for the process of asexual reproduction. Also known as a gonangium.

**resource limitation model**    A hypothesis that available larvae are too limited for fish populations to ever reach the carrying capacity of the habitat and, therefore, competitive exclusion does not occur.

**resource partitioning**    The process by which organisms share a resource.

**restoring force**    The force that causes the water in a wave to return to its undisturbed level.

**reticulopod**    A pseudopod with branches that interconnect to form a net for the capture of particles.

**rhizoid**    A rootlike anchoring structure of a nonvascular plant.

**rhizome**    A horizontal stem, growing below ground or trailing on the surface.

**rhizosphere**    The area below ground that is physically and chemically influence by the complex of roots of a vascular plant.

**ribonucleic acid (RNA)**    A usually single-stranded nucleic acid that functions in protein synthesis.

**ribosomal RNA (rRNA)**    A type of structural RNA found in organelles called ribosomes.

**ribosome**    An organelle consisting of protein and RNA. It is the site of protein synthesis.

**ridge system**    A long mountain range that forms along cracks in the ocean floor where erupting magma breaks through the earth's crust. Also known as *midocean ridge*.

**rift community**    A community of marine organisms that depend on the specialized environment found at divergence zones in the ocean floor. See also *deep-sea vent community* and *vent community*.

**rift valley**    Area of high volcanic activity that runs along portions of a midocean ridge.

**rift zone**    A region where the lithosphere splits, separates, and moves apart as new crust is formed.

**RNA**    See *ribonucleic acid*.

**roe**    An ovary with eggs.

**root hair**    A hairlike extension of an epidermal cell of a root, for absorption of water and minerals.

**rorqual**    A type of baleen whale in the family Balaenopteridae. These whales have a dorsal fin and ventral grooves on their throats.

**rough endoplasmic reticulum (RER)**    Endoplasmic reticulum that has ribosomes attached to its surface.

**salinity**    A measure of the concentration of dissolved inorganic salts in water.

**salp**    A free-swimming tunicate belonging to the class Thaliacea.

**salt gland**    One or more cells of a plant or animal that concentrate salt from sap or blood and release a salty secretion outside the body to control the body's mineral balance.

**salt wedge**    The angled boundary between saltwater and freshwater in an estuary that occurs when the rapid flow of river water prevents the saltwater from mixing with the freshwater.

**salt-wedge estuary**    An estuary in which the seawater moves in and out along an angled boundary known as a *salt wedge*.

**saxitoxin**    A class of toxins produced by certain species of dinoflagellates that cause paralytic shellfish poisoning.

**scale leaf**    Leaf that lacks a photosynthetic blade and is a protective sheath around the delicate growing tip of a stem.

**scaphopod (SKA-foh-pahd)**    A mollusc belonging to the class Scaphopoda. Scaphopods are also known as tusk shells because their shells are shaped like the tusks of elephants.

**scientific method**    The orderly pattern of gathering and analyzing information in science.

**scleractinian coral**    Coral that forms hard skeletons of calcium carbonate around its polyps.

**scyphozoan (sy-fuh-ZOH-uhn)**    An animal that belongs to the cnidarian class Scyphozoa. These are commonly called jellyfish.

**sea**    A body of saltwater that is smaller than an ocean and is more or less landlocked.

**sea anemone**    Large, heavy, complex polyps that belong to the cnidarian class Anthozoa.

**sea ranching**    The process of raising young fish in hatcheries and returning the adults to the sea. Also known as *ocean ranching*.

**seafloor spreading**    The process by which magma driven by convection currents is turned back by the lithosphere, moves laterally, and then descends, causing lateral movement of the earth's crust.

**seagrass**    A lilylike flowering plant that grows and reproduces while submerged in seawater.

**seamount**    A steep-sided formation that rises sharply from the ocean bottom.

**seaweed**    Any multicellular marine alga visible to the naked eye.

**secondary consumer**    A carnivore that feeds on herbivores. Also known as a *second-order consumer*.

**secondary metabolite**    Chemical made by cells for purposes other than routine metabolism.

**second-order consumer**    A carnivore that feeds on herbivores. Also known as a *secondary consumer*.

**sedentary polychaete (PAHL-eh-keet)**    A sessile polychaete that usually forms some sort of tube to cover its body.

**sediment**    Loose particles of inorganic and organic material.

**sedimentation**    The settlement of particles from suspension in water, usually accumulating on the seafloor.

**seed**    A specialized structure found in some species of plant. The seed contains the plant embryo and a supply of nutrients surrounded by a protective outer layer.

**seismic sea wave**    A large wave that increases in amplitude as it approaches the shore. Sometimes referred to as a tidal wave. Also known as a *tsunami*.

**selective deposit feeder**    Animals that separate organic material from minerals in sediments and ingest only the organic material.

**selective forces**    The physical and biological characteristics of the environment, such as temperature, salinity, predation, and food availability, that favor the survival of one species over another.

**semidiurnal tide**    The condition of having two high tides and two low tides each day.

**senescence**    The process of aging.

**sepia (SEE-pee-uh)**    A dark fluid produced in the ink glands of cephalopods.

**septa (singular, septum)**    A divider or partition.

**sequential hermaphrodite**    Individual possessing both functional male and female reproductive organs at different times in its life.

**sessile**    A term applied to animals that spend most of their time fixed in one place.

**seston**    Particles, living or dead, suspended in seawater.

**setae (SEE-tee)**    Bristles on the bodies of annelid worms.

**sex cell**    Cell that functions in reproduction, such as sperm, eggs, and pollen. Also known as a *gamete*.

**sextant**    A device used to measure the angle of a star with respect to the horizon.

**sexual dimorphism (dy-MOR-fiz-uhm)**    The condition in that males and females look different.

**sexual reproduction**    The process by which two parent organisms produce an offspring by the fusion of sex cells produced by each parent.

**shallow-water wave**    A deepwater wave that enters shallow water.

**sheath**    The basal, sometimes nonphotosynthetic part of the leaf of many vascular plants, usually wrapping around the stem.

**shelf break**    An abrupt change in the underwater landscape that occurs where the continental shelf ends and the continental slope begins.

**shelf zone**    The part of the ocean bottom that extends from the line of lowest tide to the edge of the continental shelf.

**shellfish**    An invertebrate animal with a shell, usually molluscs and crustaceans.

**shipworm**    A bivalve mollusc with a wormlike body belonging to the genus *Teredo* or *Bankia* that can burrow into wood.

**signature whistle**    A unique sound made by a dolphin that identifies the animal that is vocalizing, gives its location, and may relay other information as well.

**siliceous ooze**    Sediment composed of 30% or more of the remains of organisms that produce shells made of silica.

**silicoflagellate (sil-i-koh-FLAJ-uh-layt)**    A tiny photosynthetic protist that belongs to the phylum Chrysophyta. Silicoflagellates possess one or two flagella and internal shells composed of silica.

**sinus venosus**    A thin-walled chamber that collects blood from the veins and conveys it to the atrium of the heart in fishes, amphibians, and reptiles.

**siphon**    Tubular structure in some invertebrates that directs the flow of water in and out of the animal's body.

**siphuncle (SY-fuhn-kuhl)**    A cord of tissue that runs through the chambers of a nautilus shell and functions in the removal of seawater from the chambers and in the regulation of the gas content of the chambers.

**sipunculid**    Solitary worms belonging to the phylum Sipuncula.

**sirenian (sy-REE-nee-uhn)**    An animal belonging to the mammalian order Sirenia, which includes manatees and dugongs.

**slack water**    The period during a change in tide when tidal currents slow and then reverse.

**smooth endoplasmic reticulum (SER)**    A type of endoplasmic reticulum that lacks ribosomes on its surface.

**sneak attack**    Shark attack mode in which the victim is attacked without warning.

**solute**    A dissolved substance.

**solvent**    A medium for dissolving other substances.

**southeast trade winds**    The surface winds between the equator and 30 degrees S latitude.

**speciation (spee-see-AY-shun)**    The process by which new species are formed.

**species**    One or more populations of potentially interbreeding organisms that are reproductively isolated from other such groups.

**species epithet (EHP-i-thet)**    The second part of the two-part Latin name for an organism.

**specific heat**   The amount of heat energy required to increase the temperature of 1 gram of a substance by 1°C. Also known as the *thermal capacity*.

**spermaceti (spur-meh-SEH-tee)**   A thin, colorless, transparent oil that forms a waxy material when it comes into contact with air. Spermaceti is found in a cavity in the head of sperm whales.

**spermatophore**   A package of sperm.

**spicule (SPIK-yool)**   A support structure of sponges. A spicule can be composed of calcium carbonate, silica, or the protein spongin.

**spiller**   A type of wave in which the energy is dissipated gradually as the wave moves over shallow bottom.

**spiracle (SPEER-uh-kuhl)**   One of a pair of openings on the head of some cartilaginous fishes that functions in the passing of water to the gills.

**spiral valve**   A valve shaped like a spiral staircase found in the intestine of sharks.

**splash zone**   The uppermost area of a rocky shore that is covered by only the highest tides and usually is just dampened by the spray of crashing waves. Also known as the *supralittoral fringe*.

**spongin (SPUN-jin)**   A structural protein in some sponge spicules.

**spongocoel (SPUN-joh-seel)**   A spacious, water-filled cavity within the body of a sponge.

**sporangium**   The part of a seaweed or plant that produces and usually releases spores for asexual reproduction.

**sporophyte**   The multicellular stage in the life cycle of a seaweed or plant producing haploid or diploid spores that germinate into a subsequent stage without fertilization.

**spring tide**   Tide that exhibits the greatest change between the high and low tide marks. Spring tide occurs when the earth, sun, and moon are in line with each other.

**spur-and-groove formation**   The usual shape of a reef front, characterized by fingerlike projections of coral that protrude seaward.

**spy hopping**   A cetacean behavior in which a whale sticks its head straight up out of the water and surveys its surroundings.

**squalene (SKWAY-leen)**   An oily material produced by the liver of sharks.

**starch**   Polymer of glucose used by plants and some other organisms to store energy reserves.

**steroid**   A type of lipid that contains complex ring structures. Many function as important chemical messengers in the bodies of animals.

**stigma**   The part of the female flower that receives the pollen.

**stilt root**   Any aerial root (drop or prop root) that holds up a mangrove tree.

**stipe**   A stemlike structure in seaweeds, particularly brown algae.

**stock**   Separate population of commercial fishes or shellfishes within a species' geographic range that is assumed to be reproductively isolated from other stocks.

**stoma (STOH-ma; plural, stomata, stoh-MAH-tuh)**   An opening in a plant leaf that allows water to escape and gases to enter.

**stony (or true) corals**   The primary organisms depositing the calcium carbonate ($CaCO_3$) that builds the structure of coral reefs. Members of the phylum Cnidaria, class Anthozoa, order Scleractinia.

**stramenopile**   A group of organisms characterized by heterokont flagella with specialized mastigonemes and including diatoms, silicoflagellates, and brown algae.

**stroma**   The liquid that contains enzymes for carbohydrate synthesis and surrounds the grana inside chloroplasts.

**stromatolite**   Coral-like community of microbes that forms a thin layer of living cells and filaments over an accumulated mass of dead and lithified material.

**structural color**   Iridescent color produced by light reflecting from crystals located in specialized cells.

**stylet**   A hard, sharp point on the end of an organ.

**subduction zone**   Region of the ocean floor where old crust sinks to the earth's core and is recycled.

**submarine canyon**   Underwater canyon similar to canyons on land that are found on some continental slopes.

**subtidal zone**   The region of the shore that is covered by water even during low tide. It is also known as the *infralittoral zone*.

**succulent**   The condition of enlarged water-filled cells in the tissues of plants, making leaves or stems appear thicker than those of nonsucculent plants; a condition typical of plants in deserts or along marine shorelines.

**sucrose**   A disaccharide composed of glucose and fructose.

**sulcus**   The groove on the surface of a dinoflagellate that extends from the cingulum and holds one of its two flagella.

**supralittoral fringe**   Uppermost area of a rocky shore that is partially covered by only the highest (spring) tides and is usually just dampened by the spray of crashing waves. Also known as the *splash zone*, it overlaps the boundary of the supralittoral and littoral zones.

**supralittoral zone**   Zone of rocky shore that lies above the extreme high water of spring tides.

**surf zone**   The area of a coast where waves slow down before striking the shore.

**surface tension**   A property of liquids characterized by the clinging together of molecules at the exposed surface as the result of cohesive forces.

**surge channel**   Groove in the buttress zone of a coral reef that penetrates the algal ridge.

**surimi**   A product made from the flesh of the Alaskan pollock (*Theragra chalcogramma*). It is flavored to produce artificial crab, shrimp, and lobster.

**suspension feeder**   An organism that feeds on material suspended in the water.

**sustainable yield**   The number of fishes and shellfishes that can be caught over several years without stressing the population.

**sverdrup (sv)**   A measure of the volume of water transported in an ocean current. One sverdrup is equal to a flow of 1 million cubic meters of water per second.

**swarm**   A large mass of krill.

**swarming**   A behavior exhibited by some errant polychaetes in which the males and females congregate at the surface of the water to reproduce.

**swash** The water running up a beach after a wave breaks.

**sweeper tentacles** Very long, nematocyst-containing tentacles of certain corals that sweep over and kill polyps on competitive coral species.

**swell** A long-period, uniform wave that appears as a regular pattern of wave crests on the ocean's surface.

**swim bladder** A gas-filled sac in some bony fishes that allows them to maintain neutral buoyancy.

**swimmeret** An appendage modified for swimming in some crustaceans.

**syconoid (SY-kuh-noyd)** A sponge with a single spongocoel that has many invaginations.

**symbiosis (sim-by-OH-sis)** An intimate living arrangement between two different kinds of organisms.

**synchronous hermaphrodite** Individual possessing both functional male and female reproductive organs at the same time in its life.

**synchronous spawners** Organisms that spawn during a brief annual event.

**tagging** A procedure for monitoring the distribution and movements of fisheries stocks in which fish that are caught are marked with identification tags and then released to be caught again.

**tail lobbing** A cetacean behavior in which a whale lifts its tail high above the water and slaps it down on the surface with such force as to make a huge splash and produce a loud noise. Also known as *tail slapping*.

**tail slapping** A cetacean behavior in which a whale lifts its tail high above the water and slaps it down on the surface with such force as to make a huge splash and produce a loud noise. Also known as *tail lobbing*.

**tan** Individual section of a drift net.

**tannin** A class of compounds produced by plants having properties that reduce microbial infection and herbivory.

**tapeworm** A parasitic flatworm belonging to the class Cestoda. Tapeworms are found in the intestines of vertebrate animals.

**taxonomy (tak-SAHN-uh-mee)** The science of naming and classifying organisms.

**tectonic estuary** An estuary that forms when earthquakes cause the land at the mouth of a river to sink, allowing seawater to cover it.

**telson** A long spike attached to the posterior end of a horseshoe crab's abdomen that is used for steering and defense.

**temporal isolation** The prevention of interbreeding between different species in nature due to their mating at different times of day or different times of the year.

**ten percent rule** A rule in ecology that states that, on average, only approximately 10% of the energy available at one trophic level is passed along to the next trophic level.

**tentacle** Armlike or fingerlike structure that projects from an animal's body, usually the head.

**terrigenous sediment** Sediments that originate on land and are washed into the sea.

**tertiary consumer** A carnivore that feeds on carnivores that eat herbivores. Also known as a *third-order consumer*.

**test** A term used for the external covering of foraminiferans.

**tetraspore** The haploid, nonmotile, dispersal stage released by the tetrasporophyte of a red alga.

**tetrasporophyte** A multicellular diploid stage in the life cycle of a red alga growing parasitically on the female gametophyte and releasing diploid spores.

**thallus (THAL-uhs)** The body of an alga.

**thecosome pteropod** An herbivorous oceanic gastropod that has a highly reduced shell and swims by flapping lateral extensions of its foot.

**theory** A body of observations and their experimental supports that have stood the test of time.

**theory of plate tectonics** The theory that states that the movement of continental masses is the result of the movement of the rigid slabs, or plates, on which they rest.

**thermal capacity** The amount of heat energy required to increase the temperature of 1 gram of a substance by 1°C. Also known as the *specific heat*.

**thermocline (THER-moh-klyn)** A zone in the ocean that is characterized by a rapid change in temperature with increasing depth.

**thermohaline circulation** Vertical mixing that results from density changes due to changes in temperature and salinity of the water.

**third-order consumer** A carnivore that feeds on carnivores that eat herbivores. Also known as a *tertiary consumer*.

**thraustochytrids** Labyrinthomorphs that are abundant decomposers in planktonic and benthic communities.

**thylakoid** Membrane-bound structure that contains chlorophyll and other photosynthetic pigments. They are found within chloroplasts.

**tidal current** Current that occurs as the result of changing tides.

**tidal flat** An area of estuary that is exposed at low tide and covered at high tide.

**tidal overmixing** A mixing action that occurs in some estuaries at high tide due to seawater at the surface moving upstream more quickly than the bottom water from the river, causing the denser surface water to sink to the bottom and displace lighter freshwater, which moves to the surface.

**tide** The changes in sea level that occur as the result of the gravitational pull of the moon and the sun on the water in the oceans.

**tide pool** A depression in intertidal rocks or in the intertidal zone of sandy beaches that continues to hold water during a low tide.

**tiller** A secondary stem that grows from the base of a main stem (culm) in a grass and gives the plant a tufted growth if arising in multiples.

**tissue** A group of similar cells that serve a specific function.

**top-down factors** Factors such as competition, herbivory, or predation that affect the structure of communities by acting on lower trophic levels.

**trace elements**  Elements dissolved in seawater that are present in concentrations of less than one part per million.

**transfer RNA (tRNA)**  A type of RNA that carries amino acids to ribosomes for their incorporation into protein.

**transform fault**  A fault occurring in regions where lithospheric plates move past each other, particularly sections of the midocean ridge.

**transverse currents**  Currents that connect eastern- and western-boundary currents within a gyre.

**trash fish**  The noncommercial fishes that are killed during fishing for commercial species.

**trawl**  Large net dragged along the bottom or in midwater, depending on the catch, by vessels called trawlers.

**triglyceride**  A type of lipid composed of glycerol and three fatty acids.

**tripton**  Particles of dead organic matter suspended in seawater; largely synonymous with *detritus* and *POM*.

**trochophore larva (TROHK-eh-fohr)**  A free-swimming ciliated larva in the life cycles of many marine molluscs as well as some other marine organisms.

**trophic level (TROH-fik)**  An energy-storing level in a food chain.

**Tropic of Cancer**  The latitude 23.5 degrees N of the equator. It marks the northernmost extent of the tropics.

**Tropic of Capricorn**  The latitude 23.5 degrees S of the equator. It marks the southernmost extent of the tropics.

**trumpet cell**  A type of cell in the tissue of kelp carrying products of photosynthesis to parts of the thallus that lie in deeper water.

**tsunami (soo-NAH-mee)**  A large seismic sea wave that increases in amplitude as it approaches a shore. Sometimes referred to as a tidal wave. Also known as a *seismic sea wave*.

**tube feet**  Tubular structures in echinoderms that function in locomotion and feeding.

**tunic**  The body covering of animals known as tunicates, phylum Urochordata, which is composed of a molecule similar to cellulose.

**tunicate (TOO-ni-kayt)**  Animals belonging to the subphylum Urochordata. Tunicates are named for their body covering: a tunic composed of a substance similar to cellulose.

**turbellarian flatworm**  A free-living (nonparasitic) type of flatworm.

**turbidity**  The cloudy condition of seawater that results from the presence of living or dead suspended particles.

**turbidity current**  A swift, underwater avalanche of sediment and water.

**turtle exclusion device**  A grate that is placed in shrimp trawls to direct sea turtles out of the net so they won't drown.

**twilight zone**  The region of ocean between 150 and 450 meters deep where there is not enough light to power photosynthesis. Also known as the *disphotic zone*.

**type specimen**  A museum specimen that is considered to be representative of a species for the purpose of identification.

**typological definition of species**  The defining of a species based on morphology or appearance.

**umbo**  The area around the hinge of a bivalve's shell. This area represents the oldest part of the shell.

**unarmored dinoflagellate**  Dinoflagellate with a few thin layers of cellulose in its alveoli, giving the appearance of having no protective cell covering.

**understory**  The layer of low vegetation on the floor of a kelp forest.

**univalve**  A shell composed of one piece.

**upwelling**  The process by which a combination of wind and ocean currents bring nutrient-laden material from the ocean bottom into the photic zone.

**urea**  Waste product of protein metabolism that is stored in the bodies of cartilaginous fish and a few other animals to maintain osmotic balance with the sea. In vertebrates it is formed in the liver and excreted by the kidneys.

**vacuole**  Structure surrounded by a plasma membrane that may contain food, wastes, or water.

**valve (diatom)**  One half of the two-part cell wall, or frustule, of a diatom.

**valve (mollusc and brachiopod)**  The two jointed halves of a shell that surround the bodies of bivalves and lamp shells.

**vascular tissue**  Layers of cells that conduct water, minerals, and the products of photosynthesis to distant parts of a plant and also provide structural support for the plant body.

**vegetative growth**  A form of asexual reproduction in seaweeds and plants in which growth regions produce additional units of the thallus or plant body of identical genetic makeup.

**veliger larva (VEL-uh-jer)**  A free-swimming larval stage that develops from the trochophore larva of some molluscs.

**vent community**  The populations of organisms that live around a deep-sea vent, also known as a *rift community*.

**ventricle**  Thick-walled and muscular pumping chamber of the vertebrate heart.

**vertebra (plural, vertebrae)**  Segment of the spinal column of vertebrates.

**vertebrate**  An animal that possesses an internal skeletal rod (commonly called a backbone) composed of units known as *vertebrae*.

**vertical mixing**  The mixing of water in a water column that occurs when the surface water becomes more dense than the water beneath it.

**vertical overturn**  The change-over in the water of an unstable water column.

**viral replication**  The process by which the parts of many new viruses are manufactured and assembled within a cell under the control of the infecting virus.

**virion**  The mature infective viral particle released by a host cell.

**viriplankton**  Planktonic virus particles.

**virologist**  A scientist who studies viruses.

**virology**    The scientific study of viruses.

**visceral mass**    The dorsal region of the gastropod body that contains the circulatory, digestive, respiratory, excretory, and reproductive systems.

**viviparity**    Reproductive mode characterized by females producing eggs that are retained and nourished in the reproductive system until the young are mature enough to be released to the outside.

**warning (aposematic) coloration**    Bright and conspicuous coloration possessed by animals that are distasteful or dangerous to their attackers. Such colors are presumed to warn away attackers.

**water column**    The ocean water in an ocean basin.

**water vascular system**    A hydraulic system unique to echinoderms that functions in locomotion, feeding, gas exchange, and excretion.

**wave refraction**    The horizontal bending of waves that enter shallow water at an angle to the shore caused by one end of the wave slowing as it drags on the bottom.

**wave shock**    The force of waves as they crash against the rocks and the organisms that live on the rocks.

**well-mixed estuary**    An estuary in which river flow is low and tidal currents play a major role in the circulation of the water, resulting in a seaward flow of water and a uniform salinity at all depths.

**westerlies**    Winds that occur between 30 and 60 degrees N latitudes that blow from the south and the west and that occur between 30 and 60 degrees S that blow from the north and the west.

**western-boundary currents**    The currents that flow along the western edge of ocean basins, moving warm water toward the poles.

**wetland**    A basically terrestrial area that contains a large amount of water, such as a salt marsh or swamp.

**white zone**    The upper zone of the supralittoral fringe of tropical rocky shores. This zone is the true border between the land and the sea. No fully marine organisms occur here.

**whorls**    The turns of a gastropod's spiral shell.

**wind-induced vertical circulation**    The vertical movement of water that sometimes occurs when wind moves surface water horizontally, causing deeper water to move vertically.

**world ocean**    A continuous mass of water that covers nearly 70.8% of the earth's surface.

**xanthophyll (ZAN-thuh-fil)**    A group of accessory pigments that function in the process of photosynthesis in some organisms.

**xylem**    The kind of vascular tissue in plants that carries water and minerals from roots to parts of the plant that need them.

**yeast**    A fungus that consists of a single cell.

**yellow zone**    The upper part of the midlittoral zone of tropical rocky shores. The typical yellow or green color of the rock is due to microscopic boring algae.

**yolk sac**    A saclike structure in amniotic eggs that contains a supply of food in the form of yolk.

**zoea larva**    The planktonic larval stage of decapod crustaceans.

**zonation**    The separation of organisms in a habitat into definite zones or bands.

**zone of drying sand**    The uppermost vertical zone in the high intertidal region of a sandy shore characterized by the presence of moisture only during the highest tides.

**zone of intolerance**    A region that is so far removed from an organism's optimal range for an environmental variable that the organism cannot survive.

**zone of resurgence**    A vertical zone of the middle and low intertidal regions of a sandy shore that retains water at low tide.

**zone of retention**    A vertical zone in the high intertidal region of a sandy shore that retains water during low tide as the result of capillary action.

**zone of saturation**    The vertical zone of the sandy intertidal region that is constantly moist.

**zone of stress**    A region above or below an organism's optimal range for an environmental variable in which the organism must expend more energy than normal to maintain homeostasis.

**zooid**    An individual member of a bryozoan colony.

**zooplankton**    Components of the plankton that feed by ingestion of detritus or of other zooplankton.

**zooxanthella (ZOH-oh-zan-thel-ah)**    Dinoflagellate that is an important symbiont of jellyfishes, corals, and molluscs.

**zygote**    The diploid and genetically unique cell that is formed upon fusion of two haploid cells or gametes of different strains or, in the extreme case, when the egg is fertilized by a sperm.

# Photo Credits

This page constitutes an extension of the copyright page. We have made every effort to trace the ownership of all copyrighted material and to secure permission from copyright holders. In the event of any question arising as to the use of any material, we will be pleased to make the necessary corrections in future printings. Thanks are due to the following authors, publishers, and agents for permission to use the material indicated.

iv: NOAA v: NOAA viii: NOAA

**Chapter 1.** 1: NOAA 1, inset: Larry Ulrich/Stone/Getty Images 2, left, right: NOAA 3, top: Richard N. Mariscal 3, bottom: Stuart Westmorland/Stone/Getty Images 4: Copyright Library, Academy of Natural Sciences, Philadelphia 5, top: Copyright President and Fellows of Harvard College/Museum of Comparitve Zoology 5, left: Roland Birke/Peter Arnold, Inc. 5, right: Copyright President and Fellows of Harvard College/Museum of Comparitive Zoology 6, right: George Whitely/Photo Researchers, Inc. 6, right: Godfrey Argent Ltd. 7: R. DeGoursey/Visuals Unlimited

**Chapter 2.** 12: NOAA 12, inset: Hal Beral/Visuals Unlimited 13: NOAA 17: NOAA/TAO Project Office 18: Hal Beral/Visuals Unlimited 19: Adam Jones/Dembinsky Photo Associates 20: James McClintock, Bill Baker. *American Scientist,* Vol. 86, No. 3, P. 254, DOI: 10.1511/1998.3.254, used with permission 21, left: Hal Beral/Visuals Unlimited 21, center: Jeffrey Rotman 21, right: Roxanna Smolowitz, D.V. M/University of Pennsylvania

**Chapter 3.** 33: NOAA 33, inset: Woods Hole Oceanographic Institute 35: Jeffrey Bada, Department of Earth and Space Sciences, UCSD 42: WHOI/D. Foster/Visuals Unlimited 44, 45: Copyright by Marie Tharp 1977/2003. Reproduced by permission of Marie Tharp, Oceanographic Cartographer, One Washington Ave., South Nyack, New York 10960 48: Les Watling, Darling Marine Center, University of Maine 49: Leslie R. Sautter/Ocean Explorer/NOAA

**Chapter 4.** 56: NOAA 56, inset: Darrell Gulin/Dembinsky Photo Asssociates 58: Dennis Drenner 66: G. Jacobs, Stennis Space Center/Geosphere Project/SPL/Photo Researchers, Inc. 78: Reuters/Bettmann/Corbis

**Chapter 5.** 84: NOAA 84, inset: Fred McConnaughey 1985/Photo Researchers, Inc. 85, top and bottom: NOAA 86, top: Connie Toops 86, bottom: NOAA 88, 89: NOAA 91: Ester R. Angert, Ph.D./Phototake 92: Paul Johnson/Biological Photo Service 94, top: Dr. Jeremy Burgess/Photo Researchers, Inc. 94, bottom: Keith R. Porter 95: Lisa Starr with PDB file courtesy of Dr. Christina A. Bailey, Department of Chemistry & Biochemistry, California Polytechnic State University, San Luis Obispo, CA 96: Fred McConnaughey 1988/Photo Researchers, Inc. 101 top, and bottom left: Fred McConnaughey 1984/Photo Researchers, Inc. 101, bottom right: Nancy Sefton/Photo Researchers, Inc. 101, bottom: Jon L. Hawker 104, top: Jon L. Hawker 104, bottom: Fred McConnaughey 1985/Photo Researchers, Inc. 105, top left: NOAA 105, top right: Woods Hole Oceanographic Institute 105, bottom left: Paul Johnson/Biological Photo Service 105, bottom left center: Tokura, R. 1982 Arenicolous Marine Fungi from Japanese beaches. Transaction of the Mycological Society of Japan 23:423-433 105, bottom right center: Connie Toops 105, bottom right: © 1995 Peter Howorth/Mo Yung Productions/Norbert Wu

**Chapter 6.** 109: NOAA 109, inset: Dr. Dennis Kunkel/Visuals Unlimited 112, top: Ester R. Angert, Ph.D./Phototake 112, bottom: Woods Hole Oceanographic Institute 113: Dennis Drenner 114: Fred Bavendam/Peter Arnold, Inc. 118: John B. Corliss 120: Tokura, R. 1982 Arenicolous Marine Fungi from Japanese beaches. Transaction of the Mycological Society of Japan 23:423-433 122, all: Courtesy of S. Sharnoff 124: Phillip Harrington/The Stock Market/Corbis 125, 126: Dr. Elizabeth Venrick/Scripps Institute of Oceanography 126: John D. Dodge 128: Sanford Berry/Visuals Unlimited 129: Paul Johnson/Biological Photo Service 130, left: John Clegg/Ardea, London 130, right: Neal Ericsson 131, top: Peter Parks/Peter Arnold, Inc. 131, bottom: Dr. Richard Kessel & Dr. Gene Shih/Visuals Unlimited

**Chapter 7.** 136: NOAA 136, inset, 137, 138, all: Richard N. Mariscal 138, top right: Richard Tankersley, Florida Institute of Technology 140: Richard N. Mariscal 142, top and left center: Richard N. Mariscal 142, right center: Doug Wechsler/Animals Animals 142, bottom: Roger Steene/imagequestmarine.com 143, top left: Mike Guiry/AlgaeBase 143, top center: John Huisman/AlgaeBase 143, top right: Courtesy of Martin Landau 143, bottom left : John Huisman/AlgaeBase 143, bottom right: Mike Guiry/AlgaeBase 145, left: NOAA 145, right: Stephen Frink/Corbis 146, top left, center left: Richard N. Mariscal 146, center right: Colin Bates/AlgaeBase 146, top right: Mike Guiry/AlgaeBase 146, bottom left: Herb Segars/Animals Animals 146, bottom right: COLOR-PIC/Animals Animals 147: Larry Lipsky /Tom Stack & Associates 150, top: Dr. David Campbell 150, center: Florida Center for Environmental Studies 150, bottom: John Huisman/AlgaeBase 154, top: Connie Toops 154, top center: Richard N. Mariscal 154, bottom center: Rick Poley/Visuals Unlimited 154, bottom: Courtesy of Richard Turner 155, left: Jon Hawker 155, right: Richard N. Mariscal 156: Richard Herman/Visuals Unlimited 157, left: Jon Hawker 157, right: Laurie Petrone, Florida Institute of Technology

**Chapter 8.** 162: NOAA 162, inset: Fred Bavendam/Peter Arnold, Inc. 166: © George Billiris, Sponge Merchant International 167, top left: Richard N. Mariscal 167, bottom: American Museum of Natural History 167, top right: © 1995 Joshua Singer 169, top: Richard N. Mariscal 169, bottom: Courtesy J. H. Farmes. From Rees, W. J., ed. 1966. *The Cnidaria and Their Evolution,* Zoological Society of London 170: Zig Leszczynski/Animals Animals 171, top left: Claudia E. Mills/Biological Photo Service 171, center and bottom: Richard N. Mariscal 171, top right: David J. Wrobel/Biological Photo Service 172, all: Richard N. Mariscal 173: Richard N. Mariscal 175, all: Richard N. Mariscal 176, top: Richard N. Mariscal 176, bottom: National Geographic/Getty Images 177: Stan Elms/Visuals Unlimited 178: Robert Brons/Biological Photo Service 179, both: Richard N. Mariscal 180, top: Roger Steene/imagequestmarine.com 180, bottom: Richard N. Mariscal 181, all: Richard N. Mariscal 182: South Carolina Department of Natural Resources 183: Richard N. Mariscal 184, top left: Kjell B. Sandved 184, bottom left: Kathie Atkinson 1991/Oxford Scientific Films/Animals Animals 184, top right: Richard N. Mariscal 184, bottom right: C. R. Wyttenbach, University of Kansas/Biological Photo Service

**Chapter 9.** 190: NOAA 190, inset: Fred Bavendam 192: Richard N. Mariscal 194: W. Gregory Brown/Animals Animals 195, top left: NOAA 195, top right: Patti Murray/Animals Animals 195, bottom: Richard N. Mariscal 196: Robert Brons/Biological Photo Service 197, left: Tom McHugh/Photo Researchers, Inc. 197, center: Herb Segars/Animals Animals 197, right: Maryilyn and Maris Kaznesrs/Dembinsky Photo Associates 198: William Jorgensen/Visuals Unlimited 199: W. Gregory Brown 1991/Animals Animals 200, top: Biological Photo Service 200, bottom: Richard N. Mariscal 201, left: Jeffrey Rotman 201, right: Fred Bavendam 203: Richard N. Mariscal 204: Doug Allan/Oxford Scientific Films/Animals, Animals 205, top left: Neg. Trans. No. 312007. Courtesy of the Dept. of Library Services, American Museum of Natural History 205, all: Richard N. Mariscal 206, top: OSF/Peter Parks/Animals Animals 206, center: Wim van Egmond/Visuals

# Index

*Italics are used to indicate figures and illustrations. Bold text indicates tables.*